UNITED STATES AIR FORCE
SEARCH AND RESCUE
SURVIVAL TRAINING

AF REGULATION 64-4

UNITED STATES AIR FORCE
SEARCH AND RESCUE
SURVIVAL TRAINING

AF REGULATION 64-4

REPRINT OF DEPARTMENT OF THE AIR FORCE FIELD MANUAL

MetroBooks

MetroBooks

An Imprint of the Michael Friedman Publishing Group, Inc.

This edition published in 2002 by Michael Friedman Publishing Group, Inc.

This reproduction of the original Search and Rescue manual, AF Regulation 64-4,
originally issued by the United States' Department of the Air Force has been updated
with new images and accompanying appendix by the publishers.
All efforts were made to obtain all necessary permissions.
Any oversight is accidental and will be rectified in future printings.

Library of Congress Cataloging-in-Publication Data available upon request.

ISBN 1-58663-722-3

Printed in the United States of America

1 3 5 7 9 10 8 6 4 2

For bulk purchases and special sales, please contact:
Michael Friedman Publishing Group, Inc.
Attention: Sales Department
230 Fifth Avenue
New York, NY 10001
212/685-6610 FAX 212/685-3916

Visit our website:
www.metrobooks.com

DEPARTMENT OF THE AIR FORCE
Headquarters US Air Force
Washington DC 20330-5000

AF REGULATION 64-4
VOLUME I
15 July 1985

Search and Rescue

SURVIVAL TRAINING

This regulation describes the various environmental conditions affecting human survival, and describes individual activities necessary to enable that survival. This regulation is for instructor and student use in formal and USAF survival and survival continuation training. This regulation also applies to US Air Force Reserve and Air National Guard units and members. Sources used to compile this regulation are listed in the bibliography.

	Paragraph	Page

Part One

THE ELEMENTS OF SURVIVING

Chapter 1

MISSION

1-1. Introduction. An ejection sequence, a bailout, or crash landing ends one mission for the crew but starts another—to successfully return from a survival situation. Are they prepared? Can they handle the new mission, not knowing what it entails? Unfortunately, many aircrew members are not fully aware of their new mission or are not fully prepared to carry it out. All instructors teaching aircrew survival must prepare the aircrew member to face and successfully complete this new mission. (Figure 1-1 shows situations a member might encounter.)

1-2. Aircrew Mission. The moment an aircrew member leaves the aircraft and encounters a survival situation, the assigned mission is to: "return to friendly control without giving aid or comfort to the enemy, to return early and in good physical and mental condition."

a. On first impressions, "friendly control" seems to relate to a combat situation. Even in peacetime, however, the environment may be quite hostile. Imagine parachuting into the arctic when it's minus 40°F. Would an aircrew member consider this "friendly"? No. If the aircraft is forced to crash-land in the desert where temperatures may soar above 120°F, would this be agreeable? Hardly. The possibilities for encountering hostile conditions affecting human survival are endless. Crewmembers who egress an aircraft may confront situations difficult to endure.

b. The second segment of the mission, "without giving aid or comfort to the enemy," is directly related to a combat environment. This part of the mission may be most effectively fulfilled by following the moral guide—the Code of Conduct. Remember, however, that the Code of Conduct is useful to a survivor at all times and in all situations. Moral obligations apply to the peacetime situation as well as to the wartime situation.

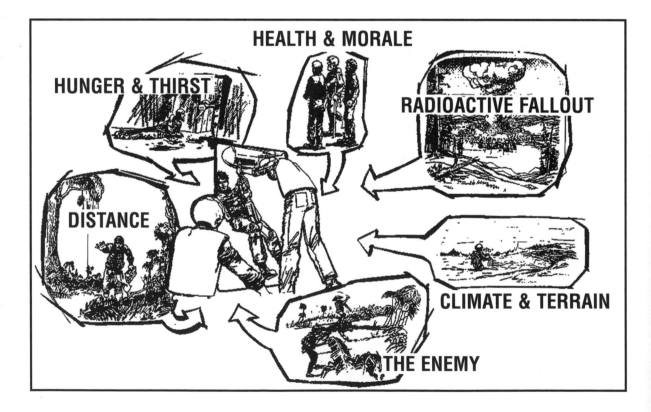

Figure 1-1. Elements of Surviving.

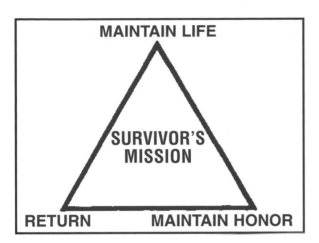

Figure 1-2. Survival Triangle.

c. The final phase of the mission is "to return early and in good physical and mental condition." A key factor in successful completion of this part of the mission may be the *will to survive*. This will is present, in varying degrees, in all human beings. Although successful survival is based on many factors, those who maintain this important attribute will increase their chance of success.

1-3. Goals. Categorizing this mission into organizational components, the three goals or duties of a survivor are to maintain life, maintain honor, and return. Survival training instructors and formal survival training courses provide training in the skills, knowledge, and attitudes necessary for an aircrew member to successfully perform the fundamental survival duties shown in figure 1-2.

1-4. Survival. Surviving is extremely stressful and difficult. The survivor may be constantly faced with hazardous and difficult situations. The stresses, hardships, and hazards (typical of a survival episode) are caused by the cumulative effects of existing conditions. (See chapter 2 pertaining to conditions affecting survival.) Maintaining life and honor and returning, regardless of the conditions, may make surviving difficult or unpleasant. The survivor's mission forms the basis for identifying and organizing the major needs of a survivor. (See survivor's needs in chapter 3.)

1-5. Decisions. The decisions survivors make and the actions taken in order to survive determine their prognosis for surviving.

1-6. Elements. The three primary elements of the survivor's mission are: the conditions affecting survival, the survivor's needs, and the means for surviving.

Chapter 2

CONDITIONS AFFECTING SURVIVAL

2-1. Introduction. Five basic conditions affect every survival situation (figure 2-1). These conditions may vary in importance or degree of influence from one situation to another and from individual to individual. At the onset, these conditions can be considered to be neutral—being neither for nor against the survivor, and should be looked upon as neither an advantage nor a disadvantage. The aircrew member may succumb to their effects—or use them to best advantage. These conditions exist in each survival episode, and they will have great bearing on the survivor's every need, decision, and action.

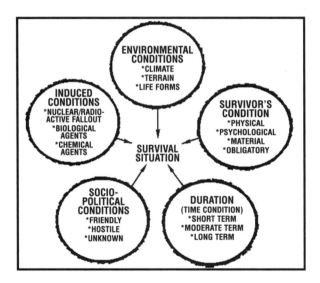

Figure 2-1. Five Basic Conditions.

2-2. Environmental Conditions. Climate, terrain, and life forms are the basic components of all environments. These components can present special problems for the survivor. Each component can be used to the survivor's advantage. Knowledge of these conditions may very well contribute to the success of the survival mission.

a. Climate. Temperature, moisture, and wind are the basic climatic elements. Extreme cold or hot temperatures, complicated by moisture (rain, humidity, dew, snow, etc.) or lack of moisture, and the possibility of wind, may have a life-threatening impact on the survivor's needs, decisions, and actions. The primary concern, resulting from the effects of climate, is the need for personal protection. Climatic conditions also have a significant impact on other aspects of survival (for example, the availability of water and food, the need and ability to travel, recovery capabilities, physical and psychological problems, etc.; figure 2-2).

b. Terrain. Mountains, prairies, hills, and lowlands are only a few examples of the infinite variety of land forms which describe "terrain." Each of the land forms have a different effect on a survivor's needs, decisions, and actions. A survivor may find a combination of several terrain forms in a given situation. The existing terrain will affect the survivor's needs and activities in such areas as travel, recovery, sustenance, and, to a lesser extent, personal protection. Depending on its form, terrain may afford security and concealment for an evader; cause travel to be easy or difficult; provide protection from cold, heat, moisture, wind, or nuclear, biological, chemical (NBC) conditions; or make surviving a *seemingly* impossible task (figure 2-3).

c. Life Forms. For survival and survival training purposes, there are two basic life forms—plant life and animal life (other than human). NOTE: The special relationship and effects of people on the survival episode are covered separately. Geographic areas are often identified in terms of the abundance of life (or lack thereof). For example, the barren arctic or desert, primary (or secondary) forests, the tropical rain forest, the polar ice cap, etc., all produce images regarding the quantities of life forms. These examples can have special meaning not only in terms of the hazards or needs they create, but also in how a survivor can use available life forms (figure 2-4).

(1) Plant Life. There are hundreds of thousands of different types and species of plant life. In some instances, geographic areas are identified by the dominant types of plant life within that area. Examples of this are savannas, tundra, deciduous forests, etc. Some species of plant life can be used advantageously by a survivor—if not for the food or the water, then for improvising camouflage, shelter, or providing for other needs.

(2) Animal Life. Reptiles, amphibians, birds, fish, insects, and mammals are life forms which directly affect a survivor. These creatures affect the survivor by posing hazards (which must be taken into consideration), or by satisfying needs.

2-3. The Survivor's Condition. The survivor's condition and the influence it has in each survival episode is often overlooked. The primary factors which constitute the survivor's condition can best be described by the four categories shown in figure 2-5. Aircrew members must prepare themselves in each of these areas before each mission, and be in a state of "constant readiness" for the possibility of a "survival mission." Crewmembers must be aware of the role a survivor's condition plays both before and during the survival episode.

a. Physical. The physical condition and the fitness level of the survivor are major factors affecting

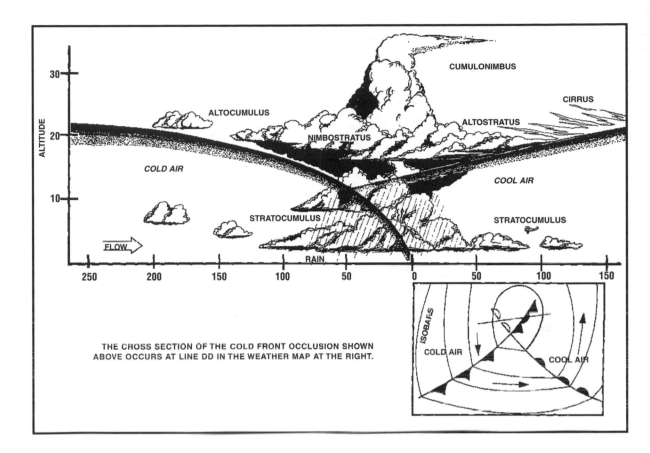

Figure 2-2. Cold Front Occlusion.

survivability. Aircrew members who are physically fit will be better prepared to face survival episodes than those who are not. Further, a survivor's physical condition (injured or uninjured) during the initial phase of a survival episode will be a direct result of circumstances surrounding the ejection, bailout, parachute landing, or crash landing. In short, high levels of physical fitness and good post-egress physical condition will enhance a survivor's ability to cope with such diverse variables as: (1) temperature extremes, (2) rest or lack of it, (3) water availability, (4) food availability, and (5) extended survival episodes. In the last instance, physical weakness may increase as a result of nutritional deficiencies, disease, etc.

    b. Psychological. Survivors' psychological state greatly influences their ability to successfully return from a survival situation.

    (1) Psychological effectiveness in a survival episode (including captivity) results from effectively coping with the following factors:

    (a) Initial shock - Finding oneself in a survival situation following the stress of ejection bailout or crash landing.

    (b) Pain - Naturally occurring or induced by coercive manipulation.

    (c) Hunger - Naturally occurring or induced by coercive manipulation.

    (d) Thirst - Naturally occurring or induced by coercive manipulation.

    (e) Cold or Heat - Naturally occurring or induced by coercive manipulation.

    (f) Frustration - Naturally occurring or induced by coercive manipulation.

    (g) Fatigue (including Sleep Deprivation) - Naturally occurring or induced by coercive manipulation.

    (h) Isolation - Includes forced (captivity) and the extended duration of any episode.

    (i) Insecurity - Induced by anxiety and self-doubts.

    (j) Loss of self-esteem - Most often induced by coercive manipulation.

    (k) Loss of self-determination - Most often induced by coercive manipulation.

    (l) Depression - Mental "lows."

    (2) A survivor may experience emotional reactions during a survival episode due to the previously stated

Emotional reactions commonly occurring in survival (including captivity) situations are:

   (a) Boredom - sometimes combined with loneliness.

   (b) Loneliness.

   (c) Impatience.

   (d) Dependency.

   (e) Humiliation.

   (f) Resentment.

   (g) Anger - sometimes included as a sub-element of hate.

   (h) Hate.

   (i) Anxiety.

   (j) Fear - often included as a part of panic or anxiety.

   (k) Panic.

Figure 2-4. Life Forms.

   (3) Psychologically survival episodes may be divided into "crisis" phases and "coping" phases. The initial crisis period will occur at the onset of the survival situation. During this initial period, "thinking" as well as "emotional control" may be disorganized. Judgment is impaired, and behavior may be irrational (possibly to the point of panic). Once the initial crisis is under control, the coping phase begins and the survivor is able to respond positively to the situation. Crisis periods may well recur, especially during extended situations (captivity). A survivor must strive to control if avoidance is impossible.

   (4) The most important psychological tool that will affect the outcome of a survival situation is the *will to survive*. Without it, the survivor is surely doomed to failure—a strong will is the best assurance of survival.

   c. Material. At the beginning of a survival episode, the clothing and equipment in the aircrew member's possession, the contents of available survival kits, and salvageable resources from the parachute or aircraft are the sum total of the survivor's material assets. Adequate premission preparations are required (must be stressed during training). Once the survival episode has started, special attention must be given to the care, use, and storage of all materials to ensure they continue to be

Figure 2-3. Terrain.

factors, previous (life) experiences (including training), and the survivor's psychological tendencies.

SURVIVOR'S CONDITION

○ PHYSICAL

○ PSYCHOLOGICAL

○ MATERIAL

○ OBLIGATORY

Figure 2-5. Survivor's Condition.

serviceable and available. Items of clothing and equipment should be selectively augmented with improvised items.

(1) Clothing appropriate to anticipated environmental conditions (on the ground) should be worn or carried as aircraft space and mission permit.

(2) The equipment available to a survivor affects all decisions, needs, and actions. The survivor's ability to improvise may provide ways to meet some needs.

d. Legal and Moral Obligations. A survivor has both legal and moral obligations or responsibilities. Whether in peacetime or combat, the survivor's responsibilities as a member of the military service continues. Legal obligations are expressly identified in the Geneva Conventions, Uniform Code of Military Justice (UCMJ), and Air Force directives and policies. Moral obligations are expressed in the Code of Conduct. (See figure 2-6.)

(1) Other responsibilities influence behavior during survival episodes and influence the *will to survive*. Examples include feelings of obligation or responsibilities to family, self, and (or) spiritual beliefs.

(2) A survivor's individual perception of responsibilities influence survival needs, and affect the psychological state of the individual both during and after the survival episode. These perceptions will be reconciled either consciously through rational thought or subconsciously through attitude changes. Training specifically structured to foster and maintain positive attitudes provides a key asset to survival.

2-4. Duration—The Time Condition. The duration of the survival episode has a major effect upon the aircrew member's needs. Every decision and action will

be driven in part by an assessment of when recovery or return is probable. Air superiority, rescue capabilities, the distances involved, climatic conditions, the ability to locate the survivor, or captivity are major factors which directly influence the duration (time condition) of the survival episode. A survivor can never be certain that rescue is imminent.

2-5. Sociopolitical Condition. The people a survivor contacts, their social customs, cultural heritage, and political attitudes will affect the survivor's status. Warfare is one type of sociopolitical condition, and people of different cultures are another. Due to these sociopolitical differences, the interpersonal relationship between the survivor and any people with whom contact is established is crucial to surviving. To a survivor, the attitude of the people contacted will be friendly, hostile, or unknown.

a. Friendly People. The survivor who comes into contact with friendly people, or at least those willing (to some degree) to provide aid, is indeed fortunate. Immediate return to home, family, or home station, however, may be delayed. When in direct association with even the friendliest of people, it is essential to maintain their friendship. These people may be of a completely different culture in which a commonplace American habit may be a gross and serious insult. In other instances, the friendly people may be active insurgents in their country and constantly in fear of discovery. Every survivor action, in these instances, must be appropriate and acceptable to ensure continued assistance.

b. Hostile People. A state of war need not exist for a survivor to encounter hostility in people. With few exceptions, any contact with hostile people must be avoided. If captured, regardless of the political or social reasons, the survivor must make all efforts to adhere to the Code of Conduct and the legal obligations of the UCMJ, the Geneva Conventions, and USAF policy.

c. Unknown People. The survivor should consider all factors before contacting unknown people. Some primitive cultures and closed societies still exist in which outsiders are considered a threat. In other areas of the world, differing political and social attitudes can place a survivor "at risk" in contacting unknown people.

2-6. Induced Conditions. Any form of warlike activity results in "induced conditions." Three comparatively new induced conditions may occur during combat operations. Nuclear warfare and the resultant residual radiation, biological warfare, and chemical warfare (NBC) create life-threatening conditions from which a survivor needs immediate protection. The longevity of NBC conditions further complicates a survivor's other needs, decisions, and actions (figure 2-7).

U.S. FIGHTING MAN'S
CODE OF CONDUCT

I

I am an American fighting man, I serve in the forces which guard my country and our way of life, I am prepared to give my life in their defense.

II

I will never surrender of my own free will. If in command, I will never surrender my men while they still have the means to resist.

III

If I am captured, I will continue to resist by all means available, I will make every effort to escape and aid others to escape. I will accept neither parole nor special favors from the enemy.

IV

If I become a prisoner of war, I will keep faith with my fellow prisoners. I will give no information or take part in any action which might be harmful to my comrades. If I am senior, I will take command. If not, I will obey the lawful orders of those appointed over me and will back them up in every way.

V

When questioned, should I become a prisoner of war, I am required to give name, rank, service number, and date of birth. I will evade answering further questions to the utmost of my ability. I will make no oral or written statements disloyal to my country and its allies or harmful to their cause.

VI

I will never forget that I am an American fighting man, responsible for my actions, and dedicated to the principles which made my country free. I will trust in my God and in the United States of America.

Figure 2-6. Code of Conduct.

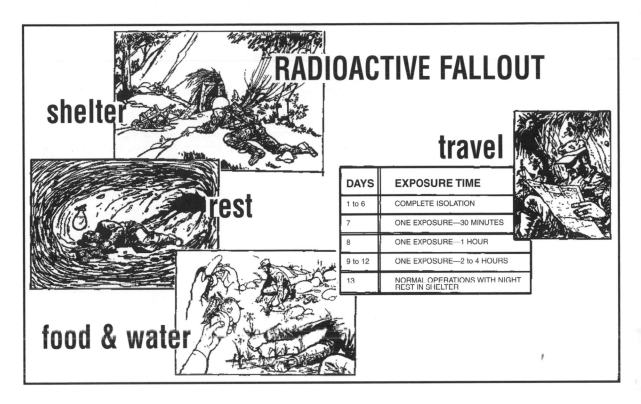

Figure 2-7. Induced Conditions.

Chapter 3

THE SURVIVOR'S NEEDS

3-1. Introduction. The three fundamental goals of a survivor—to maintain life, maintain honor, and return—may be further divided into eight basic needs. In a non-combatant situation, these needs include: personal protection, sustenance, health, travel, and communications (signaling for recovery). During combat additional needs must be fulfilled. They are: evasion, resistance if captured, and escape if captured. Meeting the individual's needs during the survival episode is essential to achieving the survivor's fundamental goals (figure 3-1).

Figure 3-1. Survivor's Needs.

3-2. Maintaining Life. Three elementary needs of a survivor in any situation which are categorized as the integral components of maintaining life are: personal protection, sustenance, and health.

a. Personal Protection. The human body is comparatively fragile. Without protection, the effects of environmental conditions (climate, terrain, and life forms) and of induced conditions (radiological, biological agents, and chemical agents) may be fatal. The survivor's primary defenses against the effects of the environment are clothing, equipment, shelter, and fire. Additionally, clothing, equipment, and shelter are the primary defenses against some of the effects of induced conditions (figure 3-2).

(1) The need for adequate clothing and its proper care and use cannot be overemphasized. The human body's tolerance for temperature extremes is very limited. However, its ability to regulate heating and cooling is extraordinary. The availability of clothing and its proper use is extremely important to a survivor in using these abilities of the body. Clothing also provides excellent

protection against the external effects of alpha and beta radiation, and may serve as a shield against the external effects of some chemical or biological agents.

(2) Survival equipment is designed to aid survivors throughout their episode. It must be cared for to maintain its effectiveness. Items found in a survival kit or aircraft can be used to help satisfy the eight basic needs. Quite often, however, a survivor must improvise to overcome an equipment shortage or deficiency.

(3) The survivor's need for shelter is twofold—as a place to rest and for protection from the effects of the environmental and (or) induced conditions (NBC). The duration of the survival episode will have some effect on shelter choice. In areas that are warm and dry, the survivor's need is easily satisfied using natural resting places. In cold climates, the criticality of shelter can be measured in minutes, and rest is of little immediate concern. Similarly, in areas of residual radiation, the criticality of shelter may also be measured in minutes (figure 3-3).

(4) Fire serves many survivor needs: purifying water, cooking and preserving food, signaling, and providing a source of heat to warm the body and dry clothing (figure 3-4).

b. Sustenance. Survivors need food and water to maintain normal body functions and to provide strength, energy, and endurance to overcome the physical stresses of survival.

(1) Water. The survivor must be constantly aware of the body's continuing need for water (figure 3-5).

(2) Food. During the first hours of a survival situation, the need for food receives little attention. During the first 2 or 3 days, hunger becomes a nagging aggravation which a survivor can overcome. The first major food crisis occurs when the loss of energy, stamina, and strength begin to affect the survivor's physical capabilities. The second major food crisis is more insidious. A marked increase in irritability and other attitudes may occur as the starvation process continues. Early and continuous attention must be given to obtaining and using any and all available food. Most people have food preferences. The natural tendency to avoid certain types of food is a major problem that must be overcome early in the survival situation. The starvation process ultimately overcomes all food aversions. The successful survivor overcomes these aversions before physical or psychological deterioration sets in (figure 3-6).

c. Health (Physical and Psychological). The survivor must be the doctor, nurse, corpsman, psychologist, and cheerleader. Self-aid is the survivor's sole recourse.

(1) Prevention. The need for preventive medicine and safety cannot be overemphasized. Attention to sanitation

Figure 3-2. Personal Protection.

and personal hygiene is a major factor in preventing physical, morale, and attitudinal problems.

(a) The need for cleanliness in the treatment of injuries and illness is self-evident. The prisoner of war (PW) who used maggots to eat away rotting flesh caused

by infection is a dramatic example. Prevention is much more preferred than such drastic procedures.

(b) Safety must be foremost in the mind of the survivor; carelessness is caused by ignorance and (or) poor judgment or bad luck. One miscalculation with a knife or axe can result in self-inflicted injury or death.

(2) Self-Aid.

(a) Injuries frequently occur during ejection, bailout, parachute landing, or ditching. Other postegress factors may also cause injury. In the event of injury, the survivor's existence may depend on the ability to perform self-aid. In many instances, common first aid procedures will suffice; in others, more primitive techniques will be required (figure 3-7).

(b) Illness and the need to treat it is more commonly associated with long-term situations such as an extended evasion episode or captivity. When preventive techniques have failed, the survivor must treat symptoms of disease in the absence of professional medical care.

(3) Psychological Health. Perhaps the survivor's greatest need is the need for emotional stability and a positive, optimistic attitude. An individual's ability to cope with psychological stresses will enhance successful survival. Optimism, determination, dedication, and humor, as well as many other psychological attributes, are all helpful for a survivor to overcome psychological stresses (figure 3-8).

3-3. Maintaining Honor. Three elementary needs that a survivor may experience during combat survival situations are categorized as integral components of maintaining honor. These three elementary needs are: (a) avoiding capture or evading, (b) resisting (if captured), and (c) escaping (if captured).

a. Avoiding Capture. Evasion will be one of the most difficult and hazardous situations a survivor will face. However difficult and hazardous evasion may be, captivity is always worse. During an evasion episode, the survivor has two fundamental tasks. The first is to use concealment techniques. The second is to use evasion movement techniques. The effective use of camouflage is common to both of these activities.

(1) Hiding oneself and all signs of presence are the evader's greatest needs. Experience indicates that the survivor who uses effective concealment techniques has a better chance of evading capture. Capture results most frequently when the evader is moving.

(2) The evader's need to move depends on a variety of needs such as recovery, food, water, better shelter, etc. Evasion movement is more successful when proven techniques are used.

b. Resisting. The PW's need to resist is self-evident. This need is both a moral and a legal obligation. Resistance is much more than refusing to divulge some bit of classified information. Fundamentally, resisting consists of two distinctly separate behaviors expected of the prisoner:

Figure 3-3. Shelters.

(1) Complying with legal and authorized requirements only.

(2) Disrupting enemy activity through resisting, subtle harassment, and tying up enemy guards who could be used on the front lines.

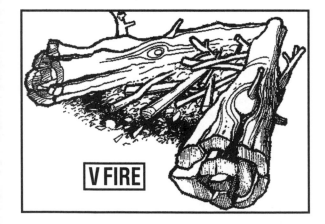

Figure 3-4. Fire.

c. Escaping (When Possible and Authorized). Escape is neither easy nor without danger. The Code of Conduct states a survivor should make every effort to escape and aid others to escape.

3-4. Returning. The need to return is satisfied by successful completion of one or both of the basic tasks confronting the survivor: aiding with recovery and traveling (on land or water).

a. Aiding With Recovery. For survivors or evaders to effectively aid in recovery, they must be able to make their position and the situation on the ground known. This is done either electronically, visually, or both (figure 3-9).

(1) Electronic signaling covers a wide spectrum of techniques. As problems such as security and safety during combat become significant factors, procedures for using electronic signaling to facilitate recovery become increasingly complex.

(2) Visual signaling is primarily the technique for attracting attention and pinpointing an exact location for rescuers. Simple messages or information may also be transmitted with visual signals.

b. Travel On Land. A survivor may need to move on land for a variety of reasons, ranging from going for

Figure 3-5. Water.

Figure 3-7. Self-Aid.

water to attempting to walk out of the situation. In any survival episode, the survivor must weigh the need to travel against capabilities and (or) safety (figure 3-10). Factors to consider may include:

restricted activity. A survivor who loses the mobility, due to injury, to obtain food, water, or shelter, can face death. There is a safe and effective way to travel across almost any type of terrain.

Figure 3-6. Food.

Figure 3-8. Health and Morale.

(1) The ability to walk or traverse existing terrain. In a nonsurvival situation, a twisted or sprained ankle is an inconvenience accompanied by some temporary pain and

(2) The need to transport personal possessions (burden carrying). There are numerous documented instances of survivors abandoning equipment and clothing, simply because carrying was a bother. Later, the abandoned materials were not available when needed to save life, limb, or aid in rescue. Burden carrying need not be difficult or physically stressful. There are many simple ways for a survivor to carry the necessities of life (figure 3-11).

Figure 3-9. Recovery.

(3) The ability to determine present position. Maps, compasses, star charts, Weems plotters, etc., permit accurate determination of position during extended travel. Yet, the knowledgeable, skillful, and alert survivor can do well without a full complement of these aids. Constant awareness, logic, and training in nature's clues to navigation may allow a survivor to determine general location even in the absence of detailed navigation aids.

(4) Restrictions or limitations to select and maintain a course of travel. The tools used in determining position are the tools used to maintain a course of travel. A straight line course to a destination is usually the simplest, but may not always be the best course for travel. Travel courses may need to be varied for diverse reasons, such as to get food or water, to enhance covert travels or to avoid hazardous or impassable obstacles or terrain. Careful planning and route selection before and during travel is essential.

c. Travel On Water. Two differing circumstances may require survivors to travel on water. First, those who crash-land or parachute into the open sea are confronted with one type of situation. Second, survivors who find a river or stream which leads in a desirable direction are faced with a different situation. In each instance, however, a common element is to stay afloat.

(1) The survivor's initial problems on the open sea are often directly related to the winds and size of the waves. Simply getting into a liferaft and staying there are often very difficult tasks. On the open sea, the winds and ocean currents have a significant effect on the direction of travel. As the survivor comes closer to shore, the direction in which the tide is flowing also becomes a factor. There are some techniques a survivor can use to aid with stabilizing the raft, controlling the direction and rate of travel, and increasing safety.

Figure 3-10. Travel.

(2) Survivors using rivers or streams for travel face both hazards and advantages as compared to overland travel. First, floating with the current is far less difficult than traveling overland. An abundance of food and water are usually readily available. Even in densely forested areas, effective signaling sites are generally available along streambeds. A survivor must use care and caution to avoid drowning, the most serious hazard associated with river travel.

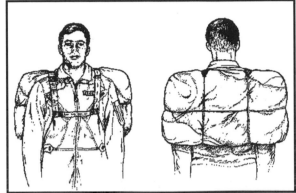

Figure 3-11. Burden Carrying.

Part Two

PSYCHOLOGICAL ASPECTS OF SURVIVAL

Chapter 4

CONTRIBUTING FACTORS

4-1. Introduction. Aircrew members in a survival situation must recognize that coping with the psychological aspects of survival is at least as important as handling the environmental factors. In virtually any survival episode, the aircrew will be in an environment that can support human life. The survivors' problems will be compounded because they never really expected to bail out or crash-land in the jungle, over the ocean, or anywhere else. No matter how well prepared, aircrews probably will never completely convince themselves that "it can happen to them." However, the records show it can happen. Before aircrew members learn about the physical aspects of survival, they must first understand that psychological problems may occur and that solutions to those problems must be found if the survival episode is to reach a successful conclusion (figure 4-1).

4-2. Survival Stresses:

a. The emotional aspects associated with survival must be completely understood just as survival conditions and equipment are understood. An important factor bearing on success or failure in a survival episode is the individual's psychological state. Maintaining an even, positive psychological state or outlook depends on the individual's ability to cope with many factors. Some include:

(1) Understanding how various physiological and emotional signs, feelings, and expressions affect one's bodily needs and mental attitude.

(2) Managing physical and emotional reactions to stressful situations.

(3) Knowing individual tolerance limits, both psychological and physical.

(4) Exerting a positive influence on companions.

b. Nature has endowed everyone with biological mechanisms that aid in adapting to stress. The bodily changes resulting from fear and anger, for example, tend to increase alertness and provide extra energy to either run away or fight. These and other mechanisms can hinder a person under survival conditions. For instance, a survivor in a raft could cast aside reason and drink seawater to quench a thirst; or, evaders in enemy territory, driven by hunger pangs, could expose themselves to capture when searching for food. These examples illustrate how "normal" reactions to stress could create problems for a survivor.

c. Two of the gravest threats to successful survival are concessions to comfort and apathy. Both threats represent attitudes which must be avoided. To survive, a person

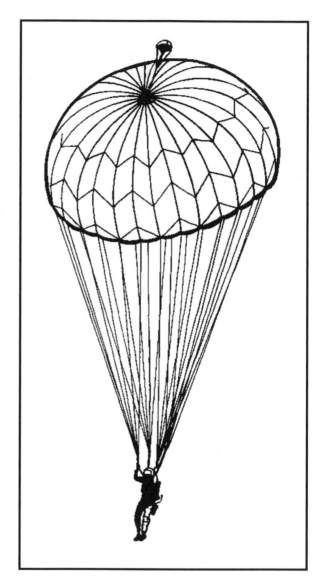

Figure 4-1. Psychological Aspects.

must focus planning and effort on fundamental needs.

(1) Many people consider comfort their greatest need. Yet, comfort is not essential to human survival. Survivors must value life more than comfort, and be willing to tolerate heat, hunger, dirt, itching, pain, and any other discomfort. Recognizing discomfort as temporary

Figure 4-2. Pain.

will help survivors concentrate on effective action.

(2) As the will to keep trying lessens, drowsiness, mental numbness, and indifference will result in apathy. This apathy usually builds up slowly, but ultimately takes over and leaves a survivor helpless. Physical factors can contribute to apathy. Exhaustion due to prolonged exposure to the elements, loss of body fluids (dehydration), fatigue, weakness, or injury are all conditions which can contribute to apathy. Proper planning and sound decisions can help a survivor avoid these conditions. Finally, survivors must watch for signs of apathy in companions and help prevent it. The first signs are resignation, quietness, lack of communication, loss of appetite, and withdrawal from the group. Preventive measures could include maintaining group morale by planning, activity, and getting the organized participation of all members.

d. Many common stresses cause reactions which can be recognized and dealt with appropriately in survival situations. A survivor must understand that stresses and reactions often occur at the same time. Although survivors will face many stresses, the following common stresses will occur in virtually all survival episodes: pain, thirst, cold and heat, hunger, frustration, fatigue, sleep deprivation, isolation, insecurity, loss of self-esteem, loss of self-determination, and depression.

4-3. Pain:

a. Pain, like fever, is a warning signal calling attention to an injury or damage to some part of the body. Pain is discomforting but is not, in itself, harmful or dangerous.

Pain can be controlled, and in an extremely grave situation, survival must take priority over giving in to pain (figure 4-2).

b. The biological function of pain is to protect an injured part by warning the individual to rest it or avoid using it. In a survival situation, the normal pain warnings may have to be ignored in order to meet more critical needs. People have been known to complete a fight with a fractured hand, to run on a fractured or sprained ankle, to land an aircraft despite severely burned hands, and to ignore pain during periods of intense concentration and determined effort. Concentration and intense effort can actually stop or reduce feelings of pain. Sometimes this concentration may be all that is needed to survive.

c. A survivor must understand the following facts about pain:

(1) Despite pain, a survivor can move in order to live.

(2) Pain can be reduced by:

(a) Understanding its source and nature.

(b) Recognizing pain as a discomfort to be tolerated.

(c) Concentrating on necessities like thinking, planning, and keeping busy.

(d) Developing confidence and self-respect. When personal goals are maintaining life, honor, and returning, and these goals are valued highly enough, a survivor can tolerate almost anything.

4-4. Thirst and Dehydration:

a. The lack of water and its accompanying problems of thirst and dehydration are among the most critical problems facing survivors. Thirst, like fear and pain, can be tolerated if the will to carry on, supported by calm, purposeful activity, is strong. Although thirst indicates the body's need for water, it does not indicate how much water is needed. If a person drinks only enough to satisfy thirst, it is still possible to slowly dehydrate. Prevention of thirst and the more debilitating dehydration is possible if survivors drink plenty of water any time it is available, and especially when eating (figure 4-3).

b. When the body's water balance is not maintained, thirst and discomfort result. Ultimately, a water imbalance will result in dehydration. The need for water may be increased if the person:

(1) Has a fever.

(2) Is fearful.

(3) Perspires unnecessarily.

(4) Rations water rather than sweat.

c. Dehydration decreases the body's efficiency or ability to function. Minor degrees of dehydration may not noticeably affect a survivor's performance, but as it becomes more severe, body functioning will become increasingly impaired. Slight dehydration and thirst can

Figure 4-3. Thirst.

also cause irrational behavior. One survivor described it:

"The next thing I remember was being awakened by an unforgettable sensation of thirst. I began to move about aimlessly and finally found a pool of water. We finally found water. In the water were two dead deer with horns locked. We went down to the water and drank away. It was the best damned drink of water I ever had in my life. I didn't taste the stench of the deer at all."

While prevention is the best way to avoid dehydration, virtually any degree of dehydration is reversible simply by drinking water.

4-5. Cold and Heat. The average normal body temperature for a person is 98.6°F. Victims have survived a body temperature as low as 20 degrees below normal, but consciousness is clouded and thinking numbed at a much smaller drop. An increase of 6 to 8 degrees above normal for any prolonged period may prove fatal. Any deviation from normal temperature, even by as little as 1 or 2 degrees, reduces efficiency.

a. Cold is a serious stress since even in mild degrees it lowers efficiency. Extreme cold numbs the mind and dulls the will to do anything except get warm again. Cold numbs the body by lowering the flow of blood to the extremities, and results in sleepiness. Survivors have endured prolonged cold and dampness through exercise, proper hygiene procedures, shelter, and food. Wearing proper clothing and having the proper climatic survival equipment when flying in cold weather areas are essential to enhance survivability (figure 4-4).

(1) One survivor described cold and its effect:

"Because of the cold water, my energy was going rapidly and all I could do was to hook my left arm over one side of the raft, hang on, and watch the low-flying planes as they buzzed me...As time progressed, the numbing increased...and even seemed to impair my thinking."

(2) Another survivor remembered survival training and acted accordingly:

"About this time, my feet began getting cold. I remembered part of the briefing I had received about feet freezing so I immediately took action. I thought about my shoes, and with my jack knife, cut off the bottom of my Mark II immersion suit and put them over my shoes. My feet immediately felt warmer and the rubber feet of the immersion suit kept the soles of my shoes dry."

b. Just as "numbness" is the principal symptom of cold, "weakness" is the principle symptom of heat. Most people can adjust to high temperatures, whether in the hold of a ship or in a harvest field on the Kansas prairie. It may take from 2 days to a week before circulation, breathing, heart action, and sweat glands are all adjusted to a hot climate. Heat stress also accentuates dehydration, which was discussed earlier. In addition to the problem of water, there are many other sources of discomfort and impaired efficiency which are directly attributable to heat or to the environmental conditions in hot climates. Extreme temperature changes, from extremely hot days to very cold nights, are experienced in desert and plains areas. Proper use of clothing and shelters can decrease the adverse effects of such extremes (figure 4-5).

c. Bright sun has a tremendous effect on eyes and any exposed skin. Direct sunlight or rays reflecting off the terrain require dark glasses or improvised eye protectors.

Figure 4-4. Cold.

Previous suntanning provides little protection; protective clothing is important.

d. Blowing wind, in hot summer, has been reported to get on some survivors' nerves. Wind can constitute an

Figure 4-5. Heat.

additional source of discomfort and difficulty in desert areas when it carries particles of sand and dirt. Protection against sand and dirt can be provided by tying a cloth around the head after cutting slits for vision.

e. Acute fear has been experienced among survivors in sandstorms and snowstorms. This fear results from both the terrific impact of the storm itself and its obliteration of landmarks showing direction of travel. Finding or improving shelter for protection from the storm itself is important.

f. Loss of moisture, drying of the mouth and mucous membranes, and accelerate dehydration can be caused by breathing through the mouth and talking. Survivors must learn to keep their mouth shut in desert winds as well as in cold weather.

g. Mirages and illusions of many kinds are common in desert areas. These illusions not only distort visual perception but sometimes account for serious incidents. In the desert, distances are usually greater than they appear and, under certain conditions, mirages obstruct accurate vision. Inverted reflections are a common occurrence.

4-6. Hunger. A considerable amount of edible material (which survivors may not initially regard as food) may be available under survival conditions. Hunger and semistarvation are more commonly experienced among survivors than thirst and dehydration. Research has revealed no evidence of permanent damage nor any decrease in mental efficiency from short periods of total fasting (figure 4-6).

Figure 4-6. Hunger.

a. The prolonged and rigorous Minnesota semistarvation studies during World War II revealed the following behavioral changes:

(1) Dominance of the hunger drive over other drives.

(2) Lack of spontaneous activity.

(3) Tired and weak feeling.

(4) Inability to do physical tasks.

(5) Dislike of being touched or caressed in any way.

(6) Quick susceptibility to cold.

(7) Dullness of all emotional responses (fear, shame, love, etc.).

(8) Lack of interest in others—apathy.

(9) Dullness and boredom.

(10) Limited patience and self-control.

(11) Lack of a sense of humor.

(12) Moodiness—reaction of resignation.

b. Frequently, in the excitement of some survival, evasion, and escape episodes, hunger is forgotten. Survivors have gone for considerable lengths of time without food or awareness of hunger pains. An early effort should be made to procure and consume food to reduce the stresses brought on by food deprivation. Both the physical and psychological effects described are reversed when food and a protective environment are restored. Return to normal is slow and the time necessary for the return increases with the severity of starvation. If food deprivation is complete and only water is ingested, the pangs of hunger disappear in a few days,

but even then the mood changes of depression and irritability occur. The individual tendency is still to search for food to prevent starvation and such efforts might continue as long as strength and self-control permit. When the food supply is limited, even strong friendships are threatened.

c. Food aversion may result in hunger. Adverse group opinion may discourage those who might try foods unfamiliar to them. In some groups, the barrier would be broken by someone eating the particular food rather than starving. The solitary individual has only personal prejudices to overcome and will often try strange foods.

d. Controlling hunger during survival episodes is relatively easy if the survivor can adjust to discomfort and adapt to primitive conditions. This man would rather survive than be fussy:

> "Some men would almost starve before eating the food. There was a soup made of lamb's head with the lamb's eyes floating around in it...When there was a new prisoner, I would try to find a seat next to him so I could eat the food he refused."

4-7. Frustration. Frustration occurs when one's efforts are stopped, either by obstacles blocking progress toward a goal or by not having a realistic goal. It can also occur if the feeling of self-worth or self-respect is lost (figure 4-7).

a. A wide range of obstacles, both environmental and internal, can lead to frustration. Frustrating conditions often create anger, accompanied by a tendency to attack and remove the obstacles to goals.

b. Frustration must be controlled by channeling energies into a positive and worthwhile obtainable goal. The survivor should complete the easier tasks before attempting more challenging ones. This will not only instill self-confidence, but also relieve frustration.

4-8. Fatigue. In a survival episode, a survivor must continually cope with fatigue and avoid the accompanying strain and loss of efficiency. A survivor must be aware of the dangers of over-exertion. In many cases, a survivor may already be experiencing strain and reduced efficiency as a result of other stresses such as heat or cold, dehydration, hunger, or fear. A survivor must judge capacity to walk, carry, lift, or do necessary work, and plan and act accordingly. During an emergency, considerable exertion may be necessary to cope with the situation. If an individual understands fatigue and the attitudes and feelings generated by various kinds of effort, that individual should be able to call on available reserves of energy when they are needed (figure 4-8).

a. A survivor must avoid complete exhaustion which may lead to physical and psychological changes. A survivor should be able to distinguish between exhaustion and being uncomfortably tired. Although a person

Figure 4-7. Frustration.

should avoid working to complete exhaustion, in emergencies certain tasks must be done in spite of fatigue.

(1) Rest is a basic factor for recovery from fatigue and is also important in resisting further fatigue. It is

Figure 4-8. Fatigue.

essential that the rest (following fatiguing effort) be sufficient to permit complete recovery; otherwise, the residual fatigue will accumulate and require longer periods of rest to recover from subsequent effort. During the early stages of fatigue proper rest provides a rapid recovery. This is true of muscular fatigue as well as mental fatigue. Sleep is the most complete form of rest available and is basic to recovery from fatigue.

(2) Short rest breaks during extended stress periods can improve total output. There are five ways in which rest breaks are beneficial:

(a) They provide opportunities for partial recovery from fatigue.

(b) They help reduce energy expenditure.

(c) They increase efficiency by enabling a person to take maximum advantage of planned rest.

(d) They relieve boredom by breaking up the uniformity and monotony of the task.

(e) They increase morale and motivation.

(3) Survivors should rest before output shows a definite decline. If rest breaks are longer, fewer may be required. When efforts are highly strenuous or monotonous rest breaks should be more frequent. Rest breaks providing relaxation are the most effective. In mental work, mild exercise may be more relaxing. When work is monotonous, changes of activity, conversation, and humor are effective relaxants. In deciding on the amount and frequency of rest periods, the loss of efficiency resulting from longer hours of effort must be weighed against the absolute requirements of the survival situation.

(4) Fatigue can be reduced by working "smarter" A survivor can do this in two practical ways:

(a) Adjust the pace of the effort. Balance the load, the rate, and the time period. For example, walking at a normal rate is a more economical effort than fast walking.

(b) Adjust the technique of work. The way in which work is done has a great bearing on reducing fatigue. Economy of effort is most important. Rhythmic movements suited to the task are best.

(5) Mutual group support, cooperation and competent leadership are important factors in maintaining group morale and efficiency, thereby reducing stress and fatigue. A survivor usually feels tired and weary before the physiological limit is reached. In addition, other stresses experienced at the same time, such as cold, hunger, fear or despair, can intensify fatigue. The feeling of fatigue involves not only the physical reaction to effort, but also subtle changes in attitudes and motivation. Remember, a person has reserves of energy to cope with an important emergency even when feeling very tired.

b. As in the case of other stresses, even a moderate amount of fatigue reduces efficiency. To control fatigue, it is wise to observe a program of periodic rest. Because of the main objective—to establish contact with friendly forces—survivors may overestimate their strength and risk exhaustion. On the other hand, neither an isolated individual nor a group leader should underestimate the capacity of the individual or the group on the basis of fatigue. The only sound basis for judgment must be gained from training and past experience. In training, a person should form an opinion of individual capacity based on actual experience. Likewise, a group leader must form an opinion of the capacities of fellow aircrew members. This group didn't think:

"By nightfall, we were completely bushed...We decided to wrap ourselves in the 'chute instead of making a shelter. We were too tired even to build a fire. We just cut some pine boughs, rolled ourselves in the nylon and went to sleep...and so, of course, it rained, and not lightly. We stood it until we were soaked, and then we struggled out and made a shelter. Since it was pitch dark, we didn't get the sags out of the canopy, so the water didn't all run off. Just a hell of a lot of it came through. Our hip and leg joints ached as though we had acute rheumatism. Being wet and cold accentuated the pain. We changed positions every 10 minutes, after gritting our teeth to stay put that long."

Figure 4-9. Sleep Deprivation.

4-9. Sleep Deprivation. The effects of sleep loss are closely related to those of fatigue. Sleeping at unaccustomed times, sleeping under strange circumstances (in a strange place, in noise, in light, or with other distractions), or missing part or all of the accustomed amount of sleep will cause a person to react with feelings of weariness, irritability, emotional tension, and some loss of efficiency. The extent of an individual's reaction depends on the amount of disturbance and on other stress factors which may be present at the same time (figure 4-9).

a. Strong motivation is one of the principal factors in helping to compensate for the impairing effects of sleep loss. Superior physical and mental conditioning, opportunities to rest, food and water, and companions help in enduring sleep deprivation. If a person is in reasonably good physical and mental condition, sleep deprivation can be endured 5 days or more without damage, although efficiency during the latter stages may be poor. A person must learn to get as much sleep and rest as possible. Restorative effects of sleep are felt even after "catnaps." In some instances, survivors may need to stay awake. Activity, movement, conversation, eating, and drinking are some of the ways a person can stimulate the body to stay awake.

b. When one is deprived of sleep, sleepiness usually comes in waves. A person may suddenly be sleepy immediately after a period of feeling wide awake. If this can be controlled, the feeling will soon pass and the person will be wide awake again until the next wave appears. As the duration of sleep deprivation increases, these periods between waves of sleepiness become shorter. The need to sleep may be so strong in some people after a long period of deprivation that they become desperate and do careless or dangerous things in order to escape this stress.

4-10. Isolation. Loneliness, helplessness, and despair, which are experienced by survivors when they are isolated, are among the most severe survival stresses. People often take their associations with family, friends,

military colleagues and others for granted. But survivors soon begin to miss the daily interaction with other people. However, these, like the other stresses already discussed, can be conquered. Isolation can be controlled and overcome by knowledge, understanding, deliberate countermeasures, and a determined will to resist it (figure 4-10).

Figure 4-10. Isolation.

4-11. Insecurity. Insecurity is the survivor's feeling of helplessness or inadequacy resulting from varied stresses and anxieties. These anxieties may be caused by uncertainty regarding individual goals, abilities, and the future in a survival situation. Feelings of insecurity may have

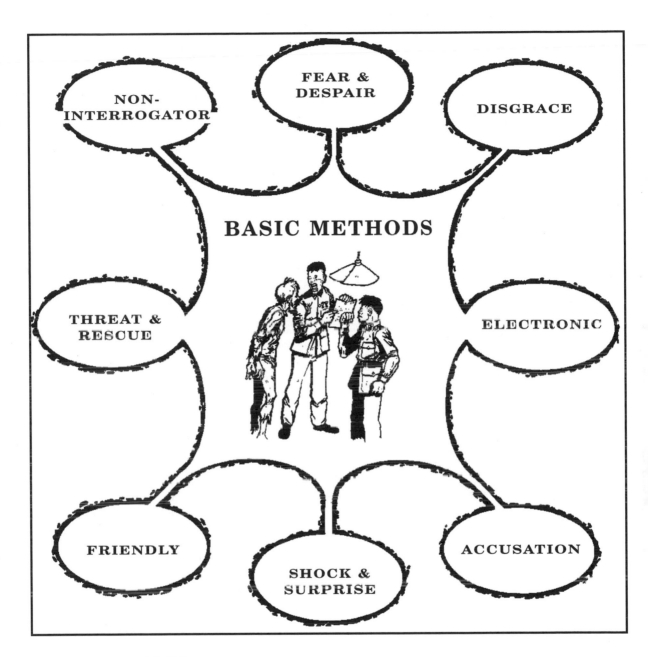

Figure 4-11. Loss of Self-Esteem.

widely different effects on the survivor's behavior. A survivor should establish challenging but attainable goals. The better a survivor feels about individual abilities to achieve goals and adequately meet personal needs, the less insecure the survivor will feel.

4-12. Loss of Self-Esteem. Self-esteem is the state or quality of having personal self-respect and pride. Lack of (or loss of) self-esteem in a survivor may bring on depression and a change in perspective and goals. A loss of self-esteem may occur in individuals in captivity. Humiliation and other factors brought on by the captor may cause them to doubt their own worth. Humiliation comes from the feeling of losing pride or self-respect by being disgraced or dishonored, and is associated with the loss of self-esteem. Prisoners must maintain their pride and not become ashamed either because they are PWs or because of the things that happen to them as a result of being a PW. The survivor who "loses face" (both personally and with the enemy) becomes more vulnerable to captor exploitation attempts. To solve this problem, survivors should try to maintain proper perspective about both the situation and themselves. Their feelings of self-worth may be bolstered if they recall the implied commitment in the Code of Conduct—PWs will not be forgotten (figure 4-11).

4-13. Loss of Self-Determination. A self-determined person is relatively free from external controls or influences over his or her actions. In everyday society, these "controls and influences" are the laws and customs of our society and of the self-imposed elements of our personalities. In a survival situation, the "controls and influences" can be very different. Survivors may feel as if events, circumstances, and (in some cases) other people, are in control of the situation. Some factors which may cause individuals to feel they have lost the power of self-determination are a harsh captor, captivity, bad weather, or rescue forces that make time or movement

Figure 4-12. Depression.

demands. This lack of self-determination is more perceived than actual. Survivors must decide how unpleasant factors will be allowed to affect their mental state. They must have the self-confidence, fostered by experience and training, to live with their feelings and decisions, and to accept responsibility for both the way they feel and how they let those feelings affect them.

4-14. Depression. As a survivor, depression is the biggest psychological problem that has to be conquered. It should be acknowledged that everyone has mental "highs" as well as mental "lows." People experiencing long periods of sadness or other negative feelings are suffering from depression. A normal mood associated with the sadness, grief, disappointment, or loneliness that everyone experiences at times is also described as depression. Most of the emotional changes in mood are temporary and do not become chronic. Depressed survivors may feel fearful, guilty, or helpless. They may lose interest in the basic needs of life. Many cases of depression also involve pain, fatigue, loss of appetite, or other physical ailments. Some depressed survivors try to injure or kill themselves (figure 4-12).

a. Psychiatrists have several theories as to the cause of depression. Some feel a person who, in everyday life and under normal conditions, experiences many periods of depression would probably have a difficult time in a survival situation. The main reason depression is a most difficult problem is that it can affect a wide range of psychological responses. The factors can become mutually reinforcing. For example, fatigue may lead to a feeling of depression. Depression may increase the feeling of fatigue, and this, in turn, leads to deeper depression, and so on.

b. Depression usually begins after a survivor has met the basic needs for sustaining life, such as water, shelter, and food. Once the survivor's basic needs are met, there is often too much time for that person to dwell on the past, the present predicament, and on future problems. The survivor must be aware of the necessity to keep the mind and body active to eliminate the feeling of depression. One way to keep busy (daily) is by checking and improving shelters, signals, and food supply.

Chapter 5

EMOTIONAL REACTIONS

5-1. Introduction. Survivors may depend more upon their emotional reactions to a situation than upon calm, careful analysis of potential danger—the enemy, the weather, the terrain, the nature of the in-flight emergency, etc. Whether they will panic from fear, or use it as a stimulant for greater sharpness, is more dependent on the survivor's reactions to the situation than on the situation itself. Although there are many reactions to stress, the following are the most common and will be discussed in detail: fear, anxiety, panic, hate, resentment, anger, impatience, dependency, loneliness, boredom, and hopelessness.

5-2. Fear. Fear can SAVE A LIFE—or it can COST ONE. Some people are at their best when they are scared. Many downed fliers faced with survival emergencies have been surprised at how well they remembered their training, how quickly they could think and react, and what strength they had. The experience gave them a new confidence in themselves. On the other hand, some people become paralyzed when faced with the simplest survival situation. Some of them have been able "to snap themselves out of it" before it was too late. In other cases, a fellow aircrew member was on hand to assist them. However, others have not been so fortunate. They are not listed among the survivors (figure 5-1).

Figure 5-1. Fear.

a. How a person will react to fear depends more upon the individual than it does upon the situation. This has been demonstrated both in actual survival situations and in laboratory experiments. It isn't always the physically strong or the happy-go-lucky people who handle fear most effectively. Timid and anxious people have met emergencies with remarkable coolness and strength.

b. Anyone who faces life-threatening emergencies experiences fear. Fear is conscious when it results from a recognized situation (such as an immediate prospect of bailout) or when experienced as apprehension of impending disaster. Fear also occurs at a subconscious level and creates feelings of uneasiness, general discomfort, worry or depression. Fear may vary widely in intensity, duration, and frequency of occurrence, and affect behavior across the spectrum from mild uneasiness to complete disorganization and panic. People have many fears; some are learned through personal experiences, and others are deliberately taught to them. Fear in children is directed through negative learning, as they are taught to be afraid of the dark, of animals, of noise, or of teachers. These fears may control behavior, and a survivor may react to feelings and imagination rather than to the problem causing fear.

c. When fantasy distorts a moderate danger into a major catastrophe, or vice versa, behavior can become abnormal. There is a general tendency to underestimate and this leads to reckless, foolhardy behavior. The principal means of fighting fear (in this case) is to pretend that it does not exist. There are no sharp lines between recklessness and bravery. It is necessary to check behavior constantly to maintain proper control.

d. One or more of the following signs or symptoms may occur in those who are afraid. However, they may also appear in circumstances other than fear.

(1) Quickening of pulse; trembling.

(2) Dilation of pupils.

(3) Increased muscular tension and fatigue.

(4) Perspiration of palms of hands, soles of feet, and armpits.

(5) Dryness of mouth and throat; higher pitch of voice; stammering.

(6) Feeling of "butterflies in the stomach," emptiness of the stomach, faintness, and nausea.

e. Accompanying these physical symptoms are the following common psychological symptoms:

(1) Irritability; increased hostility.

(2) Talkativeness in early stages, leading finally to speechlessness.

(3) Confusion, forgetfulness, and inability to concentrate.

(4) Feelings of unreality, flight, panic, or stupor.

f. Throughout military history, many people have coped successfully with the most strenuous odds. In adapting to fear, they have found support in previous training and experience. There is no limit to human control of fear. Survivors must take action to control fear. They cannot run away from fear. Appropriate actions should be to:

(1) Understand fear.

(2) Admit that it exists.

(3) Accept fear as reality.

g. Training can help survivors recognize what individual reactions may be. Using prior training, survivors should learn to think, plan, and act logically, even when afraid.

h. To effectively cope with fear, a survivor must:

(1) Develop confidence. Use training opportunities; increase capabilities by keeping physically and mentally fit; know what equipment is available and how to use it; learn as much as possible about all aspects of survival.

(2) Be prepared. Accept the possibility that "it can happen to me." Be properly equipped and clothed at all times; have a plan ready. Hope for the best, but be prepared to cope with the worst.

(3) Keep informed. Listen carefully and pay attention to all briefings. Know when danger threatens and be prepared if it comes; increase knowledge of survival environments to reduce the "unknown."

(4) Keep busy at all times. Prevent hunger, thirst, fatigue, idleness, and ignorance about the situation, since these increase fear.

(5) Know how fellow crewmembers react to fear. Learn to work together in emergencies—to live, work, plan, and help one another as a team.

(6) Practice religion. Don't be ashamed of having spiritual faith.

(7) Cultivate "good" survival attitudes. Keep the mind on a main goal and keep everything else in perspective. Learn to tolerate discomfort. Don't exert energy to satisfy minor desires which may conflict with the overall goal—to survive.

(8) Cultivate mutual support. The greatest support under severe stress may come from a tightly knit group. Teamwork reduces fear while making the efforts of every person more effective.

(9) Exercise leadership. The most important test of leadership and perhaps its greatest value lies in the stress situation.

(10) Practice discipline. Attitudes and habits of discipline developed in training carry over into other situations. A disciplined group has a better chance of survival than an undisciplined group.

(11) Lead by example. Calm behavior and demonstration of control are contagious. Both reduce fear and inspire courage.

i. Every person has goals and desires. The greatest values exercise the greatest influence. Because of strong religious, moral or patriotic values, people have been known to face torture and death calmly rather than reveal information or compromise a principle. Fear can kill or it can save lives. It is a normal reaction to danger. By understanding and controlling fear through training,

knowledge, and effective group action, fear can be overcome.

5-3. Anxiety:

a. Anxiety is a universal human reaction. Its presence can be felt when changes occur which affect an individual's safety, plans, or methods of living. It is generally felt when individuals perceive something bad is about to happen. A common description of anxiety is "butterflies in the stomach." Anxiety creates feelings of uneasiness, general discomfort, worry, or depression. Anxiety and fear differ mainly in intensity. Anxiety is a milder reaction and the specific cause(s) may not be readily apparent, whereas fear is a strong reaction to a specific, known cause. Common characteristics of anxiety are: fear of the future, indecision, feelings of helplessness, resentment (figure 5-2).

Figure 5-2. Anxiety.

b. To overcome anxiety, the individual must take positive action by adopting a simple plan. It is essential to keep your mind off of your injuries and do something constructive. For example, one PW began to try and teach English to the Chinese and to learn Chinese from them.

5-4. Panic.
In the face of danger, a person may panic or "freeze" and cease to function in an organized manner. A person experiencing panic may have no conscious control over individual actions. Uncontrollable, irrational behavior is common in emergency situations. Anybody can panic, but some people go to pieces more easily than others. Panic is brought on by a sudden overwhelming fear, and can often spread quickly through a group of people. Every effort must be made

to bolster morale and calm the panic with leadership and discipline. Panic has the same signs as fear and should be controlled in the same manner as fear. This survivor allowed pain to panic him:

"His parachute caught in the tree, and he found himself suspended about five feet above the ground...one leg strap was released while he balanced in this aerial position, and he immediately slipped toward the ground. In doing so,

Figure 5-3. Panic.

his left leg caught in the webbing and he was suspended by one leg with his head down. Unfortunately, the pilot's head touched an anthill and biting ants immediately swarmed over him. Apparently, in desperation, the flier pulled his gun and fired five rounds into the webbing holding his foot. When he did not succeed in breaking the harness by shooting at it, he placed the last shot in his head and thus took his own life. It was obvious from the discover-

er's report that if the pilot had even tried to turn around or to swing himself from his inverted position, he could have reached either the aerial roots or the latticed trunk of the tree. With these branches, he should have been able to pull himself from the harness...The fact that his head was in a nest of stinging ants only added to his panic, which led to the action that took his life." (See figure 5-3.)

5-5. Hate. Hate—feelings of intense dislike, extreme aversion, or hostility—is a powerful emotion which can have both positive and negative effects on a survivor. An understanding of the emotion and its causes is the key to learning to control it. Hate is an acquired emotion rooted in a person's knowledge or perceptions. The accuracy or inaccuracy of the information is irrelevant to learning to hate.

a. Any person, any object, or anything that may be understood intellectually, such as political concepts or religious dogma, can promote feelings of hate. Feelings of hate (usually accompanied with a desire for vengeance, revenge, or retribution) have sustained former prisoners of war through their harsh ordeals. If an individual loses perspective while under the influence of hate and reacts emotionally, rational solutions to problems may be overlooked, and the survivor may be endangered.

b. To effectively deal with this emotional reaction, the survivor must first examine the reasons why the feeling of hate is present. Once that has been determined, survivors should then decide what to do about those feelings. Whatever approach is selected, it should be as constructive as possible. Survivors must not allow hate to control them.

5-6. Resentment. Resentment is the experiencing of an emotional state of displeasure or indignation toward some act, remark, or person that has been regarded as causing personal insult or injury. Luck and fate may play a role in any survival situation. A hapless survivor may feel jealous resentment toward a fellow PW, travel partner, etc., if that other person is perceived to be enjoying a success or advantage not presently experienced by the observer. The survivor must understand that events cannot always go as expected. It is detrimental to morale and could affect survival chances if feelings of resentment over another's attainments become too strong. Imagined slights or insults are common. The survivor should try to maintain a sense of humor and perspective about ongoing events and realize that stress and lack of self-confidence play roles in bringing on feelings of resentment.

5-7. Anger. Anger is a strong feeling of displeasure and belligerence aroused by a real or supposed wrong. People become angry when they cannot fulfill a basic need or desire that seems important to them. When anger is not relieved, it may turn into a more enduring attitude of hostility, characterized by a desire to hurt or destroy the person or thing causing the frustration. When anger is intense, the survivor loses control over the situation, resulting in impulsive behavior which may be destructive in nature. Anger is a normal response which can serve a useful purpose when carefully controlled. If the situation warrants and there is no threat to survival, one could yell or scream, take a walk, do some vigorous exercise, or just get away from the source of the anger, even if only for a few minutes. Here is a man who couldn't hold it:

"I tried patiently to operate it (the radio) in every way I had been shown. Growing more angry and disappointed at its failure, I tore the aerial off, threw the cord away, beat the battery on the rocks, then threw the pieces all over the hillside. I was sure disappointed." (See figure 5-4.)

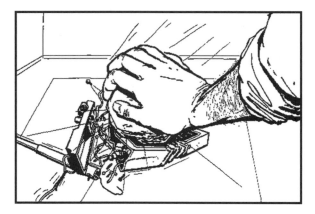

Figure 5-4. Anger.

5-8. Impatience:

a. The psychological stresses brought about by feelings of impatience can quickly manifest themselves in physical ways. Internally, the effects of impatience can cause changes in physical and mental well-being. Survivors who allow impatience to control their behavior may find that their efforts prove to be counterproductive and possibly dangerous. For example, evaders who don't have the ability or willingness to suppress annoyance when confronted with delay may expose themselves to capture or injury.

b. Potential survivors must understand they have to bear pain, misfortune, and annoyance without complaint. In the past, many survivors have displayed tremendous endurance, both mental and physical, in times of distress or misfortune. While not every survivor will be able to

display such strength of character in all situations, each person should learn to recognize the things which may make them impatient to avoid acting unwisely. This survivor couldn't wait:

"I became very impatient. I had planned to wait until night to travel but I just couldn't wait. I left the ditch about noon and walked for about two hours until I was caught."

5-9. Dependence. The captivity environment is the prime area where a survivor may experience feelings of dependency. The captor will try to develop in prisoners feelings of need, support, and trust for the captor. By regulating the availability of basic needs like food, water, clothing, social contact, and medical care, captors show their power and control over the prisoners' fate. Through emphasis on the prisoners' inability to meet their own basic needs, captors seek to establish strong feelings of prisoner dependency. This dependency can make prisoners extremely vulnerable to captor exploitation—a major captor objective. PW recognition of this captor tactic is key to countering it. Survivors must understand that, despite captor controls, they do control their own lives. Meeting even one physical or mental need can provide a PW with a "victory" and provide the foundation for continued resistance against exploitation (figure 5-5).

Figure 5-5. Dependence.

5-10. Loneliness. Loneliness can be very debilitating during a survival episode. Some people learn to control and manipulate their environment and become more self-sufficient while adapting to changes. Others rely on protective persons, routines, and familiarity of surroundings to function and obtain satisfaction (figure 5-6).

a. The ability to combat feelings of loneliness during a survival episode must be developed long before the episode occurs. Self-confidence and self-sufficiency are key factors in coping with loneliness. People develop these attributes by developing and demonstrating competence in performing tasks. As the degree of competence increases, so does self-confidence and self-sufficiency. Military training, more specifically survival training, is designed to provide individuals with the competence and self-sufficiency to cope with and adapt to survival living.

with efficient performance of the job. The ungratifying nature of a task can be counteracted by clearing up its meaning, objectives, and, in some cases, its relation to the total plan.

a. This survivor couldn't think of anything to do:
"The underground representative took me to a house to wait for another member of the underground to pick me up. This was the worst part of the whole experience—this waiting. I just sat in the house and waited for two weeks. I thought I would go mad." (See figure 5-7.)

b. This survivor invented something to do:
"Not knowing what to do, I decided to kill all the bugs. There were a lot of spiders, the big ones that do not hurt humans, so I killed the flies and gave them to the spiders to eat."

Figure 5-6. Loneliness.

Figure 5-7. Boredom.

b. In a survival situation, the countermeasure to conquer loneliness is to be active, to plan and think purposely. Development of self-sufficiency is the primary protection since all countermeasures in survival require the survivor to have the ability to practice self-control.

6-11. Boredom. Boredom and fatigue are related and frequently confused. Boredom is accompanied by a lack of interest and may include feelings of strain, anxiety, or depression, particularly when no relief is in sight and the person is frustrated. Relief from boredom must be based on correction of the two basic sources, repetitiveness and uniformity. Boredom can be relieved by a variety of methods—rotation of duties, broadening the scope of a particular task or job, taking rest breaks, or other techniques of diversification which may actually interfere

5-12. Hopelessness. Hopelessness stems from negative feelings—regardless of actions taken, success is impossible—or the certainty that future events will turn out for the worst no matter what a person tries to do. Feelings of hopelessness can occur at virtually any time during a survival episode. Survivors have experienced loss of hope in trying to maintain health due to an inability to care for sickness, broken bones, or injuries; considering their chances of returning home alive; seeing their loved ones again; or believing in their physical or mental ability to deal with the situation, for example, evade long distances or not give information to an interrogator (figure 5-8).

Figure 5-8. Hopelessness.

a. During situations in which physical exhaustion or exposure to the elements affects the mind, a person may begin to lose hope. The term "give-up-itis" was coined in Korea to describe the feeling of "hopelessness." During captivity, deaths occurred for no apparent reason. These individuals actually willed themselves to die or at least did not will themselves to live. The original premise (in the minds of such people) is that they are going to die. To them, the situation seemed totally futile and they had passively abandoned themselves to fate. It was possible to follow the process step by step. The people who died withdrew themselves from the group, became despondent, then lay down and gave up. In some cases, death followed rapidly.

b. One way to treat hopelessness is to eliminate the cause of the stress. Rest, comfort, and morale-building activities can help eliminate this psychological problem. Another method used in Korea was to make the person so angry that the person wanted to get up and attack the tormentors. A positive attitude has a powerful influence on morale and combating the feeling of hopelessness.

c. Since many stress situations cannot be dealt with successfully by either withdrawal or direct attack, it may be necessary to work out a compromise solution. The action may entail changing a survivor's method of operation or accepting substitute goals.

d. Evaders faced with starvation may compromise with their conscience and steal "just this one time." They may ignore their food aversion and eat worms, bugs, or even human flesh. A related form of compromise is acceptance of substitute means to achieve the same goals.

5-13. Summary. All the psychological factors may be overcome by survivors if they can recognize the problem, work out alternative solutions, decide on an appropriate course of action, take action, and evaluate the results. Perhaps the most difficult step in this sequence is deciding on an appropriate course of action. Survivors may face either one or several psychological problems. These problems are quite dangerous and must be effectively controlled or countered for survival to continue.

Chapter 6

THE WILL TO SURVIVE

6-1. Introduction. The will to survive is defined as the desire to live despite seemingly insurmountable mental and (or) physical obstacles. The tools for survival are furnished by the military, the individual, and the environment. The training for survival comes from survival training publications, instruction, and the individual's own efforts. But tools and training are not enough without a *will to survive*. In fact, the records prove that "will" alone has been the deciding factor in many survival cases. While these accounts are not classic examples of "how to survive," they illustrate that a single-minded survivor with a powerful *will to survive* can overcome most hardships. There are cases in which people have eaten their belts for nourishment, boiled water in their boots to drink as broth, or have eaten human flesh—though this certainly wasn't their cultural instinct.

a. One incident where the *will to survive* was the deciding factor between life and death involved a man stranded in the Arizona desert for 8 days without food and water. He traveled more than 150 miles during searing daytime temperatures, losing 25 percent of his body weight due to the lack of water (usually 10 percent loss causes death). His blood became so thick that the lacerations he received could not bleed until he had been rescued and received large quantities of water. When he started on that journey, something must have clicked in his mind telling him to live, regardless of any obstacles which might confront him. And live he did—on guts and will alone! (See figure 6-1.)

b. Let's flip a coin and check the other side of "will." Our location is the Canadian wilderness. A pilot ran into engine trouble and chose to deadstick his plane onto a frozen lake rather than punch out. He did a beautiful job and slid to a stop in the middle of the lake. He left the aircraft and examined it for damage. After surveying the area, he noticed a wooded shoreline only 200 yards away where food and shelter could be provided—he decided to go there. Approximately halfway there, he changed his mind and returned to the cockpit of his aircraft where he smoked a cigar, took out his pistol, and blew his brains out. Less than 24 hours later, a rescue team found him. Why did he give up? Why was he unable to survive? Why did he take his own life? On the other hand, why do people eat their belts or drink broth from their boots? No one really knows, but it's all related to the *will to survive*.

6-2. Overcoming Stress. The ability of the mind to overcome stress and hardship becomes most apparent when there appears to be little chance of a person surviving. When there appears to be no escape from the situation, the "will" enables a person to begin to win

Figure 6-1. Will to Survive.

"the battle of the mind." This mental attitude can bridge the gap between the crisis period and the coping period.

6-3. Crisis Period:

a. The crisis period is the point at which the person realizes the gravity of the situation and understands that the problem will not go away. At this stage, action is needed. Most people will experience shock in this stage as a result of not being ready to face this new challenge. Most will recover control of their faculties, especially if they have been prepared through knowledge and training.

b. Shock during a crisis is normally a response to being overcome with anxiety. Thinking will be disorganized. At this stage, direction will be required because the individual is being controlled by the environment. The person's center of control is external. In a group-survival episode, a natural leader may appear who will direct and reassure the others. But if the situation continues to control the individual or the group, the response may be panic, behavior may be irrational, and judgment is impaired. In a lone-survivor episode, the individual must gain control of the situation and respond constructively. In either case, survivors must evaluate the situation and develop a plan of action.

During the evaluation, the survivor must determine the most critical needs to improve the chance of living and being rescued.

6-4. The Coping Period. The coping period begins after the survivor recognizes the gravity of the situation and resolves to endure it rather than succumb. The survivor must tolerate the effects of physical and emotional stresses. These stresses can cause anxiety, which becomes the greatest obstacle to self-control and solving problems. Coping with the situation requires considerable internal control. For example, the survivor must often subdue urgent desires to travel when that would be counterproductive and dangerous. A person must have patience to sit in an emergency-action shelter while confronted with an empty stomach, aching muscles, numb toes, and suppressed feelings of depression and hopelessness Those who fail to think constructively may panic. This could begin a series of mistakes that result in further exhaustion, injury, and sometimes death. Death comes not from hunger pains but from the inability to manage or control emotions and thought processes.

6-5. Attitude. The survivor's attitude is the most important element of the *will to survive*. With the proper attitude, almost anything is possible. The desire to live is sometimes based on the feelings toward another person and (or) thing. Love and hatred are two emotional extremes that have moved people to do exceptional things physically and mentally. The lack of a *will to survive* can sometimes be identified by the individual's motivation to meet essential survival needs, emotional control resulting in reckless, panic-like behavior, and self-esteem.

a. It is essential to strengthen the *will to survive* during an emergency. The first step is to avoid a tendency to panic or "fly off the handle." Sit down, relax, and analyze the situation rationally. Once thoughts are collected and thinking is clear, the next step is to make decisions. In normal living, people can avoid decisions and let others do their planning. But in a survival situation, this will seldom work. Failure to decide on a course of action is

actually a decision for inaction. This lack of decision-making may even result in death. However, decisiveness must be tempered with flexibility and planning for unforeseen circumstances. As an example, an aircrew member down in an arctic nontactical situation decides to construct a shelter for protection from the elements. The planning and actions must allow sufficient flexibility so the aircrew can monitor the area for indications of rescuers and be prepared to make contact visually, electronically, etc.—with potential rescuers.

b. Tolerance is the next topic of concern. A survivor or evader will have to deal with many physical and psychological discomforts, such as unfamiliar animals, insects, loneliness, and depression. Aircrew members are trained to tolerate uncomfortable situations. That training must be applied to deal with the stress of environments.

c. Survivors in both tactical and nontactical situations must face and overcome fears to strengthen the *will to survive*. These fears may be founded or unfounded, be generated by the survivor's uncertainty or lack of confidence, or be based on the proximity of enemy forces. Indeed, fear may be caused by a wide variety of real and imagined dangers. Despite the source of the fear, survivors must recognize fear and make a conscious effort to overcome it.

6-6. Optimism. One of a survivor's key assets is optimism—hope and faith. Survivors must maintain a positive, optimistic outlook on their circumstances and how well they are doing. Prayer or meditation can be helpful. How a survivor maintains optimism is not so important as its use.

6-7. Summary. Survivors do not choose or welcome their fate and would escape it if they could. They are trapped in a world of seemingly total domination—a world hostile to life and any sign of dignity or resistance. The survival mission is not an easy one, but it is one in which success can be achieved. This has been an introduction to the concepts and ideas that can help an aircrew member return. Having the *will to survive* is what it's all about!

Part Three

BASIC SURVIVAL MEDICINE

Chapter 7

SURVIVAL MEDICINE

7-1. Introduction:

a. Foremost, among the many things that can compromise a survivor's ability to return are medical problems encountered during ejection, parachute descent, and (or) parachute landing. In the Southeast Asian conflict, some 30 percent of approximately 1,000 US Air Force survivors, including 322 returned PWs, were injured by the time they disentangled themselves from their parachutes. The most frequently reported injuries were fractures, strains, sprains, and dislocations, as well as burns and other types of wounds (figure 7-1).

Figure 7-1. Survival Medicine.

b. Injuries and illnesses peculiar to certain environments can reduce survival expectancy. In cold climates, and often in an open-sea survival situation, exposure to extreme cold can produce serious tissue trauma, such as frostbite, or death from hypothermia. Exposure to heat in warm climates, and in certain areas on the open seas, can produce heat cramps, heat exhaustion, or life-threatening heatstroke.

c. Illnesses contracted during evasion or in a captivity environment can interfere with successful survival. Among these are gastrointestinal disorders, respiratory diseases, skin infections and infestations, malaria, typhus, cholera, etc.

d. A review of the survival experiences from World War II, Korea, and Southeast Asia indicates that, while US military personnel generally knew how to administer first aid to others, there was a marked inability to administer self-aid. Furthermore, only the most basic medical care had been taught to most military people. Lastly, it was repeatedly emphasized that even minor injuries or ailments, when ignored, became major problems in a survival situation. Thus, prompt attention to the most minor medical problem is essential in a survival episode. Applying principles of survival medicine should enable military members to maintain health and well-being in a hostile or nonhostile environment until rescued and returned to friendly control.

e. Information in this chapter and chapter 8 is a basic reference to self-aid techniques used by PWs in captivity and techniques found in folk medicine. The information describes procedures that can maintain health in medically austere situations. It includes items used to prevent and treat injuries and illnesses. Because there is no "typical" survival situation, the approach to self-aid must be flexible, placing emphasis on using what is available to treat the injury or illness. Further, survivors recognize that medical treatment offered by people of other cultures may be far different from our own. For example, in the rural areas of Vietnam, a poultice of python meat was and is used to treat internal lower-back pain. Such treatment may be repugnant to some US military personnel; however, medical aid offered to survivors in non-US cultures may be the best available in the given circumstance.

f. The procedures in this chapter and chapter 8 must be viewed in the reality of a true survival situation. The results of treatment may be substandard compared with present medical standards. However, these procedures will not compromise professional medical care that becomes available following rescue. Moreover, in the context of a survival situation, they may represent the best available treatment to extend the individual's survival expectancy.

7-2. Procedures and Expedients. Survival medicine encompasses procedures and expedients that are:

a. Required and available for the preservation of health and the prevention, improvement, or treatment of injuries and illnesses encountered during survival.

b. Suitable for application by nonmedical personnel to themselves or comrades in the circumstances of the survival situation.

(1) Survival medicine is more than first aid in the conventional sense. It approaches final definitive treatment in that it is not dependent upon the availability of technical medical assistance within a reasonable period of time.

(2) To avoid duplication of information generally available, the basic principles of first aid will not be repeated, nor will the psychological factors affecting survival that were covered in part two.

7-3. Hygiene. In a survival situation, cleanliness is essential to prevent infection. Adequate personal cleanliness will not only protect against disease germs that are present in the individual's surroundings but will also protect the group by reducing the spread of these germs (figure 7-2).

Figure 7-2. Hygiene.

a. Washing, particularly the face, hands, and feet, reduces the chances of infection from small scratches and abrasions. A daily bath or shower with hot water and soap is ideal. If no tub or shower is available, the body should be cleaned with a cloth and soapy water, paying particular attention to the body creases (armpits, groin, etc.), face, ears, hands, and feet. After this type of "bath," the body should be rinsed thoroughly with clear water to remove all traces of soap, which could cause irritation.

b. Soap, although an aid, is not essential to keeping clean. Ashes, sand, loamy soil, and other expedients may be used to clean the body and cooking utensils.

c. When water is in short supply, the survivor should take an "air bath." All clothing should be removed and the body simply exposed to the air. Exposure to sunshine is ideal, but even on an overcast day or indoors, a 2-hour exposure of the naked body to the air will refresh the body. Care should be taken to avoid sunburn when bathing in this manner. Exposure in the shade, shelter, sleeping bag, etc., will help if the weather conditions do not permit direct exposure.

d. Hair should be kept trimmed, preferably 2 inches or less in length, and the face should be clean-shaven. Hair provides a surface for the attachment of parasites and the growth of bacteria. Keeping the hair short and the face clean-shaven will provide less habitat for these organisms. At least once a week, the hair should be washed with soap and water. When water is in short supply, the hair should be combed or brushed thoroughly and covered to keep it clean. It should be inspected weekly for fleas, lice, and other parasites. When parasites are discovered, they should be removed.

e. The principal means of infecting food and open wounds is contact with unclean hands. Hands should be washed with soap and water, if available, after handling any material that is likely to carry germs. This is especially important after each visit to the latrine, when caring for the sick and injured, and before handling food, food utensils, or drinking water. The fingers should be kept out of the mouth and the fingernails kept closely trimmed and clean. A scratch from a long fingernail could develop into a serious infection.

7-4. Care of the Mouth and Teeth. Application of the following fundamentals of oral hygiene will prevent tooth decay and gum disease:

a. The mouth and teeth should be cleansed thoroughly with a toothbrush and dentifrice at least once each day. When a toothbrush is not available, a "chewing stick" can be fashioned from a twig. The twig is washed, then chewed on one end until it is frayed and brushlike. The teeth can then be brushed very thoroughly with the stick, taking care to clean all tooth surfaces. If necessary, a clean strip of cloth can be wrapped around the finger and rubbed on the teeth to wipe away food particles that have collected on them. When neither toothpaste nor toothpowder are available, salt, soap, or baking soda can be used as substitute dentifrices. Parachute inner core can be used by separating the filaments of the inner core and using this as a dental floss. Gargling with willow bark tea will help protect the teeth.

b. Food debris that has accumulated between the teeth should be removed by using dental floss or toothpicks. The latter can be fashioned from small twigs.

c. Gum tissues should be stimulated by rubbing them vigorously with a clean finger each day.

d. Use as much care cleaning dentures and other dental appliances, removable or fixed, as when cleaning natural teeth. Dentures and removable bridges should be removed and cleaned with a denture brush or "chew stick" at least once each day. The tissue under the dentures should be brushed or rubbed regularly for proper stimulation. Removable dental appliances should be removed at night or for a 2- to 3-hour period during the day.

7-5. Care of the Feet. Proper care of the feet is of utmost importance in a survival situation, especially if the survivor has to travel. Serious foot trouble can be prevented by observing the following simple rules:

a. The feet should be washed, dried thoroughly, and massaged each day. If water is in short supply, the feet should be "air cleaned" along with the rest of the body (figure 7-3).

b. Toenails should be trimmed straight across to prevent the development of ingrown toenails.

c. Boots should be broken in before wearing them on any mission. They should fit properly, neither so tight that they bind and cause pressure spots nor so loose that they permit the foot to slide forward and backward when walking. Insoles should be improvised to reduce any friction spots inside the shoes.

Figure 7-3. Care of Feet.

d. Socks should be large enough to allow the toes to move freely but not so loose that they wrinkle. Wool socks should be at least one size larger than cotton socks to allow for shrinkage. Socks with holes should be properly darned before they are worn. Wearing socks with holes or socks that are poorly repaired may cause blisters. Clots of wool on the inside and outside should be removed from wool socks because they may cause blisters. Socks should be changed and washed thoroughly with soap and water each day. Woolen socks should be washed in cool water to lessen shrinkage. In camp, freshly laundered socks should be stretched to facilitate drying by hanging in the sun or in an air current. While traveling, a damp pair of socks can be dried by placing them inside layers of clothing or hanging them on the outside of the pack. If socks become damp, they should be exchanged for dry ones at the first opportunity.

e. When traveling, the feet should be examined regularly to see whether there are any red spots or blisters. If detected in the early stages of development, tender areas should be covered with adhesive tape to prevent blister formation.

7-6. Clothing and Bedding. Clothing and bedding become contaminated with any disease germs that may be present on the skin, in the stool, in the urine, or in secretions of the nose and throat. Therefore, keeping clothing and bedding as clean as possible will decrease the chances of skin infection and decrease the possibility of parasite infestation. Outer clothing should be washed with soap and water when it becomes soiled. Underclothing and socks should be changed daily. If water is in short supply, clothing should be "air cleaned." For air cleaning, the clothing is shaken out of doors, then aired and sunned for 2 hours. Clothing cleaned in this manner should be worn in rotation. Sleeping bags should be turned inside out, fluffed, and aired after each use. Bed linen should be changed at least once a week, and the blankets, pillows, and mattresses should be aired and sunned (figure 7-4).

7-7. Rest. Rest is necessary for the survivor because it not only restores physical and mental vigor, but also promotes healing during an illness or after an injury.

a. In the initial stage of the survival episode, rest is particularly important. After those tasks requiring immediate attention are done, the survivor should inventory available resources, decide upon a plan of action, and even have a meal. This "planning session" will provide a rest period without the survivor having a feeling of "doing nothing."

b. If possible, regular rest periods should be planned in each day's activities. The amount of time allotted for rest will depend on a number of factors, including the survivor's physical condition, the presence of hostile forces, etc., but usually, 10 minutes each hour is sufficient. During these rest periods, the survivor should change either from physical activity to complete rest or from mental activity to physical activity as the case may be. The survivor must learn to become comfortable and to rest under less than ideal conditions.

Figure 7-4. Bedding.

7-8. Rules for Avoiding Illness. In a survival situation, whether short-term or long-term, the dangers of disease are multiplied. Application of the following simple guidelines regarding personal hygiene will enable the survivor to safeguard personal health and the health of others:

a. ALL water obtained from natural sources should be purified before consumption.

b. The ground in the camp area should not be soiled with urine or feces. Latrines should be used, if available. When no latrines are available, individuals should dig "cat holes" and cover their waste.

c. Fingers and other contaminated objects should never be put into the mouth. Hands should be washed before handling any food or drinking water, before using the fingers in the care of the mouth and teeth, before and after caring for the sick and injured, and after handling any material likely to carry disease germs.

d. After each meal, all eating utensils should be cleaned and disinfected in boiling water.

e. The mouth and teeth should be cleansed thoroughly at least once each day. Most dental problems associated with long-term survival episodes can be prevented by using a toothbrush and toothpaste to remove accumulated food debris. If necessary, devices for cleaning the teeth should be improvised.

f. Bites and insects can be avoided by keeping the body clean, by wearing proper protective clothing, and by using a head net, improvised bed nets, and insect repellents.

g. Wet clothing should be exchanged for dry clothing as soon as possible to avoid unnecessary body-heat loss.

h. Personal items such as canteens, pipes, towels,

toothbrushes, handkerchiefs, and shaving items should not be shared with others.

i. All food scraps, cans, and refuse should be removed from the camp area and buried.

j. If possible, a survivor should get 7 or 8 hours of sleep each night.

k. Aircrew members should keep all immunization shots current.

7-9. General Management of Injuries:

a. Bleeding. Control of bleeding is most important in survival situations where replacement transfusions are not possible. Immediate steps should be taken to stop the flow of blood, regardless of its source. The method used should be commensurate with the type and degree of bleeding. The tourniquet, when required and properly used, will save a life. If improperly used, it may cost the life of the survivor. The basic characteristics of a tourniquet and the methods of its use are well covered in standard first aid texts; however, certain points merit emphasis in the survival situation. A tourniquet should be used only after every alternate method has been attempted. If unable to get to medical aid within 2 hours, after 20 minutes, gradually loosen the tourniquet. If bleeding has stopped, remove the tourniquet; if bleeding continues, reapply and leave in place. The tourniquet should be applied as near to the site of the bleeding as possible, between the wound and the heart, to reduce the amount of tissue lost.

b. Pain:

(1) Control of Pain. The control of pain accompanying disease or injury under survival situations is both difficult and essential. In addition to its morale-breaking discomfort, pain contributes to shock and makes the survivor more vulnerable to enemy influences. Ideally, pain should be eliminated by the removal of the cause. However, this is not always immediately possible, hence measures for the control of pain are beneficial.

(2) Position, Heat, and Cold. The part of the body that is hurting should be put at rest, or at least its activity restricted as much as possible. The position selected should be the one giving the most comfort, and be the easiest to maintain. Splints and bandages may be necessary to maintain the immobilization. Elevation of the injured part, with immobilization, is particularly beneficial in such throbbing type pain as is typical of the "mashed" finger. Open wounds should be cleansed, foreign bodies removed, and a clean dressing applied to protect the wound from the air and chance contacts with environmental objects. Generally, the application of warmth reduces pain—toothache, bursitis, etc. However, in some conditions, application of cold has the same effect—strains and sprains. Warmth or cold is best applied by using water due to its high specific heat, and the survivor can try both to determine which is most beneficial.

hes. With companions, the use of improvised litters
be possible.

(4) Reduction of dislocated joints is done similar to
of fractures. Gentle, but firm, traction is applied and
xtremity is manipulated until it "snaps" back into
:. If the survivor is alone, the problem is complicated
iot impossible. Traction can still be applied by using
ity. The distal portion of the extremity is tied to (or
ged) into the fork of a tree or similar point of fixa-
. The weight of the body is then allowed to exert the
:ssary traction, with the joint being manipulated until
dislocation is reduced.

. Infection:

(1) Infection is a serious threat to the survivor.
inevitable delay in definite medical treatment and
reality of the survival situation increases the chances
vound infection. Antibiotics may not be available in
icient amounts in the survival situation. In survival
dicine, one must place more emphasis on the preven-
i and control of infection by applying techniques used
ore the advent of antibiotics.

(2) Unfortunately, survivors have little control over
amount and type of infection introduced at the time
injury. However, they can help control the infection
wearing clean clothes. Use care to prevent additional
ection of wounds. Wounds, regardless of the type
severity, should not be touched with dirty hands or
jects. One exception to this rule is the essential control
arterial bleeding. Clothing should be removed from
ounds to avoid contamination surrounding skin areas.

(3) All wounds should be promptly cleansed. Water
s the most universally available cleaning agent, and
should be (preferably) sterile. At sea level, sterilize water
by placing it in a covered container and boiling it for 10
minutes. Above 3,000 feet, water should be boiled for
1 hour (in a covered container) to ensure adequate sterilization. The water will remain sterile and can be stored
indefinitely as long as it is covered.

(a) Irrigate wounds rather than scrubbing to minimize additional damage to the tissue. Foreign material
should be washed from the wound to remove sources of
continued infection. The skin adjacent to wounds should
be washed thoroughly before bandaging. When water is
not available for cleaning wounds, the survivor should
consider the use of urine. Urine may well be the most
nearly sterile of all fluids available and, in some cultures,
is preferred for cleaning wounds. Survivors should use
urine from the midstream of the urine flow.

(b) While soap is not essential to clean wounds, a
bar of medicated soap placed in a personal survival kit
and used routinely would do much to prevent the infection of seemingly inconsequential injuries. External antiseptics are best used for cleaning abrasions, scratches,
and the skin areas adjacent to lacerations. Used in deep,
larger wounds, antiseptics produce further tissue damage.

(c) Nature also provides antiseptics that can be
used for wound care. The American mountain ash is
found from Newfoundland south to North Carolina, and
its inner bark has antiseptic properties. The red berries
contain ascorbic acid and have been eaten to cure scurvy.
The Sweet Gum bark is still officially recognized as
being an antiseptic agent. Water from boiled Sweet Gum
leaves can also be used as antiseptic for wounds.

f. The "Open Treatment" Method. This is the only
safe way to manage survival wounds. No effort should
be made to close open wounds by suturing or by other
procedure. In fact, it may be necessary to open the wound
even more to avoid entrapment or infection and to promote drainage. The term "open" does not mean that
dressings should not be used. Good surgery requires that
although wounds are not "closed," nerves, bone, and
blood vessels should be covered with tissue. Such judgment may be beyond the capability of the aircrew member, but protection of vital structures will aid in the
recovery and ultimate function. A notable exception to
"open treatment" is the early closure of facial wounds
that interfere with breathing, eating, or drinking. Wounds,
left open, heal by formation of infection-resistant granulation tissue (proud flesh). This tissue is easily recognized by its moist red granular appearance, a good sign
in any wound.

g. Dressings and Bandages. After cleansing, all
wounds should be covered with a clean dressing. The
dressing should be sterile; however, in the survival situation, any clean cloth will help to protect the wound from
further infection A proper bandage will anchor the dressing to the wound and afford further protection. Bandages
should be snug enough to prevent slippage, yet not constrictive. Slight pressure will reduce discomfort in most
wounds and help stop bleeding. Once in place, dressings
should not be changed too frequently unless required.
External soiling does not reduce the effectiveness of a
dressing, and pain and some tissue damage will accompany any removal. In addition, changing dressings increases
the danger of infection.

h. Physiological "Logistics." Despite all precautions, some degree of infection is almost universal in survival wounds. This is the primary reason for the "open"
treatment advocated above. The human body has a
tremendous capacity for combating infections if it is permitted to do so. The importance of proper rest and nutrition to wound healing and control of infection has been
mentioned. In addition, the "logistics" of the injured part
should be improved. The injury should be immobilized in
a position to favor adequate circulation, both to and from
the wound. Avoid constrictive clothing or bandages.
Applying heat to an infected wound further aids in mobilizing local body-defense measures. Lukewarm saltwater
soaks will help draw out infection and promote oozing of
fluids from the wound, thereby removing toxic products.

(3) Painkillers. Drugs are very effective in reducing pain; however, they are not likely to be available in the survival situation. Hence, the importance of the above "natural" procedures. Aspirin, APCs, and such tablets are primarily intended to combat the discomforts of colds and upper-respiratory diseases, and, at best, will just take the edge off severe pain. They should be taken, however, if available. If no aspirin is available, there are some parts of vegetation that can be used. For example, most of the willows have been used for their pain-relieving and fever-lowering properties for hundreds of years. The fresh bark contains salicin, which probably decomposes into salicyclic acid in the human body. Wintergreen, also known as checkerberry, was used by some American Indians for body aches and pains. The leaves are made into a tea. The boiled bark of the magnolia tree helps relieve internal pains and fever, and has been known to stop dysentery. To be really effective in control of pain, stronger narcotic drugs such as codeine and morphine are required. During active hostilities, morphine may be available in aircraft and individual first aid kits.

c. Shock:

(1) Circulatory Reaction. Shock in some degree accompanies all injuries to the body, and frequently it is the most serious consequence of the injury. In essence, shock is a circulatory reaction of the body (as a whole) to an injury (mechanical or emotional). While the changes to the circulatory system initially favor body resistance to the injury (by ensuring adequate blood supply to vital structures), they may progress to the point of circulatory failure and death. All aircrew members should be familiar with the signs and symptoms of shock so that the condition may be anticipated, recognized, and dealt with effectively. However, the best survival approach is to treat ALL moderate and severe injuries for shock. No harm will be done, and such treatment will speed recovery.

(2) Fluids. Normally, fluids administered by mouth are generally prohibited in the treatment of shock following severe injury. Such fluids are poorly absorbed when given by mouth, and they may interfere with later administration of anesthesia for surgery. In survival medicine, however, the situation is different in that the treatment being given is the final treatment. Survivors cannot be deprived of water for long periods just because they have been injured; in fact, their recovery depends upon adequate hydration. Small amounts of warm water, warm tea, or warm coffee given frequently early in shock are beneficial if the patient is conscious, can swallow, and has no internal injuries. In later shock, fluids by mouth are less effective as they are not absorbed from the intestines. Burns, particularly, require large amounts of water to replace fluid lost from injured areas. Alcohol should never be given to a person in shock or who may go into shock.

(3) Psychogenic Shock. Psychogenic s____ quently noted during the period immediatel___ an emergency; for example, bailout. Psycho____ which occurs even without injury, requires a____ limit it, both in degree and duration. The deg___ post-impact shock varies widely among indi____ occurrence is almost universal. In reality, the____ passed through two major emergencies almo____ ously; the aircraft incident leading to the surv____ and the situation itself. Should the survivor be____ the majority of them are), a third emergency i____ not uncommon, then, that some psychogenic r____ circulatory implications occurs. Resistance to____ shock depends upon the individual's personalit____ amount of training previously received. Treatm___ of stopping all activities (when possible), relaxi___ ing the situation, and formulating a plan of acti___ the survival situation begins.

d. Fractures:

(1) Proper immobilization of fractures, dis____ and sprains is even more important in survival ____ than in conventional first aid. Rather than merel____ the patient comfortable during transport to even____ ment, in survival medicine, the initial immobiliz___ part of the ultimate treatment. Immobilizing bod____ to help control pain was discussed earlier. In add___ immobilization in proper position hastens healing____ tures and improves the ultimate functional result.____ survival situation, the immobilization must suffice____ relatively long period of time and permit the patie___ maintain a fairly high degree of mobility. Materials____ splinting and bandaging are available in most sur____ situations, and proper techniques are detailed in mo___ first aid manuals.

(2) The reduction of fractures is normally beyo___ the scope of first aid; however, in the prolonged sur___ situation, the correction of bone deformities is neces___ to hasten healing and obtain the greatest functional____ result. The best time for manipulation of a fracture i___ in the period immediately following the injury, before____ painful muscle spasms ensue. Traction is applied until____ overriding fragments of bone are brought into line (check against the other limb), and the extremity is firmly immobilized. Frequently, it is advantageous to____ continue traction after reduction to ensure the proper alignment of the bones.

(3) As plaster casts are not available in the survival situation, improvising an immobilization device is necessary. This may be done by using several parallel, pliable willow branches, woven together with vines or parachute lines. Use care so that the extremity is not constricted when swelling follows the injury. In an escape and evasion situation, it may be necessary to preserve the mobility of the survivor after reduction of the fracture. This is difficult in fractures of the lower extremities, although tree limbs may be improvised as

(3) Painkillers. Drugs are very effective in reducing pain; however, they are not likely to be available in the survival situation. Hence, the importance of the above "natural" procedures. Aspirin, APCs, and such tablets are primarily intended to combat the discomforts of colds and upper-respiratory diseases, and, at best, will just take the edge off severe pain. They should be taken, however, if available. If no aspirin is available, there are some parts of vegetation that can be used. For example, most of the willows have been used for their pain-relieving and fever-lowering properties for hundreds of years. The fresh bark contains salicin, which probably decomposes into salicyclic acid in the human body. Wintergreen, also known as checkerberry, was used by some American Indians for body aches and pains. The leaves are made into a tea. The boiled bark of the magnolia tree helps relieve internal pains and fever, and has been known to stop dysentery. To be really effective in control of pain, stronger narcotic drugs such as codeine and morphine are required. During active hostilities, morphine may be available in aircraft and individual first aid kits.

c. Shock:

(1) Circulatory Reaction. Shock in some degree accompanies all injuries to the body, and frequently it is the most serious consequence of the injury. In essence, shock is a circulatory reaction of the body (as a whole) to an injury (mechanical or emotional). While the changes to the circulatory system initially favor body resistance to the injury (by ensuring adequate blood supply to vital structures), they may progress to the point of circulatory failure and death. All aircrew members should be familiar with the signs and symptoms of shock so that the condition may be anticipated, recognized, and dealt with effectively. However, the best survival approach is to treat ALL moderate and severe injuries for shock. No harm will be done, and such treatment will speed recovery.

(2) Fluids. Normally, fluids administered by mouth are generally prohibited in the treatment of shock following severe injury. Such fluids are poorly absorbed when given by mouth, and they may interfere with later administration of anesthesia for surgery. In survival medicine, however, the situation is different in that the treatment being given is the final treatment. Survivors cannot be deprived of water for long periods just because they have been injured; in fact, their recovery depends upon adequate hydration. Small amounts of warm water, warm tea, or warm coffee given frequently early in shock are beneficial if the patient is conscious, can swallow, and has no internal injuries. In later shock, fluids by mouth are less effective as they are not absorbed from the intestines. Burns, particularly, require large amounts of water to replace fluid lost from injured areas. Alcohol should never be given to a person in shock or who may go into shock.

(3) Psychogenic Shock. Psychogenic shock is frequently noted during the period immediately following an emergency; for example, bailout. Psychogenic shock, which occurs even without injury, requires attention to limit it, both in degree and duration. The degree of this post-impact shock varies widely among individuals, but its occurrence is almost universal. In reality, the survivor has passed through two major emergencies almost simultaneously; the aircraft incident leading to the survival situation, and the situation itself. Should the survivor be injured (and the majority of them are), a third emergency is added. It is not uncommon, then, that some psychogenic reaction with circulatory implications occurs. Resistance to this type of shock depends upon the individual's personality and the amount of training previously received. Treatment consists of stopping all activities (when possible), relaxing, evaluating the situation, and formulating a plan of action before the survival situation begins.

d. Fractures:

(1) Proper immobilization of fractures, dislocations, and sprains is even more important in survival medicine than in conventional first aid. Rather than merely making the patient comfortable during transport to eventual treatment, in survival medicine, the initial immobilization is part of the ultimate treatment. Immobilizing body parts to help control pain was discussed earlier. In addition, immobilization in proper position hastens healing of fractures and improves the ultimate functional result. In the survival situation, the immobilization must suffice for a relatively long period of time and permit the patient to maintain a fairly high degree of mobility. Materials for splinting and bandaging are available in most survival situations, and proper techniques are detailed in most first aid manuals.

(2) The reduction of fractures is normally beyond the scope of first aid; however, in the prolonged survival situation, the correction of bone deformities is necessary to hasten healing and obtain the greatest functional result. The best time for manipulation of a fracture is in the period immediately following the injury, before painful muscle spasms ensue. Traction is applied until overriding fragments of bone are brought into line (check against the other limb), and the extremity is firmly immobilized. Frequently, it is advantageous to continue traction after reduction to ensure the proper alignment of the bones.

(3) As plaster casts are not available in the survival situation, improvising an immobilization device is necessary. This may be done by using several parallel, pliable willow branches, woven together with vines or parachute lines. Use care so that the extremity is not constricted when swelling follows the injury. In an escape and evasion situation, it may be necessary to preserve the mobility of the survivor after reduction of the fracture. This is difficult in fractures of the lower extremities, although tree limbs may be improvised as

crutches. With companions, the use of improvised litters may be possible.

(4) Reduction of dislocated joints is done similar to that of fractures. Gentle, but firm, traction is applied and the extremity is manipulated until it "snaps" back into place. If the survivor is alone, the problem is complicated but not impossible. Traction can still be applied by using gravity. The distal portion of the extremity is tied to (or wedged) into the fork of a tree or similar point of fixation. The weight of the body is then allowed to exert the necessary traction, with the joint being manipulated until the dislocation is reduced.

e. Infection:

(1) Infection is a serious threat to the survivor. The inevitable delay in definite medical treatment and the reality of the survival situation increases the chances of wound infection. Antibiotics may not be available in sufficient amounts in the survival situation. In survival medicine, one must place more emphasis on the prevention and control of infection by applying techniques used before the advent of antibiotics.

(2) Unfortunately, survivors have little control over the amount and type of infection introduced at the time of injury. However, they can help control the infection by wearing clean clothes. Use care to prevent additional infection of wounds. Wounds, regardless of the type or severity, should not be touched with dirty hands or objects. One exception to this rule is the essential control of arterial bleeding. Clothing should be removed from wounds to avoid contamination surrounding skin areas.

(3) All wounds should be promptly cleansed. Water is the most universally available cleaning agent, and should be (preferably) sterile. At sea level, sterilize water by placing it in a covered container and boiling it for 10 minutes. Above 3,000 feet, water should be boiled for 1 hour (in a covered container) to ensure adequate sterilization. The water will remain sterile and can be stored indefinitely as long as it is covered.

(a) Irrigate wounds rather than scrubbing to minimize additional damage to the tissue. Foreign material should be washed from the wound to remove sources of continued infection. The skin adjacent to wounds should be washed thoroughly before bandaging. When water is not available for cleaning wounds, the survivor should consider the use of urine. Urine may well be the most nearly sterile of all fluids available and, in some cultures, is preferred for cleaning wounds. Survivors should use urine from the midstream of the urine flow.

(b) While soap is not essential to clean wounds, a bar of medicated soap placed in a personal survival kit and used routinely would do much to prevent the infection of seemingly inconsequential injuries. External antiseptics are best used for cleaning abrasions, scratches, and the skin areas adjacent to lacerations. Used in deep, larger wounds, antiseptics produce further tissue damage.

(c) Nature also provides antiseptics that can be used for wound care. The American mountain ash is found from Newfoundland south to North Carolina, and its inner bark has antiseptic properties. The red berries contain ascorbic acid and have been eaten to cure scurvy. The Sweet Gum bark is still officially recognized as being an antiseptic agent. Water from boiled Sweet Gum leaves can also be used as antiseptic for wounds.

f. The "Open Treatment" Method. This is the only safe way to manage survival wounds. No effort should be made to close open wounds by suturing or by other procedure. In fact, it may be necessary to open the wound even more to avoid entrapment or infection and to promote drainage. The term "open" does not mean that dressings should not be used. Good surgery requires that although wounds are not "closed," nerves, bone, and blood vessels should be covered with tissue. Such judgment may be beyond the capability of the aircrew member, but protection of vital structures will aid in the recovery and ultimate function. A notable exception to "open treatment" is the early closure of facial wounds that interfere with breathing, eating, or drinking. Wounds, left open, heal by formation of infection-resistant granulation tissue (proud flesh). This tissue is easily recognized by its moist red granular appearance, a good sign in any wound.

g. Dressings and Bandages. After cleansing, all wounds should be covered with a clean dressing. The dressing should be sterile; however, in the survival situation, any clean cloth will help to protect the wound from further infection A proper bandage will anchor the dressing to the wound and afford further protection. Bandages should be snug enough to prevent slippage, yet not constrictive. Slight pressure will reduce discomfort in most wounds and help stop bleeding. Once in place, dressings should not be changed too frequently unless required. External soiling does not reduce the effectiveness of a dressing, and pain and some tissue damage will accompany any removal. In addition, changing dressings increases the danger of infection.

h. Physiological "Logistics." Despite all precautions, some degree of infection is almost universal in survival wounds. This is the primary reason for the "open" treatment advocated above. The human body has a tremendous capacity for combating infections if it is permitted to do so. The importance of proper rest and nutrition to wound healing and control of infection has been mentioned. In addition, the "logistics" of the injured part should be improved. The injury should be immobilized in a position to favor adequate circulation, both to and from the wound. Avoid constrictive clothing or bandages. Applying heat to an infected wound further aids in mobilizing local body-defense measures. Lukewarm saltwater soaks will help draw out infection and promote oozing of fluids from the wound, thereby removing toxic products.

Poultices, made of clean clay, shredded bark of most trees, ground grass seed, etc., do the same thing.

i. Drainage. Adequate natural drainage of infected areas promotes healing. Generally, wicks or drains are unnecessary. On occasion, however, it may be better to remove an accumulation of pus (abscess) and insert light, loose packing to ensure continuous drainage. The knife or other instrument used in making the incision for drainage must be sterilized to avoid introducing other types of organisms. The best way to sterilize in the field is with heat, dry or moist.

j. Antibiotics. Antibiotics, when available, should be taken for the control of infection. Consensus is that the drug should be of the so-called "broad spectrum type"; that is, be effective against any micro-organisms rather than specific for just one or two types. The exact amount to be included in survival kits will vary with the drug and basic assumptions as to the number and types of infections to be expected. Remember that antibiotics are potency-dated items (shelf life about 4 years), and including them in survival kits requires kit inspection and drug replacement with active medical stocks.

k. Debridement. (The surgical removal of lacerated, devitalized, or contaminated tissue.) The debridement of severe wounds may be necessary to minimize infection (particularly of the gas gangrene type) and to reduce septic (toxic) shock. In essence, debridement is the removal of foreign material and dead or dying tissue. The procedure requires skill and should only be done by nonmedical personnel in cases of dire emergency. If required, follow these general rules. Dead skin must be cut away. Muscle may be trimmed back to a point where bleeding starts and gross discoloration ceases. Fat that is damaged tends to die and should be cut away. Bone and nerves should be conserved where possible and protected from further damage. Provide ample natural drainage for the potentially infected wound and delay final closure of the wound.

l. Burns:

(1) Burns, frequently encountered in aircraft accidents and subsequent survival episodes, pose serious problems. Burns cause severe pain, increase the probability of shock and infection, and offer an avenue for the loss of considerable body fluids and salts. Direct initial treatment toward relieving pain and preventing infection. Covering the wound with a clean dressing of any type reduces the pain and chance for infection. Further, such protection enhances the mobility of the patient and the capability for performing other vital survival functions. In burns about the face and neck, ensure that the victim has an open airway. If necessary, cricothyroidotemy should be done before the patient develops extreme difficulties. Burns of the face and hands are particularly serious in a survival situation as they interfere with the capability of survivors to meet their own needs. Soaking certain barks (willow, oak, maple) in water soothes and protects burns by astringent action. This is a function of the acid content of the bark used.

(2) Maintenance of body fluids and salts is essential to recover from burns. The only way to administer fluids in a survival situation is by mouth; hence the casualty should ingest sufficient water early before the nausea and vomiting of toxicity intervenes. Consuming the eyes and blood (both cooked) of animals can help restore electrolyte levels if salt tablets are not available. NOTE: The survivor may also pack salt in personal survival kits to replace electrolytes (1/4 teaspoon per quart of water).

m. Lacerations: Lacerations (cuts) are best left open due to the probability of infection. Clean thoroughly, remove foreign material, and apply a protective dressing. Frequently, immobilization will hasten the healing of major lacerations. On occasion (tactical), it may be necessary to close (cover) the wound, despite the danger of infection, in order to control bleeding or increase the mobility of the patient. If a needle is available, thread may be procured from parachute lines, fabric, or clothing, and the wound closed by suturing. If suturing is required, place the stitches individually, and far enough apart to permit drainage of underlying parts. Do not worry about the cosmetic effect, just approximate the tissue. For scalp wounds, hair may be used to close after the wound is cleansed. Infection is less a danger in this area due to the rich blood supply.

n. Head Injuries. Injuries to the head pose additional problems related to brain damage as well as interfering with breathing and eating. Bleeding is more profuse in the face and head area, but infections have more difficulty taking hold. This makes it somewhat safer to close such wounds earlier to maintain function. Cricothyroidotemy may be necessary if breathing becomes difficult due to obstruction of the upper airways. In the event of unconsciousness, watch the patient closely and keep him or her still. Even in the face of mild or impending shock, keep the head level or even slightly elevated if there is reason to expect brain damage. Do not give fluids or morphine to unconscious persons.

o. Abdominal Wounds. Wounds of the abdomen are particularly serious in the survival situation. Such wounds, without immediate and adequate surgery, have an extremely high mortality rate and render patients totally unable to care for themselves. If intestines are not extruded through the wound, a secure bandage should be applied to keep this from occurring. If intestine is extruded, do not replace it due to the almost certain threat of fatal peritonitis. Cover the extruded bowel with a large dressing and keep the dressing wet with any fluid that is fit to drink, or urine. The patient should lie on the back and avoid any motions that increase intra-abdominal pressure, which might extrude more bowel. Keep the survivor in an immobile state, or move on a litter. "Nature" will eventually take care of the problem, either through death, or walling-off of the damaged area.

p. Chest Injuries. Injuries of the chest are common, painful, and disabling. Severe bruises of the chest or fractures of the ribs require that the chest be immobilized to prevent large painful movements of the chest wall. The bandage is applied while the patient deeply exhales. In the survival situation, it may be necessary for survivors to wrap their own chest. This is more difficult but can be done by attaching one end of the long bandage (parachute material) to a tree or other fixed object, holding the other end in the hand, and slowly rolling body toward the tree, keeping enough counterpressure on the bandage to ensure a tight fit.

q. Sucking Chest Wounds. These wounds are easily recognized by the sucking noise and appearance of foam or bubbles in the wound. These wounds must be closed immediately before serious respiratory and circulatory complications occur. Ideally, the patient should attempt to exhale while holding the mouth and nose closed (Valsalva) as the wound is closed. This inflates the lungs and reduces the air trapped in the pleural cavity. Frequently, a taped, airtight dressing is all that is needed, but sometimes it is necessary to put in a stitch or two to make sure the wound is closed.

r. Eye Injuries. Eye injuries are quite serious in a survival situation due to pain and interference with other survival functions. The techniques for removing foreign bodies and for treating snow blindness are covered in standard first aid manuals. More serious eye injuries involving disruption of the contents of the orbit may require that the lids of the affected eye be taped closed or covered to prevent infection.

s. Thorns and Splinters. Thorns and splinters are frequently encountered in survival situations. Reduce their danger by wearing gloves and proper footgear. Their prompt removal is quite important to prevent infection. Wounds made by these agents are quite deep compared to their width, which increases chances of infection by those organisms (such as tetanus) that grow best in the absence of oxygen. Removal of splinters is aided by the availability of a sharp instrument (needle or knife), needle-nose pliers, or tweezers. Take care to get all of the foreign body out; sometimes it is best to open the wound sufficiently to properly cleanse it and to allow air to enter the wound. When cleaned, treat as any other wound.

t. Blisters and Abrasions. Care for blisters and abrasions promptly. Foot care is extremely important in the survival situation. If redness or pain is noted, the survivor should stop (if at all possible) to find and correct the cause. Frequently, a protective dressing or bandage and (or) adhesive will be sufficient to prevent a blister. If a blister occurs, do not remove the top. Apply a sterile (or clean) dressing. Small abrasions should receive attention to prevent infection. Using soap with a mild antiseptic will minimize the infection of small abrasions, which may not come to the attention of the survivor.

u. Insect Bites. Bites of insects, leeches, ticks, chiggers, etc., pose several hazards. Many of these organisms transmit diseases, and the bite itself is likely to become infected, especially if it itches and the survivor scratches it. The body should be inspected frequently for ticks, leeches, etc., and these should be removed immediately. If appropriate and possible, the survivor should avoid infested areas. These parasites can best be removed by applying heat or other irritant to them to encourage a relaxation of their hold on the host. Then the entire organism may be gently detached from the skin, without leaving parts of the head imbedded. Treat such wounds as any other wound. Applying cold, wet dressings will reduce itching, scratching, and swelling.

7-10. Illnesses. Many illnesses that are minor in a normal medical environment become major in a survival situation when the individual is alone without medications or medical care. Survivors should use standard methods (treat symptoms) to prevent expected diseases since treatment in a survival situation is so difficult. Key preventive methods are to maintain a current immunization record, maintain a proper diet, and exercise.

a. Food Poisoning. Food poisoning is a significant threat to survivors. Due to sporadic food availability, excess foods must be preserved and saved for future consumption. Methods for food preservation vary with the global area and situation. Bacterial contamination of food sources has historically caused much more difficulty in survival situations than the ingestion of so-called poisonous plants and animals. Similarly, dysentery or waterborne diseases can be controlled by proper sanitation and personal hygiene.

b. Treatment of Food Poisoning. If the food poisoning is due to preformed toxin, staphylococcus, botulism, etc. (acute symptoms of nausea, vomiting, and diarrhea soon after ingestion of the contaminated food), supportive treatment is best. Keep the patient quiet and lying down, and ensure the patient drinks substantial quantities of water. If the poisoning is due to ingestion of bacteria that grow within the body (delayed gradual onset of same symptoms), take antibiotics (if available). In both cases, symptoms may be alleviated by frequently eating small amounts of fine, clean charcoal. In PW situations, if chalk is available, reduce it to powder, and eat to coat and soothe the intestines. Proper sanitation and personal hygiene will help prevent spreading infection to others in the party or continuing reinfection of the patient.

Chapter 8

PW MEDICINE

8-1. Introduction:

a. Imprisoned PWs are, in the physical sense at least, under the control of their captors. Thus, the application of survival medicine principles will depend on the amount of medical service and supplies the captors can, and will, give to their prisoners. An enemy may both withhold supplies and confiscate survivors' supplies. Some potential enemies (even if they wanted to provide PW medical support) have such low standards of medical practice that their best efforts could jeopardize the recovery of the patient (figure 8-1).

b. An interesting and important sociological problem arises in getting medical care for PWs. How far should prisoners go in their efforts to get adequate rations and medicines for themselves or those for whom they are responsible? The Code of Conduct is quite specific concerning consorting with the enemy. Individuals must use considerable judgment in deciding whether to forget the welfare of fellow prisoners in order to follow the letter of the Code. Even more questionable is the individual who will offer such a justification for personal actions. Again,

these questions involve more than purely medical consideration. In combat, there are apt to be frequent situations in which medical considerations are outweighed by more important ones.

8-2. History:

a. As in past wars, there were professional medical personnel among the captives in North Vietnam; however, these personnel were not allowed to care for the sick and injured as in the past. Medical care and assistance from the captors were limited and generally below comparable standards of the United States. Yet 566 men returned, most in good physical and psychological condition, having relied to a large extent on their own ingenuity, knowledge, and common sense in treating wounds and diseases. They were able to recall childhood first aid, to learn by trial and error, and to use available resources. Despite their measures of success in this respect, many released personnel felt that with some prior training, considerable improvement in self-help techniques was possible even in the most primitive conditions.

Figure 8-1. PW Medicine.

b. To determine how the services could help and to assist future PWs to care for themselves if the situation required it, the Medical Section of the Air Force Intelligence Service, with the Surgeon General of the Air Force, sponsored a 5-day seminar to examine the pertinent medical experiences of captivity and to recommend appropriate additions and changes in training techniques. As a basis for seminar discussion, Air Training Command provided data on the major diseases, wounds, and ailments, and the treatment methods used by the captives in Southeast Asia. Transcripts (325) of debriefing material were screened for medical data. Significant disease categories were established for analysis simplification based on the frequency of the problems encountered.

PROBLEMS	MAJOR CATEGORIES ESTABLISHED
Dysentery	Trauma (lacerations, burns, fractures)
Fungus	Gastrointestinal problems
Dental problems	Communicable diseases
Intestinal problems	Nutritional diseases
Fractures	Dermatological ailments
Lacerations	Dental problems
Respiratory ailments	
Burns	

In examining these major categories, attention was focused on those medical problems considered significant by the prisoners themselves in evaluating their primitive practices (self-help).

8-3. Trauma:

a. Most of the prisoners began their captivity experience with precapture injuries—burns, wounds, fractures, and lacerations. Other injuries were the result of physical abuse while a prisoner. Most of these individuals, upon their return, expressed a need to know more about managing their injuries in captivity and also what to expect about the long-term effects of injures. It was not evident to them that the practice of a few simple rules will generally lead to acceptable results in wound treatment, and that much can be done after repatriation to correct cosmetic and functional defects.

b. The groundwork for management of injuries should begin well before an individual enters the captivity or survival environment. The treatment of injuries in survival or captivity depends primarily on providing the body the best possible circumstances to "repair itself."

It is vital, therefore, to have the body in the best possible physical condition before exposure to survival or captivity. This means good cardiovascular conditioning, good muscle strength and tone, and good nutritional status. Physiological and nutritional status will markedly influence the rate and degree of healing in response to injury. The opportunity for maintaining the best possible physical conditioning and nutritional status in captivity will be greatly reduced. (Once in a captivity or survival setting, it is important to do everything possible to maintain a good physiological and nutritional status.)

8-4. Gastrointestinal Problems:

a. **Diarrhea.** This was a common ailment in the prison environment, not only in Vietnam, but also in WW II and Korea. It plagued the forces of North Vietnam and the allied forces. This was the second most frequent malady afflicting the Viet Cong forces. The causative factors of this almost epidemic state were varied. A variety of infectious agents gaining access to the body by use of contaminated food and water certainly contributed to the problem. Equally important as causative agents were the low level of sanitation and hygiene practices within the camps. Psychogenic responses to unappetizing diet, nutritional disturbances, and viral manifestations also contributed.

(1) Captor Therapy. This consisted primarily of local or imported antidiarrheal agents, antibiotics, and vitamins. Appropriate diet therapy was instituted.

(2) Captive Self-Therapy. After instituting diet restrictions (solid food denial and increased liquid intake), afflicted personnel were administered "concoctions" of banana skins, charcoal, chalk, or tree bark tea.

(3) Treatment Evaluation. The accepted therapy for diarrhea focuses on the causative factors, which in the captivity experience were largely neglected. From a symptomatic perspective, the principles of self-treatment are simple to master: restrict intake to nonirritating foods (avoid vegetables and fruits), establish hygienic standards, increase fluid intake, and, when available, use antidiarrheal agents. The prisoners often resorted to a more exotic therapeutic regimen consisting of banana skins, charcoal, chalk, salt restriction, rice, or coffee. Charcoal, chalk, and the juice of tree barks have a scientific basis for their therapeutic success. Inasmuch as diarrhea was a source of concern and a disability for the North Vietnamese as well as the captives, therapy was often offered on request and was appropriate and successful.

(4) Conclusion. Diarrhea was frequent among PWs during captivity. Seldom fatal, it was disabling and a source of concern to those afflicted. Most captives were treated on demand and improved. This condition lends itself to some form of self-therapy through an understanding of its physiological derangements. The PW responded with intelligence, common sense, and a reasonably effective self-help regimen.

b. Dysentery. From a symptomatic perspective, dysentery is a severe form of diarrhea with passage of mucous and blood. Treatment and conclusions are similar to those for diarrhea.

c. Worms and Intestinal Parasites. Worms were extremely common among the captives. Twenty-eight percent of the released prisoners indicated worms as a significant medical problem during captivity. Worms often caused gastrointestinal problems similar to those resulting from a variety of other causes. The pin worm appears to have been the primary cause. This is not surprising, as its distribution is worldwide and the most common cause of helminthic infection of people in the United States. It requires no intermediate host; hence, infection is more rapidly acquired under poor hygienic conditions so commonly found in warm climates and conditions similar to the captivity environment. Seldom fatal, worms are significant, as they can lower the general resistance of the patient and may have an adverse affect on any intercurrent illness.

(1) Captor Therapy. This consisted of antihelminthic agents (worm medicine) dispensed without regularity, but with satisfactory results.

(2) Captive Therapy. The nuisance and irritating aspects of worms led to severe rectal itching, insomnia, and restlessness. This motivated the prisoner to find some form of successful self-therapy. Prevention was a simple and readily obtainable goal. Shoes were worn when possible; hands were washed after defecation; and fingernails were trimmed close and frequently. Peppers, popular throughout the centuries in medicine, contain certain substances chemically similar to morphine. They are effective as a counter-irritant for decreasing bowel activity. Other "house remedies" popular among the captives included drinking saltwater (a glass of water with 4 tablespoons of salt added), eating tobacco from cigarettes (chewing up to two or three cigarettes and swallowing them), and infrequently drinking various amounts of kerosene. All of these remedies have some degree of therapeutic effectiveness, but are not without danger and therefore deserve further comment. Saltwater alters the environment in the gastrointestinal tract and can cause diarrhea and vomiting. Too large an amount can have harmful effects on body-fluid mechanisms and can lead to respiratory complications and death. Tobacco contains nicotine and historically was popular in the 19th century as an emetic expectorant and was used for the treatment of intestinal parasites. Nicotine is, however, one of the most toxic of all drugs and can cause death when more than 60 mg is ingested. A single cigarette contains about 30 mg of nicotine, so the captives who ate two or more cigarettes had been using a cure more dangerous than the disease. Kerosene is also toxic, with 3 to 4 ounces capable of causing death. It is particularly destructive to the lungs, and if through vomiting it were to make its way into the trachea and eventually the

lungs, the complications would then again be far worse than the presence of worms.

(3) Treatment Evaluation. The antihelminthics therapy used by the captors was extremely effective. The problem during confinement was the nonavailability of such medication on demand. In addition, the inability to practice proper hygienic standards assured the continuation of, and reinfection with, worms.

(4) Conclusion. Worm infection in confinement is common and expected. It is seldom fatal, but contributes to general disability and mental depression due to its nuisance symptoms. Under certain circumstances, worms can assist in the spread of other diseases. The principle to follow in self-care is simple—use as high a hygienic standard as possible, and use medication causing bowel peristalsis and worm expulsion. Substances that interfere with the environment of the worms will aid in their expulsion. The toxic "house remedies" must be weighed against their possible complications.

8-5. Hepatitis. Infection of the liver was fairly common in some camps and present among the prison population throughout the captivity experience. Diagnosis was usually made on the basis of change of skin color to yellow (jaundice).

a. Captor Treatment. The Vietnamese seemed to have followed the standard therapy of rest, dietary management, and vitamin supplementation. They also displayed a heightened fear of the disease and avoided direct contact, when possible, with those afflicted.

b. Captive Therapy. For the most part, it parallels the therapy of the captors. This disease allows for little ingenuity or inventiveness of therapy.

c. Comments. Hepatitis is worldwide. Presumably most cases of hepatitis in captivity were viral in origin and easily disseminated to fellow prisoners. Conditions of poor sanitation and hygiene with close communal living foster its spread. Prevention through proper hygienic practices is the most effective tool. Equally important is an understanding of the disease characteristics. The majority of the cases recovered completely and less than 1 percent succumbed to this disease.

8-6. Nutritional Deficiencies. Symptoms attributed to malnutrition were frequent in the early years of confinement and continued up through 1964. The use of polished rice and the lack of fresh fruits and vegetables contributed to vitamin and protein deficiencies. From 1969 through 1973, food supplements were provided, and by release time, few obvious manifestations of diseases were present among those returning. The primary problems during the early years were vitamin deficiencies.

a. Vitamin B Deficiency (Beriberi). Presumably present among several PWs (especially those confined to the Briarpatch [Xom Ap Lo], about 15 miles west of

Sontay), it was rarely diagnosed on return. Its primary manifestation was pain in the feet described by the captives as "like a minor frostbite that turned to shooting pains."

(1) Captor Therapy. Prisoners were treated with vitamin injections and increased caloric content.

(2) Captive Therapy. Increasing caloric intake by eating anything of value. No specific self-care program existed for this malady.

(3) Comments. Beriberi is a nutritional disease resulting from a deficiency of vitamin B (thiamine). It is widespread in the Orient and in tropical areas where polished rice is a basic dietary staple. Of the various forms of the disease, dry beriberi would seem most important to the confinement condition. Early signs and symptoms of the disease include muscle weakness and atrophy, loss of vibratory sensation over parts of the extremities, numbness, and tingling in the feet. From the comments of the PWs, it is difficult to formulate a diagnosis. Modern therapy consists of vitamin B or sources of the vitamin in food (such as green peas, cereal grains, and unpolished rice).

(4) Conclusion. In Vietnam, the possible early onset of the dry form of beriberi was encountered. This is supported by the symptoms described and by the existence of dietary shortages of vitamin B and other nutritional deficiencies.

b. Vitamin A Deficiency. There were several reported cases of decreased vision (primarily at night) attributed to vitamin A deficiency. This problem usually occurred during periods of punishment or politically provoked action when food was withheld as part of the discipline. The condition responded well to increased caloric intake and deserves little special mention. An understanding of the transient nature of this problem and its remedial response to therapy is important.

8-7. Communicable Diseases. Some communicable diseases were endemic in North Vietnam and certainly responsible for large-scale disability among the personnel of the enemy forces. Plague, cholera, and malaria are frequent and a serious public health menace. Thanks to the immunization practices of the American forces, these diseases were of little concern to Americans during their captivity.

8-8. Skin Diseases:

a. Lesions. Dermatological lesions were common to the various prison experiences. Their importance lies not in their lethality (as they apparently did not cause any deaths), but for their irritant quality and the debilitating and grating effect on morale and mental health. Boils, fungi, heat rash, and insect bites appeared frequently and remained a problem throughout the captivity experience.

b. Boils and Blisters. A deep-seated infection that usually involves the hair follicles and adjacent subcutaneous tissue, especially parts exposed to constant irritation.

(1) Captor Treatment. Prisoner complaints about the presence of boils usually brought about some action by the captors. Treatment varied considerably and obviously depended on the knowledge of medics, doctors treating their prisoners, the availability of medical supplies, and the current camp policy. For the most part, systemic antibiotics, sulfa, and tetracycline were administered. In other instances, the boils were lanced or excised and treated with topical astringents.

(2) Captive Treatment. As the medics normally responded to pleas about boils, self-treatment was practiced primarily when there was distrust of captor techniques. Prisoners would attempt to lance the boil with any sharp instrument, such as needles, wires, splinters, etc., and exude their contents by applying pressure. The area was then covered with toothpaste and, when available, iodine.

(3) Comments. As noted above, the boil is an infection of hair follicles. It is more frequent in warm weather and aggravated by sweat, which provides ideal conditions for the bacteria. Boils seldom appear singularly. Once present, they are disseminated by fingers, clothing, and discharges from the nose, throat, and groin. Modern therapy consists of hot compresses to hasten localization, and then conservative incision and drainage. Topical antibiotics and systemic antibiotics are then used. Boils increase in frequency with a decrease in resistance as seen in malnutrition and exhaustion states in a tropical environment. This almost mimics the prison conditions.

(4) Conclusions. Self-help treatment is limited. Of importance here is sterility when handling the boils, cleanliness, exposure to sunlight, keeping the skin dry, and getting adequate nutrition. The disease is self-limiting and not fatal. The application of any material or medication with a detergent effect may be used (soaks in saline, soap, iodine, and topical antibiotics).

8-9. Fungal Infections. Fungal infections were also a common skin problem for those in Southeast Asian captivity. As with other skin lesions, they are significant for their noxious characteristics and weakening effect on morale and mental health. Superficial fungal infections of the skin are widespread throughout the world. Their frequency among PWs reflects the favorable circumstances of captivity for cultivating fungal infections.

a. Captor Therapy. Treatment consisted of medication described by many PWs as iodine and the occasional use of sulfa powder.

b. Captive Therapy. Treatment (often the result of memory of childhood experiences and trial-and-error observations) consisted of the removal of body hair (to prevent or improve symptoms in the case of heat rash), exposure to sunlight to dry out fungal lesions, and development of effective techniques to foster body cooling and to decrease heat generation. Considerable effort was directed at keeping the body clean.

c. Comments. Superficial dermatoses (skin lesions) due to fungi were common. Their invasive powers are at best uniformly weak, and because of this, infections are limited to the superficial portions of the skin and seldom by themselves fatal. Modern therapy since 1958 has relied heavily on an oral antifungal agent effective against many superficial fungi. This drug is expensive and not available in many parts of the world. Several lotions and emulsions can be used with some success. Elemental iodine is widely used as a germicide and fungicide. It is an effective antiseptic and obviously found favor in North Vietnam because of its availability. Without professional therapy, self-help, although limited in scope, can be effective. The principle of wet soaks for dry lesions and dry soaks for wet lesions is a fairly reliable guide. The use of the sun as a drying agent can also be very effective.

d. Conclusions. Skin problems are common to the captivity environment. More important, extreme personal discomfort, accompanied by infection, was detrimental to the physical and mental well-being of the prisoner.

8-10. Dental Problems. These were common among all captives, not only during confinement but also before capture. They were secondary to facial injury during egress, or caused by physical abuse during interrogation. Periodontitis (inflammation of tissue surrounding the tooth), pyorrhea (discharge of pus), and damage to teeth consistent with poor hygiene and "wear and tear" were also present.

a. Specific Complaints. Pain associated with the common toothache represented one of the most distressing problems faced by the PW. It affected the PW's nutrition and robbed the PW of the physical pleasure of eating (a highlight of isolated captivity). The inability of the PW to adequately deal with this problem caused persistent anxiety and decreased the ability to practice successful resistance techniques. In a few isolated instances, PWs actually considered collaboration with the captors in exchange for treatment and relief from tooth pain.

b. Captor Treatment. Treatment varied considerably and was no doubt influenced by political considerations. "Dentists" were infrequently available in camps before 1969. Cavities were filled, although usually inadequately with subsequent loss of the filling. Use of local anesthesia also varied depending on the dentist providing care.

c. Captive Therapy. The PWs often chose to treat themselves rather than seek or accept prison dentistry when it was available. Abscesses were lanced with sharp instruments made locally out of wood, bamboo, or whatever was available. Brushing was excessive, again using

whatever was available; chew sticks common to Asia were widely used. Aspirin (ASA), when available, was applied directly to the tooth or cavity.

d. Comments. The self-help practices noted above had many positive aspects. The basic principle of maintaining a well-planned cleaning program using fiber, brushes, or branches certainly contributed to the relatively low incidence of cavities and infection among the prisoners. The lancing of abscesses using bamboo sticks, although not a professional maneuver, has merit insofar as the pressure is relieved and the tendency to develop into cellulitis (widespread infection) decreased. The application of aspirin directly into the cavity should be discouraged as might the application of any other substance not directly produced for this purpose.

e. Conclusions. The most effective tool against dental complications in captivity is proper preventive dentistry. The present program of the three services, if adhered to, is adequate to ensure a high state of dental hygiene while captive.

8-11. Burns. Burns were an extremely frequent injury among PWs. Severity ranged from first through third degree and occurred frequently on hands and arms.

a. Captor Treatment. For the most part, burns were treated by captors by cleaning the burns and applying antiseptics and bandages. The results obtained were, by and large, inadequate, with frequent infections and long-term debilitation.

b. Captive Treatment. No specific treatment was developed among the PWs for burns. Reliance on some form of therapy was almost completely left to the captor.

c. Comments. Burns are extremely painful and can severely interfere with the ability to escape or to survive in captivity. The basic principle here is prevention.

d. Prevention. Adequate protection of exposed surfaces while flying (flame-retardant suits, gloves, boots, and helmet with visor down) is the best preventive action.

8-12. Lacerations and Infections:

a. Treatment. Captor treatment for lacerations and infections reflected the medical standards in North Vietnam and their domestic priorities. Wound and infection treatment varied considerably from being adequate to substandard and malpracticed. Obviously, the availability of trained physicians, a changing political climate, and difficulty in obtaining sophisticated medical supplies and equipment dictated and influenced the quality of the care delivered. The prisoners could do little professionally with this type of injury. As with diseases, the maintenance of good nutritional standards, cleanliness, and "buddy self-care" were the basic treatments.

b. Comments. When soft tissue is split, torn, or cut, there are three primary concerns—bleeding, infection, and healing of the wound.

(1) Bleeding is the first concern and must be controlled as soon as possible. Most bleeding can be controlled by direct pressure on the wound, and that should be the first treatment used. If that fails, the next line of defense would be the use of classic pressure points to stop hemorrhaging. And the last method for controlling hemorrhage would be the tourniquet. The tourniquet should be used only as a last resort. Even in more favorable circumstances in which the tourniquet can be applied as a first aid measure and left in place until trained medical personnel remove it, the tourniquet may result in the loss of the limb. The tourniquet should be used only when all other measures have failed, and it is a life or death matter. To control bleeding by direct pressure on the wound, sufficient pressure must be exerted to stop the bleeding, and that pressure must be maintained long enough to "seal off" the bleeding surfaces. Alternate pressing and then releasing to see whether the wound is still bleeding is not desirable. It is best to apply the pressure and keep it in place for up to 20 minutes. Oozing blood from a wound of an extremity can be slowed or stopped by elevating the wound above the level of the heart.

(2) The next concern is infection. In survival or captivity, consider all breaks in the skin due to mechanical trauma contaminated, and treat appropriately. Even superficial scratches should be cleaned with soap and water and treated with antiseptics, if available. Antiseptics should generally not be used in wounds that go beneath the skin's surface since they may produce tissue damage that will delay healing. Open wounds must be thoroughly cleansed with boiled water. Bits of debris such as clothing, plant materials, etc., should be rinsed out of wounds by pouring large amounts of water into the wounds and ensuring that even the deepest parts are clean. In a fresh wound where bleeding has been a problem, care must be taken not to irrigate so vigorously that clots are washed away and the bleeding resumes. Allow a period of an hour or so after the bleeding has been stopped before beginning irrigation with the boiled water. Begin gently at first, removing unhealthy tissue, increasing the vigor of the irrigation over a period of time. If the wound must be cleaned, use great care to avoid doing additional damage to the wound. The wound should be left open to promote cleansing and drainage of infection. In captivity, frequently deep, open wounds will become infested with maggots. The natural tendency is to remove these maggots, but actually, they do a good job of cleansing a wound by removing dead tissue. Maggots may, however, damage healthy tissue after the dead tissue is removed. So the maggots should be removed if they start to affect healthy tissue. Remember that it is imperative that the wound be left open and allowed to drain.

(3) An open wound will heal by a process known as secondary intention or granulation. During the healing phase, the wound should be kept as clean and dry as possible. For protection, the wound may be covered with clean dressings to absorb the drainage and to prevent additional trauma to the wound. These dressings may be loosely held in place with bandages (clean parachute material may be used for dressings and bandages). The bandages should not be tight enough to close the wound or to impair circulation. At the time of dressing change, boiled water may be used to gently rinse the wound. The wound may then be air-dried and a clean dressing applied. (The old dressing may be boiled, dried, and reused.) Nutritional status is interrelated with the healing process, and it is important to consume all foods available to provide the best possible opportunity for healing.

c. Conclusions. Obviously the PW is at a distinct disadvantage in treating wounds, lacerations, and infections without modern medicine. Yet, knowledge of the basic principles mentioned above, locally available equipment and resources, and optimism and common sense can help a survivor to maintain life.

8-13. Fractures and Sprains. Fractures and sprains often occurred during shootdown and (or) egress from the aircraft. They also occurred during evasion attempts.

a. Captor Treatment. As with other treatment, treatment of sprains and fractures varied considerably depending on the severity of the injury and the resources available for treatment. Even after immediate treatment or surgical procedures, there was little follow-up therapy. Prisoners were usually returned to camp to care for themselves or to rely on the help of fellow prisoners.

b. Captive Treatment. Captive therapy was primarily that of helping one another to exercise or immobilize the injured area, and in severe cases, to provide nursing care.

c. Comments:

(1) An acute nonpenetrating injury to a muscle or joint can best be managed by applying cold as soon as possible after the injury. Icepacks or cold compresses should be used intermittently for up to 48 hours following the injury. This will minimize hemorrhage and disability. Be careful not to use snow or ice to the point where frostbite or cold injury occurs. As the injured part begins to become numb, the ice should be removed to permit rewarming of the tissues. Then the ice can be reapplied. Following a period of 48 to 72 hours, the cold treatment can be replaced by warm packs to the affected part. A "sprain or strain" may involve a wide variety of damage ranging from a simple bruise to deep hemorrhage or actual tearing of muscle fibers, ligaments, or tendons. While it is difficult to establish specific guidelines for treatment in the absence of a specific diagnosis, in general, injuries of this type require some period of rest (immobilization) to allow healing. The period of rest is followed by a period of rehabilitation (massage and exercise) to restore function. For what appears to be a simple

superficial muscle problem, a period of 5 to 10 days' rest followed by a gradual progressive increase in exercise is desirable. Pain should be a limiting factor. If exercise produces significant pain, the exercise program should be reduced or discontinued. In captivity, it is probably safest to treat severe injuries to a major joint like a fracture with immobilization (splint, cast) for a period of 4 to 6 weeks before beginning movement of the joint.

(2) Bone fractures are of two general types, open and closed. The open fracture is associated with a break in the skin over the fracture site that may range all the way from a broken bone protruding through the skin to a simple puncture from a bone splinter. The general goals of fracture management are: restore the fracture to a functional alignment; immobilize the fracture to permit healing of the bone; and rehabilitation. Restoring or reducing the fracture simply means realigning the pieces of bones, putting the broken ends together as close to the original position as possible. The natural ability of the body to heal a broken bone is remarkable and it is not necessary that an extremity fracture be completely straight for satisfactory healing to occur. In general, however, it is better if the broken bone ends are approximated so that they do not override. Fractures are almost always associated with muscle spasms, which become stronger with time. The force of these muscle spasms tends to cause the ends of the broken bones to override one another, so the fracture should be reduced as soon as possible. To overcome the muscle spasm, force must be exerted to reestablish the length of the extremity. Once the ends of the bone are realigned, the force of the muscle spasm tends to hold the bones together. At this point, closed fractures are ready to be immobilized, but open fractures require treatment of the soft tissue injury in the manner outlined earlier. In other words, the wound must be cleansed and dressed, then the extremity should be immobilized. The immobilization preserves the alignment of the fracture and prevents movement of the fractured parts, which would delay healing. For fractures of long bones of the body, it becomes important to immobilize the joints above and below the fracture site to prevent movement of the bone ends. In a fracture of the mid-forearm, for example, both the wrist and the elbow should be immobilized. In immobilizing a joint, it should be fixed in a "neutral" or functional position. That is, neither completely straight nor completely flexed or bent, but in a position about midway between. In splinting a finger, for example, the finger should be curved to about the same position the finger would naturally assume at rest.

(3) A splint of any rigid material, such as boards, branches, bamboo, metal boot insoles, or even tightly rolled newspaper may be almost as effective as plaster or mud casts. In conditions such as continuous exposure to wetness, the splint can be cared for more effectively than the plaster or mud cast. In cases in which there is a soft-tissue wound in close proximity to the fracture, the splint method of immobilization is more desirable than a closed cast because it permits change of dressing, cleaning, and monitoring of the soft-tissue injury. The fracture site should be loosely wrapped with parachute cloth or soft plant fibers; then the splints can be tied in place extending at least the entire length of the broken bone and preferably fashioned in such a way as to immobilize the joint above and below the fracture site. The splints should not be fastened so tightly to the extremity that circulation is impaired. Since swelling is likely to occur, the bindings of the splint will have to be loosened periodically to prevent the shutting off of the blood supply.

(4) The time required for immobilization to ensure complete healing is very difficult to estimate. In captivity, it must be assumed that healing time will, in general, be prolonged. This means that for a fracture of the upper extremity of a "non-weight-bearing bone," immobilization might have to be maintained for 8 weeks or more to ensure complete healing. For a fracture of the lower extremity or a "weight-bearing bone," it might require 10 or more weeks of immobilization.

(5) Following the period of immobilization and fracture healing, a program of rehabilitation is required to restore normal functioning. Muscle tone must be reestablished and the range of motion of immobilized joints must be restored. In cases where joints have been immobilized, the rehabilitation program should be started with a "passive range of motion exercises." This means moving the joint through a range of motion without using the muscles that are normally used to move that joint. For example, if the left wrist has been immobilized, a person would begin the rehabilitation program by using the right hand to passively move the left wrist through a range of motion that can be tolerated without pain. When some freedom of motion of the joint has been achieved, the individual should begin actively increasing that range of motion using the muscles of the joint involved. Do not be overly forceful in the exercise program—use pain as a guideline—the exercise should not produce more than minimal discomfort. Over a period of time, the joint movement should get progressively greater until the full range of motion is restored. Also, exercises should be started to restore the tone and strength of muscles that have been immobilized. Again, pain should be the limiting point of the program and progression should not be so rapid as to produce more than a minimal amount of discomfort.

8-14. Summary. Common sense and basic understanding of the type of injuries are most helpful in avoiding complication and debilitation. Adequate nourishment and maintenance of physical condition will materially assist in healing of burns, fractures, lacerations, and other injuries—the body will repair itself.

8-15. Conclusions. In the management of trauma and burns in captivity or survival, remember that the body will do the healing or repair, and the purpose of the "treater" is to provide the body with the best possible atmosphere to conduct that self-repair. Some general principles are:

a. Be in the best possible physical, emotional, and nutritional status before being exposed to the potential survival or captivity setting.

b. Minimize the risk of injury at the time of survival or captivity by following appropriate safety procedures and properly using protective equipment.

c. Maintain the best possible nutritional status while in captivity or the survival setting.

d. Don't overtreat! Overly vigorous treatment can do more harm than good.

e. Use cold applications for relief of pain and to minimize disability from burns and soft-tissue strains or sprains.

f. Clean all wounds with gentle irrigation using large amounts of the cleanest water available.

g. Leave wounds open.

h. Splint fractures in a functional position.

i. After the bone has healed, begin an exercise program to restore function.

j. Remember that even improperly healed wounds or fractures may be improved with cosmetic or rehabilitative surgery and treatment upon rescue or repatriation.

Part Four

FACTS AND CONDITIONS AFFECTING A SURVIVOR

Chapter 9

WEATHER

9-1. Introduction. History records many attempts by people of ancient cultures to understand the heavens. In the primitive past, the ability to predict weather was of primary importance. It was by observing the stars and other celestial bodies that these early societies could predict the coming of the seasons and therefore the weather patterns that played such a large role in their survivability. When people today are forced to live under the primitive conditions of a survival situation, they are no different from those who have struggled before them against those same conditions.

Figure 9-1. Structure of the Atmosphere.

9-2. Knowledge of Weather. However, today's participants in the age-old struggle against nature must (out of necessity) still be concerned with the effect of weather.

It is still true that weather cannot be controlled, but the person who is prepared (through knowledge) will be more successful.

a. Weather is not the same as climate. Weather is the state of the atmosphere, with respect to wind, temperature, cloudiness, moisture, pressure, humidity, etc. Climate, on the other hand, is the type of weather condition generally prevailing over a region throughout the year, averaged over a series of years.

b. The atmosphere extends upward from the surface of the Earth for a great many miles, gradually thinning as it approaches its upper limit. Near the Earth's surface, the air is relatively warm due to contact with the Earth. As altitude increases, the temperature decreases by about 3.5°F for every 1,000 feet until air temperature reaches about 67°F below zero at 7 miles above the Earth.

c. To understand where the weather patterns originate, a brief familiarization of the "layers" or structure of the atmosphere is needed (figure 9-l). The atmosphere is divided into two layers. The upper layer is the "stratosphere," where the temperature remains constant. The lower layer is the "troposphere," where the temperature changes. Nearly all weather occurs in this lower layer, which begins at the Earth's surface and extends upward for 6 to 10 miles.

9-3. Elements Affecting Weather. Weather conditions in the troposphere and on the Earth are affected by four elements: temperature, air pressure, wind, and moisture.

a. Temperature is the measure of the warmth or coldness of an object or substance and, for this discussion, the various parts of the atmosphere. The sunlight entering the atmosphere reaches the Earth's surface and warms both the ground and the seas. Heat from the ground and the seas then warm the atmosphere. The atmosphere absorbs the heat and prevents it from escaping into space. This process is called the greenhouse effect because it resembles the way a greenhouse works. Once the Sun sets, the ground cools more slowly than the air because it is a better conductor of heat. At night the ground is warmer than the air, especially under a clear, dry sky. The ground cools more slowly than humid night air. Temperature changes near the ground for other reasons. Dark surfaces are warmer than light-colored surfaces. Evening air settles in low areas and valleys creating spots colder than higher elevations.

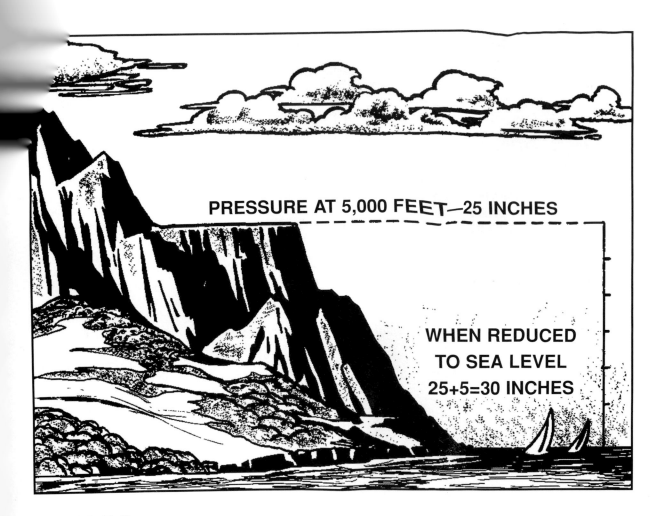

PRESSURE AT 5,000 FEET—25 INCHES

WHEN REDUCED TO SEA LEVEL 25+5=30 INCHES

Figure 9-2. Air Pressure.

Seas, lakes, and ponds retain heat and create warmer temperatures at night near shore. The opposite is true during the day, especially in the spring, when lakes are cold. On the beach, daytime temperatures will be cooler than the temperatures on land farther from shore. Knowing this will help determine where a survivor should build a shelter.

b. Air pressure is the force of the atmosphere pushing on the Earth. The air pressure is greatly affected by temperature. Cool air weighs more than warm air. As a result, warm air puts less pressure on the Earth than cool air. A low-pressure area is formed by warm air whereas cool air forms a high-pressure area (figure 9-2).

c. Wind is the movement of air from a high-pressure area to a low-pressure area. The larger the difference in pressure, the stronger the wind. On a global scale, the air around the Equator is replaced by the colder air around the poles (figure 9-3). This same convection of air on a smaller scale causes valley winds to blow upslope during the day and down the mountainside at night. Cool air blows in from the ocean during the day due to the heating and rising of the air above the land and reverses at night (figures 9-4 and 9-5). This movement of air creates winds throughout the world. When cool air moves into a low-pressure area, it forces the air that was already there to move upward. The rising air expands and cools.

d. Moisture enters the atmosphere in the form of water vapor. Great quantities of water evaporate each day from the land and oceans causing vapor in the air called humidity. The higher the humidity, the higher the moisture content of the air. Air holding as much moisture as possible is saturated. The temperature at which the air becomes saturated is called the dew point. When the temperature falls below the dew point, moisture in the air condenses into drops of water. Low clouds called fog may develop when warm, moist air near the ground is cooled to its dew point. A cooling of the air may also

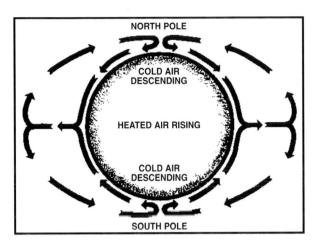

Figure 9-3. Wind Movement.

cause moisture to fall to the Earth as precipitation (rain, snow, sleet, or hail).

9-4. Circulation of the Atmosphere. If the Earth did not rotate, wind would move directly from the high-pressure areas of the poles to the low-pressure areas of the Equator. The movement of air between the poles and the Equator would go on constantly.

a. The rotation of the Earth prevents winds from the poles and the Equator from moving directly north or south. The Earth rotates from west to east, and as a result, winds moving toward the Equator seem to curve toward the west. Winds moving away from the Equator seem to curve toward the east. This is known as the Coriolis effect (figure 9-6). This effect results in winds circling the Earth in wide bands. These prevailing winds are divided into six belts that are known as the trade winds, the prevailing westerlies, and the polar easterlies; all three are found in both the Northern and Southern Hemispheres (figure 9-7).

(1) The winds blowing toward the Equator are known as the trade winds. The air above the Equator is so hot it is always rising. The north and south trade winds move in to take the place of the rising air. The Coriolis effect makes the trade winds appear to move from the east. The weather in the region of the trade winds moves from east to west because of the Earth's rotation. The doldrums is the region where the trade winds from the north and south meet near the Equator. The doldrums is usually calm, but it is quite rainy and may have periods of gusty winds.

(2) The prevailing westerlies blow away from the Equator. They occur north of the trade winds in the Northern Hemisphere and south of the trade winds in the Southern Hemisphere. The prevailing westerlies seem to move from the west because of the Coriolis effect. The weather in the region of these winds blows

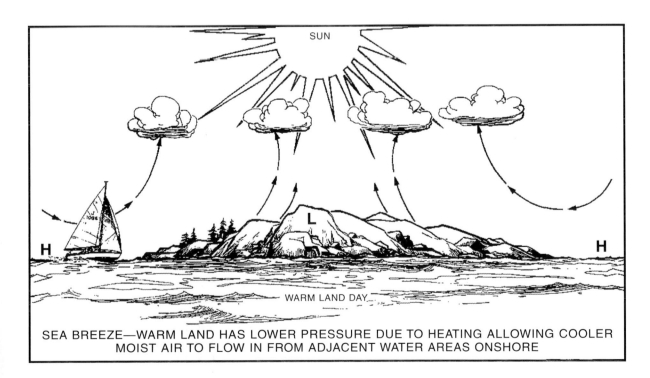

SEA BREEZE—WARM LAND HAS LOWER PRESSURE DUE TO HEATING ALLOWING COOLER MOIST AIR TO FLOW IN FROM ADJACENT WATER AREAS ONSHORE

Figure 9-4. Air Transfer (Daytime).

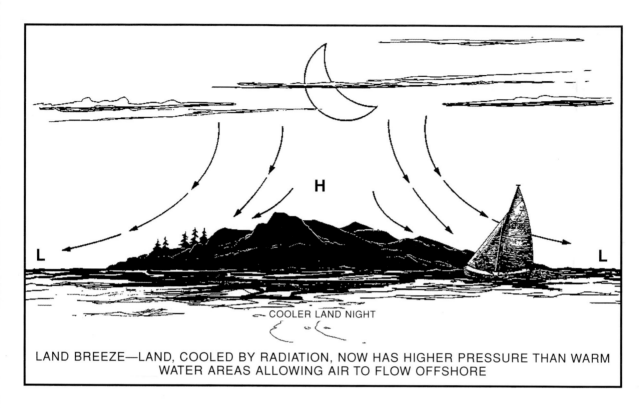

COOLER LAND NIGHT

LAND BREEZE—LAND, COOLED BY RADIATION, NOW HAS HIGHER PRESSURE THAN WARM WATER AREAS ALLOWING AIR TO FLOW OFFSHORE

Figure 9-5. Air Transfer (Nighttime).

from west to east. The prevailing westerlies move across most of the United States and Canada, and are divided from the trade winds in a region called the horse latitudes. The air in the horse latitudes blows downward to fill the space that was left between the prevailing westerlies and the trade winds. The winds are very light in the horse latitudes.

(3) The winds from the North and South Poles are known as the polar easterlies. Because the air is so cold, making it heavy, the air above the poles sinks downward. The air spreads out when it reaches the ground and moves toward the Equator. The weather in the region of the polar easterlies moves from east to west with the Coriolis effect making the winds seem to blow from the east. The polar front is the meeting place of the polar easterlies and the prevailing westerlies and is a cloudy, rainy region. Above the polar front is a band of west winds called the jet stream. The jet stream occurs about 5 to 7 miles above the ground. Its winds may exceed 200 miles per hour.

b. Pressure systems are highs or lows covering areas as large as 1 million square miles. Most pressure systems found in the United States and Canada develop along the polar front. There, the cold winds of the polar easterlies and the warmer winds of the prevailing westerlies move past one another and create swirling winds

called eddies. These eddies are carried eastward across the United States and Canada by the prevailing westerlies. There are two kinds of eddies: cyclones and anticyclones (figure 9-8).

(1) Cyclones formed by eddies are not the same as the storms known as cyclones. The winds of the eddies that create cyclones swirl inward toward a center of low pressure A low-pressure system is formed by the cyclone and its low-pressure region. Because of the rotation of the Earth, cyclones that build north of the Equator blow in a counterclockwise direction. Cyclones that form south of the Equator move in a clockwise direction. Cyclones in North America generally approach on brisk winds, bringing cloudy skies and usually rain or snow.

(2) Anticyclones swirl outward around a center of high pressure, forming a high-pressure system. Anticyclones move in a clockwise direction north of the Equator and counterclockwise south of the Equator. Anticyclones come after cyclones, bringing dry, clearing weather and light winds.

c. Air masses depend largely on the temperature and moisture of the areas in which they originate. Air masses may cover 5 million square miles. As they move away from their source regions and pass over land and sea, the air masses are constantly being modified through heating

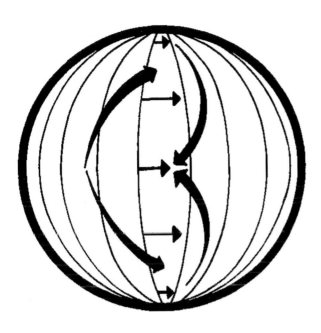

Figure 9-6. Coriolis Effect.

or cooling from below, lifting or subsiding, absorbing or losing moisture. In general, however, they retain some of their original characteristics and can be recognized and identified.

(1) There are four major types of air masses:

(a) Continental polar - cold and dry.

(b) Continental tropical - hot and dry.

(c) Maritime polar - cool and moist.

(d) Maritime tropical - warm and moist.

(2) In North America, the continental polar air mass over northern Canada blows cold, dry air into southern Canada and the United States. Maritime polar air masses off the northeast and northwest coasts of North America bring cool, damp weather to the continent. Maritime tropical air masses from the southeast and southwest coasts bring warm, muggy weather. The polar air masses are strongest in the winter, and the tropical air masses are strongest in the summer. During the winter, a cold arctic air mass from the North Pole also influences the weather of North America. A continental tropical air mass forms over the southwest United States during the warm months but disappears in the winter.

d. When two different air masses meet, they do not ordinarily mix (unless their temperatures, pressures, and relative humidities happen to be very similar). Instead, they set up boundaries called frontal zones, or "fronts." The colder air mass moves under the warmer air mass in the form of a wedge. If the boundary is not moving, it is termed a stationary front. Usually, however, the boundary moves along the Earth's surface, and as one air mass withdraws from a given area, it is replaced by another air

mass. This action creates a moving front. If warmer air is replacing colder air, the front is called "warm"; if colder air is replacing warmer air, the front is called "cold." Most changes in the weather occur along fronts. The movement of fronts depends on the formation of pressure systems. Cyclones push fronts along at speeds of 20 to 30 miles per hour. Anticyclones blow into an area after a front has passed.

(1) When a warm front moves forward, the warm air slides up over the wedge of colder air lying ahead of it (figure 9-9). This warm air usually has high humidity. As this warm air is lifted, its temperature is lowered. As the lifting process continues, condensation occurs, low nimbostratus and stratus clouds form, from which rain develops. The rain falls through the cooler air below, increasing its moisture content. Any reduction of temperature in the colder air, which might be caused by upslope motion or cooling of the ground after sunset, may result in extensive fog. As the warm air progresses up the slope, with constantly falling temperature, clouds appear at increasing heights in the form of altostratus and cirrostratus, if the warm air is stable. If the warm air is unstable, cumulonimbus clouds and altocumulus clouds will form and frequently produce thunderstorms. Finally, the air is forced up near the stratosphere and in the freezing temperatures at that level, the condensation appears as thin wisps of cirrus clouds. The upslope movement is very gradual, rising about 1,000 feet every 20 miles. Thus, the cirrus clouds, forming at perhaps 25,000 feet altitude, may appear as far as 500 miles in advance of the point on the ground that marks the position of the front. Warm fronts produce more gradual changes in the weather than do cold fronts. The changes depend chiefly on the humidity of the advancing warm air mass. If the air is dry, cirrus clouds may form and there will be little or no precipitation. If the air is humid, light, steady rain or snow may fall for several days. Warm fronts usually have light winds. The passing of a warm front brings a sharp rise in temperature, clearing skies, and an increase in humidity.

(2) When the cold front moves forward, it acts like a snowplow, sliding under the warmer air and tossing it aloft. This causes sudden changes in the weather. In fast-moving cold fronts, friction retards the front near the ground, which brings about a steeper frontal surface. This steep frontal surface results in a narrower band of weather concentrated along the forward edge of the front. If the warm air is stable, an overcast sky may occur for some distance ahead of the front, accompanied by general rain. If the warm air is conditionally unstable, scattered thunderstorms and showers may form in the warm air. In some cases, an almost continuous line of thunderstorms is formed and called a "squall line." Behind the fast-moving cold front there is usually rapid clearing, with gusty and turbulent surface winds and colder temperatures. The slope of a cold front is much steeper than that of a warm front and the progress

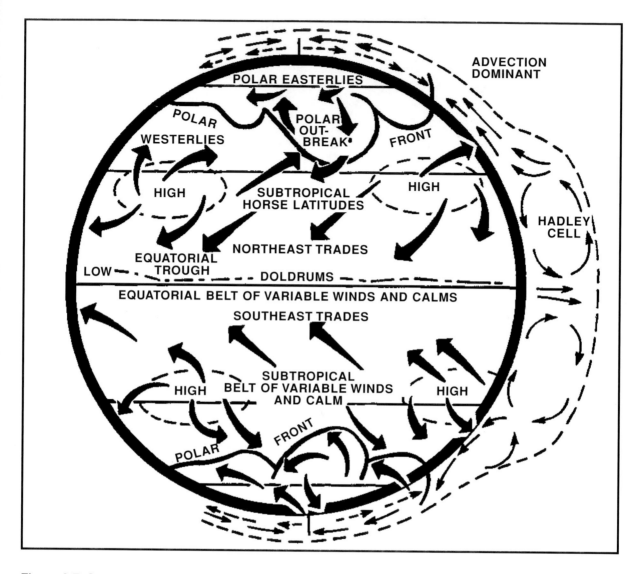

Figure 9-7. Atmosphere Circulation.

is generally more rapid—usually from 20 to 35 miles an hour; although in extreme cases, cold fronts have been known to move at 60 miles per hour (figure 9-10). Weather activity is more violent and usually takes place directly at the front instead of in advance of the front. However, especially in late afternoon during the warm season, a squall line will frequently develop as many as 50 to 200 miles in advance of the actual cold front. Whereas warm-front dangers lie in low ceilings and visibilities, cold-front dangers lie chiefly in sudden storms with high and gusty winds. Unlike the warm front, the cold front arrives almost unannounced, makes a complete change in the weather within the space of a few hours, and passes on. The squall line is ordinarily quite narrow—50 to 100 miles in width—but is likely to extend for hundreds of miles in length, frequently

lying across the entire United States in a line running from northeast to southwest. Altostratus clouds sometimes form slightly ahead of the front, but these are seldom more than 100 miles in advance. After the front has passed, the weather clears rapidly with cooler, drier air.

(3) One other form of front with which the survivor should become familiar is the "occluded front" (figure 9-11). Cold fronts travel about twice as fast as warm fronts. As a result, cold fronts often catch up to warm fronts. When a cold front reaches a warm front, an occluded front develops. Meteorologists subdivide occlusions into two types: cold-front occlusions and warm-front occlusions. In a cold-front occlusion, the air behind the cold front is colder than the air ahead of the warm front. The weather of a cold-front occlusion

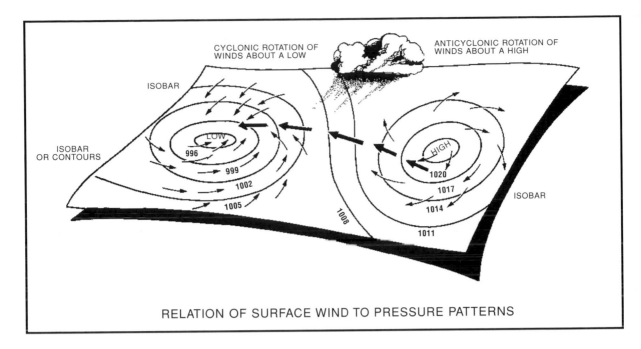

RELATION OF SURFACE WIND TO PRESSURE PATTERNS

Figure 9-8. Cyclonic and Anticyclonic Rotation.

resembles that of a cold front. When the air behind the cold front is warmer than the air ahead of the warm front, it is known as a warm-front occlusion. Warm-front occlusion is similar to a warm front.

These fronts produce milder weather than do cold or warm fronts.

(4) Stationary fronts are another type of front that occurs when air masses meet but move very slowly.

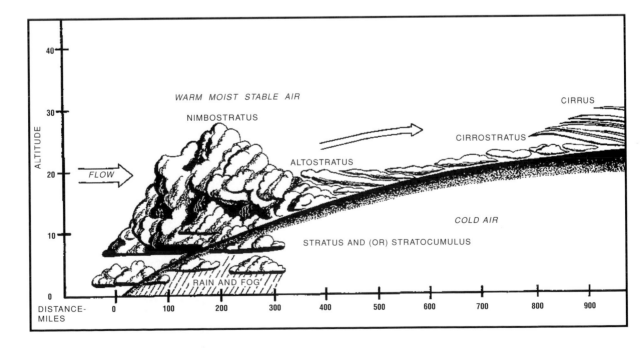

Figure 9-9. Stable Air Warm Front.

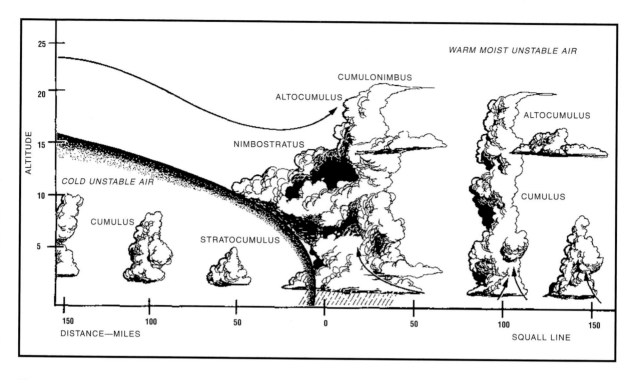

Figure 9-10. Fast Moving Cold Front.

It may remain over an area for several days bringing moderate weather.

9-5. Storms. The four main types of violent weather a person should be familiar with are thunderstorms, winter storms, tornadoes, and hurricanes.

a. Thunderstorms are the most frequent kinds of storms. As many as 50,000 thunderstorms occur throughout the world each day. Under some conditions, the rapid lifting of moist, warm air results in thunderstorms and dramatic cloud formations (figure 9-12). They develop from tall, puffy cumulonimbus clouds. Clouds may tower 5 to 10 miles high during hot, humid days. The temperatures inside the clouds are well below freezing. The air currents inside the clouds move up and down as fast as 5,000 feet per minute. Heavy rain is common because water vapor condenses rapidly in the air. Lightning and thunder occur during the life of a thunderstorm. When the sound of thunder is heard, a survivor should seek shelter immediately. Lightning causes more fatalities than any other type of weather phenomenon. In the United States alone more than 200 lightning deaths occur each year. Another reason for seeking shelter immediately is to escape the hail that sometimes accompanies the thunderstorm. Hail, which can grow as large as baseballs, is most noted for

damaging crops, but a powerful storm can bring injures, even fatalities, to survivors if shelter is not available.

b. Tornadoes are the most violent form of thunderstorms. Under certain conditions, violent thunderstorms will generate winds swirling in a funnel shape with rotational speeds of up to 400 miles per hour that extend out of the bottom of the thunderstorm. When this funnel-shaped cloud touches the surface, it can cause major destruction. The path of a tornado is narrow, usually not more than a couple of hundred yards wide. Tornadoes form in advance of a cold front and are usually accompanied by heavy rain and thunder in southern areas of the United States.

c. Winter storms include ice storms and blizzards. An ice storm may occur when the temperature is just below freezing. During this storm, precipitation falls as rain but freezes on contact with the ground. A coating of ice forms on the ground and makes it very hazardous to the traveler. Snowstorms with high winds and low temperatures are called blizzards. The winds blow at 35 miles per hour or more during a blizzard, and the temperature may be 10°F or less. Blowing snow makes it impossible to travel because of low visibility and drifting.

d. A hurricane, or typhoon, the most feared of storms, has a far more widespread pattern than a tornado. The

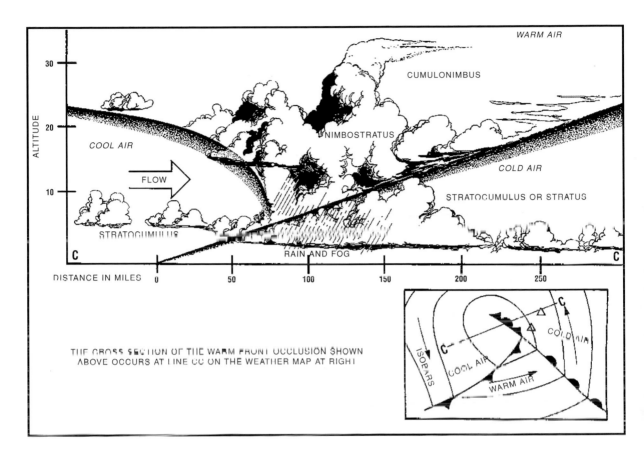

Figure 9-11. A Warm Front Occlusion.

storm forms near the Equator over the oceans and is a large, low-pressure area, about 500 miles in diameter. Winds swirl around the center (eye) of the storm at speeds of more than 75 miles per hour and can reach 190 miles per hour. Hurricanes break up over land and often bring destructive winds and floods. Thunderstorms often form within hurricanes and can produce tornadoes. Most hurricanes occurring in the United States sweep over the West Indies and strike the southeastern coast of the country. An early indication of a hurricane is a wind from an unusual direction, like the replacement of the normal flow of the trade winds from an easterly direction. The arrival of high waves and swells at sea coming from an unusual direction may also give some warning. The high waves and swells are moving faster than the storm and may give several days' warning.

9-6. Weather Forecasting. Weather forecasting enables survivors to make plans based on probable changes in the weather. Forecasts help survivors decide what clothes to wear and what type of shelter to build. During an evasion situation, it may help survivors

determine when to travel. While accurate weather prediction or forecasting normally requires special instruments, an awareness of changing weather patterns and attention to existing conditions can help a survivor or evader prepare for and, when appropriate, use changing weather conditions to enhance their survivability. The following are some elementary weather indicators that could help predict the weather and help save lives.

a. Clouds that move higher are good signs of fair weather. Lower clouds indicate an increase in humidity, which in all probability means precipitation (figure 9-13).

b. The Moon, Sun, and stars are all weather indicators. A ring around the Moon or Sun means rain (figure 9-14). The ring is created when tiny ice particles in fine cirrus clouds scatter the light of the Moon and the Sun in different directions. When stars appear to twinkle, it indicates that strong winds are not far off, and will become strong surface winds within a few hours. Also, a large number of stars in the heavens show clear visibility with a good chance of frost or dew.

c. "Low-hanging" clouds over mountains mean a weather change (figure 9-13). If they get larger during

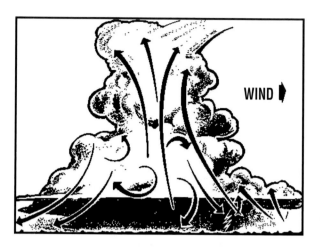

Figure 9-12. Thunderstorms.

the daytime, bad weather will arrive shortly. Diminishing clouds mean dry weather is on its way. Storms are often preceded by high, thin cirrus clouds arriving from the west. When these thicken and are obscured by lower clouds, the chances increase for the arrival of rain or snow.

d. The old saying "red skies at night, sailor's delight; red skies at morning, sailors take warning" has validity. The morning Sun turning the eastern sky crimson often signals the arrival of stormy weather. As the storm moves east, clouds may turn red as a clearing western sky opens for the setting Sun.

e. "The farther the sight, the nearer the rain," is a seaman's chant. When bad weather is near, the air pressure goes down and the atmosphere becomes clearer. High atmospheric pressure with stable and dusty air means fair weather.

f. A cold front arriving in the mountains during the summer usually means several hours of rain and thunderstorms. However, the passing of a cold front means several days of clear, dry weather.

g. A morning rainbow is often followed by a squall. An afternoon rainbow means unsettled weather, while an evening rainbow marks a passing storm. A faint rainbow around the Sun precedes colder weather.

h. Stormy weather will probably follow within hours when flowers seem to be much more fragrant.

i. People say "when sounds are clear, rain is near" because sound travels farther before storms.

j. Even birds can help predict the weather. Water birds fly low across the water when a storm is approaching. Birds will huddle close together before a storm.

k. The flowers of many plants, like the dandelion, will close as humidity increases and rain is on the horizon.

l. As humidity increases, the rocks in high mountain areas will "sweat" and provide an indication of forthcoming rain.

m. Lightning can tell survivors something by noting the color and compass direction. If the lightning looks white when seen through clear air and is located in the west or northwest, survivors would know the storm is headed toward them. Storms to the south or east will normally pass to the east. Red or colored lightning is seen at a distance in storms that will pass to the north or south.

n. Smoke, rising from a fire then sinking low to the ground, can indicate that a storm is approaching.

9-7. Summary. Even with the modern equipment available, forecasting tomorrow's weather is often difficult. This chapter provided background information and tips to use to teach survivors to predict weather. By understanding the basic characteristics and actions of weather, the survivor can better prepare for its effects.

Figure 9-13. Low Clouds.

Figure 9-14. Rings Around Moon or Sun.

Chapter 10

GEOGRAPHIC PRINCIPLES

10-1. Introduction. Geographic principles bring together the three major components of the environment (terrain, life forms, and climate) for the purpose of understanding the relationship between survivors and the physical environment around them. The more survivors know about these environmental components, the better they can help themselves in a survival situation. This chapter provides a brief introduction to these complex topics.

10-2. Components of the Environment:
a. Terrain is defined as a geographic area consisting of land and its features. The landmass of the Earth is covered with a variety of topography, including mountains, valleys, plateaus, and plains.

(1) The mountains vary greatly in size, structure, and steepness of slopes. For example, there is as much contrast between the large volcanic Cascade Mountains and those of the Rocky Mountains as there is between the Rocky Mountains and the Appalachian Mountains. Most major mountain systems will have corresponding foothills (figure l0-1).

(2) With two exceptions, valleys are formed as mountains are pushed up. The exceptions are massive gorges formed by glacial action and valleys carved out by wind and water erosion (figure 10-2).

(3) Plateaus are elevated and comparatively large, level expanses of land. Throughout the southwest, examples of the typical plateau can be seen. These plateaus were formed when a volcano deposited either lava or ash over a softer, sedimentary area. Through years of erosion, the volcanic "cap" broke loose in places and allowed the softer ground to be carried away. This type of plateau is the least common; however, it is the largest. The Columbia Plateau of Washington State is one example that covers 200,000 square miles (figure 10-3).

(4) The water forms of the Earth include oceans, seas, lakes, rivers, streams, ponds, and ice.

(a) Oceans comprise approximately 70 percent of the Earth's surface. The major oceans include the Pacific, Atlantic, Indian, and Arctic. Oceans have an enormous effect on land, not only in their physical contact but in their effect on weather. In most cases, lakes today are descended from much larger lakes or seas.

(b) Ice covers 10 percent of the Earth's surface. This permanent ice is found in two forms—pack ice and glaciers. Pack ice (normally 7 to 15 feet thick) is frozen seawater and may be as many as 150 feet thick. Those pieces that break off form ice islands. The two permanent icepacks on Earth are found near the North and South Poles—Arctic and Antarctic. The polar regions, which are thousands of feet thick, partially, but never completely, thaw. An icecap is a combination of pack ice and ice sheets. The term is usually applied to an ice plate limited to high mountain and plateau areas. During glacial periods, an icecap will spread over the surrounding lowlands (figure 10-4).

b. Life forms can best be described in terms of vegetation and animal life, with special emphasis on humans (which are covered later).

(1) There are hundreds of plant species on Earth. An in-depth study is obviously impossible. To understand the plant kingdom better, it is important to understand basic plant functions and adaptations they have made to exist in diverse environments. Vegetation will be categorized into either trees or plants.

(a) Of all the variety in species and types, trees can be divided simply into two types: coniferous or deciduous. Conifers are generally considered to be cone-bearing, evergreen trees. Some examples of conifers are pine, fir, and spruce. Deciduous trees are those which lose their leaves in winter and are generally considered as "hardwood." Some examples are maple, aspen, oak, and alder.

(b) For discussion, we will divide plants into two categories: annuals and perennials. Annuals complete their life cycle in 1 year. They produce many seeds and regenerate from seed. Climatic conditions may not be conducive for growth the following year, so seeds may remain dormant for many years. A classic example is the 1977 desert bloom in Death Valley. Plants bloomed for the first time in 80 years. Perennials are plants that last year after year without regeneration from seed.

(2) As with plants, the discussion of animal life has to be limited. Animals will be classified as either warmblooded or coldblooded. Using this division as a basic, it will be easier to describe animal adaptations to extreme climatic conditions. Warmblooded animals are generally recognized as cold-adapted animals and include all birds and mammals. Obviously, humans are a part of this classification because they are cold-adapted. Coldblooded animals gain heat from the environment. These are animals adapted for life in warm or moderate climates (lizards, snakes, etc.).

c. Climate can be described as an average condition of the weather at any given place. However, this description must be expanded to include the seasonal variations and extremes as well as the averages in terms of the climatic elements. In some areas, the climate is so domineering that the corresponding biome is named either in part or as a whole by the climate. Examples are the cloud forests and rain forests. The climate can only be described in terms of its various elements—temperature, moisture, and wind.

Figure 10-1. Composite of Mountains.

Figure 10-2. Gorges and Valleys.

(1) The atmosphere gains only about 20 percent of its temperature from the direct rays of the Sun. Most of the atmospheric heat gain comes from the Earth radiating that heat (energy) back into the atmosphere and being trapped. This is known as the greenhouse effect.

Figure 10-3. Plateau.

(2) Thinking of the atmosphere as a greenhouse, it is easier to understand the relationship water has in this "closed system." As water evaporates, the amount of water vapor the atmosphere can absorb depends solely upon temperature. The dew point is achieved when the amount of water vapor in the air equals the maximum volume the air will hold at a given air temperature. Lowering of air temperature creates condensation. Condensation appears in the form of clouds, fog, and dew. Any additional temperature reduction results in precipitation, such as rain. If the temperature of the dew point is below freezing, precipitation may appear in the form of hail or snow.

(3) Variation in air pressure is the primary cause of wind. When air is heated, it creates an area of lower pressure. As air cools, the pressure increases. Air movement occurs as the pressure tries to equalize, thus creating wind. Because wind is also a control of climate, people need to know why and how it affects climate. Let's look at wind in two aspects: localized wind (low altitude) and upper-air wind (high altitude). Localized wind is formed at low altitude, occurring due to dynamic topographical features and fluctuating air temperature

and pressure. High-altitude winds surrounding the Earth are bands of stable high- and low-pressure areas (cells). Predictable winds move off these cells, which are referred to as jet streams. These high-altitude winds control weather.

10-3. Effects of Climate on Terrain. The major effect climate has on terrain is erosion. Erosion can occur directly from heavy precipitation or indirectly through the accumulation of snow on snowpacks and glaciers. Wind and temperature both have erosion potential.

a. Heavy precipitation or melting water from icepacks and glaciers can create deep ravines by cutting into mountainous areas. Broad flood basins along major rivers can also aid in the development of river deltas in lakes, oceans, and deep fjords. The action of glaciers throughout the years has carved out deep, broad valleys with steep valley walls (figure 10-5).

b. The effects of wind erosion are greatest in barren, dry areas. The Great Arches National Park has some of the most dramatic examples of the effect of wind erosion. This type of erosion is caused by the wind driving sand and dust particles against an exposed rock or soil

Figure 10-4. Pack Ice.

surface, causing it to be worn away by the impact of the particles in an abrasive action (figure 10-6). Another form of wind erosion involves the movement of loose particles lying upon the ground surface that may be lifted into the air or rolled along the ground. Dry riverbeds, beaches, areas of recently formed glacial deposits, and dry areas of sandy or rocky ground are highly susceptible to this type of erosion. Sand dunes are attributed to this phenomenon (figure 10-7).

c. Frost action will have a weathering or eroding effect on rock land formations and ground surfaces. The frost action is the repeated growth and melting of ice crystals in the pore space or fractures of soil and rock. The tremendous force of growing ice crystals can exert a pressure great enough to pry apart rock. Many scree and talus slopes are caused by this action. Where soil water freezes, it tends to form ice layers parallel with the ground surface, heaving the soil upward unevenly. The peat-moss mounds of the tundra are an example of this action. The net effect of frost action will be dependent on the amount of surface moisture.

10-4. Effects of Terrain on Climate. The effect of terrain on climate is not nearly as subtle as the effect of climate on terrain. Three major factors exist that must he considered when studying the effects of terrain on weather.

a. Moisture for most major weather systems comes from the evaporation of the oceans of the world. The temperature, location, and flow of ocean currents, combined with the prevailing winds, will affect how much water will evaporate into the atmosphere. The warmer the ocean and corresponding current, the greater the rate of evaporation. Since the currents are deflected by landmasses, many warm currents flow parallel to major continents. When this moisture is blown inland by the prevailing winds, the net effect is the creation of a wet maritime climate, such as that found along the west coasts of Canada, Washington, Oregon, and Central Europe. If the temperature of the ocean and currents is cold, very little moisture will be yielded to the atmosphere. Examples of this occur along the Pacific coastline of Peru and Chile and along the Atlantic coastline of Angola and Southwestern Africa.

b. The interiors of large continent masses are dry because of the distance that isolates them from the effects of maritime climates. The large continents of the Northern Hemisphere create dry, high-pressure cells that isolate the interiors from the lower pressure, moist air cells and keep them from having much effect. The climate is referred to as the Continental Climate. The concept will be explained in the next chapter—Environmental Characteristics.

c. Mountains serve as moisture barriers, separating the maritime-influenced climates from the continental-influenced climates. The barrier effect of mountains on weather will be dependent on the height, length, and

Figure 10-5. Valleys.

width of the range and the severity of the weather fronts. In many cases, a lack of precipitation will extend for several hundred miles beyond the mountains. An example of this phenomenon occurs in the western states. The Cascade and Sierra Mountains block a great deal of Pacific Ocean moisture from the inland deserts of Washington, Oregon, and Nevada. The Rocky Mountains further block most of the moisture that is left in the atmosphere. Only the high cirrus clouds escape the barrier effect of these mountains. Another example can be seen in Asia. The Himalayan Mountains serve as a very effective barrier, blocking the Asiatic monsoon from central interior Asia, which helps create the Gobi Desert.

10-5. Effects of Climate and Terrain on Life Forms:

a. Since plants require water and light, climate will greatly affect the type and number of plants in an area.

(1) In areas with a great deal of rainfall, plants will be plentiful. In these areas, plants must compete for available sunlight. In areas where the primary vegetation has been knocked down (by clear cutting, landslides, or along flood basins of rivers), a thick secondary growth will occur. In time the secondary growth, if undisturbed, will become a climax forest. Some of the trees in these areas may grow to 300 feet. Because of the shade, vines and shade-tolerant perennials may sparsely cover the ground.

Figure 10-6. Erosion.

(2) By contrast, in areas where the amount of rainfall is limited, the plants must compete for the available water. The number of plants will also be sparse. Due to the harsh climatic and soil conditions, plant life is typically hardy. Through millions of years of evolution, plants have developed the following survival characteristics:

(a) Production of many seeds that germinate when water does come.

(b) Shallow root systems gather water quickly when they can.

(c) Ability to store water (cacti and other succulents).

(d) Rough, textured leaves (transpiration).

(e) Production of toxins (kill off competing plants).

b. Vegetation is also affected by the terrain. In mountainous regions, the clouds begin to lose moisture as they pass over the tops of the mountains. The result is more water is available for the growth of vegetation. However, with any increase in elevation, the temperature becomes colder. This exposure to colder temperature has a drastic effect on plant life.

c. Generally, animal life is mostly dependent on two factors: water and vegetation. The greater the rainfall, the greater the number of animals. Conversely, the drier an area, the less vegetation there will be to support animals. The location of small animals is determined by the secondary growth and ground cover, used for protection.

d. Temperature also affects the habits of animals. For example, animals may burrow to protect themselves from extreme heat or extreme cold or will be more active at night in hot, dry regions. Animals also respond physiologically to temperature extremes. During extreme cold, some species of mammals enter a state of winter dormancy (hibernation). It is a special case of temperature regulation in which animals lower the setting of their "thermostats" to maintain lower than normal body temperatures in order to save energy while maintaining minimum body functions essential for survival. This is important since their normal food supply is not always available during the winter. During periods of excessive heat, some species of fish, reptiles, mammals, and amphibians will enter a "summer sleep" called estivation. Estivation is a state in which the animal's body functions and activities are greatly reduced. Estivation and hibernation are not merely a result of temperature regulation but rather are methods by which the organisms survive unfavorable periods.

Figure 10-7. Sand Dunes.

Chapter 11

ENVIRONMENTAL CHARACTERISTICS

11-1. Introduction. Most survivors will not have a choice as to where they have to survive. The ease or difficulty in maintaining life and honor and returning are dependent on the types and extremes of the climate, terrain, and life forms in the immediate area. The Köppen-Geiger System of Climate Classification will be used in this chapter as the basis for organizing the discussion of environmental characteristics. This system has become the most widely used of climatic classifications for geographical purposes.

11-2. The Köppen-Geiger System. This system defines each climate according to fixed values of temperature and precipitation, computed according to the averages of the year or of individual months. A climate system based on these data has a great advantage in that the area covered by each subtype of climate can be outlined for large parts of the world. The five major climate groups are designated as (figure 11-1):

a. Tropical Climates. Average temperature of each month is above 64.4°F. These climates have no winter season. Annual rainfall is large and exceeds annual evaporation.

b. Dry Climates. Potential evaporation exceeds precipitation on the average throughout the year. No water surplus; hence, no permanent streams originate in dry climate zones.

c. Warm Temperate Climates. Coldest month has an average temperature below 64.4°F, but above 26.6°F. The warm temperate climates thus have both a summer and winter season.

d. Snow Climates. Coldest month average temperature below 26.6°F. Average temperature of warmest month above 50°F.

e. Ice Climates. Average temperature of warmest month below 50°F. These climates have no true summer.

11-3. Tropical Climates:

a. Tropics. Some people think of the tropics as an enormous and forbidding tropical rain forest through which every step taken must be hacked out and where every inch of the way is crawling with danger. Actually, much of the tropics is not rain forest. What rain forest there is must be traveled with some labor and difficulty. The tropical area may be rain forest, mangrove or other swamps, open grassy plains, or semi-dry brushland. The tropical area may also have deserts or cold mountainous districts. There is, in fact, a variety of tropical climates. Each region, while subject to the general climatic condition of its own zone, may show special modifications locally. Each general climate is a whole range of basic minor climates. For all their diversity, the climates of the tropics have the following in common:

(1) An almost constant length of day and night, a length that varies by no more than half an hour at the Equator to 1 hour at the limits of the tropics. The plant life thus has an evenly distributed period of daylight throughout the year.

(2) Temperature variation throughout the tropics is minimal—9°F to 18°F.

(3) There is no systematic pattern of major tropical landforms. There are high rugged mountains, such as the Andes of South America; karst formations as in Southeast Asia; plateaus like the Deccan of India; hilly lands like those which back the Republic of Guinea in Africa; and both large and small plains like the extensive one of the upper Amazon River or the restricted plain of the Irrawaddy River in Burma. The arrangement of all these landforms is part of the pattern of the larger landmasses, not of the tropics alone.

b. Vegetation:

(1) The jungles in South America, Asia, and Africa are more correctly called tropical rain forests. These forests form a belt around the entire globe, bisected somewhat equally by the Equator. However, the tropical rain forest belt is not a continuous one, even in any of the various regions in which it occurs. Usually it is broken by mountain ranges, plateaus, and even by small semi-desert areas, according to the irregular pattern of climate that regulates the actual distribution of rain forest.

(2) Some of the leading characteristics of the tropical rain forest common to those areas in South America, in Asia, and in Africa are:

(a) Temperatures average close to 80°F for every month.

(b) Vegetation consists of three stories.

(c) High rainfall (80 inches or more) distributed fairly evenly throughout the year.

(d) Areas of occurrence lie between 23.5 north and 23.5 south latitudes.

(e) Evergreen trees predominate, many of large girth up to 10 feet in diameter, with thick, leathery leaves.

(f) Vines (lianas) and air plants (epiphytes) are abundant.

(g) Herbs, grasses, and bushes are rare in the understory.

(h) Uniformity.

(i) Tree bark thin, green, smooth, and usually lacking fissures.

(3) The majority of plants that grow in the forest of the rainy tropics are woody and of the dimensions of trees. Trees form the principal elements of the vegetation.

LEGEND

Tropical Climates - Yellow
Dry Climates - Tan
Warm Temperate Climates - Green
Snow Climates - Blue
Ice Climates - Red
Highlands - Gray

Figure 11-1. Köppen-Geiger System.

LEGEND

Tropical Climates - Yellow
Dry Climates - Tan
Warm Temperate Climates - Green
Snow Climates - Blue
Ice Climates - Red
Highlands - Gray

Figure 11-2. Köppen-Geiger System. (cont)

The vines and air plants that grow on the trunks and branches of trees are woody. Grasses and herbs, which are common in the temperate woods of the United States, are rare in the tropical rain forest. The undergrowth consists of woody plants—seedling and sapling trees, shrubs, and young, woody climbers. The bamboos, which are really grasses, grow to giant proportions, 20 to 80 feet high in some cases. Bamboo thickets in parts of some rain forests are very difficult to penetrate. The plants that produce edible parts in the jungle are often scattered, and require searching to find several of the same kind. A tropical rain forest (figure 11-2) has a wider variety of trees than any other area in the world. Scientists have counted 179 species in one 8.5-acre area in South America. An area this size in a forest in the United States would have fewer than seven species of trees.

(4) The average height of the taller trees in the rain forest is rarely more than 150 to 180 feet. Old giants of the tropical rain forest attain 300 feet in height, but this is extremely rare. Trees more than 10 feet in diameter are also rare in the jungle. The trunks are, as a rule, straight and slender and do not branch until near the top. The base of many trees is provided with plank buttresses, flag-like outgrowths that are common in all tropical forests. The majority of mature tropical trees have large, leathery, dark-green leaves that resemble laurel leaves in size, shape, and texture. The general appearance is monotonous, and large and strikingly colored flowers are uncommon. Most of the trees and shrubs have inconspicuous flowers, often greenish or whitish.

(5) Travel books often give a misleading impression of the density of tropical forests. On riverbanks or in clearings, where much light reaches the ground, there is a dense growth that is often quite impenetrable. But in the interior of an old, undisturbed forest, it is not difficult to walk in any direction. Photographs give an exaggerated notion of the density of the undergrowth. It is usually possible to see another person at least 60 feet away.

(6) The abundance of climbing plants is one of the characteristic features of rain forest vegetation. The great majority of these climbers are woody and many have stems of great length and thickness. Stems as thick as a man's thigh are not uncommon. Some lianas cling closely to the trees that support them, but most ascend to the forest canopy like cables or hang down in loops or

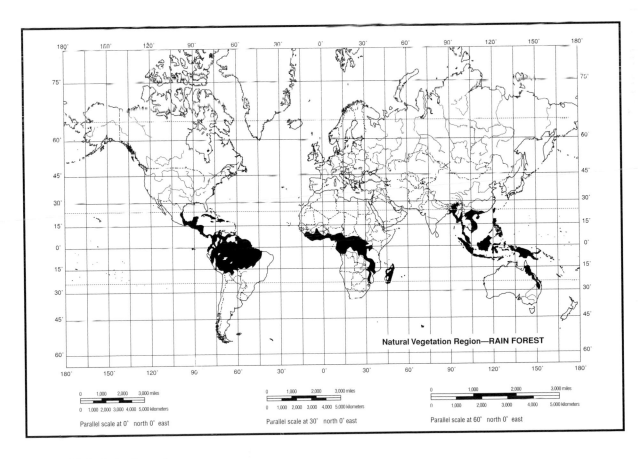

Figure 11-2. Rain Forest.

festoons. In the rain forest, there is no winter or spring, only perpetual midsummer. The appearance of the vegetation is much the same at any time of year. There are seasons of maximum flowering during which more species bloom than at any other time, and also seasons of maximum production of young leaves, plant growth, and reproduction is continuous and some flowers can be found at any time. The margins of a tropical rain forest clearing and areas around abandoned dwellings abound in edible plants. However, in the center of the virgin rain forest, trying to kind food is more difficult due to accessibility. The lofty trees are so tall that fruits and nuts are generally out of reach.

c. Distribution of Tropical Rain Forests:

(1) In the Americas, the largest continuous mass of rain forest is found in the basin of the Amazon River. This extends west to the lower slopes of the Andes and east to the Atlantic coast of the continent; it is broken only by relatively small areas of savanna and deciduous forest. This great South American rain forest extends south into the region of the Gran Chaco (south-central

South America) and north along the eastern side of Central America into southern Mexico and into the Antilles chain of the West Indies. In the extreme northwest of South America (Ecuador, Colombia), there is a narrow belt of rain forest, separated from the Amazonian forest by a wide expanse of deciduous forest, extending from about latitude 6 degrees South to a little beyond the Tropic of Capricorn. The distribution of rain forest in Central America is perhaps less well known than any other major tropical region. The main areas are below the 500-foot elevation (figure 11-3).

Figure 11-4. Rain Forest (Borneo).

Figure 11-3. Rain Forest (Central America).

(2) In Africa, the largest area of rain forest lies in the Congo basin and extends westward into the Republic of Cameroon. As a narrow strip, the forest continues still farther west, parallel to the Gulf of Guinea, through Nigeria and the Gold Coast of Liberia and Guinea. Southward from the Congo basin, the forest extends toward Rhodesia.

(3) In the eastern tropics, the rain forest extends from Ceylon and Western India to Southeast Asia and the Philippines, as well as through the Malay

Archipelago to New Guinea. The largest continuous areas are in New Guinea, the Malay Peninsula, and the adjoining islands of Sumatra and Borneo, where the Indo-Malayan rain forest reaches its greatest luxuriance and floral wealth (figure 11-4).

(a) In India, the area of rain forest is not large, but it is found locally in the western and eastern Ghats (coastal ranges) and, more extensively, in the lower part of the eastern Himalayas, the Khasia hills, and Assam.

(b) In Burma and Southeast Asia, the rain forest is developed only locally, the principal vegetation being the monsoon forest. The monsoon forest is a tropical type that is partly leafless in certain seasons.

(c) In the eastern Sunda Islands from western Java to New Guinea, the seasonal drought (due to the dry, east monsoon from Australia) is too severe for the development of a rain forest, except in locally favorable situations (figure 11-5).

(d) In Australia, the tropical rain forest of Indo-Malaya is continued south as a narrow strip along the eastern coast of Queensland. Rain forest also extends into the islands of the western Pacific (Solomons, New Hebrides, Fiji, Samoa, etc.). (See figure 11-6.)

d. Food Plants. Some of the available food plants in the rain forest include:

(1) Indian or Tropical Almond.
(2) Rose Apple.
(3) East Indian Arrowroot.
(4) Bullock's Heart.
(5) Sugar Cane.
(6) Cattail.
(7) Bael Fruit.
(8) Water Chestnut (Trapa Nut).
(9) Bamboo.
(10) Chufa (Nut Grass).
(11) Goa Bean.
(12) Luffa Sponge (Wild Gourd).
(13) Yam Bean.
(14) Wild Fig.
(15) Bignay.
(16) Wild Grape.
(17) Lotus Lilly.
(18) Water Lettuce.
(19) Breadfruit.
(20) Canna Lily.
(21) Bracken (Fern).
(22) Sego Palm.
(23) Tree Fern.
(24) Palm Sugar.
(25) Mango.
(26) Papaya.
(27) Italian Millet.
(28) Screw Pine.
(29) Pearl Millet
(30) Plantain.
(31) Mulberry.

Figure 11-5. Rain Forest (New Zealand).

(32) Batolo Plum.
(33) Cashew Nut.
(34) Pokeweed.
(35) Buri Palm.
(36) Polypody.
(37) Fishtail Palm.
(38) Air Potato (ubi tuber).
(39) Coconut Palm.
(40) Purslane.
(41) Nipa Palm.
(42) Rice.
(43) Rattan Palm.
(44) Soursop.
(45) Ceylon Spinach.
(46) Water Lilly.
(47) Sterculia.
(48) Sweetsop.
(49) Tamarind.
(50) Taro.
(51) Ti Plant.
(52) Horseradish Tree.
(53) Tropical Yam.

Figure 11-6. Rain Forest (Tahiti).

e. Semi-Evergreen Seasonal Forest:

(1) In character, the semi-evergreen seasonal forest in Central and South America and Africa corresponds essentially to the monsoon forest of Asia. Characteristics of the semi-evergreen forest are:

(a) Two stories of tree strata—upper story 60 to 80 feet high; lower story 20 to 45 feet high.

(b) Large trees are rare: average diameter about 2 feet.

(c) Seasonal drought causes leaf fall; more in dry years.

(2) The peculiar distribution of the rainy season and the dry season that occurs in the countries bordering the Bay of Bengal in southeastern Asia brings on the monsoon climate. The monsoons of India, Burma, and Southeast Asia are of two types. The dry monsoon occurs from November to April, when the dry northern winds from central Asia bring long periods of clear weather with only intermittent rain. The wet monsoon occurs from May to October, when the southern winds from the Bay of Bengal bring rain, usually in torrents, that lasts for days and often weeks at a time. During the dry season, most leaves drop completely off, giving the landscape a wintry appearance, but as soon as the monsoon rains begin, the foliage reappear immediately.

f. Plant Foods of the Semi-Evergreen Seasonal Forest. These foods include:

(1) Agave (Century Plant).
(2) Amaranth.
(3) Bael Fruit.
(4) Banana.
(5) Tropical or Indian Almond.
(6) Rose Apple.
(7) Bamboo.
(8) Goa Bean.
(9) Yam Bean.
(10) Mango.
(11) Purslane.
(12) Mulberry.
(13) Bignay.
(14) Italian Millet.
(15) Soursop.
(16) Cashew Nut.
(17) Breadfruit.
(18) Pearl Millet.
(19) Ceylon Spinach.
(20) Sterculia.
(21) Sugar Cane.
(22) English Acorns (Oak).
(23) Luffa Sponge (Wild Gourd).
(24) Cattail.
(25) Buri Palm.
(26) Sweetsop.
(27) Chestnut.
(28) Water Chestnut (Trapa Nut).
(29) Rattan Palm.
(30) Tamarind.
(31) Chufa (Nut Grass).
(32) Papaya.
(33) Taro.
(34) Ti Plant.
(35) Wild Fig.
(36) Screw Pine.
(37) Horseradish Tree.
(38) Tree Fern.
(39) Plantain.
(40) Tropical Yam.
(41) Wild Grape.
(42) Pokeweed.
(43) Water Lettuce.
(44) Polypody.
(45) Canna Lily.
(46) Air Potato (ubi tuber).
(47) Lilly Lotus.
(48) Water Lily.

g. Tropical Scrub and Thorn Forest (figure 11-7):

(1) Chief Characteristics of the Tropical Scrub and Thorn Forest:

(a) Definite dry season, with wet season varying in length from year to year. Rains appear mainly as downpours from thunderstorms.

(b) Trees are leafless during dry season; average height is 20 to 30 feet with tangled undergrowth in places (figure 11-8).

(c) Ground is bare except for a few tufted plants in bunches; grasses are not common.

(d) Plants with thorns are predominant.

(e) Fires occur at intervals.

(2) Food Plants:

(a) Within the tropical scrub and thorn forest areas, the survivor will find it difficult to get food plants in the dry season. During the height of the drought period, the primary kinds of foods come from the following plant parts:

-1. Tubers.
-2. Bulbs.
-3. Pitch.
-4. Nuts.
-5. Rootstalks.
-6. Corms.
-7. Gums and Resins.
-8. Seeds and Grains.

(b) During the rainy season in the tropical scrub and thorn forest, plant food is considerably more abundant. At this time, the survivor should look for the following edible plants:

-1. Sweet Acacia.
-2. Wild Chicory.
-3. St. John's Bread.
-4. Wild Caper.
-5. Agave (Century Plant).
-6. Wild Fig.
-7. Almond.
-8. Cashew Nut.
-9. Baobab.
-10. Juniper.
-11. Tamarind.
-12. Tropical Yam.
-13. Sea Orach.
-14. Prickly Pear.
-15. Wild Pistachio.
-16. Air Potato (ubi tuber).

h. Tropical Savanna (figure 11-9):

(1) General Characteristics of the Savanna:

(a) Savannas lie wholly within the tropical zone in South America and Africa.

(b) The savanna looks like a broad, grassy meadow with trees spaced at wide intervals.

(c) The grasses of the tropical savanna often exceed the height of a man. However, none of the savanna grasses are sod-forming in the manner of lawn grasses, but are bunch grasses with a definite space between each grass plant.

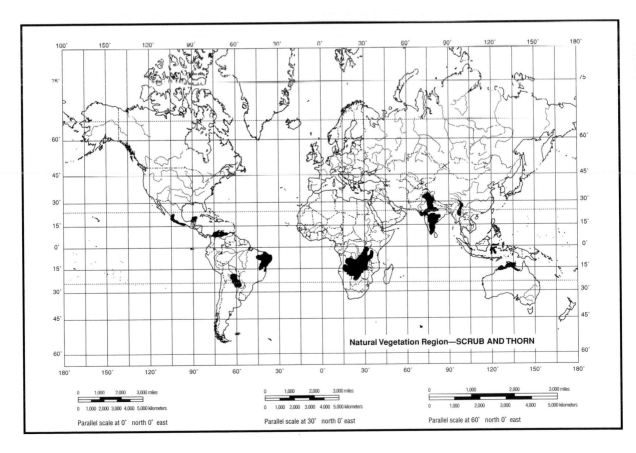

Figure 11-7. Shrub and Thorn Forest.

Figure 11-8. Tropical Shrub or Thorn Forest (Kenya).

(d) The soil in the savanna is frequently red.

(e) The scattered trees usually appear stunted and gnarled like old apple trees.

(f) Palms may be found on savannas.

(2) Savanna of South America. For the most part, the vegetation is of the bunch-grass type. A long dry season alternates with a rainy season. In these areas, both high and low grasses are present. Bright-colored flowers appear between the grass bunches during the rainy season. The grains from the numerous grasses are useful as survival food, as are the underground parts of the many seasonal plants that appear with and following the rains.

(3) Savanna in Africa. The high grass tropical savanna of Africa is dominated by very tall, coarse grasses that grow from 5 to 15 feet high. Unless the natives burn the grass during the dry season, the savanna becomes almost impenetrable. This type of savanna occurs in a broad belt surrounding the tropical rain forest and extends from western Africa eastward beyond the Nile River. From the Nile, it extends southward and westward. The tropical bunch-grass savanna comprises the greatest part of the

African savanna, consisting of grasses about 3 feet tall. The African savanna has both dwarf and large trees. The most renowned of these large trees is the monkeybread, or baobab (figure 11-10).

(4) Food Plants. The food plants found on the savanna are also found in other vegetation areas.

(a) Amaranth.	(g) Wild Chicory.
(b) Wild Crabapple.	(h) Wild Fig.
(c) Purslane.	(i) Tamarind.
(d) Wild Apple.	(j) Chufa (Nut Grass).
(e) Wild Dock.	(k) Water Plantain.
(f) Wild Sorrel.	(l) Water Lily.

i. Animal Life. The tropics abound in animal life. The tremendous varieties of animal species found in tropical areas throughout the world preclude discussions of each animal. It is essential that the survivor realize that just as people have an inherent fear of some animals, most animals also fear people. With some exceptions, animals of the tropics will withdraw from any encounter with humans. Being primarily nocturnal animals, most will never be seen by the survivor. By becoming familiar with

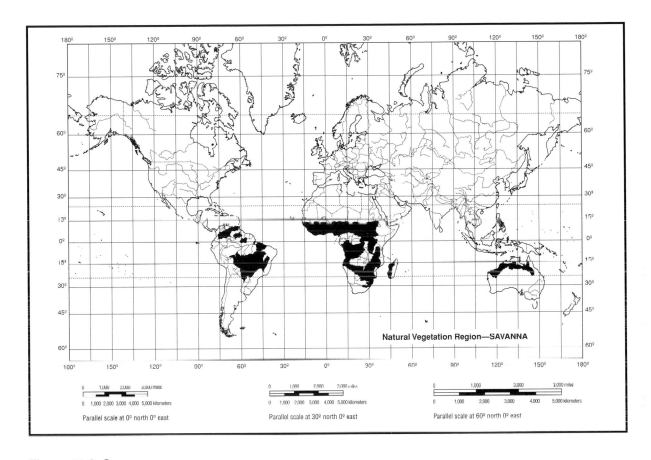

Figure 11-9. Savanna.

the wild inhabitants of the tropics, the survivor will better understand this type of environment and will respect, not fear, the surroundings in which survival takes place.

(1) All tropical areas have members of the pig family. By habit, pigs are gregarious and are omnivorous in diet. They will eat any small animals they can kill, although they feed mainly on roots, tubers, and other vegetable substances. The most common species found in the Old World tropics are the peccary, the Indian wild boar, the Babirussa of Celebes, and the Central African Giant Forest Hog. In Central and South American tropics, peccaries are common. These pigs are represented by two species, the "white-lipped" peccary and the "collared" peccary. Both are a grizzled black color, distinguishable by markings from which they derive their names. The white-lipped peccary, the larger of the two (height of approximately 18 inches), is black with white under the snout, and has the reputation of being the more ferocious. The collared peccary, reaching a height of 14 inches, is identified by the white or gray band around the body where the neck joins the shoulder. The collared peccary often travels in groups of 5 to 15. While alone, they are not particularly dangerous, but a pack can effectively repel any enemy and can make short work of a jaguar, cougar, or human. Both types of peccaries have musk glands that are located on the spine 4 inches up from the tail. This gland must be removed soon after the animal is killed, otherwise the flesh will become tainted and unfit for consumption.

(2) Tropic areas harbor many species of reptiles and amphibians. Most of them are edible when skinned and cooked. Hazards from these animals are mostly imagined; however, some are venomous or dangerous if encountered. Individual species of the crocodylidae family (alligators, crocodiles, caiman, and gavials) are usually abundant only in remote areas away from humans. Most dangerous are the saltwater crocodiles of the Far East and the Nile crocodiles in Africa. Poisonous snakes, while numerous in the tropics, are rarely seen and pose little danger to the wary survivor. There are no known poisonous lizards in the tropics. Several species of frogs and toads contain poisonous skin secretions.

Figure 11-10. Savanna (Africa).

The large, pan-tropic toad, Bufo Marimum, exudes a particularly irritating secretion if handled roughly. Aside from skin irritations, these amphibians pose little danger to humans, unless the secretion gets into the eyes, where it may cause blurred vision, intense burning, and possible blindness.

(3) All tropical areas of the world abound in monkeys. They are very curious animals, and this fact may be used to the survivor's advantage in trying to procure one for food. Only the very large species of monkey, such as the mandrill or baboon, could constitute any danger to humans.

(4) Tropical areas also have a large number of the various species of mice, rats, squirrels, and rabbits.

(5) Members of the cat family are found in jungles throughout the world. The ocelot abounds in the jungles of Central and South America. It is a small, lean, savage cat whose coloring closely resembles that of the jaguar, and it will attain a weight of approximately 40 pounds and a length of 3 feet when fully grown. Cats such as the leopard are found in the tropics of the Old World. The leopard is one of the most wary of beasts, becoming a powerful fighter when wounded or cornered. Unlike lions and tigers, the leopard can climb trees with ease; therefore caution should be used when hunting this animal.

(6) Deer are found in most jungle areas; however, their population is normally small. In the Asian jungles, several species of deer frequent the low, marshy areas

BLACK WIDOW | **BLACK WIDOW** | **BROWN RECLUSE**

Figure 11-11. Black Widow and Brown Recluse Spiders.

adjacent to rivers. In the Central and South American jungles, two species of deer are most common. The jungle species is found in thick, upland forests. It is much smaller than the North American species and seldom attains a weight of more than 80 pounds. Another deer found in the Central and South American jungles is the "brocket," or "jungle deer." This small, reddish-brown deer, which attains a height of about 23 inches, is extremely shy and is found mostly in dense cover since it has no defense against other animals.

(7) The real dangers lie in the insects located in the jungle, which can pass on diseases or parasites.

(a) Malaria may be the worst enemy. It is transmitted by mosquitoes, which are normally encountered from late afternoon until early morning. Guard against bites by camping away from swamps on high land and sleep under mosquito netting, if available; otherwise, use mud on the face as a protection against insects. Wear full clothing, especially at night, and tuck pants into the tops of socks or shoes. Wear a mosquito head net and gloves. Take antimalaria tablets (if available) according to directions.

(b) The greatest number of ant species is found in the jungle regions of the world. Nesting sites may be in the ground or in the trees. Ants can be a considerable nuisance, especially if near a campsite. They inflict pain by biting, stinging, or squirting a spray of formic acid. Before selecting a campsite, a close check of the area should be made for any nests or trails of ants.

(c) Ticks may be numerous, especially in grassy places. Use a protected area and undress often, inspecting all parts of the body for ticks, leeches, bed bugs, and other pests. If there are several people in the group, examine one another.

(d) Fleas are common in dry, dusty buildings. The females will burrow under the toenails or into the skin to lay their eggs. Remove them as soon as possible. In India and southern China, bubonic plague is a constant threat. Rat fleas carry this disease and discovery of dead rats usually means a plague epidemic in the rat population. Fleas may also transmit typhus fever and in many parts of the tropics, rats also carry parasites that cause jaundice and other fevers. Keep food in rat-proof containers or in rodent-proof caches. Do not sleep with any food in the shelter!

(e) In many parts of the Far East, a type of typhus fever is carried by tiny red mites. These mites resemble the chiggers of the southern and southwestern United States. They live in the soil and are common in tall grass, cut-over jungle, or stream banks. When a person lies or sits on the ground, the mites emerge from the soil, crawl through clothes, and bite. Usually people don't know they have been bitten, as the bite is painless and does not itch. Mite typhus is a serious disease and the survivor should take preventive measures to avoid this pest. The survivor should clear the camping ground and burn it off, sleep above the ground, and treat clothing with insect repellent.

(f) Leeches are primarily aquatic and their dependence on moisture largely determines their distribution. The aquatic leeches are normally found in still, freshwater lakes, ponds, and waterholes. They are attracted by disturbances in the water and by a chemical sense.

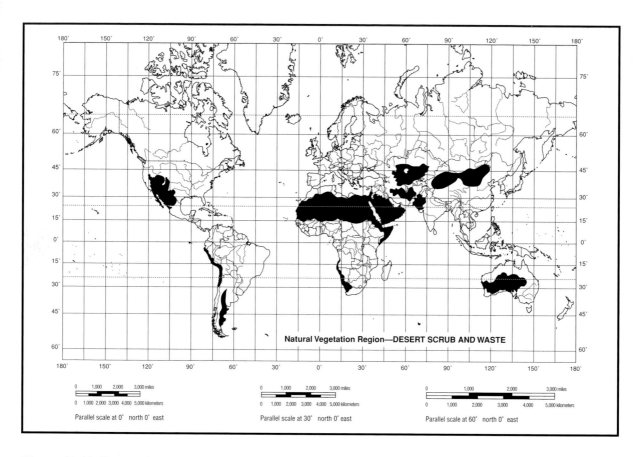

Figure 11-12. Desert Scrub and Waste Areas.

Land leeches are quite bloodthirsty and easily aroused by a combination of odor, light, temperature, and mechanical sense. These leeches are the most feared of all since they might enter air passages from which they cannot escape once they have fed and become distended. Normally, there is little pain when leeches attach themselves, and after they fill with blood, they drop off unnoticed. Some leeches, living in springs and wells, may enter the mouth or nostrils when drinking and may cause bleeding and obstruction.

    (g) Spiders, scorpions, hairy caterpillars, and centipedes are often abundant. The survivor should shake out shoes, socks, and clothing and inspect bedding morning and evening. A few spiders have poisonous bites that may cause severe pain. The black widow and the brown recluse spiders are venomous and should be considered very dangerous (figure 11-11). The large spiders called tarantulas rarely bite, but if touched, the short, hard hairs that cover them may come off and irritate the skin. Centipedes bite if touched and their bite is like that of a wasp's sting. Avoid all types of many-legged insects.

Scorpions are real pests, as they like to hide in clothing, bedding, or shoes, and strike without being touched. Their sting can cause illness or death.

    j. Population. Density of human population varies with the climate. Cultivation is difficult in areas of tropical rain forests along the Equator. The torrential rains leach out the soil and weeds grow rapidly. Consequently, cultivated food sources must be supplemented by game and other products of the forest. Villages are usually scattered along rivers since movement is easier by water than through the dense forest. Numerous people are also located along coastal areas where farming takes place and people can obtain food from the sea.

11-4. Dry Climates. Dry climates are generally thought of as hot, barren areas that receive scanty rainfall. Rainfall is limited, but dry climates are not barren wastelands and many kinds of plants and animals thrive (figure 11-12).

    a. Deserts:

    (1) Most deserts are located between the latitudes of 15 and 35 degrees on each side of the Equator and are

Figure 11-13. The Sahara.

dry regions where the annual evaporation rate exceeds the annual precipitation rate (generally less than 10 inches of rain annually). Extremes of temperature are as characteristic of deserts as is lack of rain and great distances. Hot days and cool nights are usual. A daily low-high of 45°F in the Sahara Desert and a 25°F to 35°F difference in the Gobi Desert is the rule. The difference between summer and winter temperatures is also extreme.

(2) Deserts occupy nearly 20 percent of the Earth's land surface, but only about 4 percent of the world's population lives there. The term "desert" is applied to a variety of areas. There are alkali deserts, rock deserts, and sand deserts. Some are barren gravel plains without a spear of grass, a bush, or cactus for a hundred miles. In other deserts there are grasses and thorny bushes where camels, goats, or even sheep find a subsistent diet. Anywhere they are found, deserts are places of extremes. They can be extremely dry, hot, cold, and often devoid of plants, trees, lakes, or rivers. Most important to the survivor is the extremely long time (distance) between water sources. The vast deserts of North

Africa, Arabia, Iran, West Pakistan, the Mojave Desert of the southwestern United States and northern Mexico, the interior Kalahari Desert of South Africa, and the Australian Desert are major examples of this type.

(3) Desert areas and climatic characteristics and seasonal variations of world deserts are as follows:

(a) Sahara Desert. The Sahara is the largest desert in the world. It stretches across North Africa from the Atlantic Ocean to the Red Sea and from the Mediterranean and the Sahara Atlas Mountains in the north to the Niger River in tropical Africa. It consists of 3 million square miles of level plains and jagged mountains, rocky plateaus, and graceful sand dunes. There are thousands of square miles where there is not a spear of grass, not a bush or tree, nor a sign of any vegetation. But Sahara oases—low spots in the desert where water can be reached for irrigation—are among the most densely populated areas in the world. Date groves and garden patches supporting 1,000 people per square mile are surrounded by barren plains devoid of life. Only 10 percent of the Sahara is sandy. The greater part of the desert is flat gravel plain from which the sand has been blown away and accumulated in limited areas, forming dunes. There are rocky mountains rising 11,000 feet above sea level and there are a few depressions 50 to 100 feet below sea level. The change from plain to mountain is abrupt in the Sahara. Mountains generally go straight up from the plain like jagged skyscrapers from a city street. Sharp-rising mountains on a level plain are especially noticeable in many desert landscapes because there is no vegetation to modify that abruptness. The lack of trees or bushes makes even occasional foothills appear more abrupt than in temperate climates (figure 11-13).

(b) Arabian Desert. Some geographers consider the Arabian Desert as a continuation of the Sahara. Half-a-million square miles in area, the Arabian Desert covers most of the Arabian Peninsula except for fertile fringes along the Mediterranean Sea, Red Sea, Arabian Sea, and the valleys of the Tigris-Euphrates Rivers. Along much of the Arabian coastline, the desert meets the sea. There is more sand in the Arabian Desert than in the Sahara and there are fewer date-grove oases. These are on the east side of the desert at Gatif, Hofuf, and Medina. Also, there is some rain in Arabia each year, in contrast to the decades in the Sahara that pass without a drop. Arabia has more widespread vegetation, but nomads find scanty pasture for their flocks of sheep and goats and must depend on wells for water. Oil is carried across the desert

in pipelines that are regularly patrolled by aircraft. Pumping stations are located at intervals. All these evidences of modern civilization have increased the well-being of the desert people and, as a result, chances for a safe journey afoot. However, the desert of Arabia is rugged, and native Arabs still get lost and die from dehydration.

(c) Gobi Desert - "Waterless Place." As used here, "Gobi" means only the 125,000 square mile basin or saucer-like plateau north of China that includes Inner and Outer Mongolia. On all sides of the Gobi, mountains form the rim of the basin. The basin itself slopes so gently that much of it appears to be a level plain. The Gobi has rocks, buttes, and numerous badlands, or deeply gullied areas (figure 11-14). For a hundred miles or so around the rim of the Gobi, there is a band of grassland. In average years, the Chinese find this to be a productive farmland. In drought years, agriculture retreats. Moving toward the center of the Gobi, there is less and less rainfall; soil becomes thinner, and grass grows in scattered bunches. This is the home of the Mongol herdsman. His wealth is chiefly horses, but he also raises sheep, goats, camels, and a few cattle. Beyond the rich grassland, the Gobi floor is a mosaic of tiny pebbles that often glisten in sunlight. These pebbles were once mixed with the sand and soil of the area, but in the course of centuries, the soil has been washed or blown away and the pebbles have been left behind as loose pavement. What rain there is in the Gobi drains

Figure 11-14. The Gobi.

Figure 11-15. Composite of American Deserts.

toward the basin; almost none of it cuts through the mountain rim to the ocean. There are some distinct and well-channeled watercourses, but these are usually dry. Many are remnants of prehistoric drainage systems. In the east, numerous shallow salt lakes are scattered over the plain. They vary in size and number with the changes of rainfall in the area. Sand dunes are found in the eastern and western Gobi, but these features are not as pronounced as they are in certain sections of the Sahara Desert. The Gobi is not a starkly barren waste-land like the great African Desert. Grass grows every-where, although it is often scanty. Mongols live in collective-type farm systems and habitations instead of being concentrated in oases (figure 11-14).

    (d) Australian Desert. More than one-third of Australia's total area is desert. Rainfall in the area is unpredictable, with an average of less than 10 inches per year. There are three connecting deserts that occupy western Australia and one desert located in the center of the continent. They are the Great Sandy Desert, Great Victorian Desert, Gibson Desert, and Simpon or Arunta Desert. The three largest deserts, Sandy, Victoria, and Arunta are of the sandy type, held in place by vegetation. The Gibson in the western portion is a stone-type desert. Most of the deserts of Australia have elevations of 1,000 to 2,000 feet.

    (e) Atacama-Peruvian Desert of South America. Generally, there are two regions of desert in South America. The first, and by far the largest, is along the west coast, beginning in the southern part of Ecuador, extending the entire coastline of Peru, and reaching nearly as far south as Valparaiso in Chile. This region, of about 2,000 miles in length and approximately 100 miles in width, is classified as true desert. Even so, along the shoreline of Peru as far south as Africa and inland a few miles, there is often a low-cloud or misty-fog layer. The layer is approxi-mately 1,000 feet thick and produces a fine drizzle. Because of this frequent cloud cover and other phenomena, the temperature along the coastal desert is remarkably cool, averaging about 72°F in the summer daytime and about 55°F in the winter daytime. From about 30 degrees south, the cloud cover does not exist and this region may truly be called rainless. The rare and uncertain showers are valueless for cultivated vegetation. Behind the coastal ranges in the higher elevations, the dryness is at a maxi-mum. In the nitrate fields of the Great Atacama Desert, the air is very dry and the slightest shower is very rare. Here the summer daytime temperatures range from 85°F to 90°F. The second desert region is entirely in Argentina (east of the Andes), extending in a finger-like strip from about 30 degrees south, southwest to about 50 degrees south. This region is approximately 1,200 miles long and

100 miles wide. In this highly dissected plateau region, the temperature ranges from a yearly average of 63°F in the north to 47°F in the south. The average annual rainfall pattern ranges from about 4 inches in the north to about 6 inches in the south.

(f) Southwest Deserts of the United States and Mexico:

-1. These desert areas have four major subdivisions:

-a. Great Basin—the basin between the Rocky Mountains and the Sierra Nevada-Cascade Ranges of southern Nevada and western Utah.

-b. Mojave Desert—Southwestern California.

-c. Sonoran Desert—Southeastern California across southern Arizona into the southwest corner of New Mexico and from Sonora and Baja, California, into Mexico.

-d. Chihuahuan Desert—lies to the east of the Great Sierra Madre Occidental, spreading north into southwest Texas, southern New Mexico, and the southeast corner of Arizona (figure 11-15).

-2. The flat plains with scanty vegetation and abruptly rising buttes of our Southwest are reminders of both the Gobi and Sahara. But the spectacular rock-walled canyons found in the Southwest have few counterparts in the deserts of Africa and Asia. The gullied badlands of the Gobi resemble similar formations in both the Southwest and the Dakotas, but our desert rivers—the lower Colorado, lower Rio Grande, and tributaries such as the Gila and the Peccos—have a more regular supply of water than is found in Old World deserts. The Nile and Niger are, in part, desert rivers but get their water from tropical Africa. They are desert immigrant rivers (like the Colorado, which collects the melting snows of the southern Rockies) and gain sufficient volume to carry them through the desert country. In general, the Southwest deserts have more varied vegetation, greater variety of scenery, and more rugged landscape than either the Gobi or the Sahara. In all three areas, it is often a long time (distance) between water sources. Death Valley, a part of the Mojave lying in southern California, probably has more waterholes and more vegetation than exist in vast stretches of the Sahara. The evil reputation of the Valley appears to have been started by unwise travelers who were too terrified to make intelligent searches for food and water. The dryness of the Death Valley atmosphere is unquestioned, but it lacks the vast barren plains stretching from horizon to horizon present in the Sahara. Compared to the Sahara, the desert country of the southwestern United States and Mexico sometimes looks like a luxuriant garden. There are many kinds of cactus plants in the desert, but these are not found in either the Sahara or Gobi.

(g) Kalahari Desert. This desert is located in the southern part of Africa. The wasteland covers about 200,000 square miles and lies about 3,000 feet above sea level. Some parts are largely covered with grassland and scrubby trees. The climate is similar to that of the Atacama-Peruvian Desert of South America.

b. Vegetation. The following are the more common xerophytic plants (those plants that can live with a limited water supply), which are found in the major deserts of the world.

(1) Cactus Family (Cactacae). Most cactus fruits and leaves have spines to protect them from birds and

Figure 11-16. Prickly Pear.

Figure 11-17. Barrel Cactus.

animals seeking water stored in the stems and leaves. Flat cactus leaves can be boiled and eaten, and the flowering fruit is edible.

(a) Prickly Pear. Is native and most abundant in the American deserts, but has been introduced into the Gobi, Sahara, and Australian Deserts, and other parts of the world (figure 11-16).

(b) Barrel Cactus. Found in many places, but is only native to the North American deserts. It grows to 5 or 6 feet high (figure 11-17).

(c) Suguaro (Giant) Cactus. Abundant in southern Arizona and in Sonora, Mexico. Can grow up to 50 feet tall.

(2) Wild Onion. Found in the Great Basin of the Southwest and in the Gobi Desert. The bulbs are edible if they look, smell, and taste like an onion or garlic.

(3) Wild Tulip. Found in the Gobi and Sahara Deserts. The bulbs can be eaten.

(4) Shrubs:

(a) Abal. Grows to about 4 feet tall in sandy deserts. The fresh flowers can be eaten. The dry twigs

can be crushed and used as a tea substitute. It is found in the Sahara and Arabian Deserts.

(b) Acacia. Most common in the Sahara, Gobi, and Australian Desert regions, and in the warmer and drier parts of America. The beans can be crushed and cooked as porridge. It is spiny with many branches and grows to 10 feet tall; roots yield water 4 to 5 feet from the tree trunk (figure 11-18).

(c) Saxaul. Found in the salt deserts of the Gobi Desert. The bark acts as water storage and is a good water source.

(d) St John's Bread. Found along the border of the Mediterranean coast of the Sahara and across the Arabian Desert. Grows 40 to 50 feet tall, and seeds can be pulverized and cooked as porridge (most nutritious plant food in the Middle East).

(e) Juniper. Found in the mountainous areas of the American deserts.

(f) Vines. Wild desert gourds are found in the Sahara and Arabian Deserts. They have a vine that grows from 8 to 30 feet long, and produce a melon-like poisonous fruit. The seed can be eaten when roasted or boiled. Flowers are also edible (figure 11-19).

(g) Succulent Plants. They are filled with juices and store moisture to survive. The surface is covered with a layer of wax or a blanket of fine hairs for protection against the heat. The moisture is contained in a tough cellulos that is not digestible and must be manually broken down to release the water.

(h) Creosote Bush. This is the most widespread and successful plant of the American deserts (height from 2 to 10 feet).

(5) Dates. Occur in groves around desert oases of the lower areas of the Sahara Desert.

c. Animal Life. There are more than 5,000 species of birds, reptiles, mammals, and insects found in desert areas. The raven, dove, woodpecker, owl, and hawk are common bird species. Reptiles such as lizards and snakes are numerous due to their adaptation to desert areas and ability to conserve body fluids. Many types of mammals live in desert areas and are primarily found near water sources.

d. Human Population. Humans are greatly influenced by the presence of water, and they live close to rivers, wells, cisterns, or oases. For example, in the Sahara Desert the 2 million people inhabiting this area are located near some 50 desert oases and small coastal cities. In the Gobi Desert, the Mongols live in scattered camps and move from one well to the next as they travel. In the southwest deserts of America, the population is greater along the Colorado and Rio Grande rivers.

11-5. Warm Temperate Climates:

a. The temperate zone is the area or region between the Tropic of Cancer and the Artic Circle and between the Tropic of Capricorn and the Antarctic Circle. The

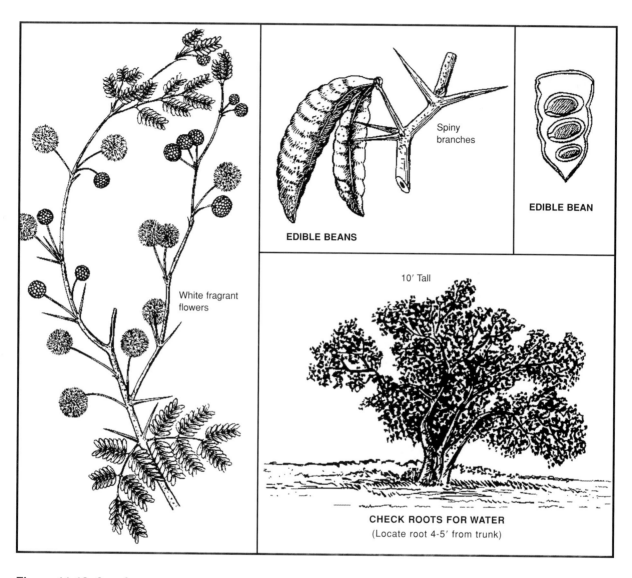

Spiny
branches

EDIBLE BEANS

EDIBLE BEAN

White fragrant
flowers

10' Tall

CHECK ROOTS FOR WATER
(Locate root 4-5' from trunk)

Figure 11-18. Acacia.

latitudes that comprise the temperate zone are 23 1/2°
north latitude to 66 1/2° north latitude and 23 1/2° south
latitude to 66 1/2° south latitude.

b. There are two main types of climate that comprise
the temperate group—the mild type, dominated by oceanic or marine climate; and a more severe one called continental climate.

(1) The temperate oceanic climate is the result of
warm ocean currents where the westerly winds carrying moisture have a warming effect on the landmass.
This oceanic type climate cannot develop over an
extensive area on the eastern or leeward side of large
continents in the middle latitudes. The extended effect
of the ocean climate can be limited by mountain

ranges. Such is the case with the Olympic, Cascade,
and Rocky Mountains. As the oceanic weather system
moves across the Olympic Mountains, it drops nearly
300 inches of precipitation annually. On the windward
side of the Cascades, the annual precipitation ranges
from 80 to 120 inches annually. In contrast, the region
from the leeward side of the Cascades is a relatively
dry area, receiving between 10 to 20 inches of precipitation annually. As the system moves across the
Rocky Mountains, most of the remaining moisture
is lost.

(2) The temperate continental climate is a land-
controlled climate, which is a product of broad middle-
latitude continents. Because of this, the continental

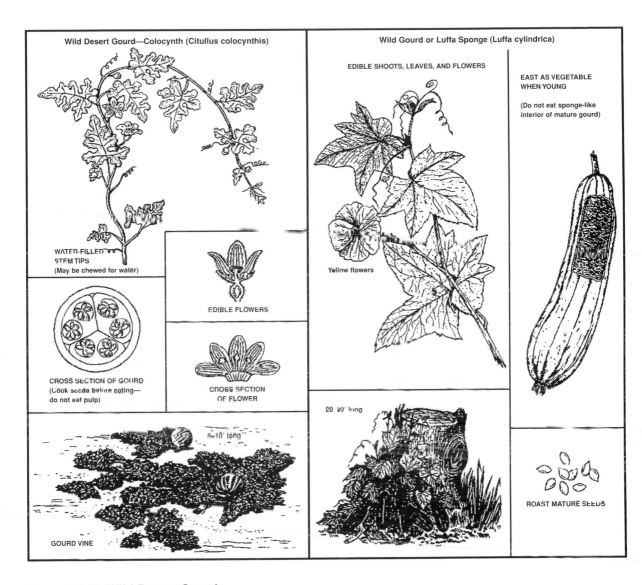

Figure 11-19. Wild Desert Gourd.

climate is not found in the Southern Hemisphere. This type of climate is very characteristic of the leeward side of mountain barriers and eastern North America and Asia. These areas are associated with dry interiors since there are few major warm-water sources available for formation of water systems. The average temperature in the winter and summer are not only extreme but also variable from one year to the next. The severe winter temperature is caused by the polar airflow toward the Equator, and neither winter nor summer temperatures are moderated by the effects of large water masses (oceans).

c. The climate within the temperate zone varies greatly in temperature, precipitation, and wind. The temperate (mid-latitude) zone is divided into four major climate zones that are controlled by both tropic and polar air masses.

(1) The humid subtropical zone is located generally between 20 and 30 degrees north and south latitude. This climate also tends to occur on the east coast of the continents that are at these latitudes. An example of this zone in the United States is the area between Missouri to lower New York and east Texas to Florida. The temperature ranges from 75°F to 80°F in the summer months to 27°F to 50°F in the winter months. The total average precipitation is 30 to 60 inches or more. During the summer months, convectional rainfall is common and thunderstorms frequent. In the winter, the rain is

Figure 11-20. Marine West Coast Climate (Olympic Peninsula).

more widespread and is usually associated with passing mid-latitude cyclones. The wind has a great influence in this area. The area is affected by both the prevailing westerlies and eastern trade winds. During the summer, the winds are influenced by eastern moist maritime air-mass flows. The weather is also influenced by low latitudes. The equatorial current that turns poleward forms warm currents (Gulf Stream and Japanese and Brazilian) that parallel the coasts.

(2) The marine west coast climate (figure 11-20) is sometimes referred to as the temperate oceanic climate. This climate is generally between 40 and 60 degrees north and south latitudes, on the west side of the continent. Examples are the west coasts of Washington to Alaska, Chile, nearly all of Europe, and New Zealand. The summer months are cool, with average temperatures of 60°F to 70°F. The winter months are mild, with temperatures averaging 27°F to 50°F. The total average rainfall ranges from 20 to 200 inches. Since the maritime climates are under the influence of the westerly winds all year, rainfall is nearly uniform from season to season. These climates are probably more cloudy than any other. They are characterized by widespread stratus and

nimbostratus clouds and frequent fog. One of the main reasons for the tremendous rainfall in these climatic areas is the warm ocean currents. These currents yield moisture to the air that is blown inland by the westerly winds (figures 11-21 and 11-22).

(3) Middle-latitude desert and steppe climates of complex origins are found generally between latitudes 35 to 50 degrees and in the interior of Asia and North America. Mountain ranges serve as barriers to the moist maritime air masses, thus resulting in low levels of precipitation. In summer, these interiors generate tropical air masses, while in winter they are overrun by polar air masses originating in Canada and Siberia. Deserts are also characterized by considerable differences between the average summer and winter temperatures. Of greater importance are the vast semi-dry steppes. Their annual precipitation of 10 to 20 inches supports short-grass vegetation. They comprise the great sheep and cattle ranges of the world; for example, the fields of South Africa and the American Great Plains support vast numbers of animals.

(4) The Mediterranean climate is sometimes referred to as subtropical dry summer climate. It is generally located from 30 to 45 degrees north and south

Figure 11-21. Maritime Climate (Southern Dahomey, Africa).

latitudes. Examples of this climate occur in the Mediterranean region, most notably in Spain, Italy, and Greece. Summer temperatures usually average 75°F to 80°F; but in coastal locations near cool currents, the average is 5°F to 10°F lower. Typical temperature averages for the coldest months are 45°F to 55°F. Coastal locations are usually somewhat warmer in the winter than inland locations. Total annual rainfall is normally 15 to 30 inches along the equatorial margins and increases poleward. This climate is a transitional zone between the dry west coast desert and the wet west coast climate. The westerly winds and cold ocean currents are the controlling influences of the Mediterranean climate. An example of a cold current that affects climates is the Humbolt Current (Peru Current) along the coast of Chile, Peru, and California.

d. Major topographical characteristics found in temperate regions are:

(1) Mountains. Areas of steep slopes with local relief of more than 2,000 feet. Examples of this landform are the Rocky Mountains of North America, the Andes Mountains of South America, and the Himalayan Mountains of Asia.

(2) High Tablelands. Upland surfaces more than 5,000 feet in elevation and having local relief of less than 1,000 feet, except where cut by widely separated canyons such as the High Tableland of the Wyoming Basin.

(3) Hills and Low Tablelands. Hill areas having local relief of more then 325 feet but less than 2,000 feet. At the ocean shoreland, however, local relief may be as low as 200 feet. A low tableland is an area less than 5,000 feet in elevation with local relief less than 325 feet, but that (unlike plains) either does not reach the sea, or where it does, terminates in a bluff overlooking a low coastal plain. Examples of this terrain can be found in the Appalachian Mountains, Quebec, and southern Argentina.

(4) Plains. Surfaces with local relief of less than 325 feet. On the marine side, the surface slopes gently to the sea. Plains rising continuously inland may attain elevations of high plains—more than 2,000 feet. The greatest expanses of plains occur in the center of the North American continent, Eastern Europe, and Western Asia (figure 11-23).

(5) Depressions. Basins surrounded by mountains, hills, or tablelands that abruptly outline the basins. Examples of depressions can be found in the southwestern United States.

e. There are several biomes of plants and animal life within the temperate zone, and the characteristic life forms are dependent upon climatic characteristics within a specific area. The biomes are named for the plants most plentiful in the area.

(1) Coniferous Forests (figure 11-24). These occur in a broad band across the northern portions of the continents of North America, Europe, and Asia. The northern boundary is the tundra and the southern limits are generally around 50 degrees north latitude. However, this zone extends down to 35 degrees north latitude in the mountainous regions of the western North American continent. This biome corresponds with the humid continental climate, except in the mountainous portions of North America below 50 degrees north latitude. The main life forms in this zone are the conifers or needle leaf, cone-producing trees, such as pines, firs, spruces, and hemlock. In these areas, the trees may grow closer together, not being severely limited by a need for sunlight, and are subject to frequent fires caused by lightning. When this occurs, the ecological succession is reversed, allowing low shrubs to spring up in the burned-over areas. Although the conifer is the predominant tree, there is more subclimax or secondary growth in these biomes than in climax forests (mature or primary forests). In these areas, the pines, alders, aspen, and poplars are the dominant trees. The dominant shrubs are heather, small maples, and yews. If forced to survive in these areas for long periods, especially in winter, the survivor will find that edible food plants are scarce (figure 11-25).

Figure 11-22. Red Mangrove and Swamps.

(a) In winter, the primary edible food plants available are:

-1. Rootstalks.
-2. Bulbs.
-3. Roots.
-4. Seeds.
-5. Resins (from pines).
-6. Infusion (teas) from evergreen needles.
-7. Bark (inner part).

(b) During the summer months, many more plants are available for food, including:

-1. Nuts.
-2. Sweet Acacia.
-3. Water Plantain.
-4. Shoots (potherbs).
-5. Wild Apple.
-6. Polypody.
-7. Leaves (potherbs).
-8. Baobab.
-9. Wild Rhubarb.
-10. Pollen (cattail).
-11. Beechnut.
-12. Flowering Bush.
-13. Flowers.
-14. Braken (Fern).
-15. Wild Sorrel.
-16. Fruits (dessert).
-17. Wild Calla.
-18. Cattail (Water Arum).
-19. Fiddleheads (Ferns).
-20. Chufa (Nut Grass).
-21. Chestnut.
-22. Wild Crabapple.
-23. Wild Dock.
-24. Chicory.
-25. Wild Filbert (Hazlenut).
-26. Wild Grape.
-27. Juniper.
-28. Common Jujube.
-29. Pine Nuts.
-30. Spreading Wood Fern.
-31. Wild Lily.
-32. English Oak (acorn).
-33. Tree Fern.
-34. Water Lily (temperate zone).

(c) In many places, the ground is thickly covered with mosses and there may be a few varieties of early flowering plants and many berry-bearing shrubs that invite birds and mammals into the open areas. Some of the largest herbivores (plant eaters) live in these evergreen forests—caribou, reindeer, moose, and deer. The small herbivores may include porcupines, several species of squirrels, mice, and rabbits. The carnivores (flesh eaters) that feed upon the smaller animals include black bear, gray wolf, lynx, wolverine, red fox, and weasel. Multitudes of insects provide food for the birds. (These insects may also present a menace to the survivor.) A large variety of birds feed not only on insects but also on plants.

(2) Deciduous Forests. Decodipis (broad leaved) forests are found extensively in the eastern portion of the United States; in Europe, between 40 to 50 degrees north latitude; and also in eastern portions of the USSR, China, Korea, and Japan from 35 to 50 degrees north latitude. This biome corresponds with the subtropic and humid continental climatic zones; the area in which any deciduous forest group determines the predominant trees or climax vegetation found there. Here are a few examples: In north central United States, Beech and Maple trees assume the dominant role; in Wisconsin and Minnesota, it is Basswood and Maple; in the eastern and southern regions, the dominant trees are Oak and

Figure 11-23. Plains.

Hickory (figure 11-26). There are also spots in this biome where pines and broadleaf evergreens grow.

(3) Deciduous and Mixed Deciduous-Coniferous Forest:

(a) Deciduous and mixed deciduous-coniferous forests manifest the following characteristics:

-1. Warm summer with rain; winters cold and drier, short drought periods.

-2. Only three stories of vegetation (trees, scrubs, herbs).

-3. Broadleaf trees without leaves in winter.

-4. Mature trees, uniform in height.

-5. Unimpeded view into interior of forest.

-6. Few herbs, ferns, mosses in summer, and abundance of edible fungi in spring and autumn.

-7. Trunks of trees covered with thick-fissured, dark-colored bark.

-8. Resting buds enclosed in hard, scaly protecting leaves frequently covered with gum or resin.

-9. For the most part, leaves are thin and delicate, rarely thick and leathery like those of tropical rain forest trees.

(b) The deciduous and mixed deciduous-coniferous forests that predominate over much of eastern United States are typical of this vegetation type.

The deciduous forest is wholly temperate in character. In contrast with the tropical evergreen forest with its richly shaded but chiefly dark, glossy green canopy, the broad-leaved temperate forest extends in a uniformly bright-green expanse. The temperate deciduous and mixed deciduous-coniferous forest-vegetation type occupies extensive areas in several parts of the world (figure 11-27).

-1. North America. Eastern United States.

-2. South America. Southern Chile, southeastern Brazil.

-3. Europe. Western and northern Europe, southern Scandanavia, southeastern Europe (Balkans).

-4. Asia. South central Siberia, southeastern Siberia and part of Manchuria, Korea (throughout), Japan (throughout), China (throughout except the extreme south and extreme north).

-5. Oceania. New Zealand.

(c) A general characteristic of a climax forest is the stratification of layers of plant growth similar to the canopy systems in the tropical rain forest. In a climax forest, there are usually a limited number of flowering plants, ferns, and shrubs for ground cover. The edible food plants in the vegetation zone are numerous, and a large array of edible species are available, including:

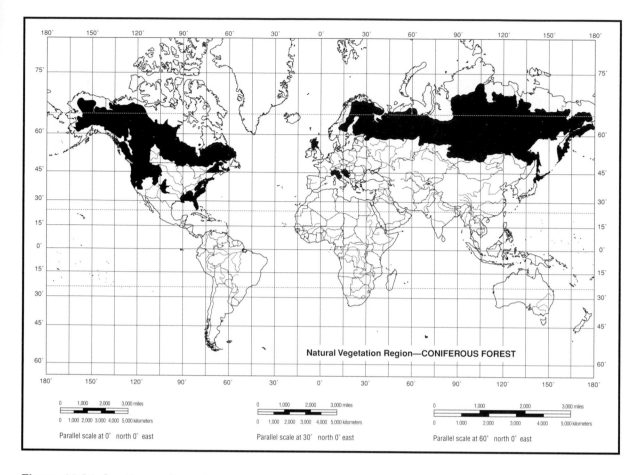

Figure 11-24. Coniferous Forest.

-1. Amarath.
-2. Water Chestnut
(Trapa Nut).
-3. Wild Lily.
-4. Wild Apple.
-5. Chicory.
-6. Lotus Lily.
-7. Beechnut.
-8. Chufa (Nut Grass).
-9. Mulberry.
-10. Braken (Fern).
-11. Wild Filbert
(Hazlenut).
-12. English Oak
(Acorn).
-13. Wild Calla
(Water Arum).
-14. Wild Grape.
-15. Wild Onion.
-16. Cattail.

-17. Common Jujube.
-18. Pine Nuts.
-19. Chestnut.
-20. Juniper.
-21. Polypody.
-22. Air Potato
(ubi tuber).
-23. Tree Fern.
-24. Wild Dock.
-25. Purslane.
-26. Wild Tulip.
-27. Water Plantain.
-28. Wild Rhubarb.
-29. Walnut.
-30. Pokeweed.
-31. Flowering Rush.
-32. Water Lily.
-33. Tropical Yam.
-34. Wild Sorrel.
-35. Wild Crabapple.

(d) Animal life associated with deciduous forests is more varied and plentiful than in evergreen forests,

though some animals, such as certain species of deer, squirrels, martins, lynx, and wildcats are common in both areas. Wolves, foxes, and other small carnivores (flesh-eating animals) feed mainly on small rodents. Some forest dwellers, such as rodents, dig their dens below the ground while other dens are dug near streams where food and shelter are found. In the aquatic environment, the beaver builds dams for food and shelter. Muskrat, otter, and mink also seek the water's edge, while snakes, turtles, and frogs are found in the streams or lakes.

(4) Steppes (figure 11-28) and Prairies:

(a) The part of Russia extending from the Volga River through central Asia to the Gobi Desert has been referred to as the steppes. However, as a vegetation type, the steppe grasslands occur in many other parts of the world. The rainfall in steppe areas averages 15 to 30 inches per year, as compared to prairie areas, which average 30 to 40 inches per year. The general aspect of a steppe area, like the prairie, is a broad, treeless expanse of open countryside that may be quite rolling in places. The principal steppe areas are:

Figure 11-25. Coniferous Forest.

Figure 11-26. Deciduous Forest.

-1. North America. Western Great Plains of the United States.

-2. South America. Argentina.

-3. Africa. Narrow belt extending across Africa on the southern rim of the Sahara, and parts of Ethiopia.

-4. Europe. Southeastern Russia.

-5. Asia, Turkey, Iran, Baluchistan, Pakistan, Turkestan, and a broad belt through central Asia.

-6. Australia. Fringes of the great central desert, especially in eastern Australia.

(b) The prairie and steppe areas are very closely related. However, the true prairie supports a somewhat different flora than the steppe areas, and for this reason, it is important that they be discussed separately. The chief distinction between prairies and steppes is the seasonal distribution of rainfall.

	PRAIRIE	STEPPE
Rainfall per year	30–40 inches	15–30 inches
Subsoil	Permanently moist	Permanently dry

In both, the precipitation comes during the short growing season (spring). Summers are hot, with intermittent showers. The primary prairie regions of the world are:

-1. North America. South central Canada, and east central United States.

-2. South America. Northeastern Argentina, Uruguay, Paraguay, and Brazil.

-3. Africa. Union of South Africa.

-4. Europe. Parts of Hungary, Rumania, and Russia (Ukraine and in a belt extending through central Russia to the Urals).

-5. Asia. Manchuria.

(c) The main plants in these biomes are grasses. Due to different conditions various characteristic grasses grow in specific areas on the prairies. The tall grasses are found near the edges of deciduous forests where larger amounts of water are available. The mid-grasses grow farther west, close to the Great Basin within the United States, with short grasses growing in the rain shadows of the mountains. Wild flowers and other annuals are found throughout these regions. On the fringes of the desert, desert plants may have moved into the grasslands. The following are food plants of the steppes:

-1. Sweet Acacia.   -9. Wild Tulip.
-2. Wild Chicory.   -10. Wild Calla
-3. Wild Rose.   (Water Arum).
-4. Amaranth.   -11. Wild Onion.
-5. Chufa (Nut Grass).   -12. Water Lily.
-6. Wild Sorrel.   -13. Cattail.
-7. Baobab.   -14. Sea Orach.
-8. Wild Dock.

(d) Common herbivores of the prairies are ground squirrels, prairie dogs, rabbits, gophers, and a great many species of mice. These are preyed upon by

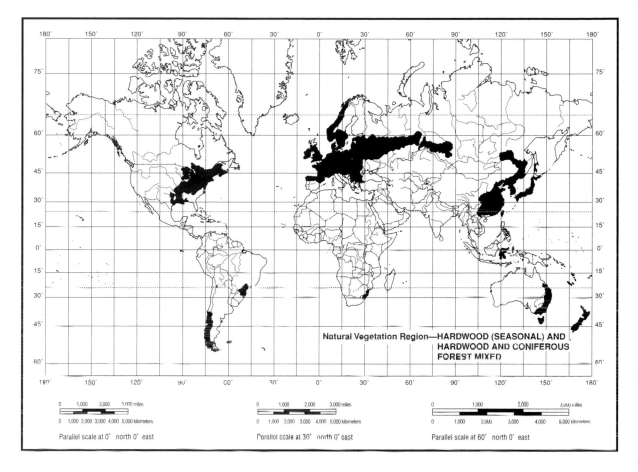

Natural Vegetation Region—HARDWOOD (SEASONAL) AND HARDWOOD AND CONIFEROUS FOREST MIXED

Figure 11-27. Hardwood (Seasonal).

badgers, coyotes, foxes, skunks, and hawks. Prairie animals travel in packs or herds that serve both to protect their individual members and assist in hunting prey. They typically have excellent vision and sense of smell, but their hearing, though keen, may be impaired by the noise of the pack or herd.

(e) A number of birds nest among the grasses. These include the meadowlark, prairie chicken, and grouse. During the dry season, some of these birds migrate to places better suited to raising their young.

(f) Insects like grasshoppers are well adapted to a grassland environment. The natural enemies of such insects are birds and reptiles that in turn become the prey of owls and hawks.

(5) Evergreen Scrub Forests. These biomes occur in southern California, in countries around the Mediterranean Sea, and in southern portions of Australia and correspond with the Mediterranean climate (figure 11-29).

(a) The major life form in this area is vegetation composed of broad-leaved evergreen shrubs, bushes, and trees usually less than 8 feet tall. This vegetation generally forms thickets. Sage and evergreen oaks are the dominant plants in North America in areas with rainfall between 20 and 30 inches. Areas with less rainfall or poorer soil have fewer, more drought-resistant shrubs such as manzanita. Scrub-forest vegetation becomes extremely dry by late summer. The hot, quick fires that commonly occur during this period are necessary for germination of many shrub seeds and also serve to clear away dense ground cover. This ground cover is difficult for the survivor to penetrate. The branches are tough, wiry, and difficult to bend. Trees are usually widely scattered, except where they occur in groves near a stream. Usually, both trees and shrubs have undivided leaves. Grasses and brightly colored spring-flowering bulbs and other flowers may also be found. The survivor will find relatively few kinds of edible plant food within the scrub forest. During the growing season—usually only the spring months—the following kinds of plant foods are available:

-1. Agave (Century   -2. Wild Dock.
    Plant). (See figure   -3. Wild Pistachio.
    11-30.)

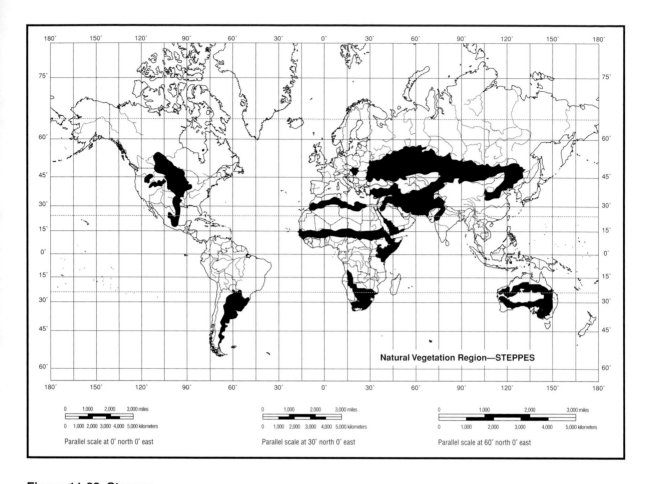

Figure 11-28. Steppes.

-4. Almond.

-5. Wild Grape.

-6. Wild Sorrel.

-7. Wild Apple.

-8. Juniper.

-9. Wild Tulip.

-10. Chicory.

-11. English Oak (Acorn).

-12. Walnut.

-13. Wild Crabapple.

-14. Wild Onion.

(b) Deer and birds usually inhabit these forests only during the wet season, which is the growth period for most scrub-forest plants. Small, dull-colored animals such as lizards, rabbits, chipmunks, and quail are year-round residents.

11-6. Snow Climates. "Snow climates" are defined as the interior continental areas of the two great landmasses of North America and Eurasia that lie between 35 and 70 degrees north latitude. The tree line provides the best natural boundary for a topographical description of the snow-climate areas. There are definite differences between the forest area to the south and the tundra to the north in snow-cover characteristics, wind conditions, animal types, and vegetation. Snow climates are comprised of two separate climate types: continental subarctic and humid continental.

a. The continental subarctic climate is one of vast extremes. The temperature may range from -108°F to 110°F. Temperatures may also fluctuate 40 to 50 degrees within a few hours. This area includes several climate subtypes. The largest areas run from Alaska to Labrador and Scandinavia to Siberia. They are cold, snowy forest climates, moist all year, with cool, short summers. A colder climate is found in northern Siberia, which has very cold winters with an average cold temperature of -36°F. Another area is found in northeastern Asia where the climate is a cold, snowy forest climate with dry winters. Winter is the dominant season of the continental subarctic climate. Because freezing temperatures occur for 6 to 7 months, all moisture in the ground is frozen to a depth of many feet.

b. The humid continental climates are generally located between 35 and 60 degrees north latitude. For the most part, these climates are located in central and eastern parts of continents of the middle latitudes. These climates are a battle zone of polar and tropical air masses.

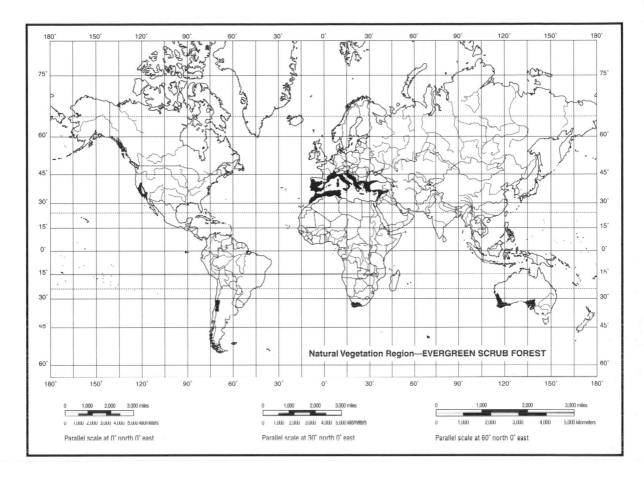

Natural Vegetation Region—EVERGREEN SCRUB FOREST

0 1,000 2,000 3,000 miles
0 1,000 2,000 3,000 4,000 5,000 kilometers
Parallel scale at 0° north 0° east

0 1,000 2,000 3,000 miles
0 1,000 2,000 3,000 4,000 5,000 kilometers
Parallel scale at 30° north 0° east

0 1,000 2,000 3,000 miles
0 1,000 2,000 3,000 4,000 5,000 kilometers
Parallel scale at 60° north 0° east

Figure 11-29. Evergreen Scrub Forest.

Seasonal contrasts are strong and the weather is highly variable. In North America, this climate extends from New England westward beyond the Great Lakes region into the Great Plains and into the prairie provinces of Canada. This climate can also be found in central Asia. The summers are cooler and shorter than in any other climate in the temperate zone, with the exception of the highland (Alpine) subarctic climate. The summer temperatures range from 60°F to 70°F. The winter temperatures range from -15°F to 26°F. The precipitation for the year varies from 10 to 40 inches. A higher percentage of the precipitation is snow, with less snow occurring in areas along the coasts. The weather is influenced by the polar easterly winds and the subtropical westerly winds. The effect of ocean currents on this continental climate is minimal. This climate is dominated by the high- or low-pressure cells centered in interiors of the continent.

c. Both climate regions have seasonal extremes of daylight and darkness resulting from the tilt of the Earth's axis (figure 11-31). Snow-climate nights are long, even continuous in winter; conversely, north of the Arctic Circle, the Sun is visible at midnight at least once a year. Darkness presents a number of problems to the survivor. No heat is received directly from the Sun in midwinter, thus the cold reaches extremes. Outside activities are limited to necessity, although the light from the Moon, stars, and auroras, shining on a light ground surface, is of some help. Confinement to cramped quarters adds boredom to discomfort, and depression becomes a dominant mood as time drags on. Fortunately, the period of complete darkness does not last long.

d. The terrain of the snow climate areas coincides with a great belt of needle-leaf forests. This region is found in the higher middle latitudes. Its poleward side usually borders on tundra and its southern margin usually adjoins continental temperate climates. This area is like the tundra because it has poor drainage. As a result, there are an abundance of lakes and swamps. The coastlines vary from gentle plains sweeping down to the ocean to steep, rugged cliffs. Glaciers are a predominant feature of the high altitudes (6,000-feet elevation or above). These glaciers flow down to lower elevations or terminate at the ocean.

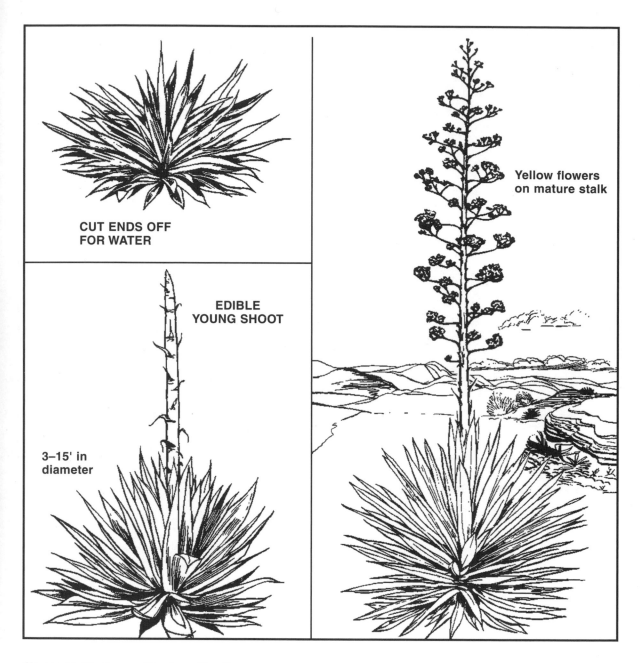

CUT ENDS OFF FOR WATER

EDIBLE YOUNG SHOOT

3–15' in diameter

Yellow flowers on mature stalk

Figure 11-30. Agave (Century Plant).

e. The vegetation is similar to that found in more temperate zones; however, the cold temperatures have caused variations in the physical appearance of the plants. Dark evergreen forests thrive south of the tree line. They consist mainly of cedar, spruce, fir, and pine, mingled with birch. These subarctic forests are called taigas. A transitional zone lies between the taiga and the tundra. In this zone, the trees are sparse and seldom grow more than 40 feet tall. Dwarf willow, birch, and alder mix with evergreens, and reindeer moss sometimes forms a thick carpet (figure 11-32).

f. Depending on the time of year and the place, chances for obtaining animal food vary considerably. Shorelines are normally scraped clean of all animals and plants by winter ice. Inland animals are migratory.

(1) Large Arctic Game. Caribou and reindeer migrate throughout northern Canada and Alaska (figure 11-33). In northern Siberia, they migrate inland to almost

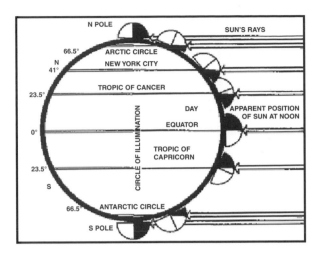

Figure 11-31. Circle of Illumination.

Musk oxen may be found in northern Greenland and on the islands of the Canadian archipelago. Sheep descend to lower elevations and to valley feeding grounds in the winter. Wolves usually run in pairs or groups. Foxes are solitary and are seen most frequently when mice and lemmings are abundant. Bears are dangerous, especially when wounded, startled, or with their young. They generally shun areas of human habitation.

(2) Small Land Game. Tundra animals include snowshoe and arctic hares, lemmings, mice, and ground squirrels. They may be trapped or shot in the winter or summer. Most prefer some cover and can be found in shallow ravines or in groves of short willows. Ground squirrels and marmots hibernate in the winter. In summer, ground squirrels are abundant along sandy banks of large streams. Marmots live in the mountains, among the rocks, usually near the edge of a meadow or in deep soil—much like woodchucks. To find the burrow in rocky areas, look for a large patch of orange-colored lichen on rocks. This plant grows best on animal or bird dung, and the marmot always seeks relief in the same spot, not far from a well-hidden entrance.

g. The Arctic is the breeding ground for many birds. In summer, ducks, geese, loons, and swans build their nests near ponds on the coastal plains or bordering lakes

50 degrees north latitude. Some are found in west Greenland. All move close to the sea or into the high mountains in summer. In winter, they feed on the tundra.

Figure 11-32. Subarctic Forest.

FOX

MOUNTAIN GOAT

CARIBOU

MUSK OX

BEAR

Figure 11-33. Arctic Game.

or rivers of the low tundra. A few ducks on a small pond usually indicates that setting birds may be found and flushed from the surrounding shores. Swans and loons normally nest on small, grassy islands in the lakes. Geese crowd together near large rivers or lakes. Smaller wading birds customarily fly from pond to pond. Grouse and ptarmigan are common in the swampy forest regions of Siberia. Sea birds may be found on cliffs or small islands off the coast. Their nesting areas can often be located by their flights to and from their feeding grounds. Jaeger gulls are common over the tundra, and frequently rest on higher hillocks. In the winter, fewer birds are available because of migratory patterns. Ravens, grouse, ptarmigan, and owls are the primary birds available. Ptarmigan are seen in pairs or flocks, feeding along grassy or willow-covered slopes.

h. Arctic and tom cod, sculpin, eelpout, and other fish may be caught in the ocean. The inland lakes and rivers of the surrounding coastal tundra generally have plenty of fish that are easily caught during the warmer season. In the North Pacific and in the North Atlantic extending slightly northward into the Arctic Sea, the coastal waters are rich in all seafoods. Varieties include fish, crawfish, snails, clams, oysters, and one of the world's largest and meatiest crabs—the king crab of the Aleutian Islands and Bering Sea areas. In the spring (breeding season), this crab comes close to shore and may be caught on fish lines set in deep water or by lowering baited lines through holes cut in the ice. Do not eat shellfish that are not covered at high tide. Never eat any type of shellfish that is dead when found, or any that do not close tightly when touched. Poisonous fish are rarer in the Arctic than in the tropics. Some fish, such as sculpins, lay poisonous eggs; but eggs of the salmon, herring, or freshwater sturgeon are safe to eat. In arctic or subarctic areas, the black mussel may be very poisonous. If mussels are the only available food, select only those in deep inlets far from the coast. Remove the black meat (liver) and eat the white meat. Arctic shark meat is also poisonous (high concentration of vitamin A).

11-7. Ice Climates. There are three separate climates in the category of ice climates: marine subarctic climate, tundra climate, and icecap climate (figure 11-34).

   a. Marine Subarctic Climate. Key characteristics of this climate are the persistence of cloudy skies and strong winds (sometimes in excess of 100 miles per hour), and a high percentage of days with precipitation. The region lies between 50 and 60 degrees north latitude and 45 to 60 degrees south latitude. The marine subarctic climate is found on the windward coasts, on islands, and over wide expanses of ocean in the Bering Sea and the North Atlantic, touching points of Greenland, Iceland, and Norway. In the Southern Hemisphere this climate is found on small landmasses.

b. Tundra Climate. The tundra region lies north of 55 degrees north latitude and south of 50 degrees south latitude. The average temperature of the warmest month is below 50°F. Proximity to the ocean and persistent cloud cover keep summer air temperatures down despite abundant solar energy at this latitude near the summer solstice (figures 11-35 and 11-36).

c. Icecap Climate. There are three vast regions of ice on the Earth. They are the Greenland and Antarctic continental icecaps and the larger area of floating sea ice in the Arctic Ocean. The continental icecaps differ in various ways, both physically and climatically, from the polar sea ice and can be treated separately (figure 11-37).

(1) Greenland. The largest island in the world is Greenland. Most of the island lies north of the Arctic Circle and ice covers about 85 percent of it. The warmest region of the island is on the southwestern coast. The average summer temperature is 50°F. The coldest region is the center of the icecap. The temperature there averages -53°F in the winter.

(2) The Antarctic. The Antarctic lies in a unique triangle formed by South America, Africa, and Australia. Surrounding the continent are portions of the Atlantic,

Pacific, and Indian Oceans. The area is almost entirely enclosed by the Antarctic Circle. The climate is considered one of the harshest in the world. The average temperature remains below 0°F all year. In the winter months, the mean temperature ranges from -40°F to -80°F. Winter temperatures inland often drop below -100°F. Great storms and blizzards (with accompanying high winds) range over the entire area due to both the continent's great elevation and being completely surrounded by warm ocean water.

(3) Sea Ice on the Arctic Ocean. Ice on the Arctic Ocean includes frozen seawater and icebergs that have broken off glaciers. This ice remains frozen near the North Pole year round. Near the coast, the sea ice melts during the summer. Currents, tides, and winds may cause it to fold and form high ridges called pressure edges. One piece of ice may slide over another, causing a formation called rafted ice. When the ice breaks into sections separated by water, these sections are called leads. Great explosions and rolling thunder are caused by the breaking and folding of the ice.

d. Terrain. The terrain of the true ice climates encompasses nearly every variation known. Much of the landmass is composed of tundra. In its true form, the tundra

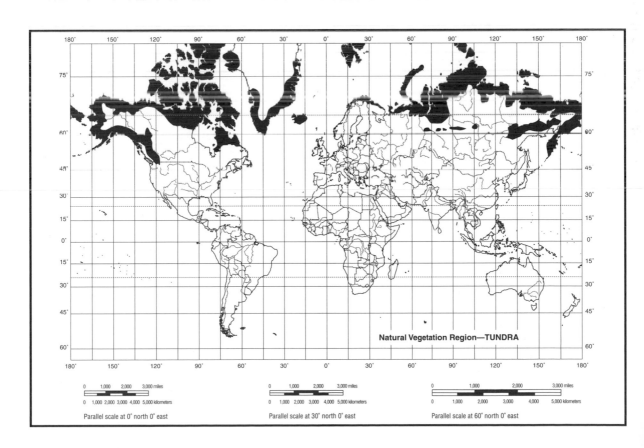

Natural Vegetation Region—TUNDRA

Figure 11-34. Tundra.

Figure 11-35. Winter Tundra.

is treeless. Vast rugged mountain ranges are found in the area and rise several thousand feet above the surrounding areas. Steep terrain, snow and ice fields, glaciers, and very high wind conditions make this area a very desolate place. Continental glaciers, such as the icecaps covering Greenland and the Antarctic continent, are large expanses of windswept ice moving slowly toward the sea. Ice thickness in continental-glacier areas can exceed 10,000 feet.

e. Vegetation: (1) Shrub Tundra. In Russia, the area surrounding the Lena River is known as a typical shrub-tundra environment. Shrubs, herbs, and mosses occur in this zone. Arctic birch predominates, but other shrubs occur and several may be useful as supplementary food, such as the crystal tea ledum (Labrador tea), willows, and the bog bilberry. In this same shrub zone a lower herbaceous layer occurs that is composed of black crowberry, several grasses, and the cowberry. On the ground, mosses and lichens are present in abundance. The shrubs on the open tundra reach a height of only 3 to 4 feet, but in valleys and along the rivers, the same shrubs may reach the height of a person.

(2) Wooded Tundra. The region immediately adjoining the treeless tundra is an extension of the coniferous areas of the south. These subarctic wooded areas include a variety of tree species of which the genus Picea (Spruce) predominates. On the Kola Peninsula of northeastern Scandinavia, these northernmost forests are birch. Siberian spruce occurs between the White Sea and the Urals. Siberian larch occurs between the Urals and the Pyasina River. Dahurian larch occurs between the Pyasina River and the upper reaches of the Anadyr River. In extreme northeastern Asia, Mongolian poplar, Korean willow, and birch are found along the rivers. The trees extending into the tundra are distinguished by their stunted growth (except in river valleys, where they reach 18 to 24 feet) and sparseness. Permanent ground frost, or permafrost, penetrates most parts of the true tundra, and the northern limits of the forest belt closely coincide with the southern limits of permafrost. A few different plants will cover very large areas so extensive stands of a single variety of plant are common in the arctic tundra. All tundra plants are small in stature compared to the plants in the warmer climates of more southerly latitudes. The

Figure 11-36. Summer Tundra.

arctic willow and birch, for instance, spread along the ground in the tundra to form large mats. Stunted growth in all the woody plants is the rule, although there are many evergreen plants and hardy bulbous or tuberous plants. Lichens, especially reindeer moss, are widespread in the tundra. As mentioned before, the plant life of the tundra is remarkably uniform in its distribution. Some species are common to all three areas, but other species are more restricted in their distribution. The tundra also contains many species of vegetation found in the forest regions to the south (figure 11-38).

(3) Bogs. The tundra has often been classified as a continuous bog, but this is far from the truth. Many bogs do exist. There are also many hilly and even mountainous areas with considerably drier soil. The moss or sphagnum bog is less common than the sedge bog. A characteristic of more southern tundras is the development of large peat mounds 9 to 15 feet high and 15 to 75 feet in diameter. These mounds have been formed by ground upheavals caused by freezing water. Many edible plants grow on these bog mounds, such as the cloudberry, dwarf arctic birch, bog bilberry, black crowberry, crystal tea ledum, sheathed cotton sedge, cowberry, and others.

f. Animal Life. Compared to other parts of the world, animal life is poor in species but rich in numbers. Large animals such as caribou, reindeer, and musk oxen migrate through the tundra areas. Carnivores—wolves, foxes, lynx, wolverines, and bears—range through the landmass

area, and polar bears, seals, walruses, and foxes are found far out on the sea ice. Small animals are the most abundant animal life found, and include hares, lemmings, marmots, mink, fishers, and porcupines.

(1) Bird life is very limited during the winter months—mainly owls and ptarmigan—but during the summer months millions of migratory waterfowl nest in the arctic tundra. Species include ducks, geese, cranes, loons, and swans, nesting in and around the swamps, bogs, and lakes of the tundra. The coastal areas are home to many species of sea birds during the summer months. The coastal waters and iceflows are rich in a variety of marine life, such as seals, walruses, whales, crustaceans, and fish.

(2) The freshwater rivers, lakes, and streams are teeming with many varieties of fish—salmon, trout, and grayling. Due to the amount of surface water in the tundra area, there are a large variety of insects. Some 40 to 60 species of mosquitoes, flies, and gnats inhabit the area.

(3) In the Antarctic, animals are virtually nonexistent. Only the lowest forms of animal life can live mainly on mosses and lichens. Marine animals, particularly whales, seals, and penguins, are found along the coastal regions. Sea birds are abundant in the summer and nest in the coastal regions and on the islands. There are a few species of wingless insects, lice, ticks, mites, etc., which live off the bird population.

Figure 11-37. Ice Climate.

11-8. Open Seas. The Köppen-Geiger system of climate classification has been used to describe the environmental characteristics of the landmasses. However, this system is not used to categorize the largest area of the world—the oceans. They are simply divided by their names and locations. All limits of oceans, seas, etc., are arbitrary, as there is only one global sea. The terms "sea" and "ocean" are often used interchangeably in reference to saltwater. However, from a geographic point of view, a sea is a body of water that is substantially smaller than an ocean or is part of an ocean.

a. The seas cover 70.8 percent of the Earth's surface. The waters are not evenly distributed, covering 61 percent of the surface in the Northern Hemisphere and 81 percent in the Southern Hemisphere. Traditionally, the seas are divided into four oceans: Atlantic, Pacific, Indian, and Arctic. These, with their fringing gulfs and smaller seas, make up the world's seas. If the land features of the Earth were smoothed out, the seas would cover the entire globe to a depth of 12,000 feet. Mount Everest, the tallest mountain

peak (29,028 feet), would disappear without a trace in the 37,800-foot deep Marinas Trench in the western Pacific Ocean. The sea floor is made up of mountains, valleys, great plains, and deep trenches. The deepest trenches and tallest mountains are found in the north Pacific. The sea-floor features do, to some extent, influence the surface properties of the seas, that is, currents, waves, and tides.

b. The average salinity of the seas is usually taken as 3.5 percent. Higher values occur at or near the surface in areas where high temperatures and strong, dry winds favor evaporation. The highest salinities occur in semi-landlocked seas at mid-latitudes, such as the Red Sea, the Persian Gulf, and the Mediterranean Sea. The Pacific Ocean is the largest and is more than twice the size of the Atlantic or Indian Oceans. Its size allows for greater climatic variations and more widespread influence. Due to similar latitudinal references the Atlantic and Pacific Oceans have many similar characteristics. The Indian Ocean is slightly smaller than the Atlantic, but is more significantly influenced by a continental landmass than

Figure 11-38. Arctic Zone Composite.

any other ocean. The Arctic Ocean is generally recognized as that body of water that lies north of 75 degrees latitude and is nearly enclosed by landmasses.

c. Within each of these four major oceans, numerous subdivisions, known as seas, may be geographically aligned along indistinct boundaries (island chains, geography of ocean floor). Examples are:

(1) The Coral Sea is an arm of the South Pacific Ocean lying east of Queensland, Australia, and west of New Hebrides and New Caledonia. It extends from the Solomon Islands in the north to the Chesterfield Islands in the south.

(2) The Bering Sea is located between Alaska and Eastern Siberia, with its southern boundary formed by the arc of the Alaskan Peninsula and the Aleutian Islands. The Bering Strait connects it with the Arctic Ocean to the north.

d. Many water bodies are partially enclosed by land and are known as gulfs. An example would be the Gulf of Mexico.

11-9. Ocean Currents. The ocean has a complex circulation system made up of a variety of currents and countercurrents. These currents move at a rate from barely measurable to about 5.75 miles per hour. They may be relatively cold or warm currents and influence the climate and environment that exists on land and over the ocean. There is a constant movement of water from areas of high density, salinity, concentration, and

pressure to areas of low density, salinity, concentration, and pressure in an attempt to establish equilibrium. These factors influence the movement of ocean currents. However, the primary influence on ocean currents is the wind. They may also be diverted by the Coriolis force and continental deflection (figure 11-39).

11-10. Climatic Conditions. To fully understand the general climatic conditions and seasonal variations that exist over the global sea, each major ocean must be examined separately, with the exception of the Atlantic and Pacific, whose similar latitudinal references result in like characteristics (exceptions will be noted). The two physical phenomena that have the greatest impact on climate are currents and systems of high and low air pressures.

a. Currents, with their basic characteristics of being either warm or cold and their inevitable convergence, influence the environment of the open seas. As equally significant as the ocean influence on typical weather sequences (for example, temperature, wind, precipitation, and storms) are semi- and quasi-permanent centers of high and low atmospheric pressures. To observe their effect on climate, imagine a hypothetical voyage from the Pole to the Equator. The southern limit of the solidly frozen arctic icepack varies in latitude from about 65 to 75 degrees between February and August. In the winter, brief periods of calm, clear weather with a mean temperature of -5°F are interspersed between passages of

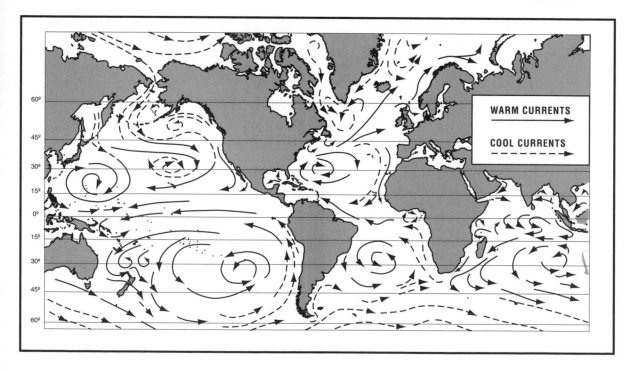

Figure 11-39. Ocean Currents.

cyclonic storms characterized by snow, winds 30 to 40 mph, temperatures from -20°F to -30°F, and gale-force winds 30 percent of the time. In the summer, frequent periods of several days of calm or light variable winds with temperatures in the mid-40s may be experienced. Skies are uniformly overcast with layers of stratus or nimbostratus clouds. Dense fogbanks are prevalent during calms. Rain or drizzle may continue for weeks at a time. One of the stormiest regions in the hemisphere is in the middle of the prevailing westerlies at 50 degrees north latitude. In the winter, calms are rare with winds of 15 to 20 knots and temperatures near freezing. Every 2 to 3 days, a pale sun and scattered clouds give way to cumulostratus clouds and rain squalls. Wind intensity may reach 50 to 60 knots with temperatures dropping to -10°F to -15°F as the rain turns to sleet, soft snow, or hail. In the summer, protracted periods of fog, low stratus clouds, and drizzle exist with moderate breezes. The weather improves in the fall, with a week or so of calm, clear weather in late September. As we move south to 40 degrees north latitude and the horse latitudes, the semi-permanent high-pressure centers result in generally fair, clear weather with a tendency toward dryness. In winter, temperatures hover near 50°F, and summer brings temperatures into the 70s with calms existing one-fourth of the time. Below 25 degrees north latitude, in the heart of the trade-wind belt, winds of 5 to 15 mph are normal. Endless bands of cumulus clouds and clear sky exist with little difference between summer and winter. Daytime temperatures range from 70°F to 80°F.

b. In the Atlantic, Pacific, and Indian Oceans between 5 degrees north latitude and 5 degrees south latitude, an equatorial trough of low pressure forms a belt where no prevailing surface winds exist and is known as the doldrums. Instead, the lack of extreme pressure gradients results in shifting winds and calms that exist as much as one-third of the time. Intense solar heating results in violent thunderstorms associated with strong squall winds. The convergence of these equatorial winds and trade winds from the intertropical front can be seen at a great distance because of towering cumulus clouds rising to 30,000 feet.

c. In the vicinity of the intertropical front, heavy convective showers are quite common. Across the Atlantic and eastern Pacific, the front is usually north of the Equator. Over the western Pacific, west of 180 degrees longitude, the doldrum belt oscillates considerably. Areas north of the Equator receive their heavy rainfall from June to September. Areas south of the Equator receive their heaviest precipitation between December and March. The meteorological sequence described above may be interrupted by periods of extreme weather centered around low pressure.

d. Waterspouts are the marine equivalent of tornadoes attached to the base of a cumulus or cumulonimbus cloud. They are common off the Atlantic and Gulf coasts of the United States and along the coasts of Japan and China during any season. They are usually seen around noon when solar heating is the greatest. They are small in diameter (10 to 100 yards) and short in duration (10 to 15 minutes). Waterspouts generally exhibit less intensity than overland tornadoes.

e. Hurricanes and typhoons are synonyms for tropical cyclones whose maximum winds exceed 75 mph. They occur in the warm, western sectors of all oceans during summer and fall. Winds may reach 170 to 230 mph. The lifespan of tropical cyclones ranges from 1 to 2 weeks. In the middle and high latitudes, extratropical cyclones contrast with tropical cyclones in several ways. There is no warm, clear eye, but rather a cold region of heavy precipitation. Sustained winds are more moderate (70 to 80 mph). Extratropical cyclones may persist for 2 to 3 days at a fixed location.

f. All ocean currents have a profound influence on climate since the properties of the surface largely determine the properties of the various air masses. The following are a few examples.

(1) The cold water of the Peru or Humbolt currents has a tremendous affect on the climate of Peru and Chile. The cold air that lies over the current is warmed as it reaches land, increasing its capacity to hold moisture. The warm air does not give up the moisture until it passes over the high Andes Mountains. This accounts for the dry climate of the coast of Chili and Peru and a more temperate climate toward the Equator than is usually found in the lower latitudes.

(2) Where the Labrador current contacts the warm gulf stream, fog prevails and steep temperature gradients are present. The northeast coast of North America has much colder climates than the west coast of Europe at the same latitude.

(3) The warm gulf stream current accounts for the continually warm and pleasant weather in the Caribbean Sea and the Gulf of Mexico.

(4) The winds blowing off the warm water of the Norwegian and east and west Greenland currents account for the unusually mild climates in northern Europe. At the same latitude elsewhere, the temperatures are usually much colder.

11-11. Life Forms. Life forms in the seas range from one-celled animals (protozoan) to complex aquatic mammals. The fish and aquatic mammals rule the sea and are of the most concern to anyone in a survival situation on the open seas. The majority of fish and mammals can be used as food sources, but some must be considered as a hazard to life, such as sharks, whales, barracudas, eels, sea snakes, rays, and jellyfish.

a. Sharks. (See figure 11-40.)

(1) Most sharks are scavengers, continuously on the move for food. If none is available, they lose interest and swim on. Even in warm oceans where most attacks

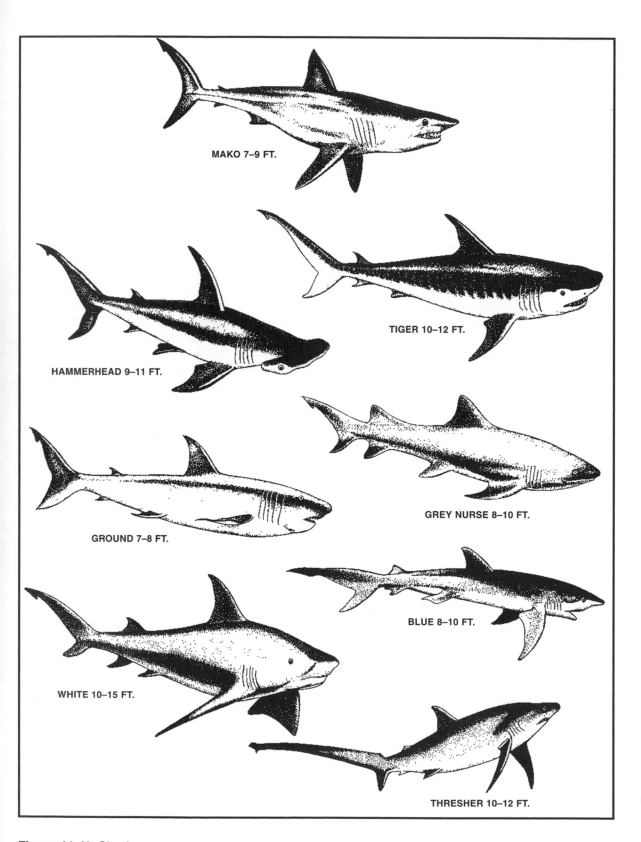

MAKO 7–9 FT.

HAMMERHEAD 9–11 FT.

TIGER 10–12 FT.

GROUND 7–8 FT.

GREY NURSE 8–10 FT.

BLUE 8–10 FT.

WHITE 10–15 FT.

THRESHER 10–12 FT.

Figure 11-40. Sharks.

occur, the risk can be reduced by knowing what to do and how to do it. Sharks live in almost all oceans, seas, and in river mouths. Normally, there isn't a shark problem in areas of colder water due to the temperature of the water decreasing swim activities. Sharks vary greatly in size, but there is no close relationship between the size of a shark and the risk of attack.

(2) Hungry sharks sometimes follow fish up to the surface and into shallow waters along the shore. When sharks explore such waters, they are more likely to come in contact with people. Sharks seem to feed most actively during the night, and particularly at dusk and dawn. After dark, they show an increased tendency to move toward the surface and into shore waters. Evidence indicates that a shark first locates food by smell or sound. Such things as garbage, body wastes, and blood probably stimulate the desire for food. A shark is also attracted by weak, fluttery movements similar to those of a wounded fish. While a shark will investigate any large, floating object as a possible food source, it probably will not attack a human unless it is hungry. Often the shark will swim away after investigating. At other times, it may approach and circle the object once or twice, or it may swim close and nudge the object with its snout. When swimming, a shark cannot stop suddenly or turn quickly in a tight circle. A shark rarely jumps out of the water to take food; however, it may grasp its prey near the surface. For this reason, people on rafts are relatively safe unless they dangle their hands, arms, feet or legs in the water.

(3) Individuals on or in the water must keep a sharp lookout for sharks. Clothing and shoes should be worn. If sharks have been noticed, survivors must be especially careful of the methods in which body wastes are eliminated and must avoid dumping blood and garbage. Vomiting, when it cannot be prevented, should be done in a container or hand and thrown as far away as possible.

(a) If a group in the water is threatened or attacked by a shark, they should bunch together, form a tight circle, and face outward so an approaching shark can be seen. Ward off attacks by kicking or stiff-arming the shark. Striking with the bare hand should be used only as a last resort; instead, survivors should use a hard and heavy object.

(b) Individuals should stay as quiet as possible and float to save energy. If it is necessary to swim, they should use strong, regular strokes, not frantic irregular movements.

(c) When alone, swimmers should stay away from schools of fish. If a single shark threatens at close range, the swimmer should use strong, regular swimming movements. Feinting toward the shark may scare it away.

(d) The survivor should not swim directly away from the shark, but face the shark and swim to one side, with strong rhythmic movements.

GIANT RAY OR MANTA

SHARK

Comparison of jumping form of porpoise and shark

PORPOISE

Figure 11-41. Animals Sometimes Mistaken for Sharks.

(e) If a shark threatens to attack or damage a raft, jabbing the snout or gills with an oar may discourage it. Check for sharks around and under the raft before going into the water.

(4) Other animals are sometimes mistaken for sharks.

(a) A school of porpoises or dolphins gracefully breaking the surface, blowing and grunting, may look alarming. Actually, it should be a reassuring sight, because porpoises and dolphins are enemies of sharks. Porpoises and dolphins are harmless to humans (figure 11-41).

(b) Giant rays or mantas, which also appear in tropical waters, may be mistaken for sharks. A swimming ray curls up the tips of its fins, and when seen from water level, the fins somewhat resemble the fins on the backs of two sharks swimming side by side. In deep water, all rays are harmless to swimmers; however, some are dangerous if stepped on in shallow waters (figure 11-41).

b. Grouper or Sea Bass. These fish do not constitute the same degree of hazard as sharks; however, these carnivorous fish are curious, bold, and have a never-ending appetite. Sea bass are most commonly found around rocks, caverns, old wrecks, and caves. Stay away from these areas.

c. Killer Whales. The killer whale has the reputation of being a fearless, ruthless, and ferocious creature. These fast swimmers are found in all oceans and seas, from the tropics to both polar regions. If encountered, a survivor can be assured there are others nearby since they hunt in packs of up to 40 creatures. They have been known to attack anything that swims or floats. If an initial attack is survived, get out of the water. The raft may afford some protection, but they have been known to come up under iceflows and knock other animals into the water. Stay out of the water. On thin ice, do not stand near seals, etc., as the whale may mistake the human form for a seal. However, the probability of being attacked by a killer whale is slim. If an aircrew member is attacked, it will probably be due to the fact that this intelligent whale simply mistook the person for its regular diet.

d. Barracuda. There are 20-odd species of barracuda; some are more feared in certain parts of the world than are sharks. If survivors come down in any tropical or subtropical sea, they may encounter this fish. Barracuda are attracted by anything that enters the water, and they seem to be particularly curious about bright objects. Accordingly, survivors should avoid dangling dogtags or other shiny pieces of equipment in the water. Dark-colored clothing is also best to wear in the water if no raft is available.

e. Moray Eels. If attacked by some species of moray eel, the survivor may have to cut off its head since some eels will retain their sharp, crushing grip until dead. The knife used to do this should be very sharp since their skin is tough and difficult to cut. Their bodies are very slippery and hard to hold. A survivor is most likely to come in contact with a moray eel when poking into holes and crevices around or under coral reefs. Use caution in these areas.

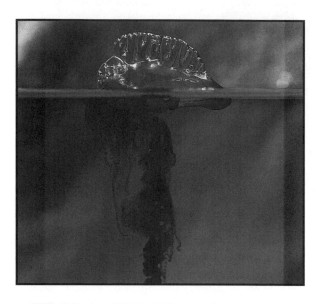

Figure 11-42. Portuguese Man-of-War.

f. Poisonous and Venomous Marine Animals (Invertebrates). There are many marine animals that have no backbone and can inflict injuries by stinging. Three major categories of invertebrates are important to the survivor.

(1) Coelenterates. This group includes jellyfish, hydroids, sea anemones, and corals. Coelenterates are all simple, many-celled organisms. They all possess tentacles equipped with stinging cells or nematocysts in addition to other technical characteristics. The family of coelenterates is divided into three major classes.

(a) Hydrozoan Class. Two of the more common members of this class are:

-1. Stinging or Fire Coral. This false coral can be found in areas of true coral reefs in warm waters.

-2. Portuguese Man-of-War or Blue Bottle. This hydroid is frequently mistaken for a true jellyfish. It is almost always found floating at the surface of the water (figure 11-42). Its stinging tentacles may extend several yards below the surface. Their float is 5 to 10 inches in length. Each tentacle may contain thousands of stinging cells. When one considers the large number of such tentacles, it is apparent that the fishing filaments of the Blue Bottle are quite a formidable venom apparatus.

(b) Anthozoa Class:

-1. Corals. Elkhorn coral and stony coral are very adaptable and have a real immunity to predators. This helps explain why they tend to dominate reef communities. Corals are carnivores, and with the use of small tentacles, capture and consume living zooplankton. Survivors should treat coral cuts by thoroughly cleaning the wound and removing any coral particles. Some coral cuts have been helped by painting them with an antiseptic solution of tincture of iodine.

-2. Sea Anemones. The sea anemone is one of the most plentiful marine creatures, with well over 1,000 species. They can be found from tide level to depths of more than 7,900 fathoms in all seas. Their size ranges from very small (less than an inch), to more than 2 feet in diameter. They eat fish, mollusks, crustaceans, and other invertebrates. Most of the stinging cells of the sea anemones are located on the outer ring of the tentacles.

(c) Scyphozoa Class—Jellyfish. There are many and varied species of jellyfish distributed throughout all seas. Their size ranges from extremely small to a diameter of 6 feet with tentacles hanging below to a depth of 100 feet. All are carnivorous. Some are transparent and glassy while others are brilliantly colored. Regardless of their size and color, they are very fragile creatures that, for the most part, depend on wind and tidal currents to help them move. Most adults can swim but this ability is weak. Whether they stay on the surface or below the surface, and to what depth, varies with each species. The stinging cells of jellyfish are located in the tentacles.

(d) Venom Apparatus of Coelenterates:

-1. All of the coelenterates have stinging cells or nemotocysts located on the tentacles. Each of these cells is like a capsule. If the survivor comes into contact with the capsule, part of it springs open and a very sharp, extremely small "thread" type tube appears. The sharp tip of the tube penetrates the skin and the venom is injected. When coming in contact with the tentacles of any coelenterate, the survivor brushes up against literally thousands of these small stinging organs.

-2. The symptoms produced by coelenterate stings will vary according to species, where the sting is located, and the physical condition of the survivor. In general, though, the sting caused by hydroids and hydroid corals is primarily skin irritations of a local nature. Stings of the Portuguese Man-of-War may be very painful. True corals and sea anemones produce a similar reaction. Some of the sting of these organisms may be hardly noticeable, while others may cause death in 3 to 8 minutes. Symptoms common to all of these may vary from an immediate mild prickly or stinging sensation, like that of touching a nettle, to a burning, throbbing, shooting-type pain that may cause the survivor to become unconscious. In some cases, the pain may be localized, while in others, it may spread to the groin, armpits, or abdomen. The area in which contact was made will usually become red, followed by severe inflammation, rash, swelling, blistering, skin hemorrhages, and sometimes ulceration. In severe cases of reaction, in

addition to shock, the person may experience one or more of the following: muscular cramps, lack of touch and temperature sensations, nausea, vomiting, backache, loss of speech, constriction of the throat, frothing at the mouth, delirium, paralysis, convulsions, and death. Since some of these traits appear quickly, the victim should try to get out of the water if at all possible to avoid drowning.

-3. One of the deadliest jellyfish is the sea wasp (an uncommon creature that is found in tropical southern Pacific waters). This animal can cause death anywhere from 30 seconds to 3 hours after contact. Most deaths occur within 15 minutes. The pain is said to be excruciating. The sea wasp can be recognized by the long tentacles that hang down from the four corners of its squarish body.

-a. Relieve pain. Tentacles or other matter on the skin should be removed immediately. This is important because as long as this matter is on the skin, additional stinging cells may be discharged. Use clothing, seaweed, or any other available material to remove the matter. Morphine is effective in relieving pain. DO NOT rub the wound with anything, especially sand, as this may cause the stinging cells to be activated. DO NOT suck the wound.

-b. Alleviate poison effects. Suntan lotion, oil, and alcohol should be applied to the area to stop further stinging. The following local remedies have been used in various parts of the world with varying degrees of success: papain (protein destroying enzyme), sodium bicarbonate, olive oil, sugar, soap, vinegar, lemon juice, diluted ammonia solution, papaya latex, plant juices, boric acid solution, flour, baking powder, etc. (Urine— with its ammonia content—may be the only source of relief available to a survivor.)

-c. Artificial respiration and cardiopulmonary resuscitation may be required. There are no known specific antidotes for most coelenterate strings; however, there is one antivenin for the sea wasp, which is papain, a proteolytic enzyme in the juice of the green fruit of the papaya. Even if the survivor is in an area where the antivenin is available, it may be too late to obtain and use it. The venom acts so quickly that medical help is often too late.

-4. Jellyfish should be given a wide berth since in some species the tentacles may trail 50 feet or more from the body. After a storm in tropical areas where large numbers of jellyfish are present, the survivor may be injured by pieces of floating tentacles that have been removed from the animals during the storm. Jellyfish washed up on the beach may appear dead but can still, in some cases, inflict painful injuries. The best prevention is to stay out of the water by getting into a raft or onto shore. If in a raft, do not let arms and legs trail over the side. The clothing (antiexposure suit) that the survivor wears should cover as much of the body as possible. Flight clothing items currently available should provide adequate protection.

(2) Mollusks. Octopus, squid, and univalve shellfish are in this category. Mollusks make up the largest single group of biotoxic marine invertebrates of direct importance to the survivor. The phylum of mollusks is generally divided into five classes. Stinging or venomous mollusks that concern the survivor fall mainly into two categories:

(a) Gastropoda (Stomache Footers):

-1. Mollusks. These in general are unsegmented invertebrates. Sometimes their soft bodies will secrete a calcereous shell. They have a muscular foot that serves a variety of functions. Some breathe by means of a type of siphon while others use gills. Some types have jaws. In those that don't have jaws, food is obtained by a rasp-like device called a radula. In the cone shells, the radula is a barb or tooth more like a hollow, needle-like structure.

-2. Gastropods. These univalves include marine snails and slugs as well as land and freshwater snails. It is estimated that there are more than 33,000 living species of gastropods; however, only members of the genus conus are of concern to the survivor. Of these cone shells, there are more than 400 species, but they will be discussed only in general terms with the emphasis placed on the more dangerous species. With few exceptions, these attractive shellfish are located in tropical or subtropical areas. All of these shells have a very highly developed venom apparatus designed for vertebrate or invertebrate creatures and are found from shallow tidal areas to depths of many hundreds of feet. The area in which the survivor may come into contact with these shellfish is in coral reefs and sandy or rubble habitat. All cone-shaped shells in these areas should be avoided. Cone shells are usually nocturnal. During the daytime, they burrow and hide in the sand, rocks, or coral; they feed at night on worms, octopus, other gastropods, and small fish. Several of these shells have caused death in humans. The venom apparatus lies within a body cavity of the animal, and the animal is capable of thrusting and injecting the poison via the barb into the flesh of the victim. The cone shell is able to inflict its wound only when the head of the animal is out of the shell.

-a. Complications. The sting made by a cone shell is a puncture-type wound. The area around the wound may exhibit one or more of the following: turning blue, swelling, numbness, stinging, or burning sensation. The amount of pain will vary from person to person. Some say the pain is like a bee sting, while others find it unbearable. The numbness and tingling sensations around the site of the wound may spread rapidly, involving the whole body, especially around the lips and mouth. Complete general muscle paralysis may occur. Coma may ensue and death is usually the result of cardiac failure.

-b. Treatment. The pain comes from the injection of venom, slime, and other irritating foreign matter into the wound site. The treatment is primarily symptomatic because there is no specific treatment. Applying hot towels or soaking the affected area in hot water may relieve some of the pain. Artificial respiration may be needed.

(b) Cephalpods. This group includes the nautilus, squid, cuttlefish, and octopus. Since the octopus is the marine animal most likely to be encountered by a survivor, it is the only one that will be discussed. The head of this animal is large and contains well-developed eyes. The mouth is surrounded by eight legs equipped with many suckers. It can move rapidly by expelling water from its body cavity, though it usually glides or creeps over the bottom. Most octopuses live in water ranging from very shallow to depths of more than 100 fathoms. All are carnivorous and feed on crabs, and other mollusks. Octopuses like to hide in holes or underwater caves—avoid these areas.

-1. Complications. The sharp, parrot-like beak of the octopus makes two small puncture wounds into which a toxic solution or venom is injected. Pain is usually felt immediately in the form of a burning, itching, or stinging sensation. Bleeding from the wound is usually very profuse, which may indicate that the venom contains an anticoagulant. The area around the wound, and in some cases the entire appendage, may swell, turn red, and feel hot. There has been one report of a fatal octopus bite. This death was attributed to the blue ringed octopus (Octopus Maculosus, figure 11-43). This small octopus is usually only 3 or 4 inches across although some may be slightly larger. Found throughout the Indo-Pacific area, this octopus is not aggressive toward humans. Because its bite is so dangerous, it should not be handled at any time. When this animal is disturbed, the intensity of its blue rings varies rapidly on a light yellow or cream to brown background.

Figure 11-43. Blue Ringed Octopus.

-2. Treatment. Treat for shock, stop bleeding, clean the wound area since more venomous saliva could be in the area, and treat symptoms as they arise. There is no known cure for the venom of the blue ringed octopus.

(3) Echinoderms. Sea cucumbers, starfish, and sea urchins are members of this group. Sea urchins comprise the most dangerous type of echinoderms. Sea urchins have rounded, egg-shaped, or flattened bodies. They have hard shells that carry spines. In some species, the spines are venomous and present a hazard if stepped on or handled. Some urchins are nocturnal. They all tend to be omnivorous, eating algae, mollusks, and other small organisms. They can be found in tidal pools or in areas of great depth in many parts of the world. Sea urchins are not good food sources. At certain times of the year, certain species can be poisonous.

(a) Complications. The needle-sharp points of sea urchin spines are able to penetrate the flesh easily. These spines are also very brittle and tend to break off while still attached to the wound and are very difficult to withdraw. Stepping on one of these spines produces an immediate and very intense burning sensation. The area of pain will also swell, turn red, and ache. Numbness and muscular paralysis, swelling of the face, and a change in the pulse have also been reported. Secondary infection usually sets in. While some deaths have been reported, other victims have experienced loss of speech, respiratory distress, and paralysis. The paralysis will last from 15 minutes to 6 hours.

(b) Treatment. Spines (pedicellaria) that are detached from the animal will continue to secrete venom into the wound. The spines of some species will be easily dislodged whereas others must be surgically removed. There will also be some discoloration due to a dye the animal secretes—do not be disturbed by this. Some experts say to apply grease to allow the spines to be scraped off. Others advise leaving them alone since some of the spines will dissolve in the wound within 24 to 48 hours. Still other experts say to apply citrus juice, if available, or soak the area in vinegar several times a day to dissolve them.

(c) Prevention. No sea urchin should be handled. The spines can penetrate leather and canvas with ease.

g. Venomous and Poisonous Marine Animals (Vertebrates). These fish can be divided into two general groups—fish that sting and fish that are poisonous to eat (figure 11-44).

(1) Venomous Spine Fish (Fish that Sting):

(a) Types of fish in this group are:

-1. Spiny dogfish.

-2. Stingrays. Includes whiprays, batrays, butterfly rays, cow-nosed rays, and round stingrays.

-3. Rat fish.

-4. Weever fish.

-5. Catfish.

-6. Toad fish.

-7. Scorpion fish.

-8. Surgeon fish.

-9. Rabbit fish.

-10. Star gazers.

NOTE: For all wounds from these types of fish, aid should be directed to three areas: alleviating the pain of the sting, trying to halt the effects of the venom, and preventing infection.

(b) Certain types of these fish have up to 18 spines. The pain caused by the sting of one of these spines is so great in some species that the victim may scream and thrash about wildly. In one case, a man stung in the face by a weever fish begged for bystanders to shoot him, even after two shots of morphine sulfate. Many of these fish are bottom dwellers that will not move out of the way when being approached by humans. Instead, they will lie quietly camouflaged, put up their spines, and simply wait for the unlucky individual to step on them. Other people have been injured while trying to remove them from fishing nets and fishing lines. In cases where humans are stung by stingrays, the barbs on the sharp spines may cause severe lacerations as well as introduce poison. These wounds should be irrigated without delay. Puncture wounds from the fish are small and make removal of the poison a difficult process. It may be necessary to remove the barb. A procedure that is fairly successful is to make a small cut across the wound (debride) and then apply suction. Even if no incision is made, suction should be tried since it is important to remove as much of the venom as possible. The more poison removed, the better. Morphine does not relieve the pain of some of these venoms. Most doctors agree that the injured part should be soaked in hot water from 30 minutes to 1 hour. The

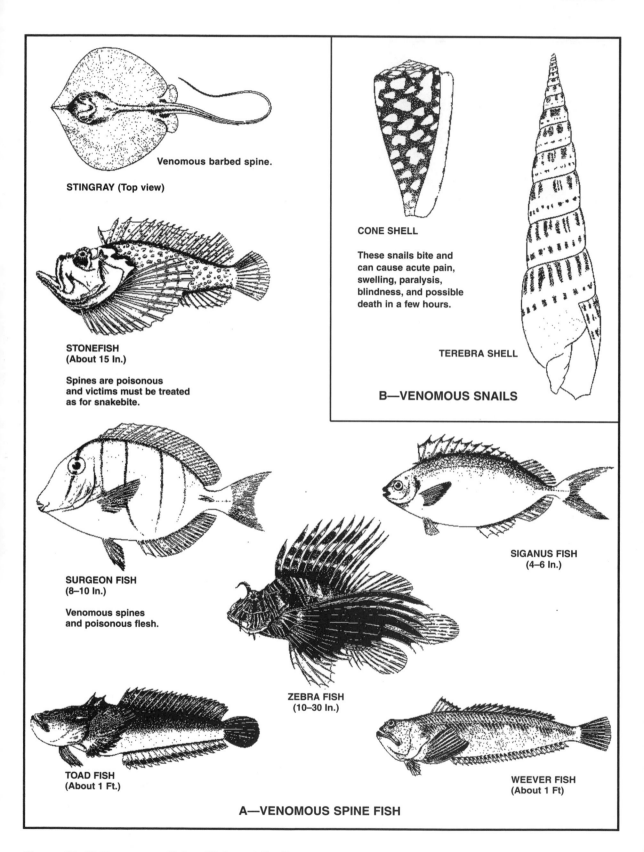

STINGRAY (Top view)
Venomous barbed spine.

STONEFISH
(About 15 In.)

Spines are poisonous
and victims must be treated
as for snakebite.

CONE SHELL

These snails bite and
can cause acute pain,
swelling, paralysis,
blindness, and possible
death in a few hours.

TEREBRA SHELL

B—VENOMOUS SNAILS

SURGEON FISH
(8–10 In.)

Venomous spines
and poisonous flesh.

SIGANUS FISH
(4–6 In.)

ZEBRA FISH
(10–30 In.)

TOAD FISH
(About 1 Ft.)

WEEVER FISH
(About 1 Ft)

A—VENOMOUS SPINE FISH

Figure 11-44. Venomous Spine Fish and Snails.

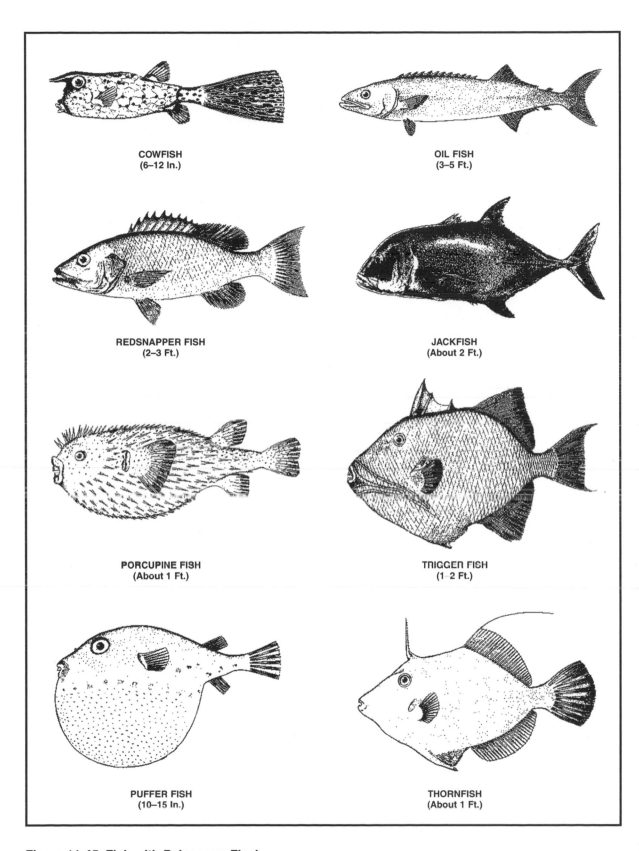

Figure 11-45. Fish with Poisonous Flesh.

temperature of the water should be as hot as the patient can stand without injury. If the wound is on the face or body, hot, moist cloth compresses can be used. The use of heat in this manner may weaken the effect of the poison in some cases. After soaking the wound, clean it again, if necessary. Cover the area of the wound with antiseptic and a clean sterile dressing. If antibiotics are available, it may be advisable to use them to help prevent infection. Treatment for shock is wise. Artificial respiration may be needed since some venoms may cause cardiac failure, convulsions, or respiratory distress.

(c) For fish that are poisonous to eat, see figure 11-45.

-1. There is no known way to detect a poisonous fish merely by its appearance. Fish that are poisonous in one area may be safe to eat in another. In general, bottom dwellers and feeders, especially those associated with coral reefs, should be suspect. Also, unusually large predator-type fish should be eaten with caution. The internal organs and roe of all tropical marine fish should never be eaten, as those parts contain a higher concentration of poison.

-2. Under certain conditions where the survivor may be required to eat questionable fish, rules should be followed. A fish will be safer if it can be caught away from reefs or entrances to lagoons. Once the fish has been secured, the "marine animal edibility test" should be used. The fish should be cut into thin strips and boiled in successive changes of water for an hour or more. This may help since some, but not all, of the toxins are water-soluble. Further, it should be noted that normal cooking techniques and temperatures will not weaken or destroy poisons.

-3. If boiling is not possible, cut the meat into thin strips and soak in changes of seawater for an hour or so, squeezing the meat juices out as thoroughly as possible. A survivor should eat only a small portion of the flesh and wait 12 hours to see whether any symptoms arise (if the fish will not spoil). Remember that the degree of poisoning is directly related to how much fish is eaten. If in doubt, do not eat it. The advice of native people on eating tropical marine fish may not be valid. In many instances they check edibility by first feeding fish portions to their dogs and cats.

-4. Treatment. As soon as any symptoms arise, vomiting should be induced by administering warm saltwater or the whites of eggs. If these procedures don't work, try sticking a finger down the person's throat. A laxative should also be given to the victim if one is available. The victim may have to be protected from injury during convulsions. If the victim starts to foam at the mouth and exhibits signs of respiratory distress, a cricothyroidotemy may have to be performed. Morphine may help relieve pain in some cases. If the victim complains of severe itching, cool showers may give some relief. Treat any other symptoms as they arise.

(2) Poisonous Marine Turtles:

(a) Species. There are more than 265 species of marine turtles. Of these, only five have been reported as poisonous and dangerous to the survivor. Many of these species are commonly eaten, but for some unknown reason, these same turtles become extremely toxic under certain conditions. Basically, the main species to be concerned with are the green, the hawksbill, and the leatherback turtles. These turtles are found mainly in tropical and subtropical seas but can also be found in temperate waters.

(b) Origin. The origin of turtle poison is unknown, but some investigators suggest it comes from the poisonous marine algae eaten by the turtles. It should be noted that a species of turtle might be safe to eat in one area but deadly in another. There is absolutely no way a survivor can distinguish between a poisonous and nonpoisonous sea turtle just by looking at it or by examining any part of it. Toxicity may occur at any time of the year; however, the most dangerous months appear to be the warmer months. The degree of freshness also has nothing to do with how poisonous the turtle is.

(c) Complications. The symptoms will vary with the amount of turtle ingested. Symptoms will develop within a few hours to a few days after eating the food. These symptoms include nausea, vomiting, diarrhea, pain, sweating, coldness in the extremities, vertigo, dry and burning lips and tongue, tightness of the chest, drooling, and difficulty in swallowing. Other victims reported a heavy feeling of the head, a white coating on the tongue, diminished reflexes, coma, and sleepiness. About 44 percent of victims poisoned by marine turtles die.

(d) Treatment. There is no known antidote for this kind of poisoning. There is no specific treatment—treat symptomatically.

(e) Prevention. If there is the slightest suspicion about the edibility of a marine turtle, it should not be eaten, or at least the marine animal edibility test should be used. Turtle liver is especially dangerous to eat because of its high vitamin A content.

h. Birds. There are roughly 260 species of sea birds. Most of the birds travel only a few miles out to sea, but the albatross ranges across the seas far from any landmasses.

i. Red Tide. Red tide is a name used to describe the reddish or brownish coloration in saltwater, resulting from tiny plants and organisms called plankton, which suddenly increase tremendously in numbers. Red tides appear in waters worldwide. In the United States, they are most common off the coasts of Florida, Texas, and southern California. Although most red tides are harmless, some may kill fish and other water creatures. Still other types of red tides do not kill sea life, but cause the shellfish feeding on them to be poisonous. Some of these creatures secrete poisons that can paralyze and kill fish, or can kill fish by using nearly all of the oxygen in the water. Although the exact reason for the sudden increase of the plankton is unknown, there is evidence that shows favorable food, temperature, sunlight, water currents, and salt in the water will increase the population. It is not unusual for it to remain from a few hours to several months. A survivor should not eat any fish that are found dead.

Chapter 12

LOCAL PEOPLE

Figure 12-1. Local People.

12-1. Introduction. One evader concluded with the following advice: "My advice is, 'When in Rome, do as the Romans do!' Show interest in their country, and they will go overboard to help you!" One of the most frequently given bits of advice is to accept, respect, and adapt to the ways of the people among whom survivors find themselves. This is good advice, but there are a number of important problems involved in putting this advice into practice (figure 12-l).

12-2. Contact with People. The survivor must give serious consideration to people. Are they people with a primitive culture? Are they farmers, fishermen, friendly people, or enemies? To the survivor, "cross-cultural contact" can vary quite radically in scope. It could mean interpersonal relationships with people of an extremely different (primitive) culture, or contacts with people who are culturally modern by our standards. A culture is identified by standards of behavior that are considered proper and acceptable for the members and may or may not conform to our idea of propriety. Regardless of who these people are, the survivor can expect that they will have different laws, social and economic values, and political and religious beliefs.

a. People will be friendly, unfriendly, or choose to ignore the survivor. Their attitude may be unknown. If the people are known to be friendly, the survivor must make every attempt to keep them that way by being courteous and respecting the religion, politics, social customs, habits, and all other aspects of their culture.

If the people are known to be enemies or are unknowns, the survivor should make every effort to avoid any contact and leave no sign of presence. Therefore, a basic knowledge of the daily habits of the local people can be extremely important in this attempt. An exception might be, if after careful and covert observation it is determined an unknown people are friendly, contact might be made if assistance is absolutely necessary.

b. Generally, there is little to fear and everything to gain from thoughtful contact with the local peoples of friendly or neutral countries. Familiarity with local customs, displaying common decency, and most important, showing respect for their customs should help a survivor avoid trouble and possibly gain needed assistance. To make contact, a survivor should wait until only one person is near and, if possible, let that person make the initial approach. Most people will be willing to help a survivor who appears to be in need; however, political attitudes and training or propaganda efforts can change the attitudes of otherwise friendly people. Conversely, in nominally unfriendly countries, many people, particularly in remote areas, may feel abused or ignored by their politicians, and may be friendlier toward outsiders.

c. The key to successful contact with local peoples is to be friendly, courteous, and patient. Displaying fear, displaying weapons, and making sudden or threatening movements can cause a local person to fear a survivor and can, in turn, prompt a hostile response. When attempting contact, smile frequently. Many local peoples may be shy and seem unapproachable or they may ignore the survivor. Approach them slowly and don't rush matters.

12-3. Survivor's Behavior:

a. Salt, tobacco, silver money, and similar items should be used discreetly in trade. Paper money is well known worldwide. Don't overpay; it may lead later to embarrassment and even danger. Treat people with respect and do not laugh at or bully them.

b. Sign language or acting out needs or questions can be very effective. Many people are accustomed to it and communicate using nonverbal sign language. Aircrew members should learn a few words and phrases of the local language in and around their area of operations. Attempting to speak someone's language is an excellent way to show respect for his or her culture. Since English is widely used, some of the local people may understand a few words of English.

c. Certain areas may be taboo. They range from religious or sacred places to diseased or dangerous areas. In some areas, certain animals must not be killed.

A survivor must learn what the rules are and follow them. The survivor must be observant and learn as much as possible. This will not only help in strengthening relations, but new knowledge and skills may be very important later. The downed aircrew member should seek advice on local hazards and find out from friendly people where there are hostile people. Keep in mind, though, that frequently, people, as in our culture, insist others are hostile because they also do not understand different cultures and distant peoples. The people that generally can be trusted, in their opinion, are their immediate neighbors—much the same as in our own neighborhood. Local people, like us, suffer from diseases that are contagious. The survivor should build a separate dwelling, if possible, and avoid physical contact without seeming to do so. Personal preparation of food and drink is desirable if it can be done without giving offense. Frequently, the use of "personal or religious custom" as an explanation for isolationist behavior will be accepted by the local people.

d. Trading or barter is common in more primitive societies. Hard coin is usually good, whether for its exchange value or as jewelry or trinkets. In isolated places, matches, tobacco, salt, razor blades, empty containers, or cloth may be worth more than any form of money.

e. The survivor must be very cautious when touching people. Many people consider "touching" taboo and such actions may be dangerous. Sexual contact should be avoided.

f. Hospitality among some people is such a strong cultural trait that they may seriously reduce their own supplies to make certain a stranger or visitor is fed. What is offered should be accepted and shared equally with all present. The survivor should eat in the same way they eat and, most importantly, attempt to eat all that is offered. If any promises are made, they must be kept. Personal property and local customs and manners, even if they seem odd, must be respected. Some kind of payment for food, supplies, etc., should be made.

g. Privacy must be respected and a survivor should not enter a house unless invited.

12-4. Changing Political Allegiance. In today's world of fast-paced international politics and "shuttle diplomacy," political attitudes and commitments within nations are subject to rapid change. The population of many countries, especially politically hostile countries, must not be considered friendly just because they do not demonstrate open hostility. Unless briefed to the contrary, avoid all contact with such people.

Part Five

PERSONAL PROTECTION

Chapter 13

PROPER BODY TEMPERATURE

13-1. Introduction. In a survival situation the two key requirements for personal protection are maintenance of proper body temperature and prevention of injury. The means for providing personal protection are many and varied. They include the following general categories: clothing, shelter, equipment, and fire. These individual items are not necessary for survival in every situation; however, all four will be essential in some

WIND SPEED		COOLING POWER OF WIND EXPRESSED AS "EQUIVALENT CHILL TEMPERATURE"																				
KNOTS	MPH	TEMPERATURE (°F)																				
Calm	Calm	40	35	30	25	20	15	10	5	0	-5	-10	-15	-20	-25	-30	-35	-40	-45	-50	-55	-60
		EQUIVALENT CHILL TEMPERATURE																				
3 - 6	5	35	30	25	20	15	10	5	0	-5	-10	-15	-20	-25	-30	-35	-40	-45	-50	-55	-65	-70
7 - 10	10	30	20	15	10	5	0	-10	-15	-20	-25	-35	-40	-45	-50	-60	-65	-70	-75	-80	-90	-95
11 - 15	15	25	15	10	0	-5	-10	-20	-25	-30	-40	-45	-50	-60	-65	-70	-80	-85	-90	-100	-105	-110
16 - 19	20	20	10	5	0	-10	-15	-25	-30	-35	-45	-50	-60	-65	-75	-80	-85	-95	-100	-110	-115	-120
20 - 23	25	15	10	0	-5	-15	-20	-30	-35	-45	-50	-60	-65	-75	-80	-90	-95	-105	-110	-120	-125	-135
24 - 28	30	10	5	0	-10	-20	-25	-30	-40	-50	-55	-65	-70	-80	-85	-95	-100	-110	-115	-125	-130	-140
29 - 32	35	10	5	-5	-10	-20	-30	-35	-40	-50	-60	-65	-75	-80	-90	-100	-105	-115	-120	-130	-135	-145
33 - 36	40	10	0	-5	-15	-20	-30	-35	-45	-55	-60	-70	-75	-85	-95	-100	-110	-115	-125	-130	-140	-150

WINDS ABOVE 40 HAVE LITTLE ADDITIONAL EFFECT

LITTLE DANGER

INCREASING DANGER (Flesh may freeze within 1 minute)

GREAT DANGER (Flesh may freeze within 30 seconds)

DANGER OF FREEZING EXPOSED FLESH FOR PROPERLY CLOTHED PERSONS

INSTRUCTIONS

MEASURE LOCAL TEMPERATURE AND WIND SPEED IF POSSIBLE; IF KNOT, ESTIMATE. ENTER TABLE AT CLOSEST 5°F INTERVAL ALONG THE TOP AND WITH APPROPRIATE WIND SPEED ALONG LEFT SIDE. INTERSECTION GIVES APPROXIMATE EQUIVALENT CHILL TEMPERATURE. THAT IS, THE TEMPERATURE THAT WOULD CAUSE THE SAME RATE OF COOLING UNDER CALM CONDITIONS.

NOTES

WIND
1. THIS TABLE WAS CONSTRUCTED USING MILES PER HOUR (MPH), HOWEVER, A SCALE GIVING THE EQUIVALENT RANGE IN KNOTS HAS BEEN INCLUDED ON THE CHART TO FACILITATE ITS USE WITH EITHER UNIT.
2. WIND MAY BE CALM BUT FREEZING DANGER GREAT IF PERSON IS EXPOSED IN MOVING VEHICLE, UNDER HELICOPTER ROTORS, IN PROPELLER BLAST, ETC. IT IS THE RATE OF RELATIVE AIR MOVEMENTS THAT COUNTS AND THE COOLING EFFECT IS THE SAME WHETHER YOU ARE MOVING THROUGH THE AIR OR IT IS BLOWING PAST YOU.
3. EFFECT OF WIND WILL BE LESS IF PERSON HAS EVEN SLIGHT PROTECTION FOR EXPOSED PARTS. LIGHT GLOVES ON HANDS, PARKA HOOD SHIELDING FACE, ETC.

ACTIVITY
DANGER IS LESS IF SUBJECT IS ACTIVE. A PERSON PRODUCES ABOUT 100 WATTS (341 BTUs) OF HEAT STANDING STILL BUT UP TO 1000 WATTS (3412 BTUs) IN VIGOROUS ACTIVITY LIKE CROSS-COUNTRY SKIING.

PROPER USE OF CLOTHING AND ADEQUATE DIET ARE BOTH IMPORTANT

COMMON SENSE
THERE IS NO SUBSTITUTE FOR IT. THE TABLE SERVES ONLY AS A GUIDE TO THE COOLING EFFECT OF THE WIND ON BARE FLESH WHEN THE PERSON IS FIRST EXPOSED. GENERAL BODY COOLING AND MANY OTHER FACTORS AFFECT THE RISK OF FREEZING INJURY.

Figure 13-1. Windchill Chart.

environments. In this part of the regulation, the conditions that affect the body temperature, the physical principles of heat transfer, and the methods of coping with these conditions will be covered.

13-2. Body Temperature. The body functions best when core temperatures range from 96°F to 102°F. Preventing too much heat loss or gain should be a primary concern for survivors. Factors causing changes in body core temperature (excluding illness) are the climatic conditions of temperature, wind, and moisture.

  a. Temperature. As a general rule, exposure to extreme temperatures can result in substantial decreases in physical efficiency. In the worst case, incapacitation and death can result.

  b. Wind. Wind increases the chill effect (figure 13-l), causes dissipation of heat, and accelerates loss of body moisture.

  c. Moisture—Precipitation, Ground Moisture, or Immersion. Water provides an extremely effective way to transfer heat to and from the body. When a person is hot, the whole body may be immersed in a stream or other body of water to be cooled. On the other hand, in the winter, a hot bath can be used to warm the body. When water is around the body, it tends to bring the "body" to the temperature of the liquid. An example is

when a hand is burned and then placed in cold water to dissipate the heat. One way to lower body temperature is by applying water to clothing and exposing the clothed body to the wind. This action causes the heat to leave the body 25 times faster than when wearing dry clothing. This rapid heat transfer is the reason survivors must always guard against getting wet in cold environments. Consider the result of a body totally submerged in water at a temperature of 50°F and determine how long a person could survive (figures 13-2 and 13-3).

13-3. Heat Transfer. There are five ways body heat can be transferred. They are radiation, conduction, convection, evaporation, and respiration.

  a. Radiation. Radiation is the primary cause of heat loss. It is defined as the transfer of heat waves from the body to the environment and (or) from the environment back to the body. For example, at a temperature of 50°F, 50 percent of the body's total heat loss can occur through an exposed head and neck. As the temperature drops, the situation gets worse. At 5°F, the loss can be 75 percent under the same circumstances. Not only can heat be lost from the head, but also from the other extremities of the body. The hands and feet radiate heat at a phenomenal rate due to the large number of capillaries present at the surface of the skin. These three areas

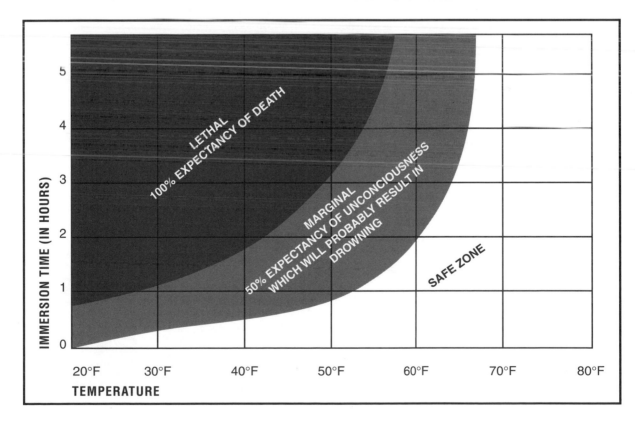

Figure 13-2. Life Expectancy Following Cold-Water Immersion.

of the body must be given particular attention during all periods of exposure to temperature extremes.

b. Conduction:

(1) Conduction is defined as the movement of heat from one molecule to another molecule within a solid

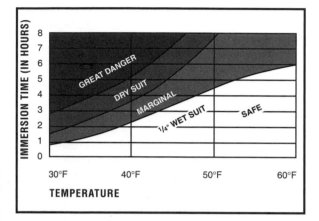

Figure 13-3. Life Expectancy Following Cold-Water Immersion (Exposure Suit).

object. Extreme examples of how heat is lost and gained quickly are deep frostbite and third-degree burns, both gained from touching the same piece of metal at opposite extremes of cold and heat. Heat is also lost from the body in this manner by touching objects in the cold with bare hands, by sitting on a cold log, or by kneeling on snow to build a shelter. These are practices that survivors should avoid because they can lead to over-chilling the body.

(2) Especially dangerous is the handling of liquid fuel at low temperatures. Unlike water, which freezes at 32°F, fuel exposed to the outside temperatures will reach the same temperature as the air. The temperature of the fuel may be 10° to 30° below zero or colder. Spilling the fluid on exposed skin will cause instant frostbite, not only from the conduction of heat by the cold fluid, but by the further cooling effects of rapid evaporation of the liquid as it hits the skin.

c. Convection. Heat movement by means of air or wind to or from an object or body is known as convection. The human body is always warming a thin layer of air next to the skin by radiation and conduction. The temperature of this layer of air is nearly equal to that of the skin. The body stays warm when this layer of warm air remains close to the body. However, when this warm layer of air is removed by convection, the body cools down. A major function of clothing is to keep the warm layer of air close to the body; however, by removing or disturbing this warm air layer, wind can reduce body temperature. Therefore, wind can provide beneficial

cooling in dry, hot conditions, or be a hazard in cold, wet conditions.

d. Evaporation. Evaporation is a process by which liquid changes into vapor, and during this process, heat within the liquid escapes to the environment. An example of this process is how a "desert water bag" works on the front of a jeep while driving in the hot desert. The wind created by the jeep helps to accelerate evaporation and causes the water in the bag to be cooled. The body also uses this method to regulate core temperature when it perspires and air circulates around the body. The evaporation method works anytime the body perspires, regardless of the climate. For this reason, it is essential that people wear fabrics that "breathe" in cold climates. If water vapor cannot evaporate through the clothing, it will condense, freeze, and reduce the insulation value of the clothing and cause the body temperature to go down.

e. Respiration. The respiration of air in the lungs is also a way of transferring heat. It works on the combined processes of convection, evaporation, and radiation. When breathing, the air inhaled is rarely the same temperature as the lungs. Consequently, heat is either inhaled or expelled with each breath. A person's breath can be seen in the cold as heat is lost to the outside. Because this method is so efficient at transferring heat, warm, moist oxygen is used to treat hypothermia patients in a clinical environment. Understanding how heat is transferred and the methods by which that transfer can be controlled can help survivors keep the body's core temperature in the 96°F to 102°F range. (See figure 13-4.)

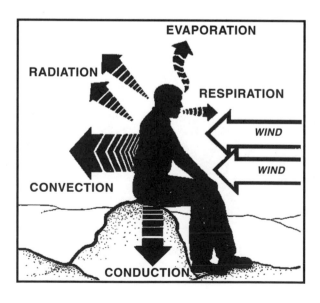

Figure 13-4. Heat Transfer.

Chapter 14

CLOTHING

14-1. Introduction. Every time people go outside they probably neglect to think about one of the most important survival-oriented assets—clothing. Clothing is often taken for granted; people tend to neglect those things that should be the most familiar to them. Clothing is an important asset to survivors and is the most immediate form of shelter. Clothing is important in staying alive, especially if food, water, shelter, and fire are limited or unobtainable. This is especially true in the first stages of an emergency situation because survivors must work to satisfy other needs. If survivors are not properly clothed, they may not survive long enough to build a fire or shelter, to find food, or to be rescued.

14-2. Protection:

a. People have worn clothing for protection since they first put on animal skins, feathers, or other coverings. In most parts of the world, people need clothing for protection from harsh climates. In snow or ice climates, people wear clothing made of fur, wool, or closely woven fabrics. They also wear warm footwear.

b. In dry climates, people wear clothing made of lightweight materials, such as cotton or linen, which have an open weave. These materials absorb perspiration and allow air to circulate around the body. People in dry climates sometimes wear white or light-colored clothes to reflect the sun's rays. They may also wear sandals, which are cooler and more comfortable than shoes. To protect the head and neck, people wear hats as sunshades.

c. Clothing also provides protection from physical injuries caused by vegetation, terrain features, and animal life, which may cause bites, stings, and cuts.

14-3. Clothing Materials:

a. Clothing is made from a variety of materials, such as nylon, wool, cotton, etc. The type of material used has a significant effect on protection. Potential survivors must be aware of both the environmental conditions and the effectiveness of these different materials in order to select the best type of clothing for a particular region.

b. Clothing materials include many natural and synthetic fibers. As material is woven together, a "dead air" space is created between the material fibers. When two or three layers of material are worn, a layer of air is trapped between each layer of material creating another layer of "dead air" or insulation. The ability of these different fibers to hold "dead air" is responsible for changing insulation values.

14-4. Natural Materials. They include fur, leather, and cloth made from plant and animal fibers.

a. Fur and leather are made into some of the warmest and most durable clothing. Fur is used mainly for coats and coat linings. Leather has to be treated to make it soft and flexible and to prevent it from rotting.

b. Wool is somewhat different because it contains natural lanolin oils. Although wool is somewhat absorbent, it retains most of its insulating qualities when wet.

c. Cotton is a common plant fiber widely used to manufacture clothing. It absorbs moisture quickly and, with heat radiated from the body, will allow the moisture to pass away from the body. It does not offer much insulation when wet. It's used as an inner layer against the skin and as an outer layer with insulation (for example, wool, Dacron pile, synthetic batting) sandwiched in between. The cotton protects the insulation and therefore provides warmth.

14-5. Synthetic Materials. Clothing manufacturers are using more and more of these materials. Many synthetic materials are stronger, more shrink-resistant, and less expensive than natural materials. Most synthetic fibers are derived from petroleum in the form of long fibers that consist of different lengths, diameters, and strengths, and sometimes have hollow cores. These fibers, woven into materials such as nylon, Dacron, and polyester, make very strong long-lasting clothing, tarps, tents, etc. Some fibers are spun into a batting-type material with air space between the fibers, providing excellent insulation used inside clothing.

a. Many fabrics are blends of natural and synthetic fibers. For example, fabrics could be a mixture of cotton and polyester or wool and nylon. Nylon covered with rubber is durable and waterproof but is also heavy. There are other coverings on nylon that are waterproof but somewhat lighter and less durable. However, most coated nylon has one drawback—it will not allow for the evaporation of perspiration. Therefore, individuals may have to change the design of the garment to permit adequate ventilation (for example, wearing the garment partially unzipped).

b. Synthetic fibers are generally lighter in weight than most natural materials and have much the same insulating qualities. They work well when partially wet and dry out easily; however, they generally do not compress as well as down.

14-6. Types of Insulation:
  a. Natural:

(1) Down is the soft plumage found between the skin and the contour feather of birds. Ducks and geese are good sources for down. If used as insulation in clothing, remember that down will absorb moisture (either precipitation or perspiration) quite readily. Because of

the light weight and compressibility of down, it has wide application in cold-weather clothing and equipment. It is one of the warmest natural materials available when kept clean and dry. It provides excellent protection in cold environments; however, if the down gets wet it tends to get lumpy and loses its insulating value.

(2) Cattail plants have worldwide distribution, with the exception of the forested regions of the far north. The cattail is a marshland plant found along lakes, ponds, and the backwaters of rivers. The fuzz on the tops of the stalks forms dead-air spaces and makes a good down-like insulation when placed between two pieces of material.

(3) Leaves from deciduous trees (those that lose their leaves each autumn) also make good insulation. To create dead-air space, leaves should be placed between two layers of material.

(4) Grasses, mosses, and other natural materials can also be used as insulation when placed between two pieces of material.

b. Synthetic:

(1) Synthetic filaments such as polyesters and acrylics absorb very little water and dry quickly. Spun synthetic filament is lighter than an equal thickness of wool and unlike down does not collapse when wet; it is also an excellent replacement for down in clothing.

(2) The nylon material in a parachute insulates well if used in the layer system because of the dead-air space. Survivors must use caution when using the parachute in cold climates. Nylon may become "cold soaked"; that is, the nylon will take on the temperature of the surrounding air. People have been known to receive frostbite when placing cold nylon against bare skin.

14-7. Insulation Measurement:

a. The next area to be considered is how well these fibers insulate against heat or cold. The most scientific way to consider the insulating value of these fibers is to use an established criterion. The commonly accepted measurement used is a comfort level of clothing, called a "CLo" factor.

b. The CLo factor is defined as the amount of insulation that maintains normal skin temperature when the outside ambient air temperature is 70°F with a light breeze. However, the CLo factor alone is not sufficient to determine the amount of clothing required. Such variables as metabolic rates, wind conditions, and the physical makeup of the individual must be considered.

c. The body's rate of burning or metabolizing food and producing heat varies among individuals. Therefore, some may need more insulation than others even though food intake is equal, and consequently the required CLo value must be increased. Physical activity also causes an increase in the metabolic rate and the rate of blood circulation through the body. When a person is physically active, less clothing or insulation is needed than when

standing still or sitting. The effect of the wind, as shown on the windchill chart, must be considered (figure 13-1). When the combination of temperature and wind drops the chill factor to minus 100°F or lower, the prescribed CLo for protecting the body may be inapplicable (over a long period of time) without relief from the wind. For example, when the temperature is minus 60°F, the wind is blowing 60 to 70 miles per hour, and the resultant chill factor exceeds minus 150°F, clothing alone is inadequate to sustain life. Shelter is essential.

d. The physical build of a person also affects the amount of heat and cold that can be endured. For example, a very thin person will not be able to endure as low a temperature as one who has a layer of fat below the skin. Conversely, heavy people will not be able to endure extreme heat as effectively as thinner people.

e. In the Air Force clothing inventory, there are many items that fulfill the need for insulating the body. They are made of the different fibers previously mentioned, and when worn in layers, provide varying degrees of insulative CLo value. The following average zone temperature chart is a guide in determining the best combination of clothing to wear.

TEMPERATURE RANGE	CLo REQUIRED
86 to 68°F	1-Lightweight
68 to 50°F	2-Intermediate Weight
50 to 32°F	3-Intermediate Weight
32 to 14°F	3.5-Heavyweight
14 to -4°F	4.0-Heavyweight
-4 to -40°F	4.0-Heavyweight

The amount of CLo value per layer of fabric is determined by the loft (distance between the inner and outer surfaces) and the amount of dead air held within the fabric. Some examples of the CLo factors of some items of clothing are:

LAYERS:		
1 - Aramid underwear (1 layer)	0.6 CLo
2 - Aramid underwear (2 layers)	1.5 CLo
3 - Quilted liners	1.9 CLo
4 - Nomex coveralls	0.6 CLo
5 - Winter coveralls	1.2 CLo
6 - Nomex jacket	1.9 CLo

This total amount of insulation should keep the average person warm at a low temperature. When comparing items 1 and 2 in the above example, it shows that when doubling the layer of underwear, the CLo value more than doubles. This is true not only of the No. 1 item but between all layers of any clothing system. Therefore, one gains added protection by using several very thin layers of insulation rather than two thick layers. The air held between these thin layers increases the insulation value.

f. The use of many thin layers also provides (through removal of desired number of layers) the ability to closely regulate the amount of heat retained inside the clothing. The ability to regulate body temperature helps to alleviate the problem of overheating and sweating, and preserves the effectiveness of the insulation.

g. The principle of using many thin layers of clothing can also be applied to the "sleeping system" (sleeping bag, liner, and bed). This system uses many layers of synthetic material, one inside the other, to form the amount of dead air needed to keep warm. To improve this system, a survivor should wear clean and dry clothing in layers (the layer system) in cold climates. While discussing the layer system, it is important to define the "COLDER" principle. This acronym is used to aid in remembering how to use and take care of clothing.

   C - Keep clothing *C*lean.

   O - Avoid *O*verheating.

   L - Wear clothing *L*oose and in *L*ayers.

   D - Keep clothing *D*ry.

   E - *E*xamine clothing for defects or wear.

   R - Keep clothing *R*epaired.

(1) Clean. Dirt and other materials inside fabrics will cause the insulation to be ineffective, abrade and cut the fibers that make up the fabric, and cause holes. Washing clothing in the field may be impractical; therefore, survivors should concentrate on using proper techniques to prevent soiling clothing.

(2) Overheating. Clothing best serves the purpose of preserving body heat when worn in layers as follows: absorbent material next to the body, insulating layers, and outer garments to protect against wind and rain. Because of the rapid change in temperature, wind, and physical exertion, garments should allow donning and removal quickly and easily. Ventilation is essential when working because enclosing the body in an airtight layer system results in perspiration that wets clothing, thus reducing its insulating qualities.

(3) Loose. Garments should be loose-fitting to avoid reducing blood circulation and restricting body movement. Additionally, the garment should overhang the waist, wrists, ankles, and neck to reduce body heat loss.

(4) Dry. Keep clothing dry since a small amount of moisture in the insulation fibers will cause heat losses up to 25 times faster than dry clothing. Internally produced moisture is as damaging as is externally dampened clothing. The outer layer should protect the inner layers from moisture as well as from abrasion of fibers; for example, wool rubbing on logs or rocks, etc. The outer shell keeps dirt and other contaminants out of the clothing. Clothing can be dried in many ways. Fires are often used; however, take care to avoid burning the items. The "bare hand" test is very effective. Place one hand near the fire in the approximate place the wet items will be and count to three slowly. If this can be done without feeling excessive

heat, it should be safe to dry items there. Never leave any item unattended while it is drying. Leather boots, gloves, and mitten shells require extreme care to prevent shrinkage, stiffening, and cracking. The best way to dry boots is upright beside the fire (not upside down on sticks because the moisture does not escape the boot), or simply walk them dry in the milder climates. The sun and wind can be used to dry clothing with little supervision except for checking occasionally on the incoming weather and to make sure the article is secure. Freeze-drying is used in subzero temperatures with great success. Survivors let water freeze on or inside the item and then shake, bend, or beat it to cause the ice particles to fall free from the material. Tightly woven materials work better with this method than do open fibers.

(5) Examine. All clothing items should be inspected regularly for signs of damage or soil.

(6) Repair. Eskimos set an excellent example in the meticulous care they provide for their clothing. When damage is detected, immediately repair it.

h. The neck, head, hands, armpits, groin, and feet lose more heat than other parts of the body and require greater protection. Work with infrared film shows tremendous heat loss in those areas when not properly clothed. Survivors in a cold environment are in a real emergency situation without proper clothing. Figure 14-1 shows some examples of how military clothing works to hold body heat.

i. Models wearing samples of aircrew attire appear as spectral figures in a thermogram, an image revealing differences in infrared heat radiated from their clothing and exposed skin. White is warmest; red, yellow, green, blue, and magenta form a declining temperature scale spanning about 15 degrees; while black represents all lower temperatures. Almost the entire scale is seen on the model in boxer shorts. Warm, white spots appear on the underarm and neck. Only the shorts block radiation from the groin. Temperatures cool along the arm to dark blue fingertips far from the heat-producing torso. The addition of the next layer of clothing (Aramid long underwear) prevents heat loss except where it is tight against the body. As more layers are added, it is easy to see the areas of greatest concern are the head, hands, and feet. These areas are difficult for crewmembers to properly insulate while flying an aircraft. Mittens are ineffective due to the degraded manual dexterity. Likewise, it is difficult to feel the rudder-pedal action while wearing bulky, warm boots. These problems require inclusion of warm hats, mittens, and footgear (mukluk type) in survival kits during cold weather operation. Research has shown that when a CLo value of 10 is used to insulate the head, hands, and feet, and the rest of the body is protected by only one CLo, the average individual can be exposed to low temperatures (-10°F) comfortably for a reasonable period of time (30 to 40 minutes). When the amount of CLo value placed on the

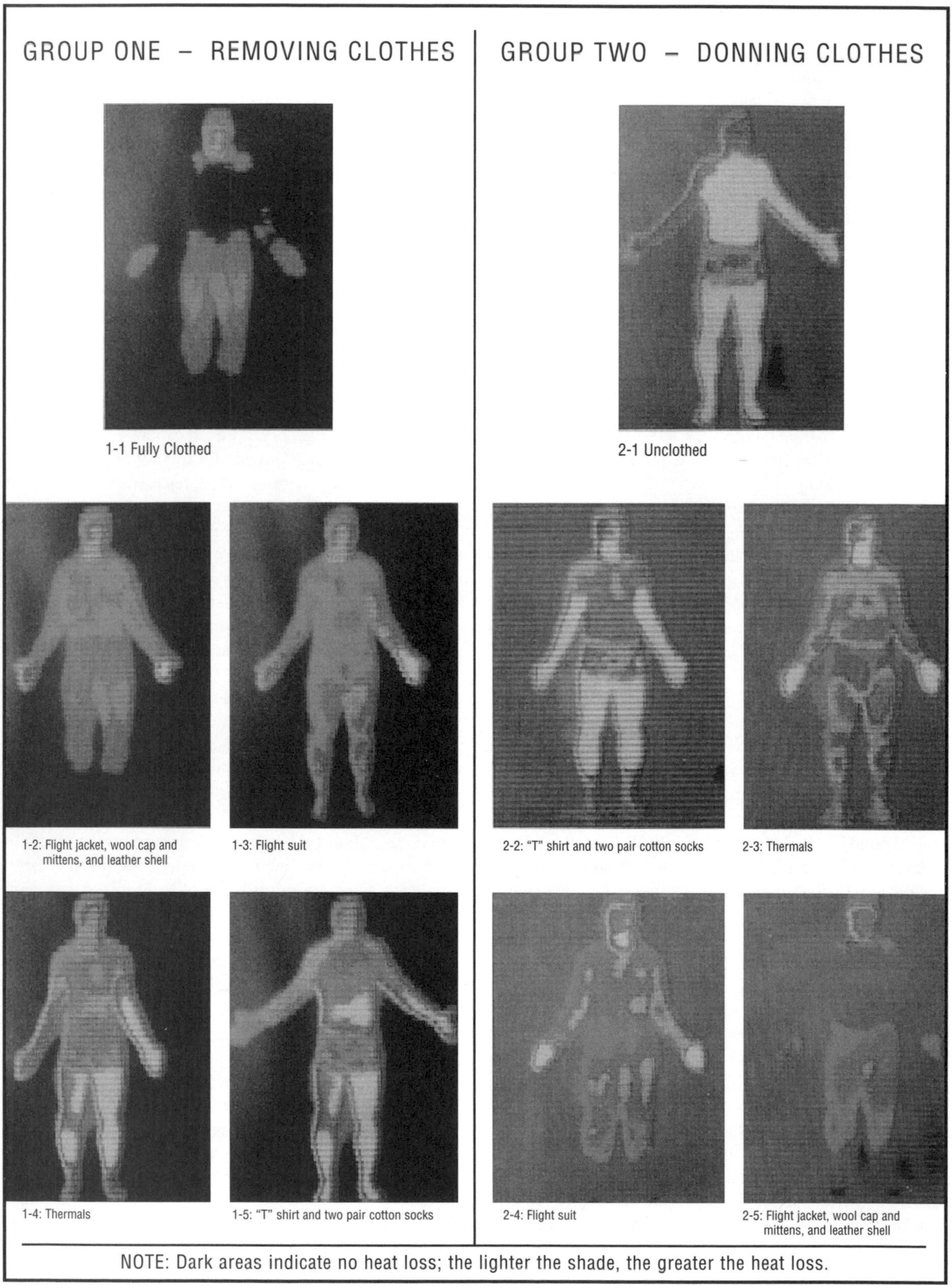

GROUP ONE – REMOVING CLOTHES

1-1 Fully Clothed

1-2: Flight jacket, wool cap and mittens, and leather shell

1-3: Flight suit

1-4: Thermals

1-5: "T" shirt and two pair cotton socks

GROUP TWO – DONNING CLOTHES

2-1 Unclothed

2-2: "T" shirt and two pair cotton socks

2-3: Thermals

2-4: Flight suit

2-5: Flight jacket, wool cap and mittens, and leather shell

NOTE: Dark areas indicate no heat loss; the lighter the shade, the greater the heat loss.

Figure 14-1. Thermogram of Body Heat Loss.

individual is reversed, the amount of time a survivor can spend in cold weather is greatly reduced due to the heat loss from their extremities. This same principle works in reverse in hot parts of the world—if one submerges the head, bands, or feet in cold water, it lets the most vascular parts of the body lose heat quickly.

14-8. Clothing Wear in Snow and Ice Areas:

a. The survivor should:

(1) Avoid restricting the circulation. Clothing should not be worn so tight that it restricts the flow of blood, which distributes the body heat and helps prevent frostbite. When wearing more than one pair of socks or gloves, ensure that each succeeding pair is large enough to fit comfortably over the other. Don't wear three or four pairs of socks in a shoe fitted for only one or two pairs. Release any restriction caused by twisted clothing or a tight parachute harness.

(2) Keep the head and ears covered. Survivors will lose as much as 50 percent of their total body heat from an unprotected head at 50°F.

(3) When exerting the body, prevent perspiration by opening clothing at the neck and wrists and loosening it at the waist. If the body is still warm, comfort can be obtained by taking off outer layers of clothing, one layer at a time. When work stops, the individual should put the clothing on again to prevent chilling.

(4) If boots are big enough, use dry grass, moss, or other material for added insulation around the feet. Footgear can be improvised by wrapping parachute cloth or other fabric lined with dry grass or moss for insulation.

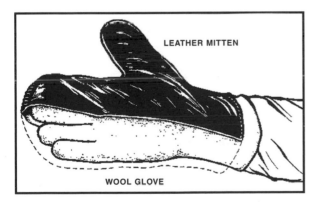

Figure 14-2. Layer System for Hands.

b. Felt booties and mukluks with the proper socks and insoles are best for dry, cold weather. Rubber-bottomed boot shoepacs with leather tops are best for wet weather. Mukluks should not be worn in wet weather. The vapor-barrier rubber boots can be worn under both conditions and are best at extremely low temperatures. The air-

release valve should be closed at ground level. These valves are designed to release pressure when airborne. Air should not be blown into the valves as the moisture could decrease insulation.

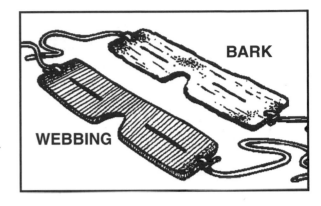

Figure 14-3. Improvised Goggles.

c. Clothing should be kept as dry as possible. Snow must be brushed from clothing before entering a shelter or going near a fire. The survivors should beat the frost out of garments before warming them, and dry them on a rack near a fire. Socks should be dried thoroughly.

d. One or two pairs of wool gloves and (or) mittens should be worn inside a waterproof shell (figure 14-2). If survivors have to expose their hands, they should warm them inside their clothing.

e. To help prevent sun or snow blindness, a survivor should wear sun or snow goggles or improvise a shield with a small horizontal slit opening (figure 14-3)

f. In strong wind or extreme cold, as a last resort, a survivor should wrap up in parachute material, if available, and get into some type of shelter or behind a windbreak. Extreme care should be taken with hard materials, such as synthetics, as they may become cold-soaked and require more time to warm.

g. At night, survivors should arrange dry, spare clothing loosely around and under the shoulders and hips to help keep the body warm. Wet clothes should never be worn into the sleeping bag. The moisture destroys the insulation value of the bag.

h. If survivors fall into water, they should roll in dry snow to blot up moisture, brush off the snow, and roll again until most of the water is absorbed. They should not remove footwear until they are in a shelter or beside a fire.

i. All clothing made of wool offers good protection when used as an inner layer. When wool is used next to the face and neck, survivors should be cautioned that moisture from the breath will condense on the surface and cause the insulating value to decrease. The use of a wool scarf wrapped around the mouth and nose is an excellent way to prevent cold injury, but it needs to be

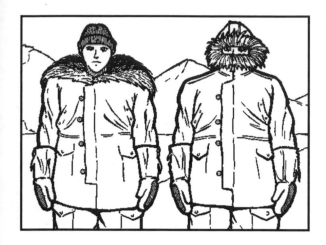

Figure 14-4. Proper Wear of Parka.

de-iced on a regular basis to prevent freezing flesh adjacent to it. An extra shell is generally worn over the warming layers to protect them and to act as a windbreak.

j. Other headgear includes the pile cap and hood. These items are most effective when used with a covering for the face in extreme cold. The pile cap is extremely warm where it is insulated, but it offers little protection for the face and back of the neck.

k. The hood is designed to funnel the radiant heat rising from the rest of the body and to recycle it to keep the neck, head, and face warm (figure 14-4). The individual's ability to tolerate cold should dictate the size of the front opening of the hood. The "tunnel" of a parka hood is usually lined with fur of some kind to act as a protecting device for the face. This same fur also helps to

protect the hood from the moisture expelled during breathing. The closed tunnel holds heat close to the face longer, the open one allows the heat to escape more freely. As the frost settles on the hair of the fur, it should be shaken from time to time to keep it free of ice buildup.

l. Sleeping systems (sleeping bag, liner, and bed) are the transition "clothing" used between normal daytime activities and sleep (figure 14-5).

m. The insulating material in the sleeping bag may be synthetic or it may be down and feathers. (Feathers and down lining require extra protection from moisture.) However, the covering is nylon. Survivors must realize that sleeping bags are compressed when packed and must be fluffed before use to restore insulation value. Clean and dry socks, mittens, and other clothing can be used to provide additional insulation.

n. Footgear is critical in a survival situation because walking is the only means of mobility. Therefore, care of footgear is essential both before and during a survival situation. Recommendations for care are:

(1) Ensure footgear is properly "broken-in" before flying.

(2) "Treat" footgear to ensure water-repellency (follow manufacturer's recommendations).

(3) Keep leather boots as dry as possible.

o. Mukluks have been around for thousands of years and have proven their worth in extremely cold weather. The Air Force mukluks are made of cotton duck with rubber-cleated soles and heels. (See figure 14-6.) They have slide fasteners from instep to collar, laces at instep and collar, and are 18 inches high. They are used by flying and ground personnel operating under *dry,* cold conditions in temperatures below 15°F. Survivors should change liners daily when possible.

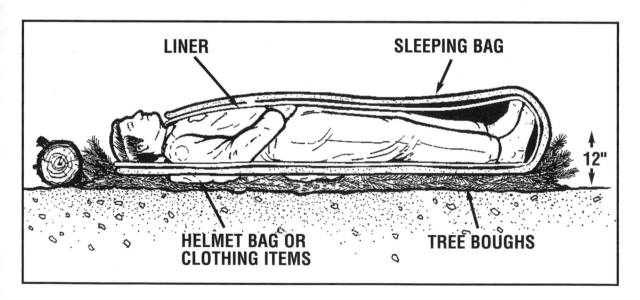

Figure 14-5. Sleeping System.

Figure 14-6. Issued Mukluks.

14-9. Care of the Feet. Foot care is critical in a survival situation. Improvising footgear may be essential to caring for feet.

a. Moose Hock Shoe. The hock skin of a moose or caribou will provide a suitable pair of shoes (figure 14-7). Cut skin around leg at A and B. Separate from the leg and pull it over the hoof. Shape and sew up small end C. Slit skin from A to B; bore holes on each side of cut for lacing; turn inside out, and lace with rawhide, suspension line, or other suitable material.

b. Grass Insoles. Used extensively by northern natives to construct inner soles. Grass is a good insulator and will collect moisture from the feet. The survivor should use the following procedure to prepare grass for use as inner soles: Grasp a sheaf of tall grass, about one-half inch in diameter, with both hands. Rotate the hands in opposite directions. The grass will break up or "fluff" into a soft mass. Form this fluff into oblong shapes and spread it evenly throughout the shoes. The inner soles should be about an inch thick. Remove these inner soles at night and make new ones the following day.

c. Hudson Bay Duffel. A triangular piece of material used as a foot covering. To improvise this foot covering, a survivor can use the following procedures:

(1) Cut two to four layers of parachute cloth into a 30-inch square.

(2) Fold this square to form a triangle.

(3) Place the foot on this triangle with the toes pointing at one corner.

(4) Fold the front cover up over the toes.

(5) Fold the side corners, one at a time, over the instep. This completes the foot wrap. (See figure 14-8.)

d. Gaiters. Made from parachute-cloth webbing or canvas. Gaiters help keep sand and snow out of shoes and protect the legs from bites and scratches (figure 14-9).

e. Double Socks. Cushion padding, feathers, dry grass, or fur stuffed between layers of socks. Wrap parachute or aircraft fabric around the feet and tie above the ankles. A combination of two or more types of improvised footwear may be more desirable and more efficient than any single type (figure 14-10).

14-10. Clothing In the Summer Arctic:

a. In the summer arctic, there are clouds of mosquitoes and black flies so thick a person can scarcely see through them. Survivors can protect themselves by wearing proper clothing to ensure no bare skin is exposed. A good head net and gloves should be worn.

b. Head nets must stand out from the face so they won't touch the skin. Issued head nets are either black or green. If one needs to be improvised they can be sewn to the brim of the hat or can be attached with an elastic band that fits around the crown. Black is the best color, as it can be seen through more easily than green or white. A heavy tape encasing a drawstring should be attached to the bottom of the head net for tying snugly at the collar. Hoops of wire fastened on the inside will make the net stand out from the face and at the same time allow it to be packed flat. The larger they are, the better the ventilation. But very large nets will not be as effective in wooded country, where they may become snagged on brush.

c. Gloves are hot, but are a necessity where flies are found in swamps. Kid gloves with a 6-inch gauntlet closing the gap at the wrist and ending with an elastic band halfway to the elbow are best. For fine work, kid gloves with the fingers cut off are good. Cotton/Nomex work gloves are better than no protection at all, but mosquitoes will bite through them. Treating the gloves with insect repellent will help. Smoky clothing may also help to keep insects away. (See figure 14-11.)

d. A survivor should remember that mosquitoes do not often bite through two layers of cloth; therefore, a lightweight undershirt and long underwear will help.

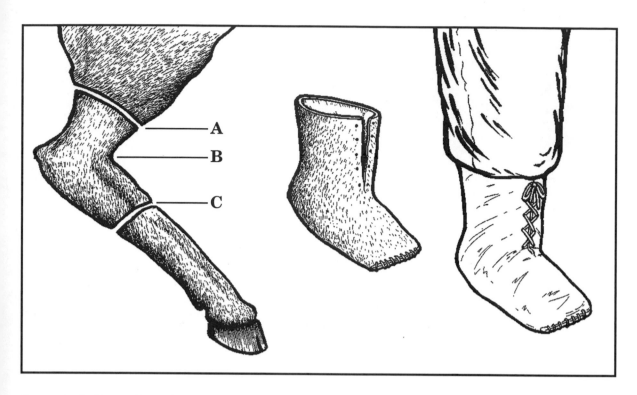

Figure 14-7. Moose Hock Shoes.

To protect ankles, blouse the bottoms of trousers around boots, or wear some type of leggings (gaiters).

e. If the head net is lost or none is available, make the best of a bad situation by wearing sunglasses with improvised screened sides, plugging ears lightly with cotton, and tying a handkerchief around the neck. Treat clothing with insect repellent at night.

14-11. Clothing at Sea. In cold oceans, survivors must try to stay dry and keep warm. If wet, they should use a windscreen to decrease the cooling effects of the wind. They should also remove, wring out, and replace outer

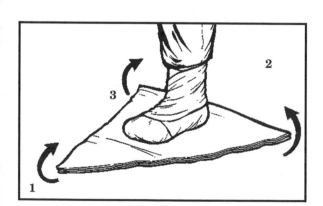

Figure 14-8. Hudson Bay Duffel.

Figure 14-9. Gaiters.

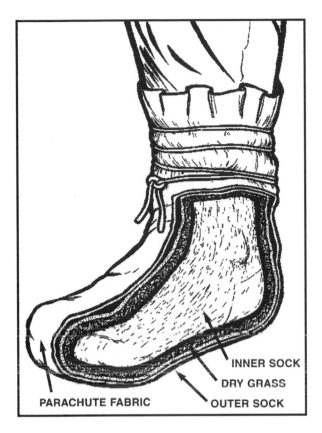

Figure 14-10. Double Socks.

garments or change into dry clothing. Hats, socks, and gloves should also be dried. If any survivors are dry, they should share extra clothes with those who are wet. Wet personnel should be given the most sheltered positions in the rain. Let them warm their hands and feet against those who are dry. Survivors should put on any extra clothing available. If no antiexposure suits are provided, they can drape extra clothing around their shoulders and over their heads. Clothes should be loose and comfortable. Also, survivors should attempt to keep the floor of the raft dry. For insulation, covering the floor with any available material will help. Survivors should huddle together on the floor of the raft and spread extra tarpaulin, sail, or parachute material over the group. If in a 20- or 25-man raft, canopy sides can be lowered. Performing mild exercises to restore circulation may be helpful. Survivors should exercise fingers, toes, shoulders, and buttock muscles. Mild exercise will help keep the body warm, stave off muscle spasms, and possibly prevent medical problems. Survivors should warm hands under armpits and periodically raise feet slightly and hold them up for a minute or two. They should also move face muscles frequently to prevent frostbite. Shivering is the body's way of quickly generating heat and is considered normal. However, persistent

Figure 14-11. Insect Protection.

shivering may lead to uncontrollable muscle spasms. They can be avoided by exercising muscles. If water is available, additional rations should be given to those suffering from exposure to cold. Survivors should eat small amounts frequently rather than one large meal.

14-12. Antiexposure Garments:

   a. Assemblies. The antiexposure assemblies, both quick donning and constant wear, are designed for personnel participating in over-water flights where unprotected or prolonged exposure to the climatic conditions

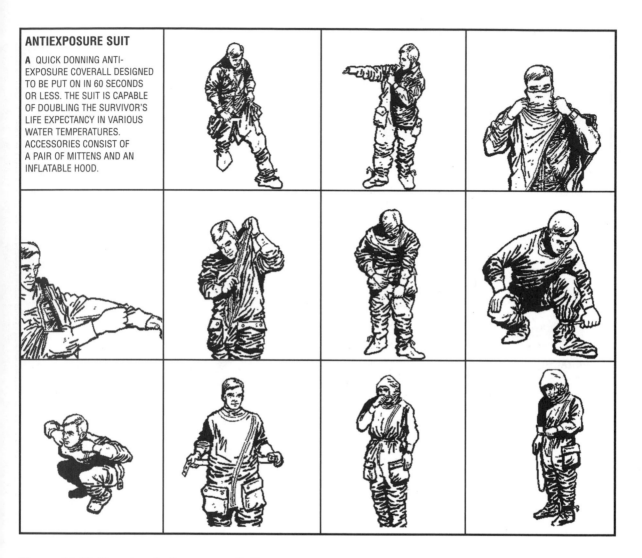

ANTIEXPOSURE SUIT

A QUICK DONNING ANTI-EXPOSURE COVERALL DESIGNED TO BE PUT ON IN 60 SECONDS OR LESS. THE SUIT IS CAPABLE OF DOUBLING THE SURVIVOR'S LIFE EXPECTANCY IN VARIOUS WATER TEMPERATURES. ACCESSORIES CONSIST OF A PAIR OF MITTENS AND AN INFLATABLE HOOD.

Figure 14-12. Donning Antiexposure Suit.

of cold air and (or) cold water (as a result of ditching or abandoning an aircraft) would be dangerous or could prove fatal. The suit provides protection from the wind and insulation against the chill of the ocean. The result of exposure in the water is illustrated in figures 13-2 and 13-3. Exposure time varies depending on the particular antiexposure assembly worn, the cold sensitivity of the person, and survival procedures used.

b. Quick-Donning Antiexposure Flying Coverall. Some antiexposure coveralls are designed for quick donning (approximately 1 minute) before emergency ditching. After ditching the aircraft, the coverall protects the wearer from exposure while swimming in cold water, and from exposure to wind, spray, and rain when adrift in a liferaft.

(1) The coverall is a one-size garment made from chloroprene-coated nylon cloth. It has two expandable-type patch pockets, an adjustable waist belt, and attached boots with adjustable ankle straps. One pair of insulated, adjustable wrist-strap mittens, each with a strap attached to a pocket, is provided. A hood, also attached with a strap, is in the left pocket. A carrying case with instructions and a snap-fastener closure is furnished for stowing in the aircraft.

(2) To use the coverall, personnel should wear it over regular flight clothing. It is large enough to wear over the usual flight gear. The gloves and hood are stowed in the pockets of the coverall and are normally worn after boarding the liferaft.

(3) The survivor should be extremely careful when

donning the coverall to prevent damage by snagging, tearing, or puncturing it on projecting objects. After donning the coverall, the waistband and boot ankle straps should be adjusted to take up fullness. If possible, crewmembers should stoop while pulling the neck seal to expel air trapped in the suit (figure 14-12). When jumping into the water, they should leap feet first with hands and arms close to sides or brought together above the head. Note that there is a constant-wear exposure suit designed to be worn continuously during over-water flights where the water temperature is 60 degrees or below. The Command may waiver it to 51 degrees.

14-13. Warm Oceans. Protection against the sun and securing drinking water are the most important problems. A survivor should keep the body covered as much as possible to avoid sunburn. A sunshade can be improvised out of any materials available, or the canopy provided with the raft may be used. If the heat becomes too intense, survivors may dampen clothing with seawater to promote evaporation and cooling. The use of sunburn preventive cream or a Chapstick is advisable. Remember, the body must be kept covered completely. Exposure to the sun increases thirst, wastes precious water, reduces the body's water content, and causes serious burns. Survivors should roll down their sleeves, pull up their socks, close their collars, wear a hat or improvised headgear, use a piece of cloth as a shield for the back of the neck, and wear sunglasses or improvise eye covers.

14-14. Tropical Climates:

a. In tropical areas, the body should be kept covered for prevention of insect bites, scratches, and sunburn.

b. When moving through vegetation, survivors should roll down their sleeves, wear gloves, and blouse the legs of their pants or tie them over their boot tops. Improvised puttees (gaiters) can be made from parachute material or any available fabric. This will protect legs from ticks and leeches.

c. Loosely worn clothing will keep survivors cooler, especially when subjected to the direct rays of the sun.

d. Survivors should wear a head net or tie material around the head for protection against insects. The most active time for insects is at dawn and dusk. An insect repellent should be used at these times.

e. In open country or in high grass, survivors should wear a neck cloth or improvised head covering for protection from sunburn and (or) dust. They should also move carefully through tall grass, as some sharp-edged grasses can cut clothing to shreds. Survivors should dry clothing before nightfall. If an extra change of clothing is available, efforts should be made to keep it clean and dry.

14-15. Dry Climates:

a. In the dry climates of the world, clothing will be needed for protection against sunburn, heat, sand, and

Figure 14-13. Protective Desert Clothing.

insects. Survivors should not discard any clothing. They should keep their head and body covered and blouse the legs of pants over the tops of footwear during the day. Survivors should not roll up sleeves, but keep them rolled down and loose at the cuff to stay cool.

b. Survivors should keep in mind that the people who live in the hot, dry areas of the world usually wear heavy, white, flowing robes that protect almost every inch of their body. The only areas open to the sun are the face and the eyes. This produces an area of higher humidity between the body and the clothing, which helps keep them cooler and conserves their perspiration (figure 14-13). The white clothing also reflects the sunlight.

c. Survivors should wear a cloth neckpiece to cover the back of the neck and protect it from the sun. A T-shirt makes an excellent neck drape, with the extra material used as padding under the cap. If hats are not available, survivors can make headpieces like those worn by the Arabs, as shown in figure 14-13. During dust storms, they should wear a covering for the mouth and nose; parachute cloth will work.

d. If shoes are lost or if they wear out, survivors can improvise footgear. One example of this is the "Russian Sock." Parachute material can be used to improvise these socks. The parachute material is cut into strips approximately 2 feet long and 4 inches wide. These strips are wrapped bandage fashion around the feet and ankles. Socks made in this fashion will provide comfort and protection for the feet.

Chapter 15

SHELTER

15-1. Introduction. Shelter is anything that protects a survivor from the environmental hazards. The information in this chapter describes how the environment influences shelter site selection and factors that survivors must consider before constructing an adequate shelter. The techniques and procedures for constructing shelters for various types of protection are also presented.

15-2. Shelter Considerations. The location and type of shelter built by survivors vary with each survival situation. There are many things to consider when picking a site. Survivors should consider the time and energy required to establish an adequate camp, weather conditions, life forms (human, plant, and animal), terrain, and time of day. Every effort should be made to use as little energy as possible and yet attain maximum protection from the environment.

a. Time. Late afternoon is not the best time to look for a site that will meet that day's shelter requirements. If survivors wait until the last minute, they may be forced to use poor materials in unfavorable conditions. They must constantly be thinking of ways to satisfy their needs for protection from environmental hazards.

b. Weather. Weather conditions are a key consideration when selecting a shelter site. Failure to consider the weather could have disastrous results. Some major weather factors that can influence the survivor's choice of shelter type and site selection are temperature, wind, and precipitation.

(1) Temperature. Temperatures can vary considerably within a given area. Situating a campsite in low areas such as a valley in cold regions can expose survivors to low night temperatures and windchill factors. Colder temperatures are found along valley floors, which are sometimes referred to as "cold air sumps." It may be advantageous to situate campsites to take advantage of the sun. Survivors could place their shelters in open areas during the colder months for added warmth, and in shaded areas for protection from the sun during periods of hotter weather. In some areas a compromise may have to be made. For example, in many deserts the daytime temperatures can be very high while low temperatures at night can turn water to ice. Protection from both heat and cold are needed in these areas. Shelter type and location should be chosen to provide protection from the existing temperature conditions.

(2) Wind. Wind can be either an advantage or a disadvantage depending upon the temperature of the area and the velocity of the wind. During the summer or on warm days, survivors can take advantage of the cool breezes and protection the wind provides from insects by locating their camps on knolls or spits of land. Conversely, wind can become an annoyance or even a hazard as blowing sand, dust, or snow can cause skin and eye irritation and damage to clothing and equipment. On cold days or during winter months, survivors should seek shelter sites that are protected from the effects of windchill and drifting snow.

(3) Precipitation. The many forms of precipitation (rain, sleet, hail, or snow) can also present problems for survivors. Shelter sites should be out of major drainages and other low areas to provide protection from flash floods or mudslides resulting from heavy rains. Snow can also be a great danger if shelters are placed in potential avalanche areas.

c. Life Forms. All life forms (plant, human, and animal) must be considered when selecting the campsite and the type of shelter that will be used. The "human" factor may mean the enemy or other groups from whom survivors wish to remain undetected. Information regarding this aspect of shelters and shelter site selection is in part nine of this regulation (Evasion). For a shelter to be adequate, certain factors must be considered, especially if extended survival is expected.

(1) Insect life can cause personal discomfort, disease, and injury. By locating shelters on knolls, ridges, or any other area that has a breeze or steady wind, survivors can reduce the number of flying insects in their area. Staying away from standing-water sources will help to avoid mosquitoes, bees, wasps, and hornets. Ants can be a major problem; some species will vigorously defend their territories with painful stings or bites or particularly distressing pungent odors.

(2) Large and small animals can also be a problem, especially if the camp is situated near their trails or waterholes.

(3) Dead trees that are standing, and trees with dead branches should be avoided. Wind may cause them to fall, causing injuries or death. Poisonous plants, such as poison oak or poison ivy, must also be avoided when locating a shelter.

d. Terrain. Terrain hazards may not be as apparent as weather and animal-life hazards, but they can be many times more dangerous. Avalanche, rock, dry streambeds, or mudslide areas should be avoided. These areas can be recognized by either a clear path or a path of secondary vegetation, such as 1- to 15-foot tall vegetation or other new growth that extends from the top to the bottom of a hill or mountain. Survivors should not choose shelter sites at the bottom of steep slopes, which may be prone to slides. Likewise, there is a danger in camping at the bottom of steep scree or talus slopes. Additionally, rock overhang must be checked for safety before using it as a shelter.

15-3. Location:

a. Four prerequisites must be satisfied when selecting a shelter location.

(1) The first is being near water, food, fuel, and a signal or recovery site.

(2) The second is that the area be safe, providing natural protection from environmental hazards.

(3) The third is that sufficient materials be available to construct the shelter. In some cases, the "shelter" may already be present. Survivors seriously limit themselves if they assume that shelters *must* be a fabricated framework having predetermined dimensions and a cover of parachute material or a signal paulin. More appropriately, survivors should consider using sheltered *places* already in existence in the immediate area. This does not rule out shelters with a fabricated framework and parachute or other manufactured material covering; it simply enlarges the scope of what can be used as a survival shelter.

(4) Finally, the area chosen must be both large enough and level enough for the survivor to lie down. Personal comfort is an important fundamental for survivors to consider. An adequate shelter provides physical and mental well-being for sound rest. Adequate rest is extremely vital if survivors are to make sound decisions. Their need for rest becomes more critical as time passes and rescue or return is delayed. Before actually constructing a shelter, survivors must determine the specific purpose of the shelter. The following factors influence the type of shelter to be fabricated.

(a) Rain or other precipitation.

(b) Cold.

(c) Heat.

(d) Insects.

(e) Available materials nearby (manufactured or natural).

(f) Length of expected stay.

(g) Enemy presence in the area—evasion "shelters" are covered in part nine of the regulation (Evasion).

(h) Number and physical condition of survivors.

b. If possible, survivors should try to find a shelter that needs little work to be adequate. Using what is already there, so that complete construction of a shelter is not necessary, saves time and energy. For example, rock overhangs, caves, large crevices, fallen logs, root buttresses, or snow banks can all be modified to provide adequate shelter. Modifications may include adding snow blocks to finish off an existing tree-well shelter, increasing the insulation of the shelter by using vegetation or parachute material, etc., or building a reflector fire in front of a rock overhang or cave. Survivors must consider the amount of energy required to build the shelter. It is not really wise to spend a great deal of time and energy in constructing a shelter if nature has provided a natural shelter nearby that will satisfy the

survivor's needs. See figure 15-1 for an example of a naturally occurring shelter.

c. The size limitations of a shelter are important only if there is either a lack of material on hand or if it is

Figure 15-1. Natural Shelter.

cold. Otherwise, the shelter should be large enough to be comfortable yet not so large as to cause an excessive amount of work. Any shelter, naturally occurring or otherwise, in which a fire is to be built must have a ventilation system that will provide fresh air and allow smoke and carbon monoxide to escape. Even if a fire does not produce visible smoke (such as heat tabs), the shelter must still be vented. See figure 15-27 for placement of ventilation holes in a snow cave. If a fire is to be placed outside the shelter, the opening of the shelter should be placed 90 degrees to the prevailing wind. This will reduce the chances of sparks and smoke being blown into the shelter if the wind should reverse direction in the morning and evening. This frequently occurs in mountainous areas. The best fire-to-shelter distance is approximately 3 feet. One place where it would not be wise to build a fire is near the aircraft wreckage, especially if it is being used as a shelter. The possibility of igniting spilled lubricants or fuels is great. Survivors may decide instead to use materials from the aircraft to add to a shelter located a safe distance from the crash site.

15-4. Immediate Action Shelters. The first type of shelter that survivors may consider using, or the first type they may be forced to use, is an immediate action shelter. An immediate action shelter is one that can be erected quickly with minimum effort; for example, raft, aircraft parts, parachutes, paulin, and plastic bags. Natural formations can also shield survivors from the elements immediately, and include overhanging ledges, fallen logs, caves, and tree wells (figure 15-2). It isn't necessary to be concerned with exact shelter dimensions. Survivors should remember that if shelter is needed, use an existing shelter if at all possible. They should improvise on natural shelters or construct new shelters only if necessary. Regardless of type, the shelter must provide whatever protection is needed and, with a little ingenuity, it should be possible for survivors to protect themselves and do so quickly. In many instances, the immediate action shelters may have to serve as permanent shelters for aircrew members. For example, many aircrew members fly without parachutes, large cutting implements (axes), and entrenching tools; therefore, multiperson lifecrafts may be the only immediate or long-term shelter available. In this situation, multiperson liferafts must be deployed in the quickest manner possible to ensure maximum advantages are attained from the following shelter principles:

a. Set up in areas that afford maximum protection from precipitation and wind and use the basic shelter principles in paragraphs 15-2 and 15-3.

b. Anchor the raft for retention during high winds.

c. Use additional boughs, grasses, etc., for ground insulation.

15-5. Improvised Shelters. Shelters of this type should be easy to construct and (or) dismantle in a short period of time. However, these shelters usually require more time to construct then an immediate action shelter. For this reason, survivors should only consider this type of shelter when they aren't immediately concerned with getting out of the elements. Shelters of this type include the following:

a. The "A-frame" design is adaptable to all environments, as it can be easily modified; for example, tropical parahammock, temperate area "A-frame," arctic thermal "A-frame," and fighter trench.

b. Simple shade shelter; these are useful in dry areas.

c. Various paratepees.

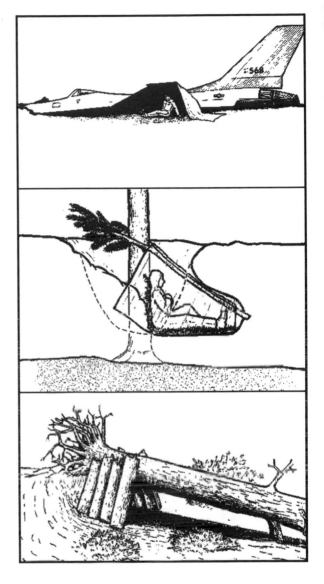

Figure 15-2. Immediate Action Shelters.

d. Snow shelters; includes tree-pit shelters.

e. All other variations of the above shelter types; sod shelters, etc.

15-6. Shelters for Warm Temperature Areas:

a. If survivors are to use parachute material, they should remember that "pitch and tightness" apply to shelters designed to shed rain or snow. Parachute material is porous and will not shed moisture unless it is stretched tightly at an angle of sufficient pitch, which will encourage runoff instead of penetration. An angle of 40 to 60 degrees is recommended for the "pitch" of the shelter. The material stretched over the framework should be wrinkle-free and tight. Survivors should not touch the material when water is running over it as this will break the surface tension at that point and allow water to drip into the shelter. Two layers of parachute material, 4 to 6 inches apart, will create a more effective water-repellent covering. Even during hard rain, the outer layer only lets a mist penetrate if it is pulled tight. The inner layer will then channel off any moisture that may penetrate. This layering of parachute material also creates a dead-air space that covers the shelter. This is especially beneficial in cold areas when the shelter is enclosed. Adequate insulation can also be provided by boughs, aircraft parts, snow, etc. These will be discussed in more depth in the area of cold-climate shelters. A double layering of parachute material helps to trap body heat, radiating heat from the Earth's surface, and other heating sources.

b. The first step is deciding the type of shelter required. No matter which shelter is selected, the building or improvising process should be planned and orderly, following proven procedures and techniques. The second step is to select, collect, and prepare all materials needed before the actual construction; this includes framework, covering, bedding, or insulation, and implements used to secure the shelter ("dead-men" lines, stakes, etc.).

(1) For shelter that use a wooden framework, the poles or wood selected should have all the rough edges and stubs removed. Not only will this reduce the chances of the parachute fabric being ripped, but it will eliminate the chances of injury to survivors.

(2) On the outer side of a tree selected as natural shelter, some or all of the branches may be left in place as they will make a good support structure for the rest of the shelter parts.

(3) In addition to the parachute, there are many other materials that can be used as framework coverings. Some of the following are both framework and

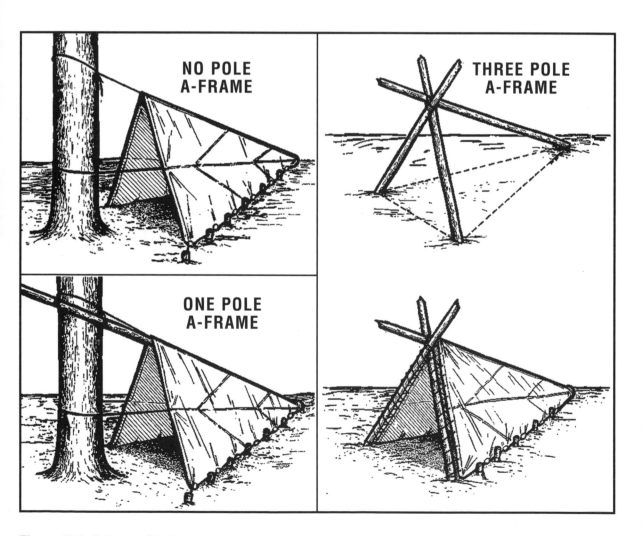

Figure 15-3. A-Frame Shelters.

covering all in one:

(a) Bark peeled off dead trees.

(b) Boughs cut off trees.

(c) Bamboo, palm, grasses, and other vegetation cut or woven into desired patterns.

(4) If parachute material is to be used alone or in combination with natural materials, it must be changed slightly. Survivors should remove all of the lines from the parachute and then cut it to size. This will eliminate bunching and wrinkling and reduce leakage.

c. The third step in the process of shelter construction is site preparation. This includes brushing away rocks and twigs from the sleeping area and cutting back overhanging vegetation.

d. The fourth step is to actually construct the shelter, beginning with the framework. The framework is very important. It must be strong enough to support the weight of the covering and precipitation buildup of snow. It must also be sturdy enough to resist strong wind gusts.

(1) Construct the framework in one of two ways. For natural shelters, branches may be securely placed against trees or other natural objects. For parachute shelters, poles may be lashed to trees or to other poles. The support poles or branches can then be laid and (or) attached depending on their function.

(2) The pitch of the shelter is determined by the framework. A 60-degree pitch is the optimum for shedding precipitation and providing shelter room.

(3) The size of the shelter is controlled by the framework. The shelter should be large enough for survivors to sit up, with adequate room to lie down and to store all personal equipment.

(4) After the basic framework has been completed, survivors can apply and secure the framework covering. The care and techniques used to apply the covering will determine the effectiveness of the shelter in shedding precipitation.

(5) When using parachute material on shelters, survivors should remove all suspension lines from the material. (Excess line can be used for lashing, sewing, etc.) Next, stretch the center seam tight; then work from the back of the shelter to the front, alternating sides and securing the material to stakes or framework by using buttons and lines. When stretching the material tight, survivors should pull the material 90 degrees to the wrinkles. If material is not stretched tight, any moisture will pool in the wrinkles and leak into the shelter.

(6) If natural materials are to be used for the covering, the shingle method should be used. Starting at the bottom and working toward the top of the shelter, the bottom of each piece should overlap the top of the preceding piece. This will allow water to drain off. The material should be placed on the shelter in sufficient quality so survivors in the shelter cannot see through it.

15-7. Maintenance and Improvements. Once a shelter is constructed, it must be maintained. Additional modifications may make the shelter more effective and comfortable. Indian lacing (lacing the front of the shelter to the tripod) will tighten the shelter. A door may help block the wind and keep insects out. Other modifications may include a fire reflector, porch or work area, or another whole addition such as an opposing lean-to.

15-8. Construction of Specific Shelters:

a. A-Frame. The following is one way to build an A-frame shelter in a warm temperate environment using parachute material for the covering. There are as many variations of this shelter as there are builders. The procedures here will, if followed carefully, result in the completion of a safe shelter that will meet survivors' needs. For an example of this and other A-frame shelters, see figure 15-3.

(1) Materials Needed:

(a) One 12- to 18- foot long sturdy ridge pole with all projections cleaned off.

(b) Two bipod poles, approximately 7 feet long.

(c) Parachute material, normally 5 or 6 gores.

(d) Suspension lines.

(e) "Buttons," small objects placed behind gathers of material to provide a secure way of affixing suspension line to the parachute material.

(f) Approximately 14 stakes, approximately 10 inches long.

(2) Assembling the Framework:

(a) Lash (See chapter 17—Equipment) the two bipod poles together at eye-level height.

(b) Place the ridge pole, with the large end on the ground, into the bipod formed by the poles and secure with a square lash.

(c) The bipod structure should be 90 degrees to the ridge pole, and the bipod poles should be spread out to an approximate equilateral triangle of a 60-degree pitch. A piece of line can be used to measure this.

(3) Application of Fabric:

(a) Tie off about 2 feet of the apex in a knot and tuck this under the butt end of the ridge pole. Use half hitches and clove hitches to secure the material to the base of the pole.

(b) Place the center radial seam of the parachute piece (or the center of the fabric) on the ridge pole. After pulling the material taut, use half hitches and clove hitches to secure the fabric to the front of the ridge pole.

(c) Scribe or draw a line on the ground from the butt of the ridge pole to each of bipod poles. Stake the fabric down, starting at the rear of the shelter and alternately staking from side to side to the shelter front. Use a sufficient number of stakes to ensure the parachute material is wrinkle-free.

(d) Stakes should be slanted or inclined away from the direction of pull. When tying off with a clove hitch, the line should pass in front of the stake first and then pass under itself to allow the button and line to be pulled 90 degrees to the wrinkle.

(e) Indian lacing is the sewing or lacing of the lower lateral band with inner core or line that is secured to the bipod poles. This will remove the remaining wrinkles and further tighten the material.

(f) A rain fly, bed, and other refinements can now be added.

b. Lean-To:

(1) Materials Needed:

(a) A sturdy, smooth ridge pole (longer than the builder's body) long enough to span the distance between two sturdy trees.

(b) Support poles, 10 feet long.

(c) Stakes, suspension lines, and buttons.

(d) Parachute material (minimum of four gores).

(2) Assembling the Framework:

(a) Lash the ridge pole (between two suitable trees) about chest or shoulder high.

(b) Lay the roof support poles on the ridge pole so the roof-support poles and the ground are at approximately a 60-degree angle. Lash the roof-support poles to the ridge pole.

(3) Application of Fabric:

(a) Place the middle seam of the fabric on the middle support pole with lower lateral band along the ridge pole.

(b) Tie off the middle and both sides of the lower lateral band approximately 8 to 10 inches from the ridge pole.

(c) Stake the middle of the rear of the shelter first, then alternate from side to side.

(d) The stakes that go up the sides to the front should point to the front of the shelter.

(e) Pull the lower lateral band closer to the ridge pole by Indian lacing.

(f) Add bed and other refinements (reflector fire, bed logs, rain fly, etc.). See figure 15-4 for lean-to examples.

c. Paratepee, 9-Pole. The paratepee is an excellent shelter for protection from wind, rain, cold, and insects. Cooking, eating, sleeping, resting, signaling, and washing can all be done without going outdoors. The paratepee, whether 9-pole, 1-pole, or no-pole, is the only improvised shelter that provides adequate ventilation to build an inside fire. With a small fire inside, the shelter also serves as a signal at night.

(1) Materials Needed:

(a) Suspension line.

(b) Parachute material, normally 14 gores are suitable.

-1. Spread out the 14-gore section of parachute and cut off all lines at the lower lateral band, leaving about 18 inches of line attached. All other suspension lines should be stripped from the parachute.

Figure 15-4. Lean-To Shelters.

-2. Sew two smoke flaps, made from two large panels of parachute material, at the apex of the 14-gore section on the outside seams. Attach suspension line with a bowline in the end to each smoke flap. The ends of the smoke-flap poles will be inserted in these (see figure 15-5).

(c) Stakes.

(d) Although any number of poles may be used, 11 poles, smoothed off, each about 20 feet long, will normally provide adequate support.

(2) Assembling the Framework. (Assume 11 poles are used. Adjust instructions if different numbers are used.)

(a) Lay three poles on the ground with the butts even. Stretch the canopy along the poles. The lower lateral band should be 4 to 6 inches from the bottoms of the poles before the stretching takes place. Mark one of the poles at the apex point.

(b) Lash the three poles together, 5 to 10 inches above the marked area. (A shear lash is effective for this purpose.) These poles will form the tripod (figure 15-5).

(c) Scribe a circle approximately 12 feet in diameter in the shelter area and set the tripod so the butts of the poles are evenly spaced on the circle. Five of the remaining eight poles should be placed so the butts are evenly spaced around the 12-foot circle, and the tops are laid in the apex of the tripod to form the smallest apex possible (figure 15-5).

(3) Application of Fabric:

(a) Stretch the parachute material along the tie pole. Using the suspension line attached to the middle radial seam, tie the lower lateral band to the tie pole 6 inches from the butt end. Stretch the parachute material along the middle radial seam and tie it to the tie pole using the suspension line at the apex. Lay the tie pole onto the shelter frame with the butt along the 12-foot circle and the top in the apex formed by the other poles. The tie pole should be placed directly opposite the proposed door.

(b) Move the canopy material (both sides of it) from the tie pole around the framework and tie the lower lateral band together and stake it at the door. The front can now be sewn or pegged closed, leaving 3 to 4 feet for a door. A sewing "ladder" can be made by lashing steps up the front of the tepee (figure 15-5).

(c) Enter the shelter and move the butts of the poles outward to form a more perfect circle and until the fabric is relatively tight and smooth.

(d) Tighten the fabric and remove remaining wrinkles. Start staking directly opposite the door and alternate from side to side, pulling the material down and to the front of the shelter. Use clove hitches or similar knots to secure material to the stakes.

(e) Insert the final two poles into the loops on the smoke flaps. The paratepee is now finished (figure 15-5).

(f) One improvement that could be made to the paratepee is the installation of a liner. This will allow a draft for a fire without making the occupants cold, since there may be a slight gap between the lower lateral band and the ground. A liner can be affixed to the inside of the paratepee by taking the remaining 14-gore piece of material and firmly staking the lower lateral band directly to the ground all the way around, leaving room for the door. The area where the liner and door meet may be sewn up. The rest of the material is brought up the inside walls and affixed to the poles with buttons (figure 15-5).

d. Paratepee, 1-Pole:

(1) Materials Needed:

(a) Normally use a 14-gore section of canopy, strip the shroud lines, leaving 16- to 18-inch lengths at the lower lateral band.

(b) Stakes.

(c) Inner core and needle.

(2) Construction of the 1-Pole Paratepee:

(a) Select a shelter site and scribe a circle about 14 feet in diameter on the ground.

(b) The parachute material is staked to the ground using the lines attached at the lower lateral band. After deciding where the shelter door will be located, stake the first line (from the lower band) down securely. Proceed around the scribed line and stake down all the lines from the lateral band, making sure the parachute material is stretched taut before the line is staked down.

(c) Once all the lines are staked down, loosely attach the center pole, and, through trial and error, determine the point at which the parachute material will be pulled tight once the center pole is placed upright—securely attach the material at this point.

(d) Using a suspension line (or innercore), sew the end gores together, leaving 3 or 4 feet for a door (figure 15-6).

e. Paratepee, No-Pole. For this shelter, the 14 gores of material are prepared the same way. A line is attached to the apex and thrown over a tree limb, etc., and tied off. The lower lateral band is then staked down starting opposite the door around a 12- to 14-foot circle. (See figure 15-7 for a no-pole paratepee example.)

f. Sod Shelter. A framework covered with sod provides a shelter that is warm in cold weather and one that is easily made waterproof and insect-proof in the summer. The framework for a sod shelter must be strong, and it can be made of driftwood, poles, willow, etc. (Some natives use whale bones.) Sod, with a heavy growth of grass or weeds, should be used since the roots tend to hold the soil together. Cutting about 2 inches of soil along with the grass is sufficient. The size of the blocks is determined by the strength of the individual. A sod house is strong and fireproof.

15-9. Shelters for Tropical Areas. Basic considerations for shelter in tropical areas are as follows:

a. In tropical areas, especially moist tropical areas, the major environmental factors influencing both site selection and shelter types are:

(1) Moisture and dampness.

(2) Rain.

(3) Wet ground.

(4) Heat.

(5) Mudslide areas.

(6) Dead standing trees and limbs.

(7) Insects.

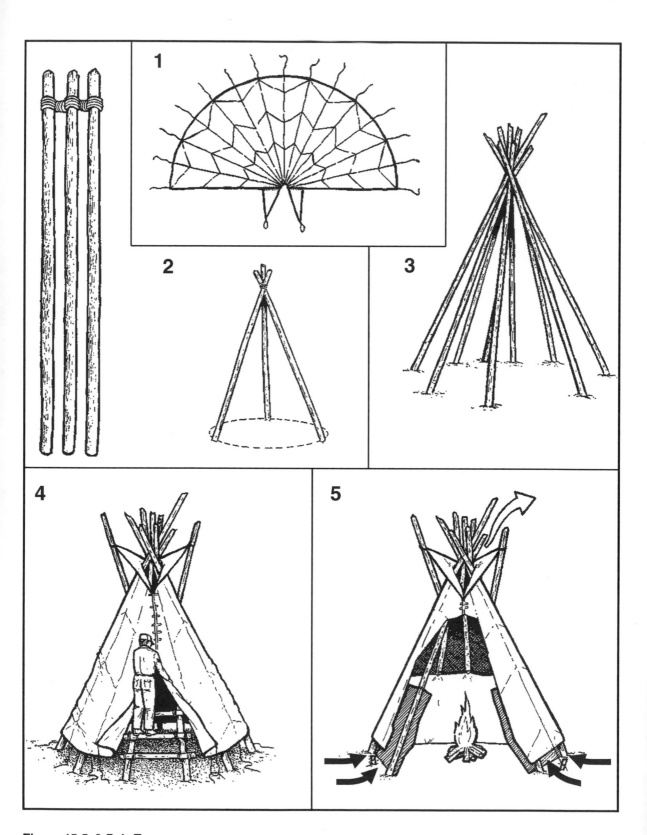

Figure 15-5. 9-Pole Tepee.

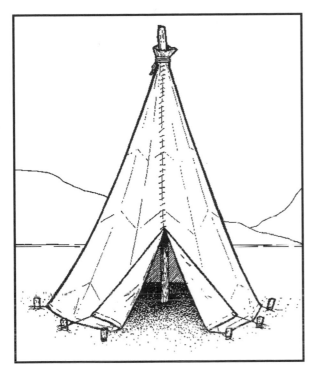

Figure 15-6. 1-Pole Tepee.

Figure 15-7. No-Pole Tepee.

b. Survivors should establish a campsite on a knoll or high spot in an open area well back from any swamps or marshy areas. The ground in these areas is drier, and there may be a breeze, which will result in fewer insects.

c. Underbrush and dead vegetation should be cleared from the shelter site. Crawling insects will not be able to approach survivors as easily due to lack of cover.

d. A thick bamboo clump or matted canopy of vines for cover reflects the smoke from the campfire and discourages insects. This cover will also keep the extremely heavy early morning dew off the bedding.

e. The easiest improvised shelter is made by draping a parachute, tarpaulin, or poncho over a rope or vine stretched between two trees. One end of the canopy should be kept higher than the other; insects are discouraged by few openings in shelters and by smudge fires. A hammock made from parachute material will keep the survivor off the ground and discourage ants, spiders, leeches, scorpions, and other pests.

f. In the wet jungle, survivors need shelter from dampness. If they stay with the aircraft, it should be used for shelter. They should try to make it mosquito-proof by covering openings with netting or parachute cloth.

g. A good rain shelter can be made by constructing an A-type framework and shingling it with a good thickness

of palm or other broad-leaf plants, pieces of bark, and mats of grass (figure 15-8).

h. Nights are cold in some mountainous tropical areas. Survivors should try to stay out of the wind and build a fire. Reflecting the heat off a rock pile or other barrier is a good idea. Some natural materials that can be used in

Figure 15-8. Banana Leaf A-Frame.

Figure 15-9. Raised Platform Shelters.

the shelters are green wood (dead wood may be too rotten), bamboo, and palm leaves. Vines can be used in place of suspension line for thatching roofs or floors, etc. Banana plant sections can be separated from the banana plant and fashioned to provide a mattress effect.

15-10. Specific Shelters for Tropical Environments:

a. Raised Platform Shelter (figure 15-9). This shelter has many variations. One example is four trees or vertical poles in a rectangular pattern that is a little longer and a little wider than the survivor, keeping in mind that the survivor will also need protection for equipment. Two long, sturdy poles are then square-lashed between the trees or vertical poles, one on each side of the intended shelter. Cross pieces can then be secured across the two horizontal poles at 6- to 12-inch intervals. This forms the platform on which a natural mattress may be constructed. Parachute material can be used as an insect net and a roof can be built over the structure using A-frame building techniques. The roof should be waterproofed with thatching laid bottom to top in a thick shingle fashion. See figure 15-9 for examples of this and other platform shelters. These shelters can also be built using three trees in a triangular pattern. At the foot of the shelter, two poles are joined to one tree.

b. Variation of Platform Shelter. A variation of the platform-type shelter is the paraplatform. A quick and comfortable bed is made by simply wrapping material around the two "frame" poles. Another method is to roll poles in the material in the same manner as for an improvised stretcher (figure 15-10).

c. Hammocks. Various parahammocks can also be made. They are more involved than a simple parachute wrapped framework and not quite as comfortable (figure 15-11).

d. Hobo Shelter. On tropical coasts and other coastal environments, if a more permanent shelter is desired as opposed to a simple shade shelter, survivors should build a "hobo" shelter. To build this shelter:

(1) Dig into the lee side of a sand dune to protect the shelter from the wind. Clear a level area large enough to lie down in and store equipment.

(2) After the area has been cleared, build a heavy driftwood framework that will support the sand.

(3) Wall sides and top with strong material (boards, driftwood, etc.) that will support the sand; leave a door opening.

(4) Slope the roof to equal the slope of the sand dune. Cover the entire shelter with parachute material to keep sand from sifting through small holes in the walls and roof.

(5) Cover with 6 to 12 inches of sand to provide protection from wind and moisture.

(6) Construct a door for the shelter (figure 15-12).

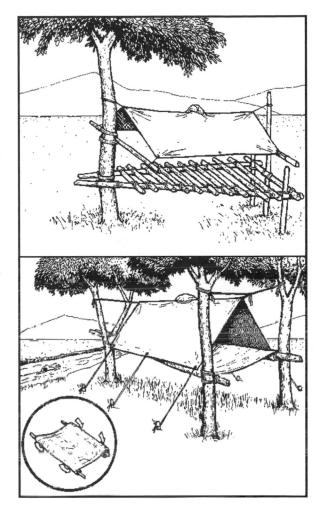

Figure 15-10. Raised Paraplatform Shelters.

15-11. Shelters for Dry Climates:

a. Natives of hot, dry areas make use of light-proof shelters with sides rolled up to take advantage of any breeze. Survivors should emulate these shade-type shelters if forced to survive in these areas. The extremes of heat *and* cold must be considered in hot areas, as most can become very cold during the night. The major problem for survivors will be escaping the heat and sun's rays.

b. Natural shelters in these areas are often limited to the shade of cliffs and the lee sides of hills, dunes, or rock formations. In some desert mountains, it is possible to find good rock shelters or cave-like protection under tumbled blocks of rocks that have fallen from cliffs. Use care to ensure that these blocks are in areas void of future rock-falling activity and free from animal hazards.

c. Vegetation, if any exists, is usually stunted and armed with thorns. It may be possible to stay in the shade by moving around the vegetation as the sun moves. The hottest part of the day may offer few shadows because the sun is directly overhead. Parachute material draped over bushes or rocks will provide some shade.

d. Materials that can be used in the construction of desert shelters include:

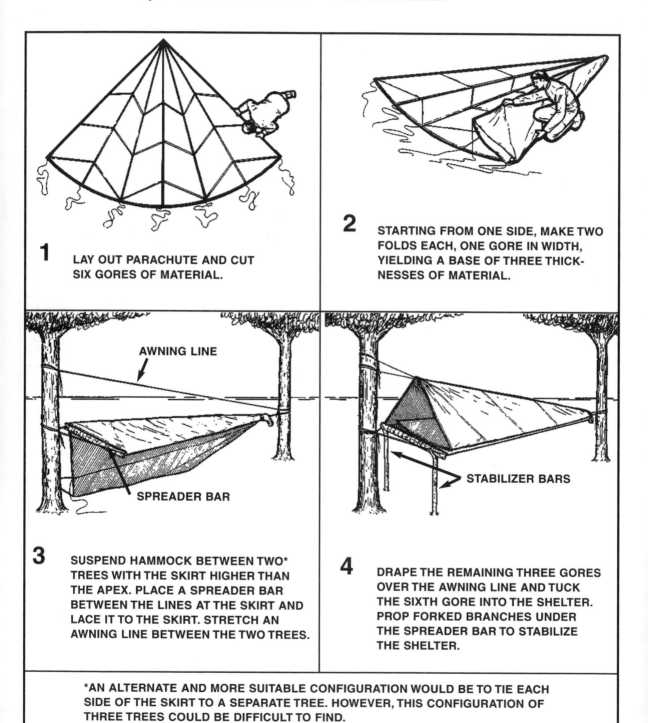

1 LAY OUT PARACHUTE AND CUT SIX GORES OF MATERIAL.

2 STARTING FROM ONE SIDE, MAKE TWO FOLDS EACH, ONE GORE IN WIDTH, YIELDING A BASE OF THREE THICK-NESSES OF MATERIAL.

AWNING LINE

SPREADER BAR

STABILIZER BARS

3 SUSPEND HAMMOCK BETWEEN TWO* TREES WITH THE SKIRT HIGHER THAN THE APEX. PLACE A SPREADER BAR BETWEEN THE LINES AT THE SKIRT AND LACE IT TO THE SKIRT. STRETCH AN AWNING LINE BETWEEN THE TWO TREES.

4 DRAPE THE REMAINING THREE GORES OVER THE AWNING LINE AND TUCK THE SIXTH GORE INTO THE SHELTER. PROP FORKED BRANCHES UNDER THE SPREADER BAR TO STABILIZE THE SHELTER.

*AN ALTERNATE AND MORE SUITABLE CONFIGURATION WOULD BE TO TIE EACH SIDE OF THE SKIRT TO A SEPARATE TREE. HOWEVER, THIS CONFIGURATION OF THREE TREES COULD BE DIFFICULT TO FIND.

Figure 15-11. Parahammock.

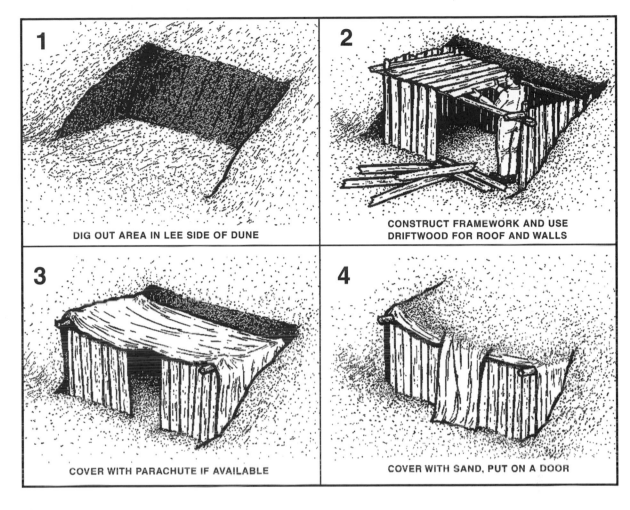

1 DIG OUT AREA IN LEE SIDE OF DUNE

2 CONSTRUCT FRAMEWORK AND USE DRIFTWOOD FOR ROOF AND WALLS

3 COVER WITH PARACHUTE IF AVAILABLE

4 COVER WITH SAND, PUT ON A DOOR

Figure 15-12. Hobo Shelter.

(1) Sand, though difficult to work with when loose, may be made into pillars by using sandbags made from parachute or any available cloth.

(2) Rock can be used in shelter construction.

(3) Vegetation such as sagebrush, creosote bushes, juniper trees, and desert gourd vines are valuable building materials.

(4) Parachute canopy and suspension lines. These are perhaps the most versatile building materials available for use by survivors. When used in layers, parachute material protects survivors from the sun's rays.

(a) The shelter should be made of dense material or have numerous layers to reduce or stop dangerous ultraviolet rays. The colors of the parachute materials used make a difference as to how much protection is provided from ultraviolet radiation. As a general rule, the order of preference should be to use as many layers as practical in the order of orange, green, tan, and white.

ULTRAVIOLET TESTS ON PARACHUTE CANOPY MATERIAL

% Ultraviolet (Short Wave 2537 A° Sunburn Rays) Blocked as compared to Direct Exposure

	1 Layer	2 Layers	3 Layers
Orange	78.2%	96.2%	99.4%
Sage Green	79.5%	96.2%	98.7%
Tan	64.1%	84.6%	93.6%
White	47.5%	61.6%	70.5%

% Ultraviolet (Long Wave 3660 A°) Blocked as Compared to Direct Exposure

	1 Layer	2 Layers	3 Layers
Orange	63.4%	92.3%	97.8%
Sage Green	60.0%	88.9%	97.8%
Tan	38.9%	66.7%	82.3%
White	28.9%	47.8%	58.9%

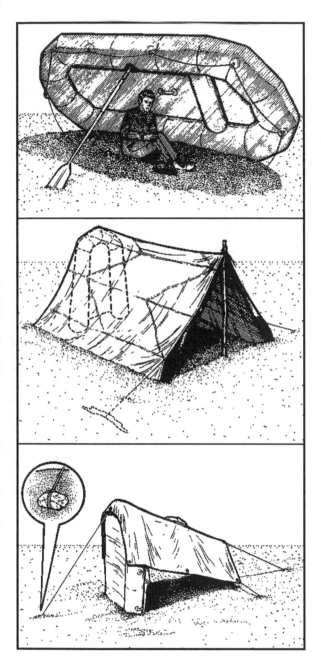

Figure 15-13. Improvised Natural Shade Shelters.

(b) The material should be kept approximately 12 to 18 inches above the individual. This allows the air to cool the underside of the material.

(c) Aircraft parts and liferafts can also be used for shade shelters. Survivors may use sections of the wing, tail, or fuselage to provide shade. However, the interior of the aircraft will quickly become superheated and should be avoided as a shelter. An inflatable raft can be tilted

against a raft paddle or natural object such as a bush or rock to provide relief from the sun (figure 15-13).

15-12. Principles of Desert Shelters:

a. The roof of a desert shelter should be multilayered so the resulting airspace reduces the inside temperature of the shelter. The layers should be separated 12 to 18 inches (figure 15-14).

b. Survivors should place the floor of the shelter about 18 inches above or below the desert surface to increase the cooling effect.

c. In warmer deserts, white parachute material should be used as an outer layer. Orange or sage green material should be used as an inner layer for protection from ultraviolet rays.

d. In cooler areas, multiple layers of parachute material should be used with sage green or orange material as the outer layer to absorb heat.

e. The sides of shelters should be movable in order to protect survivors during cold and (or) windy periods and to allow for ventilation during hot periods.

f. In a hot desert, shelters should be built away from large rocks, which store heat during the day. Survivors may need to move to the rocky areas during the evening to take advantage of the warmth heated rocks radiate.

g. Survivors should:

(1) Build shelters on the windward sides of dunes for cooling breezes.

(2) Build shelters during early morning, late evening, or at night. However, potential survivors should recall that survivors who come down in a desert area during daylight hours must be immediately concerned with protection from the sun and loss of water. In this case, parachute-canopy material can be draped over liferaft, vegetation, or a natural terrain feature for quick shelter.

15-13. Shelters for Snow and Ice Areas:

a. The differences in arctic and arctic-like environments create the need for different shelters. Basically, there are two types of environments that may require special shelter characteristics or building principles before survivors will have adequate shelter. They are:

(1) Barren lands, which include some seacoasts, icecaps, sea-ice areas, and areas above the tree line.

(2) Tree-line areas.

b. Barren lands offer a limited variety of materials for shelter construction. These are snow, small shrubs, and grasses. Ridges formed by drifting or wind-packed snow may be used for wind protection (survivors should build on the lee side). In some areas, such as sea ice, windy conditions usually exist and cause the ice to shift, forming pressure ridges. These areas of unstable ice and snow should be avoided at all times. Shelters that are suitable for barren-type areas include:

(1) Molded dome (figure 15-15).

(2) Snow cave (figure 15-16).

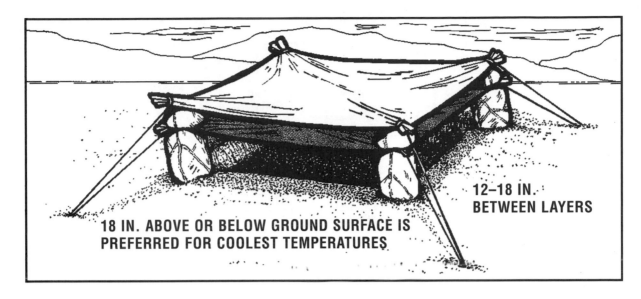

12–18 IN. BETWEEN LAYERS

18 IN. ABOVE OR BELOW GROUND SURFACE IS PREFERRED FOR COOLEST TEMPERATURES

Figure 15-14. Parachute Shade Shelter.

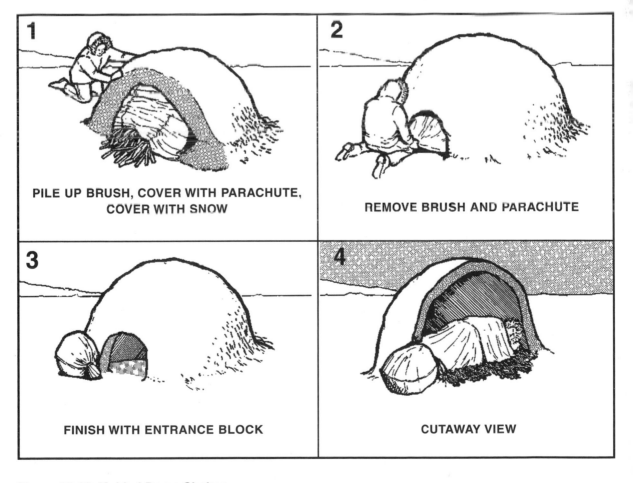

1 PILE UP BRUSH, COVER WITH PARACHUTE, COVER WITH SNOW

2 REMOVE BRUSH AND PARACHUTE

3 FINISH WITH ENTRANCE BLOCK

4 CUTAWAY VIEW

Figure 15-15. Molded Dome Shelter.

Figure 15-16. Snow Cave.

(3) Fighter trench (figure 15-17).
(4) Igloo (figure 15-18).
(5) Para-snow house (Figure 15-19).
NOTE: Of these, the ones that are quick to construct and require minimum effort and energy are the molded dome, snow cave, and fighter trench. It is important to know which of these shelters is the easiest to build since reducing or eliminating the effect of the windchill factor is essential to remaining alive.

c. In tree-covered areas, sufficient natural-shelter building materials are normally available. Caution is required. Shelters built near rivers and streams may get caught in the overflow.

d. Tree-line-area shelter types include:
(1) Thermal A-Frame construction (figure 15-20).
(2) Lean-to or wedge (figure 15-21).
(3) Double lean-to (figure 15-22).
(4) Fan (figure 15-23).
(5) Willow frame (figure 15-24).
(6) Tree well (figure 15-25).

e. Regardless of the type of shelter used, the use of thermal principles and insulation in arctic shelters is required. Heat radiates from bare ground and from ice masses over water. This means that shelter areas on land should be dug down to bare earth if possible (figure 15-26). A minimum of 8 inches of insulation above survivors is needed to retain heat. All openings except ventilation holes should be sealed to prevent heat loss. Leaving vent holes open is especially important if heat-producing devices are used. Candles, Sterno, or small oil lamps produce carbon monoxide. In addition to the ventilation hole through the roof, another may be required at the door to ensure adequate circulation of the air. (As a general rule, unless persons can see their breath, the snow shelter is too warm and should be cooled down to preclude melting and dripping.)

f. Regardless of how cold it may get outside, the temperature inside a small, well-constructed snow cave will probably not be lower than -10°F. Body heat alone can raise the temperature of a snow cave 45 degrees above the outside air. A burning candle will raise the temperature 4 degrees. Burning Sterno (small size, 2 5/8 oz.) will raise the cave temperature about 28 degrees. However, since they cannot be heated many degrees above freezing, snow shelters provide a rather rugged life. Once the inside of the shelter "glazes" over with ice, this layer of ice should be removed by chipping it off, or a new shelter built since ice reduces the insulating quality of a shelter. Maintain the old shelter until the new one is constructed. It will provide protection from the wind.

g. The aircraft should not be used as a shelter when temperatures are below freezing except in high wind conditions. Even then a thermal shelter should be constructed as soon as the conditions improve. The aircraft will not provide adequate insulation, and the floor will usually become icy and hazardous.

15-14. General Construction Techniques:

a. All thermal shelters use a layering system consisting of the frame, parachute (if available), boughs or shrubs, and snow. The framework must be sturdy enough to support the cover and insulation. A door block should be used to minimize heat loss. Insulation should be added on sleeping areas.

b. If a barren land-type shelter is being built with snow as the only material, a long knife or digging tool is a necessity. It normally takes 2 to 3 hours of hard work to dig a snow cave, and much longer for the novice to build an igloo.

c. Survivors should dress lightly while digging and working; they can easily become overheated and dampen their clothing with perspiration that will rapidly turn to ice.

d. If possible, all shelter types should have their openings 90 degrees to the prevailing wind. The entrance to the shelter should also be screened with snow blocks stacked in a L-shape.

e. Snow on the sea ice, suitable for cutting into blocks, will usually be found in the lee of pressure ridges or ice hummocks. The packed snow is often so shallow that the snow blocks have to be cut out horizontally.

f. No matter which shelter is used, survivors should take a digging tool into the shelter at night to cope with the great amount of snow that may block the door during the night.

15-15. Shelter Living:

a. Survivors should limit the number of shelter entrances to conserve heat. Fuel is generally scarce in the

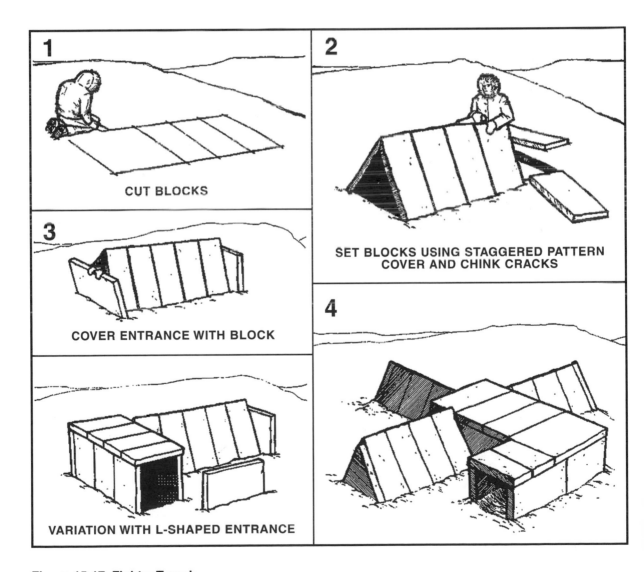

Figure 15-17. Fighter Trench.

Arctic. To conserve fuel, it is important to keep the shelter entrance sealed as much as possible (figure 15-27). When it is necessary to go outside the shelter, activities such as gathering fuel, snow, or ice for melting, etc., should be done. To expedite matters, a trash receptacle may be kept inside the door, and equipment may be stored in the entryway. Necessities that cannot be stored inside may be kept just outside the door. Any firearms (guns) the survivor may have must be stored outside the shelter to prevent condensation from building, which could cause them to malfunction.

b. A standard practice in snow-shelter living is for people to relieve themselves indoors when possible. This practice conserves body heat. If the snowdrift is large enough to dig connecting snow caves, one may be used as a toilet room. If not, tin cans may be used for urinals, and snow blocks for solid waste (fecal) matter.

c. Survivors should use thick insulation under themselves when sleeping or resting even if they have a sleeping bag. They can use a thick bough bed in shingle-fashion, seat cushions, parachute, or an inverted and inflated rubber raft.

d. Outer clothing makes good mattress material. A parka makes a good footbag. The shirt and inner trousers may be rolled up for a pillow. Socks and insoles can be separated and aired in the shelter. Drying may be completed in the sleeping bag by stowing around the hips. This drying method should be used only as a last resort.

e. Keeping the sleeping bag clean, dry, and fluffed will give maximum warmth. To dry the bag, it should be turned inside-out, frost beaten out, and warmed before the fire—taking care that it doesn't burn.

f. To keep moisture (from breath) from wetting the sleeping bag, a moisture cloth should be improvised from a piece of clothing, a towel, or parachute fabric. It can then be lightly wrapped around the head in such a way that the breath is trapped inside the cloth. A piece of fabric dries easier than a sleeping bag. If cold is experienced during the night, survivors should exercise by fluttering their feet up and down or by beating the inside of the bag with their hands. Food or hot liquids can be helpful.

g. Snow remaining in clothing will melt in a warm shelter. When the clothing is again taken outside, the water formed will turn to ice and reduce the CLo value. Brush clothes before entering the shelter. Under living conditions where drying clothing is difficult, it is easier to keep clothing from getting wet than having to dry it out later.

h. If all the snow cannot be eliminated from outer clothing, survivors should remove the clothing and store it in the entryway or on the floor away from the source of heat so it remains cold. If ice should form in clothing, it may be beaten out with a stick.

i. In the cramped quarters of any small emergency shelter, pots of food or drink can be accidentally kicked over. The cooking area, even if it is only a Sterno stove, should be located out of the way, possibly in a snow alcove.

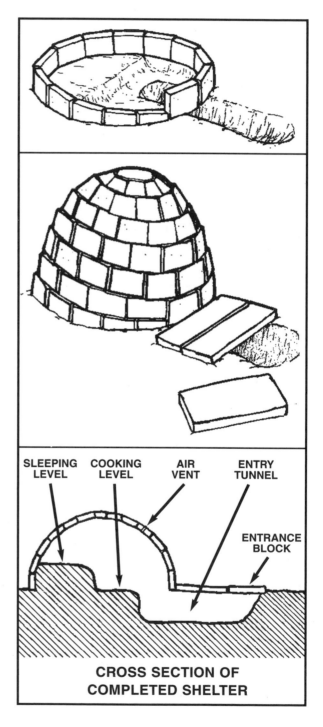

CROSS SECTION OF COMPLETED SHELTER

Figure 15-18. Igloo.

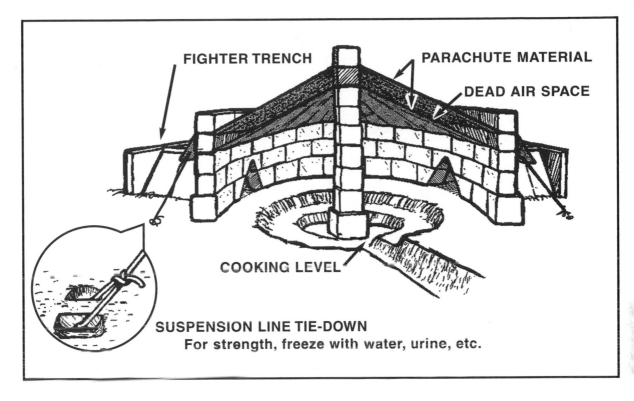

Figure 15-19. Para-Snow House.

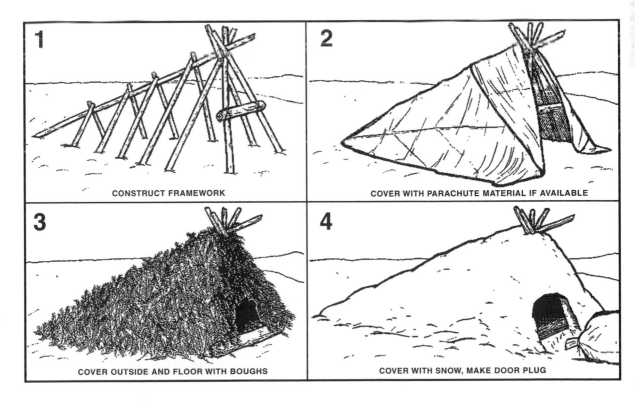

Figure 15-20. Thermal A-Frame.

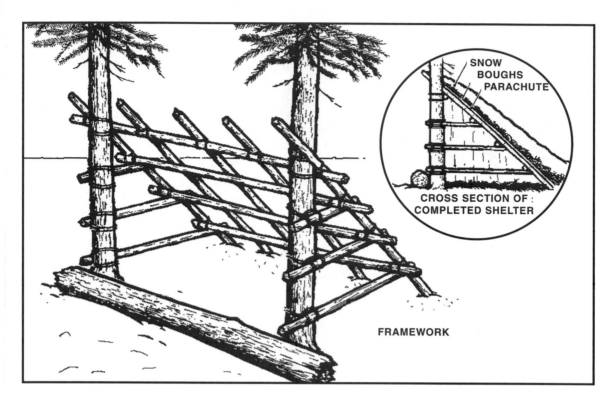

Figure 15-21. Lean-To or Wedge.

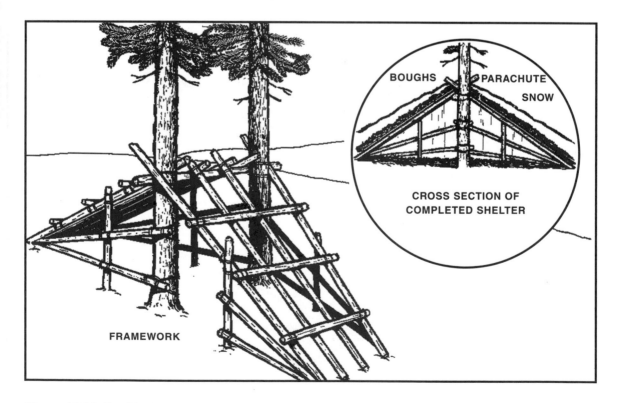

Figure 15-22. Double Lean-To.

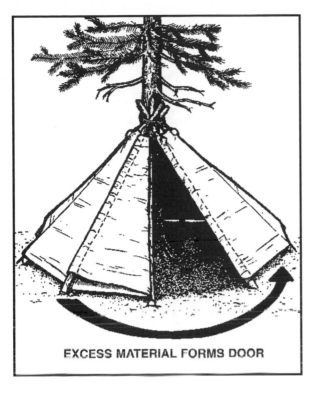

Figure 15-23. Fan Shelter.

15-16. Summer Considerations for Arctic and Arctic-Like Areas:

a. Survivors need shelter against rain and insects. They should choose a campsite near water but on high, dry ground if possible. Survivors should also stay away from thick vegetation, as mosquitoes and flies will make life miserable. A good campsite is a ridge top, cold lakeshore, or a spot that gets a breeze.

b. If survivors stay with the aircraft, it can be used for shelter during the summer. They should cover openings with netting or parachute cloth to keep insects out and cook outside to avoid carbon monoxide poisoning. Fires must be built a safe distance from the aircraft.

c. Many temperate area shelters are suitable for summer arctic conditions. The paratepee (of the 1- or no-pole variety) is especially good. It will protect from precipitation and keep insects out.

15-17. Shelter for Open Seas. Personal protection from the elements is just as important on the seas as it is anywhere else. Some rafts come equipped with insulated floors, spray shields, and canopies to protect survivors from heat, cold, and water. If rafts are not so equipped or the equipment has been lost, survivors should try to improvise these items using parachute material, clothing, or other equipment.

Figure 15-24. Willow Frame Shelter.

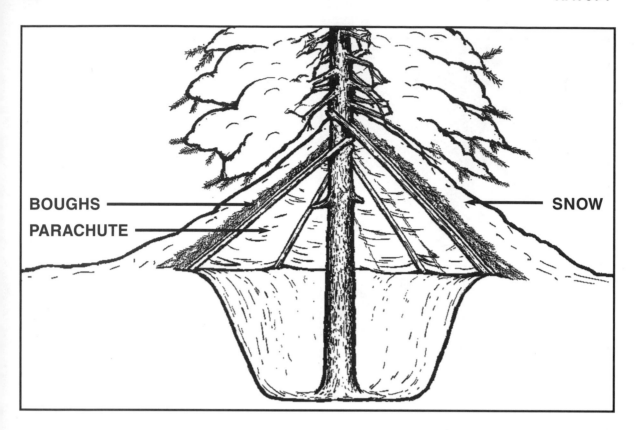

BOUGHS
PARACHUTE
SNOW

Figure 15-25. Tree Well Shelter.

Figure 15-26. Scraping Snow to Bare Earth.

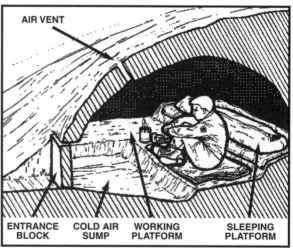

AIR VENT

ENTRANCE
BLOCK

COLD AIR
SUMP

WORKING
PLATFORM

SLEEPING
PLATFORM

Figure 15-27. Snow Cave Shelter Living.

Chapter 16

FIRECRAFT

16-1. Introduction:

a. The need for a fire should be placed high on the list of priorities. Fire is used for warmth, light, drying clothes, signaling, making tools, cooking, and water purification. When using fire for warmth, the body uses fewer calories for heat and consequently requires less food. Just having a fire to sit by is a morale booster. Smoke from a fire can be used to discourage insects.

b. Avoid building a very large fire. Small fires require less fuel, are easier to control, and their heat can be concentrated. Never leave a fire unattended unless it is banked or contained. Banking a fire is done by scraping cold ashes and dry earth onto the fire, leaving enough air coming through the dirt at the top to keep the fuel smoldering. This will keep the fire safe and allow it to be rekindled from the saved coals.

16-2. Elements of Fire:

a. The three essential elements for successful fire building are fuel, heat, and oxygen. These combined elements are referred to as the "fire triangle." By limiting fuel, only a small fire is produced. If the fire is not fed properly, there is too much or too little fire. Green fuel is difficult to ignite, and the fire must be burning well before it is used for fuel. Oxygen and heat must be accessible to ignite any fuel.

b. The survivor must take time and prepare well! Preparing all of the stages of fuel and all of the parts of the fire-starting apparatus is the key. To be successful at firecraft, one needs to practice and be patient.

c. The fuels used in building a fire normally fall into three categories (figure 16-1), relating to their size and flash point: tinder, kindling, and fuel.

(1) Tinder is any type of small material having a low flash point. It is easily ignited with a minimum of heat, even a spark. Tinder must be arranged to allow air (oxygen) between the hair-like, bone-dry fibers. The preparation of tinder for fire is one of the most important parts of firecraft. Dry tinder is so critical that pioneers used extreme care to have some in a waterproof "tinder box" at all times. It may be necessary to have two or three stages of tinder to get the flame to a useful size. Tinders include:

(a) The shredded bark from some trees and bushes.

(b) Cedar, birch bark, or palm fiber.

(c) Crushed fibers from dead plants.

(d) Fine, dry woodshavings, and straw/grasses.

(e) Resinous sawdust.

(f) Very fine pitch woodshavings (resinous wood from pine or sappy conifers).

(g) Bird or rodent nest linings.

(h) Seed down (milkweed, cattail, thistle).

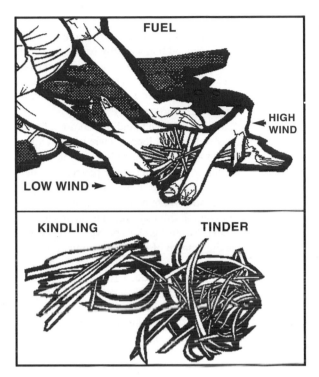

Figure 16-1. Stages of a Fire.

(i) Charred cloth.

(j) Cotton balls or lint.

(k) Steel wool.

(l) Dry powdered sap from the pine tree family (also known as pitch),

(m) Paper.

(n) Foam rubber.

(2) Kindling is the next larger stage of fuel material. It should also have a high combustible point. It is added to, or arranged over, the tinder in such a way that it ignites when the flame from the tinder reaches it. Kindling is used to bring the burning temperature up to the point where larger and less combustible fuel material can be used. Kindling includes:

(a) Dead, dry small twigs or plant fibers.

(b) Dead, dry thinly shaved pieces of wood, bamboo, or cane (always split bamboo as sections can explode).

(c) Coniferous seed cones and needles.

(d) "Squaw wood" from the underside of coniferous trees; dead, small branches next to the ground sheltered by the upper live part of the tree.

(e) Pieces of wood removed from the insides of larger pieces.

(f) Some plastics, such as the spoon from an in-flight ration.

(g) Wood that has been soaked or doused with flammable materials, that is, wax, insect repellent, petroleum fuels, and oil.

(h) Strips of petrolatum gauze from a first aid kit.

(i) Dry, split wood burns readily because it is drier inside. Also, the angular portions of the wood burn easier than the bark-covered round pieces because it exposes more surface area to the flame. The splitting of all fuels will cause them to burn more readily.

(3) Fuel, unlike tinder and kindling, does not have to be kept completely dry as long as there is enough kindling to raise the fuel to a combustible temperature. It is recommended that all fine materials be protected from moisture to prevent excessive smoke production. (Highly flammable liquids should not be poured on an existing fire. Even a smoldering fire can cause the liquids to explode and cause serious burns.) The type of fuel used will determine the amount of heat and light the fire will produce. Dry, split hardwood trees (oak, hickory, monkey pod, ash) are less likely to produce excessive smoke and will usually provide more heat than soft woods. They may also be more difficult to break into usable sizes. Pine and other conifers are fast-burning and produce smoke unless a large flame is maintained. Rotten wood is of little value since it smolders and smokes. The weather plays an important role when selecting fuel. Standing or leaning wood is usually dry inside even if it is raining. In tropical areas, avoid selecting wood from trees that grow in swampy areas or those covered with mosses. Tropical soft woods are not usually a good fuel source. Trial and error is sometimes the best method to determine which fuel is best. After identifying the burning properties of available fuel, a selection can be made of the type needed. Recommended fuel sources are:

(a) Dry, standing dead wood and dry, dead branches (those that snap when broken). Dead wood is easy to split and break. It can be pounded on a rock or wedged between other objects and bent until it breaks.

(b) The insides of fallen trees and large branches may be dry even if the outside is wet. The heartwood is usually the last to rot.

(c) Green wood that can be made to burn is found almost anywhere, especially if finely split and mixed evenly with dry dead wood.

(d) In treeless areas, other natural fuels can be found. Dry grasses can be twisted into bunches. Dead cactus and other plants are available in deserts. Dry peat moss can be found along the surface of undercut streambanks. Dried animal dung, animal fats, and sometimes even coal can be found on the surface. Oil-impregnated sand can also be used when available.

16-3. Fire Location. The location of a fire should be carefully selected. An old story is told of a mountain man

who used his last match to light a fire built under a snow-covered tree. The heat from the fire melted the snow and it slid off the tree and put out the fire. For a survivor, this type of accident can be very demoralizing or even deadly. Locate and prepare the fire site carefully.

16-4. Fire Site Preparation:

a. After a site is located, twigs, moss, grass, or duff should be cleaned away. Scrape at least a 3-foot diameter area down to bare soil for even a small fire. Larger fires require a larger area. If the fire must be built on snow, ice, or wet ground, survivors should build a platform of green logs or rocks. (Beware of wet or porous rocks, they may explode when heated.)

b. There is no need to dig a hole or make a circle of rocks in preparation for fire building. Rocks may be placed in a circle and filled with dirt, sand, or gravel to raise the fire above the moisture from wet ground. The purpose of these rocks is to hold the platform only.

c. To get the most warmth from the fire, it should be built against a rock or log reflector (figure 16-2). This will direct the heat into the shelter. Cooking fires can be walled-in by logs or stones. This will provide a platform for cooking utensils and serve as a windbreak to help keep the heat confined.

Figure 16-2. Fire Reflector.

d. After preparing the fire, all materials should be placed together and arranged by size (tinder, kindling, and fuel). As a rule of thumb, survivors should have three times the amount of tinder and kindling than is necessary for one fire. It is to their advantage to have too much rather than not enough. Having plenty of material on hand will prevent the possibility of the fire going out while additional material is gathered.

16-5. Firemaking with Matches (or Lighter):

a. Survivors should arrange a small amount of kindling in a low pyramid, close enough together so flames can jump from one piece to another. A small opening should be left for lighting and air circulation.

b. Matches can be conserved by using a "shave stick," or by using a loosely tied fagot of thin, dry twigs. The match must be shielded from wind while igniting the shave stick. The stick can then be applied to the lower windward side of the kindling.

c. Small pieces of wood or other fuel can be laid gently on the kindling before lighting or can be added as the kindling begins to burn. The survivors can then place smaller pieces first, adding larger pieces of fuel as the fire begins to burn. They should avoid smothering the fire by crushing the kindling with heavy wood.

d. Survivors have only a limited number of matches or other instant fire-starting devices. In a long-term situation, they should use these devices sparingly or carry fire with them when possible. Many primitive cultures carry fire (fire bundles) by using dry punk or fiberous barks (cedar) encased in a bark. Others use torches. Natural fire bundles also work well for holding the fire (figure 16-3).

e. The amount of oxygen must be just enough to keep the coals inside the dry punk burning slowly. This requires constant vigilance to control the rate of the burning process. The natural fire bundle is constructed in a cross section as shown in figure 16-3.

16-6. Heat Sources.

A supply of matches, lighters, and other such devices will last only a limited time. Once the supply is depleted, they cannot be used again. If possible, before the need arises, survivors should become skilled at starting fires with more primitive means, such as friction, heat, or a sparking device. It is essential that they continually practice these procedures. The need to start a fire may arise at the most inopportune times. One of the greatest aids a survivor can have for rapid fire starting is the "tinder box" previously mentioned. Using friction, heat, and sparks are very reliable methods for those who use them on a regular basis. Therefore, survivors must practice these methods. Survivors must be aware of the problems associated with the use of primitive heat sources. If the humidity is high in the immediate area, a fire may be difficult to ignite even if all other conditions are favorable. For primitive methods to be successful, the materials must be BONE DRY. The primitive people who use these ignition methods take great care to keep their tinder, kindling, and other fuels dry, even to the point of wrapping many layers of waterproof materials around them. PREPARATION, PRACTICE, and PATIENCE in the use of primitive fire-building techniques cannot be overemphasized. A key point in all primitive methods is to ensure that the tinder is not disturbed.

a. Flint and Steel:

(1) Flint and steel is one way to produce fire without matches.

(a) To use this method, survivors must hold a piece of flint in one hand above the tinder.

(b) Grasp the steel in the other hand and strike the flint with the edge of the steel with a downward glancing blow (figure 16-4).

(2) True flint is not necessary to produce sparks. Iron pyrite and quartz will also give off sparks even if they are struck against only each other. Check the area and select the best spark-producing stone as a backup for the

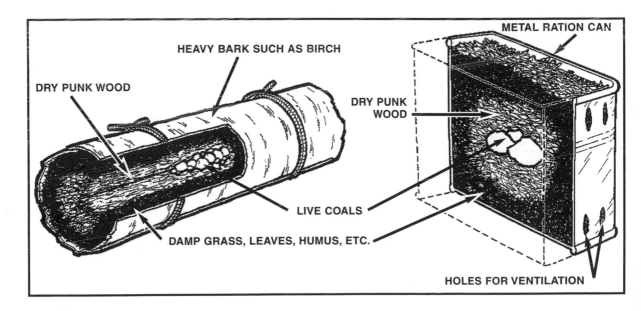

Figure 16-3. Fire Bundles.

DRY PUNK WOOD

HEAVY BARK SUCH AS BIRCH

METAL RATION CAN

DRY PUNK WOOD

LIVE COALS

DAMP GRASS, LEAVES, HUMUS, ETC.

HOLES FOR VENTILATION

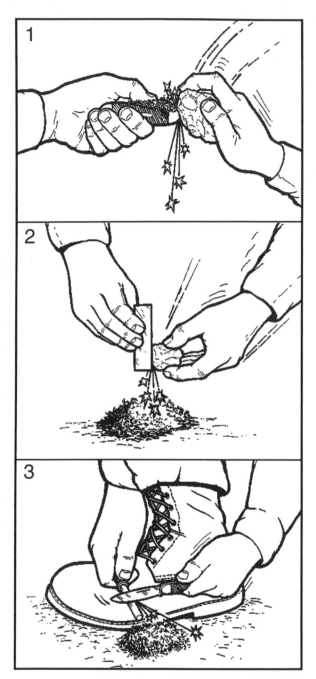

Figure 16-4. Fire Starting with Flint and Steel.

the spark is struck. The residue from the "match" burns hot and fast and will compensate for some moisture in tinder. If issued survival kits do not contain this item and the survivors choose to make one rather than buy it, lighter flints can be glued into a groove in a small piece of wood or plastic. The survivors can then practice striking a spark by scratching the flint with a knife blade. A 90-degree angle between the blade and flint works best. The device must be held close enough for the sparks to hit the tinder, but enough distance must be allowed to avoid accidentally extinguishing the fire. Cotton balls dipped in petroleum jelly make excellent tinder with flint and steel. When the tinder ignites, additional tinder, kindling, and fuel can be added.

b. Batteries:

(1) Another method of producing fire is to use the battery of the aircraft, vehicle, storage batteries, etc. Using two insulated wires, connect one end of a wire to the positive post of the battery and the end of the other wire to the negative post. Touch the two remaining ends to the ends of a piece of noninsulated wire. This will cause a short in the electrical circuit and the noninsulated wire will begin to glow and get hot. Material coming into contact with this hot wire will ignite. Survivors should use caution when attempting to start a fire with a battery. They should ensure that sparks or flames are not produced near the battery because explosive hydrogen gas is produced and can result in serious injury (figure 16-5).

Figure 16-5. Fire Starting with Batteries.

available matches. The sparks must fall on the tinder and then be blown or fanned to produce a coal and subsequent flame.

(3) Synthetic flint, such as the so-called metal match, consists of the same type of material used for flints in commercial cigarette lighters. Some contain magnesium that can be scraped into tinder and into which

(2) If fine-grade steel wool is available, a fire may be started by stretching it between the positive and negative posts until the wire itself makes a red coal.

c. Burning Glass. If survivors have sunlight and a burning glass, a fire can be started with very little physical effort (figure 16-6). Concentrate the rays of the sun on tinder by using the lens of a lensatic compass, a camera lens, or the lens of a flashlight that magnifies; even a convex piece of bottle glass may work. Hold the lens so that the brightest and smallest spot of concentrated light falls on the tinder. Once a whisp of smoke is produced, the tinder should be fanned or blown upon until the smoking coal becomes a flame. Powdered charcoal in the tinder will decrease the ignition time. Add kindling carefully as in any other type of fire. Practice will reduce the time it takes to light the tinder.

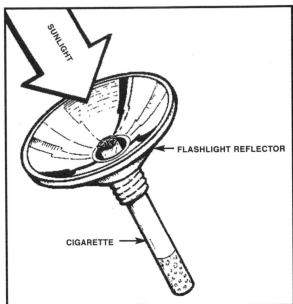

Figure 16-7. Fire Starting with Flashlight Reflector.

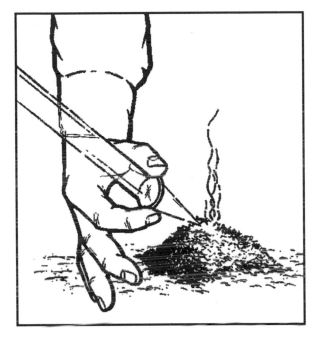

Figure 16-6. Fire Starting with Burning Glass.

d. Flashlight Reflector. A flashlight reflector can also be used to start a fire (figure 16-7). Place the tinder in the center of the reflector where the bulb is usually located. Push it up from the back of the hole until the hottest light is concentrated on the end and smoke results. If a cigarette is available, use it as tinder for this method.

e. Bamboo Fire Saw:

(1) The bamboo fire saw is constructed from a section of dry bamboo with both end joints cut off. The section of bamboo, about 12 inches in length, is split in half lengthwise. The inner wall of one of the halves (called the "running board") is scraped or shaved thin. This is done in the middle of the running board. A notch to serve as a guide is cut in the outer sheath opposite the scraped area of the inner wall. This notch runs

across the running board at a 90-degree angle (figure 16-8).

(2) The other half of the bamboo joint is further split in half lengthwise, and one of the resultant quarters is used as a "baseboard." One edge of the baseboard is shaved down to make a tapered cutting edge. The baseboard board is then firmly secured with the cutting edge up. This may be done by staking it to the ground in any manner that does not allow it to move (figure 16-8).

(3) Tinder is made by scraping the outer sheath of the remaining quarter piece of the bamboo section. The scrapings (approximately a large handful) are then rubbed between the palms of the hands until all of the wood fibers are broken down and dust-like material no longer falls from the tinder. The ball of scrapings is then fluffed to allow maximum circulation of oxygen through the mass (figure 16-8).

(4) The finely shredded and fluffed tinder is placed in the running board directly over the shaved area, opposite the outside notch. Thin strips of bamboo should be placed lengthwise in the running board to hold the tinder in place. These strips are held stationary by the hands when grasping the ends of the running board (figure 16-8).

(5) A long, very thin sliver of bamboo (called the "pick") should be prepared for future use. One end of the running board is grasped in each hand, making sure the thin strips of bamboo are held securely in place. The running board is placed over the baseboard at a right angle so that the cutting edge of the baseboard fits into

1 GUIDE NOTCH

OUTSIDE

NOTCH FOR TINDER

INSIDE

2 SCRAPE BAMBOO WITH KNIFE BLADE 90° TO THE BAMBOO

RUB BAMBOO SHAVINGS BETWEEN HANDS TO BREAK DOWN FIBERS

3 STICKS TINDER

4 PLACE GUIDE NOTCH ON SHARP EDGE OF BAMBOO BLADE; RUB BACK AND FORTH WITH INCREASING SPEED AND PRESSURE UNTIL SMOKE IS SEEN

SHARP EDGE

5 BLOW GENTLY ON COALS AND POKE THEM INTO TINDER

4 BLOW GENTLY ON THE TINDER UNTIL IT BEGINS TO FLAME BUILD UP FIRE

Figure 16-8. Bamboo Fire Saw.

the notch in the outer sheath of the running board. The running board is then slid back and forth as rapidly as possible over the cutting edge of the baseboard, with sufficient downward pressure to ensure enough friction to produce heat.

(6) As soon as "billows" of smoke rise from the tinder, the running board is picked up. The pick is used to push the glowing embers from the bottom of the running board into the mass of tinder. While the embers are being pushed into the tinder, they are gently blown upon until the tinder bursts into flame.

(7) As soon as the tinder bursts into flame, slowly add kindling in small pieces to avoid smothering the fire. Fuel is gradually added to produce the desired size fire. If the tinder is removed from the running board as soon as it flames, the running board can be reused by cutting a notch in the outer sheath next to the original notch and directly under the scraped area of the inner wall.

f. Bow and Drill:

(1) This is a friction method that has been used successfully for thousands of years. A spindle of yucca, elm, basswood, or any other straight grainwood (not softwood) should be made. The survivors should make sure that the wood is not too hard or it will create a glazed surface when friction is applied. The spindle should be 12 to 18 inches long and three-fourths inch in diameter. The sides should be octagonal, rather than round, to help create friction when spinning. Round one end and work the other end into a blunt point. The round end goes to the top upon which the socket is placed. The socket is made from a piece of hardwood large enough to hold comfortably in the palm of the hand with the curved part up and the flat side down to hold the top of the spindle. Carve or drill a hole in this side and make it smooth so it will not cause undue friction and heat production. Grease or soap can be placed in this hole to prevent friction (figure 16-9).

(2) The bow is made from a stiff branch about 3 feet long and about 1 inch in diameter. This piece should have sufficient flexibility to bend. It is similar to a bow used to shoot arrows. Tie a piece of suspension line or leather thong to both ends so that it has the same tension as that of a bow. There should be enough tension for the spindle to twist comfortably.

(3) The fireboard is made of the softwood and is about 12 inches long, three-fourths inch thick, and 3 to 6 inches wide. A small hollow should be carved in the fireboard. A V-shaped cut can then be made in from the edge of the board. This V-shape should extend into the center of the hollow where the spindle will make the hollow deeper. The object of this "V" cut is to create an angle that cuts off the edge of the spindle as it gets hot and turns to charcoal dust. This is the critical part of the fireboard and must be held steady during the operation of spinning the spindle.

(4) While kneeling on one knee, the other foot can be placed on the fireboard as shown in figure 16-9, and the tinder placed under the fireboard just beneath the V-cut. Care should be taken to avoid crushing the tinder under the fireboard. Space can be obtained by using a small, three-fourths inch diameter stick to hold up the fireboard. This allows air into the tinder where the hot powder (spindle charcoal dust) is collected.

(5) The bowstring should be twisted once around the spindle. The spindle can then be placed upright into the spindle hollow (socket). The survivor may press the socket down on the spindle and fireboard. The entire apparatus must be held steady with the hand on the socket braced against the leg or knee. The spindle should begin spinning with long even slow strokes of the bow until heavy smoke is produced. The spinning should become faster until the smoke is very thick. At this point, hot powder, which can be blown into a glowing ember, has been successfully produced. The bow and spindle can then be removed from the fireboard and

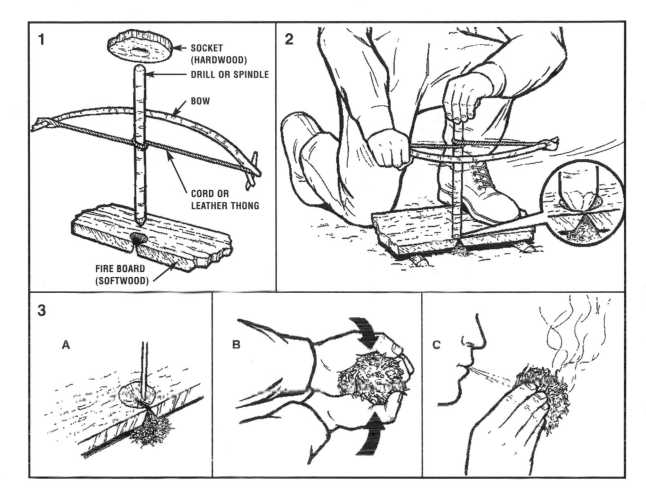

Figure 16-9. Bow and Drill.

the tinder can be placed next to the glowing ember, making sure not to extinguish it. The tinder must then be rolled gently around the burning ember, and blow into the embers, starting the tinder to burn. This part of the fire is most critical and should be done with care and planning.

(6) The burning tinder is then placed into the waiting fire "lay" containing more tinder and small kindling. At no time in this process should the survivor break concentration or change sequence. The successful use of these primitive methods of fire starting will require a great deal of patience. Success demands dedication and practice.

g. The Fire Thong. The fire thong, another friction method, is used in only those tropical regions where rattan is found. The system is simple and consists of a twisted rattan thong or other strong plant fiber, 4 to 6 feet long, less than 1 inch in diameter, and a 4-foot length of dry wood, which is softer than rattan

(deciduous wood; figure 16-10). Rub with a steady but increasing rhythm.

h. The Plow. The plow is a method used by some primitives and basically follows the principles of other friction methods. The wood used must not glaze with heat applied and must be able to produce powder with friction.

i. Ground Stake. Another variation can be constructed by driving a stake into the ground as shown in figure 16-11.

16 7. Firemaking with Special Equipment:

a. The night end of the day-night flare can be used as a fire starter. This means, however, that survivors must weigh the importance of a fire against the loss of a night flare.

b. Some emergency kits contain small fire starters, cans of special fuels, windproof matches, and other aids.

Figure 16-10. Fire Thong.

Survivors should save the fire starters for use in extreme cold and damp (moist) weather conditions.

c. The white plastic spoon (packed in various in-flight rations) may be the type that burns readily. The handle should be pushed deep enough into the ground to support the spoon in an upright position. Light the tip of the spoon. It will burn for about 10 minutes (long enough to dry out and ignite small tinder and kindling).

d. If a candle is available, it should be ignited to start a fire and thus prevent using more than one match. As soon as the fire is burning, the candle can be extinguished and saved for future use.

e. Tinder can be made more combustible by adding a few drops of flammable fuel/material. An example of this would be mixing the powder from an ammunition cartridge with the tinder. After preparing tinder in this manner, it should be stored in a waterproof container for future use. Care must be used in handling this mixture because the flash at ignition could bum the skin and clothing.

f. For thousands of years, the Eskimos and other northern peoples have relied heavily upon oils from animals to heat their homes. A fat stove or "Koodlik" is used by the Eskimos to burn this fuel.

g. Survivors can improvise a stove from a ration can and burn any flammable oil-type liquid or animal fats available. Here again, survivors should keep in mind that if there is only a limited amount of animal fat, it should be eaten to produce heat inside the body.

16-8. Burning Aircraft Fuel. On barren lands in the Arctic, aircraft fuel may be the only material survivors have available for fire.

a. A stove can be improvised to burn fuel, lubricating oil, or a mixture of both (figure 16-12). The survivor should place 1 or 2 inches of sand or fine gravel in the bottom of a can or other container and add fuel. *Care should be used when lighting the fuel because it may explode.* Slots should be cut into the top of the can to let flame and smoke out, and holes punched just above the level of the sand to provide a draft. A mixture of fuel and oil will make the fire burn longer. If no can is available, a hole can be dug and filled with sand. Fuel is then poured on the sand and ignited. The survivor should not allow the fuel to collect in puddles.

b. Lubricating oil can be burned as fuel by using a wick arrangement. The wick can be made of string, rope, rag, sphagnum moss, or even a cigarette and should be placed on the edge of a receptacle filled with oil. Rags, paper, wood, or other fuel can be soaked in oil and thrown on the fire.

c. A stove can be made of any empty waxed carton by cutting off one end and punching a hole in each side near the unopened end. Survivors can stand the carton on the closed end and loosely place the fuel inside the carton. The stove can then be lit using fuel material left hanging over the end. The stove will burn from the top down.

d. Seal blubber makes a satisfactory fire without a container if gasoline or heat tablets are available to provide an initial hot flame (figure 16-13). The heat source should be ignited on the raw side of the blubber while the fur side is on the ice. A square foot of blubber burns for several hours. Once the blubber catches fire, the heat tablets can be recovered. Eskimos light a small piece of blubber and use it to kindle increasingly larger pieces. The smoke from a blubber fire is dirty, black, and heavy. The flame is very bright and can be seen for several miles. The smoke will penetrate clothing and blacken the skin.

16-9. Useful Firecraft Hints:

a. Conserve matches by using them only on properly prepared fires. They should never be used to light cigarettes or for starting unnecessary fires.

b. Carry some dry tinder in a waterproof container. It should be exposed to the sun on dry days. Adding a little powdered charcoal will improve it. Cotton cloth is good tinder, especially if scorched or charred. It works well with a burning glass or flint and steel.

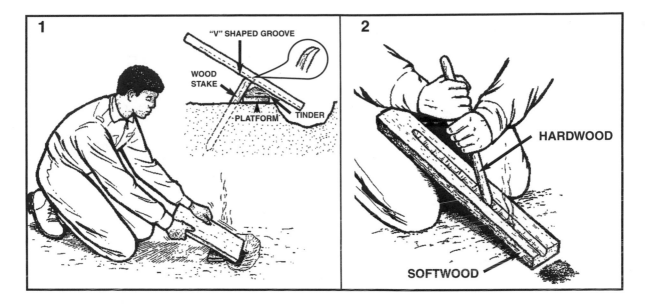

Figure 16-11. Fire Plow.

c. Remember that firemaking can be a difficult job in an arctic environment. The main problem is the availability of firemaking materials. Making a fire starts WELL before the match is lit. The fire must be protected from the wind. In wooded areas, standing timber and brush usually make a good windbreak, but in open areas, some type of windbreak may have to be constructed. A row of snow blocks, the shelter of a ridge, or a pile of brush will work as a windbreak. It must be high enough to shield the fire from the wind. It may also act as a heat reflector if it is of solid material.

d. Remember, a platform will be required to prevent the fire from melting down through the deep snow and extinguishing it. A platform is also needed if the ground is moist or swampy. The platform can be made of green logs, metal, or any material that will not burn through very readily. Care must be taken when selecting an area for fire building. If the area has a large accumulation of humus material and (or) peat, a platform is needed to avoid igniting the material as it will tend to smolder long after the flames of the fire are extinguished. A smoldering peat fire is almost impossible to put out and may burn for years.

e. In forested areas, the debris on the ground and the lichen mat should be cleared away to mineral soil, if possible, to prevent the fire from spreading.

f. The ignition source used to ignite the fire must be quick and easily operated with hand protection, such as mittens. Any number of devices will work well—matches, candle, lighter, fire starter, metal matches, etc.

16-10. Fire Lays. Most fires are built to meet specific needs or uses, either for heat, light, or preparing food and water. The following configurations are the most commonly used for fires and serve one or more needs (figure 16-14).

a. Tepee:

(1) The tepee fire can be used as a light source and has a concentrated heat point directly above the apex of the tepee, which is ideal for boiling water. To build:

(a) Place a large handful of tinder on the ground in the middle of the fire site.

(b) Push a stick into the ground, slanting over the tinder.

(c) Then lean a circle of kindling sticks against the slanting stick, like a tepee, with an opening toward the windward side for draft.

(2) To light the fire:

(a) Crouch in front of the fire lay with back to the wind.

(b) Feed the fire from the downwind side, first with thin pieces of fuel, then gradually with thicker pieces.

(c) Continue feeding until the fire has reached the desired size. The tepee fire has one big drawback: It tends to fall over easily. However, it serves as an excellent starter fire.

b. Log Cabin. As the name implies, this lay looks similar to a log cabin. Log cabin fires give off a great amount of light and heat primarily because of the amount of oxygen that enters the fire. The log cabin

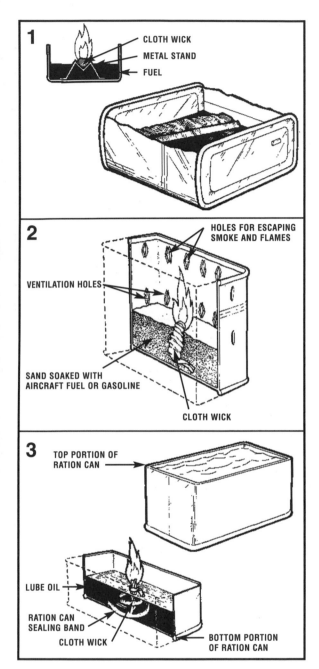

Figure 16-12. Fat or Oil Stoves.

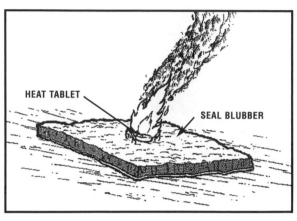

Figure 16-13. Heat Tablet/Seal Blubber.

wind. The long fire can also be built above ground by using two parallel green logs to hold the coals together. These logs should be at least 6 inches in diameter and situated so the cooking utensils will rest upon the logs. Two 1-inch thick sticks can be placed under both logs, one at each end of the long fire. This is done to allow the coals to receive more air.

d. Keyhole Fire. To construct a keyhole fire, a hole is dug in the shape of an old-style keyhole and does the same thing as the long fire.

e. Pyramid Fire. The pyramid fire looks similar to a log cabin fire except there are layers of fuel in place of a hollow framework. The advantage of a pyramid fire is that it burns for a long time resulting in a large bed of coals. This fire could possibly be used as an overnight fire when placed in front of a shelter opening.

f. Star Fire. This fire is used when conservation of fuel is necessary or a small fire is desired. It burns at the center of the "wheel" and must be constantly tended. Hardwood fuels work best with this type of fire.

g. "T" Fire. Used for large group cooking. The size of this lay may be adjusted to meet the group's cooking needs. In the top part of the "T," the fire is constructed and maintained as long as needed to provide hot coals for cooking in the bottom part of the "T" fire lay. The number of hot coals may be adjusted in the lower part of the "T" fire lay to regulate the cooking temperature.

h. "V" Fire. This fire lay is a modification of the long fire. The configuration allows a survivor to either block strong winds, or take advantage of light breezes. During high-wind conditions, the vertex of the "V"—formed by the two outside logs—is placed in the direction from which the winds are coming, thereby sheltering the tinder (kindling) for ignition. Reversing the "lay" will funnel light breezes into the tinder (kindling), thereby facilitating ease of ignition (figure 16-1).

fire creates a quick and large bed of coals and can be used for cooking or as the basis for a signal fire. If one person or a group of people are going to use the coals for cooking, the log cabin can be modified into a long fire or a keyhole fire.

c. Long Fire. The long fire begins as a trench, the length of which is layed to take advantage of existing

Figure 16-14. Fire Lays.

Chapter 17

EQUIPMENT

17-1. Introduction. Survivors in a survival situation have needs that must be met—food, water, clothing, shelters, etc. The survival kit contains equipment that can be used to satisfy these needs. Quite often, however, this equipment may not be available due to damage or loss. This chapter will address the care and use of issued equipment and improvising the needed equipment when not available. The uses of some issued items are covered in appropriate places throughout this regulation. The care and use of equipment (not covered elsewhere) will be addressed here.

17-2. Types of Kits:

a. All survival kits contain two types of equipment—mandatory and optional. The mandatory equipment for survival kits are:

(1) One-man liferaft (1 each).
(2) Compass (1 each).
(3) Smoke and illumination flares (2 each).
(4) Signal mirror (1 each).
(5) Hand-held launched flare (1 each).
(6) First aid kit (1 each).
(7) Survival radio (1 each).

NOTE: These items of equipment may not be mandatory for raft kit.

b. Optional items are authorized by the major air commands; this authority is delegated to subordinate commanders. These optional items are directly related to climatic conditions and the type of terrain that is being flown over. There are more than 40 optional items. Here are a few examples:

(1) Sleeping bag.
(2) Strobe light with lenses.
(3) Wire saw.
(4) Water container.
(5) Survival shovel.
(6) Matchbox container.

17-3. Issued Equipment. Survival equipment is designed to aid survivors throughout their survival episode. To maintain its effectiveness, the equipment must be well cared for.

a. Electronic Equipment:

(1) Electronic signaling devices are by far the survivor's most important signaling devices. Therefore, it is important for survivors to properly care for them to ensure their continued effectiveness. In cold temperatures, the electronic signaling devices must be kept warm to prevent the batteries from becoming cold soaked.

(2) In a cold environment, if survivors speak directly into the microphone, the moisture from their breath may condense and freeze on the microphone, creating communication problems.

(3) Caution must be used when using the survival radios in a cold environment. If the radio is placed against the side of the face to communicate, frostbite could result.

(4) In a wet environment, survivors should make every effort to keep their electronic signaling devices dry. In an open-sea environment, the only recourse may be to shake the water out of the microphone before transmitting.

b. Firearms:

(1) A firearm is a precision tool. It will continue functioning only as long as it is cared for. Saltwater, perspiration, dew, and humidity can all corrode or rust a firearm until it is inoperable. If immersed in saltwater, the survivor should wash the parts in freshwater and then dry and oil them. As an expedient, one way to dry the firearm is to place it in boiling water, and after removal wipe off the excess moisture. The residual heat will evaporate most of the remaining moisture. Survivors should not use uncontrolled heat to dry the firearm, as heat over 250°F can remove the temper from the springs in a short time and weaken the action.

(2) Any petroleum-based lubricants used in cold environments will stiffen or freeze, causing the firearm to become inoperative. It would be better to thoroughly clean the firearm and remove all lubricant. Metal becomes brittle from cold and is, therefore, prone to breakage.

(3) A firearm was not intended for use as a club, hammer, or pry bar. To use it for any purpose other than that for which it was designed would only result in damage to the firearm.

c. Cutting Tools:

(1) A file and sharpening stone are often packed in a survival kit. The file is normally used for axes, and the stone is normally used for knives.

(2) An old axiom states that a sharp cutting tool is a safe cutting tool. Control of a cutting tool is easier to maintain if it is sharp, and the possibility of accidental injury is reduced.

(3) One of the most valuable items in any survival situation is a knife, since it has a large number of uses. Unless the knife is kept sharp, however, it falls short of its potential.

(4) A knife should be sharpened only with a stone as repeated use of a file rapidly removes steel from the blade. In some cases, it may be necessary to use a file to remove plating from the blade before using the stone.

(5) One of two methods should be used to sharpen a knife. One method is to push the blade down the stone in

a slicing motion. Then turn the blade over and draw the blade toward the body (figure 17-1).

(6) The other method is to use a circular motion the entire length of the blade, turn the blade over and repeat

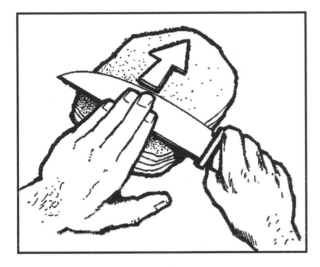

Figure 17-1. Knife Sharpening (Draw).

the process. What is done to one side of the cutting edge should also be done to the other to maintain an even cutting edge (figure 17-2).

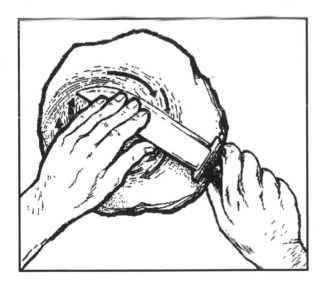

Figure 17-2. Knife Sharpening (Circular).

(7) Most sharpening stones available to survivors will be whetstones. Water should be applied to these stones. The water will help to float away the metal

removed by sharpening and make cleaning of the stone easier.

(8) If a commercial whetstone is not available, a natural whetstone can be used. Any sandstone will sharpen tools, but a gray, clay-like sandstone gives better results. Quartzite should be avoided. Survivors can recognize quartzite instantly by scratching the knife blade with it—the quartz crystals will bite into steel. If no sandstone is available, granite or crystalline rock can be used. If granite is used, two pieces of the stone should be rubbed together to smooth the surface before use.

(9) As with a knife, a sharp axe will save time and energy and be much safer.

(10) A file should be used on an axe or hatchet. Survivors should file away from the cutting edge to prevent injury if the file should slip. The file should be worked from one end of the cutting edge to the other. The opposite side should be worked to the same degree. This will ensure that the cutting edge is even. After using a file, the stone may be used to hone the axe blade (figure 17-3).

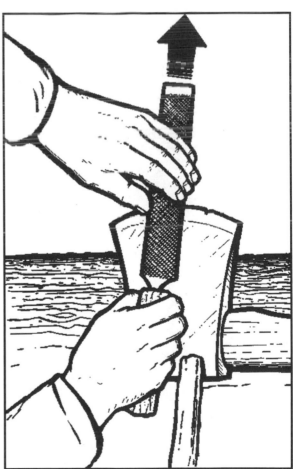

Figure 17-3. Sharpening Axe.

(11) When using an axe, don't try to cut through a tree with one blow. Rhythm and aim are more important than force. Too much power behind a swing interferes with aim. When the axe is swung properly, its weight provides all the power needed.

(12) Carving a new axe handle and mounting the axe head takes a great deal of time and effort. For this reason, a survivor should avoid actions that would require the handle to be changed. Using aim and paying attention to where the axe falls will prevent misses, which could result in a cracked or broken handle. Survivors should not use an axe as a pry bar and should avoid leaving the axe out in cold weather where the handle may become brittle.

(13) A broken handle is difficult to remove from the head of the axe. Usually the most convenient way is to burn it out (figure 17-4). For a single-bit axe, bury the bit in the ground up to the handle, and build a fire over it. For a double-bit, a survivor should dig a small trench, lay the middle of the axe head over it, cover both "bits" with earth, and build the fire. The covering of earth keeps the flame from the cutting edge of the axe and saves its temper. A little water added to the earth will further ensure this protection.

(14) When improvising a new handle, a survivor can save time and trouble by making a straight handle instead of a curved one like the original. Survivors should use a young, straight piece of hardwood without knots. The wood should be whittled roughly into shape and finished by shaving. A slot should be cut into the axe-head end of the handle. After it is fitted, a thin, dry wooden wedge can then be pounded into the slot. Survivors should use the axe awhile, pound the wedge in again, then trim it off flush with the axe. The handle must be smoothed to remove splinters. The new handle can be seasoned to prevent shrinkage by "scorching" it in the fire.

d. Whittling:

(1) Whittle means to cut, trim, or shape (a stick or piece of wood) by taking off bits with a knife. Survivors should be able to use the techniques of whittling to help save time, energy, and materials as well as to prevent injuries. They will kind that whittling is a necessity in constructing triggers for traps and snares, shuttles and spacers, and other improvised equipment.

(2) When whittling, survivors must hold the knife firmly and cut away from the body (figure 17-5). Wood should be cut with the grain. Branches should be trimmed as shown in figure 17-6.

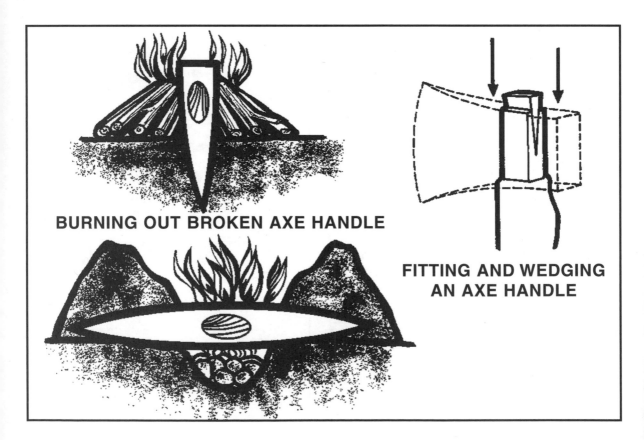

BURNING OUT BROKEN AXE HANDLE

FITTING AND WEDGING AN AXE HANDLE

Figure 17-4. Removing Broken Axe Handle.

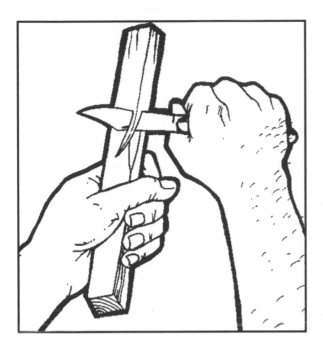

Figure 17-5. Whittling.

(3) To cut completely through a piece of wood, a series of V-cuts should be made all the way around as in figure 17-7. Once the piece of wood has been severed, the pointed end can then be trimmed.

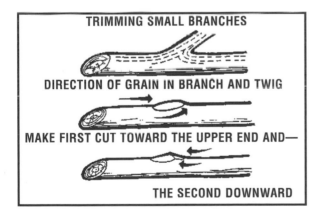

Figure 17-6. Trimming Branches.

(4) The thumb can be used to help steady the hand. Be sure to keep the thumb clear of the blade. To maintain good knife control, the right hand is steadied with the right thumb while the left thumb pushes the blade forward (figure 17-8). This method is very good for trimming.

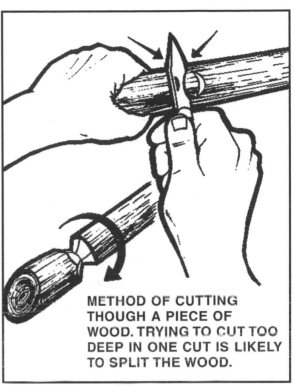

METHOD OF CUTTING THOUGH A PIECE OF WOOD. TRYING TO CUT TOO DEEP IN ONE CUT IS LIKELY TO SPLIT THE WOOD.

Figure 17-7. Cutting Through a Piece of Wood.

e. Felling Trees:

(1) To fell a tree, the survivor must first determine the direction in which the tree is to fall. It is best to fell the tree in the direction in which it is leaning. The lean of the tree can be found by using the axe as a plumb line

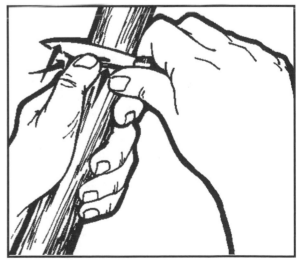

Figure 17-8. Fine Trimming.

(figure 17-9). The survivor should then clear the area around the tree from underbrush and overhanging branches to prevent injury (figure 17-10).

Figure 17-9. Using the Axe as a Plumb Line.

(2) The survivor should make two cuts. The first cut should be on the leaning side of the tree and close to the ground and the second cut on the opposite side and a little higher than the first cut (figure 17-11).

Figure 17-10. Clearing Brush from Cutting Area.

(3) Falling trees often kick back and can cause serious injury (figure 17-12), so survivors must ensure they have a clear escape route. When limbing a tree, start at

Figure 17-11. Felling Cuts.

the base of the tree and cut toward the top. This procedure will allow for easier limb removal and results in a smoother cut. For safety, the survivor should stand on one side of the trunk with the limb on the other.

(4) To prevent damage to the axe head and possible physical injury, any splitting of wood should be done on a log as in figure 17-13. The log can also be used for cutting sticks and poles (figure 17-14).

(5) To make cutting of a sapling easier, bend it over with one hand, straining grain. A slanting blow close to the ground will cut the sapling (figure 17-15).

17-4. Improvised Equipment:

a. If issued equipment is inoperative, insufficient, or nonexistent, survivors will have to rely upon their ingenuity to manufacture the needed equipment. Survivors must determine whether the need for the item outweighs the work involved to manufacture it. They will also have to evaluate their capabilities. If they have injuries, will the injuries prevent them from manufacturing the item(s)?

b. Undue haste may not only waste materials, but also waste the survivor's time and energy. Before manufacturing equipment, they should have a plan in mind.

c. The survivor's equipment needs may be met in two different ways. They may alter an existing piece of equipment to serve more than one function, or they may also construct a new piece of equipment from available

Figure 17-12. Tree Kickbacks.

materials. Since the items survivors can improvise are limited only by their ingenuity, all improvised items cannot be covered in this regulation.

Figure 17-13. Splitting Wood.

Figure 17-14. Cutting Poles.

d. The methods of manufacturing the equipment referred to in this regulation are only ideas and do not have to be strictly adhered to. Many Air Force survivors have a parachute. This device can be used to improvise a variety of needed equipment items.

e. The parachute consists of (figure 17-16):

(1) The pilot chute, which deploys first and pulls the rest of the parachute out.

(2) The parachute canopy, which consists of the apex (top) and the skirt or lower lateral band. The canopy material is divided by radial seams into 28 sections called gores. Each gore measures about 3 feet at the skirt and tapers to the apex. Each gore is further subdivided into four sections called panels. The canopy is normally divided into four colors. These colored areas are intended to aid the survivor in shelter construction, signaling, and camouflage.

(3) Fourteen suspension lines connect the canopy material to the harness assembly. Each piece of suspension line is 72 feet long from riser to riser and 22 feet long from riser to skirt and 14 feet from skirt to apex. The tensile strength of each piece of suspension line is 550 pounds. Each piece of suspension line contains seven to nine pieces of innercore with a tensile strength of 35 pounds. The harness assembly contains risers and webbing buckles, snaps, "D" rings, and other hardware that can be used in improvisation.

f. The whole parachute assembly should be considered as a resource. Every piece of material and hardware can be used.

(1) To obtain the suspension lines, a survivor should cut them at the risers or, if time and conditions permit, consider disassembling the connector links. Cut

Figure 17-15. Cutting Saplings.

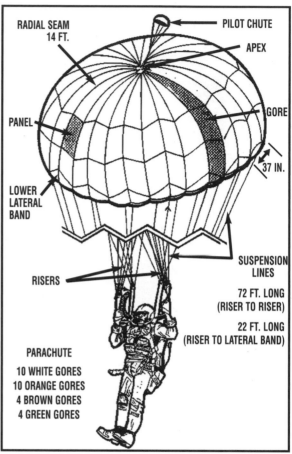

RADIAL SEAM 14 FT.

PILOT CHUTE

APEX

GORE

37 IN.

PANEL

LOWER LATERAL BAND

SUSPENSION LINES

72 FT. LONG (RISER TO RISER)

22 FT. LONG (RISER TO LATERAL BAND)

RISERS

PARACHUTE

10 WHITE GORES
10 ORANGE GORES
4 BROWN GORES
4 GREEN GORES

Figure 17-16. Parachute Diagram.

the suspension lines about 2 feet from the skirt of the canopy. When cutting suspension lines or dismantling the canopy/pack assembly, it will be necessary to maintain a sharp knife for safety and ease of cutting.

(2) Survivors should obtain all available suspension line due to its many uses. Even the line within the radial seams of the canopy should be stripped for possible use. The suspension line should be cut above the radial seam stitches next to the skirt end of the canopy (two places). The cut should not go all of the way through the radial seam (figure 17-17). At the apex of the canopy, and just below the radial seam stitching, a horizontal cut can be made and the suspension line extracted. The line can then be cut.

(3) For maximum use of the canopy, survivors must plan its disassembly. The quantity requirements for shelter, signaling, etc., should be thought out and planned for. Once these needs have been determined, the canopy may be cut up. The radial seam must be stretched tightly for ease of cutting. The radial seam can then be cut by holding the knife at an angle and following the center of the seam. With proper tension and the gentle pushing (or

pulling) of a sharp blade there will be a controlled splitting of the canopy at the seam (figure 17-17). It helps to secure the apex either to another individual or to an immobile object, such as a tree.

(4) When stripping the harness assembly, the seams of the webbing should be split so the maximum usable webbing is obtained. The harness material and webbing should not be randomly cut as it will waste much needed materials.

g. One requirement in improvising is having available material. Parachute fabric, harness, suspension lines, etc., can be used for clothing. Needles are helpful for making any type of emergency clothing. Wise survivors should always have extra sewing needles hidden somewhere on their person. A good needle or sewing awl can be made from the can-opening key from the ration tin (figure 17-18) or, as the Eskimos do, from a sliver of bone. Thread is usually available in the form of innercore. It will be to the survivor's benefit to collect small objects that may "come in handy." Wire, nails, buttons, a piece of canvas,

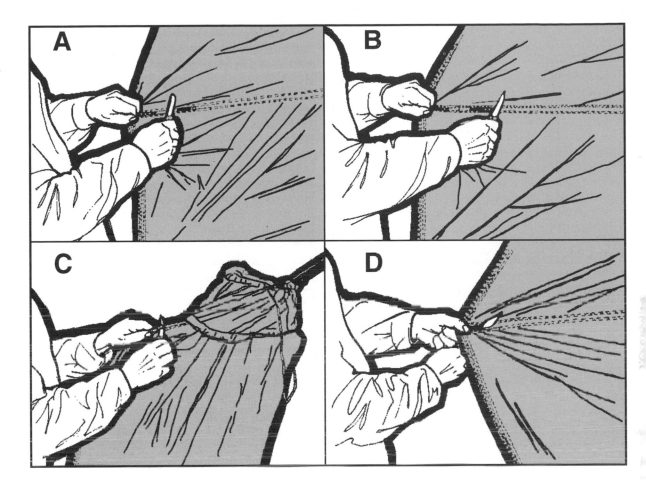

Figure 17-17. Cutting the Parachute.

or animal skin should not be discarded. Any such object may be worth its weight in gold when placed in a hip pocket or a sewing kit. Any kind of animal skin can be used for making clothing such as gloves or mittens or making a ground cover to keep the sleeping bag dry and clean. Small skins can be used for mending and for boot insoles. Mending and cleaning clothes when possible will pay dividends in health, comfort, and safety.

h. The improvised equipment survivors may need to make will probably involve sewing. The material to be sewn may be quite thick and hard to sew, and to keep from stabbing fingers and hands, a palm-type thimble can be improvised (figure 17-19). A piece of webbing, leather, or other heavy material, with a hole for the thumb, is used. A flat rock, metal, or wood is used as the thimble and this is held in place by a doughnut-shaped piece of material sewn onto the palm piece. To use, the end of the needle with the eye is placed on the thimble and the thimble is then used to push the needle through the material to be sewn.

17-5. Miscellaneous Improvised Equipment:

a. Improvised Trail-Type Snowshoes. The snowshoe frame can be made from a sapling 1 inch in diameter and 5 feet long. The sapling should be bent and spread to 12 inches at the widest point. The survivor can then include the webbing of suspension lines (figure 17-20). The foot harness, for attaching the snowshoe to the boot, is also fashioned from suspension line.

b. Improvised Bear Paw-Type Snowshoes. A sapling can be held over a heat source and bent to the shape as shown in figure 17-21. Wire from the aircraft or parachute suspension line can be used for lashing and for making webbing. Snowshoes can also be quickly improvised by cutting a few pine boughs and lashing them together at the cut ends. The lashed boughs positioned with the cut ends forward can then be tied to the feet (figure 17-22).

(1) Survivors should guard against frostbite and blistering while snowshoeing. Due to the design of the

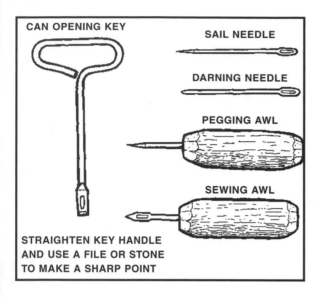

CAN OPENING KEY

SAIL NEEDLE

DARNING NEEDLE

PEGGING AWL

SEWING AWL

**STRAIGHTEN KEY HANDLE
AND USE A FILE OR STONE
TO MAKE A SHARP POINT**

Figure 17-18. Needle and Sewing Awls.

harness, the circulation of the toes is usually restricted, and the hazard of frostbite is greater. They should check the feet carefully, stop often, take off the harness, and massage the feet when they seem to be getting cold.

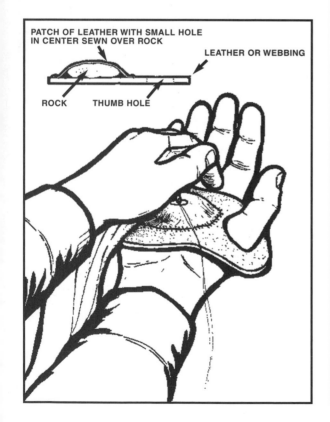

Figure 17-19. Palm Thimble.

(2) Blistering between the toes or on the ball of the foot is sometimes unavoidable in a "tenderfoot" if much snowshoeing is done. To make blisters less likely, a survivor should keep socks and insoles dry and change them regularly.

c. Sleeping Bag. Immediate action should be to use a whole parachute until conditions allow for improvising. A sleeping bag can be improvised by using four gores of parachute material or an equivalent amount of her materials (figure 17-23). The material should be folded in half lengthwise and sewn at the foot. To measure the length, the survivor should allow an extra 6 to 10 inches in addition to the individual's height. The two raw edges can then be sewn together. The two sections the bag can be filled with cattail down, goat's beard lichen, dry grass, insulation from aircraft walls, etc. The stuffed sleeping

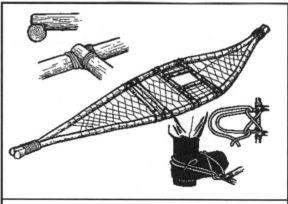

CANADIAN EMERGENCY SNOWSHOES

1. Select 6 poles 6 feet long (individual's height), ¾ inch (thumb size) at the base, ¼ inch (little finger size) at the tip. Cut 6 sticks approximately 10 inches long and ¾ inches wide and tie them in the following manner:

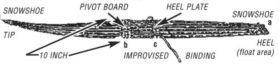

SNOWSHOE

PIVOT BOARD

HEEL PLATE

SNOWSHOE

TIP

10 INCH

IMPROVISED

BINDING

HEEL
(float area)

a. Lash one stick to the snowshoe float area (cut off excess).
b. Lash three sticks to the forward of the center of the shoe to form the pivot board. This position of the pivot board allows the float to remain on the snow and causes the tip to rise when walking.
c. Lash two sticks where your heel strikes the snowshoe to form the heel plate.
d. Tie the snowshoe tips together.

2. The snowshoe binding must be secured to the snowshoe so that the survivor's foot can pivot when walking.

Binding—make as shown from continuous length of split harness webbing or from suspension lines (braided lines preferred).

Figure 17-20. Improvised Trail Snowshoes.

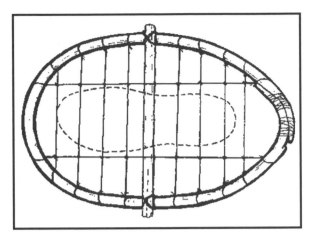

Figure 17-21. Improvised Bear Paw.

bag should then be quilted to keep the insulation from shifting. The bag can be folded in half lengthwise and the foot and open edges sewn. The length and width can be adjusted for the individual.

d. Insulating Bed:

(1) In addition to the sleeping bag, some form of ground insulation is advisable. An insulation mat will help insulate the survivor from ground moisture and the cold. Any nonpoisonous plants such as ferns and grasses will suffice. Leaves from a deciduous tree make a comfortable bed. If available, extra clothing, seat cushions, aircraft insulation, rafts, and parachute material may be used. In a coniferous forest, boughs from the trees would do well if the bed is constructed properly.

(2) The survivor should start at the foot of the proposed bed and stick the cut ends in the ground at about a 45-degree angle and very close together. The completed bed should be slightly wider and longer than the body. If the ground is frozen, a layer of dead branches can be used on the ground with the green boughs placed in the dead branches, similar to sticking them in the ground.

(3) A bough bed should be a minimum of 12 inches thick before use. This will allow sufficient insulation between the survivor and the ground once the bed is compressed. The bough bed should be fluffed up and boughs added daily to maintain its comfort and insulation capabilities.

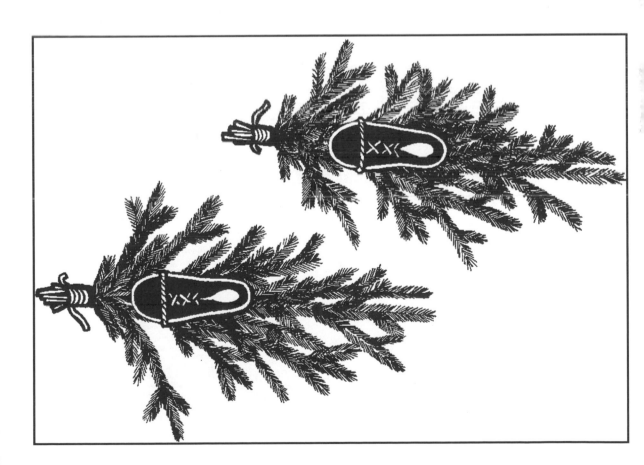

Figure 17-22. Bough Snowshoes.

(4) Spruce boughs have many sharp needles and can cause some discomfort. Also the needles on various types of pines are generally located on the ends of the boughs, and it would take an abundance of pine boughs to provide comfort and insulation. Fir boughs, on the other hand, have an abundance of needles all along the boughs and the needles are rounded. These boughs are excellent for beds, providing comfort and insulation (figure 17-24).

e. Rawhide. Rawhide is a very useful material that can be made from any animal hide. Processing it is time consuming but the material obtained is strong and very durable. It can be used for making sheaths for cutting tools, lashing materials, ropes, etc.

(1) The first step in making rawhide is to remove all of the fat and muscle tissue from the hide. The large pieces can be cut off and the remainder scraped off with a dull knife of similar instrument.

(2) The next step is to remove the hair. This can be done by applying a thick layer of wood ashes to the hair side. Ashes from a hardwood fire work best. Thoroughly sprinkle water all over the ashes. This causes lye to leach out of the ash. The lye will remove the hair. The hide should be rolled with the hair side in and stored in a cool place for several days. When the hair begins to slip (check by pulling on the hair), the hide should be unrolled and placed over a log. Remove the hair by

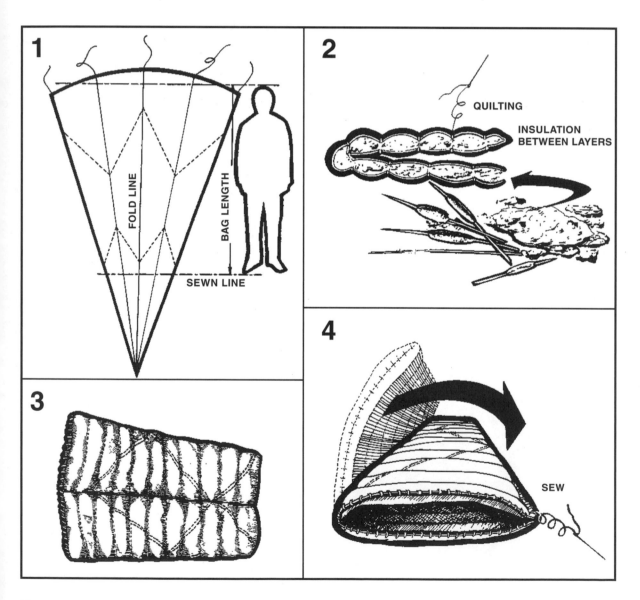

Figure 17-23. Improvised Sleeping Bag.

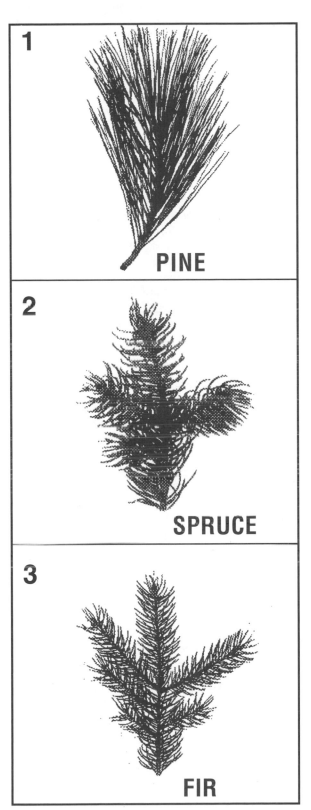

Figure 17-24. Boughs.

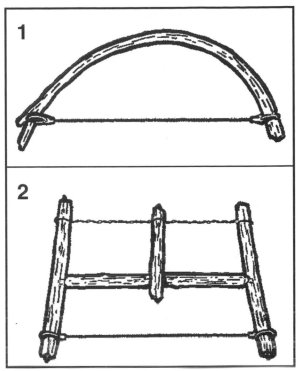

Figure 17-25. Bow Saw and Buck Saw.

scraping it off with a dull knife. Once the hair is removed, the hide should be thoroughly washed, stretched inside a frame, and allowed to dry slowly in the shade. When dry, rawhide is extremely hard. It can be softened by soaking in water.

f. Wire Saws. Wire or pieces of metal can be used to replace broken issued saws. With minor modifications, the survivor can construct a usable saw. A bow-saw

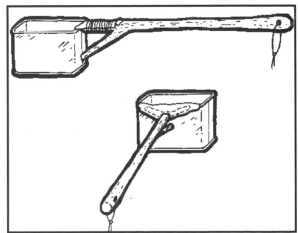

Figure 17-26. Cooking Utensils.

arrangement will help to prevent the blade from flexing. A green sapling may be used for the bow as shown in figure 17-25. If a more durable saw is required and time permits, a bucksaw may be improvised (figure 17-25). Blade tension can be maintained by use of a tightening device known as a "windlass" (figure 17-25).

g. Cooking Utensils. Ration tins can serve as adequate cooking utensils. If the end has been left intact as in figure 17-26, use a green stick long enough to prevent burning the hand while cooking. If the side has been left intact, a forked stick may be used to add support to the container (figure 17-26).

17-6. Ropes and Knots:

a. Basic Knowledge of Tying a Knot. A basic knowledge of correct rope and knot procedures will aid the survivor to do many necessary actions. Such actions as improvising equipment, building shelters, assembling packs, and providing safety devices require the use of proven techniques. Tying a knot incorrectly could result in ineffective improvised equipment, injury, or death.

b. Rope Terminology (figure 17-27):

(1) Bend. A bend (called a knot in this regulation) is used to fasten two ropes together or to fasten a rope to a ring or loop.

(2) Bight. A bight is a bend or U-shaped curve in a rope.

(3) Hitch. A hitch is used to tie a rope around a timber, pipe, or post so that it will hold temporarily but can be readily untied.

(4) Knot. A knot is an interlacement of the parts of bodies, as cordage, forming a lump or knot or any tie or fastening formed with a cord, rope, or line, including bends, hitches, and splices. It is often used as a stopper to prevent a rope from passing through an opening.

(5) Line. A line (sometimes called a rope) is a single thread, string, or cord.

(6) Loop. A loop is a fold or doubling of the rope through which another rope can be passed. A temporary loop is made by a knot or a hitch. A permanent loop is made by a splice or some other permanent means.

(7) Overhand Turn or Loop. An overhand loop is made when the running end passes over the standing part.

(8) Rope. A rope (often called a line) is made of strands of fiber twisted or braided together.

(9) Round Turn. A round turn is the same as a turn, with running end leaving the circle in the same general direction as the standing part.

(10) Running End. The running end is the free or working end of a rope.

(11) Standing End. The standing end is the balance of the rope, excluding the running end.

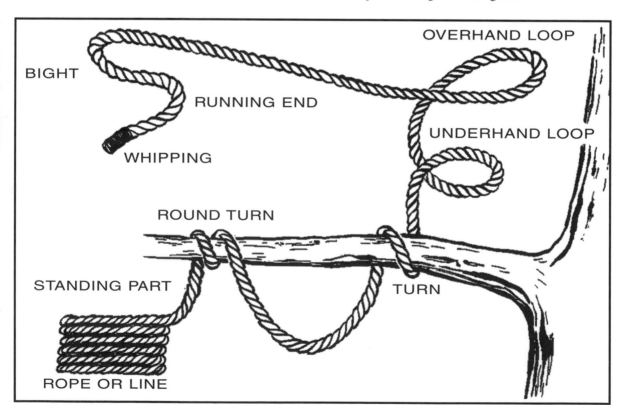

Figure 17-27. Elements of Ropes and Knots.

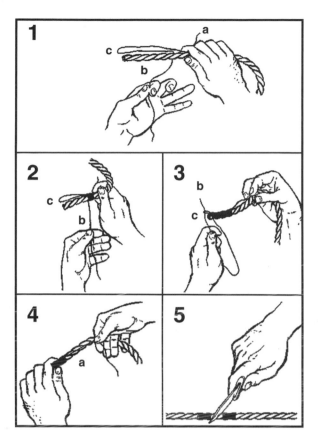

Figure 17-28. Whipping the End of a Rope.

(12) Turn. A turn describes the placing of a rope around a specific object such as a post, rail, or ring with the running end continuing in the opposite direction from the standing end.

(13) Underhand Turn or Loop. An underhand turn or loop is made when the running end passes under the standing part.

c. Whipping the Ends of a Rope. The raw, cut end of a rope has a tendency to untwist and should always be knotted or fastened in some manner. Whipping is one method of fastening the end of the rope. This method is particularly satisfactory because it does not increase the size of the rope. The whipped end of a rope will still

thread through blocks or other openings. Before cutting a rope, place two whippings on the rope 1 or 2 inches apart and make the cut between the whippings (figure 17-28-5). This will prevent the cut ends from untwisting immediately after they are cut. A rope is whipped by wrapping the end tightly with a small cord. Make a bight near one end of the cord and lay both ends of the small cord along one side of the rope (figure 17-28-1). The bight should project beyond the end of the rope about one-half inch. The running end (b) of the cord should be wrapped tightly around the rope and cord (figure 17-28-2) starting at the end of the whipping, which will be farthest from the end of the rope. The wrap should be in the same direction as the twist of the rope strands. Continue wrapping the cord around the rope, keeping it tight, to within about one-half inch of the end. At this point, slip the running end (b) through the bight of the cord (figure 17-28-3). The standing part of the cord (a) can then be pulled until the bight of the cord is pulled under the whipping and cord (b) is tightened (figure 17-28-4). The ends of cord (a and b) should be cut at the edge of the whipping, leaving the rope end whipped.

d. Knots at End of the Rope:

(1) Overhand Knot. The overhand knot (figure 17-29) is the most commonly used and the simplest of all knots. An overhand knot may be used to prevent the end of a rope from untwisting, to form a knot at the end of a rope, or as a part of another knot. To tie an overhand knot, make a loop near the end of the rope and pass the running end through the loop, pulling it tight.

(2) Figure-Eight Knot. The figure-eight knot (figure 17-30) is used to form a larger knot than would be formed by an overhand knot at the end of a rope. A figure-eight knot is used in the end of a rope to prevent the ends from slipping through a fastening or loop in another rope. To make the figure-eight knot, make a loop in the standing part, pass the running end around

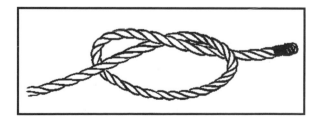

Figure 17-29. Overhand Knot.

Figure 17-30. Figure-Eight Knot.

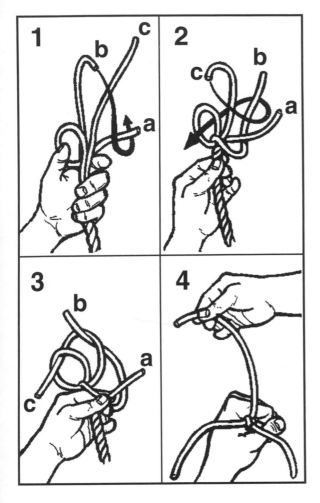

Figure 17-31. Wall Knot.

the standing part, back over one side of the loop, and down through the loop. The running end can then be pulled tight.

(3) Wall Knot. The wall knot (figure 17-31) with a crown is used to prevent the end of a rope from untwisting when an enlargement is not objectionable. It also makes a desirable knot to prevent the end of the rope from slipping through small openings, as when rope handles are used on boxes. The crown or the wall knots may be used separately. To make the wall knot, untwist the strands for about five turns of the rope. A loop in strand "a" (figure 17-32-1) should be used and strand "b" brought down (figure 17-31-2) and around strand "a." Strand "c" (figure 17-31-3) can then be brought around strand "b" and through the loop in strand "a." The knot can then be tightened (figure 17-31-4) by grasping the rope in one hand and pulling each strand tight. The strands point up or away from the rope. To make a neat, round knot, the wall knot should be crowned.

(4) Crown on Wall Knot. To crown a wall knot, the end of strand "a" (figure 17-32-1) should be moved between strands "b" and "c." Next strand "c" is passed (figure 17-32-2) between strand "b" and the loop in strand "a." Line "b" is then passed over line "a" and through the bight formed by line "c" (figure 17-32-3). The knots can then be drawn tight and the loose strands cut. When the crown is finished, strands should point down or back along the rope.

e. Knots for Joining Two Ropes:

(1) Square Knot. The square knot (figure 17-33) is used for tying two ropes of equal diameter together to prevent slippage. To tie the square knot, lay the running end of each rope together but pointing in opposite directions. The running end of one rope can be passed under the standing part of the other rope. Bring the two running

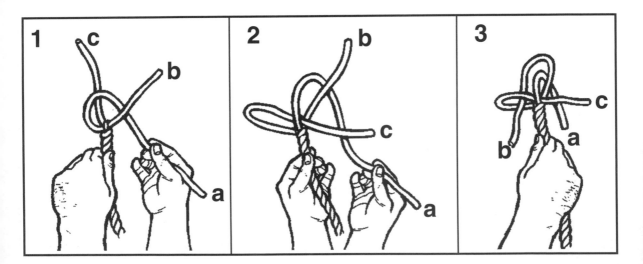

Figure 17-32. Crown on Wall Knot.

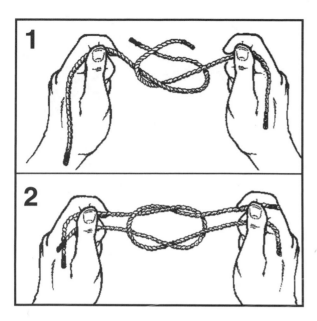

Figure 17-33. Square Knot.

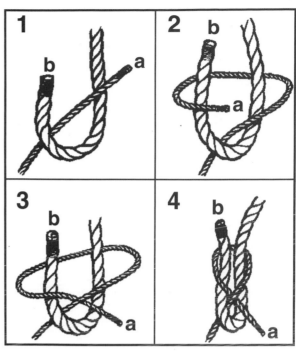

Figure 17-35. Single Sheet Bend.

ends up away from the point where they cross and crossed again (figure 17-33-1). Once each running end is parallel to its own standing part (figure 17-33-2), the two ends can be pulled tight. If each running end does not come parallel to the standing part of its own rope, the knot is called a "granny knot" (figure 17-34-1). Because it will slip under strain, the granny knot should not be used. A square knot can also be tied by making a bight in the end of one rope and feeding the running end of the other rope through and around this bight. The running end of the second rope is routed from the

standing side of the bight. If the procedure is reversed, the resulting knot will have a running end parallel to each standing part but the two running ends will not be opposite each other. This knot is called a "thief" knot (figure 17-34-2). It will slip under strain and is difficult to untie. A true square knot will draw tighter under

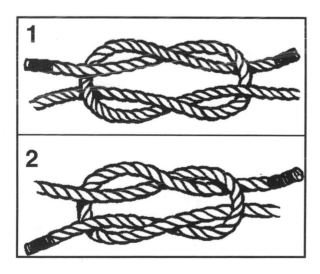

Figure 17-34. Granny and Thief Knots.

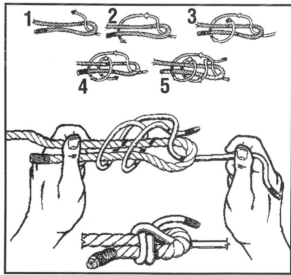

Figure 17-36. Double Sheet Bend.

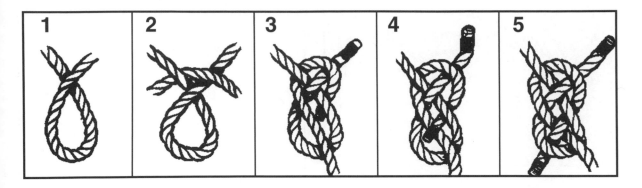

Figure 17-37. Carrick Bend.

strain. A square knot can be untied easily by grasping the bends of the two bights and pulling the knot apart.

(2) Single Sheet Bend. The use of a single sheet bend (figure 17-35), sometimes called a weaver's knot, is limited to tying together two dry ropes of unequal size. To tie the single sheet bend, the running end (a) (figure 17-35-1) of the smaller rope should pass through a bight (b) in the larger rope. The running end should continue around both parts of the larger rope (figure 17-35-2), and back under the smaller rope (figure 17-35-3). The running end can then be pulled tight (figure 17-35-4). This knot will draw tight under light loads but may loosen or slip when the tension is released.

(3) Double Sheet Bend. The double sheet bend (figure 17-36) works better than the single sheet bend for joining ropes of equal or unequal diameter, joining wet ropes, or for tying a rope to an eye. It will not slip or draw tight under heavy loads. To tie a double sheet bend, a single sheet bend is tied first. However, the running end is not pulled tight. One extra turn is taken around both sides of the bight in the larger rope with the running end for the smaller rope. Then tighten the knot.

(4) Carrick Bend. The carrick bend (figure 17-37) is used for heavy loads and for joining thin cable or heavy rope. It will not draw tight under a heavy load. To tie a carrick bend, a loop is formed (figure 17-37-1) in one rope. The running end of the other rope is passed behind the standing part (figure 17-37-2) and in front of the running part of the rope in which the loop has been formed. The running end should then be woven under one side of the loop (figure 17-37-3), through the loop, over the standing part of its own rope (figure 17-37-4), down through the loop, and under the remaining side of the loop (figure 17-37-5).

f. Knots for Making Loops:

(1) Bowline. The bowline (figure 17-38) is a useful knot for forming a loop in the end of a rope. It is also easy to untie. To tie the bowline, the running end (a) of the rope passes through the object to be affixed to the bowline and forms a loop (b) (figure 17-38-1) in the standing part of the rope. The running end (a) is then passed through the loop (figure 17-38-2) from underneath and around the standing part (figure 17-38-3) of the rope, and back through the loop from the top (figure 17-38-4). The running end passes down through the loop parallel to the rope coming up through the loop. The knot is then pulled tight.

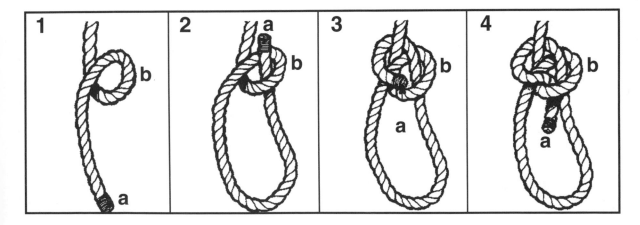

Figure 17-38. Bowline.

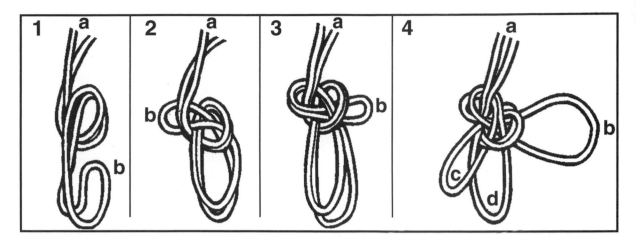

Figure 17-39. Double Bowline.

(2) Double Bowline. The double bowline (figure 17-39) with a slipknot is a rigging used by tree surgeons, who work alone in trees for extended periods. It can be made and operated by one person and is comfortable as a sling or boatswain's chair (figure 17-40). A small board with notches as a seat adds to the personal comfort of the user. To tie a double bowline, the running end (a) (figure 17-39) of a line should be bent back about 10 feet along the standing part. The bight (b) is formed as the new running end and a bowline tied as described and illustrated in figure 17-38. The new running end (b) (figure 17-39) or loop is used to support the back and the remaining two loops (c) and (d) support the legs.

(3) Rolling or Magnus Hitch (figure 17-41). A rolling or Magnus hitch is a safety knot designed to make a running end fast to a suspension line with a nonslip grip yet it can be released by hand pressure bending the knot downward. The running end (a) (figure 17-41-1) is passed around the suspension line (b) twice, making two full turns downward (figure 17-41-2). The

running end (a) (figure 17-41-3) is then turned upward over the two turns, again around the suspension line, and under itself (figure 17-41-3). This knot is excellent for fastening a rope to itself, a larger rope, a cable, a timber, or a post.

(4) Running Bowline. The running bowline (figure 17-42) is the basic air-transport rigging knot. It provides a sling of the choker type at the end of a single line and is generally used in rigging. To tie a running bowline, make a bight (b) (figure 17-42-1) with an overhand loop (c) made in the running end (a). The running end (a) is passed around the standing part, through the loop (c) (figure 17-42-2), under, then back over the side of the bight, and back through the loop (c) (figure 17-42-3).

(5) Bowline on a Bight. It is sometimes desirable to form a loop at some point in a rope other than at the end. The bowline on a bight (figure 17-43) can be used

Figure 17-40. Boatswain's Chair.

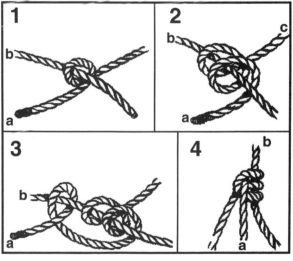

Figure 17-41. Rolling or Magnus Hitch.

Figure 17-42. Running Bowline.

place where the line is doubled or at an end which has been doubled back. The Spanish bowline is used in rescue work or to give a two-fold grip for lifting a pipe or other round object in a sling. To tie the Spanish bowline, a doubled portion of the rope is held in the left hand with the loop up and the center of the loop is turned back against the standing parts to form two loops (figure 17-44-l), or "rabbit ears." The two rabbit ears (c) and (d) (figure 17-44-2) are moved until they partly overlap each other. The top of the loop nearest the person is brought down toward the thumb of the left hand, being sure it is rolled over as it is brought down. The thumb is placed over this loop (figure 17-44-5) to hold it in position. The top of the remaining loop is grasped and brought down, rolling it over and placing it under the thumb. There are now four small loops, (c, d, e, and f) in the rope. The lower left-hand loop (c) is turned one-half turn and inserted from front to back of the upper left-hand loop (e). The lower right-hand loop (d) is turned (figure 17-44-4) and inserted through the upper right-hand loop (f). The two loops (c and d) that have been passed through are grasped and the rope pulled tight (figure 17-44-5).

for this purpose. It is easily untied and will not slip. The same knot can be tied at the end of the rope by doubling the rope for a short section. A doubled portion of the rope is used to form a loop (b) (figure 17-43-1) as in the case of the bowline. The bight end (a) of the doubled portion is passed up through the loop (b), back down (figure 17-43-2), up around the entire knot (figure 17-43-3), and tightened (figure 17-43-4).

(6) Spanish Bowline. A Spanish bowline (figure 17-44) can be tied at any point in a rope, either at a

(7) French Bowline. The French bowline (figure 17-45) is sometimes used as a sling for lifting injured people.

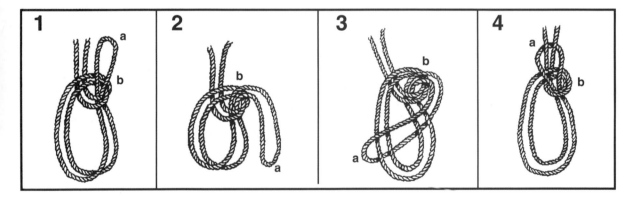

Figure 17-43. Bowline on a Bight.

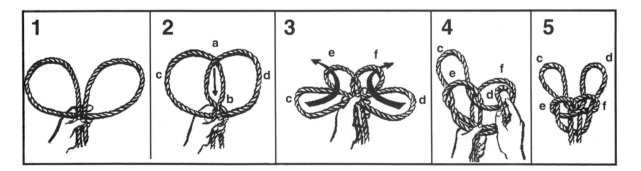

Figure 17-44. Spanish Bowline.

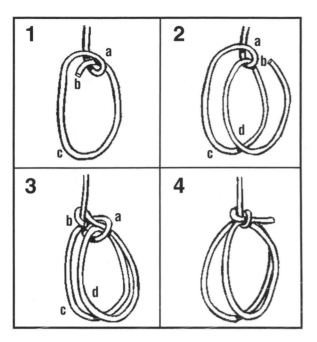

Figure 17-45. French Bowline.

When used in this manner, one loop is used as a seat and the other loop is used around the body under the arms. The weight of the injured person keeps the two loops tight so that the victim cannot fall out, and for this reason it is particularly useful as a sling for someone who is unconscious. The French bowline is started in the same way as the simple bowline. Make a loop (a) (figure 17-45-1) in the standing part of the rope. The running end (b) is passed through the loop from underneath and a separate loop (c) is made. The running end (b) is passed through the loop (a), again from underneath (figure 17-45-3), around the back of the standing part and back through the loop (a) so that it comes out parallel to the looped portion. The standing part of the rope is pulled to tighten the knot (figure 17-45-4), leaving two loops (c and d).

(8) Harness Hitch. The harness hitch (figure 17-46) is used to form a nonslipping loop in a rope. To make the harness hitch, form a bight (3) (figure 17-46-1) in the running end of the rope. Hold this bight in the left hand and form a second bight (b) in the standing part of the rope. The right hand is used to pass bight (b) over bight (a) (figure 17-46-2). Holding all loops in place with the left band, the right hand is inserted through bight (a) behind the upper part of bight (b) (figure 17-46-3). The bottom (c) of the first loop is grasped and pulled up through the entire knot (figure 17-46-4), pulling it tight.

g. Hitches:

(1) Half Hitch. The half hitch (figure 17-47-l) is used to tie a rope to a timber or to another larger rope. It is not a very secure knot or hitch and is used for temporarily securing the free end of a rope. To tie a half hitch, the rope is passed around the timber, bringing the running end around the standing part, and back under itself.

(2) Timber Hitch. The timber hitch (figure 17-47-2) is used for moving heavy timbers or poles. To make the timber hitch, a half hitch is made and similarly the running end is turned about itself at least another time. These turns must be taken around the running end itself or the knot will not tighten against the pull.

(3) Timber Hitch and Half Hitch. To get a tighter hold on heavy poles for lifting or dragging a timber hitch and half hitch are combined (figure 17-47-3). The running end is passed around the timber and back under the standing part to form a half hitch. Farther along the timber, a timber hitch is tied with the running end. The strain will come on the half hitch and the timber hitch will prevent the half hitch from slipping.

(4) Clove Hitch. A clove hitch (figure 17-47-4) is used to fasten a rope to a timber, pipe, or post. It can be tied at any point in a rope. To tie a clove hitch in the center of the rope, two turns are made in the rope close together. They are twisted so that the two loops lay back-to-back. These two loops are slipped over the timber or pipe to form the knot. To tie the clove hitch at the end of a rope, the rope is passed around the timber in two turns so that the first turn crosses the standing part and the running end comes up under itself on the second turn.

(5) Two Half Hitches. A quick method for tying a rope to a timber or pole is the use of two half hitches.

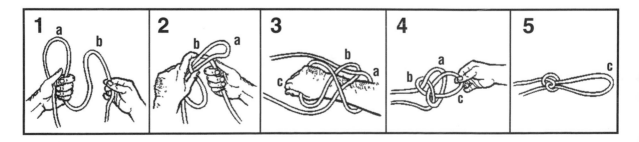

Figure 17-46. Harness Hitch.

Figure 17-47. Half Hitch, Timber Hitch, and Clove Hitch.

Figure 17-48. Round Turn and Two Half Hitches.

(6) Round Turn and Two Half Hitches. Another hitch used for fastening a rope to a pole, timber, or spar is the round turn and two half hitches (figure 17-48). The running end of the rope is passed around the pole or spar in two complete turns, and the running end is brought around the standing part and back under itself to make a half hitch. A second half hitch is made. For greater security, the running end of the rope should be secured to the standing part.

(7) Fisherman's Bend. The fisherman's bend (figure 17-49) is used to fasten a cable or rope to an anchor, or for use where there will be a slackening and tightening motion in the rope. To make this bend, the running end of the rope is passed in two complete turns through the

The running end of the rope is passed around the pole or timber, and a turn is taken around the standing part and under the running end. This is one half hitch. The running end is passed around the standing part of the rope and back under itself again.

Figure 17-49. Fisherman's Bend.

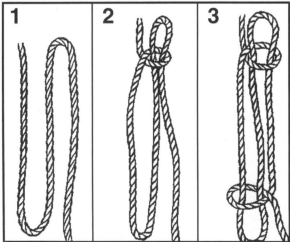

Figure 17-50. Sheepshank.

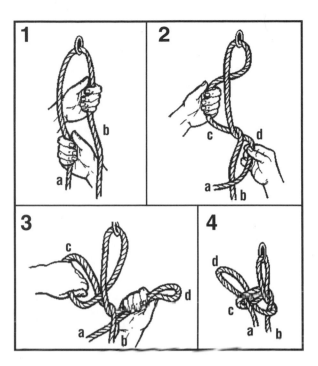

Figure 17-51. Speir Knot.

running end (a) is passed through a ring (figure 17-51-1) or around a pipe or post and brought back on the left side of the standing part (b). Both hands are placed, palms up, under both parts of the rope with the left hand higher than the right hand; grasping the standing part (b) with the left hand and the running end (a) with the right hand. The left hand is moved to the left and the right hand to the right (figure 17-51-3) to form two bights (c and d). The left hand is twisted a half turn toward the body so that bight (c) is twisted into a loop (figure 17-51-3). Pass bight (d) over the rope and down through the loop (c). The Speir knot is tightened by pulling on the bight (d) and the standing part (b) (figure 17-51-4).

(10) Rolling Hitch (Pipe or Pole). The rolling hitch (figure 17-52) is used to secure a rope to a pipe or pole so that the rope will not slip. The standing part (a) of the rope is placed along the pipe or pole (figure 17-52-1) extending in the direction opposite to the direction the pipe or pole will be moved. Two turns (b) are taken with the running end around the standing part (a) and the pole (figure 17-52-3). The standing part (a) of the rope is reversed so that it is leading off in the direction in which the pole will be moved (figure 17-52-3), and two turns are taken (c) (figure 17-52-4) with the running end (d). On the second turn around, the running end (d) is passed under the first turn (c) to secure it. To make this knot secure, a half hitch (e) (figure 17-52-6) is tied with the standing part of the rope 1 or 2 feet above the rolling hitch.

(11) Blackwall Hitch. The blackwall hitch (figure 17-53) is used for fastening a rope to a hook. To make the blackwall hitch, a bight of the rope is placed behind the hook. The running end (a) and standing part (b) are crossed through the hook so that the running end comes out at the opposite side of the hook and under the standing part.

(12) Catspaw. A catspaw can be made at the end of a rope (figure 17-54) for fastening the rope to a hook. Grasp the running end (a) of the rope in the left hand and make two bights (c and d) in the standing part (b). Hold these two bights in place with the left hand and take two turns about the junction of the two bights with the standing part of the rope. Slip the two loops (c and d) so formed over the hook.

ring or object to which it is to be secured. The running end is passed around the standing part of the rope and through the loop that has just been formed around the ring. The running end is then passed around the standing part in a half hitch. The running end should be secured to the standing part.

(8) Sheepshank. A sheepshank (figure 17-50) is a method of shortening a rope, but it may also be used to take the load off a weak spot in the rope. To make the sheepshank (which is never made at the end of a rope), two bights are made in the rope so that three parts of the rope are parallel. A half hitch is made in the standing part over the end of the bight at each end.

(9) Speir Knot. A Speir knot (figure 17-51) is used when a fixed loop, a nonslip knot, and a quick release are required. It can be tied quickly and released by a pull on the running end. To tie the Speir knot, the

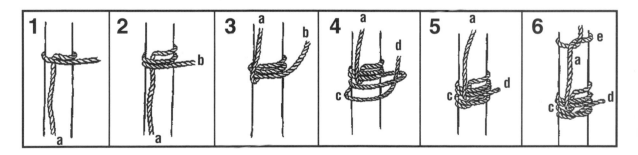

Figure 17-52. Rolling Hitch.

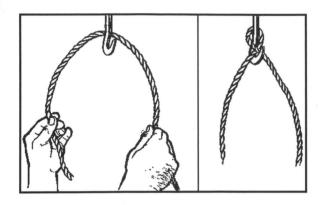

Figure 17-53. Blackwall Hitch.

(13) Scaffold Hitch. The scaffold hitch (figure 17-55) is used to support the end of a scaffold plank with a single rope. To make the scaffold hitch, the running end of the rope is laid across the top and around the plank, then up and over the standing part (figure 17-55-1). A doubled portion of the running end is brought back under the plank (figure 17-55-2) to form a bight (b) at the

opposite side of the plank. The running end is taken back across the top of the plank (figure 17-55-3) until it can be passed through the bight (b). A loop is made (c) in the standing part (figure 17-55-4) above the plank. The running end is passed through the loop (c) around the standing part, and back through the loop (c).

(14) Barrel Slings. Barrel slings can be made to hold barrels horizontally or vertically. To sling a barrel horizontally (figure 17-56), a bowline is made with a long bight. The rope at the bottom of the bight is brought up over the sides of the bight. The two "ears" are thus moved forward over the end of the barrel. To sling a barrel vertically (figure 17-57) the rope is passed under the barrel and up to the top. An overhand knot is made (a) on top (figure 17-57-1). With slight tension on the rope, the two parts (figure 17-57-2) of the overhand knot are grasped, separated, and pulled down to the center of the barrel (b and c). The rope is pulled snug and a bowline tied (d) over the top of the barrel (figure 17-57-3).

h. Lashing. There are numerous items that require lashings for construction; for example, shelters, equipment racks, and smoke generators. Three types of lashings will be discussed here—the square lash, the diagonal lash, and the shear lash.

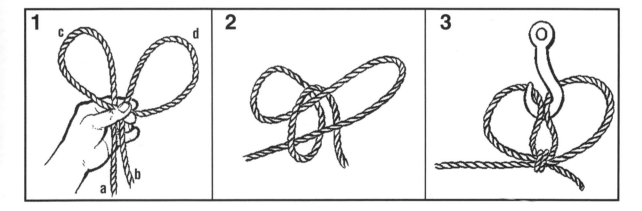

Figure 17-54. Catspaw.

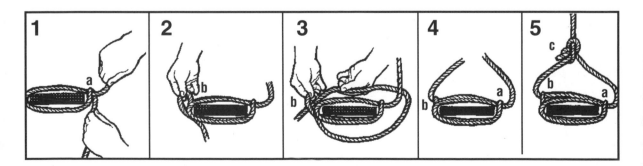

Figure 17-55. Scaffold Hitch.

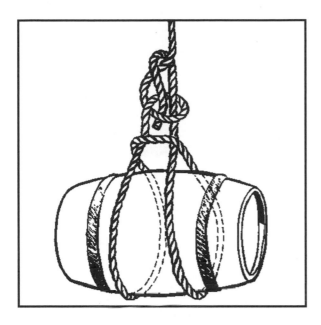

Figure 17-56. Barrel Slung Horizontally.

(1) Square Lash. Square lashing is started with a clove hitch around the log, immediately under the place where the crosspiece is to be located (figure 17-58-1). In laying the turns, the rope goes on the outside of the previous turn around the crosspiece, and on the inside of the previous turn around the log. The rope should be kept tight (figure 17-58-2). Three or four turns are necessary. Two or three "frapping" turns are made between the crosspieces (figure 17-58-3). The rope is pulled tight; this will bind the crosspiece tightly together. It is finished with a clove hitch around the same piece that the lashing was started on (figure 17-58-4). The square lash is used to secure one pole at right angles to another pole.

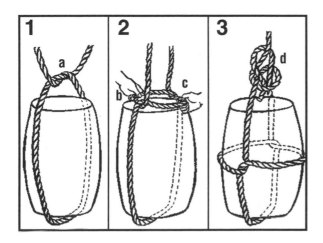

Figure 17-57. Barrel Slung Vertically.

Another lash that can be used for the same purpose is the diagonal lash.

(2) Diagonal Lash. The diagonal lash is started with a clove hitch around the two poles at the point of crossing. Three turns are taken around the two poles (figure 17-59-1). The turns lie beside each other, not on top of each other. Three more turns are made around the two poles, this time crosswise over the previous turns. The turns are pulled tight. A couple of frapping turns are made between the two poles, around the lashing turns, making sure they are tight (figure 17-59-2). The lashing is finished with a clove hitch around the same pole the lash was started on (figure 17-59-3).

(3) Shear Lash. The shear lash is used for lashing two or more poles in a series. The desired number of poles are placed parallel to one another and the lash is started with a clove hitch on an outer pole (figure 17-60-1). The poles are then lashed together, using seven or eight turns of the rope laid loosely beside each other (figure 17-60-2). Make frapping turns between each pole (figure 17-60-3). The lashing is finished with a clove hitch on the pole opposite that on which the lash was started (figure 17-60-4).

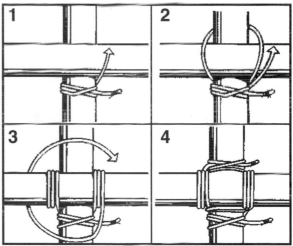

Figure 17-58. Square Lash.

I. Making Ropes and Cords. Almost any natural fibrous material can be spun into good serviceable rope or cord, and many materials that have a length of 12 to 24 inches or more can be braided. Ropes up to 3 and 4 inches in diameter can be "laid" by four people, and tensile strength for bush-made rope of 1-inch diameter range from 100 pounds to as high as 3,000 pounds.

(1) Tensile Strength. Using a three-lay rope of 1-inch diameter as the standard, the following table of tensile strengths may serve to illustrate general strengths of various materials. For safety's sake, the lowest figure should always be regarded as the tensile strength.

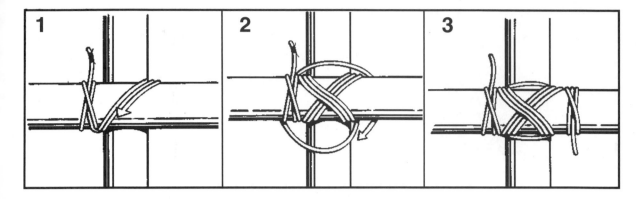

Figure 17-59. Diagonal Lash.

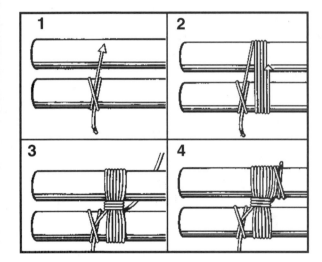

Figure 17-60. Shear Lash.

Green Grass..................…....…....100 lbs to 250 lbs
Bark Fiber..........…............…....500 lbs to 1,500 lbs
Palm Fiber..........…....…..........650 lbs to 2,000 lbs
Sedges...........…......…..........2,000 lbs to 2,500 lbs
Monkey Rope (Lianas)….…...…..560 lbs to 700 lbs
Lawyer Vine (Calamus)......¹/₂-inch diam., 1,200 lbs

NOTE: Doubling the diameter quadruples the tensile
strength. Halving the diameter reduces the tensile
strength to one-fourth.

(2) Principles of Ropemaking Materials. To ensure
that a material is suitable for ropemaking, it must have
four qualities:

(a) It must be reasonably long in the fiber.

(b) It must have "strength."

(c) It must be pliable.

(d) It must have "grip" so the fibers will "bite"
into one another.

(3) Determining Suitability of Material. There are
simple tests to determine whether a material is suitable:

(a) First, pull on a length of the material to test
for strength.

(b) Second, twist it between the fingers and
"roll" the fibers together, if it will withstand this and not
"snap" apart, an overhand knot is tied and gently tightened. If the material does not cut upon itself, but allows
the knot to be pulled taut, it is suitable for ropemaking if
the material will "bite" together and is not smooth or
slippery.

(4) Where to Find Suitable Material. These qualities
can be found in venous types of plants, in ground vines,
in most of the longer grasses, in some of the water reeds
and rushes, in the inner barks of many trees and shrubs,
and in the long hair or wool of many animals.

(5) Obtaining Fibers for Making Ropes. Some green
freshly gathered materials may be "stiff" or unyielding.
When this is the case, it should be passed through hot
flames for a few moments. The heat treatment should
cause the sap to burst through some of the cell structure,
and the material thus becomes pliable. Fibers for rope
making may be obtained from many sources, such as:

(a) Surface roots of many shrubs and trees have
strong fibrous bark.

(b) Dead inner bark of fallen branches of some
species of trees and in the new growth of many trees,
such as willows.

(c) The fibrous material of many water- and
swamp-growing plants and rushes.

(d) Many species of grass and weeds.

(e) Some seaweeds.

(f) Fibrous material from leaves, stalks, and trunks
of many palms.

(g) Many fibrous-leaved plants, such as the aloes.

(6) Gathering and Preparing Materials. There may
be a high content of vegetable gum in some plants. This
can often be removed by soaking the plants in water, by

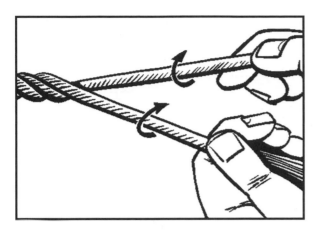

Figure 17-61. Twisting Fibers.

boiling, or by drying the material and "teasing" it into thin strips.

(a) Some of the materials have to be used green if any strength is required. The materials that should be green include the sedges, water rushes, grasses, and lianas.

(b) Palm fiber is harvested in tropical or subtropical regions. It is found at the junction of the leaf and the palm trunk, or it will be found lying on the ground beneath many palms. Palm fiber is a "natural" for making ropes and cords.

(c) Fibrous matter from the inner bark of trees and shrubs is generally more easily used if the plant is dead or half-dead. Much of the natural gum will have dried out and when the material is being teased, prior to spinning, the gum or resin will fall out in fine powder.

(7) Making a Cord by Spinning with the Fingers:

(a) Use any material with long strong threads or fibers that have been previously tested for strength and pliability. The fibers are gathered into loosely held strands of even thickness. Each of these strands is twisted clockwise. The twist will hold the fibers together. The strands should be formed one-eighth inch diameter. As a general rule, there should be about 15 to 20 fibers to a strand. Two, three, or four of these strands are later twisted together, and this twisting together or "laying" is done with a counterclockwise twist, while at the same time, the separate strands that have not yet been laid up are twisted clockwise. Each strand must be of equal twist and thickness.

(b) Figure 17-61 shows the general direction of twist and the method whereby the fibers are bonded into strands. In a similar manner, the twisted strands are put together into lays, and the lays into ropes.

(c) The person who twists the strands together is called the "layer" and must see that the twisting is even, the strands are uniform, and the tension on each strand is equal. In "laying," care must be taken to ensure each of the strands is evenly "laid up," that is, one strand does not twist around the other one.

(d) When spinning fine cords for fishing lines, snares, etc., considerable care must be taken to keep the strands uniform and the lay even. Fine, thin cords of no more than 1/32-inch thickness can be spun with the fingers and are capable of taking a breaking strain of 20 to 30 pounds or more.

(e) Normally two or more people are required to spin and lay up the strands for cord. However, many native people spin cord unaided. They twist the material by running the flat of the hand along the thigh, with the fibrous material between hand and thigh, and with the free hand, they feed in fiber for the next "spin." Using this technique, one person can make long lengths of single strands. This method of making cord or rope with the fingers is slow if any considerable length of cord is required.

(f) An easier and simpler way to rapidly make lengths of rope from 50 to 100 yards or more in length is to make a rope machine and set up multiple spinners in the form of cranks. Figure 17-62 shows the details of rope spinning.

(g) To use a rope machine, each feeder holds the material under one arm and with one free hand feeds it into the strand that is being spun by the crank. The other hand lightly holds the fibers together till they are spun. As the lightly spun strands are increased in length, they must be supported on crossbars. They should not be allowed to lie on the ground. Spin strands from 20 to 100 yards before laying up. The material should not be spun in too thickly. Thick strands do not help strength in any way, rather, they tend to make a weaker rope.

(8) Setting Up a Rope Machine:

(a) When spinning ropes of 10 yards or longer, it is necessary to set crossbars every 2 or 3 yards to carry the strands as they are spun. If crossbars are not set up, the strands or rope will sag to the ground, and some of the fibers will tangle up with grass, twigs, or dirt on the ground. Also, the twisting of the free end may either be stopped or interrupted and the strand will be unevenly twisted.

(b) The easiest way to set up crossbars for the rope machine is to drive pairs of stakes into the ground about 6 feet apart and at intervals of about 6 to 10 feet. The crossbars must be smooth and free from twigs and loose portions of bark that might twist in with the spinning strands.

(c) The crossbar (a) is supported by two uprights and pierced to take the cranks (b). These cranks can be made out of natural sticks, morticed slab, and pegs, or if available, bent wire. The connecting rod (c) enables one person to turn all cranks clockwise simultaneously. Crossbars supporting the strands as they are spun are shown (d). A similar crank handle to the previous ones (b) is supported on a forked stick at the end of the rope

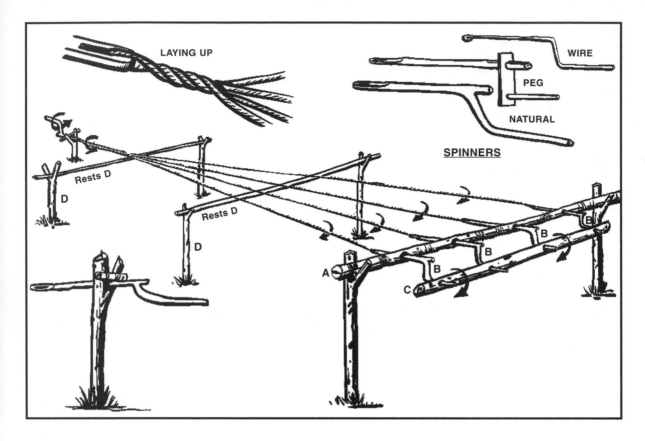

Figure 17-62. Rope Machine.

machine. This handle is turned in reverse (counterclockwise) to the cranks (c) to twist the connected strands together. These are "laid up" by one or more of the feeders.

(d) The first strand should be turned clockwise, then the laying up of the strands will be done counterclockwise and the next laying will again be clockwise. Proof that the rope is well made is that the individual fibers lay lengthways along the rope.

(e) In the process of laying up the strands, the actual twisting together or laying will take some of the original "twist" out of the strand which has not yet been laid. Therefore, it is necessary to keep twisting the strands while laying together.

(f) When making a rope too long to be spun and laid in one piece, a section is laid up and coiled on the ground at the end of the rope walk farthest from the cranks. Strands for a second length are spun, and these strands are married, or spliced, into the strands of the first section and then the laying up of the second section continues the rope.

(g) The actual "marrying" of the strands is done only in the last lay, which makes the rope when completed. The ends where the strands are married should be staggered in different places. By this means, rope can be made and extended in sections to a great length.

(h) After a complete length of rope is laid up, it should be passed through a fire to burn off the loose ends and fibers. This will make the rope smooth and more professional looking.

(9) Laying the Strands:

(a) The strands lie on the crossbar as they are spun. When the strands have been spun to the required length, which should not be more than about a hundred feet, they are joined together by being held at the far end. They are then ready for laying together. The turner, who is facing the cranks, twists the ends together counterclockwise, at the same time keeping full weight on the rope that is being laid up. The layer advances, placing the strands side by side as they turn.

(b) It is important to learn to feed the material evenly and lay up slowly, thereby getting a smooth, even rope (figure 17-63). Do not try to rush the ropemaking. Speed in ropemaking only comes with practice. At first it will take a team of three or four up to 2 hours or more to make a 50-yard length of rope of three lays, each of three strands; that is, 9 strands for a rope with a finished diameter of about 1 inch. With practice, the same three or four people will make the same rope in 15 to 20 minutes. These times do not include time for gathering material.

(c) In feeding the free ends of the strands, twist the loose material fed in by the feeder. As the feeders move backward, they must keep a slight tension on the strands.

(10) Making Rope with a Single Spinner:

(a) Using a Single Crank. Two people can make rope using a single crank. A portion of the material is fastened to the eye of the crank, as with the multiple crank. Supporting crossbars, as used in a ropewalk, are required when a length of more than 20 or 30 feet is being spun.

(b) Feeding:

-1. If the feeder is holding material under the left arm, the right hand is engaged in continuously pulling material forward to the left hand, which feeds it into the turning strand. These actions, done together as the feeder walks backward, govern the thickness of the strands. The left hand, lightly closed over the loose turning material, must "feel" the fibers "biting," or twisting together.

-2. When the free end of the turning strand, which is against the loose material under the arm, takes in too thick a tuft of material, the left hand is closed and so arrests the twist of the material between the left hand and the bundle. This allows teasing out the overall "bite" with the right hand, thus maintaining a uniform thickness of the spinning strand.

(c) Thickness of Strands. Equal thickness and twist for each of the strands throughout their length is important. The thickness should not be greater than necessary with the material being used. For a grass rope, the strand should not be more than one-fourth inch diameter; for coarse bark or palm, not more than one-eighth or three-sixteenth inch; and for fine bark, hair, or sisal fiber, not more than one-eighth inch.

(d) Common Errors in Ropemaking:

-1. There is a tendency with beginners to feed unevenly. Thin, wispy sections of strand are followed by thick portions. Such feeding degrades the quality of rope. Rope made from such strands will break with less than one-fourth of the tensile strain on the material.

-2. Beginners are wise to twist and feed slowly. Speed, with uniformity of twist and thickness, comes with practice.

-3. Thick strands do not help. It is useless to try and spin a rope from strands an inch or more in thickness. Such a rope will break with less than half the tensile strain on the material. Spinning "thick" strands does not save time in ropemaking.

(e) Lianas, Vines, and Canes. Lianas and ground vines are natural ropes, and grow in subtropical and tropical scrub and jungle. Many are of great strength and useful for braiding, tree climbing, and other purposes. The smaller ground vines, when "braided," give great strength and flexibility. Canes and stalks of palms provide excellent material if used properly. Only the outer skin is tough and strong, and this skin will split off easily if the main stalk is bent away from the skin. This principle also applies to the splitting of lawyer cane (calamus), palmleaf stalks, and all green material. If the split starts to run off, bend the material away from the inside, and it will gradually gain in size and come back to an even thickness with the other split side.

(f) Bark Fibers:

-1. The fibers in many barks that are suitable for ropemaking are located near the innermost layers. This is the bark next to the sapwood. When seeking suitable barks of green timber, cut a small section about 3 inches long and 1 inch wide. Cut this portion from the wood to the outer skin of the bark.

-2. The specimen should be peeled and the different layers tested. Green bark fibers are generally difficult to spin because of "gum," and it is better to search round for windfall dead branches and try the inner bark of these. The gum probably has leached out, and the fibers should separate easily.

-3. Many shrubs have excellent bark fiber, and here it is advisable to cut the end of a branch and peel off a strip of bark for testing. Thin bark from green shrubs is sometimes difficult to spin into fine cord and is easier to use as braid for small cords.

-4. Where it is necessary to use green bark fiber for rope spinning the gum will generally wash out when the bark is teased and soaked in water for a day or so. After removing from the water, the bark strips should allowed to dry before shredding and teasing into fiber.

(11) Braiding. One person may require a length of rope. If there is no help available to spin materials, it is necessary to find reasonably long material. With this material one person can braid and make suitable rope. The usual three-strand braid makes a flat rope, and while quite good, it does not have finish or shape, nor is as "tight" as the four-strand braid. On other occasions,

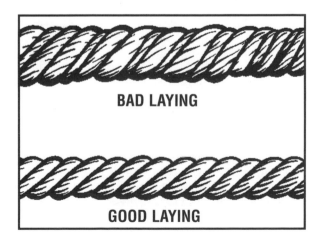

BAD LAYING

GOOD LAYING

Figure 17-63. Rope Laying.

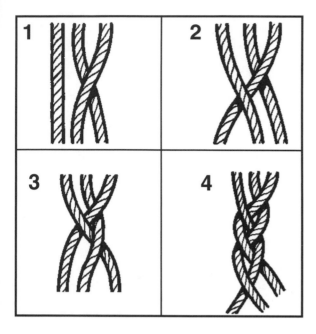

Figure 17-64. Three-Strand Braid.

it may be necessary to braid broad bands for belts or for shoulder straps. There are many fancy braids that can be developed from these, but these three are basic, and essential for practical woodcraft work. A general rule for all braids is to work from the outside into the center.

(a) Three Plait:

-1. The right-hand strand is passed over the strand to the left.

-2. The left-hand strand is passed over the strand to the right.

-3. This is repeated alternately from left to right (figure 17-64).

(b) Flat Four-Strand Braid:

-1. The four strands are placed side by side. The right-hand strand is taken (figure 17-65-1) and placed over the strand to the left.

-2. The outside left-hand strand (figure 17-65-2) is laid under the next strand to itself and over what was the first strand.

-3 The outside right-hand strand is laid over the first strand to its left (figure 17-65-3).

-4. The outside left strand is placed under and over the next two strands, respectively, moving toward the right.

5. Thereafter, the right-hand strand goes over the strand to the left, and the left hand strand under and over to the right (figure 17-65-4).

(c) Broad Braid. Six or more strands are held flat and together.

-1. A strand in the center is passed over the next strand to the left, as in figure 17-66-1.

-2. The second strand to the left of center is passed toward the right and over the first strand so that it points toward the right (figure 17-66-2).

-3. The strand next to the first one is taken over under and over (figure 17-66-3).

-4. The next strands are woven from left and right alternately toward the center (figure 17-66-4 through 6). The finished braid should be tight and close (figure 17-66-7).

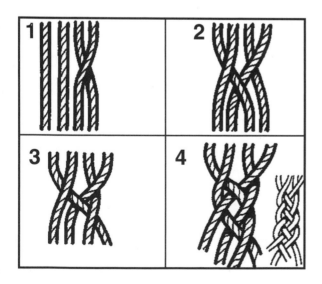

Figure 17-65. Four-Strand Braid.

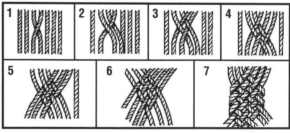

Figure 17-66. Broad Braid.

-5. To finish the broad braid:

-a. One of the center strands is laid back on itself (figure 17-67-1).

-b. Now take the first strand that it enclosed in being folded back, and weave this back upon itself (figure 17-67-2).

-c. Strand from the opposite side is laid back and woven between the strands already braided (figure 17-67-3).

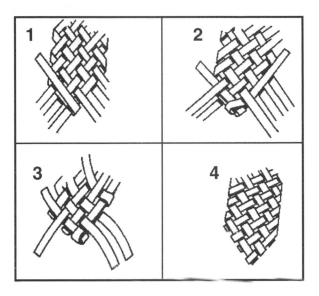

Figure 17-67. Finishing the Broad Braid.

-d. All the strands should be so woven back that no strands show an uneven pattern, and there should be a regular under-over-under of the alternating weaves (figure 17-67-4).

e. If the braid is tight, there may be difficulty in working the loose ends between the plaited strands.

-f. This can be done easily by sharpening a thin piece of wood to a chisel edge to open the strands sufficiently to allow the ends being finished to pass between the woven strands.

-g. It should be rolled under a bottle or other round object and made smooth as final finishing.

17-7. Personal Survival Kits:

a. Even though a survival kit may be available, aircrew members should consider assembling and carrying personal survival kits. Survival experiences have occurred where survivors hit the ground running, and because of shock and fear left their survival kits behind. If survivors have a personal survival kit in a pocket, it may improve their survival chances considerably.

b. A great deal of thought should go into preparing personal survival kits. The potential needs of the survivors must be a consideration, such as the impact of the environmental elements, type of mission to be flown (tactical or nontactical), availability of rescue, and how far to friendly forces (figure 17-68).

c. There are two basic ways to carry a personal survival kit. One way is to pack all items into one or two waterproof containers. The other way is to scatter the items throughout personal clothing. Any type of small container can be used to encase the contents of the personal survival kit. Plastic cigarette cases, soap dishes, and Band-aid boxes are excellent containers.

d. Examples of items that can be packed into a small container are:

(1) Matches.
(2) Safety pins (varied sizes).
(3) Fishhooks.
(4) Knife (small, multi-bladed).
(5) Button compass.
(6) Prophylactic (for water container).
(7) Bouillon cubes.
(8) Salt.
(9) Snare wire.
(10) Water purification tablets.
(11) Signal mirror.
(12) Needles.
(13) Band-aids.
(14) Aluminum foil.
(15) Insect repellent stick.
(16) Chapstick.
(17) Soap (antiseptic).

NOTE: All kits carried aboard the aircraft should be approved by the unit's life-support officer.

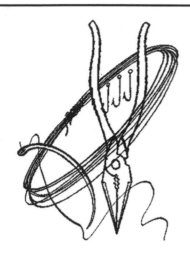

MINIMUM ESSENTIAL ITEMS

High-quality pocket knife with at least two cutting blades.

Pocket compass.

Match safe with matches.

• Plastic or metallic container.

• Waterproof kitchen-type matches (cushion heads against friction), or

• Waterproof matches rolled in paraffin-soaked muslin in an easily opened container, such as small soap box, toothbrush case, etc.

Needles—sailmakers, surgeons, and darning—at least one of each.

Assorted fishhooks in heavy foil, tin, or plastic holders.

Snare wire—small hank.

Needle-nosed pliers with side cutters, high quality.

Bar surgical soap or hand soap containing phisohex.

Small fire starter of pyrophoric metal (some plastic match cases have a strip of the metal anchored on the bottom outside of the case).

Personal medicines.

Water-purification tablets.

"Band-aids."

Insect repellent stick.

Chapstick.

GOOD TO HAVE ITEMS

*Pen-gun and flares.

*Colored cloth or scarf for signaling.

Stick-type skin dye (for camouflage).

Plastic water bottle.

*Flexible saw (wire saw).

*Sharpening stone.

Safety pins (several sizes).

Travel razor.

Small steel mirror.

6-inch flat bastard file.

Aluminum foil.

ADDITIONAL SUGGESTIONS

Toothbrush—small type.

Surgical tape.

Prophylactics (make good waterproof containers or canteens).

*Penlight with batteries.

Fishline.

*Fishline monofilament.

*Clear plastic bags.

Emergency ration can opener (can be taped shut and strung on dogtag chain).

Split shot—for fishing sinkers.

Gill net.

Small, high quality candles.

INDIVIDUAL MEDICAL KIT

Sterile gauze compress bandage.

Antibiotic ointment (Neomycin polymycin bacitracin opthalmic ointment is good).

Tincture of zephrine—skin antiseptic.

Aspirin tablets.

Salt tablets.

Additional medications may be desirable, depending upon nature of the mission and on individual's particular personal needs.

This should be discussed with and procured from your local flight surgeon.

*Especially valuable.

Figure 17-68. Personal Survival Kit Items.

Part Six

SUSTENANCE

Chapter 18

FOOD

18-1. Introduction. Except for the water they drink and the oxygen they breathe, survivors must meet their body's needs through the intake of food. This chapter will explore the relationship of proper nutrition to physical and mental efficiency. It is extremely important that survivors maintain a proper diet at all times. A nutritionally sound body stands a much better chance of surviving. Improper diet over a long period of time may lead to a lack of stamina, slower reactions, less resistance to illness, and reduced mental alertness, all of which can cost survivors their life in a survival situation. Knowledge of the body's nutritional requirements will help survivors select foods to supplement their rations.

18-2. Nutrition. Survivors and evaders expend much more energy in survival situations than they would in the course of their normal everyday jobs and life. Basal metabolism is the amount of energy expended by the body when it is in a resting state. The rate of basal metabolism will vary slightly with regard to the sex, age, weight, height, and race of a person. The basic energy expended, or number of calories consumed by the hour, will change as a person's activity level changes. A person who is simply sitting in a warm shelter, for example, may consume anywhere from 20 to 100 calories an hour, while that same person evading through thick undergrowth with a heavy pack would expand a greater amount of energy. In a survival situation, proper food can make the difference between success and failure.

a. The three major constituents of food are carbohydrates, fats, and proteins. Vitamins and minerals are also important as they keep certain essential body processes in good working order. It is also necessary for survivors to maintain proper water and salt levels in their body, as they aid in preventing certain heat disorders.

(1) Carbohydrates. Carbohydrates are composed of very simple molecules that are easily digested. Carbohydrates lose little of their energy to the process of digestion and are therefore efficient energy suppliers. Because carbohydrates supply easily used energy, many nutritionists recommend that, if possible, survivors should try to use them for up to half of their calorie intake. Examples of carbohydrates are: starches, sugars, and cellulose. These can be found in fruits, vegetables, candy, milk, cereals, legumes, and baked goods. Cellulose cannot be digested by humans, but it does provide needed roughage for the diet.

(2) Fats. Fats are more complex than carbohydrates. The energy contained in fats is released more slowly than the energy in carbohydrates. Because of this, it is a longer lasting form of energy. Fats supply certain fat-soluble vitamins. Sources of these fats and vitamins are butter, cheese, oils, nuts, egg yolks, margarine, and animal fats. If survivors eat fats before sleeping, they will sleep warmer. If fats aren't included in the diet of survivors, they can become run-down and irritable. This can lead to both physical and psychological breakdown.

(3) Protein. The digestive process breaks protein down into various amino acids. These amino acids are formed into new body-tissue protein, such as muscles. Some protein gives the body the exact amino acids required to rebuild itself. These proteins are referred to as "complete." Protein that lacks one or more of these essential amino acids is referred to as "incomplete." Incomplete protein examples are cheese, milk, cereal grains, and legumes. Incomplete protein, when eaten in combination with milk and beans for example, can supply an assortment of amino acids needed by the body. Some complete protein is found in fish, meat, poultry, and blood. No matter which type of protein is consumed, it will contain the most complex molecules of any food type listed.

(a) If possible, the recommended daily allowance of 2 1/2 to 3 ounces of complete protein should be consumed by each survivor each day. If only the incomplete protein is available, two, three, or even four types of foods may need to be eaten in combination so that enough amino acids are combined to form complete protein.

(b) If amino acids are introduced into the body in great numbers and some of them are not used for the rebuilding of muscle, they are changed into fuel or stored in the body as fat. Because protein contains the more complex molecules, over fats or carbohydrates, they supply energy after those forms of energy have been used up. A lack of protein causes malnutrition, skin and hair disorders, and muscle atrophy.

b. Vitamins occur in small quantities in many foods, and are essential for normal growth and health. Their chief function is to regulate the body processes. Vitamins can generally be placed into two groups: fat-soluble and water-soluble. The body stores only small amounts of the water-soluble type. In a long survival episode where a routinely balanced diet is not available, survivors must overcome food aversions and eat as much of a variety of vitamin-rich foods as possible. Often one or more of the four basic food groups (meat, fish, poultry; vegetables and fruits; grain and cereal; milk and milk products) are not available in the form of familiar foods, and vitamin deficiencies such as beriberi or scurvy result. If the survivor can overcome aversions to local foods high in vitamins, these diseases, as well as signs and symptoms such as depression and irritability, can be warded off.

c. Adequate minerals can also be provided by a balanced diet. Minerals build and (or) repair the skeletal system and regulate normal body functions. Minerals needed by the body include iodine, calcium, iron, and salt, to name but a few. A lack of minerals can cause problems with muscle coordination, nerves, water retention, and the ability to form or maintain healthy red blood cells.

d. For survivors to maintain their efficiency, the following number of calories per day is recommended. These figures will change because of individual differences in basal metabolism, weight, etc. During warm weather survivors should consume anywhere from 3,000 to 5,000 calories per day. In cold weather the calorie intake should rise from 4,000 to 6,000 calories per day. A familiarity with the calorie and fat amounts in foods is important for survivors to meet their nutritional needs. For example, it would take quite a few mussels and dandelion greens to meet those requirements. Survivors should attempt to be familiar enough with foods that they can select or find foods that provide a high calorie intake (figure 18-1).

(1) Survivors should also be familiar with the number of calories supplied by the food in issued rations. In most situations, rations will have to be supplemented with other foods procured by survivors. If possible, survivors should limit their activities to save energy. Rationing food is a good idea since survivors never know when their ordeal will end. They should eat when they can, keeping in mind that they should maintain at least a minimum calorie intake to satisfy their basic activity needs.

(2) Caloric and fat values of selected foods are shown in the chart, and unless otherwise specified, the foods listed are raw. Depending on how survivors cook the food, the usable food value can be increased or decreased.

18-3. Food. Survivors should be able to find something to eat wherever they are. One of the best places to find food is along the seacoast, between the high and low watermark. Other likely spots are the areas between the beach and a coral reef; marshes, mud flats, or mangrove swamps where a river flows into the ocean or into a larger river; riverbanks, inland waterholes, and shores of ponds and lakes; margins of forests, natural meadows,

FOOD	CALORIES	FAT
WHOLE LARGE DUCK EGG	177	12.0
SMALL OR LARGE MOUTH BASS—3 TO 4 OZ.	109	3.6
CLAMS—4 TO 5 LARGE	88	0.2
FRESHWATER CRAYFISH —3 TO 4 OZ.	75	0.6
EEL—3 TO 5 OZ.	240	20.0
OCTOPUS—3 TO 4 OZ.	76	0.9
ATLANTIC SALMON—4 OZ.	220	14.0
RAINBOW TROUT—4 OZ.	200	11.8
BANANA—ONE SMALL	87	0.3
BREADFRUIT—3 TO 4 OZ.	105	0.5
GUAVA—ONE MEDIUM	64	0.7
MANGO—ONE SMALL	68	0.5
WILD DUCK—4 OZ.	230	16.0
BAKED OPOSSUM—4 OZ.	235	10.6
WILD RABBIT—4 OZ.	124	4.0
VENISON—4 OZ.	128	3.1
DANDELION GREENS —ONE COOKED CUP	70	1.4
POTATO—MEDIUM	78	0.2
PRICKLY PEAR—4 OZ.	43	0.2

Figure 18-1. Food and Calorie Diagram.

protected mountain slopes, and abandoned cultivated fields.

a. Rations placed in survival kits have been developed especially to provide some of the proper sustenance needed during survival emergencies. When eaten as directed on the package, they will keep the survivor relatively efficient. If enough other food can be found, rations should be conserved for emergency use.

b. Consideration must be given to available food and water and how long the survival episode may last. Environmental conditions must also be considered. If a survivor is in a cold environment, more of the proper food will be required to provide necessary body heat. Rescue may vary from a few hours to several months, depending on the environment, operational commitments, and availability of rescue resources in that area. Available food must be rationed based on the estimated time that will elapse before being able to supplement issued rations with natural foods. If it is decided that some of the survivors should go for help, each traveler should be given twice as much food as the ones remaining behind. In this way, the survivors resting at the encampment and those walking out will stay in about the same physical condition for about the same length of time.

c. If available water is less than a quart a day, avoid dry, starchy, and highly seasoned foods and meat. Keep in mind that eating increases thirst. For water conservation, the best foods to eat are those with high carbohydrate content, such as hard candy and fruit. All work requires additional food and water. When work is being performed, the survivor must increase food and water consumption to maintain physical efficiency. If food is available, it is all right to nibble throughout the day. It is preferable, though, to have at least two meals a day, with one being hot. Cooking usually makes food safer, more digestible, and palatable. The time spent cooking will provide a good rest period. On the other hand, some food, such as sapodilla, star apple, and soursop, are not palatable unless eaten raw.

d. Native foods may be more appetizing if they are eaten alone. Rations and native foods usually do not mix well. In many countries, vegetables are often contaminated by human feces, which the natives use as fertilizer. Dysentery is transmitted in this way. If possible, survivors should try to select and prepare their own meals. If necessary to avoid offending the natives, indicate that religious beliefs or taboos require self-preparation of food.

e. Learn to overcome food prejudices. Foods that may not look good to the survivor are often a part of the natives' regular diet. Wild foods are high in mineral and vitamin content. With a few exceptions, all animals are edible when freshly killed. Avoid strange looking fish and fish with flesh that remains indented when depressed, as it is probably becoming spoiled and should not be eaten. With knowledge and the ability to overcome food prejudices, a survivor can eat and sustain life in a strange or hostile environment.

18-4. Animal Food. Animal food gives the most food value per pound. Anything that creeps, crawls, swims, or flies is a possible source of food. People eat grasshoppers, hairless caterpillars, wood-boring beetle larvae and pupae, ant eggs, spider bodies, and termites. Such insects are high in fat and should be cooked until dried. Everyone has probably eaten insects contained in flour, cornmeal, rice, beans, fruits, and greens in their daily foods.

a. Man As a Predator. To become successful at hunting, the hunter must go through a behavioral change and reorganize personal priorities. This means the one and only goal for the present is to kill an animal to eat. To kill this animal, the hunter must mentally become a predator. The hunter must be prepared to undergo stress in order to hunt down and kill an animal. Because of the type of weapons survivors are likely to have, it will be necessary to get very close to the animal to immobilize or kill it. This is going to require all the stealth and cunning

survivors can muster. In addition to stealth and cunning, knowledge of the animal being hunted is very important. If in an unfamiliar area, survivors may learn much about the animal life of the area by studying signs such as trails, droppings, and bedding areas.

b. Animal Signs. The survivor should establish the general characteristics of the animals. The size of the tracks will give a good idea of the size of the animal. The depth of the tracks will indicate the weight of the animal. The animal dung can tell the hunter much. For example, if it is still warm or slimy, it was made very recently; if there is a large amount scattered around the area, it could well be a feeding or bedding area. The droppings may indicate what the animal feeds on. Carnivores often have hair and bone in the dung; herbivores have coarse portions of the plants they have eaten. Many animals mark their territory by urinating or scraping areas on the ground or trees. These signs could indicate good trap or ambush sites. Following the signs (tracks, droppings, etc.) may reveal the feeding, watering, and resting areas. Well-worn trails will often lead to the animal's watering place. Having made a careful study of all the signs of the animal, the hunter is in a much better position to procure it, whether electing to stalk, trap, or snare it, or lie in wait to shoot it.

c. Hunting. If survivors elect to hunt, there are some basic techniques that will be helpful and improve chances of success. Wild animals rely entirely on their senses for their preservation. These senses are smell, vision, and hearing. Humans have lost the keenness of some of their senses, such as smelling, hearing, etc. To overcome this disadvantage, they have the ability to reason. As an example, some animals have a fantastic sense of smell, but this can be overcome by approaching the quarry from a downwind direction. The best times to hunt are at dawn and dusk as animals are either leaving or returning to their bedding areas. Both diurnal and nocturnal animals are active at this time. There are five basic methods of hunting:

(1) Still or Stand. This is the best method for inexperienced hunters, as it involves less skill. The main principle of this method is to wait in ambush along a well-used game trail until the quarry approaches within killing range. Morning and evening are usually the best times to "still" hunt. Care should be taken not to disturb the area; always wait downwind. Patience and self-control are necessary to remain motionless for long periods of time.

(2) Stalking. "Stalking" refers to the stealthy approach toward game. This method is normally used when an animal has been sighted, and the hunter then proceeds to close the distance using all available cover. Stalking must be done slowly so that minimum noise is made, as quick movement is easily detected by the animal. Always approach from the downwind side and move when the animal's head is down eating, drinking, or looking in another direction. The same techniques are

used in blind stalking as in the regular stalk, the main difference being that the hunter is stalking a position where the animal is expected to be while the animal is not yet in sight.

(3) Tracking. Tracking is very difficult unless conditions are ideal. This method involves reading all of the signs left behind by the animal, interpreting what the animal is doing, and how it can best be killed. The most common signs are trails, beds, urine, droppings, blood, tracks, and feeding signs.

(4) Driving. Some wild animals can be scared or driven in a direction where other hunters or traps have been set. This method is normally used where the game can be funneled; a valley or canyon is a good place to make a drive. More than one person is usually necessary to make a drive.

(5) Calling. Small predators may be called in by imitating an injured animal. Ducks and geese can be attracted by imitating their feeding calls. These noises can be made by sucking on the hand, blowing on a blade of grass or paper, sucking the lip, or using specially designed devices. Survivors should not call animals unless they know what they are doing, as strange noises may "spook" the animal.

d. Killing Implements. It is difficult to kill animals of any size without using some type of tool or weapon. As our technology has increased in complexity, so have our killing tools. If a firearm is available, a basic knowledge of shooting and hunting techniques is necessary.

(1) Learning to become proficient with primitive weapons is important. Many primitive tribes of the world are still effectively using spears, clubs, bows and arrows, slingshots, etc., to provide food for their families. One of the limiting factors in the use of firearms is the amount of ammunition on hand. Therefore, a survivor cannot afford to waste ammunition on moving game or game that is beyond the effective range of the firearm being used. Wait for a pause in the animal's motions. The shot must be placed in a vital area with any firearm. Aim for the brain, spine, lungs, or heart (figure 18-2). A hit in any of these areas is usually fatal.

(2) A full-jacketed bullet often won't immediately down a larger animal hit in a vital area such as the lungs or heart. The alternative to losing the animal is tracking it to where it falls. Often it's better to wait awhile before pursuing the animal. If not pursued, it may lay down and stiffen or perhaps bleed to death. Follow the blood trail to where the game has gone down and kill it if it is still alive. Even

Figure 18-2. Shooting Game.

though ammunition might be limited, small game may be more productive than large game. Although they present smaller targets and have less meat, they are less wary, more numerous, and travel less distance to escape if wounded. A large amount of edible meat on small game can be destroyed from a bullet wound. On rodents, most of the meat is on the hindquarters and frontquarters; on birds, it is on the breast and legs. The survivor should try to hit a vital spot that spoils the least meat.

(3) Night hunting is usually best, since most animals move at night. A flashlight or torch may be used to shine in the animal's eyes. It will be partly blinded by the light and a survivor can get much closer than in the daytime. If no gun is available, the animal can be killed with a club or a sharpened stick used as a spear.

(4) Remember that large animals, when wounded, cornered, or with their young, can be dangerous. Be sure the animal is dead, not just wounded, unconscious, or "playing possum." Animals usually die with their eyes open and glazed-over. Poke all "dead" animals in the eye with a long sharp stick before approaching them.

(5) Small freshwater turtles can often be found sunning themselves along rivers and lakeshores. If they dash into shallow water, they can still be procured with nets, clubs, etc., but watch out for mouth and claws. Frogs and snakes also sun and feed along streams. Use both hands to catch a frog—one to attract it and keep it busy while grabbing it with the other. Bright cloth on a fishhook also works. All snakes are good eating and can be killed with a long stick. Both marine and land lizards are edible. A noose, small fishhook baited with a bright cloth lure, slingshot, or club can be used. A slingshot can be made with a forked stick and the elastic from the parachute pack or surgical tubing found in some survival kits (figure 18-3). With practice, the slingshot can be very effective for killing small animals.

e. Snaring and Trapping. Snaring and trapping animals are ways in which survivors can procure animal food to supplement issued rations. Because small animals are usually more abundant than large animals, they will

probably be the survivor's main source of food. Snares should be set out on a 15:1 ratio—15 snares should be set out for every one animal expected to be caught.

(1) Using traps and snares is more advantageous then going out on foot and physically hunting the animal. The most important advantage being that traps work 24 hours a day with no assistance from the hunter. A large area can be effectively trapped, with the possibility of catching many animals within the same period of time. Survivors (generally) use much less energy maintaining a trapline than is used by hunting. This means less food is required because less energy is expended.

(2) The traps or snares should be set in areas where the game is known to live or travel. Look for signs such as tracks, droppings, feeding signs, or actual sightings of the animal. If snares are used, they should be set up to catch the animal around the neck. Therefore, the loop must allow the head to pass through but not the body. Loops will vary in size from one animal to another. When placing snares, try to find a narrow area of the game trail where the animal has no choice but to enter the loop. If a narrow area cannot be found, brush or other obstacles can be arranged to funnel the animal into the snare (figure 18-4). Do not overdo the tunneling; use as little as possible. Avoid disturbing the natural surroundings if possible. Do not walk on game trails but approach 90 degrees to the trail, set the snare, and back away. Snares may also be set over holes or burrows. All snares and traps should be set during the midday because most animals are nocturnal in nature. Check snares and traps twice daily. If possible, check after sunup and before sunset. The checks should be made from a distance so any animals moving at the time of checking will not be disturbed or frightened away.

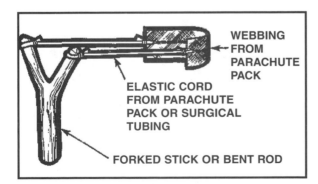

Figure 18-3. Slingshot.

WEBBING FROM PARACHUTE PACK

ELASTIC CORD FROM PARACHUTE PACK OR SURGICAL TUBING

FORKED STICK OR BENT ROD

FUNNELING

Figure 18-4. Funneling.

Figure 18-5. Mangle.

(3) There are three ways to immobilize or trap animals.

(a) Strangle. This is done by simply using a free-sliding noose, which, when tightened around the neck, will restrict circulation of air and blood. The material should be strong enough to hold the animal, for example, suspension line, string, wire, cable, or rawhide.

(b) Mangle. Mangle traps use a weight that is suspended over the animal's trail or over bait. When the animal trips the trigger, the weight (log) will descend and mangle the animal (figure 18-5).

(c) Hold. Any means of impeding the animal and detaining its progress would be considered a hold-type trap.

(4) The apache foot snare is an example of a hold-type trap. It is used for large browsers and grazers like deer (figure 18-6). It should be located along game trails where an obstruction, such as a log, blocks the trail. When animals jump over this obstruction, a very shallow depression is formed where their hooves land. The apache foot snare should be placed at this depression. The box trap for birds is another example of a hold-type trap (figure 18-7).

Figure 18-6. Apache Foot Snare.

Figure 18-7. Box Trap.

(5) The simple loop is the quickest snare to construct. All snares and traps should be simple in construction with as few moving parts as possible. This loop can be constructed from any type of bare wire, suspension line, inner core, vines, long strips of green bark, clothing strips or belt, and any other material that will not break under the strain of holding the animal. If wire is being used for snares, a figure eight or locking loop should be used (figure 18-8). Once tightened around the animal, the wire is locked into place by the figure eight, which prevents the loop from opening again. A simple-loop snare is generally placed in the opening of a den, with the end of the snare anchored to a stake or similar object (figure 18-8). The simple-loop snare can also be used when making a squirrel pole (figure 18-9), or with some types of trigger devices.

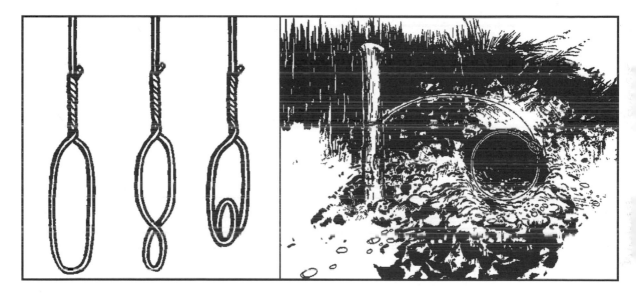

Figure 18-8. Locking Loop and Setting Noose.

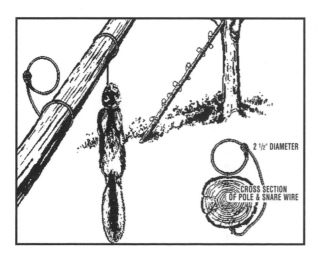

2 ½" DIAMETER

CROSS SECTION OF POLE & SNARE WIRE

Figure 18-9. Squirrel Pole.

f. Triggers. Triggers may be used with traps. The purpose of the trigger is to set the device in motion, which will eventually strangle, mangle, or hold the animal. There are many triggers. Some of the more common ones are:

(1) Two-pin toggle with a counterweight for small to medium animals, which are lifted out of the reach of predators (figure 18-10).

(2) Figure "H" with wire snare for small mammals and rodents (figure 18-11).

(3) Canadian ace for predators such as the bobcat, coyote, etc., (figure 18-12).

(4) Three-pin toggle with deadfall for medium to large animals (figure 18-13). Medium and large animals can be captured using deadfalls, but this type of trap is recommended only when big game exists in large quantities to justify the great expense of time and effort spent in constructing the trap.

Figure 18-10. Two-Pin Toggle.

Figure 18-12. Canadian Ace.

(5) The twitch-up snare, which incorporates the simple loop, can be used to catch small animals (figure 18-14). When the animal is caught, the sapling jerks it up into the air and keeps the carcass out of the reach of predators. This type of snare will not work well in cold climates, since the bent sapling will freeze in position and not spring up when released.

(6) A long, forked stick can be used as a twist stick to procure ground squirrels, rabbits, etc. A den that has signs of activity must be located. Using the long, forked stick, the survivor probes the hole with the forked end until something soft is felt, then twisting the stick will entangle the animal's hide in the stick, and the animal can be extracted (figure 18-18).

g. Birds. Birds can be caught with a gill net. The net should be set up at night vertically to the ground in some natural flyway, such as an opening in dense foliage. A small gill net on a wooden frame with a disjointed stick for a trigger can also be used. A gill net can be made by using inner core from parachute suspension line (figure 18-21).

(1) Birds can be caught on baited fishhooks (figure 18-15) or simple slipping-loop snares. Birds' nests can be a source of food. All bird eggs are edible when fresh. Large wading birds such as cranes and herons often nest in mangrove swamps or in high trees near water.

(2) During molting season, birds cannot fly because of the loss of their "flight" feathers—they can be procured by clubbing or netting.

Figure 18-11. Figure H.

Figure 18-13. Three-Pin Toggle.

Figure 18-14. Twitch Up.

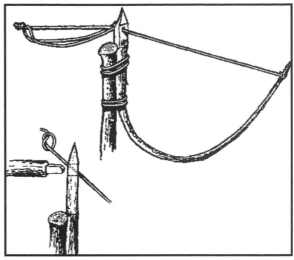

Figure 18-16. Ojibwa Bird Snare.

(3) Birds can be also caught in an Ojibwa snare. This snare is made by cutting a 1- or 2-inch thick sapling at a height of 4 ½ to 5 feet above the ground (figure 18-16). A springy branch is then whittled flat at the butt end and a rectangular hole is cut through the flattened end. One end of a ½-inch thick stick, 15 inches long, is then whittled to fit slightly loose in the hole, and the top corner of the whittled end is rounded off so the stick will easily drop away from the hole. The branch is then tied by its butt end to the top of the sapling. A length of inner core from suspension line is tied to the bottom end of the branch and the branch is bent into a bow with the line passing through the hole in the butt end. A knot is tied in the line and the 15-inch stick is then placed in the hole to lock the line in place (just behind the knot). An 8-inch

loop is made at the end of the line and laid out on the 15-inch stick (spread out as well as possible). A piece of bait is placed on top of the sapling, and when a bird comes to settle on the 15-inch stick, the stick drops from the hole causing the loop to tighten around the bird's legs.

(4) When many birds frequent a particular type of bush, some simple-loop snares may be set up throughout the bush. Make the snares as large as necessary for the particular type of birds that come to perch, feed, or roost there (figure 18-17).

(5) In wild, wooded areas, many larger species of birds such as spruce grouse and ptarmigan may be approached. The spruce grouse, which has merited the name of "fool's hen," can be approached and killed with

Figure 18-15. Baited Fishhook.

Figure 18-17. Ptarmigan or Small Game Snare.

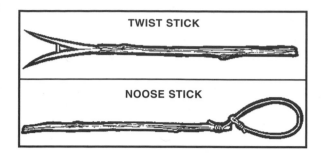

Figure 18-18. Twist Stick and Noose Stick.

a stick with little trouble. It often sits on the lower branches of trees and can be easily caught with a long stick with a loop at the end (figure 18-18).

(6) Ground-feeding birds (Quail, Hungarian Partridge, Chukar) can be trapped in a trench dug into the ground. The trench should be just wide enough for the bird to walk into, so survivors must first observe the type of ground-feeding birds in the area. The trench should be 2 to 3 feet long and about 10 to 12 inches deep at the deep end. The other end of the trench should be ramped down from the surface level. Bait is scattered along the surface into the pit and after having pecked the last piece of bait, the bird will not be able to get out of the pit because it can't fly or climb out, its feathers keep it from backing out, and it can't turn around to walk out.

(7) Perching birds may be captured using birdlime. Birdlime is a term applied to any sticky or gluey substance that is rubbed on a branch to prevent the flight of a bird that has landed on it or has flapped a wing against it. Birdlime is usually made from the sap of plants in the Euphorbia family. The common names of some of these plants are spotted spurge, cypress spurge, snow-on-the-mountain, and poinsettias. The Euphorbias have a wide range in North and Central America. The milky sap is poisonous and may cause blisters on the skin and should be handled with care. Birdlime is most effective in the desert and jungle, and it will not work in cold weather. Dust will make birdlime ineffective, so it should be used

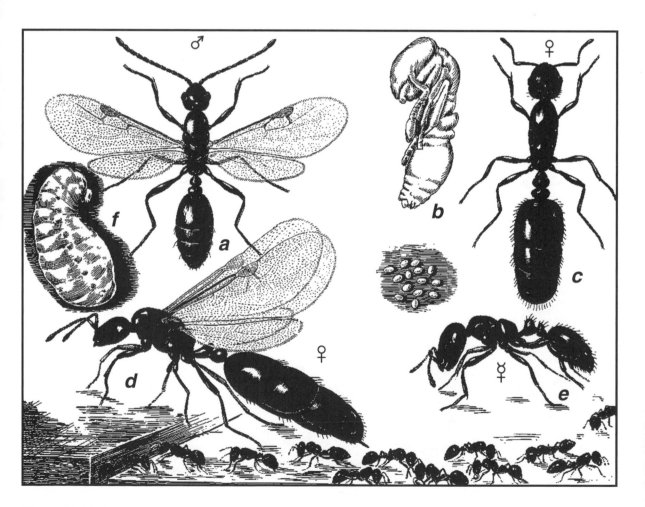

Figure 18-19. Ants.

in spots where dust is not prevalent. The sap of the bread-fruit tree makes excellent birdlime as it swells and becomes glutinous on contact with air.

h. Insects. If there is ever a time when food aversions must be overcome, it is when survivors turn to insects as a food source.

(1) Primitive peoples eat insects and consider them great delicacies. When food is limited and insects are available, they can become a valuable food source. In some places, locusts and grasshoppers, cicadas, and crickets are eaten regularly; occasionally termites, ants, and a few species of stonefly larvae are consumed. Big beetles such as the Goliath Beetle of Africa, the Giant Water Beetles, and the big Long Horns are relished the world over. Clusters, like those of the Snipefly Atherix (that overhang the water), and the windrows of Brinefly puparia are eaten. Aquatic water bugs of Mexico are grown especially for food. All stages of growth can be eaten, including the eggs, but the large insects must be cooked to kill internal parasites.

(2) Termites and white ants are also an important food source. Strangely enough, these are closely related to cockroaches. The reason they are eaten so extensively in Africa is the fact that they occur in enormous numbers and are easily collected both from their nests and during flight. They are sometimes attracted to light in unbelievable numbers, and the natives become greatly excited when the large species appear.

(3) Many American Indian tribes made a habit of eating the large carpenter ants that are sometimes pests in houses. These were eaten both raw and cooked. Even today the practice of eating them has not entirely disappeared, although they do not form an essential part of the diet of any of the inhabitants of this country (figure 18-19).

(4) It is not at all unnatural that the American Indians should have relished the honey ants in all parts of the continent where they occur. These ants are peculiar in that some of the workers become veritable storehouses for honey, their abdomens becoming more or less spherical and so greatly enlarged that they are scarcely able to move. They cluster on the ceilings and walls of their nests and disgorge part of their stored food to other inhabitants. The American Indians discovered the sweetness stored in these insects and made full use of it. At first they ate the ants alive, later gathering them in quantity and crushing them so they formed an enticing dish— one that was considered a delicacy and served to guests of distinction as a special favor. The next step in the use of the honey ant was the extraction of the pure honey by crushing the insects and straining the juices. After the honey was extracted, it was allowed to ferment, forming what is said to be a highly flavored wine.

(5) Natives of the American tropics, with a much larger ant fauna from which to choose, select the queens of the famous leafcutting or so-called umbrella ants upon which to feed, eating only the abdomens, either raw or cooked.

(6) It is natural that caterpillars—the larvae of moths and butterflies—should form a very substantial part of the food of primitive peoples because they are often of large size or occur in great abundance. In Africa, many tribes consider caterpillars choice morsels of food, and much time is spent in collecting them. Some of the native tribes recognize 20 or more different kinds of caterpillars that are edible, and they are sufficiently well acquainted with the life history of the insects to know the plants upon which they feed and the time of year when they have reached the proper stage of development for collecting. Caterpillars with hairs should be avoided. If eaten, the hairs may become lodged in the throat, causing irritation or infection. Today it is known that insects have nutritional or medicinal value. The praying mantis, for example, contains 58 percent protein, 12 percent fat, 3 percent ash, vitamin B complex, and vitamin A. The insect's outer skeleton is an interesting compound of sugar and amino acids.

(7) Bee larvae were eaten by the ancient Chinese. Today, some Chinese eat locusts, dragonflies, and bumblebees. Cockroaches and locusts are a favorite dish in Szechwan. In Kwangtun, grasshoppers, golden June beetles, crickets, wasp larvae, and silkworm larvae are used for food.

(8) Stinging insects should have their stinging apparatus removed before they are eaten.

(9) As can be seen, insects have been used as a food source for thousands of years and will undoubtedly continue to be used. If survivors cannot overcome their aversion to insects as a food source, they will miss out on a valuable and plentiful supply of food.

i. Fishing. Fishing is one way to get food throughout the year wherever water is found. There are many ways to catch fish, including hook and line, gill nets, poisons, traps, and spearing.

(1) If an emergency fishing kit is available, there will be a hook and line in it, but if a kit is not available a hook and line will have to be procured elsewhere or improvised. Hooks can be made from wire or carved from bone or wood. The line can be made by unraveling parachute suspension line or by twisting threads from clothing or plant fibers. A piece of wire between the fishing line and the hook will help prevent the fish from biting through the line. Insects, smaller fish, shellfish, worms, or meat can be used as bait. Bait can be selected by observing what the fish are eating. Artificial lures can be made from pieces of brightly colored cloth, feathers, or bits of bright metal or foil tied to a hook. If the fish will not take the bait, try to snag or hook them in any part of the body as they swim by. In freshwater, the deepest water is usually the best place to fish. In shallow streams, the best places are pools below falls, at the foot of rapids, or behind rocks. The best time to fish is usually

1 OVERHANGING BRUSH

2 UNDERCUT

3 POOL FROM BACKWASH

4 FEEDER STREAM

5 BEHIND ROCKS

6 FALLEN TREE

Figure 18-20. Fishing Places.

early morning or late evening (figure 18-20). Sometimes fishing is best at night, especially in moonlight or if a light is available to attract the fish. The survivor should be patient and fish at different depths in all kinds of water. Fishing at different times of the day and changing bait often is rewarding.

(2) The most effective fishing method is a net because it will catch fish without having to be attended (figures 18-21 and 18-22). If a gill net is used, stones can be used for anchors and wood for floats. The net should be set at a slight angle to the current to clear itself of any floating refuse that comes down the stream. The net should be checked at least twice daily (figure 18-23). A net with poles attached to each end works effectively if moved up or down a stream as rapidly as possible while moving stones and threshing the bottom or edges of the streambanks. The net should be checked every few moments so the fish cannot escape.

(3) Shrimp (prawns) live on or near the sea bottom and may be scraped up. They may be lured to the surface with light at night. A hand net made from parachute cloth or other material is excellent for catching shrimp. Lobsters are creeping crustaceans found on or near the sea bottom. A lobster trap, jig, baited hook, or dip net can be used to catch lobster. Crabs will creep,

climb, and burrow and are easily caught in shallow water with a dip net or in traps baited with fish heads or animal viscera.

(4) Fishtraps (figure 18-24) are very useful for catching both freshwater and saltwater fish, especially those that move in schools. In lakes or large streams, fish tend to approach the banks and shallows in the morning and evening. Sea fish, traveling in large schools, regularly approach the shore with the incoming tide, often moving parallel to the shore guided by obstruction in the water.

(a) A fishtrap is basically an enclosure with a blind opening where two fence-like walls extend out, like a funnel, from the entrance. The time and effort put into building a fishtrap should depend on the need for food and the length of time survivors plan to stay in one spot.

(b) The trap location should be selected at high tide and the trap built at low tide. One to 2 hours of work should do the job. Consider the location, and try to adapt natural features to reduce the labors. Natural rock pools should be used on rock shores. Natural pools on the surface of reefs should be used on coral islands by blocking the opening as the tide recedes. Sandbars, and the ditches they enclose, can be used on sandy shores. The best fishing off sandy beaches is the lee side of offshore sandbars. By watching the swimming habits of fish, a simple dam can be built that extends out into the water forming an angle with the shore. This will trap fish as they swim in their natural path. When planning a more complex brush dam, select protected bays or inlets using the narrowest area and extending one arm almost to the shore.

(c) In small, shallow streams, the fishtraps can be made with stakes or brush set into the stream bottom or weighted down with stones so the stream is blocked except for a small narrow opening into a stone or brush pen or shallow water. Wade into the stream herding the fish into the trap, and catch or club them when they get into shallow water. Mud-bottom streams can be trampled until cloudy and then netted. The fish are blinded and cannot avoid the nets. Freshwater crawfish and snails can be found under rocks, logs, overhanging bushes, or on mud bottoms.

(5) Fish may be confined in properly built enclosures and kept for days. In many cases, it may be advantageous to keep them alive until needed and thus ensure there is a fresh supply without danger of spoilage. Mangrove swamps are often good fishing grounds. At low tide, clusters of oysters and muscles are exposed on the mangrove "knees," or lower branches. Clams can be found in the mud at the base of trees. Crabs are very active among branches or roots and in the mud. Fish can be caught at high tide. Snails are found on mud and clinging to roots. Shellfish that are not covered at high tide or those from a colony containing diseased members should not be eaten. Some indications of diseased shellfish are shells gaping open at low tide, foul odor, and (or) milky juice.

(6) Throughout the warm regions of the world, there are various plants that the natives use for poisoning fish. The active poison in these plants is harmful only to cold-blooded animals. Survivors can eat fish killed by this poison without ill effects.

(a) In Southeast Asia, the derris plant is widely

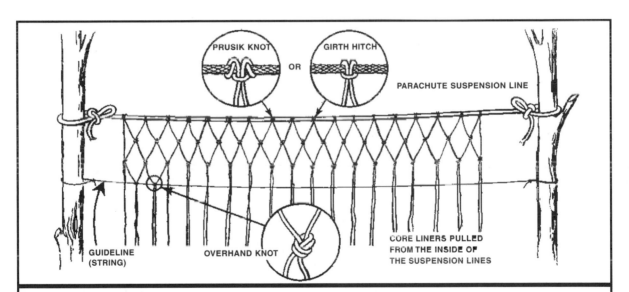

1. Suspend a suspension-line casing (from which the core liners have been pulled) between two uprights, approximately at eye level.

2. Hang core liners (an even number) from the line suspended as in 1, above. These lines should be attached with a Prusik knot or girth hitch and spaced in accordance with the mesh you desire. One-inch spacing will result in a 1-inch mesh, etc. The number of lines used will be in accordance with the width of the net desired. If more than one man is going to work on the net, the length of the net should be stretched between the uprights, thus providing room for more than one man to work. If only one man is to set up the net, the depth of the net should be stretched between the uprights and step 8, below, followed.

3. Start at left or right. Skip the first line and tie the second and third lines together with on overhand knot. Space according to mesh desired. Then tie fourth and fifth, sixth and seventh, etc. One line will remain at the end.

4. On the second row, tie the first and second, third and fourth, fifth and sixth, etc., to the end.

5. Third row, skip the first line and repeat step 3, above.

6. Repeat step 4, and so on.

7. You may want to use a guideline that can be moved down for each row of knots to ensure equal mesh. The guideline should run across the net on the side opposite the one you are working from so it will be out of your way.

8. When you have stretched the depth between the uprights and get close to ground level, move the net up by rolling it on a stick and continue until the net is the desired length.

9. String suspension-line casing along the sides when net is completed to strengthen it and make the net easier to set.

Figure 18-21. Making a Gill Net.

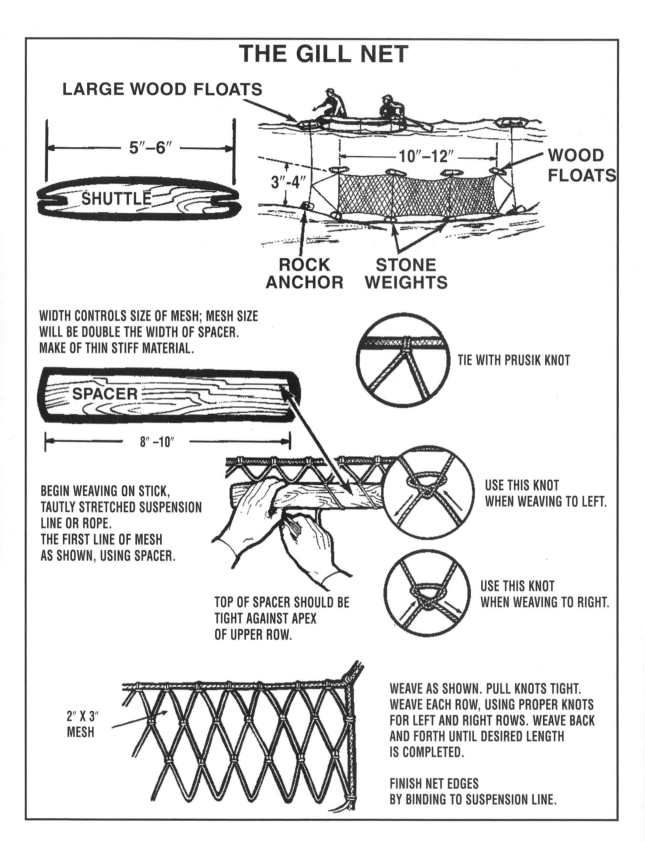

THE GILL NET

LARGE WOOD FLOATS

5"–6"

SHUTTLE

10"–12"

3"-4"

WOOD FLOATS

ROCK ANCHOR

STONE WEIGHTS

WIDTH CONTROLS SIZE OF MESH; MESH SIZE WILL BE DOUBLE THE WIDTH OF SPACER. MAKE OF THIN STIFF MATERIAL.

TIE WITH PRUSIK KNOT

SPACER

8"–10"

BEGIN WEAVING ON STICK, TAUTLY STRETCHED SUSPENSION LINE OR ROPE. THE FIRST LINE OF MESH AS SHOWN, USING SPACER.

USE THIS KNOT WHEN WEAVING TO LEFT.

TOP OF SPACER SHOULD BE TIGHT AGAINST APEX OF UPPER ROW.

USE THIS KNOT WHEN WEAVING TO RIGHT.

2" X 3" MESH

WEAVE AS SHOWN. PULL KNOTS TIGHT. WEAVE EACH ROW, USING PROPER KNOTS FOR LEFT AND RIGHT ROWS. WEAVE BACK AND FORTH UNTIL DESIRED LENGTH IS COMPLETED.

FINISH NET EDGES BY BINDING TO SUSPENSION LINE.

Figure 18-22. Making a Gill Net with Shuttle and Spacer.

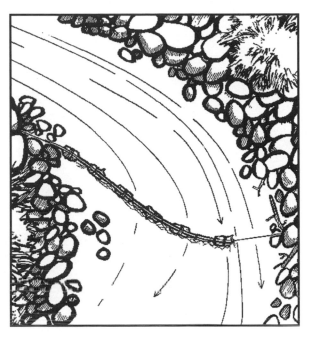

Figure 18-23. Setting the Gill Net.

used as a source of fish poison. The derris plant, a large woody vine, is also used to produce a commercial fish poison called rotenone. Commercial rotenone can be used in the same manner as crushed derris roots; it causes respiratory failure in fish, but has no ill effects on humans. However, rotenone has no effect if dusted over the surface of a pond. It should be mixed to a malted-milk consistency with a little water, and then distributed in the water. If the concentration is strong, it takes effect within 2 minutes in warm water, or it may take an hour in colder water. Fish sick enough to turn over on their back will eventually die. An ounce of 12 percent rotenone can kill every fish for half a mile down a slow-moving stream that is about 25 feet wide. A few facts to remember about the use of rotenone are:

-1. It is very swift acting in warm water at 70°F and above.

-2. It works slower in cold water and is not practical in water below 55°F.

-3. It is best applied in small ponds, streams, or tidal pools.

-4. Excess usage will be wasted. However, too little will not be effective.

(b) A small container of 12 percent rotenone (½ oz.) is a valuable addition to any emergency

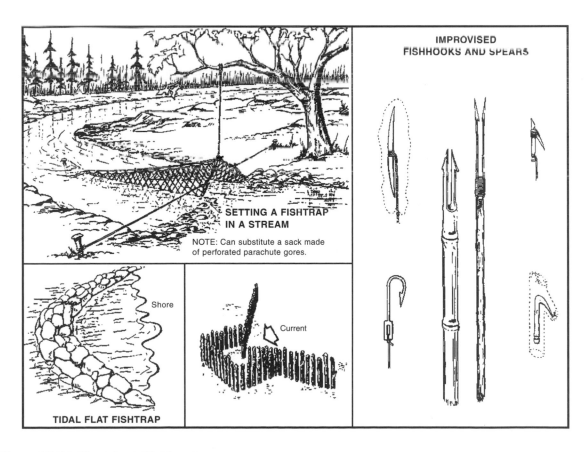

Figure 18-24. Maze-type Fishtraps.

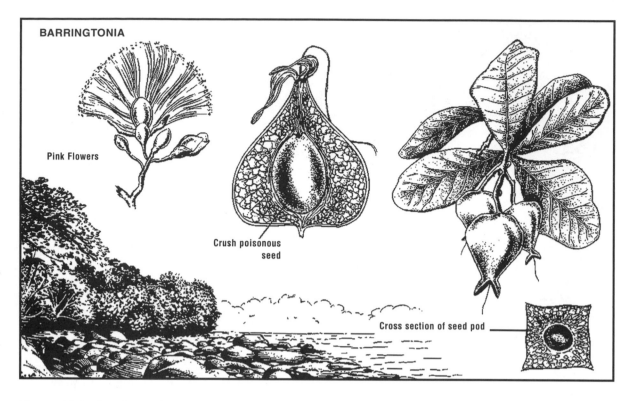

Figure 18-25. Barringtonia Plant for Poisoning Fish.

kit. Do not expose it unnecessarily to air or light; it retains its toxicity best if kept in a dark-colored vial. Lime thrown in a small pond or tidal pool will kill fish in the pool. Lime can be obtained by burning coral and seashells.

(c) The most common method of using fish-poison plants is to crush the plant parts (most often the roots) and mix them with water. Drop large quantities of the crushed plant into pools or the headwaters of small streams containing fish. Within a short time, the fish will rise in a helpless state to the surface. After putting in the poison, follow slowly downstream and pick up the fish as they come to the surface, sink to the bottom, or swim crazily to the bank. A stick dam or obstruction will aid in collecting fish as they float downstream. The husk of "green" black walnuts can be crushed and sprinkled into small sluggish streams and pools to act as a fish-stupefying agent. In the southwest Pacific, the seeds and bark from the barringtonia tree (figure 18-25) are commonly used as a source of fish poison. The barringtonia tree usually grows along the seashore.

(7) Tickling can be effective in small streams with undercut banks or in shallow ponds left by receding floodwaters. Place hands in the water and reach under the bank slowly, keeping hands close to the bottom if possible. Move fingers slightly until they make contact with a fish. Then work hands gently along its belly

until reaching its gills. Grasp the fish firmly just behind the gills and scoop it onto land. In the tropics, this type of fishing can be dangerous due to hazardous marine life in the water, such as piranhas, eels, and snakes.

18-5. Plant Food. The thought of having a diet consisting only of plant food is often distressing to stranded aircrew members. This is not the case if the survival episode is entered into with the confidence and intelligence based on knowledge or experience. If the survivors know what to look for, can identify it, and know how to prepare it properly for eating, there is no reason they shouldn't find sustenance. In many isolated regions, survivors who have had some previous training in plant identification can enjoy wild plant food.

a. Plants provide carbohydrates, which provide the body energy and calories. Carbohydrates keep weight and energy up, and include important starches and sugars.

b. A documented and authoritative example of the value of a strictly plant diet in survival can be cited in the case of a Chinese botanist who had been drafted into the Japanese army during World War II. Isolated with his company in a remote section of the Philippines, the Chinese botanist kept 60 of his fellow soldiers alive for 16 months by finding wild plants and preparing them properly. He selected six men to assist him, and then found 25 examples of edible plants in the vicinity of their camp. He acquainted the men with these samples, showing them what parts of the plants could be used for food. He then sent the men out to look for similar plants and had them separate the new plants according to the original examples to avoid any poisonous plants mingling with the edible ones. The result of this effort was impressive. Though all the men had a natural desire for ordinary food, none suffered physically from the plant-food diet. The report was especially valuable because the botanist kept a careful record of all the food used, the results, and the comments of the men. This case history reflects the same opinions as those found in questionnaires directed to American survivors during World War II.

c. Another advantage of a plant diet is availability. In many instances, a situation may present itself in which procuring animal food is out of the question because of injury, being unarmed, being in enemy territory, exhaustion, or being in an area that lacks wildlife. If convinced that vegetation can be depended on for daily food needs, the next question is "where to get what and how."

(1) Experts estimate there are about 300,000 classified plants growing on the surface of the Earth, including many that thrive on mountain tops and on the floors of the oceans. There are two considerations survivors must keep in mind when procuring plant foods. The first consideration, of course, is the plant be edible, and preferably, palatable. Next, it must be fairly abundant in the areas where it is found. If it includes an inedible or poisonous variety in its family, the edible plant must be distinguishable to the average eye from the poisonous one. Usually a plant is selected because one special part is edible, such as the stalk, the fruit, or the nut.

(2) To aid in determining plant edibility, there are general rules that should be observed and an edibility test that should be performed. In selecting plant foods, the following should be considered. Select plants resembling those cultivated by people. It is risky to rely on a plant (or parts thereof) being edible for human consumption merely because animals have been seen eating it (for example, horses eat leaves from poison ivy; some rodents eat poisonous mushroom). Monkeys will put poisonous plants and fruits in pouches of their mouth and spit them out later. When selecting an unknown plant as a possible food source, apply the following general rules:

(a) Mushrooms and fungi should not be selected. Fungi have toxic peptides, a protein-based poison that has no taste. There is no field test other than eating to determine whether an unknown mushroom is edible. Anyone gathering wild mushrooms for eating must be absolutely certain of the identity of every specimen picked. Some species of wild mushrooms are difficult even for an expert to identify. Because of the potential

for poisoning, relying on mushrooms as a viable food source is not worth the risk.

(b) Plants with umbrella-shaped flowers are to be completely avoided, although carrots, celery, dill, and parsley are members of this family. One of the most poisonous plants, poison water hemlock, is also a member of this family (figure 18-26).

(c) All of the legume family should be avoided (beans and peas). They absorb minerals from the soil and cause problems. The most common mineral absorbed is selenium. Selenium is what has given locoweed its fame. (Locoweed is a vetch.)

(d) As a general rule, all bulbs should be avoided. Examples of poisonous bulbs are tulips and death camas.

(e) White and yellow berries are to be avoided, as they are almost always poisonous. Approximately one-half of all red berries are poisonous. Blue or black berries are generally safe for consumption.

(f) Aggregated fruits and berries are always edible (for example, thimbleberry, raspberry, salmonberry, and blackberry).

(g) Single fruits on a stem are generally considered safe to eat.

(h) Plants with shiny leaves are considered to be poisonous and caution should be used.

(i) A milky sap indicates a poisonous plant.

(j) Plants that are irritants to the skin should not be eaten, such as poison ivy.

(k) A plant that grows in sufficient quantity within the local area should be selected to justify the edibility test and provide a lasting source of food if the plant proves edible.

(l) Plants growing in the water or moist soil are often the most palatable.

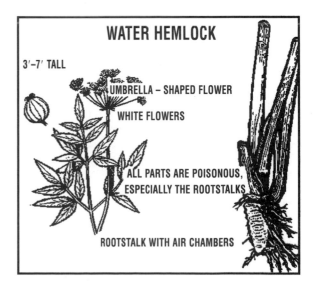

Figure 18-26. Water Hemlock.

(m) Plants are less bitter when growing in shaded areas.

(3) The previously mentioned information concerning plants is general. There are exceptions to every rule, but when selecting unknown plants for consumption, plants with these characteristics should be avoided. Plants that do not have these characteristics should be considered as possible food sources. Apply the edibility test to only one plant at a time so if some abnormality does occur, it will be obvious which plant caused the problem. Once a plant has been selected for the edibility test, proceed as follows:

(a) Crush or break part of the plant to determine the color of its sap. If the sap is clear, proceed to the next step.

(b) Touch the plant's sap or juice to the inner forearm or tip of the tongue. (A small taste of a poisonous plant will not do serious harm.) If there are no ill effects, such as a rash or burning sensation to the skin, bitterness to the taste, or numbing sensation of the tongue or lips, then proceed with the rest of the steps. (NOTE: Sometimes heavy smokers are unable to taste various poisons, such as alkaloids.)

(c) Prepare the plant or plant part for consumption by boiling in two changes of water. The toxic properties of many plants are water-soluble or destroyed by heat; cooking and discarding in two changes of water lessens the amount of poisonous material, or removes it completely. Parboiling is a process of boiling the individual plant parts in repeated changes of water to remove bitter elements. This boiling period should last about 5 minutes.

(d) Place about 1 teaspoonful of the prepared plant food in the mouth for 5 minutes and chew but do not swallow it. A burning, nauseating, or bitter taste is a warning of possible danger. If any of these ill effects occur, remove the material from the mouth at once and discard that plant as a food source. However, if no burning sensation or other unpleasant effect occurs, swallow the plant material and wait 8 hours.

(e) If after this 8-hour period there are no ill effects, such as nausea, cramps, or diarrhea, eat about 2 tablespoonfuls and wait an additional 8 hours.

(f) If no ill effects occur at the end of this 8-hour period, the plant may be considered edible.

(g) Keep in mind that any new or strange food should be eaten with restraint until the body's system has become accustomed to it. The plant may be slightly toxic and harmful when large quantities are eaten.

(4) If cooking facilities are not available, survivors will not be able to boil the plant before consumption. In this case, plant food may be prepared as follows:

(a) Leach the plant by crushing the plant material and placing it in a container. Pour large quantities of cold water over it (rinse the plant parts). Leaching removes some of the bitter elements of nontoxic plants.

EDIBLE PARTS OF PLANTS	
Underground Parts	Tubers Roots and Rootstalks Bulbs
Stems and Leaves (potherbs)	Shoots and Stems Leaves Pith Bark
Flower Parts	Flowers Pollen
Fruits	Fleshy Fruits (dessert and vegetable) Seeds and Grains Nuts Seed Pods Pulps
Gums and Resins	
Saps	

Figure 18-27. Edible Parts of Plants.

(b) If leaching is not possible, survivors should follow what steps they can in the edibility test.

d. The survivor will find some plants that are completely edible, but many plants that they may find will have only one or more identifiable parts having food and thirst-quenching value. The variety of plant component parts that might contain substance of food value is shown in figure 18-27.

(1) Underground Parts:

(a) Tubers. The potato is an example of an edible tuber. Many other kinds of plants produce tubers, such as the tropical yam, the Eskimo potato, and tropical water lilies. Tubers are usually found below the ground. Tubers are rich in starch and should be cooked by roasting in an earth oven or by boiling to break down the starch for ease in digestion. The following are some of the plants with edible tubers.

-1. East Indian Arrowroot.
-2. Taro.
-3. Cassava (Tapioca).
-4. Bean, Yam.
-5. Chufa (Nut Grass).
-6. Water Lily (Tropical).
-7. Sweet Potato (Kamote).
-8. Tropical Yam.

(b) Roots and Rootstalks. Many plants produce roots that may be eaten. Edible roots are often several feet in length. In comparison, edible rootstalks are underground portions of the plant that have become thickened, and are relatively short and jointed. Both true roots and rootstalks are storage organs rich in stored starch. The following are some of the plants with edible roots or rootstalks (rhizomes):

-1. Baobab.
-2. Screw Pine.
-3. Goa Bean.
-4. Water Plantain.
-5. Bracken.
-6. Reindeer Moss.
-7. Wild Calla (Water Arum).
-8. Rock Tripe.
-9. Polypody.
-10. Canna Lily.
-11. Flowering Rush.
-12. Cattail.
-13. Spinach, Ceylon.
-14. Chicory.
-15. Ti Plant.
-16. Horseradish.
-17. Tree Fern.
-18. Lotus Lily.
-19. Water Lily (temperate zone).
-20. Manioc.

(c) Bulbs. The most common edible bulb is the wild onion, which can easily be detected by its characteristic odor. Wild onions may be eaten uncooked, but other kinds of bulbs are more palatable if cooked. In Turkey and Central Asia, the bulb of the wild tulip may be eaten. All bulbs contain a high percentage of starch. (Some bulbs are poisonous, such as the death camas, which has white or yellow flowers.) The following are some of the plants with edible bulbs:

-1. Wild Lily.
-2. Wild Tulip.
-3. Wild Onion.
4. Blue Camas.
-5. Tiger Lily.

(2) Shoots and Leaves:

(a) Shoots (Stems). All edible shoots grow in much the same fashion as asparagus. The young shoots of ferns (fiddleheads), and especially those of bamboo and numerous kinds of palms, are desirable for food. Some kinds of shoots may be eaten raw, but most are better if boiled for 5 to 10 minutes, the water drained off, and the shoots reboiled until they are sufficiently cooked for eating (parboiled). (See figure 18-28.)

-1. Agave (Century Plant).
-2. Coconut Palm.
-3. Purslane.
-4. Reindeer Moss.
-5. Bamboo.
-6. Fishtail Palm.
-7. Goa Bean.
-8. Nipa Palm.
-9. Bracken.
-10. Rattan Palm.
-11. Wild Rhubarb.
-12. Cattail.
13. Sago Palm.
-14. Spinach, Ceylon.
-15. Rock Tripe.
-16. Colocynth.
-17. Sugar Palm.
-18. Papaya.
-19. Sugar Cane.
-20. Lotus Lily.
-21. Pokeweed (poisonous roots).
-22. Sweet Potato (Kamote).
-23. Luffa Sponge.
-24. Water Lily (Tropical).
-25. Polypody.
26. Buri Palm.
-27. Arctic Willow.

(b) Leaves. The leaves of spinach-type plants (potherbs), such as wild mustard, wild lettuce, and lamb quarters, may be eaten either raw or cooked. Prolonged cooking, however, destroys most of the vitamins. Plants that produce edible leaves are perhaps the most numerous of all edible plants. The young, tender leaves of most nonpoisonous plants are edible. The following

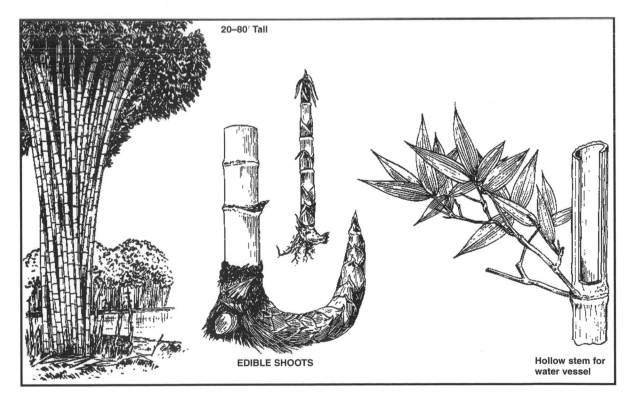

20–80′ Tall

EDIBLE SHOOTS

Hollow stem for water vessel

Figure 18-28. Bamboo.

are only some of the plants with edible leaves:

-1. Amarath.	-17. Plantain.
-2. Luffa Sponge.	-18. Pokeweed
-3. Rock Tripe.	(poisonous roots).
-4. Avocado.	-19. Sweet Potato
-5. Mango.	(Kamote).
-6. Wild Sorrel.	-20. Tamarind.
-7. Baobab.	-21. Horseradish.
-8. Sea Orach.	-22. Prickly Pear.
-9. Goa Bean.	-23. Taro (only
-10. Papaya.	after cooking).
-11. Spinach, Ceylon.	-24. Water Lettuce.
-12. Cassava.	-25. Purslane.
-13. Chickory.	-26. Ti Plant.
-14. Screw Pine.	-27. Arctic Willow.
-15. Spreading	-28. Lotus Lily.
Wood Fern.	-29. Reindeer Moss.
-16. Dock.	

(c) Pith. Some plants have edible piths in the center of the stem. The pith of some kinds of tropical plants is quite large. Pith of the sago palm is particularly valuable because of its high food value. The following are some of the palms with edible pith (starch):

-1. Buri.	-4. Coconut.
-2. Fishtail.	-5. Rattan.
-3. Sago.	-6. Sugar.

(d) Bark. The inner bark of a tree—the layer next to the wood—may be eaten raw or cooked. It is possible in northern areas to make flour from the inner bark of such trees as the cottonwood, aspen, birch, willow, and pine. The outer bark should be avoided in all cases because this part contains large amounts of bitter tannin. Pine bark is high in vitamin C. The outer bark of pines can be cut away and the inner bark stripped from the trunk and eaten fresh, dried, or cooked, or it may be pulverized into flour. Bark is most palatable when newly formed in spring. As food, bark is most useful in the arctic regions, where plant food is often scarce.

(3) Flower Parts:

(a) Flowers and Buds. Fresh flowers may be eaten as part of a salad or to supplement a stew. The hibiscus flower is commonly eaten throughout the southwest Pacific. In South America, the people of the Andes eat nasturtium flowers. In India, it is common to eat the flowers of many kinds of plants as part of a vegetable curry. Flowers of desert plants may also be eaten. The

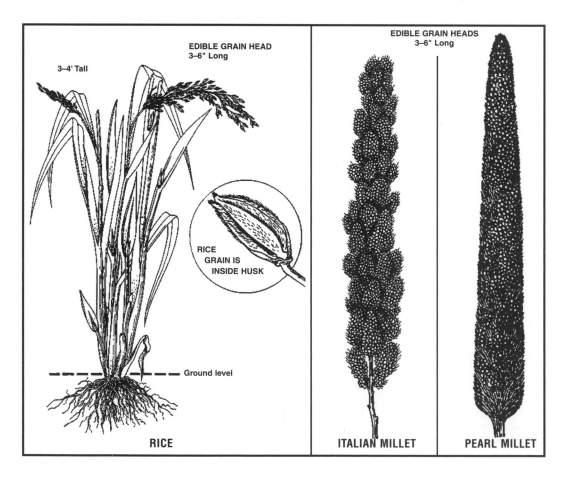

Figure 18-29. Grains.

following are plants with edible flowers:

-1. Abal.   -5. Horseradish.
-2. Colocynth.   -6. Wild Caper.
-3. Papaya.   -7. Luffa Sponge.
-4. Banana.

(b) Pollen. Pollen looks like yellow dust. All pollen is high in food value and in some plants, especially the cattail. Quantities of pollen may easily be collected and eaten as a kind of gruel.

(4) Fruits. Edible fruits can be divided into sweet and nonsweet (vegetable) types. Both are the seed-bearing parts of the plant. Sweet fruits are often plentiful in all areas of the world where plants grow. For instance, in the far north, there are blueberries and crowberries; in the temperate zones, cherries, plums, and apples; and in the American deserts, fleshy cactus fruits. Tropical areas have more kinds of edible fruit than other areas, and a list would be endless. Sweet fruits may be cooked, or for maximum vitamin content, left uncooked. Common vegetable fruits include the tomato, cucumber, and pepper.

(a) Fleshy Fruits (sweet). The following are plants with edible fruits:

1. Wild Apple.   13. Jackfruit.
-2. Bael Fruit.   14. Common Jujube.
-3. Banana.   -15. Mango.
-4. Bignay.   -16. Mulberry.
-5. Wild Blueberry.   -17. Papaya.
-6. Bullock's Heart.   -18. Batako Plum.
-7. Cloudberry.   -19. Pokeberry.
-8. Crabapple.   -20. Prickly Pear.
-9. Cranberry.   -21. Rose Apple.
-10. Wild Fig.   -22. Soursop.
-11. Wild Grape.   -23. Sweetsop.
-12. Huckleberry.

(b) Fleshy Fruits (vegetables). The following are plants with edible fruits:

-1. Breadfruit.   -4. Wild Caper.
-2. Horseradish.   -5. Luffa Sponge.
-3. Plantain.

(c) Seeds and Grains. The seeds of many plants, such as buckwheat, ragweed, amaranth, and goosefoot, contain oils and are rich in protein. The grains of all cereals and many other grasses, including millet, are also extremely valuable sources of plant protein. They may be either ground between stones, mixed with water and cooked to make porridge, parched or roasted over hot stones. In this state, they are still wholesome and may be kept for long periods without further preparation (figure 18-29). The following are some of the plants with edible seeds and grains:

-1. Amaranth.   -6. Nipa Palm.
-2. Italian Millet.   -7. Tamarind.
-3. Rice.   -8. Screw Pine.
-4. Bamboo.   -9. Coloynth.
-5. Pearl Millet.   -10. Water Lily (Tropical).

-11. Sterculia.   -16. Lotus Lily.
-12. Baobab.   -17. Purslane.
-13. Sea Orach.   -18. Water Lily
-14. St. John's Bread.   (temperate zone).
-15. Goa Bean.   -19. Luffa Sponge.

(d) Nuts. Nuts are among the most nutritious of all raw plant foods and contain an abundance of valuable protein. Plants bearing edible nuts occur in all the climatic zones of the world and on all continents except in the arctic regions. Inhabitants of the temperate zones are familiar with walnuts, filberts, almonds, hickory nuts, acorns, hazelnuts, beechnuts, and pine nuts, to name just a few. Tropical zones produce coconuts and other palm nuts, Brazil nuts, cashew nuts, and macadamia nuts (figure 18-30). Most nuts can be eaten raw but some, such as acorns, are better when cooked. The following are some of the plants with edible nuts:

-1. Almond.   -8. Filbert (Hazelnut).
-2. Water Chestnut   -9. Fishtail Palm.
   (Trapa Nut).   -10. Jackfruit Seeds.
-3. Buri Palm.   -11. English Oak (Acorn).
-4. Almond, Indian   -12. Sago Palm.
   or Tropical.   -13. Sugar Palm.
5. Mountain Chestnut.   14. Pine.
-6. Coconut Palm.   -15. Wild Pistachio.
-7. Beechnut.   -16. Walnut.

(e) Pulps. The pulp around the seeds of many fruits is the only part that can be eaten. Some fruits produce sweet pulp; others have a tasteless or even bitter pulp. Plants that produce edible pulp include the custard apple, inga pod, breadfruit, and tamarind. The pulp of breadfruit must be cooked, whereas in other plants, the pulp may be eaten uncooked. Use the edibility rules in all cases of doubt.

(5) Gums and Resins. Gum and resin are sap that collects and hardens on the outside surface of the plant. It is called gum if it is soft and soluble, and resin if it is hard and not soluble. Most people are familiar with the gum that exudes from cherry trees and the resin that seeps from pine trees. These plant byproducts are edible and are a good source of nutritious food that should not be overlooked.

(6) Saps. Vines or other plant parts may be tapped as potential sources of usable liquid. The liquid is obtained by cutting the flower stalk and letting the fluid drain into some sort of container, such as a bamboo section. Palm sap with its high sugar content is highly nutritious. The following are some plants with edible sap and drinking water:

-1. Sweet Acacia (water).   -9. Rattan Palm (water).
-2. Colocynth (water).   -10. Cactus (water).
-3. Coconut Palm (sap).   -11. Grape (water).
-4. Fishtail Palm (sap).   -12. Banana (water).
-5. Agave (water).   -13. Sago Palm (sap).
-6. Cuipo Tree (water).   -14. Sugar Palm (sap).
-7. Saxual (water).   -15. Buri Palm (sap).
-8. Nipa Palm (sap).

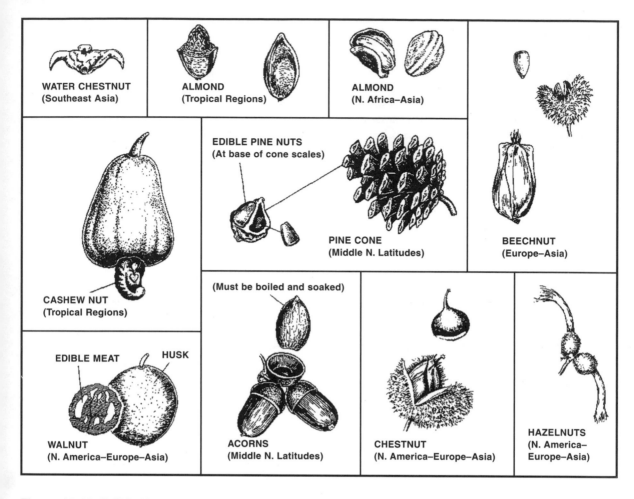

Figure 18-30. Edible Nuts.

18-6. Food in Tropical Climates. There are more types of animals in the jungles of the world than in any other region. A jungle visitor who is unaware of the lifestyle and eating habits of these animals would not observe the presence of a large number of the animals.

a. Game trails are the normal routes along which animals travel through a jungle. Some of the animals used as food are hedgehogs, porcupines, anteaters, mice, wild pigs, deer, wild cattle, bats, squirrels, rats, monkeys, snakes, and lizards.

(1) Reptiles are located in all jungles and should not be overlooked as a food source. All snakes should be considered poisonous and extreme caution used when killing the animal for a food source. All cobras should be avoided since the spitting cobra aims for the eyes; the venom can blind if not washed out immediately. Lizards are good food, but may be difficult to capture since they can be extremely fast. A good blow to the head of a reptile will usually kill it. Crocodiles and caimans are extremely dangerous on land as well as in the water.

(2) Frogs can be poisonous; all brilliantly colored frogs should be totally avoided. Some frogs and toads in the tropics secrete substances through the skin that has a pungent odor. These frogs are often poisonous.

(3) The larger, more dangerous animals such as the tiger, rhinoceros, water buffalo, and elephants are rarely seen and should be left alone. These larger animals are usually located in the open grasslands.

b. Seafood such as fish, crabs, lobsters, crayfish, and small octopi can be poked out of holes, crevices, or rock pools (figure 18-31). Survivors should be ready to spear them before they move off into deep water. If they are in deeper water, they can be teased shoreward with a baited hook or a stick.

(1) A small heap of empty oyster shells near a hole may indicate the presence of an octopus. A baited hook placed in the hole will often catch the octopus. The survivor should allow the octopus to surround the hook and line before lifting. Octopi are not scavengers like sharks, but they are hunters, fond of spiny lobsters and other crab-like fish. At night, they come into shallow water and can be easily seen and speared.

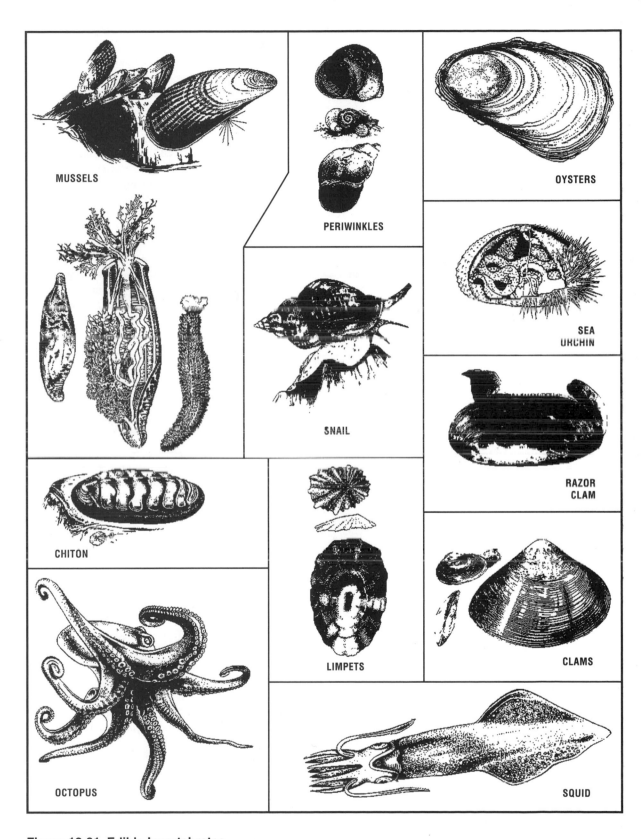

Figure 18-31. Edible Invertebrates.

(2) Snails and limpets cling to rocks and seaweed from the low watermark and above. Large snails called chitons adhere tightly to rocks just above the surf line.

(3) Mussels usually form dense colonies in rock pools, on logs, or at the bases of boulders. *Mussels are poisonous in tropical zones during the summer,* especially when seas are highly phosphorescent or reddish.

(4) Sluggish sea cucumbers and conchs (large snails) live in deep water. The sea cucumber will shoot out its stomach when excited. The stomach is not edible. The skin and the five strips of muscle can be eaten after boiling. Conchs can be boiled out of their shells and have very firm flesh. Use care when picking conchs up. The bottom of its "foot" has a boney covering that can severely cut the survivor who procures it.

(5) The safest fish to eat are those from the open sea or deep water beyond the reef. Silvery fishes, river eels, butterfly fishes, and flounders from bays and rivers are good to eat.

(6) Land crabs are common on tropical islands and are often found in coconut groves. An open coconut can be used for bait.

(7) A number of methods can be used for procuring fish.

(a) Hook-and-Line Fishing. This type of fishing on a rocky coast requires a lot of care to keep the line from becoming entangled or cut on sharp edges. Most shallow-water fish are nibblers. Unless the bait is well placed and hooked and the barb of the hook offset by bending, the bait may be lost without catching a fish. Use hermit crabs, snails, or the tough muscle of a shellfish as bait. Take the cracked shells and any other animal remains and drop them into the area to be fished. This brings the fish to the area and provides a better procurement opportunity. Examine stomach contents of the first fish caught to determine what the fish are feeding on.

(b) Jigging. A baited or spooned hook dipped repeatedly beneath the surface of the water is sometimes effective. This method may be used at night.

(c) Spearing. This method is difficult except when the stream is small and the fish are large and numerous during the spawning season, or when the fish congregate in pools. Make a spear by sharpening a long piece of wood, lashing two long thorns on a stick, or fashioning a bone spear point, and take a position on a rock over a fish run. Wait patiently and quietly for fish to swim by.

(d) Chop Fishing. Chop fishing is effective at night during low tide. This method requires a torch and a machete. The fish are attracted by the light of the torch, and then they may be stunned by slashing at them with the *back* of the machete blade. Care should be taken when swinging the machete (figure 18-32).

c. The jungle environment has uniquely favorable conditions for plant and animal life. The variety and richness of plant growth in these areas are paralleled nowhere else on the Earth. Because the rainfall is distributed throughout the year and there is a lack of cold seasons, plants in the humid regions can grow, produce leaves, and flower year round. Some plants grow very rapidly. For example, the stem of the giant bamboo may grow more than 22 inches in a single day.

(1) A survivor in search of plant food should apply some basic principles to the search. A survivor is lucky to find a plant that can readily be identified as edible. If a plant resembles a known plant, it is very likely to be of the same family and can be eaten. If a plant cannot be identified, the edibility test should be applied. A survivor will find many edible plants in the tropical forest, but chances of finding them in abundance are better in an area that has been cultivated in the past (secondary growth).

(2) Some plants a survivor might find:

(a) Citrus-fruit trees may be found in uncultivated areas but are primarily limited to areas of secondary growth. The many varieties of citrus-fruit trees and shrubs have leaves 2 to 4 inches long alternately arranged. The leaves are leathery, shiny, and evergreen. The leaf stem is often winged. Small (usually green) spines are often present by the side of the bud. The flowers are small and white to purple in color. The fruit has a leathery rind with numerous glands and is round and fleshy with several cells (fruit sections or slices) and many seeds. The great number of wild and cultivated fruits (oranges, limes, lemons, etc.) native to the tropics are eaten raw or used in beverages.

(b) Taro can be found in both secondary growth and in virgin areas. It is usually found in the damp, swampy areas in the wild, but certain varieties can be found in the forest. It can be identified by its large heart-shaped or arrowhead-shaped leaves growing at the top of a vertical stem. The stem and leaves are usually green and rise a foot or more from a tuber at the base of the stem. Taro-leaf tips point down; poisonous elephant ear points up. All varieties of taro must be cooked to break down the irritating crystals in the plant.

(c) Wild pineapple can be found in the wild, and common pineapples may be found in secondary growth areas. The wild pineapple is a coarse plant with long, clustered, sword-shaped leaves with sawtoothed edges. The leaves are spirally arranged in a rosette. Flowers are violet or reddish. The wild-pineapple fruit will not be as fully developed in the wild state as when cultivated. The seeds from the flower of the plant are edible as well as the fruit. The ripe fruit may be eaten raw, but the green fruit must be cooked to avoid irritation. (The leaf fibers make excellent lashing material and ropes can be manufactured from it.)

(d) Yams may be found cultivated or wild. There are many varieties of yam, but the most common has a vine with square-shaped cross section and two rows of heart-shaped leaves growing on opposite sides of the vine. The vine can be followed to the ground to locate

Figure 18-32. Chop Fishing.

the tuber. The tubers should be cooked to destroy the poisonous properties of the plant (figure 18-33).

(e) Ginger grows in the tropical forest and is a good source of flavoring for food. It is found in shaded areas of the primary forest. The ginger plant grows 5 to 6 feet high. It has seasonal white snapdragon-type flowers; some variations have red flowers. The leaves when crushed produce a very sweet odor and are used for seasoning or tea. The tea is used by primitive people to treat colds and fever.

(f) The coconut palm is found wild on the seacoast and in farmed areas inland. It is a tree 50 to 100 feet high, either straight or curved, and marked with ringlike leaf scars. The base of the tree is swollen and surrounded by a mass of rootlets. The leaves are leathery and reach a length of 15 to 20 feet. (The leaves make excellent sheathing for shelter.) The fruit grows in clusters at the top of the tree. Each nut is covered with a fibered, hard shell. The "heart" of the coconut palm is edible and is found at the top. (The new leaves grow out of the heart.) Cut the tree down and remove the leaves to gain access to the heart. The flower of the coconut tree is also edible and is best used as a cooked vegetable. The germinating nut is filled with a meat that can be eaten

raw or cooked. There are many other varieties of palm found in the tropics that have edible hearts and fruits (figure 18-34).

(g) The papaya is an excellent source of food and can be found in secondary growth areas. The tree grows to a height of 6 to 20 feet. The large, dark green, many-fingered, rough-edged leaves are clustered at the top of the plant. The fruit grows on the stem clustered under the leaves. The fruit is small in the wild state, but cultivated varieties may grow to 15 pounds. The peeled fruit can be eaten raw or cooked. The peeling should never be eaten. The green fruit is usually cooked. The milky sap of the green fruit is used as a meat tenderizer; care should be taken not to get it in the eyes. Always wash hands after handling fresh green papayas. If some of the sap does get in the eyes, they should be washed immediately (figure 18-35).

(h) Cassava (tapioca) can be found in secondary growth areas. It can be identified by its stalk-like leaves, which are deeply divided into numerous pointed sections or fingers. The woody (red) stem of the plant is slender and at points appears to be sectioned. When found growing wild in secondary growth areas, pull the trunks to find where a root grows. When one is found, a tuber can be dug. Tubers have been found growing around a portion of the stem that was covered with vegetation. The brown tuber of the plant is white inside and must be boiled or roasted. The tuber must also be peeled before boiling. (The green-stemmed species of cassava is poisonous and must be cooked in several changes of water before eating it.)

(i) Ferns can be found in the virgin tropical forest or in secondary growth areas. The new leaves (fiddleheads) at the top are the edible parts. They are covered with fuzzy hair that is easily removed by rubbing or washing. Some can be eaten raw but as a rule should be cooked as a vegetable (figure 18-36).

(j) Sweetsops can be found in the tropical forest. It is a small tree with simple, oblong leaves. The fruit is shaped like a blunt pinecone with thick, gray-green or yellow, brittle spines. The fruit is easily split or broken when ripe, exposing numerous dark brown seeds imbedded in the cream-colored, very sweet pulp.

(k) The star apple is common in the tropical forests. The tree grows to a height of 60 feet and can be identified by the leaves, which have shiny, silky, brown hairs on the bottom. The fruit looks like a small apple or plum with a smooth greenish or purple skin. The meat is greenish in color and milky in texture. When cut through the center, the brown, elongated seeds make a figure like a 6- or 10-pointed star. The fruit is sweet and can be eaten only when fresh. When cut the rind will, like other parts of the tree, emit a white sticky juice or latex, which is not poisonous (an exception to the milky-sap rule).

(3) Of the 300,000 different kinds of wild plants in the world, a large number of them are found in the

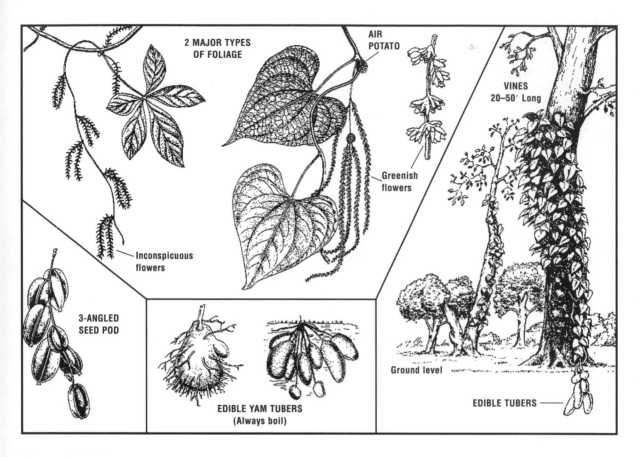

2 MAJOR TYPES OF FOLIAGE

AIR POTATO

Greenish flowers

VINES 20–50′ Long

Inconspicuous flowers

3-ANGLED SEED POD

EDIBLE YAM TUBERS
(Always boil)

Ground level

EDIBLE TUBERS

Figure 18-33. Yams.

tropics, and many of them are potentially edible. Very few are deadly when eaten in small quantities. Those that are poisonous may be detected by using the edibility rules. Only a small number of jungle plants have been discussed. It would be of great benefit to anyone flying over or passing through a tropical environment to study the plant foods available in this type of environment.

18-7. Food in Dry Climates. Although not as readily available as in the tropical climate, food is available and obtainable.

a. Plant life in the desert is varied due to the different geographical areas. It must be remembered, therefore, that available plants will depend on the actual desert, the time of year, and whether there has been any recent rainfall. Aircrew members should be familiar with plants in the area to be flown over.

(1) Date palms are located in most deserts and are cultivated by the native people around oases and irrigation ditches. They bear a nutritious, oblong, black fruit (when ripe).

(2) Fig trees are normally located in tropical and subtropical zones; however, a few species can be found in the deserts of Syria and Europe. Many kinds are cultivated. The fruit can be eaten when ripe. Most figs resemble a top or a small pear somewhat squashed in shape. Ripe figs vary greatly as to palatability. Many are hard, woody, covered with irritating hairs, and worthless as survival food. The edible varieties are soft, delectable, and almost hairless. They are green, red, or black when ripe.

(3) Millet, a grain-bearing plant, is grown by natives around oases and other water sources in the Middle Eastern deserts.

(4) The fruit of all cacti are edible. Some fruits are red, some yellow, but all are soft when ripe. Any of the flat-leaf variety, such as the prickly pear, can be boiled and eaten as greens (like spinach) if the spines are first removed. During severe droughts, cattlemen burn off the spines and use the thick leaves for fodder. Although the cactus originates in the American deserts, the prickly pear has been introduced to the desert edges in Asia, Africa, the Near East, and Australia, where it grows profusely. Natives eat the fruit as fast as it ripens.

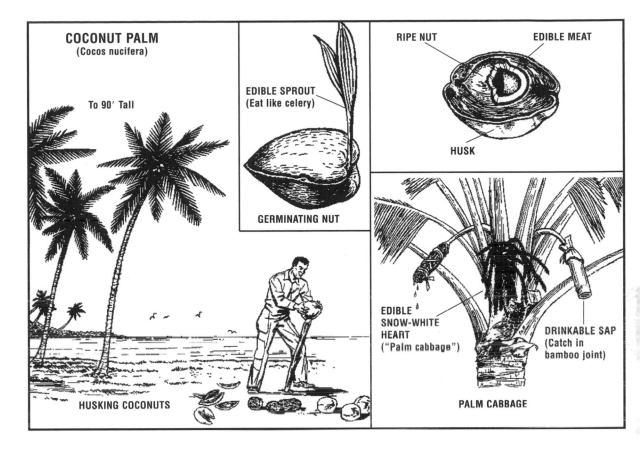

Figure 18-34. Coconut Palm.

(5) There are two types of onions in the Gobi Desert. A hot, strong, scallion-type grows in the late summer. It will improve the taste of food but should not be used as a primary food. The highland onions grow 2 to 2.5 inches in diameter. These can be eaten like apples and the greens can also be eaten raw or cooked.

(6) All desert flowers can be eaten except those with milky or colored sap.

(7) All grasses are edible. Usually the best part is the whitish tender end that shows when the grass stalk is pulled from the ground. All grass seeds are edible.

b. Animal food sources may be used to supplement diets and provide needed protein and fats. When looking at a desert area, it is sometimes difficult to visualize an abundance of animal life existing in it. There is, however, a great quantity of animal life present. Most are edible, but some may be hazardous to a survivor during the procurement stage. Some of the abundant animal life includes:

(1) At the peak of seasonal plant growth, the desert crawls and buzzes with an enormous number and variety of beetles, ants, wasps, moths, and bugs. They appear with the first good rains and generally feed during nighttime. The Ute Indians of North America have harvested crickets, and peoples of the Middle East have roasted locusts. The human diet in Mexico and of the American Indians of the Southwest frequently includes grasshoppers and caterpillars.

(2) On the Playas of the Sonora and Chihuahua deserts, several species of freshwater shrimp appear every summer in warm temporary ponds. In the Mohave Desert, where summer rains are rare, they may appear only a few times in a century.

(3) Snakes, lizards, tortoises, etc., have adapted well to the desert environment. Care must be observed when procuring them, as some are hazardous, such as the Gila monster and rattlesnake. The desert tortoise, about a foot long when full-grown, lives in some of the harshest regions of the Mohave and Sonora deserts. It is clubfooted, herbivorous, and can crawl about 20 feet per minute. The tortoise converts some of its food into water, which is stored for the hot months in two sacs under the upper shell. A pint of water lasts the dry season. In spring and fall, the tortoise browses in broad

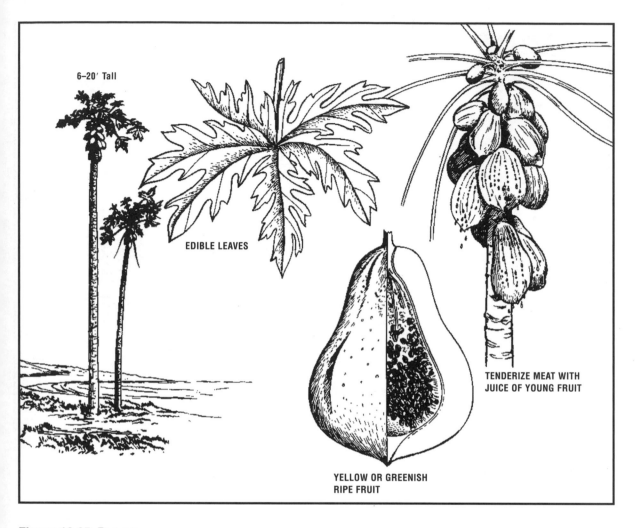

6–20' Tall

EDIBLE LEAVES

TENDERIZE MEAT WITH
JUICE OF YOUNG FRUIT

YELLOW OR GREENISH
RIPE FRUIT

Figure 18-35. Papaya.

daylight, becoming livelier as the day warms up. In the heat of the summer, it comes out of its shallow burrow in the early morning, the late evening, or not at all.

(4) In general, desert birds stay in areas of heavier vegetation and many need water daily; therefore, most will be found within short flights of some type of water source. Many birds will migrate during the drought season. If an abundance of birds is seen, insects, vegetation, and a water source can usually be found nearby.

(5) Rabbits, prairie dogs, and rats have learned to live in deserts. They remain in the shade or burrow into the ground, protecting themselves from the direct sun and heated air as well as from the hot desert surface.

(6) Larger mammals are also found in the desert. This group consists of gazelles, antelope, deer, foxes, small cats, badgers, dingos, hyenas, etc., and is amazingly abundant. Most are nocturnal and generally avoid humans. They roam at night eating smaller game and

insects; a few eat plants; and a few can be hazardous to a survivor. Any of these mammals should be approached with caution.

(7) Only a few of the available animals and plants have been discussed. If the possibility of having to survive in a desert area exists, the aircrew member should try to become familiar with the food sources available in that area.

18-8. Food in Snow and Ice Climates. In the snow and ice climates, food is more difficult to find than water. Animal life is normally more abundant during the warm months, but it can still be found in the cold months. Fish are available in most waters during the warmer months, but they congregate in deep waters, large rivers, and lakes during the cold months. Some edible plant life can be found throughout the year in most areas of the Arctic.

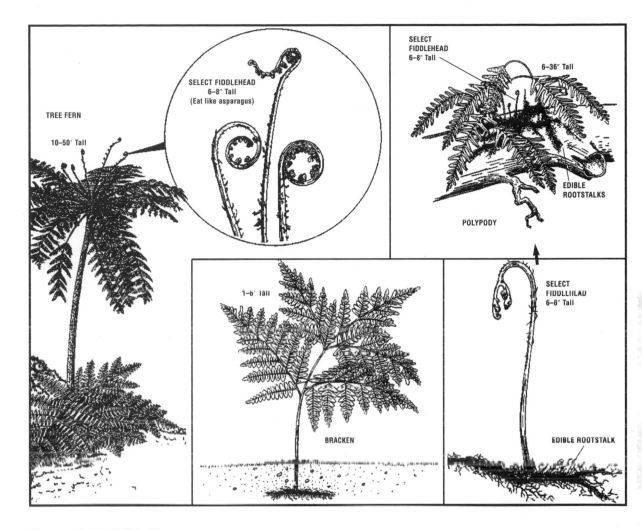

Figure 18-36. Edible Ferns.

a. All animals in the arctic regions are edible, but the livers of seals and polar bears must not be eaten because of the high concentration of vitamin A. Death could result from ingesting large quantities of the liver. On the open sea ice, game animals such as seal, walrus, polar bear, and fox are available (figure 18-37). Many types of birds can be found during the warmer months. Fish can be caught throughout the year.

(1) Seals will probably be the main source of food when stranded on the open sea ice. They can be found in open leads, areas of thin ice, or where snow has drifted over a pressure ridge forming a cave, which could have open water or very thin ice. These areas may also house polar bears, which feed primarily on seals. Polar bears should be avoided.

(a) Newborn seals have trouble staying afloat or swimming and will be found on the ice in the early summer. The seal cubs can be easily killed with a club, spear, knife, or firearm and make an excellent source of food. The meat, blubber (fat), and coagulated milk in their stomach are edible. When killing a cub, it is best to keep a lookout for the mother. She tends to protect offspring in any way possible.

(b) Seals must surface periodically to breathe. When the icepack is thin, the seals poke their noses through the ice and take a breath of air in a lead or in open water. In thick ice, the seal will chew and (or) claw a breathing hole through the ice. Normally most seals will have more than one breathing hole. While hunting seals, it is best to take a position beside a breathing hole and wait until a seal comes up to breathe, then spear or strike it on the head with a club. Seals are very sensitive to blows on or about the nose. They will often lose consciousness but not die. A hook can be suspended through the breathing hole so it hangs down at least 6 inches below the ice. When a seal comes to breathe, it can become hooked when it tries to depart the breathing hole. Seals can be recovered by gaffing or grabbing by hand, but in some cases the breathing hole might have to be enlarged to pull the body through. If the seal is killed in open water a "manak" or "grapple hook" can be used to retrieve it. All seals killed in open water or those that fall into open water should be recovered immediately. During the cold months, they will float for quite awhile, but during the warm months or when a female is nursing young, they sink rapidly. This is due to the loss of body fat.

(2) Birds are plentiful during the summer months and can be procured by spearing, clubbing, catching with a baited fishhook, or by using a weapon.

(3) In tundra areas there are large game, small game, and birds available as food sources.

(a) The large game consists of caribou, musk oxen, sheep, wolves, and bears (figure 18-38). Even though the large game animals can be a food source, they will be difficult to procure if a firearm is not available. Therefore, they should be considered a hazard to a survivor without a firearm. In the spring, bears tend to congregate along rivers and streams due to the amount of food available—normally salmon. During the fall, bears will be found feeding at berry patches. During certain seasons of the year, these areas should be avoided.

(b) Small game animals of the tundra include hares, lemmings, mice, ground squirrels, marmots, and foxes. They may be trapped or killed the entire year. When snaring, it is best to use a simple loop made of strong line or wire. The wire must be a two-strand twisted wire since metal becomes brittle in the cold and breaks very easily. Other snares and triggers will be less effective in the cold climate. A gill net can be used as a snare by spreading it across a trail so the animal will entangle itself.

(c) Surface water is generally plentiful due to the number of lakes, ponds, bogs, and marshes. Waterfowl and birds are very abundant during the warm months and include ducks, terns, geese, gulls, owls, and ptarmigans. The eggs and young birds are an excellent food source and can be easily procured (figure 18-39).

(4) As in the tundra areas, the forested areas in the Arctic and arctic-like areas abound in wildlife.

(a) The large game species include moose, deer, caribou, and bear.

Figure 18-37. Walruses.

Figure 18-38. Big Horn Sheep.

(b) Small game of the forests includes hares, squirrels, porcupine, muskrat, and beaver. They can be snared or trapped easily in winter or summer. Small-animal trails can be found in the winter with great ease. Most animals do not like to travel in deep snow so they tend to travel the same trail most of the time and this trail will look like a small super-highway—the snow packed down well below the normal snow level. Most trails will also be located in heavy cover or undergrowth or parallel to roads and open areas. The same trails will normally be used during the summer.

(5) During the summer months, the open water provides an excellent opportunity to procure all types of fish, both freshwater and saltwater, and freshwater mussels. The mussels can be handpicked off the bottom, while the fish can be netted, speared, clubbed, or caught with a hook and line. After freezeup, fishing is still possible through the ice. Shallow lakes, rivers, or ponds can freeze completely, killing off all fish life. Fish tend to congregate in the deepest water possible. A hole should be cut through the ice at the estimated deepest point. Other good locations are at outlets or where tributaries flow into lakes or ponds. The ice is normally thinner over rapid moving water and at the edges of deep streams or rivers with snowdrifts extending out from the banks. Open water is often marked by a mist or fog formed over the area by vaporizing water. All methods of procuring fish in the summer will work in the winter (figure 18-40).

(6) The ocean shores are rich hunting grounds for edible sea life such as clams, mussels, scallops, snails, limpets, sea urchins, chitons, and sea cucumbers. They can be procured most of the year wherever there is open water. Tidal pools usually contain a great number of both fish and mollusks. The fish can be netted, speared, or hand-caught. All sea life can be eaten raw, but cooking usually makes it more palatable (figure 18-41).

b. The plant life of the arctic regions is generally small and stunted due to the effects of permafrost, low mean temperatures, and a short growing season.

Figure 18-39. Swans.

(1) In the barren tundra areas, a wide variety of small edible plants and shrubs exist. During the short summer months in the tundra, Labrador tea, fireweed, coltsfoots, dwarf arctic birch, willow, and numerous other plants and berries can be found. During the winter, roots, rootstalks, and frozen berries can be found beneath the snow. Lichens and mosses are abundant but should be selected carefully, as some species are poisonous.

(2) In bog or swamp areas, many types of water sedge, cattail, dwarf birch, and berries are available. During spring and summer, many young shoots from these plants are easily collected.

(3) The wooded areas of the Arctic contain a variety of trees (birch, spruce, poplar, aspen, and others). Many berry-producing plants can be found, such as blueberries, cranberries, raspberries, cloudberries, and crowberries. Wild rose hips, Labrador tea, alder, and other shrubs are very abundant. Many wild edible plants are highly nutritious. Greens are particularly rich in carotene (vitamin A). Leafy greens, many berries, and rose hips are all rich in ascorbic acid (vitamin C). Many roots and rootstalks

contain starch and can be used as a potato substitute in stews and soups.

(4) Although there are several types of edible mushrooms, fungi, and puffballs in the Arctic, a person should avoid ingesting them because it is difficult to identify the poisonous and nonpoisonous species. During the growing season, the physical characteristics can change considerably, making positive identification even more difficult.

(5) There are many poisonous plants and a few poisonous berries in the Arctic. Very few cause death; many will cause extreme nausea, dizziness, abdominal pain, and diarrhea. Contact poisonous plants, such as poison ivy, are not found in the Arctic. The more common poisonous plants are shown in figures 18-42 through 18-49.

(6) When selecting edible plants, choose young shoots when possible as these will be the most tender. Plants should be eaten raw to obtain the most nutritive value. Some of the more common edible plants are:

(a) Dandelions generally grow with grasses but may be scattered over rather barren areas. Both leaves and roots are edible raw or cooked. The young leaves

Figure 18-40. Winter Fishing.

make good greens; the roots (when roasted) are used as a substitute for coffee.

(b) Black and white spruce are generally the northern most evergreens. These trees have short, stiff needles that grow singularly rather than in clusters like pine needles. The cones are small and have thin scales. Although the buds, needles, and stems have a strong resinous flavor, they provide essential vitamin C by chewing them raw. In spring and early summer, the inner bark can be used for food.

(c) The dwarf arctic birch is a shrub with thin tooth-edged leaves and bark that peels off in sheets. The fresh green leaves and buds are rich in vitamin C. The inner bark may also be eaten.

(d) There are many different species of willow in the arctic. Young tender shoots may be eaten as greens and the bark of the roots is also edible. They have a decidedly sour taste but contain a large amount of vitamin C (figure 18-50).

(7) Lichens are abundant and widespread in the far North and can be used as a source of emergency food. Many species are edible and rich in starch-like substances, including Iceland moss, peat moss and reindeer lichen. Beard lichen growing on trees has been used as food by American Indians. However, some of it contains a bitter acid that causes irritation of the digestive tract. If lichens are boiled, dried, and powdered, this acid is removed and the powder can then be used as flour or made into a thick soup.

18-9. Food on the Open Seas. Almost all sea life is not only edible but is also an excellent source of nutrients essential to humans. The protein is complete because it contains all the essential amino acids, and the fats are similar to those of vegetables. Seafoods are high in minerals and vitamins. The majority of life in the sea (fish, birds, plants, and aquatic animals) is edible.

a. Most seaweed is edible and is a good source of food, especially for vitamins and minerals. Some seaweed contains as much as 25 percent protein, while others are composed of more than 50 percent carbohydrates. At least 75 different species are used for food by seacoast residents around the world. For many people,

Figure 18-41. Shell Fish.

especially the Japanese, seaweed is an essential part of the diet, and the most popular varieties have been successfully farmed for hundreds of years. The high cellulose content may require gradual adaptation because of their laxative quality if they comprise a large part of the diet. As with vegetables, some species are more flavorful than others. Generally, leafy green, brown, or red seaweed can be washed and eaten raw or dried. The following list of edible seaweed gives a description of the plant, tells where it may be found, and in many cases, suggests a method of preparation:

(1) Common green seaweed (figure 18-51), often called sea lettuce (Ulva lactuca), is in abundance on both sides of the Pacific and North Atlantic oceans. After washing it in clean water, it can be used as a garden lettuce.

(2) The most common edible brown seaweed is the sugar wrack, kelp, and Irish moss (figure 18-52).

(a) The young stalks of the sugar wrack are sweet to taste. This seaweed is found on both sides of the Atlantic and on the coasts of China and Japan.

(b) Edible kelp has a short cylindrical stem and thin, wavy, olive-green or brown fronds 1 to several feet in length. It is found in the Atlantic and Pacific Oceans, usually below the high-tide line on submerged ledges and rocky bottoms. Kelp should be boiled before eating. It can be mixed with vegetables or soup.

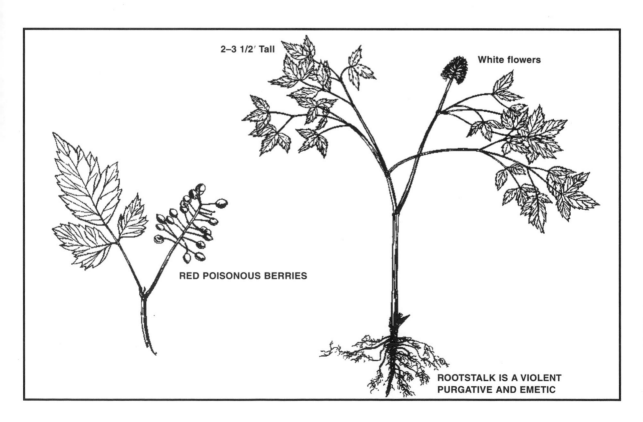

Figure 18-42. Baneberry.

(c) Irish moss, a variety of brown seaweed, is quite edible, and is often sold in marketplaces. It is found on both sides of the Atlantic Ocean and can be identified by its tough, elastic, and leathery texture; however, when dried, it becomes crisp and shrunken. It should be boiled before eating. It can be found at or just below the high-tide line. It is sometimes found cast up on the shore.

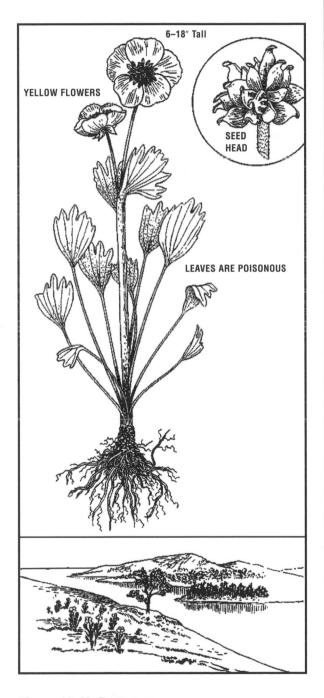

Figure 18-43. Buttercup.

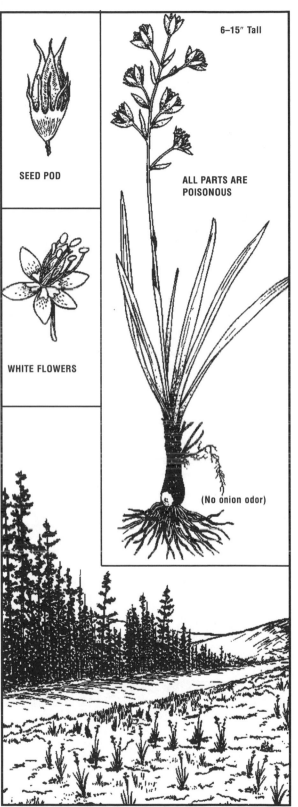

Figure 18-44. Death Camas.

(3) Red seaweed can usually be identified by its characteristic reddish tint, especially the edible varieties. The most common and edible red seaweed include the dulse, laver, and other warm-water varieties (figure 18-53).

(a) Dulse has a very short stem that quickly broadens into a thin, broad, fan-shaped expanse that is dark red and divided by several clefts into short, round-tipped lobes. The entire plant is from a few inches to a foot in length. It is found attached to rocks or coarser seaweed, usually at the low-tide level, on both sides of the Atlantic Ocean and in the Mediterranean. Dulse is leathery in consistency and is sweet to the taste. If dried and rolled, it can be used as a substitute for tobacco.

(b) Laver is usually red, dark purple, or purplish-brown, and has a satiny sheen or filmy luster. Common

to both the Atlantic and Pacific Oceans, it has been used as food for centuries. This seaweed is used as a relish, or is cleaned and then boiled gently until tender. It can also be pulverized and added to crushed grains and fried in the form of flatcakes. During World War II, laver was chewed for its thirst-quenching value by New Zealand troops. Laver is usually found on the beach at the low-tide level.

(c) A great variety of red warm-water seaweed is found in the South Pacific area. This seaweed accounts for a large portion of the native diet. When found on the open sea, bits of floating seaweed may not only be edible but often contains tiny animals that can be used for food. The small fish and crabs can be dislodged by shaking the clump of seaweed over a container.

b. Plankton includes both minute plants and animals that drift about or swim weakly in the ocean. These basic organisms in the marine food chain are generally more common near land since their occurrence depends on the nutrients dissolved in the water. Plankton can be caught by dragging a net through the water. The taste of the plankton will depend on the types of organisms predominant in the area. If the population is mostly fish larvae, the plankton will taste like fish. If the population is mostly crab or shellfish larvae, the plankton will taste like crab or shellfish. Plankton contains valuable protein, carbohydrates and fats. Because of its high chiton and cellulose content, however, plankton cannot be immediately digested in large quantities. Therefore, anyone subsisting primarily on a plankton diet must gradually increase the quantities consumed. Most of the planktonic algae (phytoplankton) are smaller than the planktonic animals (zooplankton) and, although edible, are less palatable. Some plankton algae, for example, those dinoflagellates that cause "red tides" and paralytic shellfish poisoning, are toxic to humans.

(1) If a survivor is going to use plankton as a food source, there must be a sufficient supply of freshwater for drinking. Each plankton catch should be examined to remove all stinging tentacles broken from jellyfish or Portuguese man-of-war. The primarily gelatinous species may also be selectively discarded since their tissues are composed predominantly of saltwater. When the plankton is found in subtropical waters during the summer months, and the presence of poisonous dinoflagellates is suspected (due to discoloration or high luminescence of the ocean), the edibility test should be applied before eating.

(2) The final precaution that a survivor may wish to take before ingesting plankton is to feel or touch the plankton to check for species that are especially spiny. The catch should be sorted (visually) or dried and crushed before eating if it contains large numbers of these spiny species.

c. If a fishing kit is available, the task of fishing will be made much easier. Small fish will usually gather under the shadow of the raft or in clumps of floating seaweed.

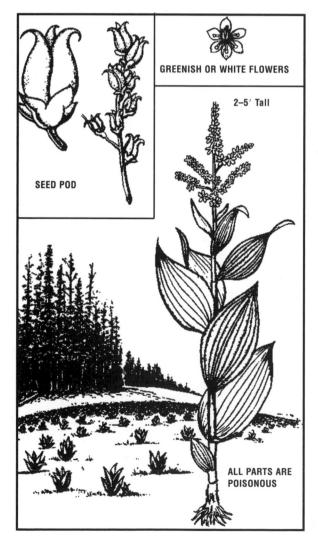

GREENISH OR WHITE FLOWERS

2–5' Tall

SEED POD

ALL PARTS ARE POISONOUS

Figure 18-45. False Hellebore.

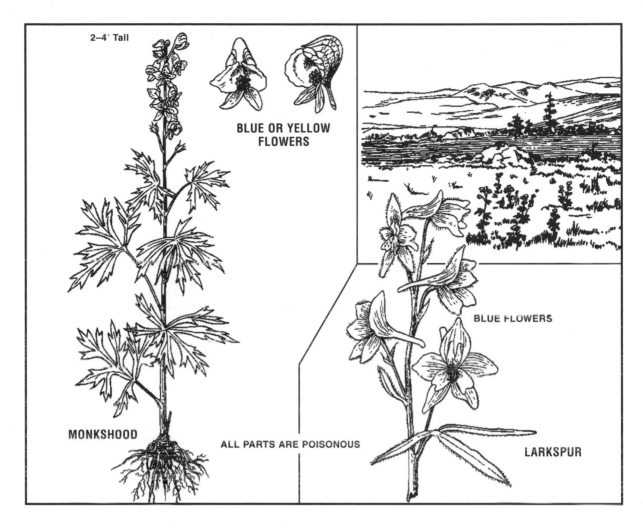

2–4′ Tall

BLUE OR YELLOW FLOWERS

MONKSHOOD

ALL PARTS ARE POISONOUS

BLUE FLOWERS

LARKSPUR

Figure 18-46. Monkshood and Larkspur.

These fish can be eaten or used as bait for larger fish. A net can also be used to procure most of sea life. Light attracts some types of fish. A flashlight or reflected moonlight can be used. It is not advisable to secure fishing lines to the body or the raft because a large fish may pull a person out of the raft or damage the raft. Fish, bait, or bright objects dangling in the water can attract large, dangerous fish. All large fish should be killed outside the raft with a blow to the head or by cutting off the head.

d. Sea birds have proven to be a useful food source that may be more easily caught than fish. Survivors have reported capturing birds by using baited hooks, by grabbing, and by shooting. Freshly killed birds should be skinned rather than plucked, to remove the oil glands. They can be eaten raw or cooked. The gullet contents can be a good food source. The flesh should be eaten or preserved immediately after cleaning. The viscera, along with any other unused parts, make good fish bait.

e. Marine mammals are rarely encountered by a person in the water, although they may be seen from a distance. Any large mammal is capable of inflicting injuries, but unless such mammals are pursued, they will generally avoid people. The killer whale (Orca) is rarely seen and, although large enough to feed on humans, has never been known to do so. Almost all sea mammals are a good source of food but are difficult to obtain. The liver, especially that of any arctic or cold-water mammal, should not be eaten because of toxic concentrations of vitamin A.

f. All sea life must be cleaned, cut up, and eaten as soon as possible to avoid spoilage. Any meat left over can be preserved by sun-drying or smoking. The internal parts can be used as bait. If any doubt exists as to the edibility of seafood, apply the "marine animal edibility test" found in chapter 11.

18-10. Preparing Animal Food. Survivors must know how to use the meat of game and fish to their advantage and how to do this with the least effort and physical exertion. Many people have died from starvation because they failed to take full advantage of a game carcass. They abandoned the carcass on the mistaken theory that they could get more game when needed.

a. If the animal is large, the first impulse is usually to pack the meat to camp. In some cases, it might be easier to move the camp to the meat. A procedure often advocated for transporting the kill is to use the skin as a sled for dragging the meat. When the entire animal is dragged, this method may prove satisfactory only on frozen lakes or rivers or over very smooth, snow-covered terrain. In rough or brush-covered country, however, it is generally more difficult to use this method, although it will work. Large mountain animals can sometimes be dragged down a snow-filled gully to the base of the mountain. If meat is the only consideration, and the survivors do not care about the condition of the skin, mountain game can sometimes be rolled for long distances. Before transporting a whole animal, it should be gutted and the incision closed. Once the bottom of the hill is reached, almost invariably the method is either to backpack the meat to camp, making several trips if no other survivors are present, or to pack the camp to the animal. Under survival conditions, home is on the back. When the weight of the meat proves excessive and moving the camp is not practical, some of the meat could be eaten at the scene. The heart, liver, and kidneys should be eaten as soon as possible to avoid spoilage.

(1) Under survival conditions, skinning and butchering must be done carefully so all edible meat can be saved. When the decision is made to discard the skin, a rough job can be done. However, considerations should be given to possible uses of the skin. A square of fresh skin, long enough to reach from the head to the knees, will not weigh much less when it is dried and is an excellent ground cloth for use under a sleeping bag on frozen ground or snow. The best time to skin and butcher an animal is immediately after the kill. However, if an animal is killed late in the day, it can be gutted immediately and the other work done the next morning. An effort to keep the carcass secure from predators should be made.

(2) When preparing meat under survival conditions, all edible fat should be saved. This is especially important in cold climates, as the diet may consist almost entirely of lean meat. Fat must be eaten in order to provide a complete diet. Rabbits lack fat, and the fact that a person will die after an extended diet consisting only of rabbit meat indicates the importance of fat in a primitive diet. The same is true of birds, such as the ptarmigan.

(3) Birds should be handled in the same manner as other animals. They should be cleaned after killing and protected from flies. Birds, with the exception of sea birds, should be plucked and cooked with the skin on. Carrion-eating birds, such as vultures, must be boiled for at least 20 minutes to kill parasites before further cooking and eating. Fish-eating birds have a strong fish-oil flavor. This may be lessened by baking them in mud or by skinning them before cooking.

3 1/2' Tall

BLUE FLOWERS

ALL PARTS ARE POISONOUS

SEED PODS

Figure 18-47. Lupine.

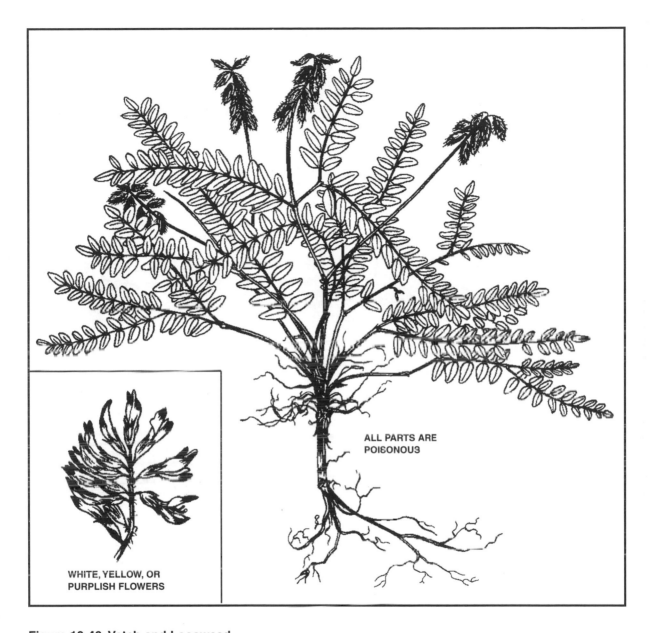

ALL PARTS ARE
POISONOUS

WHITE, YELLOW, OR
PURPLISH FLOWERS

Figure 18-48. Vetch and Locoweed.

b. There are two general ways to skin animals, depending on the size: the big-game method, or the glove-skinning method.

(1) Survivors should use the big-game method when skinning and butchering large game.

(a) The first step in skinning is to turn the animal on its back and with a sharp knife, cut through the skin in a straight line from the tailbone to a point under its neck, as illustrated in figure 18-54. In making this cut, pass around the anus and, with great care, press the skin open until the first two fingers can be inserted between the skin and the thin membrane enclosing the entrails. When the fingers can be forced forward, place the blade of the knife between the fingers, blade up, with knife held firmly. While forcing the fingers forward, palm upward, follow with the knife blade, cutting the skin but not cutting the membrane.

(b) If the animal is a male, cut the skin parallel to but not touching the penis. If the tube leading from the bladder is accidentally cut, a messy job and unclean meat will result. If the gall or urine bladders are broken, washing will help clean the meat. Otherwise, it is best

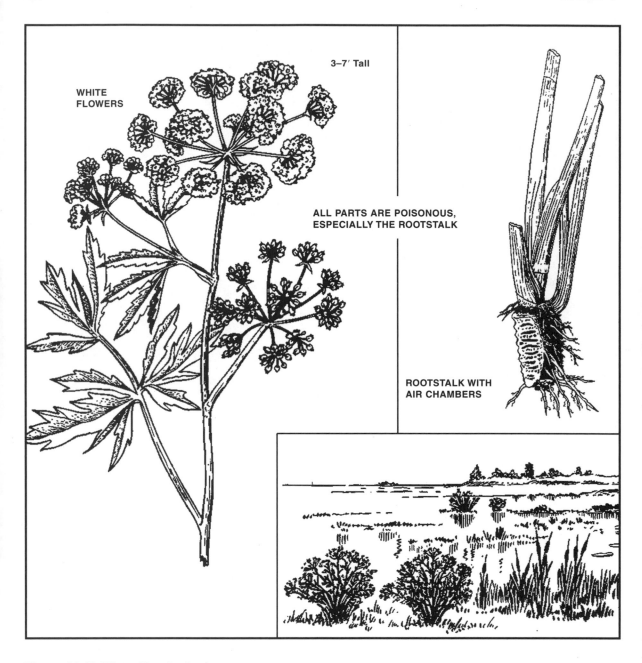

WHITE FLOWERS

3–7' Tall

ALL PARTS ARE POISONOUS, ESPECIALLY THE ROOTSTALK

ROOTSTALK WITH AIR CHAMBERS

Figure 18-49. Water Hemlock.

not to wash the meat but to allow it to form a protective glaze.

(c) On reaching the ribs, it is no longer possible to force the fingers forward, because the skin adheres more strongly to flesh and bone. Furthermore, care is no longer necessary. The cut to point C can be quickly completed by alternately forcing the knife under the skin and lifting it. With the central cut completed, make side cuts consisting of incisions through the skin, running from the central cut (A-C) up the inside of each leg to the knee and

hock joints. Then make cuts around the front legs just above the knees and around the hind legs above the hocks. Make the final crosscut at point C, and then cut completely around the neck and in back of the ears. Now is the time to begin skinning.

(d) On a small or medium-size animal, one person can skin on each side. The easiest method is to begin at the corners where the cuts meet. When the animal is large, three people can skin at the same time. However, one should remember that when it is getting dark and

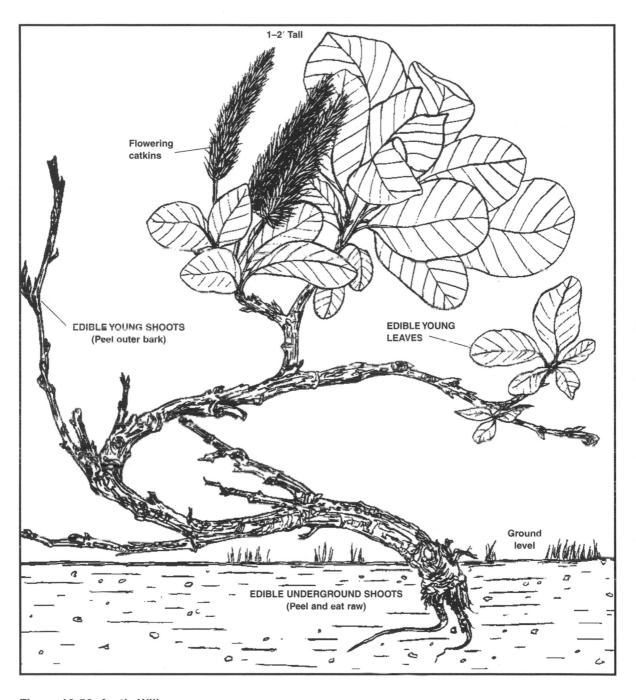

Figure 18-50. Arctic Willow.

hands are clumsy because of the cold, a sharp skinning knife can make a deep wound. After skinning down the animal's side as far as possible, roll the carcass on its side to skin the back. Then spread out the loose skin to prevent the meat from touching the ground and turn the animal on the skinned side. Follow the same procedure on the opposite side until the skin is free.

(e) In opening the membrane that encloses the entrails, follow the same procedure used in cutting the skin by using the fingers of one hand as a guard for the knife and separating the intestines from membrane. This thin membrane along the ribs and sides can be cut away in order to see better. Be careful to avoid cutting the intestines or bladder. The large intestine passes through an aperture in the pelvis. This tube must be separated from the bone surrounding it with a knife. Tie

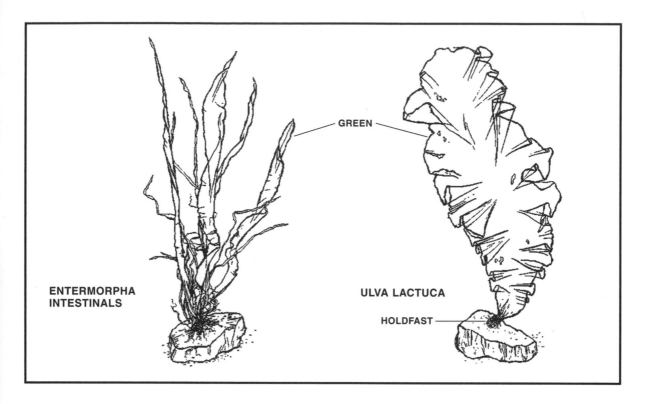

Figure 18-51. Edible Green Seaweeds.

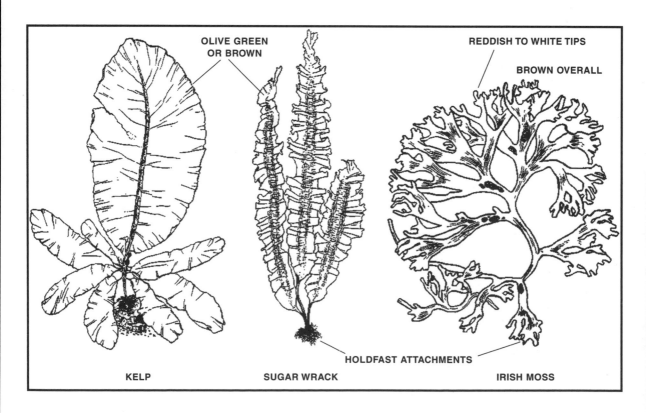

Figure 18-52. Edible Brown Seaweeds.

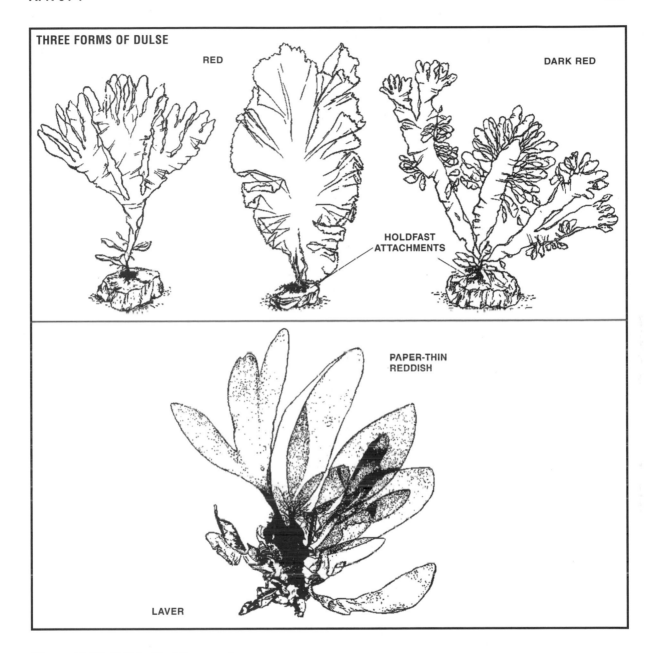

THREE FORMS OF DULSE

RED

DARK RED

HOLDFAST
ATTACHMENTS

PAPER-THIN
REDDISH

LAVER

Figure 18-53. Edible Red Seaweeds.

a knot in the bladder tube to prevent the escape of urine. With these steps completed, the entrails can be easily disengaged from the back and removed from the carcass. Another method of gutting or field dressing is shown in figure 18-55. After gutting is completed, it may be advisable to hang the animal. Figure 18-56 shows two methods. (NOTE: If it is hot, gut the animal before skinning it.)

(f) The intestines of a well-conditioned animal are covered with a lace-like layer of fat, which can be lifted off and placed on nearby bushes to dry for later use. The gall bladder, which is attached to the liver of some animals, should be carefully removed. If it should happen to rupture, the bile will taint anything it touches. Be sure to clean the knife if necessary. The kidneys are imbedded in the back, forward of the pelvis, and are covered with fat. Running forward from the kidneys on each side of the backbone are two long strips of chop meat or muscle called tenderloin or backstrap. Eat this after the liver, heart, and kidneys, as it is usually very tender. Edible meat can also be removed from the head, brisket, ribs, backbone, and pelvis.

HEAD UPHILL WHILE GUTTING
WILL HELP BODY DRAINAGE

Figure 18-54. Big Game Skinning.

(g) Large animals should be quartered. To do this, cut down between the first and second rib and then sever the backbone with an axe or machete. Cut through the brisket of the front half and then chop lengthwise through the backbone to produce the front quarters. On the rear half, cut through the pelvic bone and lengthwise through the backbone. To make the load lighter and easier to transport, a knife could be used to bone the animal, thereby eliminating the weight of the bones. Butchering is the final step and is simplified for survival purposes. The main purpose is to cut the meat into manageable-size portions (figure 18-57).

(2) Glove skinning is usually performed on small game (figure 18-58).

(a) The initial cuts are made down the insides of the back legs. The skin is then peeled back so the hindquarters are bare and the tail is severed. To remove the remaining skin, pull it down over the body in much the same way a pullover sweater is removed. The head and front feet are severed to remove the skin from the body. For one-cut skinning of small game, cut across the lower back and insert two fingers under each side of the slit. By pulling quickly in opposite directions, the hide will be easily removed (figure 18-59).

(b) To remove the internal organs, a cut should be made into the abdominal cavity without puncturing the organs. This cut must run from the anus to the neck. There are muscles that connect the internal organs to the trunk and they must be severed to allow the viscera to be removed. A rabbit may be gutted using a knife-less method with no mess and little time lost. Squeeze the entrails toward the rear, resulting in a tight, bulging abdomen. Raise the rabbit over the head and sling it down hard striking the forearms against the thighs. The momentum will expel the entrails through a tear in the vent (figure 18-60). Save the internal organs such as heart, liver, and kidneys, as they are nutritious. The liver should be checked for any white blotches and discarded if affected, as these indicate tularemia (also known as rabbit fever). The disease is transmitted by rodents but also infects humans.

c. Cold-blooded animals are generally easy to clean and prepare.

(1) Snakes and lizards are very similar in taste and they have similar skin. Like the mammals, the skin and viscera should be removed. The easiest way to do this is to sever the head and (or) legs. In the case of a lizard, peel back enough skin so it may be grasped securely and simply pull it down the length of the body turning the skin inside-out as it goes. If the skin does not come away easily, a cut down the length of the animal can be made. This will allow the skin to part from the body more easily. The entrails are then removed and the animal is ready to cook.

(2) Except for the larger amphibians such as the bullfrog, the hind legs are the largest part of the animal worth saving. To remove the hindquarters, simply cut

through the backbone with a knife, leaving the abdomen and upper body. Pull the skin from the legs and they are ready to cook. With the bullfrogs and larger amphibians, the whole body can be eaten. The head, skin, and viscera should be removed and discarded (use as bait to catch something else).

d. Most fish need little preparation before they are eaten. Scaling the fish before cooking is not necessary. A cut from the anus to the gills will expose the internal organs, which should be removed. The gills should also be removed before cooking. The black line along the inside of the backbone is the kidney and should be removed by running a thumbnail from the tail to the head. There is some meat on the head and it should not be discarded. See figure 18-61 for a method of filleting a fish.

e. All birds have feathers that can be removed in two ways: by plucking or pulling out the feathers, and by skinning. The gizzard, heart, and liver should be retained. The gizzard should be split open, as it contains partially digested food and that must be discarded before being eaten.

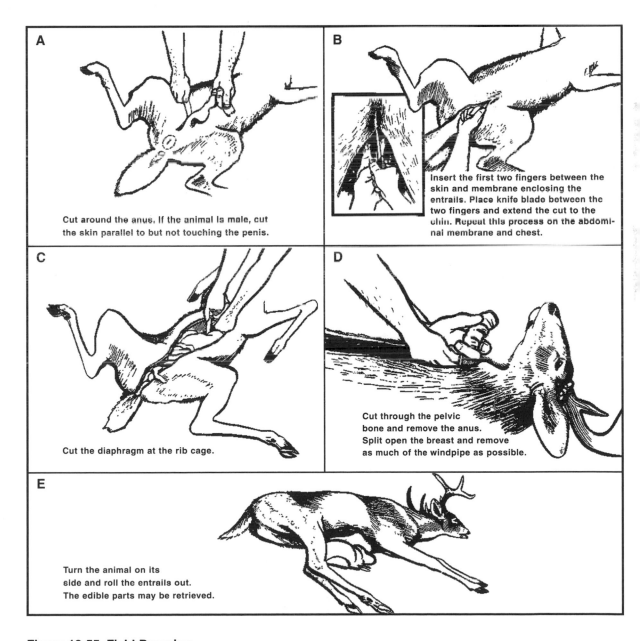

A Cut around the anus. If the animal is male, cut the skin parallel to but not touching the penis.

B Insert the first two fingers between the skin and membrane enclosing the entrails. Place knife blade between the two fingers and extend the cut to the chin. Repeat this process on the abdominal membrane and chest.

C Cut the diaphragm at the rib cage.

D Cut through the pelvic bone and remove the anus. Split open the breast and remove as much of the windpipe as possible.

E Turn the animal on its side and roll the entrails out. The edible parts may be retrieved.

Figure 18-55. Field Dressing.

Figure 18-56. Hanging Game.

f. Insects are an excellent food source and they require little or no preparation. The main point to remember is to remove all hard portions such as the hind legs of a grasshopper and the hard wing covers of beetles. The rest is edible.

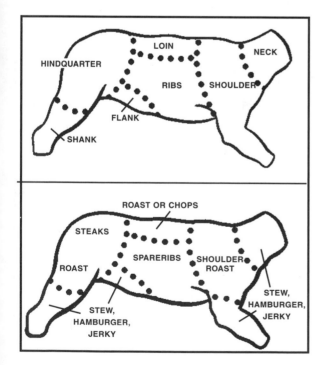

Figure 18-57. Butchering.

18-11. Cooking. All wild game, large insects (grasshoppers), freshwater fish, clams, mussels, snails, and crawfish must be thoroughly cooked to kill internal parasites. Mussels and large snails may have to be minced to make them tender.

a. Boiling is the most nutritious, simplest, and safest method of cooking (figure 18-62). Numerous containers can be used for boiling; for example, a metal container suspended above or set beside a heat source to boil foods. Green bamboo makes an excellent cooking container. Stone boiling is a method of boiling using super-heated rocks and a container that holds water but cannot be suspended over an open flame. Examples of containers are survival kit containers, a flying helmet, a hole in the ground lined with waterproof material, or a hollow log. The container is filled with food and water and then heated with super-hot stones until the water boils. Stones from a stream or damp area should not be used. The moisture in the stones may turn to steam and cause the stone to explode while the stones are being heated in the fire. The container should be covered and new stones added as the water stops boiling. The rocks can be removed with the aid of a wire secured to the rock before being put into the container, or with two sticks used in a chopstick fashion.

b. Baking is a good method of cooking as it is slow and is usually done by putting food into a container and cooking it slowly. Baking is often used with various types of ovens. Foods may be wrapped in wet leaves (figure 18-63; avoid using a type of plant that will give an unpleasant flavor to what is being cooked), placed inside a metal container, or they may be packed with mud or clay and placed

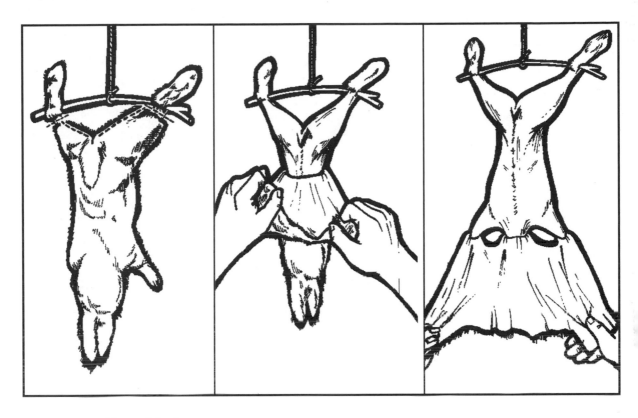

Figure 18-58. Glove Skinning.

directly on the coals. Fish and birds packed in mud and baked must not be skinned because the scales, skin, or feathers will come off the animal when the mud or clay is removed. Clambake-style baking is done by heating a number of stones in a fire and allowing the fire to burn down to coals. A layer of wet seaweed or leaves is then placed over the hot rocks. Food such as mussels and clams in their shells are placed on the wet seaweed and (or) leaves (figure 18-64). More wet seaweed and (or) leaves and soil is used as a cover. When thoroughly steamed in their own juices, clam, oyster, and mussel shells will open and may be eaten without further preparation.

c. Any type of food can be cooked in the ground in a rock oven (figure 18-65). First, a hole is dug about 2 feet deep and 2 or 3 feet square, depending on the amount of food to be cooked. The sides and bottom are then lined with rock. Next, procure several green trees about 6 inches in diameter and long enough to bridge the hole. Firewood and grass or leaves for insulation should also be gathered. A fire is started in the hole. Two or three green trees are placed over the hole and several rocks are placed on the trees. The fire must be maintained until the green trees burn through. This indicates that the fire has burned long enough to thoroughly heat the rock and the oven is ready. The fallen rocks, fire, and ash are removed from the hole and a thin layer of dirt is spread

over the bottom. The insulating material (grass, leaves, moss, etc.) is placed over the soil, then the more insulating material on top and around the food, another thin layer of soil, and the extra hot rocks are placed on top. The hole is then filled with soil up to ground level.

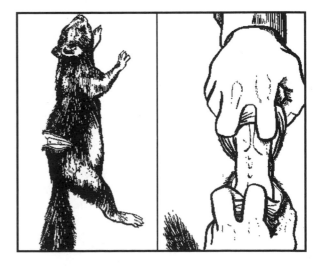

Figure 18-59. Small Animal Skinning.

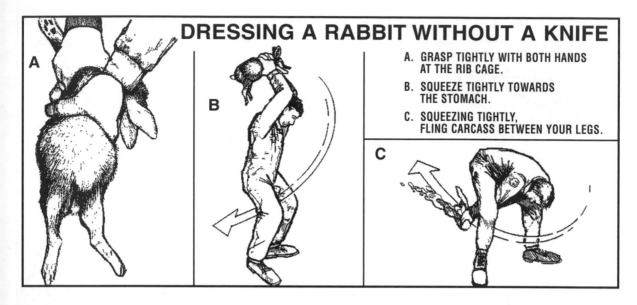

DRESSING A RABBIT WITHOUT A KNIFE

A. GRASP TIGHTLY WITH BOTH HANDS AT THE RIB CAGE.

B. SQUEEZE TIGHTLY TOWARDS THE STOMACH.

C. SQUEEZING TIGHTLY, FLING CARCASS BETWEEN YOUR LEGS.

Figure 18-60. Dressing a Rabbit Without a Knife.

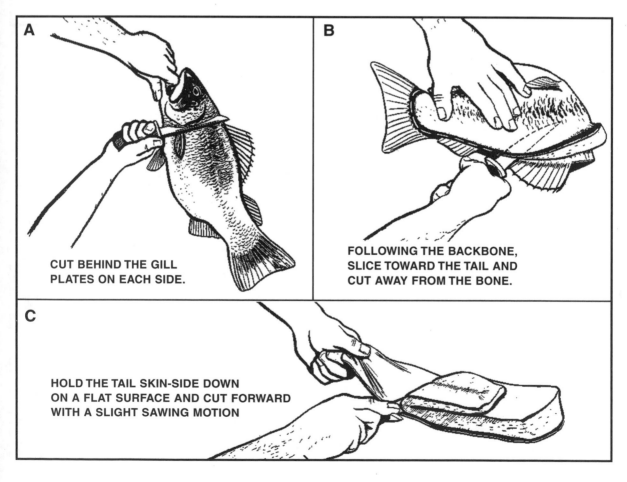

CUT BEHIND THE GILL PLATES ON EACH SIDE.

FOLLOWING THE BACKBONE, SLICE TOWARD THE TAIL AND CUT AWAY FROM THE BONE.

HOLD THE TAIL SKIN-SIDE DOWN ON A FLAT SURFACE AND CUT FORWARD WITH A SLIGHT SAWING MOTION

Figure 18-61. Filleting a Fish.

Small pieces of meat (steaks, chops, etc.) cook in 1 1/2 to 2 hours and large pieces take 5 to 6 hours.

d. Roasting is less desirable as it involves exposing the food to direct heat, which quickly destroys the nutritional properties (figure 18-66). Putting a piece of meat on a stick and holding it over the fire is considered roasting.

e. Broiling is the quickest way to prepare fish. A rock broiler may be made by placing a layer of small stones on top of hot coals, and laying the fish on the top. Scaling the fish before cooking is not necessary, and small fish need not be cleaned. Cooked in this manner, fish have a moist and delicious flavor. Crabs and lobsters may also be placed on the stones and broiled.

f. Meat may be cooked by laying it on a net board or stone (planking) that is propped up close to the fire (figure 18-67). The meat will have to be turned over at least once to allow thorough cooking. The cooking time depends on how close the meat is to the fire.

g. Frying is by far the least favorable method of preparing food. It tends to make the meat tough because most of the natural juices are cooked out of the meat. Some of the nutritional value of the meat will also be destroyed. Frying can be done on any nonporous surface that can be heated. Examples are unpainted aircraft parts, turtle shells, large seashells, flat rocks, and some survival kit parts.

18-12. Preparing Food in Enemy Areas. The problem of preparing food in a hostile area becomes acute when fire, even a small cooking fire, can bring about capture. After finding food in a hostile area, the problem of preparing the food in a manner that will not compromise the survivor's presence must be resolved. Of course, it

Figure 18-62. Boiling.

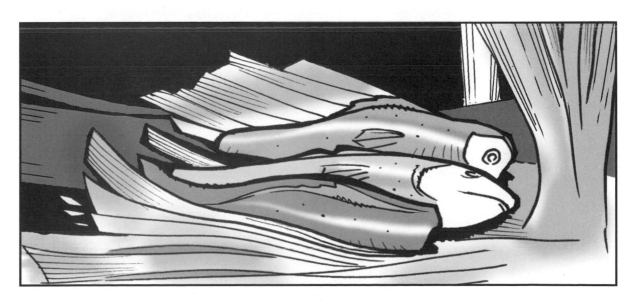

Figure 18-63. Baking.

Figure 18-64. Clam Baking.

Figure 18-65. Rock Oven.

Figure 18-66. Broiling and Roasting.

c. Assuming there will have to be a way to prepare food under hostile conditions, a survivor should be aware of some of the methods in order to achieve some degree of safety, and at the same time, improve palatability. Parasites and other organisms living in the flesh of the animals depend on the body temperature of the animals, the moisture within the flesh of the animals, and other factors to support their life. Any action that modifies these conditions (for example, freezing or thorough drying of the meat) and kills some parasites may improve the palatability.

d. If cooking is considered necessary, use extreme care in selecting the site for a fire and ensure that security considerations are favorable. The food should be prepared in very small quantities in order to keep the size of the fire as small as possible. The use of the "Dakota Hole" configuration is more appropriate for cooking food during a tactical situation (figure 16-14).

18-13. Preserving Food. Finding natural foods is an uncertain aspect of survival. The survivor must make the best use of the available food. Food, especially meat, has a tendency to spoil within a short period of time unless it is preserved. There are many ways to preserve food: some of the most common are cooking, refrigeration, freezing, and dehydration.

would be simple to state that the best solution would be to eat the food without cooking it.

a. In some respects, this would be a more reasonable solution than it might initially seem to be. From the standpoint of palatability, it is mostly a matter of adjusting the "frame of mind." Animal foods are recognized as being palatable when cooked to a very minor degree. The need for food cannot be ignored and the situation may demand that it be eaten partially cooked or even uncooked.

b. With regard to the health considerations involved, many of the reasons for cooking are recognized as a means of destroying organisms that may be present in the food and can cause sickness or ill effects if they enter the body. Under survival conditions in a hostile area, one may be forced to forego thorough cooking and accept the risk involved until their return to friendly forces where professional treatment is available.

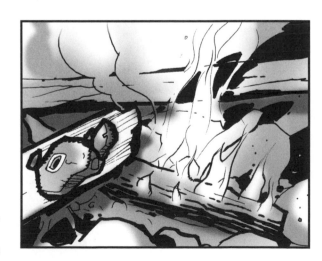

Figure 18-67. Planking.

a. Cooking will slow down the decomposition of food but will not eliminate it. This is because many bacteria are present, which work to break it down. Cooking methods that are the best for immediate consumption, such as boiling, are the least effective for preserving food. Food should be re-cooked every day until all is consumed.

b. Cooling is an effective method of storing food for short periods of time. Heat tends to accelerate the decomposition process whereas cooling retards decomposition. The colder food becomes, the less the likelihood of deterioration, until freezing eliminates decomposition. Cooling devices available to a survivor are:

(1) Food items buried in snow will maintain a temperature of approximately 32°F.

(2) Food wrapped in waterproof material and placed in streams will remain cool in summer months. Care should be taken to ensure food is secured.

(3) Earth below the surface, particularly in shady areas or along streams, remains cooler than the surface. A hole may be dug, lined with grass, and covered to form an effective cool-storage area much the same as a root cellar.

(4) When water evaporates, it tends to cool down the surrounding area. Using this fact, articles of food may be wrapped in an absorbent material such as cotton or burlap and rewetted as the water evaporates.

c. Once food is frozen, it will not decompose. Food should be frozen in meal-size portions so refreezing is avoided.

d. Drying removes all moisture and preserves the food. Drying is done by sunning, smoking, or burying it in hot sand.

(1) For sun-drying, the food should be sliced very thin and placed in direct sunlight. Meat should be cut across the grain to improve tenderness and decrease drying time. If salt is available it should be added to improve flavor and accelerate the drying process (figure 18-68).

(2) Smoking is a process done through the use of nonresinous wood such as willow or aspen and is used to produce smoke, which adds favor and dries the meat. A smoke rack is also necessary to contain the smoke (figure 18-69). The following are the procedures for drying meat by using smoke:

(a) Cut meat very thin and across the grain. If the meat is warm and difficult to slice thin, cut the meat in 1 or 2-inch cubes and beat it thin with a clean wooden mallet (improvised).

(b) Remove fat.

(c) Hang the meat on a rack so each piece is separate.

(d) Elevate meat no less than 2 feet above coals.

Figure 18-68. Sun-Drying.

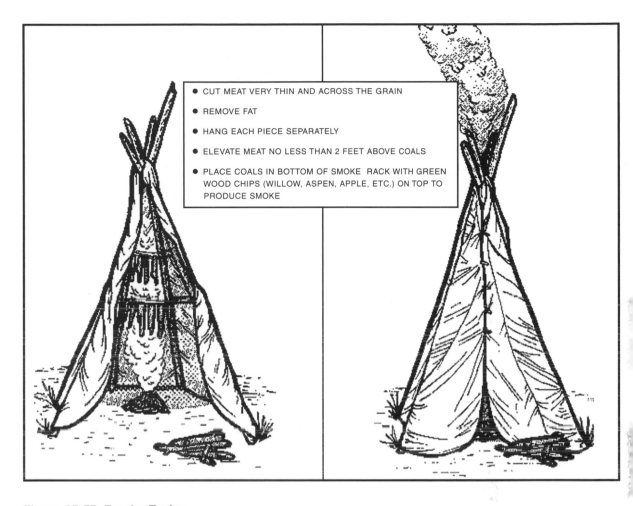

- CUT MEAT VERY THIN AND ACROSS THE GRAIN

- REMOVE FAT

- HANG EACH PIECE SEPARATELY

- ELEVATE MEAT NO LESS THAN 2 FEET ABOVE COALS

- PLACE COALS IN BOTTOM OF SMOKE RACK WITH GREEN WOOD CHIPS (WILLOW, ASPEN, APPLE, ETC.) ON TOP TO PRODUCE SMOKE

Figure 18-69. Smoke-Drying.

(e) Coals are placed in the bottom of a smoke rack with green woodchips on top to produce smoke.

e. The method used to preserve fish through warm weather is similar to that used in preserving meat (figure 18-70). When there is no danger of predatory animals disturbing the fish, the fish should be placed on available fabric and allowed to cool during the night. Early the next morning, before the air gets warm, the fish should be rolled in moist fabric (and leaves). This bundle can be placed inside the survivor's pack. During the rest periods, or when the pack is removed, it should be placed in a cool location out of the sun's rays.

(1) Fish may be dried in the same manner described for smoking meat. To prepare fish for smoking, the heads and backbone are removed and the fish are spread flat on a grill. Thin willow branches with bark removed make skewers.

(2) Fish may also be dried in the sun. They can be suspended from branches or spread on hot rocks. When

the meat has dried, seawater or salt should be used on the outside, if available.

f. In survival environments, there are many animals and insects that will devour a survivor's food if it is not correctly stored. Protecting food from insects and birds is done by wrapping it in parachute material, wrapping and tying brush around the bundle, and finally, wrapping it with another layer of material. This creates "dead air" space, making it more difficult for insects and birds to get to the food. If the outer layer is wetted, evaporation will also cool the food to some degree. In most cases, if the food is stored several feet off the ground, it will be out of reach of most animals. This can be done by hanging the food or putting it in a "cache." If the food is dehydrated, the container must be completely waterproof to prevent reabsorption. Frozen food will remain frozen only if the outside temperature remains below freezing. Burying food is a good way to store it as long as scavengers are not in the area to uncover it. Insects and small animals should also be

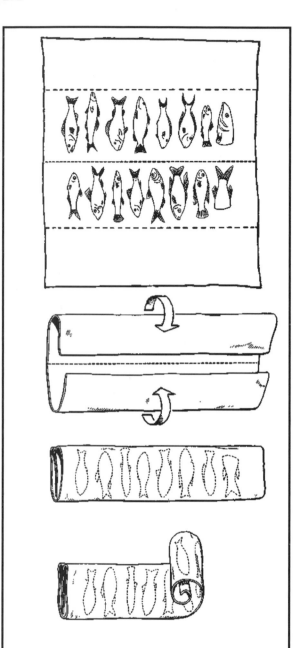

1. Arrange fish on available fabric.

2. Turn down the upper edge of wrap over the top line of fish and turn up the lower edges over the lower line.

3. Fold in the center as shown.

4. Then begin on the edge and roll the wrap. You will have a rounded roll of protected fish. This roll should be securely, but not tightly, tied and wrapped in a sleeping bag, parachute fabric, or clothing, as you would do with the meat.

Figure 18-70. Preserving Fish.

thought of when burying the food. Food should never be stored in the shelter as this may attract wild animals and could be hazardous to the survivors.

18-14. Preparing Plant Food. Preparing plant foods can be more involved than preparing animal life.

a. Some plant foods, such as acorns and tree bark, may be bitter because of tannin. These plants will require leaching by chopping up the plant parts and pouring several changes of fresh water over them. This will help wash out the tannin, making the plant more palatable. Other plants such as cassava and green papaya must be cooked before eating to break down the harmful enzymes and chemical crystals within them and make them safe to eat. Plants such as skunk cabbage must undergo this cooking process several times before it is safe to eat.

b. All starchy foods must be cooked since raw starch is difficult to digest. They are boiled, steamed, roasted, or fried and are eaten plain, or mixed with other wild foods. The manioc (cassava) is best cooked, because the bitter form (green stem) is poisonous when eaten raw. Starch is removed from sago palm, cycads, and other starch-producing trunks by splitting the trunk and pounding the soft, whitish inner parts with a pointed club. This pulp is washed with water and the white sago (pure starch) is drained into a container. It is washed a second time, and then it may be used directly as flour. One trunk of the sago palm will supply a survivor's starch needs for many weeks.

c. The fiddleheads of all ferns are the curled, young, succulent fronds, which have the same food value as cabbage or asparagus. Practically all types of fiddleheads are covered with hair, which makes them bitter. The hair can be removed by washing the fiddleheads in water. If fiddleheads are especially bitter, they should be boiled for 10 minutes and then reboiled in fresh water for 30 to 40 minutes. Wild bird eggs or meat may be cooked with the fiddleheads to form a stew.

d. Wild grasses have an abundance of seeds, which may be eaten boiled or roasted after separating the chaff from the seeds by rubbing. No known grass is poisonous. If the kernels are still soft and do not have large, stiff barbs attached, they may be used for porridge. If brown or black rust is present, the seeds should not be eaten (ergot poisoning). To gather grass seeds, a cloth is placed on the ground and the grass heads beaten with sticks.

e. Plants that grow in wet places along margins of rivers, lakes, and ponds, and those growing directly in water, are of potential value as survival food. The succulent underground parts and stems are most frequently eaten. Poisonous water plants are rare. In temperate climates, the water hemlock is the most poisonous plant found around marshes and ponds. In the tropics, the various members of the calla lily family often grow in very wet places. The leaves of the calla lily look like

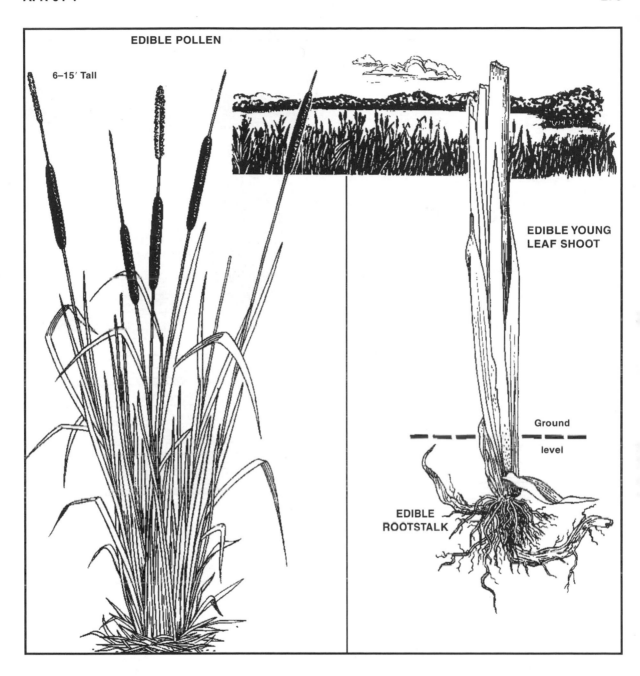

EDIBLE POLLEN

6–15′ Tall

EDIBLE YOUNG LEAF SHOOT

Ground level

EDIBLE ROOTSTALK

Figure 18-71. Cattails.

arrowheads. Jack-in-the-pulpit, calla lily, and sweet flag are members of the Arum family. To be eaten, the members of this plant family must be cooked in frequent changes of water to destroy the irritant crystals in the stems. Two kinds of marsh and water plants are the cattail and the water lily.

(1) The cattail (Typha) is found worldwide except in tundra regions of the far north (figure 18-71). Cattails can be found in the damper places in desert areas of all continents, as well as in the moist tropic and temperate zones of both hemispheres. The young shoots taste like asparagus. The spikes can be boiled or steamed when green and then eaten. The rootstalks, without the outer covering, are eaten boiled or raw. Cattail roots can be cut into thin strips, dried, and then ground into flour. They are 46 percent starch, 11 percent sugar, and the rest is fiber. While the plant is in flower, the yellow pollen is very abundant; this may be mixed with water

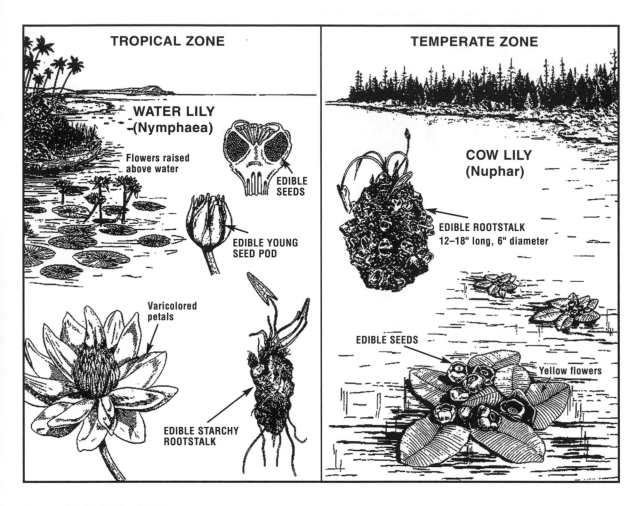

Figure 18-72. Water Lillies.

and made into small cakes and steamed as a substitute for bread.

(2) Water lilies (Nymphaea and Nuphar) occur on all the continents, but principally in southern Asia, Africa, North America, and South America (figure 18-72). Two main types are:

(a) Temperate water lilies produce enormous rootstalks and yellow or white flowers that float on the water.

(b) Tropical water lilies produce large edible tubers and flowers that are elevated above the water surface.

(3) Rootstalks or tubers may be difficult to obtain because of deep water. They are starchy and high in food value. They can be eaten either raw or boiled. Stems may be cooked in a stew. Young seedpods may be sliced and eaten as a vegetable. Seeds may be bitter, but are very nourishing. They may be parched and rubbed between stones as flour. The water lily is considered an important food source by native peoples in many parts of the world.

f. Nuts are very high in nutritional value and usually can be eaten raw. Nuts may be roasted in the fire or by shaking them in a container with hot coals from the fire. They may then be ground to make flour. If a survivor does not wish to eat a plant or plant part raw, it can be cooked using the same methods used for cooking meat—by boiling, roasting, baking, broiling, or frying.

g. If survivors have been able to procure more plant foods than can be eaten, the excess can be preserved in the same manner as animal foods. Plant foods can be dried by wind, air, sun, or fire, with or without smoke. A combination of these methods can be used. The main object is to remove the moisture. Most wild fruits can be dried. If the plant part is large, such as some tubers, it should be sliced, and then dried. Some type of protection may be necessary to prevent consumption and (or) contamination by insects. Extra fruits or berries can be carried with the survivor by wrapping them in leaves or moss.

Chapter 19

WATER

19-1. Introduction. Nearly every survival account details the need survivors had for water. Many ingenious methods of locating, procuring, purifying, and storing water are included in the recorded experiences of downed aircrew members. If survivors are located in temperate, tropic, or dry climates, water may be their first and most important need. The priority of finding water over that of obtaining food must be emphasized to potential survivors. An individual may be able to live for weeks without food, depending on the temperature and amount of energy being exerted. A person who has no water can be expected to die within days. Even in cold climate areas or places where water is abundant, survivors should attempt to keep their body fluids at a level that will maintain them in the best possible state of health. Even in relatively cold climates, the body needs 2 quarts of water per day to remain efficient (figure 19-1).

19-2. Water Requirements. Normally, with an atmospheric temperature of about 68°F, the average adult requires 2 to 3 quarts of water daily.

a. This water is necessary to replace that lost daily in the following ways:

(1) Urine. Approximately 1.4 quarts of water is lost in urine.

(2) Sweat. About 0.1 quart of water is lost in sweat.

(3) Feces. Approximately 0.2 quart of water is lost in feces.

(4) Insensible Water Loss. When the individual is unaware that water loss is actually occurring, it is referred to as insensible water loss. Insensible water loss occurs by the following mechanisms:

(a) Diffusion through the skin. Water loss through the skin occurs as a result of the actual diffusion of water molecules through the cells of the skin. The average loss of water in this manner is approximately 0.3 to 0.4 quart. Fortunately, loss of greater quantities of water by diffusion is prevented by the outermost layer of the skin, the epidermis, which acts as a barrier to this type of water loss.

(b) Evaporation through the lungs. Inhaled air initially contains very little water vapor. However, as soon as it enters the respiratory passages, the air is exposed to the fluids covering the respiratory surfaces. By the time this air enters the lungs, it has become totally saturated with moisture from these surfaces. When the air is exhaled, it is still saturated with moisture and water is lost from the body.

b. Larger quantities of water are required when water loss is increased in any one of the following circumstances:

(1) Heat Exposure. When an individual is exposed

to very high temperatures, water lost in the sweat can be increased to as much as 3.5 quarts an hour. Water loss at this increased rate can deplete the body fluids in a short time.

(2) Exercise. Physical activity increases the loss of water in two ways, as follows:

(a) The increased respiration rate causes increased water loss by evaporation through the lungs.

(b) The increased body heat causes excessive sweating.

(3) Cold Exposure. As the temperature decreases, the amount of water vapor in the air also decreases. Therefore, breathing cold air results in increased water loss by evaporation from the lungs.

(4) High Altitude. At high altitudes, increased water loss by evaporation through the lungs occurs not only as a result of breathing cooler air but also as a result of the increased respiratory efforts required.

(5) Burns. After extensive burns, the outermost layer of the skin is destroyed. When this layer is gone, there is no longer a barrier to water loss by diffusion, and the rate of water loss in this manner can increase up to 5 quarts each day.

(6) Illness. Severe vomiting or prolonged diarrhea can lead to serious water depletion.

c. Dehydration (body-fluid depletion) can occur when required body fluids are not replaced.

(1) Dehydration is accompanied by the following symptoms:

(a) Thirst.

(b) Weakness.

(c) Fatigue.

(d) Dizziness.

(e) Headache.

(f) Fever.

(g) Inelastic abdominal skin.

(h) Dry mucous membranes, that is, dry mouth and nasal passages.

(i) Infrequent and reduced volume of urination. The urine is concentrated, so it is very dark in color. In severe cases, urination may be quite painful.

(2) Companions will observe the following behavioral changes in individuals suffering from dehydration:

(a) Loss of appetite.   (e) Apathy.

(b) Lagging pace.   (f) Emotional instability.

(c) Impatience.   (g) Indistinct speech.

(d) Sleepiness.   (h) Mental confusion.

(3) Dehydration is a complication that causes decreased efficiency in the performance of even the simplest task. It also predisposes survivors to the development of severe shock following minor injuries. Constriction of blood vessels in the skin as a result of dehydration increases the danger of cold injury during

NO WALKING AT ALL	MAXIMUM DAILY TEMPERATURE (°F) IN SHADE	AVAILABLE WATER PER MAN, US QUARTS					
		0	1 Qt	2 Qts	4 Qts	10 Qts	20 Qts
		DAYS OF EXPECTED SURVIVAL					
	120°	2	2	2	2.5	3	4.5
	110	3	3	3.5	4	5	7
	100	5	5.5	6	7	9.5	13.5
	90	7	8	9	10.5	15	23
	80	9	10	11	13	19	29
	70	10	11	12	14	20.5	32
	60	10	11	12	14	21	32
	50	10	11	12	14.5	21	32

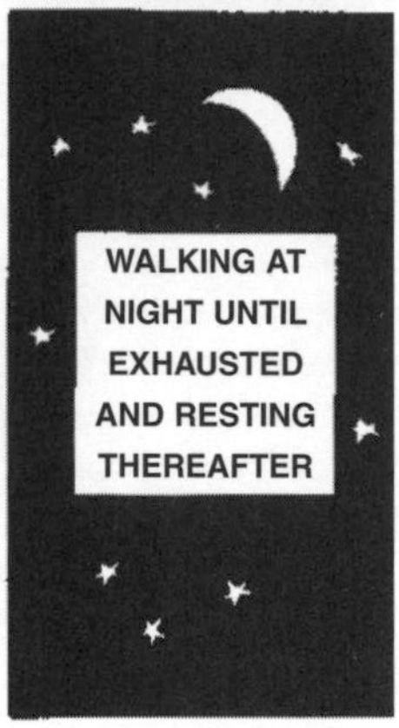

WALKING AT NIGHT UNTIL EXHAUSTED AND RESTING THEREAFTER	MAXIMUM DAILY TEMPERATURE (°F) IN SHADE	AVAILABLE WATER PER MAN, US QUARTS					
		0	1 Qt	2 Qts	4 Qts	10 Qts	20 Qts
		DAYS OF EXPECTED SURVIVAL					
	120°	1	2	2	2.5	3	
	110	2	2	2.5	3	3.5	
	100	3	3.5	3.5	4.5	5.5	
	90	5	5.5	5.5	6.5	8	
	80	7	7.5	8	9.5	11.5	
	70	7.5	8	9	10.5	13.5	
	60	8	8.5	9	11	14	
	50	8	8.5	9	11	14	

Figure 19-1. Water Requirements.

cold exposure. Failure to replace body fluids ultimately results in death.

(a) Proper treatment for dehydration is to replace lost body fluids. The oral intake of water is the most readily available means of correcting this deficiency. A severely dehydrated person will have little appetite. This person must be encouraged to drink small quantities of water at frequent intervals to replenish the body's fluid volume. Cold water should be warmed so the system will accept it easier.

(b) To prevent dehydration, water loss must be replaced by periodic intake of small quantities of water throughout the day. As activities or conditions intensify, the water intake should be increased accordingly. Water intake should be sufficient to maintain a minimum urinary output of 1 pint every 24 hours. Thirst is not an adequate stimulus for water intake, and a person often dehydrates when water is available. Therefore, water intake should be encouraged even when the person is not thirsty. Humans cannot adjust to decreased water intake for prolonged periods of time. When water is in short supply, any available water should be consumed sensibly. If sugar is available, it should be mixed with the water, and efforts should be made to find a local water source. Until a suitable water source is located, individual water losses should be limited in the following ways:

-1. Physical activity should be limited to the absolute minimum required for survival activities. All tasks should be performed slowly and deliberately with minimal expenditure of energy. Frequent rest periods should be included in the daily schedule.

-2. In hot climates, essential activity should be conducted at night or during the cooler part of the day.

-3. In hot climates, clothing should be worn at all times because it reduces the quantity of water lost by sweating. Sweat is absorbed into the clothing and evaporated from its surface in the same manner it evaporates from the body. This evaporation cools the air trapped between the clothing and the skin, causing a decrease in the activity of the sweat glands and a subsequent reduction in water loss.

-4. In hot weather, light-colored clothing should be worn rather than dark-colored clothing. Dark-colored clothing absorbs the sun's rays and converts them into heat. This heat causes an increase in body temperature, which activates the sweat glands and increases water lost through sweating. Light-colored clothing, however, reflects the sun's rays, minimizing the increase in body temperature and subsequent water loss.

19-3. Water Sources. Survivors should be aware of both the water sources available to them and the resources at their disposal for producing water.

a. Survivors may obtain water from solar stills, desalter kits, or canned water packed in various survival kits. It would be wise for personnel, who may one day have to use these methods of procuring water, to be knowledgeable of their operating instructions and the amount of water they produce.

(1) Canned water provides 10 ounces per can.

(2) Desalter kits are limited to 1 pint per chemical bar—kits contain eight chemical bars.

(3) A "sea solar still" can produce as many as 2 1/2 pints per day.

(4) "Land solar stills" produce varied amounts of water. This amount is directly proportionate to the amount of water available in the soil or placed into the still (vegetation, entrails, contaminated water, etc.) and the ambient temperature.

b. Aircrew members would be wise to carry water during their missions. Besides the fact that the initial shock of the survival experience sometimes produces feelings of thirst, having an additional water container can benefit survivors. The issued items (canned water, desalter kits, and solar stills) should be kept by survivors for times when no natural sources of freshwater are available.

c. Naturally occurring indicators of water are:

(1) Surface water, including streams, lakes, springs, ice, and snow.

(2) Precipitation, such as rain, snow, dew, sleet, etc.

(3) Subsurface water, which may not be as readily accessible as wells, cisterns, and underground springs and streams, can be difficult for survivors to locate and use.

d. Several indicators of possible water are:

(1) Presence of abundant vegetation of a different variety, such as deciduous growth in a coniferous area.

(2) Drainages and low-lying areas.

(3) Large clumps of plush grass.

(4) Animal trails that may lead to water. The "V" formed by intersecting trails often point toward water sources.

e. Survivors may locate and procure water as follows:

(1) Precipitation may be procured by laying a piece of nonporous material such as a poncho, piece of canvas,

plastic, or metal material on the ground. If rain or snow is being collected, it may be more efficient to create a bag or funnel shape with the material so the water can be easily gathered. Dew can be collected by wiping it up with a sponge or cloth first, and then wringing it into a container (figure 19-2). Consideration should be given to the possibility of contaminating the water with dyes, preservatives, or oils on the surfaces of the objects used to collect the precipitation. Ice will yield more water per given volume than snow and requires less heat to do so. If the sun is shining, snow or ice may be placed on a dark surface to melt (dark surfaces absorb heat, whereas light surfaces reflect heat). Ice can be found in the form of icicles on plants and trees, sheet ice on rivers, ponds, and lakes, or sea ice. If snow must be used, survivors should use snow closest to the ground. This snow is packed and will provide more water for the amount of snow than will the upper layers. When snow is to be melted for water, place a small amount of snow in the bottom of the container being used and place it over or near a fire. Snow can be added a little at a time. Survivors should allow water in the container bottom to become warm so that when more snow is added, the mixture remains slushy. This will prevent burning the bottom out of the container. Snow absorbs water and if packed, forms an insulating airspace at the bottom of the container. When this happens, the bottom may burn out.

(2) Several things may help survivors locate groundwater such as rivers, lakes, and streams.

(a) The presence of swarming insects indicates water is near. In some places, survivors should look for signs of animal presence. For example, in damp places, animals may have scratched depressions into the ground to obtain water; insects may also hover over these areas.

(b) In the Libyan Sahara, doughnut-shaped mounds of camel dung often surround wells or other water sources. Bird flight patterns can indicate the direction to or from water. Pigeons and doves make their way to water regularly. They fly from water in the morning and to it in the evening. Large flocks of birds may also congregate around or at areas of water.

(c) The presence of people will indicate water. The location of this water can take many forms—stored water in containers that are carried with people who are traveling, wells, irrigation systems, pools, etc. Survivors who are evaders should be extremely cautious when approaching any water source, especially if they are in dry areas; these places may be guarded or inhabited.

(3) When no surface water is available, survivors may have to tap the earth's supply of groundwater. Access to this depends upon the type of ground—rock or loose material, clay, gravel, or sand.

(a) In rocky ground, survivors should look for springs and seepages. Limestone and lava rocks will have more and larger springs than any other rocks. Most

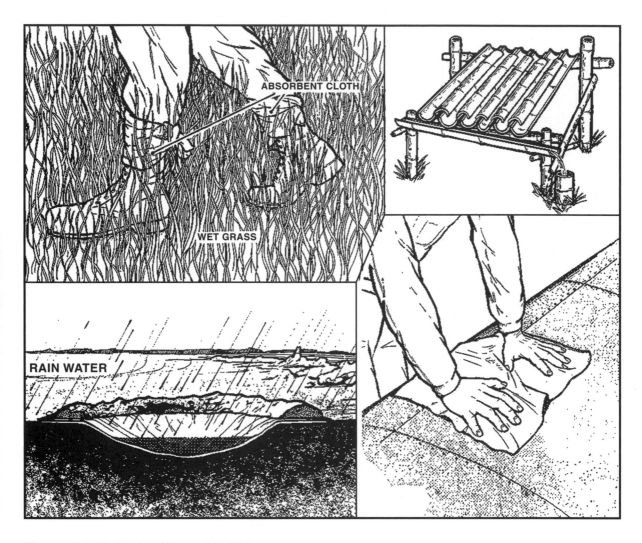

Figure 19-2. Methods of Procuring Water.

lava rocks contain millions of bubble holes; groundwater may seep through them. Survivors can also look for springs along the walls of valleys that cross a lava flow. Some flows will have no bubbles but do have "organ pipe" joints—vertical cracks that part the rocks into columns a foot or more thick and 20 feet or more high. At the foot of these joints, survivors may find water creeping out as seepage, or pouring out in springs.

(b) Most common rocks, like granite, contain water only in irregular cracks. A crack in a rock with bird dung around the outside may indicate a water source that can be reached by a piece of surgical hose used as a straw or siphon.

(c) Water is more abundant and easier to find in loose sediments than in rocks. Springs are sometimes found along valley floors or down along their sloping sides. The flat benches or terraces of land above river

valleys usually yield springs or seepages along their bases, even when the stream is dry. Survivors shouldn't waste time digging for water unless there are signs that water is available. Digging in the floor of a valley under a steep slope, especially if the bluff is cut in a terrace, can produce a water source. A lush green spot where a spring has been during the wet season is a good place to dig for water. Water moves slowly through clay, but many clays contain strips of sand that may yield springs. Survivors should look for a wet place on the surface of clay bluffs and try digging it out.

(d) Along coasts, water may be found by digging beach wells (figure 19-3). Locate the wells behind the first or second pressure ridge. Wells can be dug 3 to 5 feet deep and should be lined with driftwood to prevent sand from refilling the hole. Rocks should be used to line the bottom of the well to prevent stirring up sand when procuring the water. The average well may take as long as 2 hours to produce 4 to 5 gallons of water. (Do not be discouraged if the first try is unsuccessful—dig another.)

19-4. Water in Snow and Ice Areas. Due to the extreme cold of arctic areas, water requirements are greatly increased. Increased body metabolism, respiration of cold air, and extremely low humidity play important roles in reducing the body's water content. The processes of heat production and digestion in the body also increase the need for water in colder climatic zones. The constructing of shelters and signals and the obtaining of firewood are extremely demanding tasks for survivors. Physical exertion and heat production in extreme cold place the water requirements of a survivor close to 5 or 6 quarts per day to maintain proper hydration levels. The diet of survivors will often be dehydrated rations and high-protein food sources. For the body to digest and use these food sources effectively, increased water intake is essential.

a. Obtaining water need not be a serious problem in the Arctic because an abundant supply of water is available from streams, lakes, ponds, snow, and ice. All surface water should be purified by some means. In the summer, surface water may be discolored but is drinkable when purified. Water obtained from glacier-fed rivers and streams may contain high concentrations of dirt or silt. By letting the water stand for a period of time, most silt will settle to the bottom; the remaining water can be strained through porous material for further filtration.

b. A "water machine" can be constructed that will produce water while the survivors are doing other tasks. It can be made by placing snow on any porous material (such as parachute or cotton), gathering up the edges, and suspending the "bag" of snow from any support near the fire. Radiant heat will melt the snow and the water will drip from the lowest point on the bag. A container should be placed below this point to catch the water (figure 19-4).

c. In some arctic areas, there may be little or no fuel supplies with which to melt ice and snow for water. In this case, body heat can be used to do the job. The ice or snow can be placed in a waterproof container like a waterbag and placed between clothing layers next to the body. This cold substance should not be placed directly

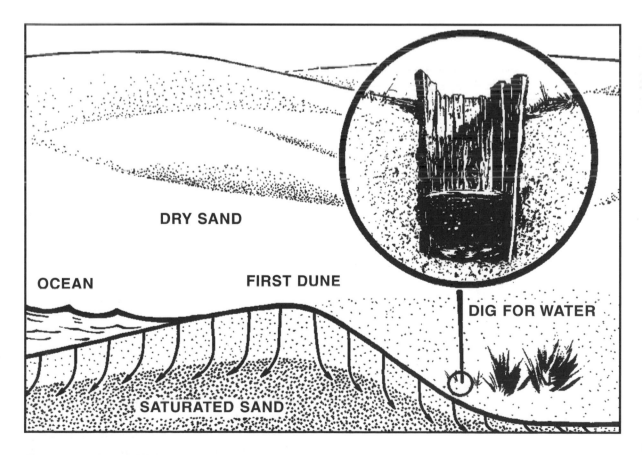

Figure 19-3. Beach Well.

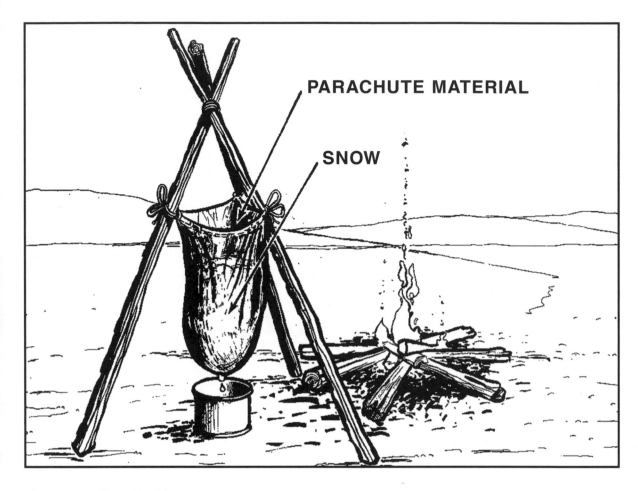

Figure 19-4. Water Machine.

next to the skin—it causes chilling and lowering of the body temperature.

d. Since icebergs are composed of freshwater, they can be a readily available source of drinking water. Survivors should use extreme caution when trying to obtain water from this source. Even large icebergs can suddenly roll over and dump survivors into the frigid seawater. If sea ice is the primary source of water, survivors should recall that like seawater itself, saltwater ice should never be ingested. To obtain water in polar regions or sea ice areas, survivors should select old sea ice—a bluish or blackish ice that shatters easily and generally has rounded corners. This ice will be almost salt-free. New sea ice is milky or gray-colored with sharp edges and angles. This type of ice will not shatter or break easily. Snow and ice may be saturated with salt from blowing spray; if it tastes salty, survivors should select different snow or ice sources.

e. The ingesting of unmelted snow or ice is not recommended. Eating snow or ice lowers the body's temperature, induces dehydration, and causes minor cold injury to lips and mouth membranes. Water consumed in cold areas should be in the form of warm or hot fluids. The ingestion of cold fluids or foods increases the body's need for water and requires more body heat to warm the substance.

19-5. Water on the Open Seas. The lack of drinkable water could be a major problem on the open seas. Seawater should never be ingested in its natural state. It will cause an individual to become violently ill in a very short period of time. When water is limited and cannot be replaced by chemical or mechanical means, it must be used efficiently. As in the desert, conserving sweat, not water, is the rule. Survivors should stay in the shade as much as possible and dampen clothing with seawater to keep cool. They should not overexert themselves but relax and sleep as much as possible.

a. If it rains, survivors can collect rainwater in available containers and store it for later use. Storage containers could be cans, plastic bags, or the bladder of a life preserver. Drinking as much rainwater as possible while it is raining is advisable. If the freshwater should become contaminated with small amounts of seawater or salt spray, it will remain safe for drinking (figure 19-5). At night and on foggy days, survivors should try to collect dew for drinking water by using a sponge, chamois, handkerchief, etc.

b. Solar stills will provide a drinkable source of water. Survivors should read the instructions immediately and set them up, using as many stills as available. (Be sure to attach them to the raft.) Desalter kits, if available, should probably be saved for the time when no other means of procuring drinking water is available. Instructions on how to use the desalter kit are on the container.

Figure 19-5. Collecting Water from Spray Shield.

c. Only water in its conventional sense should be consumed. The so-called "water substitutes" do little for the survivor, and may do much more harm than not consuming any water at all. There is no substitute for water. Fish juices and other animal fluids are of doubtful value in preventing dehydration. Fish juices contain protein that requires large amounts of water to be digested and the waste products must be excreted in the urine, which increases water loss. Survivors should never drink urine—urine is body-waste material and serves only to concentrate waste materials in the body and require more water to eliminate the additional waste.

19-6. Water in Tropical Areas. Depending on the time of the year and type of jungle, water in the tropical climates can be plentiful; however, it is necessary to know where to look for and procure it. Surface water is normally available in the form of streams, ponds, rivers, and swamps. In the savannas during the dry season, it may be necessary for the survivor to resort to digging for water in the places previously mentioned. Water obtained from these sources may need filtration and should be purified. Jungle plants can also provide survivors with water.

a. Many plants have hollow portions that can collect rainfall, dew, etc. (figure 19-6). Since there is no absolute way to tell whether this water is pure, it should be purified. The stems or the leaves of some plants have a hollow section where the stem meets the trunk. Look for water collected here. This includes any Y-shaped plants (palms or air plants). The branches of large trees often support air plants (relatives of the pineapple) whose overlapping, thickly growing leaves may hold a considerable amount of rainwater. Trees may also catch and store rainwater in natural receptacles such as cracks or hollows.

Figure 19-6. Water Collectors.

b. Pure freshwater needing no purification can be obtained from numerous plant sources. There are many varieties of vines that are potential water sources. The vines are from 50 feet to several hundred feet in length and 1 to 6 inches in diameter. They also grow as a hose along the ground and up into the trees. The leaf structure of the vine is generally high in the trees. Water

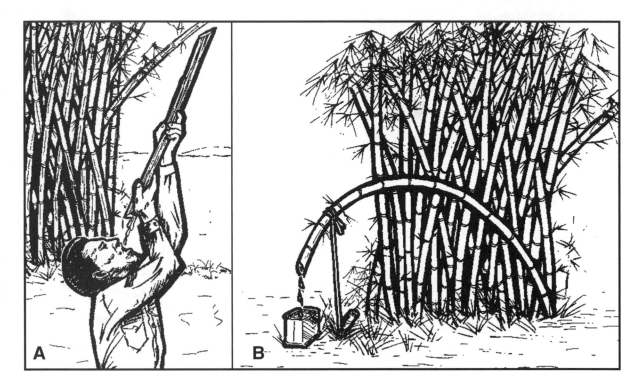

Figure 19-7. Water Vines and Bamboo.

vines are usually soft and easily cut. The smaller species may be twisted or bent easily and are usually heavy because of the water content. The water from these vines should be tested for potability. The first step in testing the water from vines is for survivors to nick the vine and watch for sap running from the cut. If milky sap is seen, the vine should be discarded; if no milky sap is observed, the vine may be a safe water vine. Survivors should cut out a section of the vine, hold that piece vertically, and observe the liquid as it flows out. If it is clear and color-less, it may be a drinkable source. If it is cloudy or milky-colored, they should discard the vine. They should let some of the liquid flow onto the palm of the hand and observe it. If the liquid does not change color, they can now taste it. If it tastes like water or has a woody or sweet taste, it should be safe for drinking. Liquid with a sour or bitter taste should be avoided. Water trapped within a vine is easily obtained by cutting out a section of the vine. The vine should first be cut high above the ground and then near to the ground. This will provide a long length of vine and, in addition, will tend to hide evi-dence of the cuts if the survivors are in an evasion situa-tion. When drinking from the vine, it should not touch the mouth as the bark may contain irritants that could affect the lips and mouth (figure 19-7). The pores in the upper end of the section of vine may re-close,

stopping the flow of water. If this occurs, survivors should cut off the end of the vine opposite the drinking end. This will reopen the pores allowing the water to flow.

c. Water from the rattan palm and spiny bamboo may be obtained in the same manner as from vines. It is not necessary to test the water if positive identification of the plant can be made. The slender stem (runner) of the rat-tan palm is an excellent water source. The joints are over-lapping in appearance, as if one section is fitted inside the next.

d. Water may be trapped within sections of green bam-boo. To determine if water is trapped within a section of bamboo, it should be shaken. If it contains water, a slosh-ing sound can be heard. An opening may be made in the section by making two 45-degree angle cuts, both on the same side of the section, and prying loose a piece of the section wall. The end of the section may be cut off and the water drunk or poured from the open end. The inside of the bamboo should be examined before consuming the water. If the inside walls are clean and white, the water will be safe to drink. If there are brown or black spots, fungus growth, or any discoloration, the water should be purified before consumption. Sometimes water can also be obtained by cutting the top off certain types of green bamboo, bending it over, and staking it to the ground (figure 19-7). A water container should be placed under it to catch the dripping water. This method has also proven effective on some vines and the rattan palm.

e. Water can also be obtained from banana plants in a couple of different ways, neither of which is satisfactory in a tactical situation. First, survivors should cut a banana plant down, then a long section should be cut off that can be easily handled. The section is taken apart by slitting from one end to the other and pulling off the layers one at a time. A strip 3 inches wide, the length of the section, and just deep enough to expose the cells, should be removed from the convex side. This section is folded toward the convex side to force the water from the cells of the plant. The layer must be squeezed gently to avoid forcing out any tannin into the water. Another technique for obtaining water from the banana plant is by making a "banana-well." This is done by making a bowl out of the plant stump, fairly close to the ground, through cutting out and removing the inner section of the stump (figure 19-8). Water that first enters the bowl may contain a concentration of tannin (an astringent that has the same effect as alum). A leaf from the banana plant or other plant should be placed over the bowl while it is filling to prevent contamination by insects, etc.

f. Water trees can also be a valuable source of water in some jungles. They can be identified by their blotched bark, which is fairly thin and smooth. The leaves are large, leathery, fuzzy, and evergreen, and may grow as large as 8 or 9 inches. The trunks may have short outgrowths with fig-like fruit on them or long tendrils with round fruit comprised of corn-kernel-shaped nuggets. In a nontactical situation, the tree can be tapped in the same manner as a rubber tree, with either a diagonal slash or a "V." When the bark is cut in to, it will exude a white sap, which if ingested causes temporary irritation of the urinary tract. This sap dries up quite rapidly and can easily be removed. The cut should be continued into the tree with a spigot (bamboo, knife, etc.) at the bottom of the tap to direct the water into a container. The water flows from the leaves back into the roots after sundown, so water can be procured from this source only after sundown or on overcast (cloudy) days. If survivors are in a tactical situation, they can obtain water from the tree and still conceal the procurement location. If the long tendrils are growing thickly, they can be separated and a hole bored into the tree. The white sap should be scraped off and a spigot placed below the tap with a water container to catch the water. Moving the tendrils back into place will conceal the container. Instead of boring into the tree, a couple of tendrils can be cut off or snapped off if no knife is available. The white sap should be allowed to dry and then be removed. The ends of the tendrils should be placed in a water container and the container concealed.

g. Coconuts contain a refreshing fluid. Where coconuts are available, they may be used as a water source. The fluid from a mature coconut contains oil, which when consumed in excess can cause diarrhea. There is little

Figure 19-8. Water from Banana Plant.

problem if used in moderation or with a meal and not on an empty stomach. Green unripe coconuts about the size of a grapefruit are the best for use because the fluid can be taken in large quantities without harmful effects. There is more fluid and less oil so there is a decreased possibility of diarrhea.

h. Water can also be obtained from liquid mud. Mud can be filtered through a piece of cloth. Water taken by this method must be purified. Rainwater can be collected from a tree by wrapping a cloth around a slanted tree and arranging the bottom end of the cloth to drip into a container (figure 19-9).

19-7. Water in Dry Areas. Locating and procuring water in a dry environment can be a formidable task. Some of the ways to find water in this environment have been

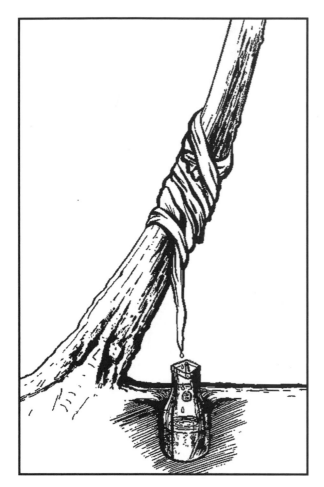

Figure 19-9. Collecting Water from Slanted Tree.

explored, such as locating a concave bend in a dry riverbed and digging for water (figure 19-10). If there is any water within a few feet of the surface, the sand will become slightly damp. Dig until water is obtained.

a. Some deserts become humid at night. The humidity may be collected in the form of dew. This dew can be collected by digging a shallow basin in the ground about 3 feet in diameter, and lining it with a piece of canvas, plastic, or other suitable material. A pyramid of stones taken from a minimum of 1 foot below the surface should then be built in this basin. Dew will collect on and between the stones and trickle down onto the lining material, where it can be collected and placed in a container.

b. Plants and trees with roots near the surface may be a source of water in dry areas. Water trees of dry Australia are a source of water, as their roots run out 40 to 80 feet at a depth of 2 to 9 inches under the surface. Survivors may obtain water from these roots by locating a root 4 to 5 feet from the trunk and cutting the root into 2- or 3-foot lengths. The bark can then be peeled off and the liquid

from each section of root drained into a container. The liquid can also be sucked out. The trees growing in hollows or depressions will have the most water in their roots. Roots that are 1 to 2 inches thick are an ideal size. Water can be carried in these roots by plugging one end with clay.

c. Cactus-like or succulent plants may be sources of water for survivors, but they should recall that no plants should be used for water procurement that have a milky sap. The barrel cactus of the United States provides a water source. To obtain it, survivors should first cut off the top of the plant. The pulpy inside portions of the plant should then be mashed to form a watery pulp. Water may ooze out and collect in the bowl, if not, the pulp may be squeezed through a cloth directly into the mouth.

d. The solar still is a method of obtaining water that uses both vegetation and ground moisture to produce water (figure 19-11). A solar still can be made from a sheet of clear plastic stretched over a hole in the ground. The moisture in the soil and from plant parts (fleshy stems and leaves) will be extracted and collected by this emergency device. Obviously, where the soil is extremely dry and no fleshy plants are available little, if any, water can be obtained from the still. The still may also be used to purify polluted water.

(1) The parts for the still are a piece of plastic about 6 feet square, a water-collector container or any waterproof material from which a collector container can be fashioned, and a piece of plastic tubing about 1/4 inch in diameter and 4 to 6 feet long. The tubing is not absolutely essential but makes the still easier to use. A container can be made from such materials as plastic, aluminum foil, ponchos, emergency ration tins, or a flight helmet. The tubing, when available, is fastened to the bottom of the inside of the container and used to remove drinking water from the container without disturbing the plastic. Some plastics work better than others, although any clear plastic should work if it is strong.

(2) If plants are available or if polluted water is to be purified, the still can be constructed in any convenient spot where it will receive direct sunlight throughout the day. Ease of digging will be the main consideration. If soil moisture is to be the only source of water, some sites will be better than others. Although sand generally does not retain as much moisture as clay, wet sand will work very well. Along the seacoast or in any inland areas where brackish or polluted water is available, any wet soil, even sand, produces usable amounts of water. On cloudy days, the yield will be reduced because direct sunlight is necessary if the still is to operate at full efficiency.

(3) Certain precautions must be kept in mind. If polluted water is used, survivors should make sure that none is spilled near the rim of the hole where the plastic touches the soil and that none comes in contact with the

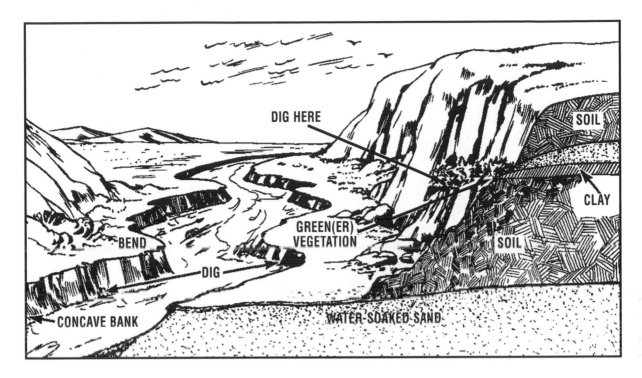

Figure 19-10. Dry Streambed.

container to prevent the freshly distilled water from becoming contaminated. Survivors should not disturb the plastic sheet during daylight "working hours" unless it is absolutely necessary. If a plastic drinking tube is not available, raise the plastic sheet and remove the container as little as possible during daylight hours. It takes half an hour for the air in the still to become resaturated and the collection of water to begin after the plastic has been disturbed. Even when placed on fairly damp soil and in an area where 8 hours of light per day is directed on the solar still, the average yield is only about 1 cup per day per still. Due to the low yields obtained from this device, survivors must give consideration to the possible danger of excessive dehydration brought about by constructing the solar still. In certain circumstances, solar still returns, even over 2- or 3-day periods, will not equal the amount of body fluid lost in construction and will actually hasten dehydration.

(4) Steps survivors should follow when constructing a solar still are: Dig a bowl-shaped hole in the soil about 40 inches in diameter and 20 inches deep. Add a smaller, deeper sump in the center bottom of the hole to accommodate the container. If polluted waters are to be purified, a small trough can be dug around the side of the hole about halfway down from the top. The trough ensures that the soil wetted by the polluted water will be exposed to the sunlight and at the same time that the

polluted water is prevented from running into the container. If plant material is used, line the sides of the hole with pieces of plant or its fleshy stems and leaves. Place the plastic over the hole and put soil on the edges to hold it in place. Place a rock no larger than a plum in the center of the plastic until it is about 15 inches below ground level. The plastic will now have the shape of a cone. Put more soil on the plastic around the rim of the hole to hold the cone securely in place and to prevent water-vapor loss. Straighten the plastic to form a neat cone with an angle of about 30 degrees so the water drops will run down and fall into the container. It takes about 1 hour for the air to become saturated and start condensing on the underside of the plastic cone.

e. The vegetation bag is a simpler method of water procurement. This method involves cutting foliage from trees or herbaceous plants, sealing it in a large, clear plastic bag, and allowing the heat of the sun to extract the fluids contained within. A large, heavy-duty clear plastic bag should be used. The bag should be filled with about 1 cubic yard of foliage, sealed, and exposed to the sun. The average yield for a bag tested was 320 ml/bag for a 5-hour day. This method is simple to set up. The vegetation-bag method of water procurement does have one primary drawback. The water produced is normally bitter to taste, caused by biological breakdown of the leaves as they lay in the water that has become super-heated

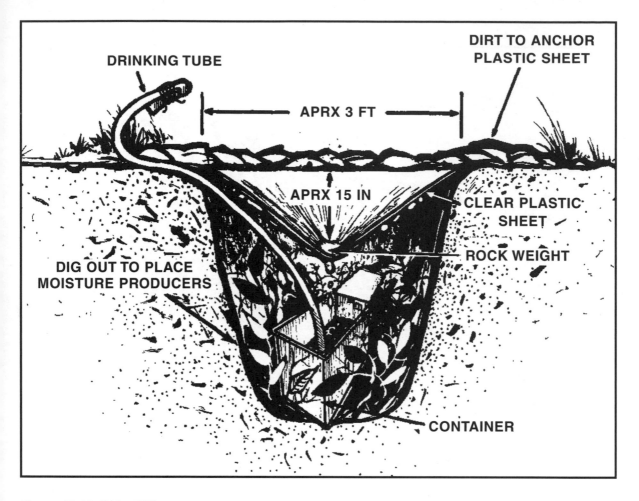

DRINKING TUBE

DIRT TO ANCHOR PLASTIC SHEET

APRX 3 FT

APRX 15 IN

CLEAR PLASTIC SHEET

ROCK WEIGHT

DIG OUT TO PLACE MOISTURE PRODUCERS

CONTAINER

Figure 19-11. Solar Still.

in the moist "hothouse" environment. This method can be readily used in a survival situation, but before the water produced by certain vegetation is consumed, it should undergo the taste test. This is to guard against ingestion of cyanide-producing substances and other harmful toxins, such as plant alkaloids. (See figure 19-12.)

f. One more method of water procurement is the water-transpiration bag, a method that is simple to use and has great potential for enhancing survival. This method is the vegetation-bag process taken one step further. A large plastic bag is placed over a living limb of a medium-size tree or large shrub. The bag opening is sealed at the branch, and the limb is then tied down to allow collected water to flow to a corner of the bag. For a diagram of the water-transpiration method, see figure 19-13.

(1) The amount of water yielded by this method will depend on the species of trees and shrubs available. During a test of this method, a transpiration bag produced approximately a gallon per day for 3 days with plastic bag on the same limb, and with no major deterioration of the branch. Other branches yielded the same amount.

Transpired water has a variety of tastes depending on whether or not the vegetation species is allowed to contact the water.

(2) The effort expended in setting up water-transpiration collectors is minimal. It takes about 5 minutes' work and requires no special skills once the method has been described or demonstrated. Collecting the water in a survival situation would necessitate survivors dismantling the plastic bag at the end of the day, draining the contents, and setting it up again the following day. The same branch may be reused (in some cases with almost similar yields); however, as a general rule, when vegetation abounds a new branch should be used each day.

(3) Without a doubt, the water-transpiration bag method surpasses other methods (solar stills, vegetation bag, cutting roots, barrel cactus) in yield, ease of assembly, and in most cases, taste. The benefits of having a simple plastic bag can't be overemphasized. As a water procurer in dry, semi-dry, or desert environments where low woodlands predominate, it can be used as a water transpirator; in scrubland, steppes, or treeless plains, as

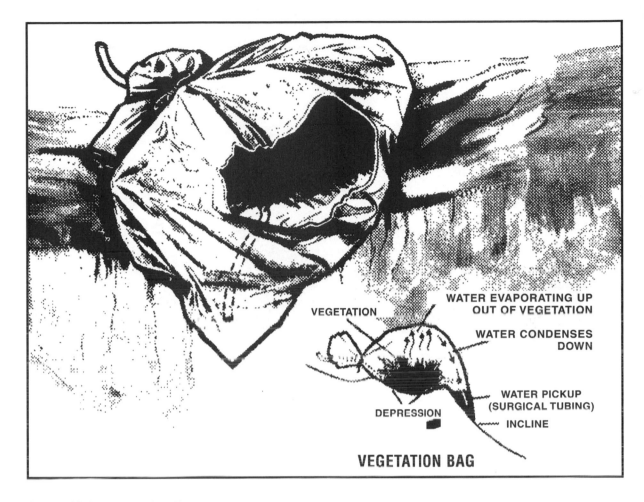

VEGETATION

WATER EVAPORATING UP
OUT OF VEGETATION

WATER CONDENSES
DOWN

WATER PICKUP
(SURGICAL TUBING)

DEPRESSION

INCLINE

VEGETATION BAG

Figure 19-12. Vegetation Bag.

a vegetation bag; in sandy areas without vegetation, it can be cut up and improvised into solar stills. Up to three large, heavy-duty bags may be needed to sustain one survivor in certain situations.

19-8. Preparation of Water for Consumption:

a. The following are ways in which survivors can possibly determine the presence of harmful agents in the water:

(1) Strong odors, foam, or bubbles in the water.

(2) Discoloration or turbid (muddy with sediment) water.

(3) Water from lakes found in desert areas is sometimes salty because the lake has been without an outlet for extended periods of time. Magnesium or alkali salts may produce a laxative effect; if not too strong, it is drinkable.

(4) If the water gags survivors or causes gastric disturbances, drinking should be discontinued.

(5) The lack of healthy green plants growing around any water source.

b. Because of survivors' potential aversion to water from natural sources, it should be rendered as potable as possible through filtration. Filtration removes only the solid particles from water—it does not purify it. One simple and quick way of filtering is to dig a sediment hole or seepage basin along a water source and allow the soil to filter the water (figure 19-14). The seepage hole should be covered when not in use. Another way is to construct a filter—layers of parachute material stretched across a tripod (figure 19-15). Charcoal is used to eliminate bad odors and foreign materials from the water. Activated charcoal (obtained from freshly burned wood) is used to filter the water. If a solid container is available for making a filter, use layers of fine to coarse sand and gravel along with charcoal and grass.

c. Purification of water may be done in a variety of ways. The method used will be dictated by the situation (such as tactical or nontactical).

(1) Boil the water for at least 10 minutes.

(2) To use purification tablets survivors should follow instructions on the bottle. One tablet per quart of clear water, two tablets if water is cloudy. Let water

Figure 19-13. Transpiration Bag.

stand for 5 minutes (allowing the tablet time to dissolve), then shake and leave standing for 15 minutes. Survivors should remember to turn the canteen over and allow a small amount of water to seep out and cover the neck part of the canteen. In an evasion situation, water-purification tablets should be used for purifying water. If these are not available, plant sources or non-stagnant, running water obtained from a location upstream from habitat should be consumed.

(3) Eight drops of 2 ½ percent iodine per quart—stir or shake and let stand for at least 10 minutes.

d. After water is found and purified, survivors may wish to store it for later consumption. The following make good containers:

(1) Waterbag.

(2) Canteen.

(3) Prophylactic inside a sock for protection of bladder.

(4) Segment of bamboo.

(5) Birch bark and pitch canteen.

(6) LPU bladder.

(7) Hood from antiexposure suit.

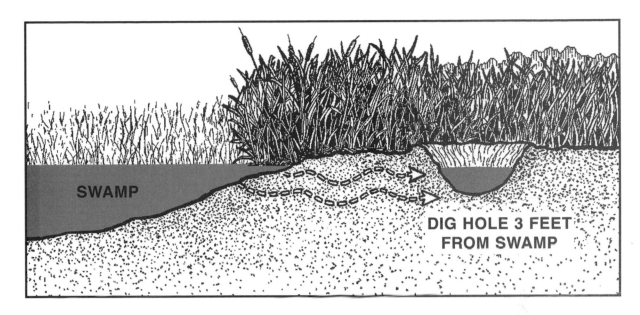

Figure 19-14. Sediment Hole.

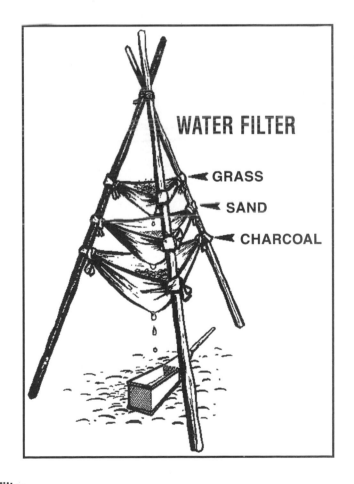

Figure 19-15. Water Filter.

Part Seven

TRAVEL

Chapter 20

LAND NAVIGATION

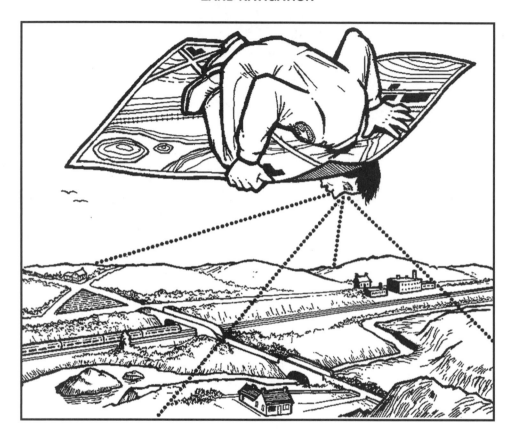

Figure 20-1. Land Navigation.

20-1. Introduction:

a. Survivors must know their location in order to intelligently decide if they should wait for rescue or if they should determine a destination and (or) route to travel. If the decision is to stay, the survivors need to know their location in order to radio the information to rescue personnel. If the decision is to travel, survivors must be able to use a map to determine the best routes of travel, location of possible food and water, and hazardous areas that they should avoid.

b. This chapter provides background information in the use of the map and compass (figure 20-1).

20-2. Maps:

a. A map is a pictorial representation of the Earth's surface drawn to scale and reproduced in two dimensions. Every map should have a title, legend, scale, north arrow,

grid system, and contour lines. With these components, a survivor can determine the portion of the Earth's surface the map covers. Survivors should be able to understand all of the markings on the map and use them to advantage. They should also be able to determine the distance between any two points on the map and be able to align the map with true north so it conforms to the actual features on the ground.

b. A map is a conceptionalized picture of the Earth's surface as seen from above, simplified to bring out important details and lettered for added identification. A map represents what is known about the Earth rather than what can be seen by an observer. However, a map is selective in that only the information that is necessary for its intended use is included on any one map. Maps also include features that are not visible on Earth, such as parallels, meridians, and political boundaries.

c. Since it is impossible to accurately portray a round object, such as the Earth, on a flat surface, all maps have some elements of distortion. Depending on the intended use, some maps sacrifice constant scale for accuracy in measurement of angles, while others sacrifice accurate measurement of angles for a constant scale. However, most maps used for ground navigation use a compromise projection in which a slight amount of distortion is introduced into the elements that a map portrays, but in which a fairly true picture is given.

d. A planimetric map presents only the horizontal positions for the features represented. It is distinguished from a topographic map by the omission of relief in a measurable form.

e. A topographic map (figure 20-2) portrays terrain and landforms in a measurable form and the horizontal positions of the features represented. The vertical positions, or relief, are normally represented by contours. On maps showing relief, the elevations and contours are measured from a specified vertical datum plane and usually mean sea level.

f. A plastic relief map is a reproduction of an aerial photograph or a photomosaic made from a series of aerial photographs upon which grid lines, marginal data, place names, route numbers, important elevations, boundaries, approximate scale, and approximate direction have been added.

g. PICTOMAP (figure 20-3) is the acronym for photographic image conversion by tonal masking procedures. It is a map on which the photographic imagery of a standard photomap has been convened into interpretable colors and symbols.

h. A photomosaic is an assembly of aerial photographs and is commonly called a mosaic in topographic usage. Mosaics are useful when time does not permit the compilation of a more accurate map. The accuracy of a mosaic depends on the method used in its preparation and may vary from simply a good pictorial effect of the ground to that of a planimetric map.

i. A military city map is a topographic map, usually 1:12,500 scale, of a city, outlining streets and showing street names, important buildings, and other urban elements of military importance, which are compatible with the scale of the map. The scales of military city maps can vary from 1:25,000 to 1:5,000, depending on the importance and size of the city, density of detail, and available intelligence information.

j. Special maps are for special purposes such as trafficability, communications, and assault. These are usually overprinted maps of scales smaller than 1:100,000 but larger than 1:1,000,000. Other types of special maps are those made from organosol or materials other than paper to meet the requirements of special climatic conditions.

k. A terrain model is a scale model of the terrain showing landforms, and in large-scale models, industrial and cultural shapes. It is designed to provide a means for

visualizing the terrain for planning or indoctrination purposes and for briefing on assault landings.

l. A special-purpose map is one that has been designed or modified to give information not covered on a standard map, or to elaborate on standard map data. Special-purpose maps are usually in the form of an overprint. Overprints may be in the form of individual sheets or combined and bound into a study of an area. A few of the subjects covered are:

    (1) Landforms.
    (2) Drainage characteristics.
    (3) Vegetation.
    (4) Climate.
    (5) Coast and landing beaches.
    (6) Railroads.
    (7) Airfields.
    (8) Urban areas.
    (9) Electric power.
    (10) Fuels.
    (11) Surface water resources.
    (12) Groundwater resources.
    (13) Natural construction materials.
    (14) Cross-country movement
    (15) Suitability for airfield construction.
    (16) Airborne operations.

20-3. Aeronautical Charts. Air navigation and planning charts are used for flight planning. Each different series of charts is constructed at a different scale and format to meet the needs of a particular type of air navigation. The air navigation and planning charts are smaller in scale and less detailed than Army maps or air target materials. The control of positional error is less critical. The following list includes the charts most commonly used in intelligence operations. They are available through the Defense Mapping Agency (DMA), Officer of Distribution Services, Washington DC. A description of each chart follows the listing:

CHART	SCALE	CODE
USAF Global Navigation and Planning Chart	1:5,000,000	GNC
USAF Jet Navigation Chart	1:3,000,000	JNC-A
USAF Operational Navigation Chart	1:1,000,000	ONC
USAF Tactical Pilotage Chart	1:500,000	TPC
USAF Jet Navigation Chart	1:2,000,000	JNC
Joint Operations Graphic	1:250,000	JOG

Figure 20-2. Topographic Map.

Figure 20-3. Pictomap.

a. Global Navigation and Planning Chart (GNC). (See figure 20-4.)

This chart is designed for general planning purposes where large areas of interest and long-distance operations are involved. It serves as a navigation chart for long-range, high-altitude, and high-speed aircraft since sheet lines have been selected on the basis of primary areas of strategic interest. Several other general planning charts are available through the DMA. Some of these charts are produced on selected areas of strategic interest; others provide wide coverage. All general planning charts are produced at a small or very small scale, which provides extensive area coverage on a single sheet.

b. USAF Jet Navigation Chart (JN/JNC-A). (See figure 20-5.)

The basic JNC is produced at a scale of 1:2,000,000. The JNC-A is produced on the north polar area and on the United States at a scale of 1:3,000,000. Both jet navigation charts are printed on 41 1/2-by-57 3/8-inch sheets.

(1) The JN chart is used for preflight planning and en route navigation by long-range jet aircraft with dead reckoning, radar, celestial, and grid navigational capabilities. The charts are designed so they can be joined to produce a strip chart, which provides the necessary navigational information for any intended course. Relief is indicated through the use of contours, spot elevations, and gradient tints. Large, level terrain areas are indicated by a symbol that consists of narrow, parallel lines with the elevation annotated within the symbol.

(2) Principal cities and towns and principal roads and rail networks are shown on the JN chart. The transportation network is shown in the immediate area of populated places. Lakes and principal drainage patterns are also pictured. The elevations of major lakes are indicated so the altitude may be determined by using the aircraft radar altimeter.

c. USAF Operational Navigation Chart (ONC). (See figure 20-6.)

(1) The ONC was developed to meet military requirements for a chart adaptable to low-altitude navigation. The ONC is used for preflight planning and en route navigation. It is also used for operational planning, intelligence briefing and plotting, and flight-planning displays.

(2) This chart covers an area of 8° of latitude and 12° of longitude. ONC sheets are identified by combining a letter and a number (figure 20-7). Letters identify 8° bands of latitude, starting at the North Pole and progressing southward. Numbers identify 12° sections of longitude from the prime meridian eastward. The successful execution of low-altitude missions depends on visual and radar identification of ground features used as checkpoints and a rapid visual association of these features with their chart counterparts. The ONC portrays, by conventional signs and symbols, cultural features, which have low-altitude checkpoint significance. Powerlines

are shown (except on cities) and are indicated by the usual line-and-pole symbol.

(3) For certain circumstances, operational requirements may be more effectively satisfied with pictorial illustrations rather than the conventional symbolization of such structures as prominent buildings, bridges, dams, towers, holding or storage tanks, stadiums, and related features. For these reasons, significant landmarks are depicted on ONCs with pictorial symbols.

(4) The ONC portrays relief in perspective so the user gets instantaneous appreciation of relative heights, slope gradients, and the forms of ground patterns. Topographic expression, illustrated basically with contours and spot elevations, is emphasized by the use of shaded relief and terrain-characteristic tints defining the overall elevation levels. ONC contour intervals and terrain-characteristic tints are selected regionally. This captures the relative significance of ground forms as a complete picture, and this feature aids preflight planning and in-flight identification.

d. USAF Tactical Pilotage Chart (TPC). (See figure 20-8.)

(1) The TPC is produced in a coordinated series at a scale of 1:500,000. Sheet sizes are the same dimensions as the ONC sheets; however, a TPC covers only one-fourth the area of an ONC sheet. The TPC breakdown on the ONC is illustrated in figure 20-9. A TPC is identified by the ONC identification and the letters "A," "B," "C," or "D."

(2) The TPC is used for detailed preflight planning and mission analysis. In designing the TPC, emphasis was placed on ground features that are significant for low-level, high-speed navigation, using visual and radar means. The selected ground features also permit immediate ground-chart orientation at predetermined checkpoints.

(3) Relief on the TPC is displayed by contours (intervals may vary between 100 feet and 1,000 feet), spot elevations, relief shading, and terrain-characteristic tints. Cultural features such as towns and cities, principal roads, railroads, power-transmission lines, boundaries, and other features of value for low-altitude visual missions are included on the TPC. Pictorial symbols are used for features that provide the best checkpoints. Other features of the TPC that enhance its tactical air navigation qualities are as follows:

(a) UTM grid overprint.

(b) Vegetation color and symbol code.

(c) Enlarged vertical-obstruction symbols.

(d) Enlarged road and railroad symbols.

(e) Emphasized radio aid to navigation symbols.

(f) Foreign place name glossary.

(g) Airdrome runway patterns to scale when information is available.

(h) Spot elevation, gradient tints, and shaded relief depicted for all elevations.

(i) The highest elevation for each 15-minute quadrangle is shown in thousands and hundreds of feet.

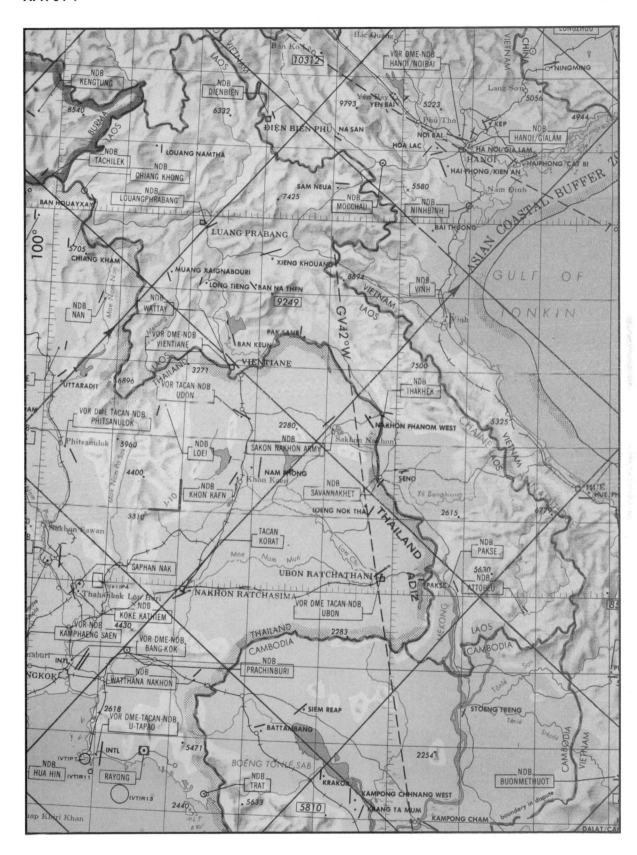

Figure 20-4. GNC Map.

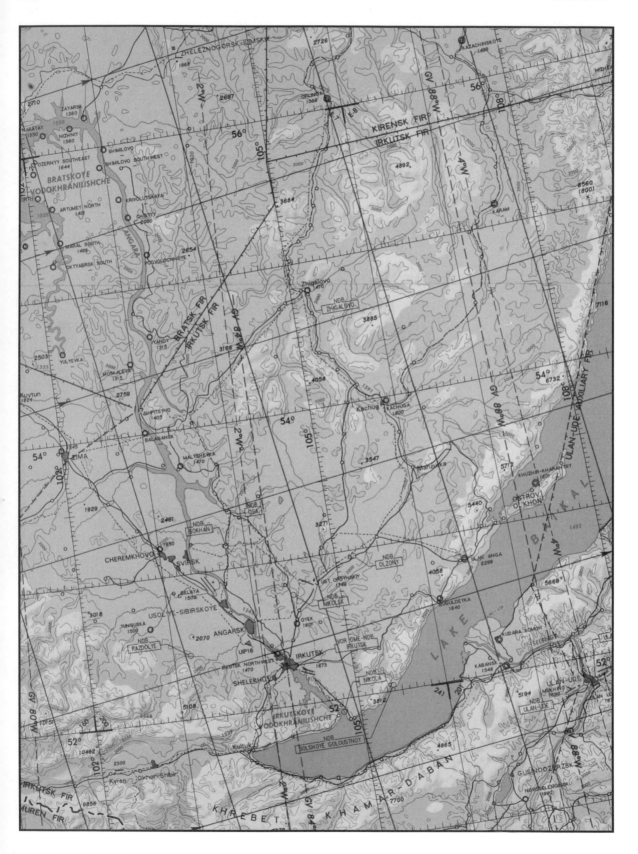

Figure 20-5. JNC Map.

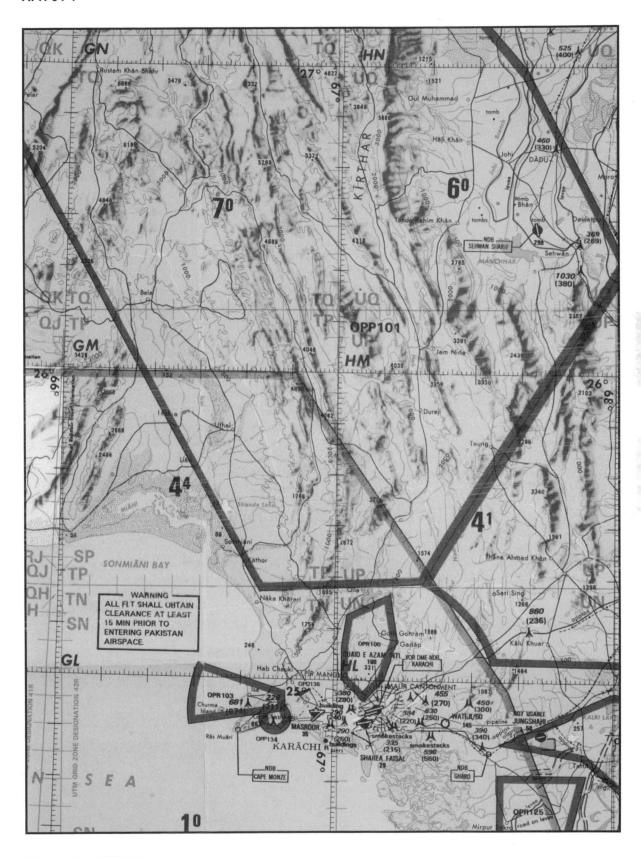

Figure 20-6. ONC Map.

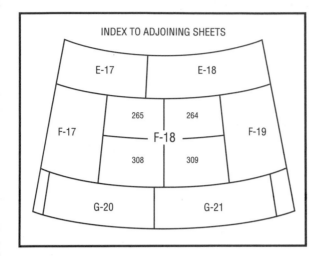

Figure 20-7. Operational Navigation Chart Index.

e. Joint Operations Graphic (JOG). (Series 1501 AIR.)

(1) JOGs (figure 20-10) are a series of 1:250,000-scale military maps designed for joint ground and air operations. The maps are published in ground and air editions. Both series emphasize the air-landing facilities, but the air series has additional symbols to identify aids and obstructions to air navigation.

(2) JOG was designed to provide a common-scale graphic for Army, Navy, and Air Force use. Air Forces make use of it for tactical air operations, close air support, and interdiction by medium- and high-speed aircraft at low altitudes. The chart may also be used for dead reckoning and visual pilotage for short-range en route navigation. Due to its large scale, it is unsuitable for local-area command planning for strategic and tactical operations.

(a) Relief on the JOG is indicated by contour lines (in feet). In some areas, the intervals may be in meters, with the approximate value in feet indicated in the margin of the chart. Spot elevations are used through all terrain levels. The ground series show elevations and contours in meters, while the air series show the same elevations and contours in feet.

(b) Relief is also shown through gradient tints, supplemented by shaded relief. The highest elevations in each 15-minute quadrangle are indicated in thousands and hundreds of feet.

(c) Cultural features such as cities, towns, roads, trails, and railroads are illustrated in detail. The locations of boundaries and power-transmission lines are also shown. Vegetation is shown by symbol. Detailed drainage patterns and water tint are used to illustrate water features, such as coastlines, oceans, lakes, rivers and streams, canals, swamps, and reefs. The JOG includes

aeronautical information such as airfields, fixed radio navigation and communication facilities, and any known obstructions more than 200 feet above ground. If the information is available, the airfield runway patterns are shown to scale by diagram.

(d) The basic numbering system of the JOG consists of two letters and a number, which identifies an area 6° in longitude-by-4° in latitude. If the chart covers an area north of the Equator, the first letter is "N"; a chart covering an area south of the Equator is identified with an initial "S." The second letter identifies the 4°-bands of latitude that are lettered north and south from the Equator. The number identifies the 6°-sections of longitude that are numbered from the 180° meridian eastward. The 6°-by-4° areas identified by two letters and a number from 1 to 60 are further broken down to either 12 or 16 sheets. Figure 20-11 illustrates how the sheets are numbered in each breakdown. The figure also indicates the respective latitudes at which the 12- and 16-sheet breakdowns are used. Charts produced in Canada use a slightly different sheet-identification system. The DOD Aeronautical Chart Catalog contains an explanation of the system.

f. DOD Evasion Charts. (See figure 20-12.)

The Defense Mapping Agency and the Aeronautical Chart and Information Center prepare DOD evasion charts. The Korea and Southeast Asia charts have been completed. The scale for these charts is 1:250,000. The charts have both longitude and latitude and the UTM grid-coordinate systems. The relief is duplicated by both contour lines and shading. The magnetic variation is shown by a compass rose superimposed on the chart. The charts also indicate the direction of seasonal ocean currents. These charts may include geographic environmental data consisting of a description of the people, climate, water, food, hazards, and vegetation. A conversion of elevation bar scale may aid in communicating with other forces. The star chart is provided to aid in night navigation.

20-4. Information Contained in Margin:

a. Before using any piece of equipment, a wise operator always reads the manufacturer's book of instructions. This is also true with maps. The instructions are placed around the outer edges of the map and are known as marginal information. All maps are not the same, so it becomes necessary each time a different map is used to carefully examine the marginal information.

b. Figure 20-13 is a large-scale (1:50,000) topographic map. The circled numbers indicate the marginal information with which the map user must be familiar. The location of the marginal information will vary with each different type of map. However, the following items are on most maps. The circled numbers correspond to the item numbers listed and described below.

(1) Sheet Name (1). The sheet name is in two places: the center of the upper margin and the right side of the lower margin. Generally, a map is named after its

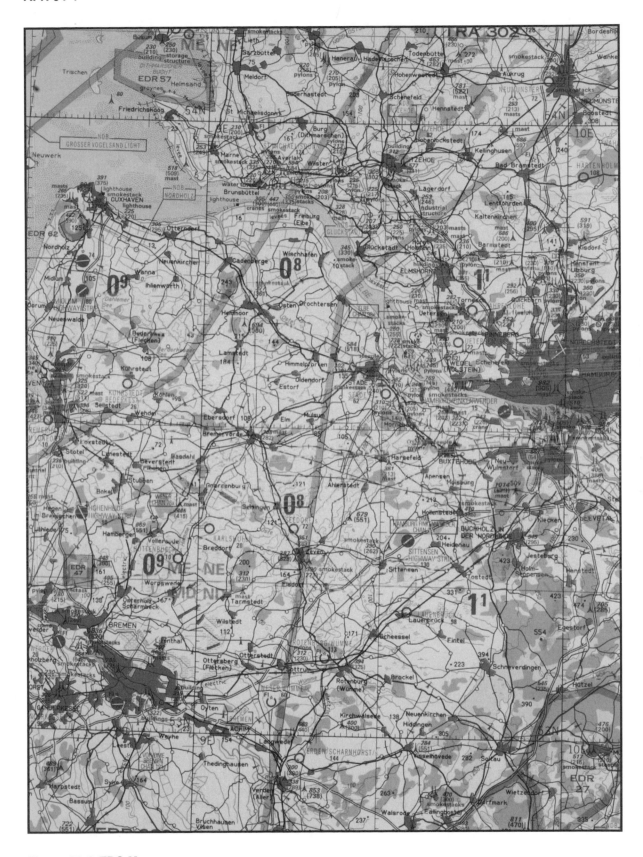

Figure 20-8. TPC Map.

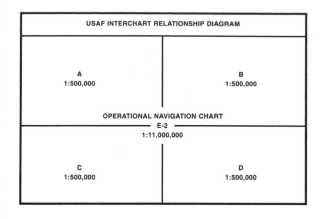

Figure 20-9. Relationship of TPC and ONC.

outstanding cultural or geographic feature. When possible, the name of the largest city on the map is used (not shown).

(2) Sheet Number (2). The sheet number is in the upper-right margin and is used as a reference number for that map sheet. For maps at 1:100,000 scale and larger, sheet numbers are based on an arbitrary system that makes possible the ready orientation of maps at scales of 1:100,000, 1:50,000, and 1:25,000 (figures 20-14 and 20-15).

(3) Series Name and Scale (3):

(a) The map-series name is in the upper-left margin. A map series usually comprises a group of similar maps at the same scale and on the same sheet lines or format designed to cover a particular geographic area. It may also be a group of maps that serve a common purpose, such as military city maps. The name given a series is of the most prominent area. The scale note is a representative fraction that gives the ratio of map distance to the corresponding distance on the Earth's surface. For example, the scale note 1:50,000 indicates that one unit of measure on the map equals 50,000 units of the same measure on the ground.

(b) Scale. The scale is expressed as a fraction and gives the ratio of map distance to ground distance. The terms "small scale," "medium scale," and "large scale" may be confusing when read with the numbers. However, if the number is viewed as a fraction, it quickly becomes apparent that 1:600,000 of something is smaller than 1:75,000 of the same thing. Hence, the larger the number after 1, the smaller the scale of the map.

-1. Small Scale. Maps at scales of 1:600,000 and smaller are used for general planning and strategical studies at the high echelons. The standard small scale is 1:1,000,000.

-2. Medium Scale. Maps at scales larger than 1:600,000 but smaller than 1:75,000 are used for

planning operations, including the movement and concentration of troops and supplies. The standard medium scale is 1:250,000.

-3. Large Scale. Maps at scales of 1:75,000 and larger are used to meet the tactical, technical, and administrative needs of field units. The standard large scale is 1:50,000.

(4) Series Number (4). The series number appears in the upper-right margin and the lower-left margin. It is a comprehensive reference expressed either as a four-digit numeral (example, 1125), or as a letter followed by a three- or four-digit numeral (example, V7915).

(5) Edition Number (5). The edition number is in the upper margin and lower-left margin. It represents the age of the map in relation to other editions of the same map and the agency responsible for its production. The latest edition will have the highest number. EDITION 1 DMATC indicates the first edition prepared by the Defense Mapping Agency Topological Center. Edition numbers run consecutively: a map bearing a higher edition number is assumed to contain more recent information than the same map bearing a lower edition number. Advancement of the edition number constitutes authority to rescind or supersede the previous edition.

(6) Bar Scales (6). The bar scales are located in the center of the lower margin. They are rulers used to convert map distance to ground distance. Maps normally have three or more bar scales, each a different unit of measure.

(7) Credit Note (7). The credit note is in the lower-left margin. It lists the producer, dates, and general methods of preparation or revision. This information is important to the map user in evaluating the reliability of the map, as it indicates when and how the map information was obtained. On some recent 1:50,000 scale maps, the map credits are shown in tabular form in the lower margin, with reliability information presented in a coverage diagram.

(8) Adjoining Sheets Diagram (8). (Not shown.) Maps at all standard scales contain a diagram that illustrates the adjoining sheets.

(a) On maps at 1:100,000 and larger scales and at 1:1,000,000 scales, the diagram is called the Index to Adjoining Sheets, and consists of as many rectangles, representing adjoining sheets, as are necessary to surround the rectangle that represents the sheet under consideration. The diagram usually contains nine rectangles, but the number or names may vary depending on the location of the adjoining sheets. All represented sheets are identified by their sheet numbers. Sheets of an adjoining series, whether published or planned, that are the same scale are represented by dashed lines. The series number of the adjoining series is indicated along the appropriate side of the division line between the series (figure 20-16).

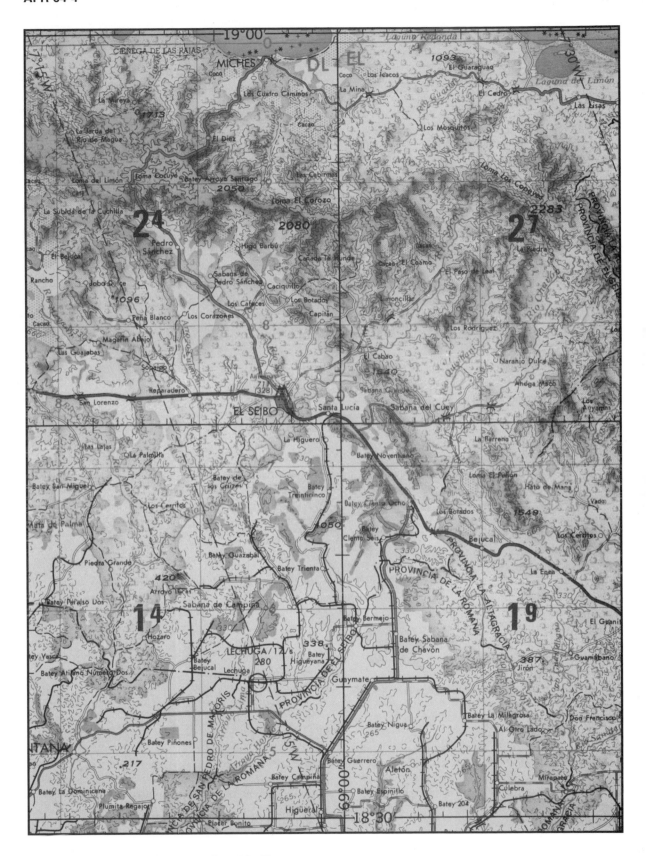

Figure 20-10. Joint Operations Graphic Map.

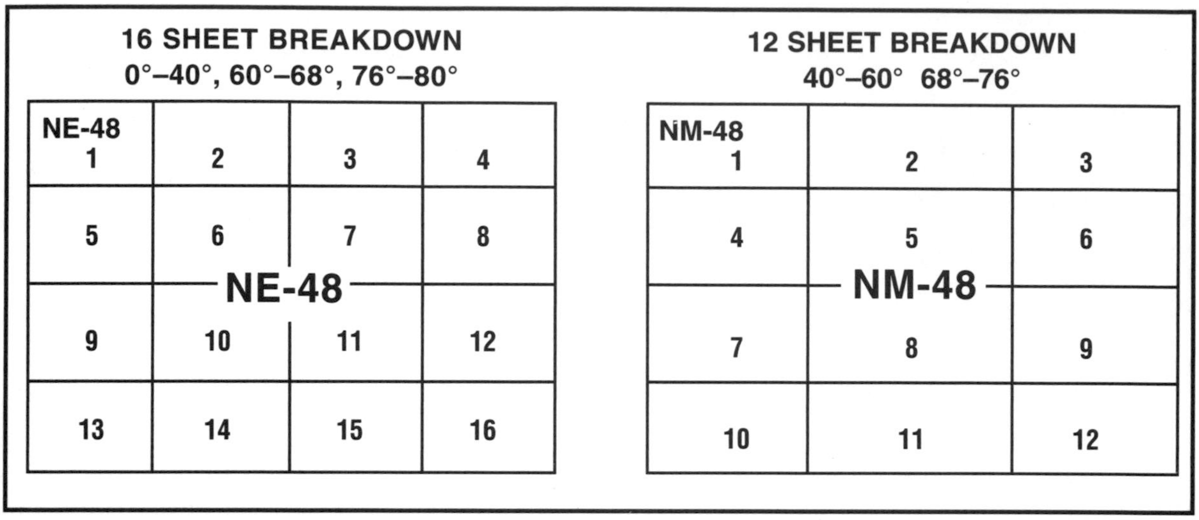

16 SHEET BREAKDOWN
0°–40°, 60°–68°, 76°–80°

NE-48 1	2	3	4
5	6	7	8
9	10	11	12
13	14	15	16

NE-48

12 SHEET BREAKDOWN
40°–60° 68°–76°

NM-48 1	2	3
4	5	6
7	8	9
10	11	12

NM-48

Figure 20-11. JOG Sheet Numbering System.

(b) On 1:50,000-scale maps, the sheet number and series number of the 1:250,000-scale map of the area are shown below the Index to Adjoining Sheets.

(c) On maps at 1:250,000 scale, the adjoining sheets are shown in the location diagram. Usually, the diagram consists of 25 rectangles, but the number may vary with the locations of the adjoining sheets.

(9) Index to Boundaries (9). The index to boundaries diagram appears in the lower-right margin of all sheets at 1:100,000 scale or larger, and at 1:1,000,000 scale. This diagram, which is a miniature of the map, shows the boundaries that occur within the map area, such as county lines and state boundaries. On 1:250,000-scale maps, the boundary information is included in the location diagram.

(10) Projection Note (10). The projection system is the framework of the map. For maps, this framework is the conformal type; that is, small areas of the surface of the Earth retain their true shapes on the projection, measured angles closely approximate true values, and the scale factor is the same in all directions from a point. The projection is identified on the map by a note in the lower margin.

(11) Grid Note (11). The grid note is in the center of the lower margin. It gives information pertaining to the grid system used, the interval of grid lines, and the number of digits omitted from the grid values. Notes pertaining to overlapping or secondary grids are also included when appropriate.

(12) Grid Reference Box (12). The grid reference box has instructions for composing a grid reference.

(13) Vertical Datum Note (13). This note is in the center of the lower margin. It designates the basis for all vertical control stations, contours, and elevations appearing on the map. On JOGs at 1:250,000 scale, the vertical datum note may appear in the reliability diagram.

(14) Horizontal Datum Note (14). This note is located in the center of the lower margin. It indicates the basis for all horizontal control stations appearing on the map. This network of stations controls the horizontal positions of all mapped features. On JOGs at 1:250,000 scale, the horizontal datum note may appear in the reliability diagram.

(15) Legend (15). The legend is located in the lower-left margin. It illustrates and identifies the topographic symbols used to depict some of the more prominent features on the map. The symbols are not always the same on every map. To avoid error in the interpretation of symbols, the legend must always be referred to when a map is read.

(16) Declination Diagram (16). The declination diagram is usually located in the lower margin of large-scale maps and indicates the angular relationships of true north, grid north, and magnetic north. On maps at 1:250,000 scale, this information is expressed as a note in the lower margin.

(17) User's Note (17). A user's note is in the center of the lower margin. It requests cooperation in correcting errors or omissions on the map. Errors should be marked and the map forwarded to the agency identified in the note.

(18) Unit Imprint (18). The unit imprint, in the lower-left margin, identifies the agency that printed the map and the printing date. The printing date should

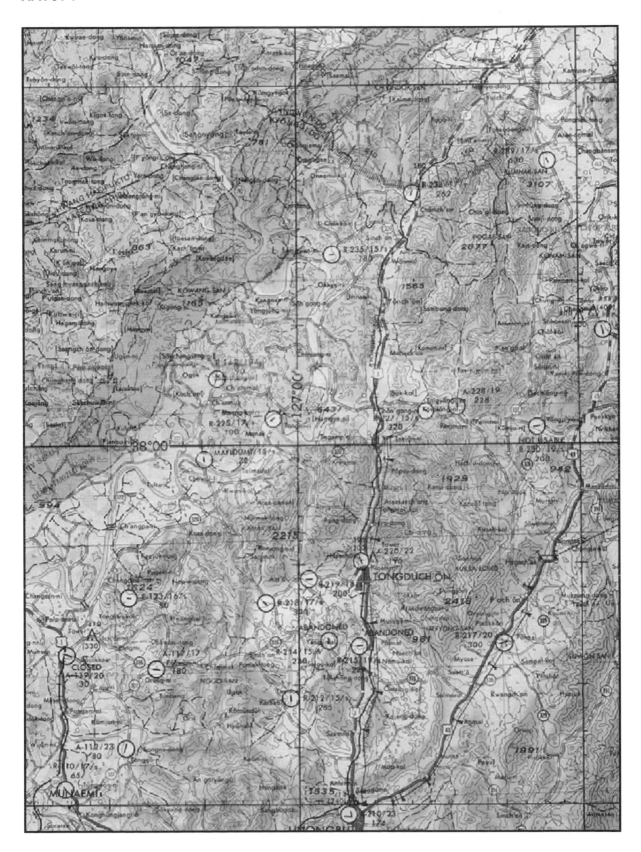

Figure 20-12. DOD Evasion Chart.

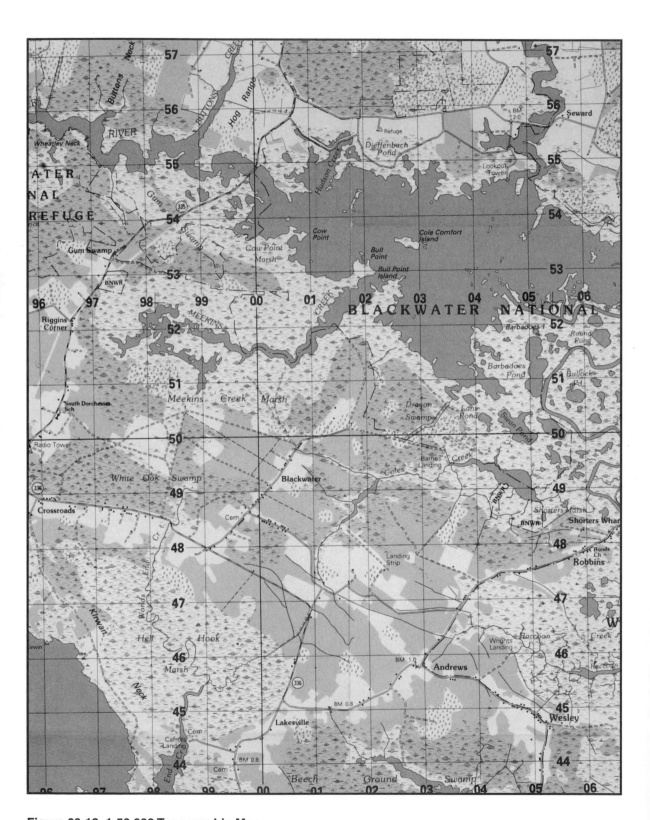

Figure 20-13. 1:50,000 Topographic Map.

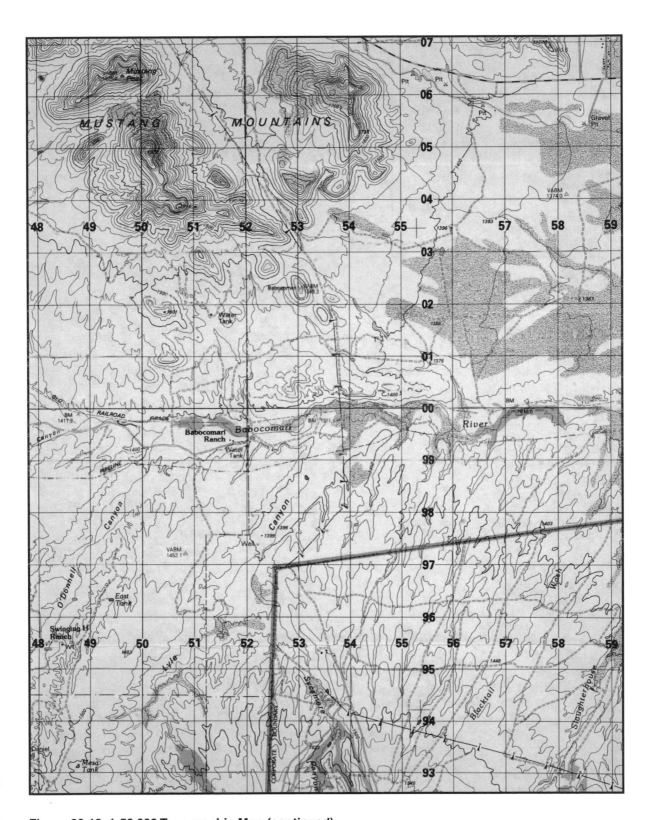

Figure 20-13. 1:50,000 Topographic Map (continued).

64	7064	7164	7264
63	6963	7063	7163	7263
62	6962	7062	7162	7262
	69	70	71	72

Figure 20-14. Basic Development, 1:100,000 Scale.

not be used to determine when the map information was obtained.

(19) Contour Interval (19). The contour interval note appears in the center of the lower margin. It states the vertical distance between adjacent contour lines on the map. When supplementary contours are used, the interval is indicated.

(20) Special Notes and Scales (20). Under certain conditions, special notes or scales may be added to the margin information to aid the map user. The following are examples:

(a) Glossary. A glossary is an explanation of technical terms or a translation of terms on maps of foreign areas where the native language is other than English.

(b) Classification. Certain maps require a note indicating the security classification. This is shown in the upper and lower margins.

(c) Protractor Scale. A protractor scale may appear in the upper margin on some maps. It is used to lay out the magnetic-grid declination of the map, which in turn is used to orient the map sheet with the aid of a magnetic compass.

(d) Coverage Diagram. A coverage diagram may be used on maps at scales of 1:100,000 and larger. It is normally in the lower or right margin and indicates the methods by which the map was made, dates of photography, and reliability of the sources. On maps at 1:250,000 scale, the coverage diagram is replaced by a reliability diagram.

(e) Elevation Guide. On some maps at scales of 1:100,000 and larger, a miniature characterization of the terrain is shown by a diagram in the lower-right margin of the map. The terrain is represented by bands

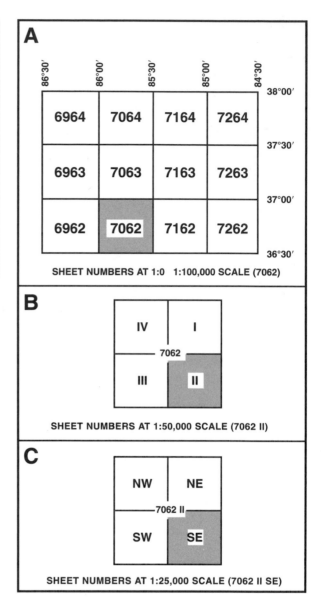

Figure 20-15. Systems for Numbering Maps.

of elevation, spot elevations, and major drainage features. The elevation guide provides the map reader with a means of rapid recognition of major landforms.

(f) Special Notes. A special note is any statement of general information that relates specifically to the mapped area. For example, rice fields are generally subject to flooding; however, they may be seasonally dry.

(21) Stock Number Identification (21). All maps published by or for the Department of the Army or Defense Mapping Agency contain stock number identifications, which are used in requisitioning map supplies. The identification consists of the words "STOCK NO." followed by a unique designation that is composed of the

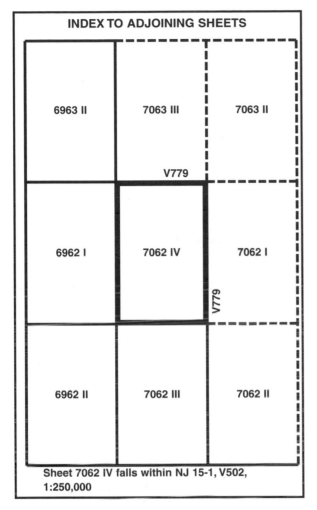

INDEX TO ADJOINING SHEETS

6963 II	7063 III	7063 II
6962 I	7062 IV	7062 I
6962 II	7062 III	7062 II

V779

V779

Sheet 7062 IV falls within NJ 15-1, V502, 1:250,000

Figure 20-16. Index to Adjoining Sheets.

series number, the sheet number of the individual map, and on recently printed sheets, the edition number.

20-5. Topographic Map Symbols and Colors:

a. The purpose of a map is to permit one to visualize an area of the Earth's surface with pertinent features properly positioned. Ideally, all the features within an area would appear on the map in their true proportion, position, and shape. This, however, is not practical because many of the features would be unimportant and others would be unrecognizable because of their reduction in size. The mapmaker has been forced to use symbols to represent the natural and manmade features of the Earth's surface. These symbols resemble, as closely as possible, the actual features as viewed from above (figures 20-17 and 20-18).

b. To facilitate identification of features on the map by providing a more natural appearance and contrast, the topographic symbols are usually printed in different colors, with each color identifying a class of features. The colors vary with different types of maps, but on a standard large-scale topographic map, the colors used and the features represented are:

(1) Black—the majority of cultural or manmade features.

(2) Blue—water features, such as lakes, rivers, and swamps.

(3) Green—vegetation, such as woods, orchards, and vineyards.

(4) Brown—all relief features, such as contours.

(5) Red—main roads, built-up areas, and special features.

(6) Occasionally, other colors may be used to show special information. (These, as a rule, are indicated in the marginal information. For example, aeronautical symbols and related information for air-ground operations are shown in purple on JOGs.)

c. In the process of making a map, everything must be reduced from its size on the ground to the size that appears on the map. For purposes of clarity, this requires some of the symbols to be exaggerated. They are positioned so the center of the symbol remains in its true location. An exception to this would be the position of a feature adjacent to a major road. If the width of the road has been exaggerated, then the feature is moved from its true position to preserve its relation to the road.

d. Army Field Manual 21-31 gives a description of topographic symbols and abbreviations authorized for use on US military maps. Figure 20-19 illustrates several of the symbols used on maps.

20-6. Coordinate Systems.
The intersections of reference lines help to locate specific points on the Earth's surface. Three of the primary reference-line systems are the geographic-coordinate system, the reference (GEOREF) system, and the universal transverse Mercator grid system (UTM). Knowing how to use these plotting systems should help a survivor to determine point locations.

a. **Coordinates.** Quantities that give position with respect to two reference lines are called coordinates. Thus, the intersection of F Street and 4th Avenue (figure 20-20) is the coordinate location of the Gridville Public Library. The coordinates of the local theater are D Street and 6th Avenue. One can see from this simplified example that coordinates are read at intersections of vertical and horizontal lines. The basic coordinate system used on maps and charts is the geographic military grid. The structure and use of the geographic coordinate system, the world geographic reference system, and the military grid reference system will be discussed and illustrated.

Figure 20-17. Area Viewed from Ground Position.

(1) Geographic Coordinates. The geographic coordinate system is a network of imaginary lines that circle the Earth. They are used to express Earth position or

Figure 20-18. Area Viewed from Ground Position—Map.

location. There are north-south lines called meridians of longitude and east-west lines called parallels of latitude. The location of any point on the Earth can be expressed in terms of the intersection of the line of latitude and the line of longitude passing through the point.

(2) Meridians of Longitude. The lines of latitude and longitude are actually great and small circles formed by imaginary planes cutting the Earth. A great circle divides the Earth into two equal parts (halves), whereas a small circle divides the Earth into two unequal parts. Study figure 20-21 and note that: (1) each north-south line is a great circle, and (2) each great circle passes through both the North and South poles. Each half of each of these great circles from one pole, in either direction, to the other pole is called a meridian of longitude. The other half of the same great circle is a second meridian of longitude.

(a) Meridian is derived from the Latin word *meridianum*, which means "lines that pass through the highest point on their course" (in this case, both the North and South Poles). Any angular distance measured east or west of the meridian is called longitudinal distance, hence the term "meridian of longitude." It is necessary, of course, to assign values to the meridians to make them meaningful. The most appropriate values to use for circles are degrees (°), minutes ('), and seconds ("). Circles are customarily divided into 360° per circle, 60' per degree, and 60" per minute.

(b) All meridians are equal in value; hence, one of them must be assigned the value of 0° (the starting point). The meridian passing through Greenwich, England, is zero degrees (0°). This meridian is also called the prime meridian (figure 20-22). The other half of the great circle on which the prime meridian is located is designated the 180th meridian. Portions of this meridian are also called the international dateline.

Other land surveys:

Township or range line ....

Section line ....

Land grant or mining claim; monument .... ☐

Fence line ....

ROADS AND RELATED FEATURES

Primary highway ....

Secondary highway ....

Light duty road ....

Unimproved road ....

Trail ....

Dual highway ....

Dual highway with median strip ....

Road under construction ....

Underpass; overpass ....

Bridge ....

Drawbridge ....

Tunnel ....

BUILDINGS AND RELATED FEATURES

Dwelling or place of employment: small; large ...

School; church ....

Barn, warehouse, etc.: small; large ....

House omission tint ....

Racetrack ....

Airport ....

Landing strip ....

Well (other than water); windmill ....

Water tank: small; large ....

Other tank: small; large ....

Covered reservoir ....

Gaging station ....

Landmark object ....

Campground; picnic area ....

Cemetery: small; large .... Cem

RAILROADS AND RELATED FEATURES

Standard gauge single track; station ....

Standard gauge multiple track ....

Abandoned ....

Under construction ....

Narrow gauge single track ....

Narrow gauge multiple track ....

Railroad in street ....

Juxtaposition ....

Roundhouse and turntable ....

TRANSMISSION LINES AND PIPELINES

Power transmission line: pole; tower ....

Telephone or telegraph line ....

Aboveground oil or gas pipeline ....

Underground oil or gas pipeline ....

CONTOURS

Topographic:

Intermediate ....

Index ....

Supplementary ....

Depression ....

Cut; fill ....

Bathymetric:

Intermediate ....

Index ....

Primary ....

Index Primary ....

Supplementary ....

MINES AND CAVES

Quarry or open pit mine .... ×

Gravel, sand, clay, or borrow pit ....

Mine tunnel or cave entrance ....

Prospect; mine shaft .... × ▪

Mine dump ....

Tailings ....

SURFACE FEATURES

Levee ....

Sand or mud area, dunes, or shifting sand ......

Intricate surface area ....

Gravel beach or glacial moraine ....

Tailings pond ....

VEGETATION

Woods ....

Scrub ....

Orchard ....

Vineyard ....

Mangrove ....

MARINE SHORELINE

Topographic maps:

Approximate mean high water ....

Indefinite or unsurveyed ....

Topographic-bathymetric maps:

Mean high water ....

Apparent (edge of vegetation) ....

COASTAL FEATURES

Foreshore flat ....

Rock or coral reef ....

Rock bare or awash ....

Group of rocks bare or awash ....

Exposed wreck ....

Depth curve; sounding 10

Breakwater, pier, jetty, or wharf ....

Seawall ....

BATHYMETRIC FEATURES

Area exposed at mean low tide; sounding datum .

Channel ....

Offshore oil or gas: well; platform .... ○ ▪

Sunken rock .... +

RIVERS, LAKES, AND CANALS

Intermittent stream ....

Intermittent river ....

Disappearing stream ....

Perennial stream ....

Perennial river ....

Small falls; small rapids ....

Large falls; large rapids ....

Masonry dam ....

Dam with lock ....

Dam carrying road ....

Intermittent lake or pond ....

Dry lake ....

Narrow wash ....

Wide wash ....

Canal, flume, or aqueduct with lock ....

Elevated aqueduct, flume, or conduit ....

Aqueduct tunnel ....

Water well; spring or seep ....

GLACIERS AND PERMANENT SNOWFIELDS

Contours and limits ....

Form lines ....

SUBMERGED AREAS AND BOGS

Marsh or swamp ....

Submerged marsh or swamp ....

Wooded marsh or swamp ....

Submerged wooded marsh or swamp ....

Rice field ....

Land subject to inundation ....

Figure 20-19. Topographic Map Symbols.

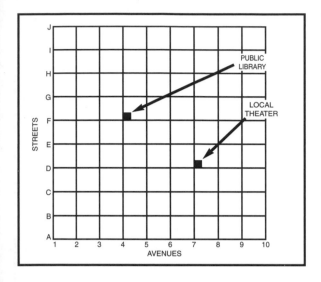

Figure 20-20. Gridville City.

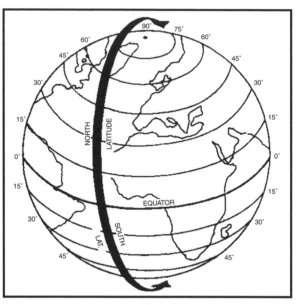

Figure 20-22. Parallels of Latitude.

(c) From the prime meridian east of the international dateline, meridians are assigned values of 0° through 180° east. Similarly, from the prime meridian

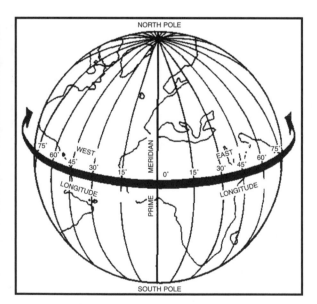

Figure 20-21. Meridian of Longitude.

west to the international dateline, meridians are assigned values of 0° through 180° west. The 0° meridian together with the 180° meridian forms a great circle that divides the Earth into the east and west longitude (or hemispheres). There are 180° of east longitude plus 180° of west longitude for 360° of longitude.

b. Parallels of Latitude. Notice in figure 20-22 that the circles running in an east-west direction are of varying diameters (sizes). Only the circle designated "Equator" is a great circle. All others are small circles. Note that all circles are parallel to the Equator and run laterally around the Earth. Hence, each circle is called a parallel of latitude. Unlike meridians, which extend only halfway around the Earth, a parallel of latitude extends all the way around the Earth; for the record, the Equator is also a parallel of latitude. Since the Equator is the only great circle of latitude, it is a natural starting point for the 0° value of latitude. The North and South poles are designated 90° north latitude and 90° south latitude, respectively. Parallels between the Equator and North Pole carry values between 0° and 90° north; parallels between the Equator and the South Pole are assigned values between 0° and 90° south.

(1) Figure 20-23 combines the lines of latitude and longitude. Lines 0° through 90° north or south latitude and 0° through 180° east or west longitude form the grid of the geographic coordinate system. Study the positions of Points A and B in figure 20-23. Determine the geographic coordinates of each in degrees. Note that point A is positioned 32° north of the Equator and 35° east of the prime meridian. The geographic position of point A, therefore, is 32° north 35° east. Point B is located 25° south of the Equator and 40° west of the prime meridian. Hence, the geographic position of point B is 25° south 40° west.

(2) Just as any point within the city of Gridville (figure 20-20) can be referenced by the intersection of

two imaginary lines, any point on the Earth's surface can be referenced by the intersection of the imaginary lines of latitude and longitude.

c. Writing Geographic Coordinates. To illustrate the proper way to write geographic coordinates, let's assume that a person needs to write the coordinates of a target. The target is located 30°20′ north of the Equator and 135°06′ east of the prime meridian. Thus, the position is located at 30°20′ north latitude and 135°06′ east longitude. By combining latitude and longitude, the position of the geographic location can be expressed as 30°20′N 135°06′E. To write these coordinates in the correct military form, eliminate the degree (°) and minute (′) symbols. Thus, the coordinates would be written 302000N1350600E.

(1) Writing geographic coordinates in the military form is necessary for wire and radio transmission of geographic coordinates. Why? The transmission equipment does not include the degree (°), minute (′), and second (″) characters on its keyboard. Coordinates are also stored in automated data-processing computers, which are programmed to handle coordinates in military characters or spaces. If the sequence of numbers and letters fed into a computer is less than 15 spaces or in error, the resulting printout will be meaningless.

(2) When a position is located that is less than 10° latitude, a zero is added to the left of the degree number. For example, 7° of latitude is written as 07. Likewise, two digits always designate minutes and two digits for seconds. Thus, 7°N becomes 07N; 7°6′N becomes 0706N; and 7°6′5″N becomes 070605N. In expressing longitude, three digits are required to indicate degrees, two digits for minutes, and two digits for seconds. Thus,

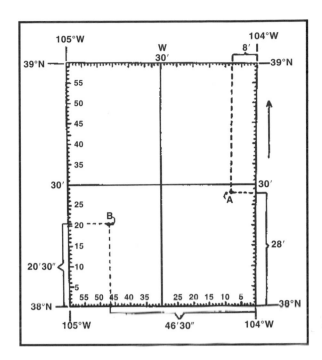

Figure 20-24. Plotting Geographic Coordinates.

8°E becomes 008E; 8°5′E becomes 00805E; and 8°5′4″E becomes 0080504E.

(3) In general, there are five rules to follow in correctly writing geographic coordinates:

(a) Write latitude first, followed by longitude.

(b) Use an even number of digits for latitude and an odd number of digits for longitude.

(c) Do not use a dash or leave a space between latitude and longitude.

(d) Use a single upper-case letter to indicate direction from the Equator and prime meridians.

(e) Omit the symbols for degrees, minutes, and seconds.

d. Plotting Geographic Coordinates:

(1) One can probably read the coordinates of point A and B in figure 20-24 rather easily; however, plotting points on maps from given coordinates must also be done. To do this, first get acquainted with the map being used. Assume that figure 20-24 is the map being used. Note that it covers an area from 38N to 39N and from 104W to 105W, an area of 1° by 1°. Also note that latitude and longitude are subdivided by 30′ division lines and then with tick marks into 5- and 1-minute subdivisions.

(2) Assume the coordinates of the point that must be plotted are 382800N1040800W. Next, follow the general procedure listed below to plot the point on the map:

(a) Locate the parallel of latitude for degrees (38°N).

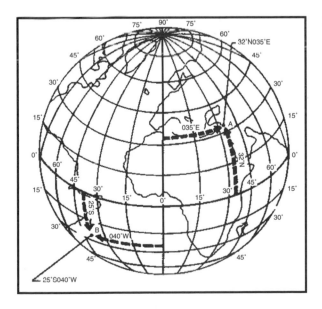

Figure 20-23. Latitudes and Longitudes.

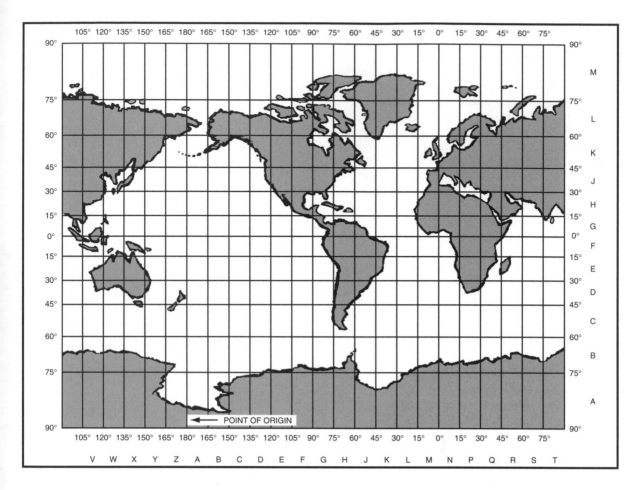

Figure 20-25. GEOREF 15-Degree Quadrants.

(b) Find the meridian of longitude for degrees (104°W).

(c) Move to the meridian (usually a tick mark) for minutes (08°W).

(d) Move to the parallel (usually a tick mark) for minutes (28′N).

(e) Plot the point on the map (point A in figure 20-24; plot at 382800N 1040800W.

(3) Recovery points, rally points, and destination positions may be plotted or identified on a map or chart to enable rescue personnel, the survivors, and evaders to locate these positions. Seconds are not shown between the 1-minute tick marks on maps and charts; they must be estimated. It is easy to estimate halfway tick marks (30 seconds); one-fourth (15 seconds) and three-fourths (45 seconds) are also reasonably easy to estimate. Then, as experience is gained, people will find that on large-scale maps they can estimate the sixths (10 seconds) and eighths (about 8 seconds). They cannot, however, accurately estimate to sixths or eighths at the scale shown in figure 20-24.

(4) To write geographic coordinates more precisely

than minutes, merely carry the coordinates out to include seconds. In the previous example, the coordinates of a target located 30°20′ north of the Equator and 135°06′ east of the prime meridian were written as 302000N1350600E. A more exact position of the target might be 30°20′05″N latitude and 135°06′16″E longitude. This more precise position is correctly written as 302005N1350616E.

e. World Geographic Reference System (GEOREF). The geographic coordinate system has several shortcomings when it is used in military operations. One objectionable feature is the large number of characters necessary to identify a location. To specify a location within 300 yards, a coordinate reading such as 241412NO141512W is necessary, with a total of 15 characters. Another objectionable feature is the diversity of directions used in applying the grid numbering system. Any particular point on a geographic grid can be north and east, north and west, south and east, or south and west. This means there are four different ways to proceed when reading various geographic coordinates.

Such a system obviously promotes errors. To overcome the disadvantages and promote speed in position reporting, other grid systems are used. We shall now examine one of these systems—that which is commonly called GEOREF. The Air Force uses the GEOREF system as a reference in the control and direction of forces engaged in large-area operations and operations of a global nature.

(1) GEOREF System Structure. The geographic coordinate grid serves as the base for the GEOREF system. The grid originates at the 180° meridian and the South Pole. Starting at the 180° meridian, it proceeds right, or eastward, around the world and back to the 180° starting point. From the South Pole, it proceeds northward to the North Pole (figure 20-25).

(a) Notice in figure 20-25 that the basic layout is subdivided into 24 east-west zones and 12 north-south zones. This forms 288 quadrangles that measure 15°-by-15°. The 24 east-west zones are lettered "A" through "Z," omitting "I" and "O." The 12 north-south zones are lettered "A" through "M" (omitting "I"). Each quadrangle is identified by two letters and is located by reading right and up. For example, the southern tip of Florida is located in GEOREF quadrangle G-H (figure 20-25).

(b) Each of the 15° quadrangles is divided into 1° quadrangles (figure 20-26). First they are divided to the right into 15 zones lettered "A" through "Q" (omitting "I" and "O"), then up into 15 zones lettered "A" through "Q" ("I" and "O" omitted).

(c) This system makes it possible to identify any quadrangle by four letters, for example, WGAN. The two letters designate the 15° grid zone, and the other letters

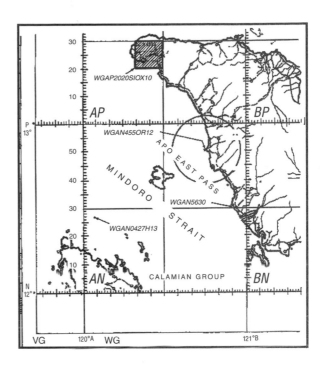

Figure 20-27. GEOREF 1-Degree Quadrants WGAN.

identify a 1° quadrangle within the 15° grid zone. In figure 20-27, WGAN refers to the quadrangle situated between 120° east longitude and 12° and 13° north latitude. Notice the 1° quadrangle WGAN is further divided by 30-minute division lines, and then with tick marks into 5- and 1-minute subdivisions.

(2) GEOREF Coordinates:

(a) Any feature within a 1° quadrangle can be located by reading the number of minutes to the right and the number of minutes up. For example, the city of Magaran (figure 20-27) can be located by proceeding as follows:

-1. 15° x 15° quadrangle identification	WG
-2. 1° x 1° quadrangle identification	WGAN
-3. Minutes to the right	WGAN 56
-4. Minutes up	WGAN 5630
-5. Full GEOREF coordinate	WGAN 5630

(b) If a reference of greater accuracy than 1 minute is required, the 1-minute tick marks may be divided into decimal values (tens or hundreds). By doing this, it is possible to locate a point within one-tenth of a minute with four letters and six numbers and within one-hundredth of a minute with four letters and eight numbers.

(3) GEOREF Special References. Another real advantage of the GEOREF system is the simplicity with

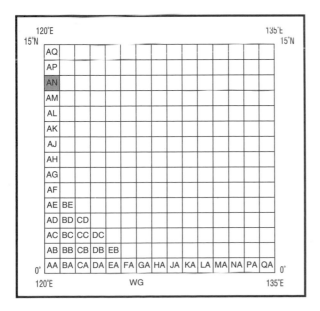

Figure 20-26. GEOREF 1-Degree Quadrants.

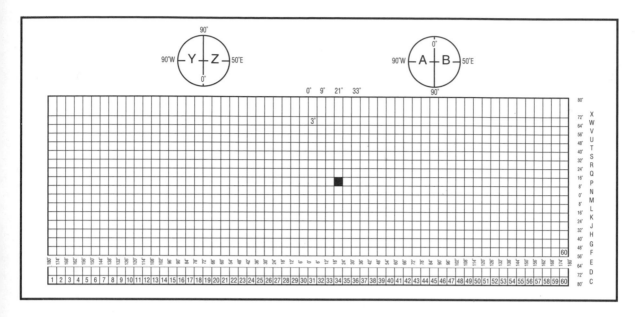

Figure 20-28. UPS Grid Zones.

which it allows a person to designate an area or indicate the elevation of a point. To designate the location and dimensions of a rectangular area, first read the GEOREF coordinates of the southwest corner of the area. Then add and "S," which denotes "side," and digits denoting the number of nautical miles that the area extends to the east. Then add an "X," denoting "by," and digits denoting the number of nautical miles that the area extends to the north. An example of such a reference is WGAP2020S10X10 (figure 20-27). Circular areas are designated in much the same manner. First, read the GEOREF coordinates of the center of the area. Then add an "R," denoting radius, and digits defining the radius in nautical miles. This is also illustrated in figure 20-27 as WGAN4550R12.

(4) Military Grid Reference System. A grid is a rectangular coordinate system superimposed on a map. It consists of two sets of equally spaced parallel lines that are mutually perpendicular and form a pattern of squares. Some maps carry more than one grid. In such cases, each grid is shown in a different color or is otherwise distinguished. The military grid reference system is comprised of two grid systems. The US Army adopted the universal transverse Mercator (UTM) grid for areas between 80°south latitude and 84°north latitude. For the polar caps, areas below 80°south latitude and above 84°north latitude, the universal polar stereographic (UPS) grid was adopted. The unit of measurement for the UTM and UPS grids is the meter, but the interval at which the grid lines are shown on the maps depends on the scale.

(5) The UTM Grid System. In the UTM system, the surface of the Earth is divided into large, quadrilateral

grid zones (figure 20-28). Beginning at the 180th meridian, 6° columns are numbered 1 through 60 eastward with each column broken down into rows. From 80°south through 72°north, each row is 8° south-north.

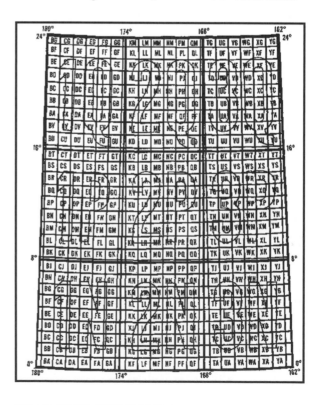

Figure 20-29. UTM Grid Zones.

The top row, 72° to 84° north is 12° south-north. The south-north rows are lettered "C" through "X" (omitting "I" and "O"), as shown in figure 20-28. The grid zones are located by reading right and up. For instance, right to column 34 and up to "P" locates grid zone 34P, which is the shaded grid zone of figure 20-28. The UPS grid zones covering the polar areas are designated by a single "A," "B," "Y," or "Z" (figure 20-28).

(a) Each UTM grid zone is divided into columns and rows to form small squares measuring 100,000 meters on each side and are called 100,000-meter squares. Each square is identified with two letters. The columns are lettered "A" through "Z" (omitting "I" and "O"), starting at the 180th meridian and progressing eastward around. The 24 letters are repeated every 18° (figure 20-29). Starting at the Equator, the horizontal-row 100,000-meter squares are lettered "A" through "V" (omitting "I" and "O") northward. From the Equator southward, the designation "V" through "A" is used. The letters are repeated periodically (figure 20-29).

(b) The Earth's meridians converge toward the poles. Therefore, the grid zones are not square or rectangular. The actual width of each grid zone decreases toward the poles. This condition causes partial squares to occur along the grid zones. In the far north and south latitudes, the grid zones become so narrow that 100,000-meter-square designations may disappear completely. However, each full or partial 100,000-meter square within a grid zone is referenced with two letters. The first letter refers to the vertical column (left to right), and the second letter identifies the horizontal row (bottom to top). Thus, a grid zone designation plus two letters identifies or designates an area 100,000 meters on each side. Furthermore, as the UTM system is set up, no two squares with the same designation are included in a grid zone or on the same map sheet.

(c) Observe grid zone 34P, which is expanded in figure 20-30 to show the 100,000-meter squares. For grid zone 34P, the columns are designated "A" through "H," and the rows are designated "K" through "T" (omitting "O"). The left column begins with "A" because, as stated earlier, columns repeat the alphabet every 18°. The bottom row begins with "K" because "A" through "J" (omitting "I") was used up in the previous 8° of north latitude.

(d) Next, note the partial squares along the left and right sides of grid zone 34P (figure 20-30). Partial squares occur because the distance east and west of the central meridian of each grid zone does not contain an even number of 100,000-meter squares. The last squares, therefore, must terminate at the meridian junctions. In the north-south direction, partial 100,000-meter squares seldom occur.

(e) Figure 20-31 shows the grid reference box for a map or chart. Note the statement in the upper-left corner of the grid reference box. It identifies the grid zones that are represented on the map sheet—52S and 53S. Thus, the full UTM coordinates of any point within the map area begin with either 52S or 53S.

(f) Still, it is not clear which area is 52S and which is 53S. Therefore, study the 100,000-meter square block identification located directly below the grid-zone designation. From the diagram a person can see that everything to the left of the center meridian is grid zone 52S and everything to the right of the same meridian is 53S.

(g) Other important information given in the grid reference block includes: (1) sample reference and, (2) step-by-step procedures for locating or writing coordinates. Each time a new map is used, identify the sample point and write its UTM coordinates to ensure the grid breakdown for that map is understood.

(h) A troublesome and sometimes confusing situation exists in which 100,000-meter squares fuse together along meridians separating grid zones. Remember, this happens every 6° around the world. Notice in figure 20-32 that the 100,000-meter squares GP and KJ are only partial squares, fusing along the center meridian (so are GQ, GN, KH, and KK). There are then full and

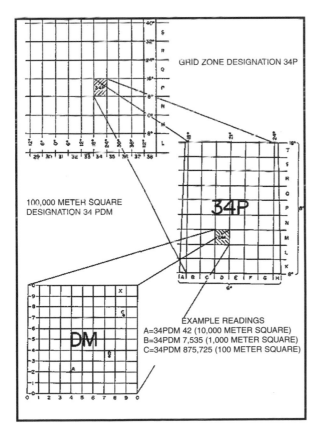

Figure 20-30. Plotting UTM Grid Coordinates.

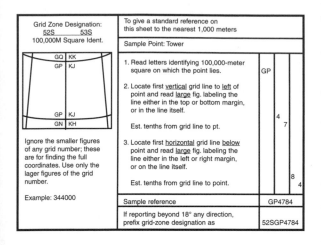

Figure 20-31. UTM Grid Reference Box.

partial 10,000-meter squares within GP and KJ. Column 7 of the GP is comprised of partial 10,000-meter squares; columns 8 and 9 are missing because of the forced fusing along the meridian; similarly, column 2 of KJ is partial; columns O and I are missing. There is no problem in reading coordinates with full 10,000-meter squares. The tower in GP (sample point in figure 20-32) is 47 right and 84 up (omitting the grid-zone and 100,000-meter square designation; all partial 100,000-meter squares are full sized in a north-south dimension). Therefore, distances up are referenced as full squares. However, partial squares, which occur in an east-west dimension, are something less than 10,000 meters long. Points within such partial squares are referenced as if the omitted part were present. That is, each partial square is imaginarily expanded into a full-size square for reference purposes.

(i) The city of Bergen is 40 up; Celle is 35 up. Celle is three-tenths of the horizontal distance between and line 7 and grid line 8—if there were a grid line 8. Thus, Celle is 13 right, and its full coordinates are 52SGP7335. Bergen is eight-tenths of the distance (reading left to right) between grid line 2—if there were a grid line 2—and grid line 3. Thus Bergen is 28 right, and its full coordinates are 53SKJ2840.

(j) Figure 20-32 depicts a UTM grid breakdown as it normally appears on 1:250,000-scale charts. The smallest physical square is 10,000 meters on each side. However, larger scale maps with grid squares of 1,000 or even 100 meters on each side may be used. If so, add the values for the smaller grid squares. As additional digits are added, more precise points on the Earth's surface can be located.

(6) The UPS Grid System. The UPS reference system is the companion system to the UTM system. It covers the area of the world above 84°north latitude and

below 80°south latitude. The UPS has a similar divisional breakdown and is read or written like the UTM system.

(a) Figure 20-33 shows the arbitrarily assigned designations for the UPS system in the North and South Pole regions. Note that from the small circles of the figure, the polar area is divided into two grid-zone divisions by the 180° and 0° meridians. The west longitude half is designated "grid zone A or Y." Also notice that no numbers are used with the letters to identify the grid zones.

(b) The two grid zones "A" and "B" of the South Pole are divided into 100,000-meter squares, as shown in the large circle of figure 20-33. Each square is identified by a two-letter designation, which is assigned so no duplication exists between the two grid zones. The letters "I" and "O" are omitted, and to avoid confusion with 100,000-meter squares in adjoining UTM zones, the letters D, E, M, N, V, and W are also omitted.

(c) The UPS system is also read right and then up. Thus, the shaded 100,000-meter square at 10 o'clock in figure 20-33 is identified as AQR. (Remember, no numbers are used in identifying the grid zone.) The shaded 100,000-meter square near the South Pole of the same drawing is identified as BBM.

(d) The UPS breakdown of the North Pole region is similar to the South Pole region. Conversion of figure 20-33 to fit the North Pole would require the following changes: Substitute grid-zone letters "Y" for "A" and "Z" for "B," and interchange the 0° and 180° positions on the common parallel at 80°south latitude. Designation of the 100,000-meter squares for the North Pole region is shown in figure 20-33.

(e) If the map scale is sufficiently large, the 100,000-meter squares can be subdivided into smaller squares of 10,000 meters on each side. Then, the 10,000-meter squares can be divided into 1,000- or even 100-meter squares. However, there is rarely a requirement in the polar regions for such a large-scale chart.

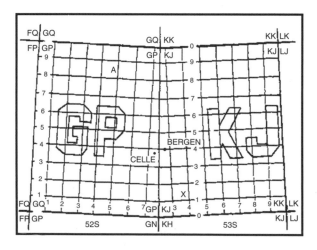

Figure 20-32. Fusion of Grid Zones 52S and 53S.

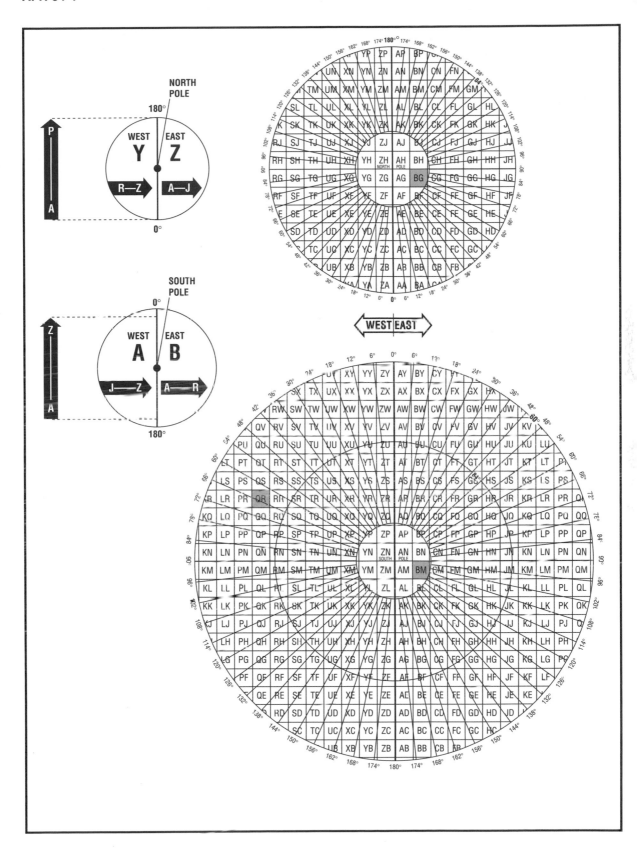

Figure 20-33. UPS Grid Zones.

Generally, a person can expect to work with small-scale charts with the grid broken down no further than 100,000-meter squares.

(7) Public Land Survey. In the western part of the United States or in areas that were not settled before the Federal Government was formed, all land is laid out in rectangular survey as established by the Government. This public land survey is based on all land being divided in relation to true north. Public land surveys all originate from six or seven initial points, which are exact locations of even latitude and longitude lines that have been established astronomically.

(a) From any one of the initial points a true north-south line, referred to as the principal meridian, is established. From the same point a true east-west line, referred to as the baseline, is established. Along this principal meridian and baseline are laid out 6-mile squares or townships. Each of these townships is numbered in relation to the initial point of survey. To the east and west of the initial point, the townships are designated by range numbers; to the north and south of the initial point, the townships are designated by township numbers. Therefore, township 2 north, range 3 east, would lie between 6 and 12 miles north of the initial point and between 12 and 18 miles east of the initial point.

(b) Each township contains 36 square miles and is divided into 36 sections. A section is 1 square mile, or 640 acres. The section layout on townships is the same throughout the Public Land Survey. The sections are numbered in rows back and forth beginning in the upper right-hand corner of the township and ending in the lower right-hand corner (figure 20-34).

(c) Each section is divided into quarters or quarter sections of 160 acres each. These quarter sections are named by the compass location in relation to the section. The upper right-hand quarter section is referred to as the northeast 1/4, the lower right-hand quarter is the southeast 1/4, the lower left-hand quarter is the southwest 1/4, and the upper left-hand quarter is the northwest 1/4.

(d) Each quarter section is further subdivided into quarters or four blocks of 40 acres each known as forties. The forties are also located by the compass directions. In locating a particular piece of property, the 40-compass quadrant is given first, followed by the quarter-section quadrant. Thus the SW-SE means the southwest 40 of the southeast-quarter section. This is the basic unit of land management and, therefore, one should become familiar with the Public Land Survey and the means of locating specific pieces of property.

20-7. Elevation and Relief. A knowledge of map symbols, grids, scale, and distance gives enough information to identify two points, locate them, measure between them, and determine how long it would take to travel between them. But what happens if there is an obstacle between the two points? The map user must become proficient in recognizing various landforms and irregularities of the Earth's surface and be able to determine the elevation and differences in height of all terrain features.

a. Datum Plane. This is the reference used for vertical measurements. The datum plane for most maps is mean or average sea level.

b. Elevation. This is defined as the height (vertical distance) of an object above or below a datum plane.

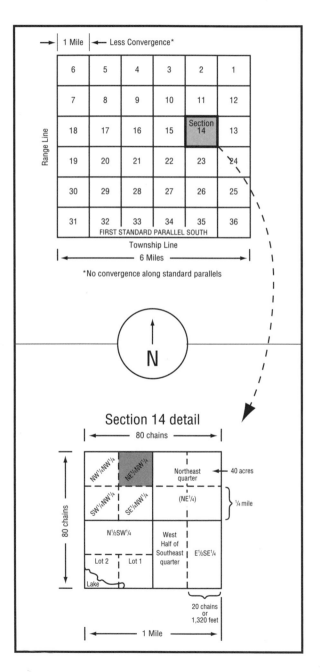

Figure 20-34. Section 14.

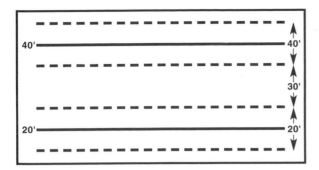

Figure 20-35. Estimating Elevation and Contour Lines.

c. Relief. Relief is the representation of the shape and height of landforms and characteristics of the Earth's surface.

d. Contour Lines:

(1) There are several ways of indicating elevation and relief on maps. The most common way is by contour lines. A contour line is an imaginary line connecting points of equal elevation. Contour lines indicate a vertical distance above or below a datum plane. Starting at sea level, each contour line represents an elevation above sea level. The vertical distance between adjacent contour lines is known as the contour interval. The

amount of contour interval is given in the marginal information. On most maps, the contour lines are printed in brown. Starting at zero elevation, every fifth contour line is drawn with a heavier line. These are known as index contours, and somewhere along each index contour the line is broken and its elevation is given. The contour lines falling between index contours are called intermediate contours. They are drawn with a finer line than the index contours and usually do not have their elevations given.

(2) Using the contour lines on a map, the elevation of any point may be determined by:

(a) Finding the contour interval of the map from the marginal information, and noting the amount and unit of measure.

(b) Finding the numbered contour line (or other given elevation) nearest the point for which elevation is being sought.

(c) Determining the direction of slope from the numbered contour line to the desired point.

(d) Counting the number of contour lines that must be crossed to go from the numbered line to the desired point and noting the direction—up or down. The number of lines crossed multiplied by the contour interval is the distance above or below the starting value. If the desired point is on a contour line, its elevation is that of the contour; for a point between contours,

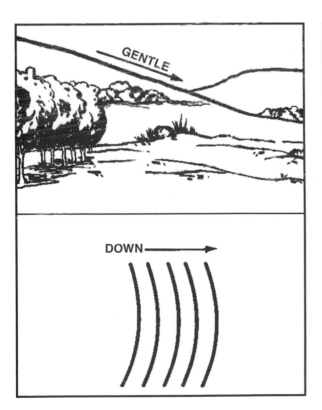

Figure 20-36. Uniform Gentle Slope.

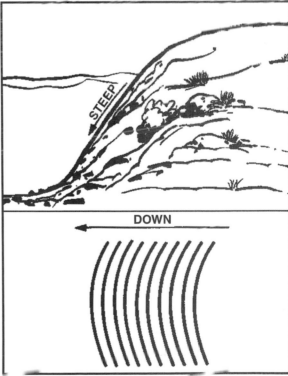

Figure 20-37. Uniform Steep Slope.

most military needs are satisfied by estimating the elevation to an accuracy of one-half the contour interval. All points less than one-fourth the distance between the lines are considered to be at the same elevation as the line. All points one-fourth to three-fourths the distance from the lower line are considered to be at an elevation one-half the contour interval above the lower line (figure 20-35).

(e) To estimate the elevation of the top of an unmarked hill, add half the contour interval to the elevation of the highest contour line around the hill. To estimate the elevation of the bottom of a depression, subtract half the contour interval from the value of the lowest contour around the depression.

(f) On maps where the index and intermediate contour lines do not show the elevation and relief in as much detail as may be needed, supplementary contour may be used. These contour lines are dashed brown lines, usually at one-half the contour interval for the map. A note in the marginal information indicates the interval used. They are used exactly as are the solid contour lines.

(g) On some maps contour lines may not meet the standards of accuracy but are sufficiently accurate in both value and interval to be shown as contour rather than as form lines. In such cases, the contours are considered approximate and are shown with a dashed symbol; elevation values are given at intervals along the heavier (index contour) dashed lines. The contour note in the map margin identifies them as approximate contours.

(h) In addition to the contour lines, benchmarks and spot elevations are used to indicate points of known elevation on the map. Benchmarks, the more accurate of the two, are symbolized by a black X, as in X BM 124. The elevation value shown in black refers to the center of the X. Spot elevations shown in brown generally are located at road junctions, on hilltops, and other prominent landforms. The symbol designates an accurate horizontal control point. When a benchmark and a horizontal control point are located at the same point, the symbol BM is used.

(i) The spacing of the contour lines indicates the nature of the slope. Contour lines evenly spaced and wide apart indicate a uniform, gentle slope (figure 20-36). Contour lines evenly spaced and close together indicate a uniform, steep slope. The closer the contour lines to one another, the steeper the slope (figure 20-37). Contour lines closely spaced at the top and widely spaced at the bottom indicate a concave slope (figure 20-38). Contour lines widely spaced at the top and closely spaced at the bottom indicate a convex slope (figure 20-39).

(j) To show the relationship of land formations to one another and how they are symbolized on a contour map, stylized panoramic sketches of the major relief

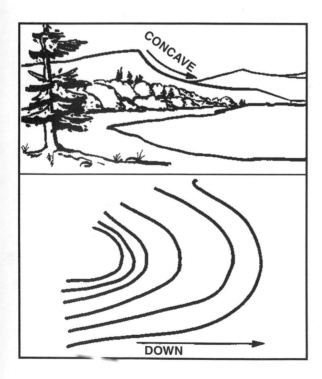

Figure 20-38. Concave Slope.

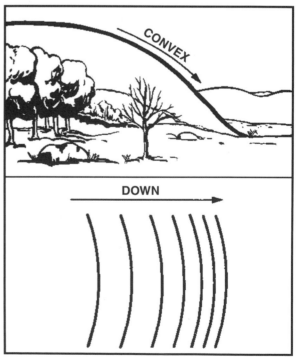

Figure 20-39. Convex Slope.

formations were drawn, and a contour map of each sketch developed. Each figure (figure 20-40 through 20-46) shows a sketch and a map with a different relief feature and its characteristic contour pattern.

(3) Hill: A point or small area of high ground (figure 20-40). When one is located on a hilltop, the ground slopes down in all directions.

(4) Valley: Usually a stream course that has at least a limited extent of reasonably level ground bordered on the sides by higher ground (figure 20-41A). A valley generally has maneuvering room within its confines. Contours indicating a valley are U-shaped and tend to parallel a major stream before crossing it. The more gradual the fall of a stream, the farther apart each contour inner part. The curve of the contour crossing always points upstream.

(5) Drainage: A less-developed stream course in which there is essentially no level ground and, therefore, little or no maneuvering room within its confines (figure 20-41B). The ground slopes upward on each side and toward the head of the drainage. Drainages occur frequently along the sides of ridges, at right angles to the valleys between the ridges. Contours indicating a drainage are V-shaped, with the point of the "V" toward the head of the drainage.

(6) Ridge: A range of hills or mountains with normally minor variations along its crest (figure 20-42A). The ridge is not simply a line of hills; all points of the ridge crest are appreciably higher than the ground on both sides of the ridge.

(7) Finger Ridge: A ridge or line of elevation projecting from or subordinate to the main body of a mountain or mountain range (figure 20-42B). A finger ridge is often formed by two roughly parallel streams cutting drainages down the side of a ridge.

(8) Saddle: A dip or low point along the crest of a ridge. A saddle is not necessarily the lower ground between two hilltops; it may simply be a dip or break along an otherwise level ridge crest (figure 20-43).

(9) Depression: A low point or sinkhole surrounded on all sides by higher ground (figure 20-44).

(10) Cuts and fills: Manmade features by which the bed of a road or railroad is graded or leveled off by

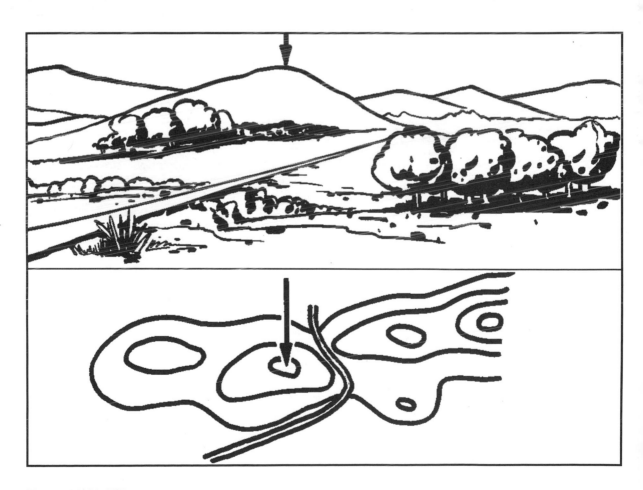

Figure 20-40. Hill.

cutting through high areas (figure 20-45A) and filling in low areas (figure 20-45B) along the right-of-way.

(11) Cliff: A vertical or near-vertical slope (figure 20-46). When a slope is so steep it cannot be shown at the contour interval without the contours fusing, it is shown by a ticked "carrying" contour(s). The ticks always point toward lower ground.

20-8. Representative Fraction (RF):

a. The numerical scale of a map expresses the ratio of horizontal distance on the map to the corresponding horizontal distance on the ground. It usually is written as a fraction, called the representative fraction (RF). The representative fraction is always written with the map distance as one (1). It is independent of any unit of measure. An RF of 1/50,000, or 1:50,000, means that one (1) unit of measure on the map is equal to 50,000 of the same units of measure on the ground.

b. The ground distance between two points is determined by measuring between the points on the map and multiplying the map measurement by the denominator of the RF.

$$\text{Example: RF} = 1:50,000 \text{ or } \frac{1}{50,000}$$
$$\text{Map distance} = 5 \text{ units (CM)}$$
$$5 \times 50,000 = 250,000 \text{ units (CM) of}$$
$$\text{ground distance (figure 20-47).}$$

c. When determining ground distance from a map, the scale of the map affects the accuracy. As the scale becomes smaller, the accuracy of measurement decreases because some of the features on the map must be exaggerated so they may be readily identified.

20-9. Graphic (Bar) Scales:

a. On most military maps, there is another method of determining ground distance. It is by means of the graphic (bar) scales. A graphic scale is a ruler printed on the map in which distances on the map may be measured as actual ground distances. To the right of the zero (0), the scale is marked in full units of measure and is called the primary scale. The part to the left of zero (0) is divided into tenths of a unit and is called the extension scale. Most maps have three or more graphic scales, each of which measures distance in a different unit of measure (figure 20-48).

b. To determine a straight-line ground distance between two points on a map, lay a straight-edged piece of paper on the map so the edge of the paper touches both points. Mark the edge of the paper at each point. Move the paper down to the graphic scale and read the ground distance between the points. Be sure to use the scale that measures in the unit of measure desired (figure 20-49).

c. To measure distance along a winding road, stream, or any other curved line, the straight edge of a piece of paper is used again. Mark one end of the paper and place it at the point from which the curved line is to be measured. Align the edge of the paper along a straight portion and mark both the map and the paper at the end of the aligned portion. Keeping both marks together, place the point of the pencil on the mark on the paper to hold it in place. Pivot the paper until another approximately straight portion is aligned and again mark on the map and the paper. Continue in this manner until measurement is complete. Then place the paper on the graphic scale and read the ground distance (figure 20-50).

d. Often, marginal notes give the road distance from the edge of the map to a town, highway, or junction of the map. If the road distance is desired from a point on the map to such a point off the map, measure the distance to the edge of the map and add the distance specified in

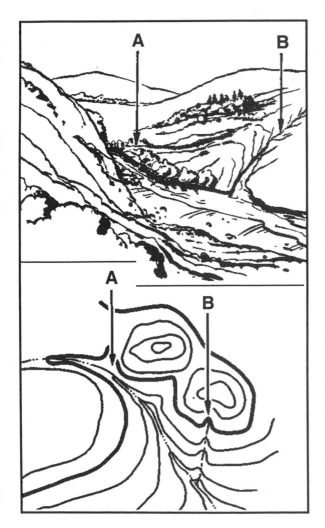

Figure 20-41. (A) Valley (B) Drainage.

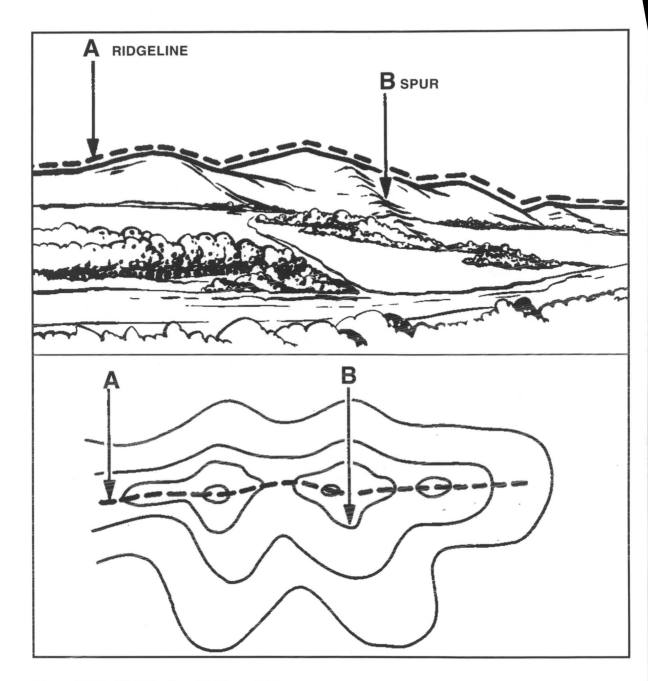

Figure 20-42. (A) Ridgeline (B) Finger Ridge.

the marginal note to that measurement. Be sure the unit of measure is the same (figure 20-51).

20-10. Using a Map and Compass, and Expressing Direction:

a. To use a map, the map must correspond to the lay of the land, and the user must have knowledge of direction and how the map relates to the cardinal directions. In essence, to use a map for land navigation, the map must be "oriented" to the lay of the land. This is usually done with a compass. On most maps, either a declination diagram or compass rose, and lines of map magnetic variations are provided to inform the user of the difference between magnetic north and true north.

b. Directions are expressed in everyday life as right, left, straight ahead, etc.; but the question arises, "To the right of what?" Military personnel require a method of expressing direction that is accurate, adaptable for use

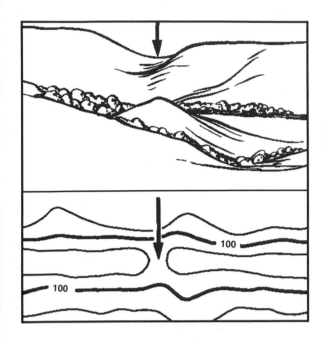

Figure 20-43. Saddle.

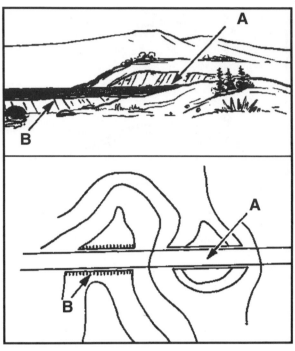

Figure 20-45. (A) Cut (B) Fill.

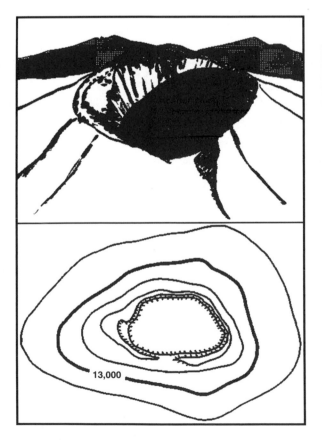

Figure 20-44. Depression.

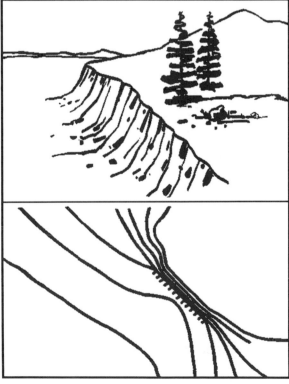

Figure 20-46. Cliff.

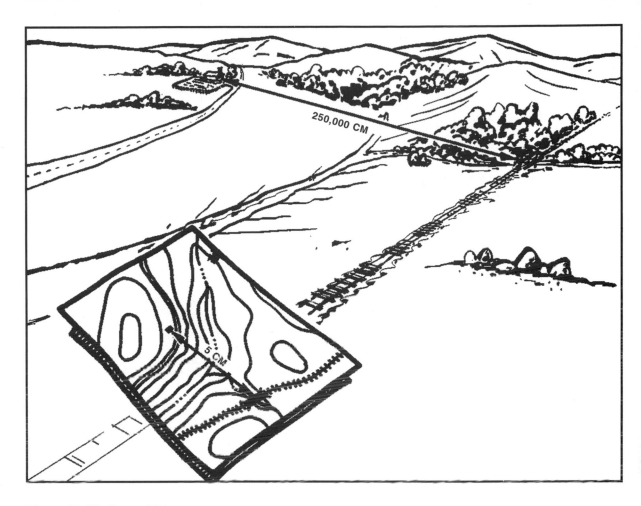

Figure 20-47. Ground Distance.

in any area of the world, and has a common unit of measure. Directions are expressed as units of angular measure. The most commonly used unit of angular measure is the degree, with its subdivisions of minutes and seconds.

(1) Baselines. To measure anything, there must always be starting point or zero measurement. To express a direction as a unit of angular measure, there must be a starting point or zero measure and a point of reference. These two points designate the base or reference

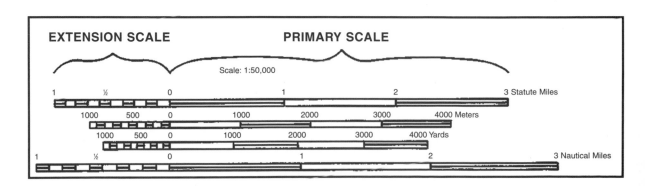

Figure 20-48. Graphic Bar Scale.

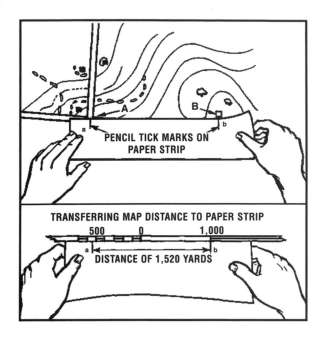

Figure 20-49. Measuring Straight Line Map Distances.

line. There are three baselines—true north, magnetic north, and grid north. Those most commonly used are magnetic and grid—the magnetic when working with a compass, the grid when working with a military map.

(a) True north—a line from any position on the Earth's surface to the North Pole. All lines of longitude are true north lines. True north is usually symbolized by a star (figure 20-52).

(b) Magnetic north—the direction to the north magnetic pole, as indicated by the north-seeking needle of a magnetic instrument. Magnetic north is usually symbolized by a half-arrowhead (figure 20-52).

(c) Grid north—the north established by the vertical grid lines on a map. Grid north may be symbolized by the letters GN or the letter Y.

(2) Azimuth and Back Azimuth:

(a) The most common method used by the military for expressing a direction is azimuths. An azimuth is defined as a horizontal angle, measured in a clockwise manner from a north baseline. When the azimuth between two points on a map is desired, the points are joined by a straight line and a protractor is used to measure the angle between north and the drawn line. This measured angle is the azimuth of the drawn line (figure 20-53). When using an azimuth, the point from which the azimuth originates is imagined to be the center of the azimuth circle (figure 20-54). Azimuths take their name from the baseline from which they are measured: true azimuths from true north, magnetic azimuths from magnetic north, and grid azimuths from grid north (figure

20-52). Therefore, any given direction can be expressed in three different ways: a grid azimuth if measured on a military map, a magnetic azimuth if measured by a compass, or a true azimuth if measured from a meridian of longitude.

(b) A back azimuth is the reverse direction of an azimuth. It is comparable to making an "about face." To obtain a back azimuth from an azimuth, add 180° if the azimuth is 180° or less, or subtract 180° if the azimuth is 180° or more (figure 20-55). The back azimuth of 180° may be stated as either 000° or 360°.

(3) Declination Diagram. A declination diagram is placed on most large-scale maps to enable the user to properly orient the map. The diagram shows the interrelationship of magnetic north, grid north, and true north (figure 20-56). On medium-scale maps, declination information is shown by a note in the map margin.

(a) Declination is the angular difference between true north and magnetic or grid north. There are two declinations, a magnetic declination (figure 20-57) and a grid declination.

(b) Grid-Magnetic (G-M) Angle is an arc indicated by a dashed line, which connects the grid north and the magnetic north prongs. The value of this arc (G-M ANGLE) states the size of the angle between grid north and magnetic north and the year it was prepared. This value is expressed to the nearest 1/2°, with mil equivalents shown to the nearest 10 mils.

(c) Grid Convergence is an arc, indicated by a dashed line, which connects the prongs for true north and grid north. The value of the angle for the center of the sheet is given to the nearest full minute with its equivalent to the nearest mil. These data are shown in the form of a grid-convergence note.

(d) Conversion notes may also appear with the diagram explaining the use of the G-M angle. One note

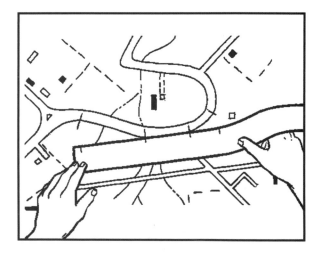

Figure 20-50. Measuring Curved Line Distances.

B-5 Conversion Factors.

One	Inches	Feet	Yards	Statute Miles	Nautical Miles	Millimeters
Inch	1	0.0833	0.0277	....		25.40
Foot	12	1	0.333	....		304.8
Yard	36	3	1	0.00056		914.4
Statute Mile	63,360	5,280	1,760	1	0.8684	....
Nautical Mile	72,963	6,080	2,026	1.1516	1	....
Millimeter	0.0394	0.0033	0.0011	....		1
Centimeter	0.3937	0.0328	0.0109	....		10
Decimeter	3.937	0.328	0.1093	....		100
Meter	39.37	3.2808	1.0936	0.0006	0.0005	1,000
Decameter	393.7	32.81	10.94	0.0062	0.0054	10,000
Hectometer	3,937	328.1	109.4	0.0621	0.0539	100,000
Kilometer	39,370	3,281	1,094	0.6214	0.5396	1,000,000
Myriameter	393,700	32,808	10,936	6.2137	5.3959	10,000,000

One	cm	dm	m	dkm	hm	km	mym
Inch	2.540	0.2540	0.0254	0.0025	0.0003		
Foot	30.48	3.048	0.3048	0.0305	0.0030	0.0003	....
Yard	91.44	9.144	0.9144	0.0914	0.0091	0.0009	....
Statute Mile	160,930	16,093	1,609	160.9	16.09	1.6093	0.1609
Nautical Mile	185,325	18,532	1,853	185.3	18.53	1.8532	0.1853
Millimeter	0.1	0.01	0.001	0.0001	....		
Centimeter	1	0.1	0.01	0.001	0.0001	....	
Decimeter	10	1	0.1	0.01	0.001	0.0001	....
Meter	100	10	1	0.1	0.01	0.001	0.0001
Decameter	1,000	100	10	1	0.1	0.01	0.001
Hectometer	10,000	1,000	100	10	1	0.1	0.01
Kilometer	100,000	10,000	1,000	100	10	1	0.1
Myriameter	1,000,000	100,000	10,000	1,000	100	10	1

Example I

Problem: Reduce 76 centimeters to (") inches

76 cm × 0.3937 = 29 inches

Answer: There are approximately 30 inches in 76 centimeters.

Example II

Problem: How many feet are there in 2.74 meters?

$$\frac{2.74}{.3048} = 9 \text{ feet}$$

Answer: There are approximately 9 feet in 2.74 meters.

Figure 20-51. Conversion Factors.

provides instructions for converting magnetic azimuth to grid azimuth; the other note provides instructions for converting grid azimuth to magnetic azimuth. The conversion (add or subtract) is governed by the direction of the magnetic-north prong relative to the grid-north prong.

(e) The grid north prong is aligned with the easting grid lines on the map, and on most maps is formed by an extension of an easting grid line into the margin. The angles between the prongs are seldom plotted exactly. The relative position of the directions is obtained from the diagram, but the numerical value should not be measured from it. For example, if the amount of declination from grid north to magnetic north is 1°, the arc shown in the diagram may be exaggerated if measured,

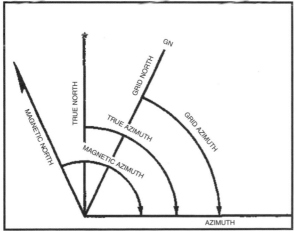

Figure 20-52. True, Grid, and Magnetic Azimuths.

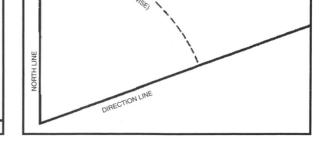

Figure 20-53. Azimuth Angle.

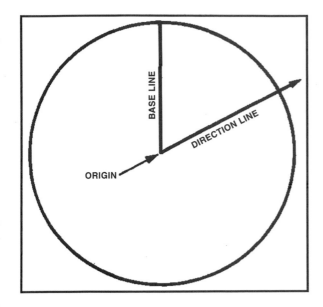

Figure 20-54. Origin of Azimuth Circle.

having an actual value of 5°. The position of the three prongs in relation to one another varies according to the declination data for each map.

(f) Some older maps have a note under the declination diagram that gives the magnetic declination for a certain year and the amount of annual change. The annual change is so small when compared to the 1/2° value of the G-M angle that it is no longer shown on standard large-scale maps.

(4) Protractors. Protractors come in several forms—full circle, half circle, square, and rectangular (figure 20-58). All of them divide a circle into units of angular measure, and regardless of their shape, consist of a scale around the outer edge and an index mark. The index mark is the center of the protractor circle from which all the direction lines radiate.

(a) To determine the grid azimuth of a line from one point to another on the map from A to B or C to D (figure 20-59), draw a line connecting the two points.

-1. Place the index of the protractor at that point where the line crosses a vertical (north-south) grid line.

-2. Keeping the index at this point, align the 0° to 180° line of the protractor on vertical grid line.

-3. Read the value of the angle from the scale; this is the grid azimuth to the point.

(b) To plot a direction line from a known point on a map (figure 20-60):

-1. Construct a north-south grid line through the known point:

-a. Generally, align the 0° to 180° line of the protractor in a north-south direction through the known point.

-b. Holding the 0° to 180° line of the protractor on the known point, slide the protractor in the north-south direction until the horizontal line of the protractor (connecting the protractor index and the 90° tick mark) is aligned on an east-west grid line.

-c. Then draw a line connecting 0°, the known point, and 180°.

-2. Holding the 0° to 180° line on the north-south line, slide the protractor index to the known point.

-3. Make a mark on the map at the required angle. (In an evasion situation, do not mark on the map.)

-4. Draw a line from the known point through the mark made on the map. This is the grid direction line.

(5) The Compass and Its Uses:

(a) The magnetic compass is the most commonly used and simplest instrument for measuring directions and angles in the field. The lensatic compass (figure 20-61) is the standard magnetic compass for military use today.

(b) The lensatic compass must always be held level and firm when sighting on an object and reading an azimuth (figure 20-62). There are several techniques for holding the compass and sighting. One way is to align the sighting slot with the hairline on the front sight in the cover and the target. The azimuth

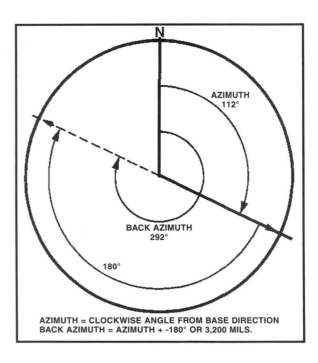

Figure 20-55. Azimuth and Back Azimuth.

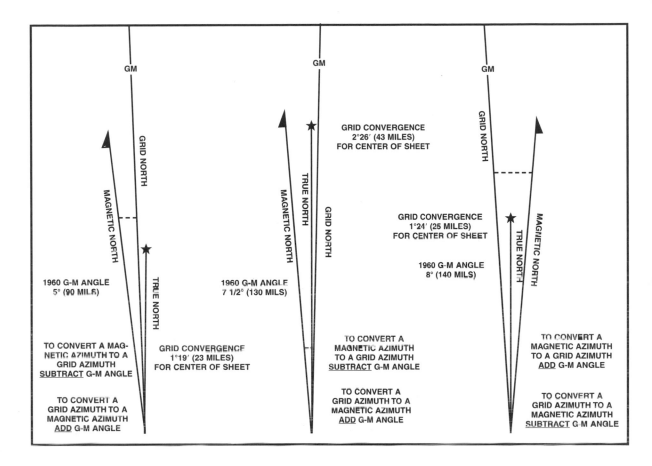

Figure 20-56. Declination Diagram (East and West).

can then be read by glancing down at the dial through the lens. This technique provides a reading precise enough to use.

(6) Night Use of the Compass. For night use, special features of the compass include the luminous markings, the bezel ring, and two luminous sighting dots. Turning the bezel ring counterclockwise causes an increase in azimuth, while turning it clockwise causes a decrease. The bezel ring has a stop and spring that allows turns at 3° intervals per click and holds it at any desired position. An accepted method for determining compass directions at night is:

(a) Rotate the bezel ring until the luminous line is over the black index line.

(b) Hold the compass with one hand and rotate the bezel ring in a counterclockwise direction with the other hand to the number of clicks required. The number of clicks is determined by dividing the value of the required azimuth by 3. For example, for an azimuth of 51°, the bezel ring would be rotated 17 clicks counterclockwise (figure 20-63).

(c) Turn the compass until the north arrow is directly under the luminous line on the bezel.

(d) Hold the compass open and level in the palm of the left hand with the thumb along the side of the compass. In this manner, the compass can be held consistently in the same position. Position the compass approximately halfway between the chin and the belt, pointing to the direct front. (Practicing in daylight will make a person proficient in pointing the compass the same way every time.) Looking directly down into the compass, turn the body until the north arrow is under the luminous line. Then proceed forward in the direction of the luminous sighting dots (figure 20-61). When the compass is to be used in darkness, an initial azimuth should be set while light is still available. With this initial azimuth as a base, any other azimuth that is a multiple of 3° can be established through use of the clicking feature of the bezel ring. The magnetic compass is a delicate instrument, especially the dial balance. The survivor should take care in its use. Compass readings should never be taken near visible masses of iron or electrical circuits.

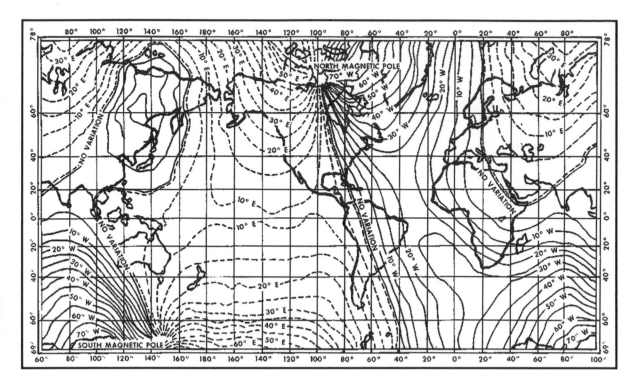

Figure 20-57. Lines of Magnetic Variation.

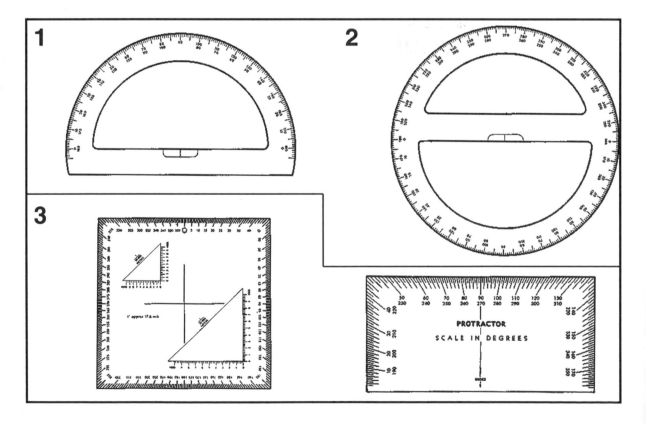

Figure 20-58. Types of Protractors.

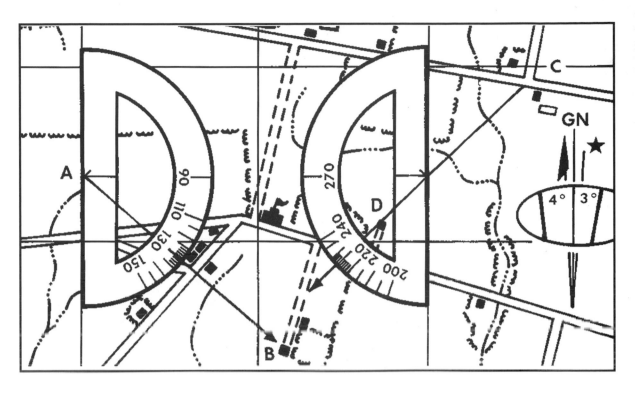

Figure 20-59. Measuring an Azimuth on a Map.

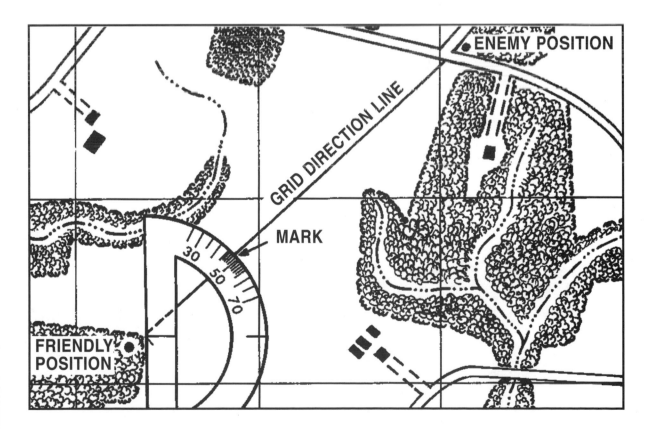

Figure 20-60. Plotting an Azimuth on a Map.

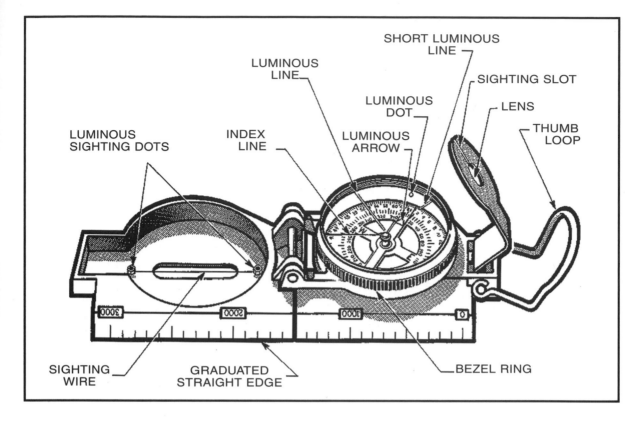

Figure 20-61. Lensatic Compass.

Figure 20-62. Holding the Compass.

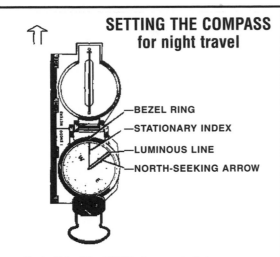

SETTING THE COMPASS
for night travel

—BEZEL RING
—STATIONARY INDEX
—LUMINOUS LINE
—NORTH-SEEKING ARROW

Each click of the BEZEL ring equals 3 degrees.

Heading between 0 and 180 degrees is divided by 3. Sum is number of clicks to the left of stationary index line. Heading between 180 and 360 degrees, subtract heading from 360 then divide sum by 3. New sum is the number of clicks to the right from stationary index line.

EXAMPLES

Heading of 027° = 9 clicks left
Heading of 300° = 20 clicks right

Figure 20-63. Night Travel.

(7) Map Orientation:

(a) A map is oriented when it is in a horizontal position with its north and south corresponding to north and south on the ground. The best way to orient a map is with a compass. (NOTE: Caution should be used to ensure nothing {metal, mine ore, etc.} in the area will alter the compass reading.)

(b) With the map in a horizontal position, the lensatic compass is placed parallel to a north-south grid line with the cover side of the compass pointing toward the top of the map. This will place the black index line on the dial of the compass parallel to and north. Since the needle on the compass points to magnetic north, a declination diagram (on the face of the compass) is formed by the index line and the compass needle.

(c) Rotate the map and compass until the directions of the declination diagram formed by the black index line and compass needle match the directions shown on the declination diagram printed in the margin of the map. The map is then oriented (grid north).

(d) If the magnetic-north arrow on the map is to the left of grid north, the compass reading will equal the G-M angle (given in the declination diagram). If the magnetic north is to the right of grid north, the compass reading will equal 360° minus the G-M angle. In figure 20-64, the declination diagram illustrates a magnetic north to the right of grid north, and the compass reading will be 360° minus 21 1/2°, or 338 1/2°.

(e) Remember to point the compass-north arrow in the same direction as the magnetic-north arrow, and the compass reading (equal to the G-M angle or the 360° minus G-M angle) will be quite apparent.

(f) In summary, if the variation is to the east of true north or the magnetic-north arrow of the declination diagram is to the east (right) of the grid-north line, subtract the degrees of variation from 360°. If it is to the left (west), add to 000°. East is least and west is best.

(g) If a grid line is not used, a true north-south line can be used. True north-south lines are longitudinal lines or lines formed by the vertical lines on a tick map (assuming the top of the map is north). The same procedure is used if magnetic variation is figured from true north—not grid north.

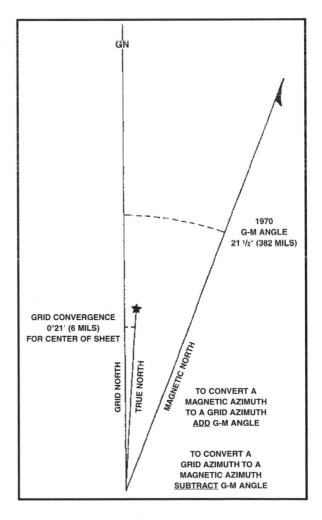

Figure 20-64. Declination Diagram.

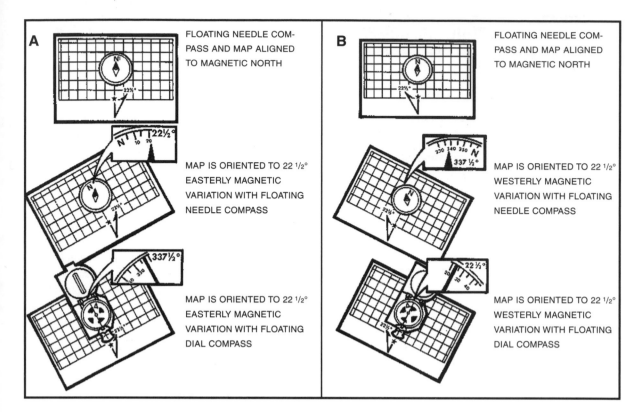

Figure 20-65. Floating Needle Compass.

(h) A floating needle compass (figures 20-65A and 20-65B) has a needle with a north direction marked on it. The degree and direction marks are stationary on the bottom inside of the compass. The button and wrist compasses may be floating dial or floating needle. To determine the heading, line up the north-seeking arrow more than 360° by rotating the compass. Then read the desired heading. Orienting a map with a floating needle compass is similar to the method used with the floating dial. The only exception is with the adjustment for magnetic variation. If magnetic variation is to the east, turn the map and the compass to the left (the north axis of the compass should be aligned with the map north) so the magnetic north-seeking arrow is pointing at the number of degrees on the compass that corresponds with the angle of declination.

(i) When a compass is not available, map orientation requires a careful examination of the map and the ground to find linear features common to both, such as roads, railroads, fence lines, power lines, etc. By aligning the feature on the map with the same feature on the ground (figure 20-66), the map is oriented. Orientation by this method must be checked to prevent the reversal of directions that may occur if only one linear feature is used. This reversal may be prevented by aligning two or more map features (terrain or manmade). If no second

linear feature is visible but the map user's position is known, a prominent object may be used. With the

Figure 20-66. Map Orientation by Inspection.

prominent object and the user's position connected with a straight line on the map, the map is rotated until the line points toward the feature.

(j) If two prominent objects are visible and plotted on the map and the position is not known, move to one of the plotted and known positions, place the straightedge or protractor on the line between the plotted positions, then turn the protractor and the map until the other plotted and visible point is seen along the edge. The map is then oriented.

(k) When a compass is not available and there are no recognizable prominent landforms or other features, a map may be oriented by any of the field-expedient methods we will now discuss.

(8) Determining Cardinal Directions Using Field Expedients:

*(a) Shadow-tip method of determining direction and time. This simple method of finding direction by the Sun consists of only three basic steps (figure 20-67).

-1. Step 1. Place a stick or branch into the ground at a fairly level spot where a distinct shadow will be cast. Mark the shadow tip with a stone, twig, or other means.

-2. Step 2. Wait until the shadow tip moves a few inches. If a 4-foot stick is being used, about 10 minutes should be sufficient. Mark the new position of the shadow tip in the same way as the first.

-3. Step 3. Draw a straight line through the two marks to obtain an approximate east-west line. If uncertain which direction is east and which is west, observe this simple rule: The Sun "rises in the east and sets in the west" (but rarely DUE east and DUE west). The shadow tip moves in just the opposite direction. Therefore, the first shadow-tip mark is always in the west direction, and the second mark in the east direction, anyplace on Earth.

(b) A line drawn at right angles to the east-west line at any point is the approximate north-south line, which will help orient a person to any desired direction of travel.

(c) Inclining the stick to obtain a more convenient shadow does not impair the accuracy of the shadow-tip method. Therefore, a traveler on sloping ground or in highly vegetated terrain need not waste valuable time looking for a large level area. A flat spot, the size of the hand, is all that is necessary for shadow-tip markings, and the base of the stick can be either above, below, or to one side of it. Also, any stationary object (the end of a tree limb or the notch where branches are jointed) serves just as well as an implanted stick because only the shadow tip is marked.

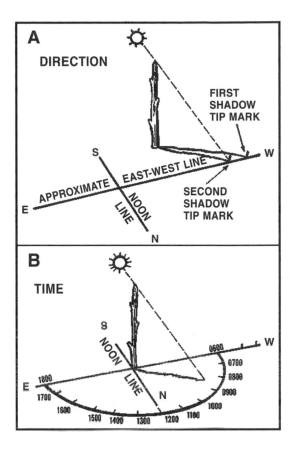

Figure 20-67. Determining Time and Direction by Shadow

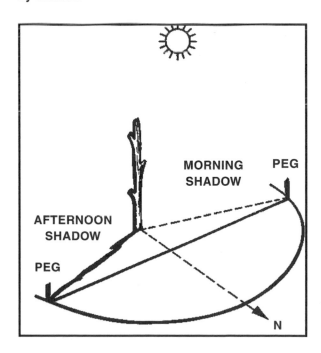

Figure 20-68. Equal Shadow Method of Determining Direction.

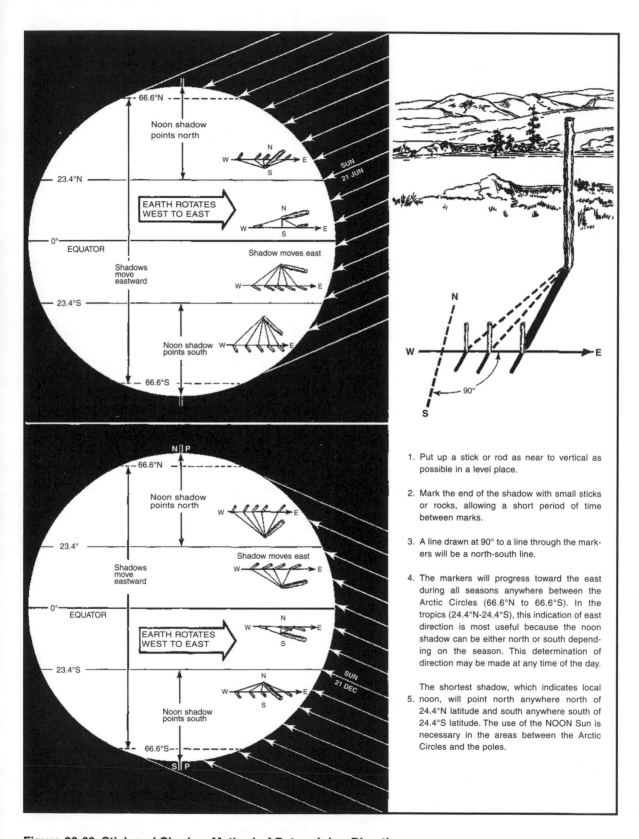

1. Put up a stick or rod as near to vertical as possible in a level place.

2. Mark the end of the shadow with small sticks or rocks, allowing a short period of time between marks.

3. A line drawn at 90° to a line through the markers will be a north-south line.

4. The markers will progress toward the east during all seasons anywhere between the Arctic Circles (66.6°N to 66.6°S). In the tropics (24.4°N-24.4°S), this indication of east direction is most useful because the noon shadow can be either north or south depending on the season. This determination of direction may be made at any time of the day.

5. The shortest shadow, which indicates local noon, will point north anywhere north of 24.4°N latitude and south anywhere south of 24.4°S latitude. The use of the NOON Sun is necessary in the areas between the Arctic Circles and the poles.

Figure 20-69. Stick and Shadow Method of Determining Direction.

(d) The shadow-tip method can also be used to find the approximate time of day (figure 20-67B).

-1. To find the time of day, move the stick to the intersection of the east-west line and the north-south line, and set it vertically in the ground. The west part of the east-west line indicates the time is 0600 and the east part is 1800, ANYWHERE on Earth, because the basic rule always applies.

-2. The north-south line now becomes the noon line. The shadow of the stick is an hour hand in the shadow clock, and with it the time can be estimated using the noon line and 6 o'clock line as the guides. Depending on the location and the season, the shadow may move either clockwise or counterclockwise, but this does not alter the manner of reading the shadow clock.

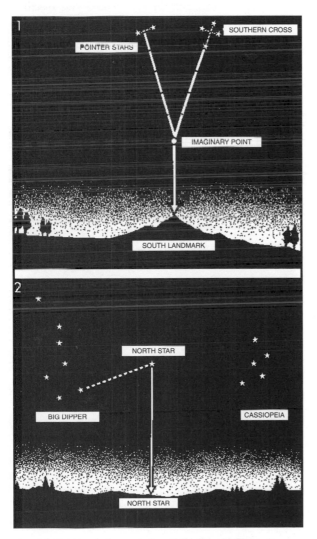

Figure 20-70. Determination of Direction Using the Stars.

-3. The shadow clock is not a timepiece in the ordinary sense. It always reads 0600 at sunrise and 1800 at sunset. However, it does provide a satisfactory means of telling time in the absence of properly set watches. Being able to establish the time of day is important for such purposes as keeping a rendezvous, prearranged concerted action by separated persons or groups, and estimating the remaining duration of daylight. Shadow-clock time is closest to conventional clock time at midday, but the spacing of the other hours, compared to conventional time, varies somewhat with the locality and date.

(e) The shadow-tip system is ineffective for use beyond 66 ½° latitude in either hemisphere due to the position of the Sun above the horizon. Whether the Sun is north or south of a survivor at midday depends on the latitude. North of 23.4°N, the Sun is always due south at local noon and the shadow points north. South of 23.4°S, the Sun is always due north at local noon and the shadow points south. In the tropics, the Sun can be either north or south at noon, depending on the date and location, but the shadow progresses to the east regardless of the date.

(f) Equal-shadow method of determining direction (figures 20-68 and 20-69). This variation of the shadow-tip method is more accurate and may be used at all latitudes less than 66° at all times of the year.

-1. Step 1. Place a stick or branch into the ground vertically at a level spot where a shadow at least 12 inches long will be cast. Mark the shadow tip with a stone, twig, or other means. This must be done 5 to 10 minutes before noon (when the Sun is at its highest point, or zenith).

-2. Step 2. Trace an arc using the shadow as the radius and the base of the stick as the center. A piece of string, shoelace, or a second stick may be used to do this.

-3. Step 3. As noon is approached, the shadow becomes shorter. After noon, the shadow lengthens until it crosses the arc. Mark the spot as soon as the shadow tip touches the arc a second time.

-4. Step 4. Draw a straight line through the two marks to obtain an east-west line.

(g) Although this is the most accurate version of the shadow-tip method, it must be performed around noon. It requires the observer to watch the shadow and complete step 3 at the exact time that the shadow tip touches the arc.

(h) At night, the stars may be used to determine the north line in the Northern Hemisphere or the south line in the Southern Hemisphere. Figure 20-70 shows how this is done.

(i) A watch can be used to determine the approximate true north or south (figure 20-71). In the Northern Hemisphere, the hour hand is pointed toward the Sun. A south line can be found midway between the hour hand

and 1200 standard time. During daylight saving time, the north-south line is midway between the hour hand and 1300. If there is any doubt as to which end of the line is north, remember that the Sun is in the east before noon and in the west in the afternoon.

(j) The watch may also be used to determine direction in the Southern Hemisphere; however, the method is different. The 1200-hour dial is pointed toward the Sun, and halfway between 1200 and the hour hand will be a north line. During daylight saving time, the north line lies midway between the hour hand and 1300.

(k) The watch method can result in error, especially in the extreme latitudes, and may cause "circling." To avoid this, make a shadow clock and set the watch to the time indicated. After traveling for an hour, take another shadow-clock reading.

(9) Determining Specific Position. When using a map and compass, the map must be oriented using the method described earlier in this chapter. Next, locate two or three known positions on the ground and the map. Using the compass, shoot an azimuth to one of the known positions (figure 20-72). Once the azimuth is

determined, recheck the orientation of the map and plot the azimuth on the map. To plot the azimuth, place the front corner of the straightedge of the compass on the corresponding point on the map. Rotate the compass until the determined azimuth is directly beneath the stationary index line. Then draw a line along the straightedge of the compass and extend the line past the estimated position on the map. Repeat this procedure for the second point (figure 20-72). If only two azimuths are used, the technique is referred to as biangulation (figure 20-72). If a third azimuth is plotted to check the accuracy of the first two, the technique is called triangulation (figure 20-72). When using three lines, a triangle of error may be formed. If the triangle is large, the work should be checked. However, if a small triangle is formed, the user should evaluate the terrain to determine the actual position. One azimuth may be used with a linear land feature such as a river, road, railroad, etc., to determine specific position (figure 20-72).

(10) Determining Specific Location without a Compass. A true north-south line determined by the stick and shadow, Sun and watch, or celestial constellation

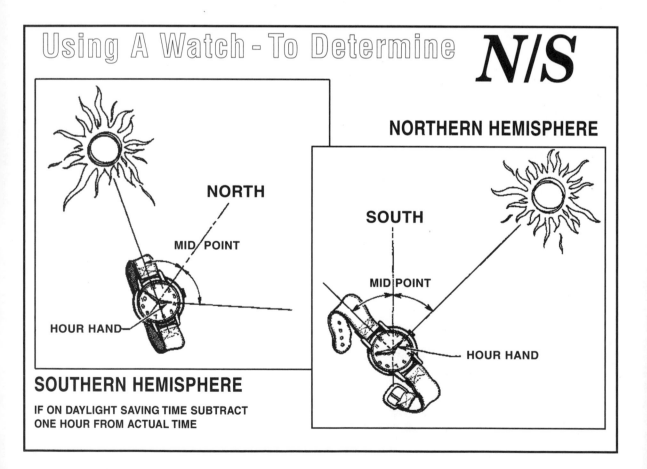

Figure 20-71. Directions Using a Watch.

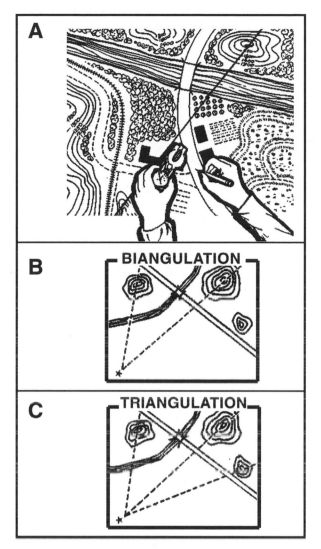

Figure 20-72. Azimuth, Biangulation, and Triangulation.

(11) Dead Reckoning:

(a) Dead reckoning is the process of locating one's position by plotting the course and distance from the last known location. In areas where maps exist, even poor ones, travel is guided by them. It is a matter of knowing one's position at all times by associating the map features with the ground features. A great portion of the globe is unmapped, or only small-scale maps are available. The survivor may be required to travel in these areas without a usable map. Although these areas could be anywhere, they are more likely to be found in frozen wastelands and deserts.

(b) For many centuries, mariners used dead reckoning to navigate their ships when they were out of sight of land or during bad weather, and it is just as applicable to navigation on land. Movement on land must be carefully planned. In military movement, the starting location and destination are known, and if a map is available, they are carefully plotted along with any known intermediate features along the route. These intermediate features, if clearly recognizable on the ground, serve as checkpoints. If a map is not available, the plotting is done on a blank sheet of paper. A scale is selected so the entire route will fit on one sheet. A north direction is clearly established. The starting point and destination are then plotted in relationship to each other. If the terrain and enemy situations permit, the ideal course is a straight line from starting point to destination. This is seldom possible or practicable. The route of

method may be used to orient the map without a compass. However, visible major land features can be used to orient the map to the lay of the land. Once the map is oriented, identify two or three landmarks and mark them on the map. Lay a straightedge on the map with the center of the straightedge at a known position as a pivot point, and rotate the straightedge until the known position of the map is aligned with present position, and draw a line. Repeat this for the second and third position. Each time a line of position is plotted, the map must still be aligned with true north and south. If three lines of position are plotted and form a small triangle, use terrain evaluation to determine present position. If they form a large triangle, recheck calculations for errors.

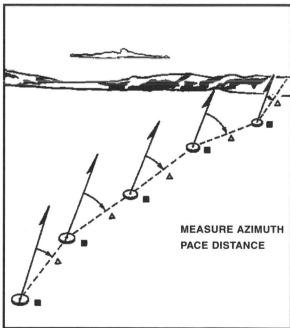

Figure 20-73. Compass Navigation on Foot.

1	2	3	4	5	6	7
ODOMETER READING AT START OF EACH COURSE	DISTANCE IN MILES	FORWARD AZIMUTH (MAGNETIC)	DECLINATION CORRECTION	DEVIATION CORRECTION	TRUE AZIMUTH	NOTES
A 4750						
	6	17°	13°	+3°	33°	
B 4756						
	9	358°	"	+2°	13°	
C 4765						
	8	341°	"	+1°	355°	
D 4773						
	1	314°	"	0°	327°	
E 4774						
	1.5	341°	"	+1°	355°	
F 4775	5					
	1.5	322°	"	0°	335°	
G 4777						
	1	312°	"	0°	325°	
H 4778						
	12	300°	12°	-1°	311°	
I 4790						
	6	341°	"	+1°	354°	
J 4796						
	6	302°	"	-1°	313°	
K 4802						
	20	319°	"	0°	331°	
4810						Wahoo River
						Crossing
4814						Cut 2 miles
L 4824			Base Camp (data)			

Figure 20-74. Sample Log.

travel usually consists of several courses, with an azimuth established at the starting point for the first course to be followed. Distance measurement begins with the departure and continues through the first course until a change in direction is made. A new azimuth is established for the second course and the distance is measured until a second change of direction is made, and so on. Records of all data are kept and all positions are plotted.

(c) A pace (for our purposes) is equal to the distance covered every time the same foot touches the ground (surveyor's paces). To measure distance, count the number of paces in a given course and convert to the map unit. Usually, paces are counted in hundreds, and hundreds can be kept track of in many ways: making

notes in a record book; counting individual fingers; placing small objects such as pebbles into an empty pocket; tying knots in a string; or using a mechanical hand counter. Distances measured this way are only approximate but with practice can become very accurate. It is important that each person who uses dead-reckoning navigation establish the length of an average pace. This is done by pacing a measured course many times and computing the mean (figure 20-73). In the field, an average pace must often be adjusted because of the following conditions:

-1. Slopes. The pace lengthens on a downgrade and shortens on an upgrade.

-2. Winds. A headwind shortens the pace, while a tailwind increases it.

-3. Surfaces. Sand, gravel, mud, and similar surface materials tend to shorten the pace.

-4. Elements. Snow, rain, or ice reduces the length of the pace.

-5. Clothing. Excess weight of clothing shortens the pace, while the type of shoes affects traction and therefore the pace length.

(d) A log (figure 20-74) should be used for navigation by dead reckoning to record all of the distances and azimuths while traveling. Often, relatively short stretches of travel cannot be traversed in a straight course because of some natural features such as a river, or a steep, rugged slope. This break in normal navigation is shown on the log to ensure proper plotting.

(e) The course of travel may be plotted directly on the face of the map, or on a separate piece of paper at the same scale as the map. If the latter method is chosen, the complete plot can be transferred to the map sheet, if at least one point of the plot is also shown on the map. The actual plotting can be done by protractor and scale. The degree of accuracy obtained depends on the quality of draftmanship, the environmental conditions, and the care taken in obtaining data while en route. Figure 20-75 illustrates a paper plot of the data

obtained for the log sample in figure 20-74. It should be noted that four of the courses from A to H are short and have been plotted as a single course, equal to the sum of the four distances and using a mean azimuth of the four. This is recommended because it saves time without a loss of accuracy. If possible, a plot should be tied in to at least one known intermediate point along the route. This is done by directing the route to pass near or over a point. If the plotted position of the intermediate point differs from its known location, discard the previous plot and start a new plot from the true location. The previous plot may be inspected to see whether there is a detectable constant error applicable to future plots; otherwise, it is of no further use.

(f) An offset is a planned magnetic deviation to the right or left of an azimuth to an objective. It is used when approaching a linear feature from the side, and a point along the linear feature (such as a road junction) is the objective. Because of errors in the compass or in map reading, one may reach the linear feature and not know whether the objective lies to the right or left. A deliberate offset by a known number of degrees in a known direction compensates for possible errors and ensures that, upon reaching the linear feature, the user knows whether to go right or left to reach the objective.

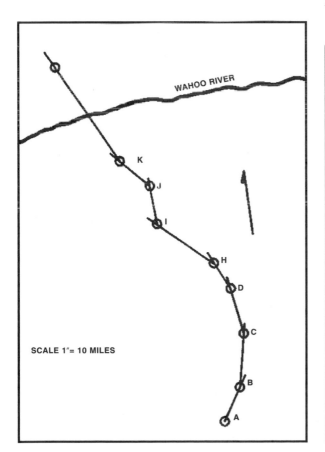

Figure 20-75. Separate Paper Plot.

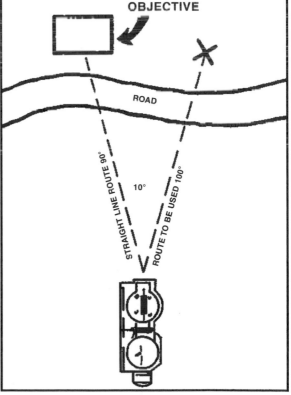

Figure 20-76. Deliberate Offset.

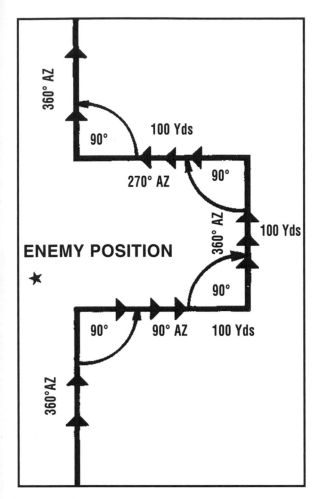

Figure 20-77. Detour Around Enemy Position.

Figure 20-76 shows an example of the use of offset to approach an objective. It should be remembered that the distance from "X" to the objective will vary directly with the distance to be traveled and the number of degrees offset. Each degree offset will move the course about 20 feet to the right or left for every 1,000 feet traveled. For example: In figure 20-76, the number of degrees offset is 10 to the right. If the distance traveled to "X" is 1,000 feet, then "X" is located about 200 feet to the right of the objective.

(g) Figure 20-77 shows an example of how to bypass enemy positions or obstacles by detouring around them and maintaining orientation by moving at right angles for specified distances; for example, moving on an azimuth 306° and wish to bypass an obstacle or position. Change direction to 90° and travel for 100 yards, change direction back to 360° and travel for 100 yards, change direction to 270° and travel for 100 yards, then change direction to 360°, and back on the original azimuth.

Bypassing an unexpected obstacle at night is done in the same way.

(12) Polar Coordinates:

(a) A point on the map may be determined or plotted from a known point by giving a direction and a distance along the direction line. This method of point location uses polar coordinates (figure 20-78). The reference direction is normally expressed as an azimuth, and the distance is determined by any convenient unit of measurement, such as meters or yards.

(b) Polar coordinates are especially useful in the field because magnetic azimuth is determined from the compass and the distance can be estimated.

(13) Position Determination:

(a) Determining Latitude. (From the Sun at sunrise and sunset.) Figure 20-79 shows the true azimuth of the rising Sun and the relative bearing of the setting Sun for all of the months in the year in the Northern and Southern Hemispheres (the table assumes a level horizon and is inaccurate in mountainous terrain).

-1. Latitude can be determined by using a compass to find the angle of the Sun at sunrise or sunset (subtracting or adding magnetic variation) and the date. According to the chart in figure 20-79, on January 26, the azimuth of the rising Sun will be 120° to the left when facing the Sun in the Northern Hemisphere (it would be 120° to the right for the setting Sun); therefore, the latitude would be 50°. If in the Southern Hemisphere, the direction of the Sun would be the opposite.

-2. The table does not list every day of the year, nor does it list every degree of latitude. If accuracy is desired within 1° of azimuth, interpolation may be

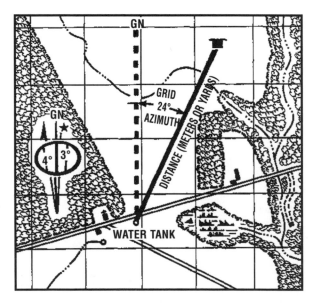

Figure 20-78. Polar Coordinates Used to Designate Position on Map.

Angle to north from the rising or setting Sun (level terrain)

LATITUDE

Date		0°	5°	10°	15°	20°	25°	30°	35°	40°	45°	50°	55°	60°
January	1	113	113	113	114	115	116	117	118	121	124	127	133	141
	6	112	113	113	113	114	115	116	118	120	123	127	132	140
	11	112	112	112	113	113	114	115	117	119	122	125	130	138
	16	111	111	111	112	112	113	114	116	118	120	124	129	136
	21	110	110	110	111	111	112	113	115	117	119	122	127	133
	26	109	109	109	109	110	110	111	112	113	115	120	124	130
February	1	107	107	108	108	108	109	110	111	113	115	117	121	126
	6	106	106	106	106	107	107	108	109	111	113	115	118	123
	11	104	104	105	105	105	106	107	108	109	110	112	116	120
	16	103	103	103	103	103	104	105	106	107	108	110	112	116
	21	101	101	101	101	101	102	102	103	104	105	107	109	112
	26	99	99	99	99	100	100	100	101	102	103	104	106	108
March	1	98	98	98	98	99	99	99	100	100	101	102	104	106
	6	96	96	96	96	96	97	97	97	98	98	99	100	102
	11	94	94	94	94	94	94	95	95	95	96	96	97	98
	16	92	92	92	92	92	92	92	92	93	93	93	93	94
	21	90	90	90	90	90	90	90	90	90	90	90	90	90
	26	88	88	88	88	88	88	88	88	87	87	87	87	86
April	1	86	86	86	86	85	85	85	85	84	84	83	82	81
	6	84	84	84	83	83	83	83	82	82	81	80	79	77
	11	82	82	82	82	81	81	81	80	80	79	77	76	74
	16	80	80	80	80	79	79	78	78	77	76	74	72	70
	21	78	78	78	78	78	77	76	76	75	73	72	69	66
	26	77	77	76	76	76	75	75	74	72	71	69	66	63
May	1	75	75	75	74	74	73	73	72	70	69	66	63	59
	6	74	74	73	73	73	72	71	70	68	67	64	61	56
	11	72	72	72	72	71	70	69	68	67	64	62	58	52
	16	71	71	71	70	70	69	68	67	65	63	60	55	49
	21	70	70	70	69	69	68	67	65	63	61	58	53	47
	26	69	69	69	68	68	67	66	64	62	60	56	51	44
June	1	68	68	68	67	66	66	64	63	61	58	54	49	41
	6	67	67	67	67	66	65	64	62	60	57	53	48	40
	11	67	67	67	66	65	64	63	62	59	56	53	47	39
	16	67	67	67	66	65	64	63	62	59	56	53	47	39
	21	67	67	67	66	65	64	63	62	59	56	53	47	39
	26	67	67	67	66	65	64	63	62	59	56	53	47	39
July	1	67	67	67	66	65	64	63	62	59	56	53	47	39
	6	67	67	67	66	66	65	64	62	60	57	53	48	40
	11	68	68	68	67	66	65	64	63	61	58	54	49	41
	16	69	68	68	68	67	66	65	64	62	59	55	50	43
	21	69	69	69	69	68	67	66	65	63	60	57	52	45
	26	70	70	70	70	69	68	67	66	64	62	59	54	48
August	1	72	72	72	71	71	70	69	68	66	64	61	57	51
	6	73	73	73	73	72	71	71	69	68	66	63	60	55
	11	75	75	74	74	74	73	72	71	70	68	66	63	58
	16	76	76	76	76	75	74	73	72	70	68	65	61	
	21	78	78	77	77	77	76	76	75	74	72	71	68	65
	26	79	79	79	79	79	78	78	77	76	75	73	71	68
September	1	82	82	82	81	81	81	80	80	79	78	77	75	73
	6	83	83	83	83	83	83	82	82	81	81	80	78	77
	11	85	85	85	85	85	85	85	84	84	83	83	82	81
	16	87	87	87	87	87	87	87	86	86	86	85	85	84
	21	89	89	89	89	89	89	89	89	89	89	88	88	88
	26	91	91	91	91	91	91	91	91	91	91	92	92	92
October	1	93	93	93	93	93	93	93	94	94	94	95	95	96
	6	95	95	95	95	95	96	96	96	97	97	98	99	100
	11	97	97	97	97	97	98	98	99	99	100	101	102	104
	16	99	99	99	99	99	100	100	101	101	102	104	105	108
	21	101	101	101	101	101	102	102	103	104	105	107	109	112
	26	102	102	103	103	103	104	104	105	106	108	109	112	115
November	1	104	104	105	105	105	106	107	108	109	110	113	116	120
	6	106	106	106	107	107	108	109	110	111	113	115	119	123
	11	107	107	108	108	108	109	110	111	113	115	117	121	126
	16	109	109	109	109	110	111	112	113	115	117	120	124	130
	21	110	110	110	111	111	112	113	114	116	119	122	126	133
	26	111	111	111	112	112	113	114	116	118	120	124	128	135
December	1	112	112	112	113	113	114	115	117	119	122	125	130	138
	6	112	112	113	113	114	115	116	118	120	123	126	132	140
	11	113	113	113	114	115	116	117	118	121	124	127	133	141
	16	113	113	113	114	115	116	117	118	121	124	127	133	141
	21	113	113	113	114	115	116	117	118	121	124	127	133	141
	26	113	113	113	114	115	116	117	118	121	124	127	133	141

NOTE. When the Sun is rising, the angle is reckoned from east to north. When the Sun is setting, the angle is reckoned from west to north.

Figure 20-79. Finding Direction from the Rising or Setting Sun.

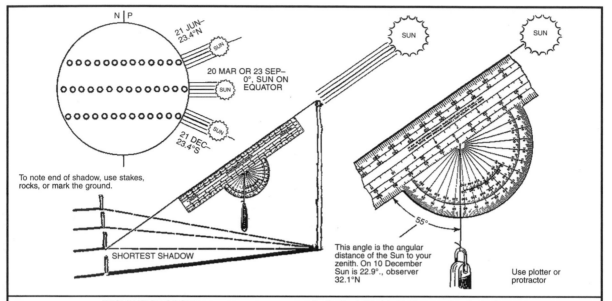

To note end of shadow, use stakes, rocks, or mark the ground.

SHORTEST SHADOW

This angle is the angular distance of the Sun to your zenith. On 10 December Sun is 22.9°., observer 32.1°N

Use plotter or protractor

DECLINATION OF SUN
(IN DEGREES AND TENTHS OF A DEGREE)

Declination is tabulated to the nearest tenth of a degree rather than to the nearest minute of arc. To convert 1/10° (0.1°) to minutes, multiply by 6 (i.e., 27.9° = 27°54')

DAY	JAN	FEB	MAR	APR	MAY	JUN	JUL	AUG	SEP	OCT	NOV	DEC
1	S 23.1	S 17.5	S 7.7	N 4.4	N 15.0	N 22.0	N 23.1	N 18.1	N 8.4	S 3.1	S 14.3	S 21.8
2	23.0	17.2	7.3	4.8	15.3	22.1	23.1	17.9	8.1	3.4	14.6	21.9
3	22.9	16.9	6.9	5.2	15.6	22.3	23.0	17.6	7.7	3.8	15.0	22.1
4	22.9	16.6	6.6	5.6	15.9	22.4	22.9	17.3	7.3	4.2	15.3	22.2
5	22.8	16.3	6.2	5.9	16.2	22.5	22.8	17.1	7.0	4.6	15.6	22.3
6	S 22.7	S 16.0	S 5.8	N 6.3	N 16.4	N 22.6	N 22.7	N 16.8	N 6.6	S 5.0	S 15.9	S 22.5
7	22.5	15.7	5.4	6.7	16.7	22.7	22.6	16.5	6.2	5.4	16.2	22.6
8	22.4	15.4	5.0	7.1	17.0	22.8	22.5	16.3	5.8	5.7	16.5	22.7
9	22.3	15.1	4.6	7.4	17.3	22.9	22.4	16.0	5.5	6.1	16.8	22.8
10	22.2	14.8	4.2	7.8	17.5	23.0	22.3	15.7	5.1	6.5	17.1	22.9
11	S 22.0	S 14.5	S 3.8	N 8.2	N 17.8	N 23.1	N 22.2	N 15.4	N 4.7	S 6.9	S 17.3	S 23.0
12	21.9	14.1	3.5	8.6	18.0	23.1	22.0	15.1	4.3	7.3	17.6	23.1
13	21.7	13.8	3.1	8.9	18.3	23.2	21.9	14.8	3.9	7.6	17.9	23.1
14	21.5	13.5	2.7	9.3	18.5	23.2	21.7	14.5	3.6	8.0	18.1	23.2
15	21.4	1.1	2.3	9.6	18.8	23.3	21.6	14.2	3.2	8.4	18.4	23.3
16	S 21.2	S 12.8	S 1.9	N 10.0	N 19.0	N 23.3	N 21.4	N 13.9	N 2.8	S 8.8	S 18.7	S 23.3
17	21.0	12.4	1.5	10.4	19.2	23.4	21.3	13.5	2.4	9.1	18.9	23.3
18	20.8	12.1	1.1	10.7	19.5	23.4	21.1	13.2	2.0	9.5	19.1	23.4
19	20.6	11.7	0.7	11.1	19.7	23.4	20.9	12.9	1.6	9.9	19.4	23.4
20	20.4	11.4	0.3	11.4	19.9	23.4	20.7	12.6	1.2	10.2	19.6	23.4
21	S 20.2	S 11.0	N 0.1	N 11.7	N 20.1	N 23.4	N 20.5	N 12.2	N 0.8	S 10.6	S 19.8	S 23.4
22	20.0	10.7	0.5	12.1	20.3	23.4	20.4	11.9	0.5	10.9	20.1	23.4
23	19.8	10.3	0.9	12.4	20.5	23.4	20.2	11.6	N 0.1	11.3	20.3	23.4
24	19.5	9.9	1.3	12.7	20.7	23.4	20.0	11.2	S 0.3	11.6	20.5	23.4
25	19.3	9.6	1.7	13.1	20.9	23.4	19.7	10.9	0.7	12.0	20.7	23.4
26	S 19.0	S 9.2	N 2.1	N 13.4	N 21.1	N 23.4	N 19.5	N 10.5	S 1.1	S 12.3	S 20.9	S 23.4
27	18.8	8.8	2.5	13.7	21.2	23.3	19.3	10.2	1.5	12.7	21.1	23.3
28	18.5	8.5	2.9	14.0	21.4	23.3	19.1	9.8	1.9	13.0	21.3	23.3
29	18.3	8.1	3.2	14.4	21.6	23.3	18.8	9.5	2.3	13.3	21.4	23.3
30	18.0	...	3.6	14.7	21.7	23.2	18.6	9.1	2.7	13.7	21.6	23.2
31	S 17.7	...	N 4.0	...	N 21.9	...	N 18.4	N 8.8	...	S 14.0	...	S 23.1

EXAMPLE: On December 10, the declination of the Sun is 22.9°S, so observers who measure the zenith distance as 0° would know that they are at latitude 22.9°S. If they measure a zenith distance of 5° with the Sun south of this zenith, they are 5° north of 22.9°S, or at latitude 17.9°S; and if the Sun is north, they are 5° south of 22.9°S, or latitude 27.9°S.

Figure 20-80. Determining Latitude by Noon Sun.

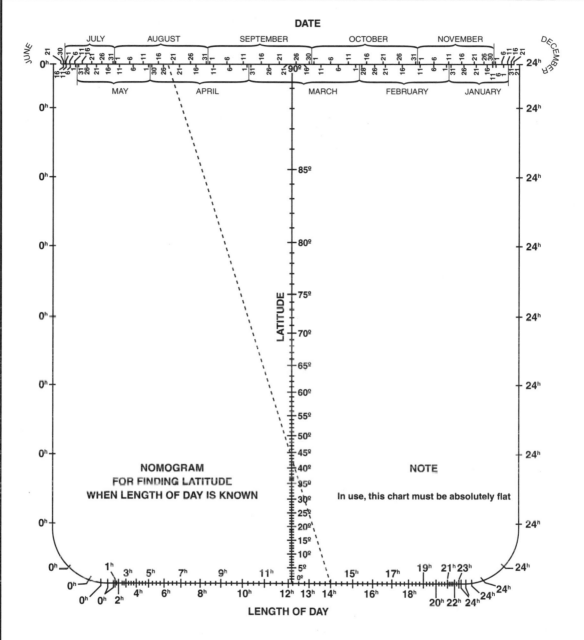

Figure 20-81. Nomogram.

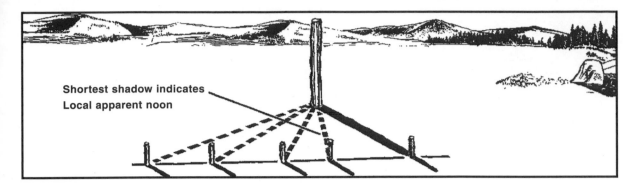

Figure 20-82. Stick and Shadow.

necessary (split the difference) between the values given in the table. For example, between 45° latitude and 50° latitude is 5°. The difference in latitudes (5°) and the difference in azimuths (3°) split (⁵/₃) is 1°²/₃′ (1°40′), so the more accurate degree of latitude would be 46°40′ latitude.

(b) Latitude by Noon Altitude of the Sun. On any given day, there is only one latitude on Earth where the Sun passes directly overhead, or through the zenith, at noon. In all latitudes north of this, the Sun passes to the south of the zenith; and in those south of it, the Sun passes to the north. For each 1° change of latitude, the zenith distance also changes by 1 degree. Figure 20-80 gives the latitude for each day of the year where the Sun is in the zenith at noon. If a Weems plotter or other protractor is available, maximum altitude of the Sun should be used to find latitude by measuring the angular distance of the Sun from the zenith at noon. Local noon can be found using the methods described earlier. Stretch a string from the top of a stick to the point where the end of the noon

shadow rested, place the plotter along the string, and drop a plumb line from the center of the plotter. The intersection of the plumb line with the outer scale of the plotter shows the angular distance of the Sun from the zenith.

(c) Latitudes by Length of Day. This method is used most effectively while on open seas. When in any latitude between 60°N and 60°S, the exact latitude within 30 nautical miles (¹/₂°) can be determined if the length of the day within 1 minute is known. This is true throughout the year except for about 10 days before and 10 after the equinoxes—approximately March 11–31 and September12–October 2. During these two periods, the day is about the same length at all latitudes. A level horizon is required to time sunrise and sunset accurately. Find the length of day from the instant the top of the Sun first appears above the ocean horizon to the instant it disappears below the horizon. This instant is often marked by a green flash. Write down the times of sunrise and sunset. Don't count on remembering them.

Date	Eq. of Time* (minutes)	Date	Eq. of Time* (minutes)	Date	Eq. of Time* (minutes)	Date	Eq. of Time* (minutes)	Date	Eq. of Time* (minutes)	Date	Eq. of Time* (minutes)
Jan. 1	-3.5	Mar. 4	-12.0	May 2	+3.0	Aug. 4	-6.0	Oct. 1	+10.0	Dec. 1	+11.0
2	-4.0	8	-11.0	14	+3.8	12	-5.0	4	+11.0	4	+10.0
4	-5.0	12	-10.0	May 28	+3.0	17	-4.0	7	+12.0	6	+9.0
7	-6.0	16	-9.0			22	-3.0	11	+13.0	9	+8.0
9	-7.0	19	-8.0	June 4	+2.0	26	-2.0	15	+14.0	11	+7.0
12	-8.0	22	-7.0	9	+1.0	Aug. 29	-1.0	20	+15.0	13	+6.0
14	-9.0	26	-6.0	14	0.0			Oct. 27	+16.0	15	+5.0
17	-10.0	Mar. 29	-5.0	19	-1.0	Sept. 1	0.0			17	+4.0
20	-11.0			23	-2.0	5	+1.0			19	+3.0
24	-12.0	Apr. 1	-4.0	June 28	-3.0	8	+2.0	Nov. 4	+16.4	21	+2.0
Jan. 28	-13.0	5	-3.0			10	+3.0	11	+16.0	23	+1.0
		8	-2.0	July 3	-4.0	13	+4.0	17	+15.0	25	0.0
Feb. 4	-14.0	12	-1.0	9	-5.0	16	+5.0	22	+14.0	27	-1.0
13	-14.3	16	0.0	18	-6.0	19	+6.0	25	+13.0	29	-2.0
19	-14.0	20	+1.0	July 27	-6.6	22	+7.0	Nov. 28	+12.0	Dec. 31	-3.0
Feb. 28	-13.0	Apr. 25	+2.0			25	+8.0				
						Sept. 28	+9.0				

*Add plus values to mean time and subtract minus values from mean time to get apparent time.

Figure 20-83. Equation of Time.

Note that only the length of day counts in the determination of latitude; a watch may have an unknown error and yet serve to determine this factor. If only one water horizon is available, as on a seacoast, find local noon by the stick-and-shadow method. The length of day is twice the interval from sunrise to noon or from noon to sunset. Knowing the length of day, latitude can be found by using the nomogram shown in figure 20-81.

(d) Longitude from Local Apparent Noon. To find longitude, a survivor must know the correct time and the rate at which a watch gains or loses time. If this rate and the time the watch was last set are known, the correct time can be computed by adding or subtracting the gain or loss. Correct the zone time on the watch to Greenwich time; for example, if the watch is on eastern standard time, add 5 hours to get Greenwich time. Longitude can be determined by timing the moment a celestial body passes the meridian. The easiest body to use is the Sun. Use one of the following methods:

-1. Stick and Shadow. Put up a stick or rod (figure 20-82) as nearly vertical as possible in a level place. Check the alignment of the stick by sighting along the line of a makeshift plumb bob. (To make a plumb bob, tie any heavy object to a string and let it hang free. The line of the string indicates the vertical.) Sometime before midday, begin marking the position of the end of the stick's shadow. Note the time for each mark. Continue marking until the shadow definitely lengthens. The time of the shortest shadow is the time when the Sun passed the local meridian or local apparent noon. A survivor will probably have to estimate the position of the shortest shadow by finding a line midway between two shadows of equal length, one before noon and one after. If the times of sunrise and sunset are accurately determined on a water horizon, local noon will be midway between these times.

-2. Double Plumb Bob. Erect two plumb bobs about 1 foot apart so both strings line up with Polaris, much the same as a gun sight. Plumb bobs should be set up when Polaris is on the meridian and has no east-west correction. The next day, when the shadows of the two plumb bobs coincide, they will indicate local apparent noon.

-3. Mark Down the Greenwich Time of Local Apparent Noon. The next step is to correct this observed time of meridian passage for the equation of time; that is, the number of minutes the real Sun is ahead of or behind the mean Sun. (The mean Sun was invented by astronomers to simplify the problems of measuring time. The mean Sun rolls along the Equator at a constant rate of 15° per hour. The real Sun is not so considerate; it changes its angular rate of travel around the Earth with the seasons.) Figure 20-83 gives the value in minutes of time to be added to or subtracted from mean (watch) time to get apparent (Sun) time.

-4. After computing the Greenwich time of local noon, the difference of longitude between the survivor's position and Greenwich can be found by converting the interval between 1200 Greenwich and the local noon from time to arc. Remember that 1 hour equals 15° of longitude, 4 minutes equal 1° of longitude, and 4 seconds equal 1' of longitude. Example: The

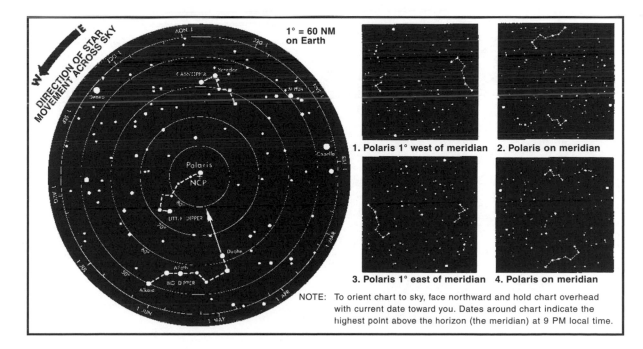

NOTE: To orient chart to sky, face northward and hold chart overhead with current date toward you. Dates around chart indicate the highest point above the horizon (the meridian) at 9 PM local time.

Figure 20-84. Finding Direction from Polaris.

survivor's watch is set to eastern standard time, and it normally loses 30 seconds a day. It hasn't been set for 4 days. The local noon is timed at 15:08 on the watch on February 4. Watch correction is 4 x 30 seconds, or plus 2 minutes. Zone time correction is plus 5 hours. Greenwich time is 15:08 plus 2 minutes plus 5 hours, or 20:10. The equation of time for February 4 is minus 14 minutes. Local noon is 20:10 minus 14 minutes or 19:56 Greenwich. The difference in time between Greenwich and present position is 19:56 minus 12:00, or 7:56. A time of 7:56 equals 119° of longitude. Since local noon is later than Greenwich noon, the survivor is west of Greenwich, and longitude is 119°W.

(e) Direction and Position Finding at Night:

-l. Direction from Polaris. In the Northern Hemisphere one star, Polaris (the Pole Star), is never more than approximately 1° from the North Celestial Pole. In other words, the line from any observer in the Northern Hemisphere to the Pole Star is never more than 1° away from true north. Find the Pole Star by locating the Big Dipper or Cassiopeia, two groups of stars that are very close to the North Celestial Pole. The two stars on the outer edge of the Big Dipper are called pointers because they point almost directly to Polaris. If the pointers are obscured by clouds, Polaris can be identified by its relationship to the constellation Cassiopeia. Figure 20-84 indicates the relation between the Big Dipper, Polaris, and Cassiopeia.

-2. Direction from the Southern Cross. In the Southern Hemisphere, Polaris is not visible. There the Southern Cross is the most distinctive constellation. When flying south, the Southern Cross appears shortly before Polaris drops from sight astern. An imaginary line through the long axis of the Southern Cross, or True Cross, points toward the South Pole. The True Cross should not be confused with a larger cross nearby known as the False Cross, which is less bright and more widely spaced. Two of the four stars in the True Cross are among the brightest stars in the heavens; they are the stars on the southern and eastern arms. Those of the northern and western arms are not as conspicuous but are bright.

-3. There is no conspicuous star above the South Pole to correspond to Polaris above the North Pole. In fact, the point where such a star would be, if one existed, lies in a region devoid of stars. This point is so dark in comparison with the rest of the sky that it is known as the Coalsack. Figure 20-85 shows the True Cross and—to the west of it—the False Cross.

(f) Finding Due East and West by Equatorial Stars. Due to the altitude of Polaris above the horizon, it may sometimes be difficult to use as a bearing. Using a point directly on the horizon may be more convenient.

-1. The celestial equator, which is a projection of the Earth's Equator onto an imaginary celestial sphere, always intersects the horizon line at the due east and west points of the compass. Therefore, any star on the celestial equator rises due east and sets due west (disallowing a small error because of atmospheric refraction). This holds true for all latitudes except those of the North and South Poles, where the celestial

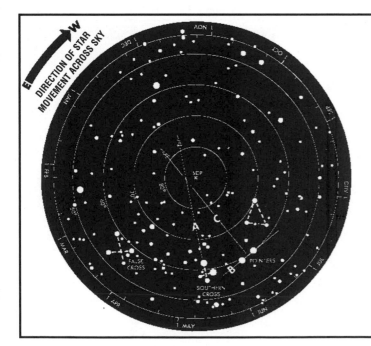

TO LOCATE THE SOUTH CELESTIAL POLE:

1. Extend an imaginary line (A) along the long axis of the True Cross to the south.

2. Join the two bright stars to the east of the Cross with an imaginary line (B). Bisect this line with one at right angles (C), and let it extend southward to intersect line (A).

3. The intersection of line (C) with the line through the Cross (A) is a few degrees from the South Celestial Pole (approximately 5 or 6 full-moon widths).

NOTE: To orient chart to sky, face southward and hold chart overhead with current date toward you. Dates around chart indicate the highest point above the horizon (the meridian) at 9 PM local time.

Figure 20-85. Finding Direction from Southern Cross.

equator and the horizon have a common plane. However, if a survivor is at the North or South Pole, it will probably be known, so this technique can be assumed to be of universal use.

-2 Certain difficulties arise in the practical use of this technique. Unless a survivor is quite familiar with the constellations, it may be difficult to spot a specific rising star as it first appears above the eastern horizon. It will probably be simpler to depend on the identification of an equatorial star before it sets in the west.

-3. Another problem is caused by atmospheric extinction. As stars near the horizon, they grow fainter in brightness because the line of sight between the observer's eyes and the star passes through a constantly thickening atmosphere. Therefore, faint stars disappear from view before they actually set. However, a fairly accurate estimate of the setting point of a star can be made some time before it actually sets. The atmospheric conditions of the area have a great effect on obstructing a star's light as it sets. Atmospheric haze, for example, is far less of a problem in deserts than along temperate-zone coastal strips.

-4. Figure 20-86 shows the brighter stars and some prominent star groups that lie along the celestial equator. There are few bright stars actually on the celestial equator. However, there are a number of stars that lie quite near it, so an approximation within a degree or so can be made. Also, a rough knowledge of the more conspicuous equatorial constellations will give the survivor a continuing checkpoint for maintaining orientation.

(g) Finding Latitude from Polaris. A survivor can find the latitude in the Northern Hemisphere north of 10°N by measuring the angular altitude of Polaris above the horizon, as shown in figure 20-87.

(h) Finding Direction (North) from Overhead Stars that are not in the General Location of the Celestial Poles:

-1. At times, survivors may not be able to locate Polaris (the North Star) due to partial cloud cover, or its position below the observer's horizon. In this situation, it would seem that survivors would be unable to locate direction. Fortunately, survivors who wish to initially find direction or who wish to check a course of

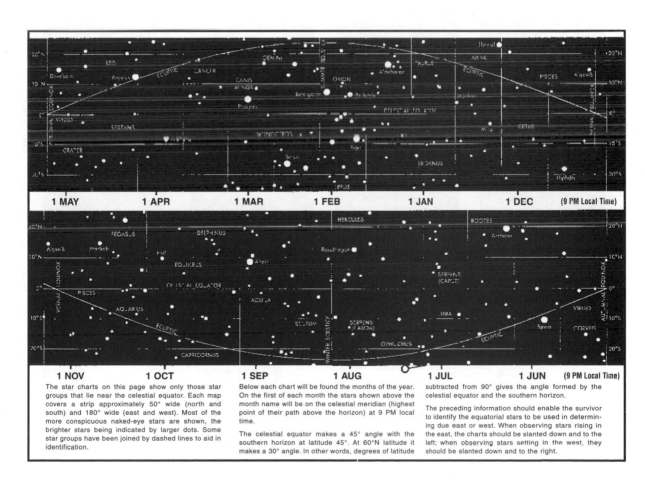

| 1 MAY | 1 APR | 1 MAR | 1 FEB | 1 JAN | 1 DEC | (9 PM Local Time) |

| 1 NOV | 1 OCT | 1 SEP | 1 AUG | 1 JUL | 1 JUN | (9 PM Local Time) |

The star charts on this page show only those star groups that lie near the celestial equator. Each map covers a strip approximately 50° wide (north and south) and 180° wide (east and west). Most of the more conspicuous naked-eye stars are shown, the brighter stars being indicated by larger dots. Some star groups have been joined by dashed lines to aid in identification.

Below each chart will be found the months of the year. On the first of each month the stars shown above the month name will be on the celestial meridian (highest point of their path above the horizon) at 9 PM local time.

The celestial equator makes a 45° angle with the southern horizon at latitude 45°. At 60°N latitude it makes a 30° angle. In other words, degrees of latitude

subtracted from 90° gives the angle formed by the celestial equator and the southern horizon.

The preceding information should enable the survivor to identify the equatorial stars to be used in determining due east or west. When observing stars rising in the east, the charts should be slanted down and to the left; when observing stars setting in the west, they should be slanted down and to the right.

Figure 20-86. Charts of Equatorial Stars.

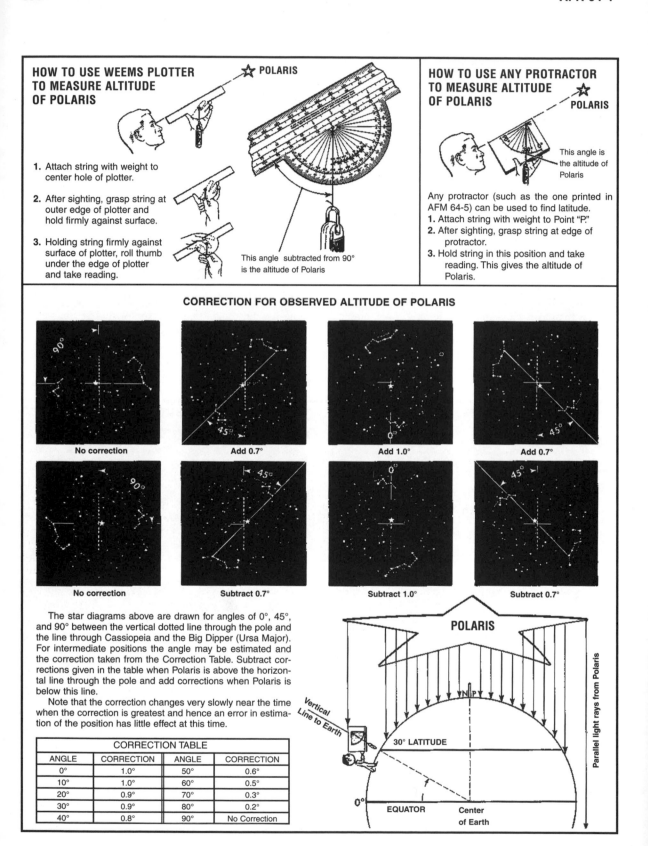

HOW TO USE WEEMS PLOTTER TO MEASURE ALTITUDE OF POLARIS

1. Attach string with weight to center hole of plotter.

2. After sighting, grasp string at outer edge of plotter and hold firmly against surface.

3. Holding string firmly against surface of plotter, roll thumb under the edge of plotter and take reading.

This angle subtracted from 90° is the altitude of Polaris

HOW TO USE ANY PROTRACTOR TO MEASURE ALTITUDE OF POLARIS

This angle is the altitude of Polaris

Any protractor (such as the one printed in AFM 64-5) can be used to find latitude.
1. Attach string with weight to Point "P."
2. After sighting, grasp string at edge of protractor.
3. Hold string in this position and take reading. This gives the altitude of Polaris.

CORRECTION FOR OBSERVED ALTITUDE OF POLARIS

No correction	Add 0.7°	Add 1.0°	Add 0.7°

No correction	Subtract 0.7°	Subtract 1.0°	Subtract 0.7°

The star diagrams above are drawn for angles of 0°, 45°, and 90° between the vertical dotted line through the pole and the line through Cassiopeia and the Big Dipper (Ursa Major). For intermediate positions the angle may be estimated and the correction taken from the Correction Table. Subtract corrections given in the table when Polaris is above the horizontal line through the pole and add corrections when Polaris is below this line.

Note that the correction changes very slowly near the time when the correction is greatest and hence an error in estimation of the position has little effect at this time.

POLARIS

Parallel light rays from Polaris

Vertical Line to Earth

30° LATITUDE

0°   EQUATOR   Center of Earth

CORRECTION TABLE			
ANGLE	CORRECTION	ANGLE	CORRECTION
0°	1.0°	50°	0.6°
10°	1.0°	60°	0.5°
20°	0.9°	70°	0.3°
30°	0.9°	80°	0.2°
40°	0.8°	90°	No Correction

Figure 20-87. Finding Latitude by Polaris.

travel during the night need not worry about being lost or unable to travel if Polaris cannot be identified.

-2. The following is an adaptation of the stick-and-shadow method of direction finding. This method is based on the principle that all the heavenly bodies (Sun, Moon, planets, and stars) rise (generally) in the east and set (generally) in the west. This technique can be used anywhere on Earth with any stars except those which are circumpolar. Circumpolar stars are those which appear to travel around Polaris instead of apparently "moving" from east to west.

-3. To use this technique, survivors should keep in mind that they may use *any star other than a circumpolar one.*

-4. Survivors who wish to know general direction must prepare a device to aid them. This can be done by placing a stick (about 5 feet in length) at a slight angle in the ground in an open area (figure 20-88). Thin material (suspension line, string, vine, braided cloth, etc.) is then attached to the tip of the stick. This material should be longer than what is required to reach the ground (figure 20-88).

-5. The survivor should lie on the back with the head next to this hanging line and pull the cord up to the temple area and hold it tautly.

-6. Next, the survivor moves around on the ground until the taut line is pointing directly at the selected bright non-circumpolar star (or planet).

-7. The taut line is now in position to simulate the star's (or planet's) shadow. Survivors should remember that this method of finding direction is an adaptation of the Sun, stick, and shadow approach. Here the more distant stars and (or) planets take the place of our Sun. Since these objects are too distant from the Earth to create shadow, the string represents the shadow.

-8. With the taut line simulating the star's shadow, survivors should mark the point on the ground where the line touches with a stick, stone, etc. The survivor should repeat this sighting on the same star (or planet) after about 15 to 20 minutes (marking the spot at which the line "shadow" touches the ground). A line scribed on the ground that connects these two points will run west-east (as the stars and planets move from east to west, the "shadow" will move in the opposite direction). The first mark will be in the west. Drawing a line perpendicular to the west-east line, a survivor will have a north-south line and be able to travel.

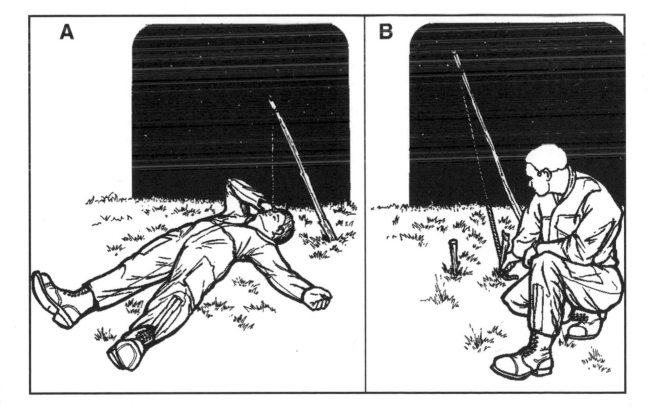

Figure 20-88. Stick and String Direction Finding.

Chapter 21

LAND TRAVEL

21-1. Introduction:

a. In any survival situation following an aircraft emergency, a decision must be made to either move or remain as close as possible to the parachute or crash site. In this chapter, land travel will be discussed, as well as the various considerations that survivors should address before determining whether travel is or is not a necessity.

b. Survivors may need to carry supplies and equipment while traveling to sustain life. For this reason, the techniques of backpacking and improvised packing are discussed to help a person do this task.

c. As a survivor, the ability to walk effectively is important in conserving energy and safety. Additionally, in rough terrain, travel may need to be done with the aid of a rope. The techniques of ascending and descending steep terrain are fundamental to understanding and performing rescue from rough terrain. These techniques, as well as techniques for snow travel, are covered. Travel may not be easy, but a knowledgeable traveler can travel safely and effectively while saving time and energy.

21-2. Decision to Stay or Travel.

In hostile areas, the decision to travel is normally automatic. To stay in the vicinity of the crash or parachute landing can lead to capture. In friendly areas, a choice exists. The best advice is to stay with the aircraft. Most rescues have been made when downed aircrews remained with the aircraft.

a. Survivors should leave the area only when they are certain of their location and know that water, shelter, food, and help can be reached, or, after having waited several days, they are convinced that rescue is not coming and they are equipped to travel.

b. Before making any decision, survivors should consider their personal physical condition and the condition of others in the party when estimating their ability to sustain travel. If people are injured, they should try to get help. If travel for help is required, they should send the people who are in the best physical and mental condition. Send two people if possible. To travel alone is dangerous. Before any decision is made, survivors should consider all of the facts.

(1) If the decision is to stay, these problems should be considered:

    (a) Environmental conditions.

    (b) Health and body care; camp sanitation.

    (c) Rest and shelter.

    (d) Water supplies.

    (e) Food.

(2) If the decision is to travel: In addition to the primary survival problems of providing food, water, and shelter, the following must be considered:

    (a) Direction of travel and why.

    (b) Travel plan.

    (c) Equipment required.

c. Before departing the site, survivors should leave information at their aircraft (nontactical situation only) stating departure time, destination, route of travel, personal condition, and available supplies.

d. From the air, it is easier to spot the aircraft than it is to spot people traveling on the ground. Someone may have seen the aircraft crash and may investigate. The aircraft or parts from it can provide shelter, signaling aids, and other equipment (cowling for reflecting signals, tubing for shelter framework, gasoline and oil for fires, etc.). Avoiding the hazards and difficulties of travel is another reason to stay with the aircraft. Rescue chances are good if survivors made radio contact, landing was made on course or near a traveled air route, and weather and air observation conditions are good.

e. Present location must be known to decide intelligently whether to wait for rescue or to determine a destination and route of travel. The survivors should try to locate their position by studying maps, landmarks, and flight data, or by taking celestial observations. Downed personnel should try to determine the nearest rescue point, the distance to it, the possible difficulties and hazards of travel, and the probable facilities and supplies en route and at the destination.

f. There are a number of other factors that should be considered when deciding to travel.

(1) The equipment and materials required for cross-country travel should be analyzed. Travel is extremely risky unless the necessities of survival are available to provide support during travel. Survivors should have sufficient water to reach the next probable water source indicated on a map or chart and enough food to last until they can procure additional food. To leave shelter to travel in adverse weather conditions is foolhardy unless in an escape and evasion situation.

(2) In addition to the basic requirements, the physical condition of the survivor must be considered in any decision to travel. If in good condition, the survivor should be able to move an appreciable distance, but if the survivor is not in good condition or is injured, the ability to travel extended distances may be reduced. Analyze all injuries received during the emergency. For example, if a leg or ankle injury occurred during landing, this must be considered before traveling.

(3) If possible, survivors should avoid making any decision immediately after the emergency. They should wait a period of time to allow for recovery from the mental—if not the physical—shock resulting from the emergency. When shock has subsided, survivors can then evaluate the situation, analyze the factors involved, and make valid decisions.

21-3. Travel. Once the survivors decide to travel, there are several considerations that apply regardless of the circumstances.

a. The ranking person must assume leadership, and the party must work as a team to ensure that all tasks are done in an equitable manner. Full use should be made of any survival experience or knowledge possessed by members of the group, and the leader is responsible for ensuring that the talents of all survivors are used.

b. Survivors should keep the body's energy output at a steady rate to reduce the effects of unaccustomed physical demands.

(1) A realistic pace should be maintained to save energy. It increases durability and keeps body temperature stable because it reduces the practice of quick starts and lengthy rests. More important, a moderate, realistic pace is essential in high altitudes to avoid the risks of lapses of judgment and hallucinations due to lack of oxygen (hypoxia). Travel speed should provide for each survivor's physical condition and daily needs, and the group's pace should be governed by the pace of the slowest group member. Additionally, rhythmic breathing should be practiced to prevent headache, nausea, loss of appetite, and irritability.

(2) Rest stops should be short since it requires added energy to begin again after cooling off. Survivors should wear their clothing in layers (layer system) and make adjustments to provide for climate, temperature, and precipitation. It is better to start with extra clothing and stop and shed a layer when beginning to warm up.

(3) Wearing loose clothing provides for air circulation, allows body moisture to evaporate, and retains body heat. Loose clothing also allows freedom of movement.

(4) Travelers should keep in mind when planning travel time and distance that the larger the group, the slower the progress will be. Time must be added for those survivors who must acclimate themselves to the climate, altitudes, and the task of backpacking. Survivors should also allow time for unexpected obstacles and problems that could occur.

(5) Proper nutrition and water are essential to building and preserving energy and strength. Several small meals a day are preferred to a couple of large ones so calories and fluids are constantly available to keep the body and mind in the best possible condition. Survivors should try to have water and a snack available while trekking, and they should eat and drink often to restore energy and prevent chills in cold temperatures. This also applies at night.

21-4. Land Travel Techniques. Land-travel techniques are based largely on experience, which is acquired through performance. However, experience can be partially replaced by the intelligent application of specialized practices that can be learned through instruction and observation. For example, travel routes may be established by observing the direction of a bird's flight, the actions of wild animals, the way a tree grows, or even the shape of a snowdrift. Bearings read from a compass, the sun, or stars will improve on these observations and confirm original headings. All observations are influenced by the location and physical characteristics of the area where they are made and by the season of the year.

a. **Route Finding.** The novice should follow a compass line, whereas the experienced person follows lines of least resistance by realizing that a curved route may be faster and easier under certain circumstances. Use game trails when they follow a projected course only. For example, trails made by migrating caribou are frequently extensive and useful. On scree or rockslides, mountain-sheep trails may be helpful. Game trails offer varying prospects, such as the chance of securing game or locating waterholes. Successful land travel requires knowledge beyond mere travel techniques. Survivors should have at least a general idea of the location of their starting place and their ultimate destination. They should also have knowledge of the people and terrain through which they will travel. If the population is hostile, they must adapt their entire method of travel and mode of living to this condition.

b. **Wilderness.** Wilderness travel requires constant awareness. A novice views a landscape from the top of a hill with care and interest, and says, "let's go." The experienced person carefully surveys the surrounding countryside. A distant blur may be mist or smoke; a faint, winding line on a far-off hill may be manmade or an animal trail, a blur in the lowlands may be a herd of caribou or cattle. People should plan travel only after carefully surveying the terrain. Study distant landmarks for characteristics that can be recognized from other locations or angles. Careful and intelligent observation will help survivors to correctly interpret the things they see, distant landmarks, or a broken twig at their feet. Before leaving a place, travelers should study their backtrail carefully. Survivors should know the route forward and backward. An error in route planning may make it necessary to backtrack in order to take a new course. For this reason, all trails should be marked (figure 21-1).

c. **Mountain Ranges.** Mountain ranges frequently affect the climate of a region and the climate in turn influences the vegetation, wildlife, and the character and number of people living in the region. For example, the ocean side of mountains has more fog, rain, and snow than the inland side of a range. Forests may grow on the ocean side, while inland, it may be semi-dry. Therefore, a complete change of survival technique may be necessary when crossing a mountain range.

(1) Travel in mountainous country is simplified by conspicuous drainage landmarks, but it is complicated by the roughness of the terrain. A mountain traveler can

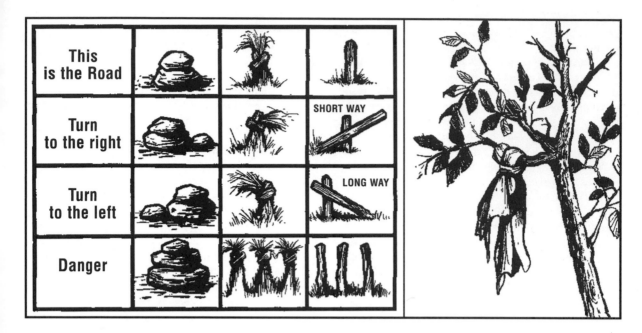

This is the Road			
Turn to the right			SHORT WAY
Turn to the left			LONG WAY
Danger			

Figure 21-1. Marking a Trail.

readily determine the direction in which rivers or streams flow; however, surveying is necessary to determine whether a river is safe for rafting, or whether a snowfield or mountainside can be traversed safely. Mountain travel differs from travel through rolling or level country, and certain cardinal rules govern climbing methods. A group descending into a valley, where descent becomes increasingly steep and walls progressively more perpendicular, may be obliged to climb up again in order to follow a ridge until an easier descent is possible. In such a situation, rappelling with a parachute-line rope may save many weary miles of travel. In mountains, travelers must avoid possible avalanches of earth, rock, and snow, as well as crevasses (deep cracks in the ice) in ice fields.

(2) In mountainous country, it may be better to travel on ridges—the snow surface is probably firmer and there is a better view of the route from above. Survivors should watch for snow and ice overhanging steep slopes. Avalanches are a hazard on steep snow-covered slopes, especially on warm days and after heavy snowfalls.

(3) Snow avalanches occur most commonly and frequently in mountainous country during wintertime, but they also occur with the warm temperatures and rainfalls of springtime. Both small and large avalanches are a serious threat to survivors traveling during winter, as they have tremendous force. The natural phenomenon of snow avalanches is complex. It is difficult to definitely predict impending avalanches, but knowing general behaviors of

avalanches and how to identify them can help people avoid avalanche-hazard areas.

(a) Snow or Sluff Avalanche. The loose snow or sluff avalanche is one kind of avalanche that starts over a small area or in one specific spot. It begins small and builds up in size as it descends. As the quantity of snow increases, the avalanche moves downward as a shapeless mass with little cohesion.

(b) Terrain Factors Affecting Avalanches:

-1. Steepness. Most commonly, avalanches occur on slopes ranging from 30 to 45 degrees (60 to 100 percent grade), but large avalanches do occur on slopes ranging from 25 to 60 degrees (40 to 173 percent grade). (See figure 21-2.)

-2. Profile. The dangerous slab avalanches have more chance of occurring on convex slopes because of the angle and the gravitational pull. Concave slopes cause a danger from slides that originate at the upper, steep part of the slope (figure 21-3).

-3. Slopes. Midwinter snowslides usually occur on north-facing slopes. This is because the north slopes do not receive the required sunlight, which would heat and stabilize the snowpack. South-facing slope slides occur on sunny, spring days when sufficient warmth melts the snow crystals, causing them to change into wet, watery, slippery snowslides. Leeward slopes are dangerous because the wind blows the snow into well-packed drifts just below the crest. If the drifts have not adhered to the snow underneath, a slab avalanche can occur. Windward slopes generally have less snow and are compact. They are usually

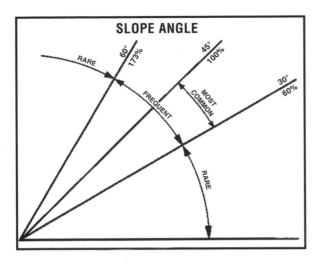

Figure 21-2. Slope Angle.

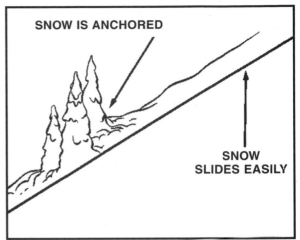

Figure 21-4. Snow Slides.

strong enough to resist movement, but avalanches may still occur with warm temperature and moisture.

-4. Surface Features. Most avalanches are common on smooth, grassy slopes that offer no resistance.

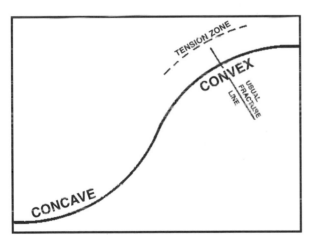

Figure 21-3. Profile of Slope.

Brush, trees, and large rocks bind and anchor the snow, but avalanches can still occur in tree areas (figure 21-4).

(c) Weather Factors:

-1. Old-snow depth covers up natural anchors (rocks, brush, and fallen trees) so the new snowfall slides easier. The type of old-snow surface is important. Sun crests or smooth surface snows are unstable, whereas a rough, jagged surface would offer stability and an anchorage. A loose-snow layer underneath is far more hazardous than a compacted one, as the upper layer of snow will slide more easily with no rough texture to

restrain it. Travelers should check the underlying snow by using a rod or stick.

-2. Winds, 15 miles per hour or more, cause the danger of avalanches to develop rapidly. Leeward slopes will collect snow that has been blown from the windward sides, forming slabs or sluffs, depending upon the temperature and moisture. Snow plumes or cornices indicate this condition (figure 21-5).

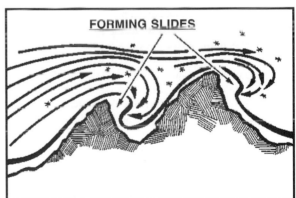

Figure 21-5. Forming Slides.

-3. A high percentage of all avalanches occur during or shortly after storms. Layers of different types of snow from different storms will cause unstable snow because the bond between layers will vary in strength. The rate of snowfall also has a significant effect on stability. A heavy snowfall spread out over several days is not as dangerous as a heavy snowfall in a few hours because slow accumulation allows time for the snow to settle and stabilize. A large amount of snow over a short period of time results in the snow constantly changing and building up, giving it little time to settle and stabilize. If the snow is light and dry, little settling or cohesion occurs, resulting in instability.

-4. Under extremely cold temperatures, snow is unstable. In temperatures around freezing or just above, the snow tends to settle and stabilize quickly. Storms, starting at low temperatures with light, dry snow that are followed by rising temperature, cause the top layer of snow to be moist and heavy, providing opportune conditions for avalanching. The light, dry snow underneath lacks the strength and elastic bondage necessary to hold the heavier, wetter snow deposited on top; therefore, the upper layer slides off. Also, extreme temperature differences between night and day cause the same problems. Rapid changes in weather conditions cause adjustments and movement within the snowpack. Survivors should be alert to rapid changes in winds, temperatures, and snowfall, which may affect snow stability.

-5. Avalanches of wet snow are more likely to occur on south slopes. Sun, rainstorms, or warmer temperatures brought on by spring weather are absorbed by the snow, causing it to become less stable.

(d) Warning Signs. Avalanches generally occur in the same area. After a path has been smoothed, it's easier for another avalanche to occur. Steep, open gullies and slopes, pushed-over small trees, or limbs broken off, and tumbled rocks, are signs of slide paths. Snowballs tumbling downhill or sliding snow is an indication of an avalanche area on leeward slopes. If the snow echoes or sounds hollow, conditions are dangerous. If the snow cracks and the cracks persist or run, the danger of a slab avalanche is imminent. The deeper the snow, the more the terrain features will be obscured. Knowledge of common terrain features can help survivors visualize what they may be up against, what to avoid, and the safest areas in which to travel. Knowing the general weather pattern for the area is helpful. Survivors should try to determine what kind of weather will be coming by observing and knowing the signs that indicate certain weather conditions.

(e) Route Selection. If avoiding the mountains and avalanche-danger areas is impossible, there are precautions survivors should take when confronting dangerous slopes. They should decide which slopes will be the safest by analyzing the factors that determine what makes one slope safe and another deadly. Study the slope terrain and keep in mind why avalanches occur.

-1. When survivors decide to cross a slope, one person at a time should cross. If all go together, they should not tie together since there is no way one person can hold another against an avalanche. Instead, they should tie a contrasting color line about 100 feet long (using suspension line or PLD tape) to each person. If they should get caught in an avalanche, the line will help identify their position if it is exposed. Survivors should

select escape routes before and throughout the climb and keep these routes in mind at all times. They should also stay on the fall line (natural path an object will follow when falling or sliding down the slope) when climbing and not zigzag or climb a different route because it seems easier. Staying on the fall line will prevent making fractures and at the same time, compact the snow, making it more stable for others who follow. If traversing, they should travel above the danger area. Survivors should travel quickly and quietly to avoid extended exposure to the probable danger of avalanches.

-2. If caught in an avalanche, any equipment that weighs survivors down should be dropped. The pack, snowshoes, and any other articles should be jettisoned.

-3. The standard rule is to use swimming motions to try to move towards the snow surface. Further, survivors should go for the sides and not try to out-swim the avalanche. If near the surface, they should try to keep one arm or hand above the surface to mark their position. If buried, a person should inhale deeply (nose down) before the snow stops moving to make room for their chest. Trapped persons should try to make a breathing space around their face. They shouldn't struggle but should relax and conserve their energy and oxygen. Only when fellow survivors or rescuers are nearby should the trapped individual shout. Rescue should be done as quickly as possible. Avalanche victims generally have a 50 percent chance of surviving after 30 minutes have passed because the snow will set up (harden).

-4. Glaciers may offer emergency travel routes across mountain ranges. Glacier crossing demands special knowledge, techniques, and equipment, such as the use of a lifeline and poles for locating crevasses. There are many places where mountain ranges can be negotiated on foot in a single day by following glaciers. Survivors must be especially careful on glaciers and watch for crevasses covered by snow. If traveling in groups of three persons or more, they should be roped together at intervals of 30 to 40 feet. Every step should be probed with a pole. Snow-bridged crevasses should be crossed at right angles to their course. The strongest part of the bridge can be found by probing. When crossing a bridged crevasse, weight should be distributed over a large area by crawling or by wearing snowshoes.

-5. When forested areas are dense, river travel and ridges usually afford the easiest travel routes. In open forests, land travel is easier and offers a better selection of travel routes; however, such forests may not offer sufficient cover or concealment for escape and evasion travel.

-6. After a fire, windstorm, or logging operation, second-growth timber usually grows thick. It is worse after it grows about 20 feet high since any space between the trees is filled in by branches and the overhead timber isn't (yet) thick enough to cause the lower branches to die from lack of sunlight.

-a. Deciduous brush also contributes to the overgrowth. Blowdowns, avalanche fans, and logging slush are difficult to negotiate. Such obstructions, even of a few hundred feet, may require major changes to the original travel route.

-b. Scrub cedar (subalpine fir) is hard to penetrate. There are tactics that can be used to make travel easier. Survivors can use fallen trees as walkways to provide a short route of travel through the scrub cedar to a clear area. Gloves should be worn when penetrating thorny vegetation. Overlaying bushes can be separated to allow passage. When land is steep, brush can be used to provide handholds if it is strong and anchored well.

-c. Brush can be dangerous. Survivors should be aware of the possibility of slipping while going downhill. Therefore, they should ensure that each step is firmly placed. Survivors should be aware of travel difficulties presented by cliffs, boulders, and ravines that are covered by brush.

-7. Do not travel through dense brush if it can be avoided.

-a. Travel on trails rather than taking shortcuts through the brush. Brush is frequently easier to travel through (over) during the winter season when it is covered by snow and when snowshoes are available (improvised).

-b. During the summer, avoid avalanche tracks because the debris may be difficult to penetrate. Traveling on the "timber cones" between avalanche paths is best when climbing a valley.

-c. The heaviest timber is the best area to travel in because little or no brush will be growing on the forest floor.

-d. Try to avoid areas near creeks and valley floors because they have more brush and trees than the valley walls and ridges. However, traveling in the stream channels may be preferable when the area is covered with dense brush and vegetation. Survivors may have to wade, but the stream may offer the best route through the brush.

-e. Traveling high above the brush at the timberline may be worthwhile if the bottom and sides of the valley look futile.

d. Snow and Ice Areas. Travel in snow and ice areas is not recommended except to move from an unsafe to a safe area, or from an area that has few natural resources to an area of greater resources (shelter material, food, and signaling area).

(1) Before traveling to a possible rescue site, town, village, or cabin, travelers should know their approximate position and the location of the desired site. The greatest hazard in snow and ice areas is the intense cold and high winds. Judging distance is difficult due to the lack of landmarks and the clear arctic air. Image distortion is a common phenomenon. "Whiteout" conditions exist and the survivor should not travel during this time. A whiteout condition occurs when there is complete snow cover and the clouds are so thick and uniform that light reflected by the snow is about the same intensity as that from the sky. If traveling during bad weather, great care must be taken to avoid becoming disoriented or falling into crevasses, over cliffs or high snow ridges, or walking into open leads. A walking stick is very useful to probe the area in the line of travel.

(2) Strong winds often sweep unchecked across tundra areas (due to the lack of vegetation) causing whiteout conditions. Because of blowing snow, fog, and lack of landmarks, a compass is a must for travel, yet it is still difficult to navigate a true course since the magnetic variation in the high latitudes (polar areas) is often extreme.

(3) During the summer months, the area is a mass of bogs, swamps, and standing water. Crossing these areas will be difficult at best. Rain and fog are common. Insects such as mosquitoes, midges, and black flies can and will cause the survivor physical discomfort and may cause travel problems. If the body is not completely covered with clothing, or if survivors do not use a head net or insect repellent, insect bites may be severe and infection can set in.

(4) In mountainous country, it is often best to travel along ridgelines because they provide a firmer walking surface and there is usually less vegetation to contend with. High winds make travel impractical if not impossible at times. Glaciers have many hidden dangers. Glacial streams may run just under the surface of the snow or ice, creating weak spots, or they may run on the surface and cause slick ice. Crevasses that run across the glacier can be a few feet to several hundred feet deep. Quite often crevasses are covered over with a thin layer of snow, making them practically invisible. Survivors could fall into crevasses and sustain severe injuries or death. If glacier travel is required, it is best to use a probe pole to test the footing ahead.

(5) Summer travel in timbered areas should not present any major problems; however, travel on ridges is preferred since the terrain is drier and there are usually fewer insects. During the cold months, snow may be deep and travel will be difficult without some type of snowshoes or skis. Travel is generally easier on frozen rivers, streams, and lakes since there is less snow or windpacked snow and they are easier to walk on.

(6) River and stream travel can be hazardous. Rivers comparatively straight are that way because of the volume of water flow and extremely fast currents. These rivers tend to have very thin ice in the winter (cold climates), especially where snowbanks extend out over the water. If an object protrudes through the ice, the immediate area will be weak and should be avoided if possible. Where two rivers and streams come together, the current is swift and the ice will be weaker than the ice on the rest of the river. Very often after freeze-up, the

source of the river or stream dries up so rapidly that air pockets are formed under the ice and can be dangerous if fallen into. During the runoff months (spring and summer), rivers and streams usually have a large volume of water that is very cold and can cause cold injuries. Wading across or down rivers and streams should be done with proper footwear and exposure protection due to the depth, swiftness, unsure footing, and coldness of the water. Generally, streams are too small and shallow for rafting. Streams are often bordered by high cliffs or banks at the headwaters. As a stream progresses, its banks are often choked with alder, devil's club, and other thick vegetation making travel very slow and difficult. Many smaller streams will simply lead the traveler to a bog or swamp where they end, causing more problems for the survivor.

(7) Sea-ice conditions vary greatly from place to place and season to season. During the winter months, there is generally little open water except between the edges of floes. Crossing from one floe to another can be done by jumping across the open-water area, but footing may be dangerous. When large floes are touching each other, the ice between is usually ground into brash ice by the action of the floes against each other, and this ground-up ice will not support a survivor's weight. Pressure ridges are long ridges in sea ice caused by the horizontal pressure of two ice floes coming together. Pressure ridges may be 100 feet high and several miles long; they may occur in a gulf or bay, or on polar seas. They must be crossed with great care because of the ruggedness of ice formations, weak ice in the area, and the possibility of open water covered with a thin layer of snow or ground-up ice. During summer months, the ice surface becomes very rough and covered with water. The ice also becomes soft and honeycombed (candlestick ice) even though the air temperature may be below freezing. Traveling over sea ice in the summer months is very dangerous.

(8) Icebergs are great masses of ice and are driven by currents and winds; about two-thirds of the iceberg is below the surface of the sea. Icebergs in open seas are always dangerous because the ice under the water will melt faster than the surface exposed to the air, upsetting the equilibrium and toppling them over. The resulting waves can throw small pieces of ice in all directions. Avoid pinnacle-shaped icebergs—low, flat-topped icebergs are safer.

e. Dry Climates. Before traveling in the desert, the decision to travel must be weighed against the environmental factors of terrain and climate, condition of survivors, possibility of rescue, and the amount of water and food required.

(1) The time of day for traveling is greatly dependent on two significant factors; the first and most apparent is temperature, and the second is type of terrain. For example, in rocky or mountainous deserts, the eroded drainages and canyons may not be seen at night and could result in a serious fall. Additionally, manmade features such as mining shafts or pits and irrigation channels could cause similar problems. If the temperature is not conducive to day travel, survivors should travel during the cooler parts of the day (in early morning or late evening). Traveling on moonlit nights is another possibility; however, survivors must be aware that moonlight can cast deceiving shadows. This problem can be decreased by scanning the ground to allow the night-sensitive portions of the eye time to pick up the slight differences in lighting. In hot desert areas where these hazards do not occur, traveling at night is a very practical solution. During the winter in the mid-latitude deserts, the cold temperatures make day travel most sensible.

(2) There are three types of deserts: mountain, rocky plateau, and sandy or dune deserts. These deserts can present difficult travel problems.

(a) Mountain deserts are characterized by scattered ranges or areas of barren hills or mountains, separated by dry, flat basins. High ground may rise gradually or abruptly from flat areas to a height of several thousand feet above sea level. Most desert rainfall occurs at high elevations and the rapid runoff causes flash floods, eroding deep gullies or ravines and depositing sand and gravel around the edges of the basins. These floods are a problem on high and low grounds. The floodwaters rapidly evaporate, leaving the land as barren as before, except for plush vegetation that rapidly becomes dormant. Basins without shallow lakes will have alkaline flats, which can cause problems with chemical burns and can destroy clothing and equipment.

(b) Rocky plateau deserts have relief interspersed by extensive flat areas of solid or broken rock at or near the surface. They may be cut by dry, steep-walled, eroded valleys known as wadis in the Middle East and arroyos or canyons in the United States and Mexico. The narrower of these valleys can be extremely dangerous to humans and material due to flash flooding. Travel in these valleys may present another problem: a survivor can lose site of reference points and travel farther than intended. The Golan Heights in Israel is an example of a rocky plateau desert.

(c) Sandy (dune) deserts are extensive, flat areas covered with sand or gravel. They are the product of ancient and modern wind erosion. "Flat" is relative in this case, as some areas contain sand dunes that are more than 1,000 feet high and 10 to 15 miles long. Travel in such terrain depends on windward or leeward gradients of the dunes and texture of sand. These dunes help a survivor determine general direction. Longitudinal dunes are continuous banks of sand at even heights that lie parallel with the dominant wind. Other areas, however, may be totally flat for distances in excess of 2 miles. Plant life varies from none to scrub reaching more than 6 feet. Examples of this desert include the ergs of the Sahara

Desert, Empty Quarter of the Arabian Desert, areas of California and New Mexico, and the Kalahari Desert in South Africa. A seif dune has forms similar to a drift behind a rock. Its length lies in the direction of the prevailing winds. Additionally, the horseshoe-shaped crescent dune has a hollow portion that faces downward. Ripples caused by wind in the sand may indicate the direction of the prevailing winds. These ripples generally lie perpendicular to the prevailing winds. In deserts, it is easier to travel on the windward side of the tops of dunes. Even though these ridges may not lie in a straight line and may wander, they offer a better route of travel than traveling in straight lines. A great deal of energy and time can be expended walking up and down dunes, especially in the loose sand on the leeward side of dunes.

(3) People who travel the desert at night orient themselves by the stars and moon. Those who traveled by day use compasses, when available, and the sun. Survivors should use all directional aids during emergency travel and each aid should be frequently cross-checked against each other. For desert travel, a compass is a valuable piece of equipment.

(a) Without a compass, landmarks must be used for local navigation. This can lead to difficulties. Mirages cause considerable trouble. Ground haze throughout the day may obscure vision. Distances are deceptive in the desert, and survivors have reported difficulty in estimating distances and the size of objects. In southern Egypt, one survivor reported that large boulders always appeared smaller than they were and in other cases small obstacles appeared insurmountable. Survivors in Saudi Arabia and in Tunisia warned that it is difficult to maintain a single landmark in navigation. Several groups reported that they found it necessary to take turns keeping an eye on a specific mountain, peak, or object that was their goal. Objects have a way of vanishing in some cases when the eye is moved for an instant, and in other cases, many peaks or hills looked alike and caused difficulties in determining the original object. In Tunisia, twin peaks are not reliable landmarks because of their frequent occurrence. (Survivors have found that after a short period of traveling they might have had up to a dozen twin peaks for reference in the same vicinity.) The Great Sand Sea (Egypt and Libya) was the emergency landing site of several groups of survivors and caused navigational difficulties. In these rolling sand hills, it is impossible to keep one object in view, and even footmarks fail to provide a reliable back-trail for determining travel directions. The extreme flatness of other stretches of desert terrain in North Africa also makes navigation difficult. With no landmarks to follow, no objectives to sight, survivors may walk in circles or large arcs before realizing their difficulties.

(b) A Marine pilot who made an emergency landing in the Arizona desert took the precaution of immediately spreading his parachute on the ground and putting rocks on the edges to ensure maximum visibility from the air. Then he decided to walk to his crashed plane, a distance he estimated to be 500 yards from his landing spot. He reached the plane and found it gutted by fire, and spent 5 days wandering the flat desert trying to find his parachute.

(c) Navigational difficulties of a different type may be experienced in Ethiopia, Kenya, and Somalia. Here the density of the thorn brush, even though it is primarily acacia with small leaves, makes it extremely hard to navigate from one point to another. In this area, survivors should follow animal trails and hope they lead to rivers or waterholes. Elephant trails seem to offer the best and clearest route.

(d) In the Sinai Desert area and in portions of Egypt, travel routes may be used; survivors can "stay put" on the trails. One survivor who made a trail encountered a camel caravan almost immediately, although he reported that it bothered him because he had not seen them approach, as they suddenly appeared out of a mirage. Another commented that it was awfully hard to be alone in his section of the desert, for in every direction, he saw wandering tribes, camel herds, or people watching him. Two survivors independently suggested that survivors pay attention to the wind as an aid in navigation. One survivor, on the Arabian Peninsula, noted that the wind blew consistently from the same direction. The other, in the Libyan Desert, made the same comment and said he was able to judge his direction of travel by the angle at which the wind blew his clothes or struck his body. Survivors in certain areas may orient themselves to the prevailing winds once it is established that these are consistent.

(4) People who walked across the North African deserts had much to say about the local environment, and little of it was complimentary. The extreme temperatures bothered them most. It was extremely hot during the day and often bitterly cold at night, especially during January and February.

(a) The bright sunlight was hard on their eyes, extremities, and exposed skin. The blinding effects of the sun reflecting off the terrain caused many persons to express concern regarding sunglasses. Several built fires and smoked their goggles to obtain protection against the glare. Lenses alone do not adequately protect the eyes from glare. Sunglasses may be improvised to reduce this problem. Light-skinned individuals tended to sunburn faster and more severely than darker-skinned individuals. Some reported that no amount of previous suntanning seemed to make any difference. The heat affected their feet and hands. Survivors reported that the surface became so hot that their feet were blistered through their shoes. Exposure of bare hands to the sunlight resulted in painful burns. Placing sunburned hands in bare armpits gave considerable relief since the armpits were one of the few places on the body where a person could find continuous perspiration to aid in cooling.

(b) The persistent winds of desert areas seem to provide no cooling effect, and several survivors found the constant blowing of the wind "got on their nerves." More significant is the fact that the constant winds usually carried an amount of sand or dirt particles. These particles got in the eyes, ears, nostrils, and mouth and caused irritations that were often severe. Additionally, this persistent wind also caused earaches. One survivor reported that the abrasions of the eyes by the particles of dust reached a point in which first the man's eyes watered so much that he could not see, then eventually the watering stopped and "emery-cloth eyelids remained," making life miserable for him.

(c) Extreme winds blew sandstorms that lasted from a few minutes to months. Generally, survivors reported that they could see the approach of such storms and were able to take proper precautions; however, sandstorms completely surprised a few groups, and they had difficulty navigating. None of the survivors who experienced sandstorms in the northern deserts underestimated the power and danger of such storms. Protection from the storms was uppermost in their minds. Most survivors used rock cairns, natural ledges, boulders, depressions, or wells for shelter. Survivors had time to dig depressions and rig a shelter from blankets, parachutes, or tarpaulins. A few wrapped themselves in their parachutes and endured the storm in a prone position.

(d) Nearly all of these people had some comment to make on orientation before, during, and after a sandstorm. They warned specifically that it is necessary to adequately mark the direction of travel before the storm. A few survivors said that when the storm was over, they had no idea which way they had been traveling, and all their landmarks were forgotten, obliterated, or indistinguishable. The general plan for marking travel routes before a sandstorm is to place a stick to indicate direction. One survivor oriented himself with a rock a few feet in front of his position. He commented that after the storm, one point was not adequate and recommended using a row of stones, sticks, or heavy gear about 10 yards in length to give adequate direction following such an emergency.

(e) Several survivors reported that they learned the hard way to keep their mouth shut in the desert. This meant that breathing through the mouth caused drying, and talking not only got on the other's nerves but also caused excessive drying of the mucous membranes.

(5) Mirages are common in desert areas. They are optical phenomena due to the refraction of light by uneven density distribution in the lower layer of the air. The most common desert mirage occurs during the heat of the day when the air close to the ground is much warmer than the air aloft. Under this condition, atmospheric refraction is less than normal, and the image of the distant low sky appears on the ground looking like a sheet of water. Distant objects may appear to be reflected in the "water." When the air close to the ground is much

colder than the air aloft, as in the early morning under a clear sky, atmospheric refraction is greater than normal. When this condition occurs, distant objects appear larger and closer than they are and objects below the normal horizon are visible. Unless the density distribution in the lower layers is such that the light rays from an object reach the observer along two or more paths, they will see a distorted image or multiple images of the object.

(a) Reports of mirages were very common in the survival episodes examined. In most cases, they were recognized as mirages and caused only minor difficulties. No survivor reported that these mirages actually represented bodies of water. While traveling, the survivor experienced problems as a result of mirages. Distances could not be judged because intermediate terrain was obscured by the mirage. Mirages hampered vision and navigation since they concealed objects. Additionally, mirages "magnified some objects and concealed others." A man hunting in the heat of the day reported that when he sighted an animal, it ran into or hid in a mirage. The lower layer of hot air that causes the mirage, commonly called desert haze, hampers vision and distorts objects. Signaling difficulties resulted from this since sighting a reflection on an object was apparently very difficult due to the low haze on the desert.

(b) Several survivors reported cases of imaginary illusions that were due to the haze or mirages. One group looked for a hill for a viewpoint so long that the entire party began to see hills in all directions. They finally held a conference to iron out their difficulties, and all settled on one hill that the group should approach. Everyone in the party saw the hill, and the group walked an estimated 9 miles looking for the hill, which never existed and which eventually disappeared into the desert flatness. Dawn and dusk illusions also occurred and were reported in the survivors' stories. One group was severely troubled with the false-dawn spectral light on the western horizon. The fact that the sun at first appeared to be rising in the west caused anxious moments. Another party had one person who claimed he saw a flashing beacon on the evening horizon. The pilot explained the illusion as the occasional refraction of bright starlight near the horizon, through the residual heat waves of the cliff before them. But one crewmember was so convinced it was a beacon that he started walking to investigate this object and was never seen again.

(6) The following are manmade characteristics of the desert:

(a) Roads and trails are scarce in the desert, as complex road systems are not needed. Most existing road systems have been used for centuries to connect centers of commerce or important religious shrines, such as Mecca and Medina in Saudi Arabia. These roads are now supplemented with routes for transporting oil or other mineral deposits to collection points. Rudimentary trails exist in many deserts for caravans and nomadic tribesmen. These trails have wells or oases approximately every 20 to 40 miles, although there are waterless stretches of more than 100 miles. These trails vary in width from a few yards to more than 800 yards. Vehicle travel in mountainous desert terrain may be severely restricted. Available routes may be easily blocked by hostile people or climatic conditions. Passes may be blocked by snow in the winter. The travel distance on foot or by animal between two points in the mountains may be less than a tenth of the distance required if vehicles are used to make the trip.

(b) Apart from nomadic tribesmen who live in tents, desert inhabitants live in thick-walled structures with small windows, usually built of masonry or a mud-and-straw (adobe) mixture. The ruins of earlier civilizations are scattered across the deserts. Ancient posts and forts invariably command important avenues of approach and frequently dominate the only available passes in difficult terrain. Control of these positions may be imperative for forces intent on dominating the immediate area.

(c) Exploration for and exploitation of natural resources, of which oil is the best known, occurs in many desert areas, especially in the Middle East. Wells, pipelines, refineries, quarries, and crushing plants may lead a survivor to rescue—or captivity. Additionally, pipelines are often elevated, making them visible from a distance.

(d) Many desert areas are irrigated for agricultural and habitation purposes. Agriculture and irrigation canals are signs that can lead a survivor to people.

f. Tropical Climates. The inexperienced person's view of jungle travel may range from difficult to nearly impossible. However, with patience and good planning, the best and least difficult route can be selected. In some cases, the easiest routes are rivers, trails, and ridgelines. However, there may be hazards associated with these routes.

(1) Rivers and streams may be overgrown, making them difficult to reach and impossible to raft. These waterways may also be infested with leeches. Trails may have traps or animal pits set on them. Trails can also lead to a dead end or into thick brush or swamps. Ridges may end abruptly at a cliff. The vegetation along a ridge may also conceal crevices or extend out past cliffs, making the cliff unnoticeable until it's too late.

(2) The machete is the best aid to survival in the jungle. However, survivors should not use it unless there is no other way. They should part the brush rather than cut it, if possible. If the machete must be used, cut at a down-and-out angle, instead of flat and level, as this method requires less effort.

(3) Survivors should take their time and not hurry. This allows them to observe their surroundings and gives better insight as to the best route of travel. Watch the ground for the best footing, as some areas may be slippery or give way easily. Avoid grabbing bushes or plants when traveling. Falling may be a painful experience, as many plants have sharp edges, thorns, or hooks. (NOTE: Wear gloves and fully button clothes for personal protection.)

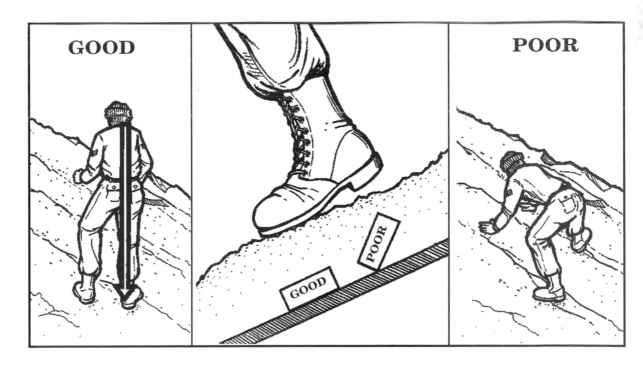

Figure 21-6. Weight of Body Over Feet.

(4) Quicksand can be a problem. In appearance, quicksand looks just like the surrounding area, with an absence of vegetation. It is usually located near the mouths of large rivers and on flat shores. The simplest description of quicksand is as a natural water tank filled with sand and supplied with water. The bottom consists of clay or other substances capable of holding water. The sand grains are rounded, as opposed to normal sharper-edged sand. This is caused by water movement, which also prevents it from settling and stabilizing. The density of this sand-water solution will support a person's body weight. The danger, if a survivor panics, may be drowning. In quicksand, the survivor should assume the spread-eagle position to help disperse the body weight to keep from sinking, and use a swimming technique to return to solid ground. (NOTE: Remember to avoid panicking and struggling, and spread out and swim or pull along the surface.)

21-5. Mountain Walking Techniques. Depending on the terrain formation, mountain walking can be divided into four different techniques—walking on hard ground, walking on grassy slopes, walking on scree slopes, and walking on talus slopes. Included in all of these techniques are two fundamental rules that must be mastered in order to expend a minimum amount of energy and time. These fundamentals are: The weight of the body must be kept directly over the feet (figure 21-6), and the

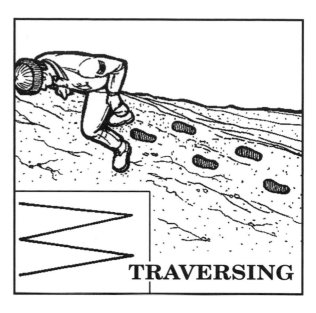

Figure 21-8. Traversing.

sole of the boot must be placed flat on the ground (figure 21-6). These fundamentals are most easily accomplished by taking small steps at a slow and steady pace. An angle of descent that is too steep should be avoided, and any indentations or protrusions of the ground, however small, should be used to advantage.

a. Hard ground is usually considered to be firmly packed dirt that will not give way under the weight of a person's step. When ascending, the above fundamentals should be applied with the addition of locking the knees on every step in order to rest the leg muscles (figure 21-7). When steep slopes are encountered, they can be traversed easier than climbed straight up. Turning at the end of each traverse should be done by stepping off in the new direction with the uphill foot (figure 21-8). This prevents crossing the feet and possible loss of balance. In traversing, the full-sole principle is done by rolling the ankle away from the hill on each step. For narrow stretches, the herringbone step may be used; that is, ascending straight up a slope with toes pointed out (figure 21-9) and using the principles stated above. Descending is usually easiest by coming straight down a slope without traversing. The back must be kept straight and the knees bent (figure 21-10) in such a manner that they take up the slack of each step. Again, remember that the weight must be directly over the feet, and the full sole must be placed on the ground with every step.

b. Grassy slopes are usually made up of small tussocks of growth rather than one continuous field. In ascending, the techniques previously mentioned are applicable; however, it is better to step on the upper side of each tussock (figure 21-11), where the ground is more level than on the lower side. Descending is best done by traversing.

Figure 21-7. Locking Knees.

Figure 21-9. Walking Uphill.

c. Scree slopes consist of small rocks and gravel that have collected below rock ridges and cliffs. The size of the scree varies from small particles to the size of a fist. Occasionally, it is a mixture of all size rocks, but normally scree slopes will be made up of rocks the same size. Ascending scree slopes is difficult, tiring, and should be avoided when possible. All principles of ascending hard ground apply, but each step must be picked carefully so the foot will not slide down when weight is placed on it. This is best done by kicking in with the toe of the upper foot so a step is formed in the scree (figure 21-12). After determining that the step is stable, carefully transfer weight from the lower foot to the upper and repeat the process. The best method for descending scree is to come straight down the slope with feet in a slight pigeon-toed position using a short shuffling step with the knees bent and back straight (figure 21-13). When several climbers descend a scree slope together, they should be as close together as possible, one behind the other, to prevent injury from dislodged rocks. Since there is a tendency to run down scree slopes, care must be taken to ensure that this is avoided and control is not lost. By leaning forward, one can obtain greater control. When a scree slope must be traversed with no gain or loss of altitude, use the hop-skip method. This is a hopping motion in which the lower foot takes all the weight and the upper foot is used for balance.

d. Talus slopes are similar in makeup to the scree slopes, except the rock pieces are larger. The technique of walking on talus is to step on top of and on the uphill side of the rocks (figure 21-14). This prevents them from tilting and rolling downhill. All other previously mentioned fundamentals are applicable. Usually, talus is easier to ascend and traverse, while scree is a more desirable avenue of descent.

21-6. Burden Carrying. Backpacking is essential when heavy loads must be carried for distances. Using a suitable harness and following certain approved packing and carrying techniques can eliminate unnecessary hardships, and help in transporting the load with greater safety and comfort. Carrying a burden initially creates mental irritation and fatigue, either of which can lower morale. Survivors should keep their mind occupied with other thoughts when packing a heavy load. Adjustments should be made during each rest stop to improve the fit and comfort of the pack. Additionally, the rate of travel should be adjusted to the weight of the pack and the environmental characteristics of the terrain being crossed.

a. Burden carrying is a task that must be done in most survival environments. Often survivors must quickly

Figure 21-10. Walking Downhill.

Figure 21-11. Walking on Grassy Slopes.

gather their equipment and move out without the assistance of a good pack. The gear may have to be carried in the arms while rapidly leaving the area. In such an instance, it would be better to fashion a roll of the gear and wear it over the shoulder, time permitting. When time is not a factor, it may be desirable to make a semi-rigid pack, such as a square pack. The convenience of being able to keep track of equipment, particularly small items, can be critical in survival situations.

(1) Packsack. A packsack can be fashioned from available survival kit containers or several layers of the fabric from the parachute canopy. The sack is sewn with inner core and a needle.

(2) Square Pack:

(a) The following instructions explain how to improvise a square pack (figure 21-15). Lay a rectangle of

Figure 21-13. Walking on Down Scree Slopes.

Figure 21-12. Walking on Scree Slopes.

Figure 21-14. Walking on Talus Slopes.

material, waterproof if available (5-by-5 feet minimum size), flat on the ground (examples are plastic, tarp, poncho, paulin, etc.; illustration A). Visualize the material divided into squares like a tic-tac-toe board. The largest piece of soft, bulky equipment (sleeping bag, parachute canopy, etc.) is placed in the center square in an "S" fold. This places the softest item in the pack against the tarp, which rests on the back while traveling. If using a poncho, place the sleeping bag just below the hood opening (illustration B). Place hard, heavy objects between the top layer and the middle layer of the "S" fold near the top of the pack. Soft items can be placed between the middle and bottom layer (illustration C).

(b) After all desired items are inside the folds, tie the inner pack in the fashion shown in illustration D. Start with a 1-inch diameter loop in the end of a long piece of parachute suspension line or other suitable line and loop it around the "S" fold laterally. Standing at the bottom of the pack, divide it into thirds and secure the running end of the line to the loop with a trucker's hitch. Both of these hitches should be at the intersection of the thirds so as to divide the pack vertically into thirds. Wrap the running ends around the pack at 90 degrees (working toward the center) to the line, and when it crosses another line, use a jam (postal) hitch to secure it and pull both ways to ensure tightness in all directions. When returning to the original figure 21-15 starting position, use the loop of the tied trucker's hitch to secure another trucker's hitch, and the inner part is complete (illustration D). The waterproof materials are then folded around the inner pack as shown in illustration E. Tie the "outer" pack in the same manner, ensuring that it is waterproof with all edges folded in securely. If a poncho is used, the head portion may be used to get into the pack, if necessary. However, it must be properly secured to ensure that the inner items are protected. With a square pack constructed in this manner, there is no reason why the equipment should get wet. (NOTE: With practice, an excellent pack can be

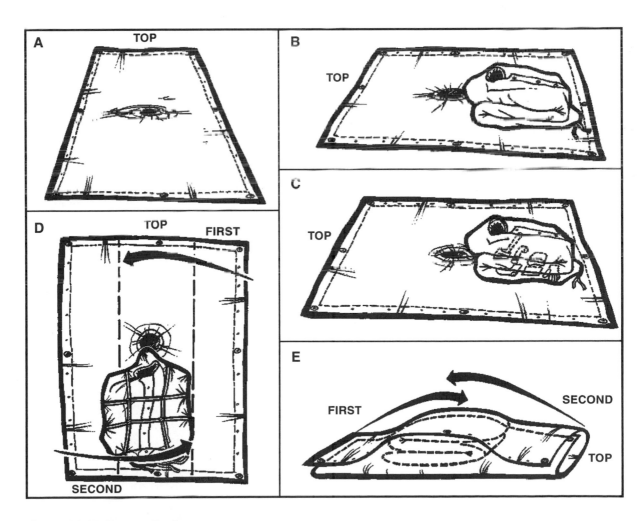

Figure 21-15. Square Pack.

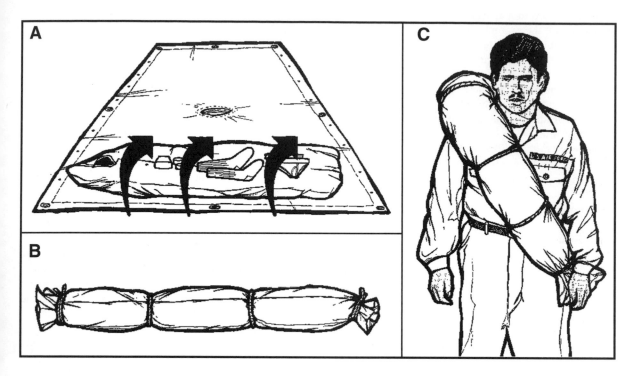

Figure 21-16. Horseshoe Pack.

constructed by tying the inner pack and outer cover simultaneously.)

(3) Horseshoe Pack. The horseshoe pack has been referred to by many names in history, including the "Johnny Reb Pack," "cigarette pack," "bedroll," and others. It is simple to make and use and relatively comfortable to carry over one shoulder. It is constructed as follows (figure 21-16): Lay the available square-shaped (preferably 5-by-5 feet) material (waterproof if available) flat on the ground (illustration A) and place all gear on the long edge of the material, leaving about 6 inches at each end. All hard items should be padded. Roll the material with the gear to the opposite edge of the square (illustration B). Tie each end securely. Place at least two or three evenly spaced ties around the roll. Bring both tied ends together and secure. This pack is compact and comfortable if all hard, heavy items are packed well inside the padding of the soft gear. If one's shoulder is injured, the pack can be carried on the other shoulder. It is easy to put on and remove (illustration C).

b. The most widely used improvised packstrap is called an Alaskan packstrap (figure 21-17). This type of packstrap can be fashioned out of any pliable and strong material. Some suitable materials for constructing the packstrap are animal skins, canvas, and parachute-harness webbing. The pack should be worn so it can be released from the strap with a single pull of the cord in the event of an emergency, such as falling into water. The knot

securing the pack should be made with an end readily available that can be pulled to drop the pack quickly: for example, a trucker's hitch with safety for normal terrain travel, and with the safety removed when in areas of danger, such as water or rough terrain.

(1) Some advantages of the Alaskan packstrap are:

(a) Small in bulk and light in weight.

(b) Easily carried in a pocket while traveling.

(c) Quickly released in an emergency.

(d) Can be adjusted to efficiently pack items of a variety of shapes and sizes.

(e) Can be used with a tumpline to help distribute the weight of the pack over the shoulders, neck, and chest, thereby eliminating sore muscles and chafed areas.

(2) Some disadvantages of the Alaskan packstrap are:

(a) Difficult to put on (without practice).

(b) Experience and ingenuity are necessary to use it with maximum efficiency.

c. The following principles should be considered when packing and carrying a pack:

(1) The pack or burden-carrying device should be adequate for the intended job.

(2) The pack or burden may be adaptable to a pack frame. The pack frame could have a bellyband to distribute the weight between the shoulders and hips and prevent undue swaying of the pack. Pack frames are also used to carry other burdens such as meat, brush, and firewood.

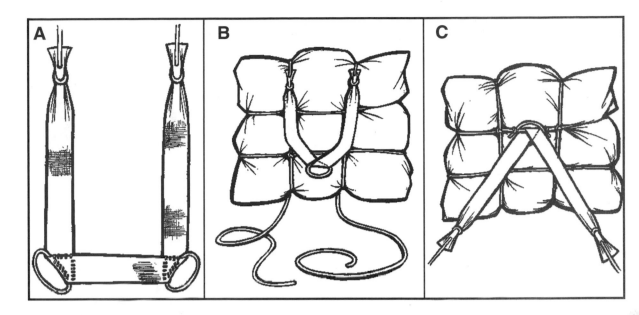

Figure 21-17. Alaskan Packstrap.

(3) Proper weight distribution is achieved by ensuring that the weight is equally apportioned on each side of the pack and as close as possible to the body's center of gravity. This enhances balance and the ability to walk in an upright position. If heavy objects are attached to the outside of the pack, the body will be forced to lean forward. A pack bundle without a frame or packboard should be carried high on the back or shoulders. For travel on level terrain, weight can be carried high. When traveling on rough terrain, weight should be carried low or midway on the back to help maintain balance and footing.

(4) Emergency and other essential items (extra and [or] protective clothing, first aid kit, radio, flashlight, etc.) should be readily available by being placed in the top of the pack.

(5) Fragile items are protected by padding them with extra clothing or some soft material and placing them in the pack where they won't shift or bounce around. Hard and (or) sharp objects cannot damage the pack or other items if cutting edges are properly sheathed, padded, and not pointed toward the bearer. Items outside the pack should be firmly secured but not protruding, where they could snag on branches and rocks.

(6) Adjust and carry the pack so that overloading or straining of muscles or muscle groups is avoided. When using a pack, the straps should be adjusted so they ride comfortably on the trapezius muscles and avoid movement when walking. Back support should be tight and placed to ensure good ventilation and support. During breaks on the trail, rest using the proper position to ease

the weight of the pack and take the strain off muscles. (See figure 21-18 for methods of resting.) A comfortable pack is adjustable to the physique of the person. A waistband will support 80 to 90 percent of the weight and is fitted relatively tight. The waistband should be cinched down around the pelvic girdle/crest area to avoid constricting circulation or restricting muscle movement.

d. A tumpline is an excellent aid to burden carrying since it transfers part of the weight of the pack to the skeletal system (figure 21-19).

(1) Tumpline Construction:

(a) Attach a soft band, which rests on the upper forehead, to the pack by using light line. Make the band out of any strong, soft material, such as animal skin with hair, tanned skin, an old sock, or parachute cloth. Make the band long enough to reach over the forehead and down to a point opposite each ear. A tumpline does not require any sewing.

(b) Adjust the tumpline to fit the head by making loops at the ends. It is difficult to reach down and behind to make necessary adjustments of the pack, but a person can easily reach up and adjust the pack by using the loops on either side of the forehead.

(c) Make mainstrings from rawhide or parachute suspension lines. Tie them to the lower corners of the pack, bring them up to the loops at the ends of the tumpline, and tie them with a slipknot. Experience is needed to estimate proper adjustments before putting on the pack; however, adjustments can always be made after the packstraps are adjusted.

Figure 21-18. Methods of Resting.

Figure 21-19. Tumpline.

(2) Tumpline Use. The tumpline should be tight enough to transfer about half of the weight of the pack/burden from the shoulders to the head. Occasionally, a heavy pack may cut off the blood circulation to the shoulders and arms. In such cases, a tumpline is of great value. By slight adjustments, most of the weight can be transferred to the head and neck, thereby loosening the shoulder straps and permitting circulation to return to numb arms. A tumpline may cause the neck muscles to be sore for the first few days due to the unusual strain placed on them; however, this discomfort soon disappears. With practice, heavy weights can be supported with only the neck and head. A tumpline can also be used to pack animal carcasses, firewood, or equipment. Since it can be rolled up and carried in a pocket, it can be a real aid to survival.

Chapter 22

ROUGH LAND TRAVEL AND EVACUATION TECHNIQUES

22-1. Introduction. There are survival situations in which traveling over rough terrain is required. However, if it is necessary to travel in such areas, specialized skills, knowledge, and equipment are required. These skills may include rope management, specialized knots, belaying, and climbing techniques, prusik climbing, rope bridges, rappelling, litter evacuation, etc. The amount and type of equipment needed will depend on the type of operation (climbing, evacuation, etc.).

22-2. Safety Considerations. A safety rope must be used when there is danger of the rescue team or climber falling. The rescue team leader will identify the members of the team who will climb together. Any member of the team may request to be roped during any mountain operation. The rope will be maintained as long as requested by any member. The rescue team leader will decide whether or not the entire team will rope up. Environmental factors may require a rapid retreat. During these circumstances, when speed is critical, it may be desirable to unrope. The rescue team leader's decision to unrope may be overridden by any team member who desires to remain roped. This option pertains to areas such as ridge climbs and low-grade climbs when roping is not required. Solo climbing in severe terrain is not recommended.

22-3. Climbing Ropes. Climbing ropes are made of synthetic fibers, such as nylon. The two essential qualities of a climbing rope are tensile strength and the ability to absorb the shock of a fall. This combination of elongation plus strength varies with the type of rope construction.

a. There are two basic types. The tensile strengths average 5,500 pounds, and the construction designs are called the Kernmantle and the Hauser Lay.

(1) Kernmantle rope consists of a core of braided or twisted strands protected by a braided sheath. They have a 6- to 8-percent stretch factor. Depending on use, Kernmantle ropes are 9 to 11 millimeters in diameter; standard lengths are 150 feet and 165 feet. Ropes for wet conditions are also available.

(2) Mountain-Lay (three-strand hard lay) ropes consist of a right-hand lay of three main strands twisted around one another. These ropes have a 9- to 13-percent stretch (twice as much as Kernmantle ropes) factor. They are constructed in 120, 150, and 165-foot lengths and are three-eighths to seven-sixteenths of an inch in diameter.

b. Climbing ropes and anchor slings require special care. Stepping on the ropes will grind abrasive dirt particles into them, which will cut the fibers of the rope. Contact between the rope and sharp corners or edges of rocks should be avoided. This may cut the rope. If the

rope must run over a sharp edge, it should be padded (figure 22-1). The ropes should be kept dry because wet ropes are not as strong and collect more dirt. Do not hang ropes on sharp edges. Do not let one rope rub against another during use. Nylon rubbing on nylon will create friction, which can burn through the rope(s). This is called weld abrasion. Smoking around ropes should be prohibited, and they should not contact any source of excessive heat. Petroleum products will accelerate deterioration of nylon. When not in use, protect the rope from the ultraviolet rays of the sun, which also have a deteriorating effect on nylon.

c. Before constructing any rope system, the ropes should be backcoiled or layed out. This is done by simply removing the rope from the coil and, starting with one end, arranging the rope into neat overlapping loops on the ground. This inspection ensures the rope is free of knots or kinks (figure 22-1).

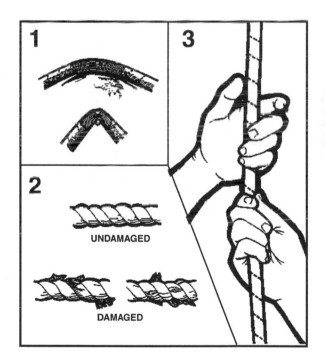

Figure 22-1. Damaged Rope.

(1) Before throwing the rope, one end should be secured. This end should be the standing end on the backcoil. Next, several small loops are taken from the working end and placed in one hand, and several more loops are placed in the other hand. The throw is done by raising the hand closet to the rope end (running end)

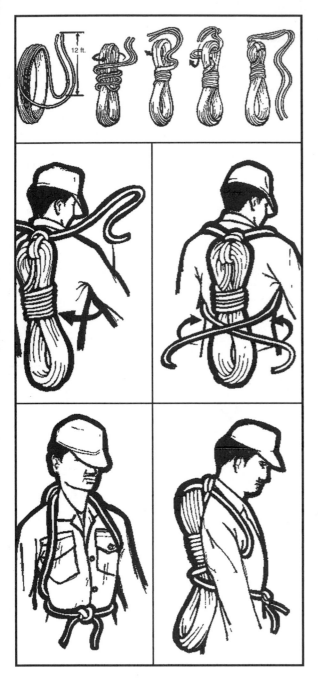

Figure 22-2. The European Coil.

be no knots in the rope or any equipment attached to the rope due to the possibility of the rope becoming entangled if it should snag on a rock, brush, etc.

(2) Two common methods are used to coil rope:

(a) The European coil is done by starting in the center of the rope, coiling the rope into 3-foot diameter loops until there is about 12 feet of rope tail left on the two ends. The ends are then wrapped three to four times around the coil, a bight is passed through the center of the coil, and the tails passed through the bight. This is then tightened down with two remaining tails of about 6 feet. The coil can now be carried by placing the coil on the back, passing one end over each shoulder, between the arm and the chest, and passing around the back to return to the front of the body. The two ends are now secured at waist level (figure 22-2).

(b) To form a mountain coil, one end of the rope is taken in one hand while the other hand is run along the rope until both hands are outstretched. The hands are then brought together forming a loop, which is transferred to one hand. This is repeated, forming uniform coils, until the entire rope is coiled. To secure the coil, a 12-inch bight is made in the standing end of the rope and laid on top of the coil. Drop the last loop on the coil and take this length of rope and wrap it around the coil and the bight. The first wrap is made at the open end of the bight to lock. Continue to wrap toward the closed end of the bight until just enough rope remains to be run through the inside of the bight. Pull the running end of the bight to secure the coil (figure 22-3). The coil may now be carried draped over a pack or over one shoulder and beneath the other.

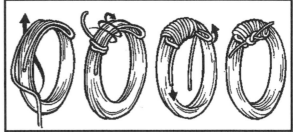

Figure 22-3. The Mountain Coil.

over the head and throwing the loops out and down in the intended direction. The loops in the other hand are allowed to drop immediately after the running end is thrown. Properly done, the remaining rope will feed out. Before throwing the rope, the word "rope" should be sounded. This will alert anyone below to the danger of falling rope. Wait for the response "clear" if there is someone below. Under normal conditions, there should

22-4. Specialized Knots for Climbing and Evacuation. Each of the following knots has a specific purpose. These knots have survived the test of time and are used in maintaining operations. They are designed to have the least effect on the fiber of a rope lock without slipping, and they are easy to untie when wet and icy. All knots reduce the strength of ropes; however, these knots reduce the strength of the rope as little as possible. Most

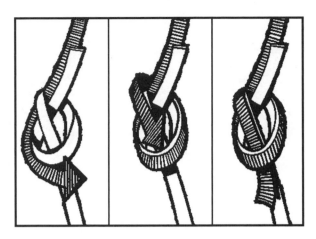

Figure 22-4. The Water Knot.

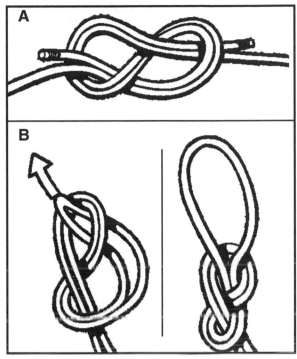

Figure 22-6. Figure-Eight Loop.

knots should be safetied with an overhand knot or two half hitches. A knot does not have to be safetied if the knot is designed for the middle of a line or if it is 13 feet or more from the end of the rope.

a. The water knot or right bend is used for joining nylon webbing (figure 22-4).

b. The double fisherman's bend is used to securely join Kernmantle or hard-lay lines (figure 22-5).

c. The double figure-eight knot is used for temporarily joining Kernmantle or hard-lay ropes (rappels, Tyrolean traverses; figure 22-6A). Carabiners should be placed between lines within the knot to prevent the knot from clinching down when loaded. The figure eight loop may be tied at the end or in the middle of a line establishing a fixed loop (figure 22-6B). If the loop is tied at the end of the line, then an overhand or single fisherman's safety knot must be used. The figure eight on a bight is used to provide two loops from a single knot to attach to two different anchors (figure 22-7).

d. The butterfly is used to make a fixed loop in the middle of a line where the direction is between 120- to 180-degree angles (figure 22-8). For angles less than 120 degrees, a figure eight in a loop will suffice. For an angle of pull greater than 120 degrees, the figure eight will become weakened and begin to split.

Figure 22-5. Double Fisherman's Bend.

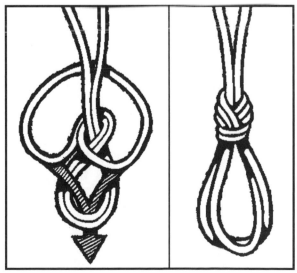

Figure 22-7. Figure-Eight with Two Loops.

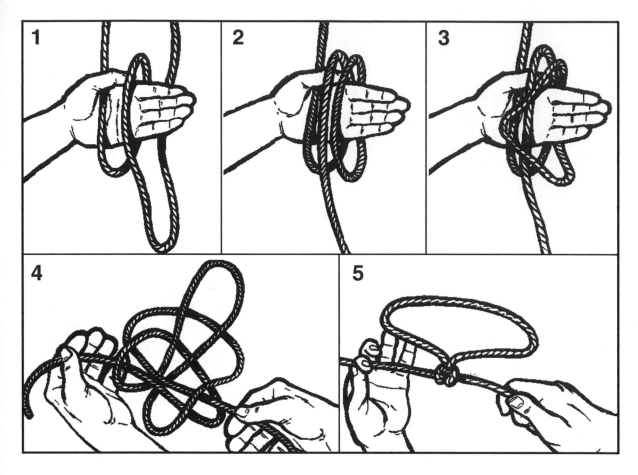

Figure 22-8. Butterfly Knot.

e. The prusik knot may be used to ascend a fixed line or safety a rappel (figure 22-9).

f. A three-loop bowline is a variation of the bowline on a bight. It is used for three anchor points or as an improvised harness (figure 22-10).

g. The mariner's knot is a combination of two knots tied with a sling and a carabiner (figure 22-11) and is

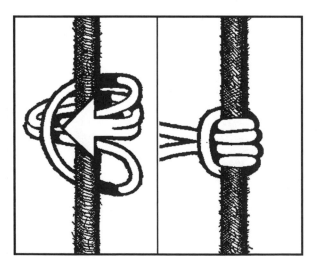

Figure 22-9. Prusik Knot.

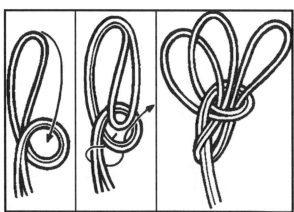

Figure 22-10. Three-Loop Bowline.

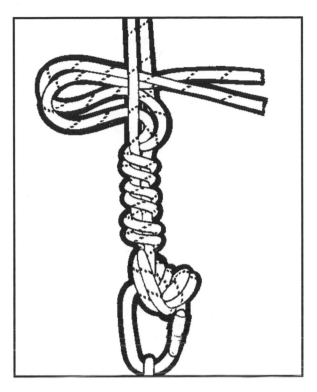

Figure 22-11. Mariner's Knot.

used to transfer loads from one system to another. With a sling, a prusik is placed on the main rope (load rope). The remaining portion of the sling is then wrapped three times through the anchor carabiner and then three times around itself next to the carabiner. The remaining tail is tucked between the two sides of the sling. Tension may be gradually released by removing the wraps.

22-5. Special Equipment. Without proper equipment, safety is jeopardized and travel is impossible in severe terrain.

a. Seat Harness. The seat harness is a safety sling that is used to attach the rope to the climber or rappeller. It must be tied correctly for safety and comfort reasons. An improvised seat harness can be made of 1-inch tubular nylon tape. The tape is placed across the back so the midpoint (center) is on the hip opposite the hand that will be used for braking during belaying or rappelling. Keep the midpoint on the appropriate hip, cross the ends of the tape in front of the body, and tie half a surgeon's knot (three or four overhand wraps) where the tapes cross (figure 22-12). The ends of the tape are brought between the legs (front to rear), around the legs, and then secured with a jam hitch to tape around the waist on both sides. The tapes are tightened by pulling down on the running ends of the tape. This must be done to prevent the tape from crossing between the legs (figure 22-12). Bring both ends around to the front and across the tape again. Then bring the tape to the opposite side of the intended brake hand and tie a square knot with an overhand knot or two half-hitch safety knots on either side of the square knot. The safety knots should be passed around as much of the tape as possible (figure 22-12). Once the seat harness has been properly tied, attach a single-locking carabiner to the harness by clipping all of the web around the waist and the web of the half-surgeon knot together. The gate of the carabiner should open on top and away from the climber.

b. Commercial Seat Harness. A commercial seat harness is a climbing harness that is worn around the pelvic girdle and is used with climbing ropes for descending and ascending devices and for rescue evacuations. The harness is easily donned, comfortable, allows freedom of movement, and evenly distributes the body

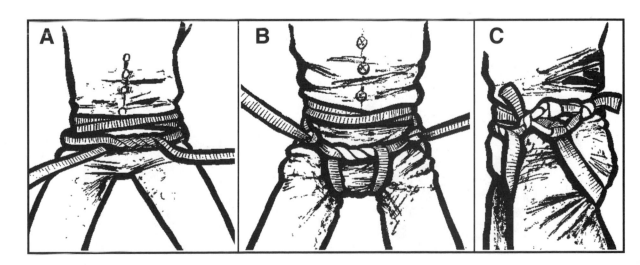

Figure 22-12. Improvised Seat Harness.

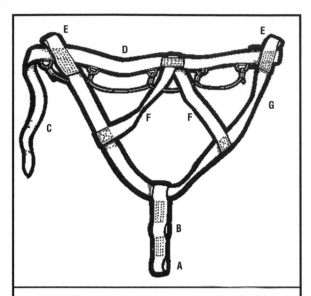

A—CROTCH STRAP – TOP LOOP

B—CROTCH STRAP – TWIN LOOP
    FOR CARABINER

C—WAIST STRAP

D—WAIST BELT

E—TIE-IN LOOPS

F—BUTTOCK STRAP

G—THIGH STRAP

Figure 22-13. Seat Harness.

weight to the pelvic region. Figure 22-13 identifies the parts of the harness. The seat-harness webbing should be cared for in the same manner as all webbing. The plastic buckle should not be subjected to direct jars since it may fracture or establish a weak point on the buckle, which can fail when used. It is primarily constructed of nylon webbing with box stitching at juncture points. Small metal rings are sewn on the lower half of the waist belt and connected with 300-pound line for equipment storage. The harness should have tie-in loops. Webbing should not have tears, rips, burns, or have been subjected to chemical exposure. The plastic buckle should not have fracture marks. All seams and threads should be intact. Harnesses showing signs of excessive wear and tear, rips, burns, or exposure to chemicals should be removed from service. Fractured buckles should be replaced.

c. Improved Chest Harness. The Parisian bandolier, or chest harness, is a secondary safety sling used to attach to a safety sling at the top of climbs or for top belaying on high-angle rappels. A Parisian bandolier is made from a continuous loop of webbing. The loop should be 3 to 4 feet long. The knot connecting the two ends of tape together is a water knot, with an overhand knot or half hitch as safeties. To don the bandolier, place one arm through the loop and bring the running end of the loop behind the back and under the opposite arm. A sheet bend is tied in the center of the chest by using the center of the tape that was passed under the climber's arm and inserted through the portion of bight formed by the tape that goes over the shoulder (figure 22-14).

d. Climbing Helmet. Climbers should wear this hard-shell helmet, which is designed to minimize injury

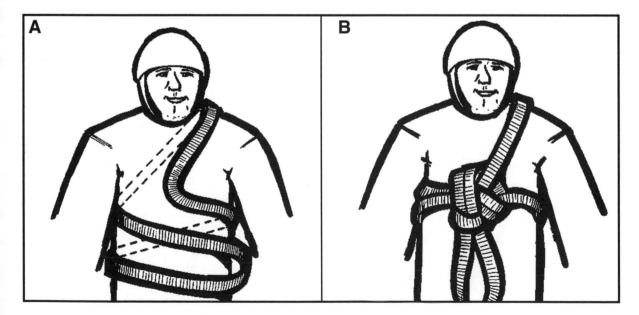

Figure 22-14. Parisian Bandolier.

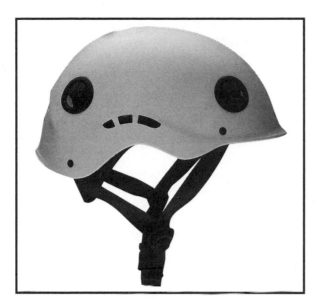

Figure 22-15. Climbing Helmet.

during falls or when struck by a small object falling from above their position. The helmet consists of a hard shell that is held away from the head by a suspension system. The suspension system absorbs a portion of the blow to the top of the helmet. Helmet fit should be such that the side of the head is protected. A "Y" style strap system of lightweight webbing holds the helmet on the head better than a single strap. The headband is adjustable

SIZE	MOUNTAIN-LAY	KERNMANTLE	NYLON WEB
5/16" OR 7 MM WIDE	2,700	2,300	
3/8" OR 9 MM WIDE	3,700	3,200	
7/16" OR 11 MM	5,500	4,800	
1/2" WIDE			1,100
9/16" WIDE			1,700
1" WIDE			4,000

Figure 22-16. Breaking Strength of Rope.

CARABINER BREAKING STRENGTH		
TYPE	MINIMUM	MAXIMUM
STANDARD OVAL, ALLOY	3,890	4,210
STANDARD D, ALLOY	4,065	5,515
LOCKING D, ALLOY	4,960	6,310
LOCKING D, STEEL	GREATER THAN 11,000 POUNDS	

Figure 22-17. Breaking Strength of Carabiners.

to allow donning over a watch cap in cold weather (figure 22-15). The hard shell should not have cracks or breaks. The suspension system must be securely riveted in place. The straps must be free of cuts, frays, and securely fastened to the helmet.

e. Gloves. Gloves should be worn when belaying or rappelling. Since these techniques are performed frequently, the gloves should be attached to the climber (rescuer) at all times.

f. Slings. Slings have many uses in adverse-terrain work—construction of anchor systems, stirrups to aid movement, attachment to chocks, improvised harnesses,

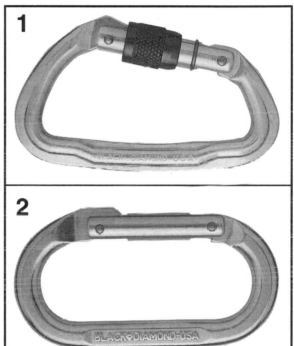

Figure 22-18. Types of Carabiners.

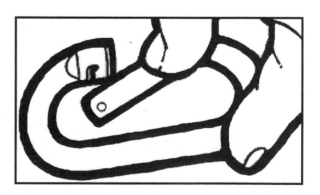

Figure 22-19. Nonlocking Carabiners.

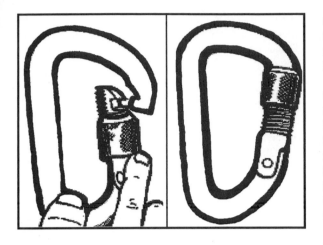

Figure 22-20. Large Locking Carabiner.

and to attach equipment to rope systems. Slings may be constructed from Mountain-Lay, Kernmantle, and flat or tubular nylon. The standard length for long slings is 13 feet, with smaller slings cut to the individual task. For climbing, slings about 4 feet long, with a circumference of 2 feet, are acceptable. Sling rope diameters for Mountain-Lay or Kernmantle are 5 mm, 7 mm, and 9 mm, while tape slings are usually one-half inch or 1 inch wide. During construction of any system or slinging of chocks, the largest possible rope diameter or tape width should be used. Mountain-Lay and Kernmantle ropes used for slings are the same as climbing ropes. Nylon webbing is constructed of multiple strands of nylon sewn in a weave pattern throughout the length. It should be constructed in flat or tubular shapes. Specific strengths

are noted in figure 22-16. Nylon webbing that is burned, torn, cut, or frayed should be removed from service. If nylon webbing is suspected to be inferior due to shock, loading, or abuse, it should not be used.

g. Carabiners. The carabiners used by climbers (rescuers) fall into two material designs—aluminum alloy and chrome vanadium steel. Both material designs may be either locking or nonlocking. Alloy metals provide the maximum strength for adverse-terrain work. As with ropes, carabiners are rated differently depending on the manufacturer. Figure 22-17 depicts the breaking strengths for various types of carabiners. Aluminum alloy and chrome vanadium steel carabiners will be oval-shaped or D-shaped (figure 22-18).

(1) Alloy standard oval (nonlocking) is shaped in an oval, with a gate to allow access into the center of the oval. The gate contains a locking pin that fits into the locking notch when the gate is closed (figure 22-19). The alloy D-shape (nonlocking) is shaped in the fashion of a D, which allows greater distribution of weight applied to the longer side. Features and operation are the same as with the standard oval.

(2) Alloy locking D or oval are both constructed the same as previously discussed, but are machined with threads on the gate and a sleeve that, when screwed clockwise, will cover the locking notch and pin for a positive lock. The steel locking D is the same construction as the alloy locking D, but is made of steel and is normally used for mountain rescue work (figure 22-20).

(3) Carabiners should not be dropped or used for anything other than the designed purpose since small fracture lines may develop and weaken the structure. Carabiners should not be used as a hammer nor loaded

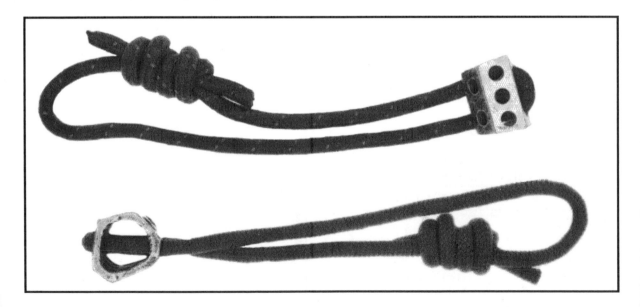

Figure 22-21. Hexagonal Chock with Kermantle Sling.

Figure 22-22. Chocks.

(stressed) beyond their maximum breaking strength. The moving parts, hinge and sleeve, of locking carabiners should be kept clean for free movement. If a carabiner "binds" do not oil it—discard it! Carabiners should not be filed, stamped, or marked with an engraving tool. Colored tape or Teflon paint may be used to identify carabiners. All moving parts (gate, locking sleeve) should operate freely, and the locking pin must properly align with the locking notch. Obvious fractures, regardless of size, are cause for condemning a carabiner.

h. Chocks:

(1) Chocks are metal alloy or copper shapes with unequal sides that are placed within cracks in rocks and serve as anchor points or parts of a protection system. Figures 22 21 and 22-22 show the common chocks used. Each is designed to fit within a variety of cracks in rocks. Sizes vary from one-sixteenth inch thick by one-fourth inch wide to several inches thick and wide. The various shapes with uneven sides allow one size to fit many rock openings. Carrying different sizes of chocks allows a person to choose the most suitable chock for the job. Depending on the style, chocks are constructed differently and require a sling for use.

(a) Copperhead. A cylindrical copper chuck ranging from three-sixteenths inch in diameter by one-half inch long to one-half inch in diameter by 1-inch long. The relatively soft head bites well into rock and has little rotation after placement. It is manufactured with a cable sling attached (figure 22-23).

(b) Hexagonal. A metal-alloy chock constructed with six sides. Each side is a different length than the opposing side (figure 22-21). The two ends are tapered,

gradually getting smaller from the back to the front of the chock. The front of the hexentric chock is the narrowest of the faces. Small hexagonal chocks are manufactured with wire slings, and larger ones have two holes bored through the front to back for threading of a Kernmantle sling. Kernmantle slings vary with the size of the chock, from 5 mm to 9 mm. The largest sling size possible will be used with each chock.

(c) Wedge. A solid alloy-metal chock shaped in the form of a wedge. All four sides decrease in size from back to front. As with small hexentrics, small wedges are manufactured with wire slings, while larger wedges require Kernmantle slings.

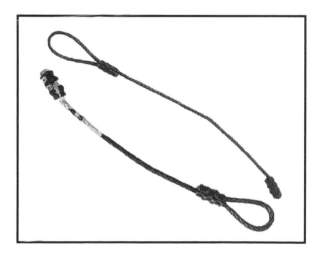

Figure 22-23. Copperheads.

(2) All chocks should be treated the same as carabiners. However, chocks will take considerable abuse since they are wedged in rock, and at times, the leverage of a hammer may be required to remove them. Chocks manufactured with wire slings and wedges should be closely monitored for security. If there is doubt about the condition of a wedge, it must not be used. Rope or tape slings on chocks must meet inspection requirements. Cracks on any surface of the chock are cause to remove it from service. Small scratches or grooves on the outer surfaces are not cause for removal from service.

i. Pitons. Pitons are metal-alloy spikes of various shapes, widths, and lengths that are hammered into cracks in rock surfaces for anchor points or as part of a protection system. The different shapes are identified by name; each name has the same shape but comes in various sizes of length and width. Figure 22-24 depicts the types of pitons. Pitons are divided into two categories—blades, whose holding power results from being wedged into cracks; and angles, which hold by wedging and blade compression.

(1) Blades comprise those pitons that have flat surfaces and range in size from knife-blade thickness to one-half inch.

(2) Angles are those that have rounded or V-shaped blades and allow the sharp edges to cut into the edge of a crack. Sizes vary from a short, "shallow angle" of one-half inch thick and 2 1/2 inches long to "bongs" (large

angles of 4 inches in width). The leeper, a special Z-shaped angle, is designed to give greater holding power due to its cutting ability and blade compression.

(3) Pitons are made of chrome-molybdenum alloy metal, which has a high strength-versus-weight ratio. Multiple holes further reduce the weight but do not compromise their strength. Pitons are virtually indestructible. However, they should be used only for the designed purpose. Damage may occur during the placement of a piton by ineffective or off-center hammering. Pitons must not have cracks or be bent from the shape of manufacture. Any piton that is suspected of cracks or bends should be removed from service.

j. Figure-Eight Clog. A mechanical friction device used for belaying, rappelling, and breaking while lowering personnel and equipment. The device is formed in a figure eight (figure 22-25) with two different size openings comprising the inner holes. The rope is passed through the larger hole of the "8" and over the connecting portion of the device. The smaller hole is attached to a carabiner. The figure-eight clog is manufactured in various sizes.

k. Protection System (Belay System). This system is commonly referred to as a belay, the securing of a climber tied to the end of a rope by a stationary second climber (figures 22-26A and 22-26B).

(1) The major components of the belay system are (figure 22-27):

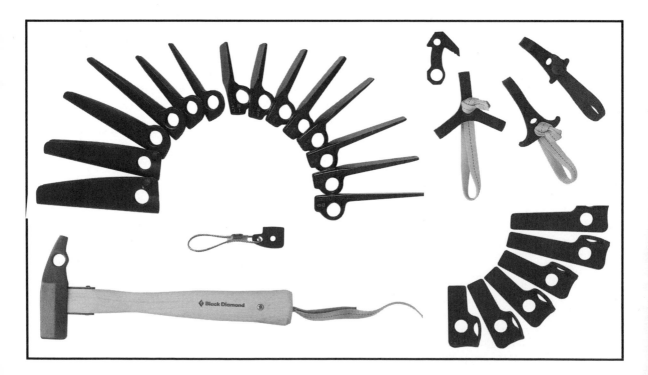

Figure 22-24. Pitons.

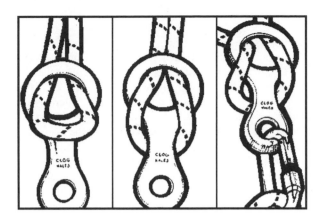

Figure 22-25. Figure-Eight Clog Direct Belay Installation.

　　(a) An anchor—secure point to which the belayer is attached.

　　(b) Belayer—individual responsible for the security of the climber.

　　(c) Intermediate protection point(s)—placement of slings, chocks, or pitons along the route of the climb by the lead climber. The climbing rope is threaded through carabiners attached to the devices and the climbing rope clipped into the carabiner.

　　(d) Climbing-rope size—11 mm Kernmantle or seven-sixteenths inch Goldline.

　　(2) The minimum equipment for a belay system is divided between rescue team and personal equipment. Rescue-team equipment is comprised of the items necessary for the climbers to reach the objective. These are the climbing ropes and climbing hardware, including chocks, pitons, slings, and carabiners. Individual equipment is comprised of items that allow each climber to perform belaying and climbing. These equipment items are climbing helmet, seat harness, climbing boots, and gloves (worn while belaying or rappelling only).

22-6. The Successful Climb. The successful accomplishment of a climb is based on the strict application of basic principles and techniques.

　a. Route Selection. Route selection can be the deciding factor in planning a climb. A direct line is seldom the proper route from a given point to the area of the survivor. Time spent at the beginning of the climbing operation in proper route selection may save a large amount of time once the operation has started. The entire route must be planned before it is carried out, with the safest route selected. Natural hazards present,

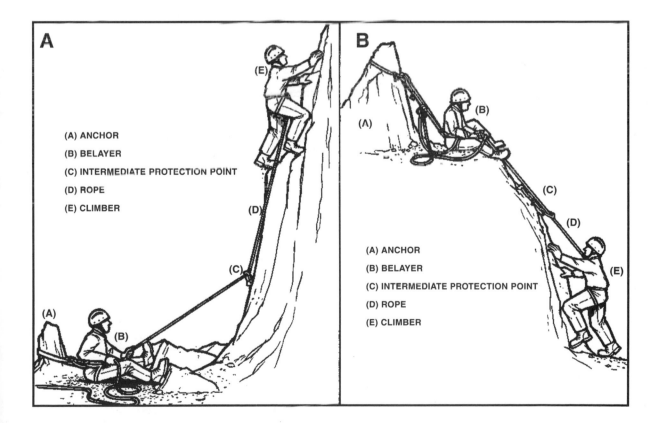

Figure 22-26. Belay System (A) Bottom (B) Top.

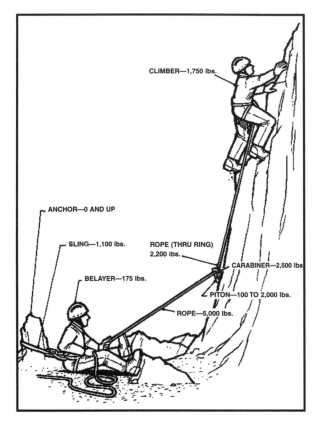

Figure 22-27. Belay Chain.

retreat routes available, time involved to perform the climb, and logistics will be major influencing factors in selecting the route.

b. Lead Climber Techniques. The lead climber is responsible for following the preplanned route of climb and altering the route as necessary. If the route of climb must be altered, consideration must be given to the experience of the climber(s). Climbing moves that are considered easy for the leader may be difficult for the second climber. Additionally, lead climbers must provide for their own personal protection during climbs to minimize the distance of a possible fall. Each intermediate protection point placed must be secure and follow as direct a line as possible.

c. Protection Placement. The basic principle to placement of protection is to decrease the distance of a fall. If the protection is not placed at frequent intervals, the climber's descent during a fall will equal twice the distance from the climber to the belayer. Slack in the belay system and rope stretch will slightly increase the distance. If a leader climbed 90 feet above the belayer without adding protection, the resultant fall will be 180 feet plus slack and stretch (figure 22-28). The belayer

is helpless to stop the leader's fall in this situation. However, providing protection at intervals (figure 22-28) of 15, 20, and 25 feet, for example, will decrease the distance of a leader's fall to a little more than 30 feet, making it easier for the belayer to stop the fall. Preferably, protection should be placed at about 10-foot intervals during all climbing operations to minimize the distance of a fall.

(1) The leader can reduce the drag on the rope by climbing in as straight a line as possible through protection points. The zigzagging of the climbing rope through protection points widely spaced or at abrupt angles (figure 22-29) will increase rope drag. Ideally, when points of protection are separated, a sling should be added to the anchor to keep the climbing rope in a straight line (figure 22-29). The attachment of a sling to an intermediate protection device is called a runner.

(2) Chocks on wire should always have a runner placed on the wire, as shown in figure 22-30. The runner reduces the direct rope movement of the wired chock, thereby reducing possible dislodgment. For all wired and roped chocks, runners should be clipped into a carabiner at the chock, and the climbing rope clipped into an additional carabiner.

(3) Pitons may also be extended in a similar manner to that used for chocks. If sufficient runners are not available on long climbs, a chock sling may be used. The climbing leader must correctly thread the climbing rope through the carabiners attached to the piton. The carabiner should open either down and out or toward the belayer. Furthermore, the rope running through the carabiner should run from the inside to the outside to prevent binding or the carabiner gate from opening.

d. Leader Belay. The length of the climbing rope or the length of the route climbed will determine when a belay will be established by the leader. If the route is longer than a rope length, a belay must be established by the leader to protect the first belayer's ascent to the leader's position. The belay chain is established as shown in figure 22-31. The only alteration necessary to the belay chain would be if the second climber continued the climb as the leader once the belayer's (first leader's) position is reached. This leap frogging of the leader is referred to as "climbing through." The sequence for the climbing-through method begins when the first leader reaches the end of the climbing-rope length. The leader selects a suitable belay position when the belayer (second climber) states that approximately 20 feet of rope remain. With a suitable site selected, one that has anchor and an area for a standing or sitting belay, the lead climber belays. The second climber then climbs up to the lead climber. Once reaching the lead climber, the second climber assumes the lead climber's role and continues the climb.

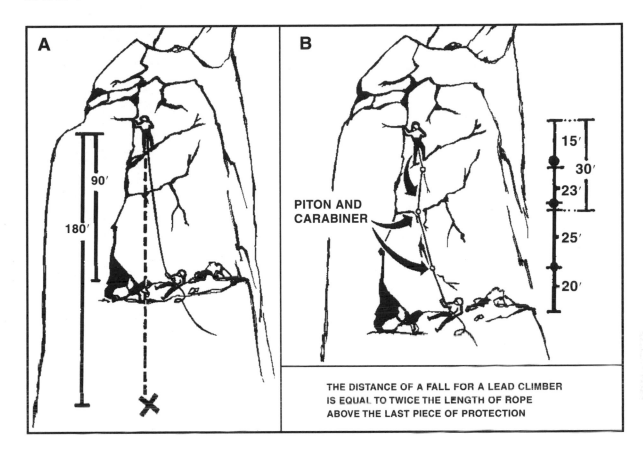

Figure 22-28. Protection Placement.

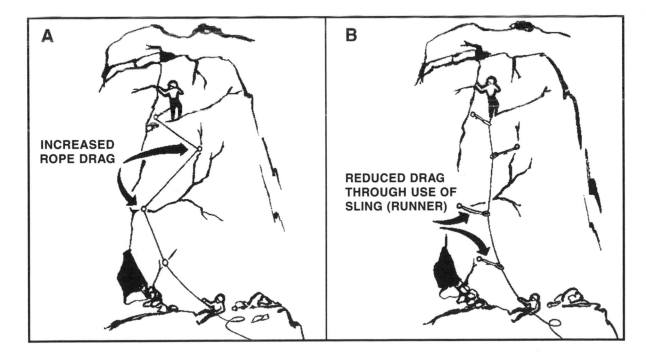

Figure 22-29. Direct Placement.

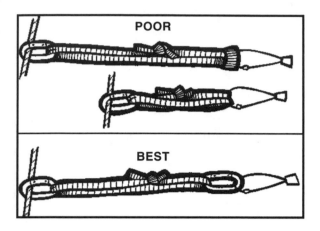

Figure 22-30. Runner Placement on Chocks.

(1) Terrain must be analyzed to find an efficient route of travel. The rescuer (climber) must make a detailed reconnaissance, noting each rock obstacle, the best approach, height, angle, type of rock, difficulty, distance between belay positions, amount of equipment, and number of trained rescuers needed to accomplish the mission on or beyond the rocks. If the strata dips toward the rescuer, holds will be difficult, as the slope will be the wrong way. However, strata sloping away from the rescuer and toward the mountain mass provides natural stairs with good holds and ledges.

(2) At least two vantage points should be used so a three-dimensional understanding of the climb can be attained. Use of early morning or late afternoon light, with its longer shadows, is helpful in this respect. Actual ground reconnaissance should be made, if possible.

e. Dangers to Avoid:

(1) On long routes, changing weather will be an important consideration. Wet or icy rock can make an otherwise easy route almost impassable; cold may reduce climbing efficiency; snow may cover holds. A weather forecast should be obtained, if possible. Smooth rock slabs are treacherous, especially when wet or iced after freezing rain. Ledges should then be sought. Rocks overgrown with moss, lichens, or grass become treacherous when wet. Under these conditions, cleated boots are by far better than composition soles.

(2) Tufts of grass and small bushes that appear firm may be growing from loosely packed and unanchored soil, all of which may give way if the grass or bush is pulled upon. Grass and bushes should be used only for balance by touch or as push holds—never as pull holds. Gently inclined but smooth slopes of rock may be covered with pebbles that may roll treacherously underfoot.

(3) Ridges can be free of loose rock but topped with unstable blocks. A route along the side of a ridge just below the top is usually best. Gullies provide the best protection and often the easiest routes, but are more subject to rockfalls. The side of the gully is relatively free from this danger. Climbing slopes of talus, moraines, or other loose rock is not only tiring to the individual, but dangerous because of the hazards of rolling rocks to others in the party. Rescuers should close up intervals when climbing simultaneously. In electrical storms, lightning

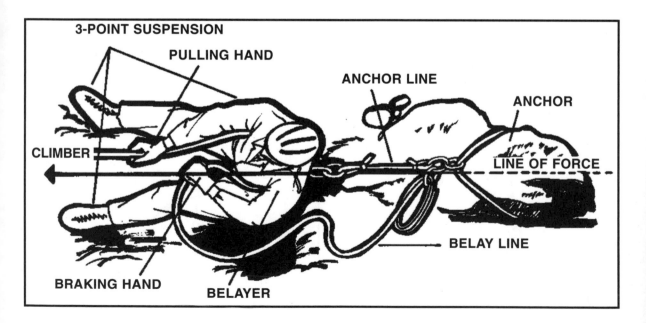

Figure 22-31. Belay Chain.

Figure 22-32. Sitting Belay.

can endanger the climber. Peaks, ridges, pinnacles, and lone trees should be avoided.

(4) Rockfalls are the most common mountaineering danger. The most frequent causes of rockfalls are other climbers, heavy rain, and extreme temperature changes in high mountains, and resultant splitting action caused by intermittent freezing and thawing. Warning of a rockfall may be the cry "ROCK," a whistling sound, a grating sound, a thunderous crashing, or sparks where the rocks strike at night. A rockfall can be a single rock or a rockslide covering a relatively large area. Rockfalls occur on all steep slopes, particularly in gullies and chutes. Areas of frequent rockfalls may be indicated by fresh scars on the rock walls, fine dust on the talus piles, or lines, grooves, and rock-strewn areas on the snow beneath cliffs. Immediate action is to seek cover, if possible. If there is not enough time to avoid the rockfall, the climber should lean in to the slope to

minimize exposure. Danger from falling rock can be minimized by careful climbing and route selection. The route selected must be commensurate with the ability of the least-experienced team member. (NOTE: Yell "rock" when equipment is dropped.)

22-7. Belaying:

a. Belaying provides the safety factor, or tension, which enables the party to climb with greater security. Without belaying skill, the use of rope in party climbing is a hazard. When climbing, a climber is belayed from above or below by another rescue team member.

(1) The belayer must run the rope through the guiding hand, which is the hand on the rope running to the climber or rescuer, and around their body to the brake hand, making certain that it will slide readily. The belayer must ensure that the remainder of the rope is laid out so it will run freely through the braking hand.

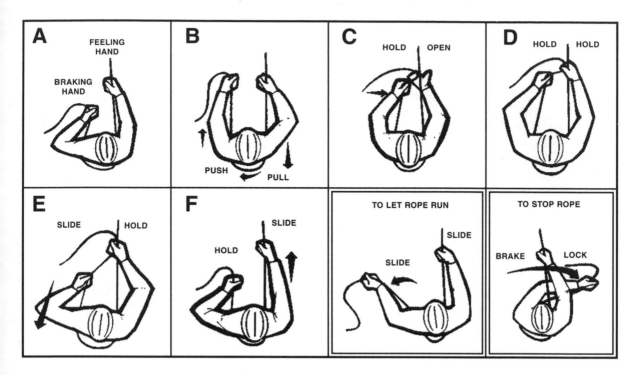

Figure 22-33. The Belay Sequence.

(2) The belayer must constantly be alert to the climber's movements in order to anticipate any needs. Avoid letting too much slack develop in the rope through constant use of the guiding hand. Keep all slack out of the rope leading to the rescuer, thus sensing any movement. If belaying a lead climber, the climber will need a constant flow of slack while climbing. If the rope is fed too slow or fast, the climber must communicate with the belayer to adjust the rate. Avoid taking up slack too suddenly to prevent throwing the climber off balance. When taking up slack, the braking hand is not brought in front of the guiding hand, but just behind the guiding hand. This allows the braking hand to slide back and to remain constantly on the rope. The braking hand is never removed from the rope during a belay.

(3) The belayer should brace well for the expected direction of pull in a fall so the force of the pull will, when possible, pull the belayer more firmly into position. A climber should neither trust nor assume a belay position that has not been personally tested.

b. The sitting belay is normally the most secure and preferred position (figure 22-32).

(1) The belayer sits and attempts to get good triangular bracing position with the legs and buttocks. Legs should be straight when possible, and the guiding hand must be on the side of the better-braced leg. The rope should run around the hips. If the belay spot is back from the edge of a cliff, friction of the rope will be greater and will simplify the holding of a fall, but the direction of pull on the belayer will be directly outward. The rope must not pass over sharp edges.

(2) Even with a good belay stance, if the rope is too high or too low on the back of a belayer, the belayer may be unable to hold a falling climber (figure 22-32). With the rope too high on the back of the belayer, on a bottom belay, the rope will ride up in the belayer's armpits. (NOTE: Rope should run under the anchor.) This will pull the belayer forward and off balance if a climber were to fall. If the rope is too low on the back of a belayer, the rope will be pulled under the buttocks of the belayer. (NOTE: Rope should run on top of the anchor.) This will force the belayer to attempt to stop the fall by trying to hold on to the rope.

(3) When necessary, seek a belay position that offers cover from a rockfall.

(4) If the climber falls, the belayer should be able to perform the following movements automatically.

(a) Relax the guiding hand.

(b) Apply immediate braking action. This is done by bringing the braking hand across the chest or in front of the body (figure 22-33).

c. When a falling climber has been brought to a stop, the belayer must hold on until the situation is relieved.

There are different ways to do this. If the climber is all right and can safely climb on to the rock or can be lowered to a secure ledge, all is well. However, the climber may be injured, or there may be no place to which the climber can be lowered. The belayer cannot continue holding the climber. The belayer must be relieved to assist the climber in the next step. This can be done by using a prusik sling. If the belayer is alone, the braking hand will hold the static climber while the other hand is free to anchor the belay line. This particular method is only good when using a belay anchor. Follow the procedure shown in figure 22-34. Climbers who are in good condition can either prusik up or pendulum across to a ledge.

22-8. Communications. Because constant communication between the belayer and climber is essential for safety, a standard group of climbing commands must be mastered.

a. Communications while climbing must be as simple as possible, and single words are preferred. Communications in the form of commands are necessary when climbing. Commands that sound alike should be avoided. The command system follows a set pattern, which leads to safe climbing and ensures that belays are used. Commands should be clear, specific, and given in a loud voice. The sequence of the system should not be broken. A review of the command system should be made before any climb. The European command system has all the necessary commands to ensure safety. Only one command has a reply; the other commands are answered by the next command in the system. Sounding off the next command is not done until the requirements of the prior command are complete.

b. Commands are generally climber initiated. The belayer, in each instance, acknowledges when commands are heard and understood. If the command is not understood, the belayer should not say anything. The silence indicates to the climber that the belayer doesn't know what is happening. The command must be repeated.

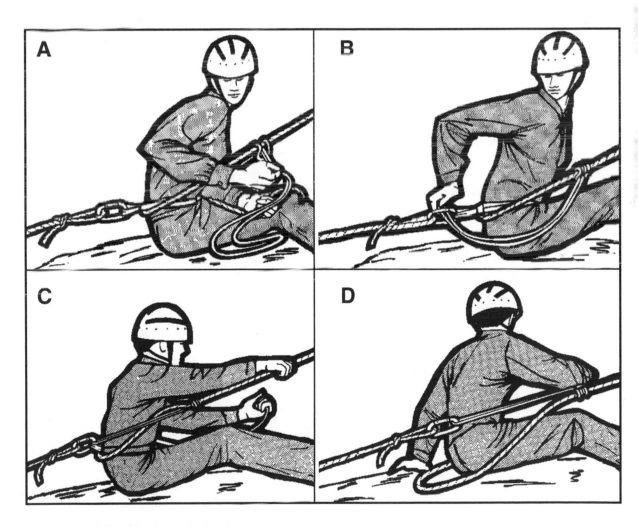

Figure 22-34. Safety Off the Belay Line.

(1) The following are the standard commands used:

(a) Initiation of a Climbing Sequence:

-1. "ON BELAY"—a signal indicating the climber is secured on the end of the rope and asks if belayer is ready.

-2. "BELAY ON"—(the belayer acknowledges the climber) means the belayer is in position and ready.

-3. "BELAY TEST"—the climber is asking to test the belay.

-4. "TEST"—signifies the belayer is ready for the climber to gradually apply weight on the rope. Releasing of the tension will indicate the test is finished.

-5. "UP ROPE"—the climber is directing the belayer to take up the slack.

-6. Belayer response: "ROPE UP."

-7. "THAT'S ME"—the climber signifies to the belayer that the tension felt on the rope is the climber.

-8. Belayer response: "THANK YOU."

-9. "CLIMBING"—the climber signifies the climber has chosen a route and is ready to climb.

-10. "CLIMB ON"—the belayer is acknowledging being ready for the climber to begin climbing. The climber will not start climbing before the command is heard.

(b) General Commands Used While Climbing:

-1. "TENSION"—the climber is telling the belayer to take up all of the slack in the rope. This command is given if the climber needs assistance or there is an impending fall.

-2. Belayer responds with: "TENSION ON."

-3. "RESTING"—the climber is in a position for rest; belayer will lock brake across waist.

-4. Belayer responds with: "REST ON."

-5. When the climber begins again, the command of "CLIMBING" will be given, and "CLIMB ON" is given when the belayer is ready.

-6. "SLACK FEET"—the climber will use this command when slack is needed in the rope to get over a difficult section or traverse. The climber should say "SLACK" and then the estimated number of feet the climber needs. Belayer will respond by repeating the number of feet of slack given. The climber will indicate that enough slack has been given by saying "THANK YOU."

-7. The command "FALLING" should be given when the climber is going to fall. This provides the belayer adequate time to apply the brake. Before continuing the climb, the climber will give the command "CLIMBING."

(c) Commands to Terminate the Climbing Sequence:

-1. "OFF BELAY"—the climber uses this command to indicate the climber is secure and finished climbing.

-2. "BELAY OFF"—the belayer uses this command to indicate the brake hand is off the rope and the climber is no longer on belay.

22-9. Anchors. Anchors are secure points in the belay chain providing protection for the belayer and the climber. Anchor systems must be able to withstand high loads. The basis for any type of anchor is strong, secure points for attachment.

a. Anchor Points. Anchor systems may be simple and consist of a single anchor point, or complex and consist of multiple anchor points. Anchor points are divided into two classes—natural and artificial. Anchor systems may be constructed entirely of one class, or a combination of the two.

(1) Natural Anchor Points (figure 22-35):

(a) Spike. A spike is a vertical projection of rock. For use as an anchor point, a sling is placed around the spike.

(b) Rock Bollard. A rock bollard is a large rock or portion of such a rock that has an angular surface enabling a sling or rope to be placed around it in such a manner that will not allow it to slip off. Care must be taken to ensure the bollard will not be pulled loose when subjected to a sudden load.

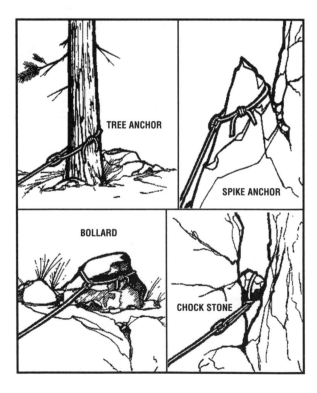

Figure 22-35. Natural Anchor Points.

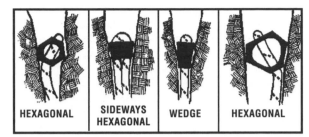

Figure 22-36. Placement of Chocks.

(c) Chockstone. A natural chockstone is a securely wedged rock providing an anchor point for a sling. In most cases, the rock is wedged within a crack.

(d) Tree. Trees often make very secure anchor points. A rope can be tied directly to the tree (or a sling doubled around the tree and connected by a carabiner). In loose or rocky soil, trees should be carefully watched, and avoided if other anchor points are available.

(2) Artificial Anchor Points. Artificial anchor points are those constructed from equipment carried by the team. These are usually the chocks or pitons placed in cracks, or bolts drilled in the rock.

(a) Chock Placement. The basic principle is to wedge the selected chock into a crack so a pull in the direction of fall will not pull the chock out. The proper method is to select a crack suitable for an anchor point and select a chock to fit that crack. The chock chosen should closely fit the widest portion of the crack. Work the chock into the crack until it is securely seated. It should be seated so the load will come on the entire chock without rotating it out of position. When the chock is in place, jerk hard on the attached sling in the direction of fall to ensure it is well seated. (Ensure the chock cannot continue to work downward to a larger area of the crack and become dislodged.) A chock may also be placed so its pull is in one direction while the second chock has a pull in the opposing direction. One sling is passed through the loop of the other chock. A force exerted downward will pull the two chocks toward each other along the axis of the crack. In vertical cracks, the lower sling should be passed through the upper for best results. The two chocks should be placed far enough apart so neither sling will reach the other chock. Figure 22-36 shows the various types of chocks

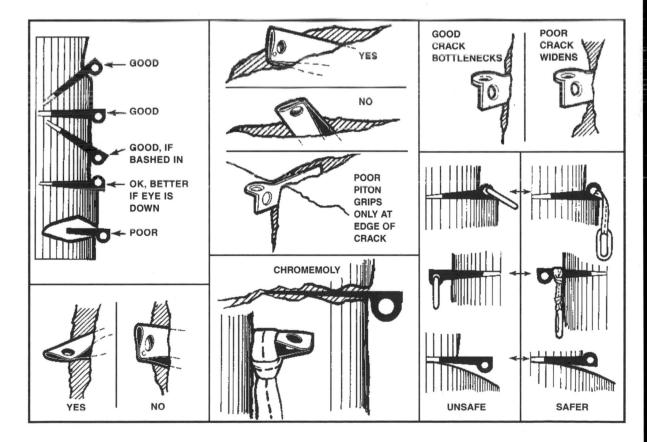

Figure 22-37. Placement of Pitons.

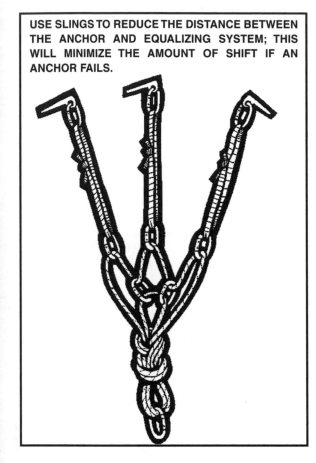

USE SLINGS TO REDUCE THE DISTANCE BETWEEN THE ANCHOR AND EQUALIZING SYSTEM; THIS WILL MINIMIZE THE AMOUNT OF SHIFT IF AN ANCHOR FAILS.

Figure 22-38. Equalizing Anchor.

and their placement. (NOTE: Chocks are preferred over pitons and bolts since they do not deface the rock.)

(b) Piton Placement. Basically, the use of pitons is a matter of locating a crack, selecting a piton that fits, driving the piton in, clipping a carabiner to the eye, and attaching the rope system to the carabiner.

-1. First look at the crack to decide the best position for driving the piton. The piton should be driven into the wider portion of the crack to reduce the likelihood of shifting or rotating under pressure. The crack must not widen or flare internally, or have a change in direction of more than 15 to 20 degrees.

-2. Before driving, the piton should fit one-half to two-thirds of the blade length into the crack. Drive until only the eye protrudes, or the piton meets resistance. Do not attempt to overdrive the piton, because it may fracture. Lightly tap the piton to test for movement or improper seating. If movement is noted, remove the piton and replace it with the next larger size,

or locate another anchor position. The proper placement of pitons in horizontal cracks is with the eye down. If the piton cannot be driven into the rock until only the eye protrudes, it may be tied off by placing a short sling around the piton, with a girth or clove hitch, as close to the rock as possible. In this case, the sling should be used as the attachment point instead of the eye of the piton. Figure 22-37 shows the proper placement of pitons.

b. Anchor System. The purpose of an anchor system is to unite weak anchor points with a strong anchor system. Systems are divided into two classes—equalizing and nonequalizing.

(1) Equalizing systems are constructed so if there is a change in direction of the load, the stress will be equally distributed to all anchor points. One major problem with this system is if one point fails, the remaining points will be shock loaded (figure 22-38).

(2) Nonequalizing systems are constructed when a change of direction is not expected. The major advantage is that the entire load is shared by the anchor points equally and would require the entire system to fail before coming loose. Any multipoint anchor system tied together so the attaching rope does not slip would be considered nonequalizing.

22-10. Climbing:

a. Balance Climbing. Balance climbing is the type of movement used to climb rock faces. It is a combination of the balance movement of a tightrope walker and the unbalanced climbing of a person ascending a tree or ladder. During the process of route selection, the climber should mentally climb the route to know what is expected. Climbers should not wear gloves when balance climbing.

(1) Body Position. The climber must keep good balance when climbing (the weight placed over the feet during movement). (See figure 22-39.) The feet, not the hands, should carry the weight (except on the steepest cliffs). The hands are for balance. The feet do not provide proper traction when the climber leans in toward the rock. With the body in balance, the climber moves with a slow, rhythmic motion. Three points of support, such as two feet and one hand, are used when possible. The preferred handholds are waist to shoulder high. Resting is necessary when climbing because tense muscles tire quickly. When resting, the arms should be kept low where circulation is not impaired. Use of small, intermediate holds is preferable to stretching and clinging to widely separated big holds. A spread-eagle position, in which a climber stretches too far (and cannot let go), should be avoided.

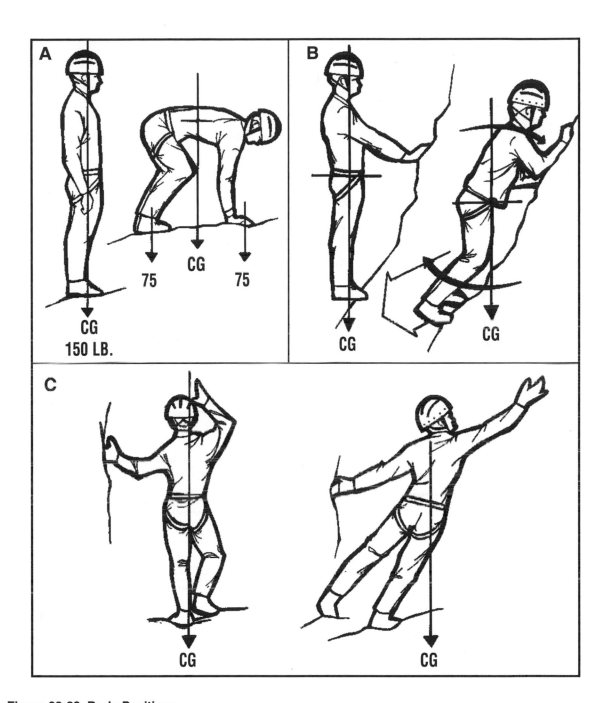

Figure 22-39. Body Positions.

(2) Types of Holds:

(a) Push Holds. Push holds (figure 22-40) are desirable because they help the climber keep the arms low; however, they are more difficult to hold on to in case of a slip. A push hold is often used to advantage in combination with a pull hold.

(b) Pull Holds. Pull holds (figure 22-41) are those that are pulled down upon and are the easiest holds to use. They are also the most likely to break out.

(c) Jam Holds. Jam holds (figure 22-42) involve jamming any part of the body or extremity into a crack. This is done by putting the hand into the crack and clenching it into a fist or by placing the arm into the crack and twisting the elbow against one side and the

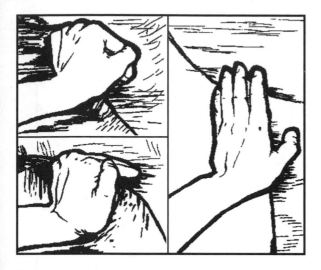

Figure 22-40. Push Holds.

-2. The lay-back (figure 22-44) is done by leaning to one side of an offset crack with the hands pulling and the feet pushing against the offset side. Lay-backing is a classic form of force or counterforce in which the hands and feet pull and push in opposite directions enabling the climber to move up in a series of shifting moves. It is very strenuous.

-3. Underclings (figure 22-45) permit pressure between hands and feet.

-4. Mantleshelving, or mantling, takes advantage of downward pressure exerted by one or both hands on a slab or shelf. By straightening and locking the arm, the body is raised, allowing a leg to be placed on a higher hold (figure 22-46).

(e) Chimney Climb. This is a body-jam hold used in very wide cracks (figure 22-47). The arms and legs are used to apply pressure against the opposite faces of the rock in a counterforce move. The

hand against the other side. When using the foot in a jam hold, care should be taken to ensure the boot is placed so it can be removed easily when climbing is continued.

(d) Combination Holds. The holds previously mentioned are considered basic and from these any number of combinations and variations can be used. The number of these variations depends only on the limit of the individual's imagination. Following are a few of the more common ones:

-1. The counterforce (figure 22-43) is attained by pinching a protruding part between the thumb and fingers and pulling outward or pressing inward with the arms.

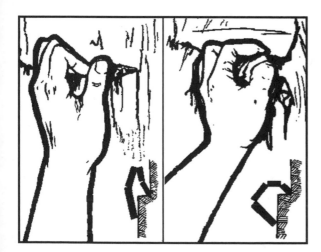

Figure 22-41. Pull Holds.

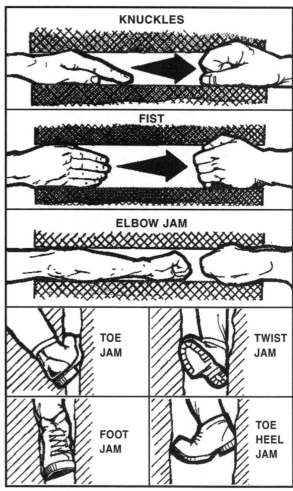

Figure 22-42. Jam Holds.

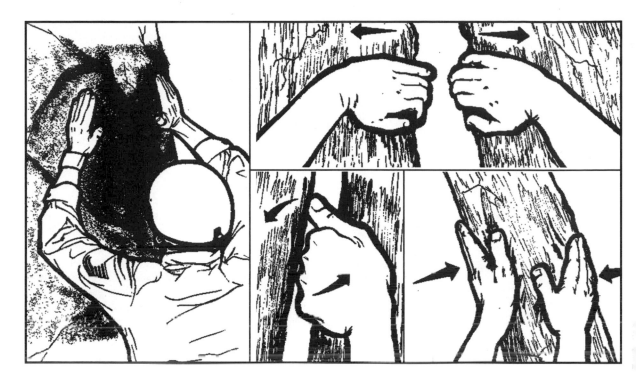

Figure 22-43. Combination Holds.

outstretched hands hold the body while the legs are drawn as high as possible. The legs are flexed, forcing the body up. This procedure is continued as necessary. Another method is to place the back against one wall and the legs and arms against the other and "worm" upward (figure 22-47).

b. Friction Climbing:

(1) A slab is a relatively smooth portion of rock lying at an angle. When traversing, the lower foot is pointed slightly downhill to increase balance and friction of the foot. All irregularities in the slope should be used for additional friction. On steep slabs, it may be necessary to squat with the body weight well over the feet, and with hands used alongside for added friction. This position may be used for ascending, traversing, or descending. A slip may result if the climber leans back or lets the buttocks down. Wet, icy, mossy or a scree-covered slab is the most dangerous.

(2) Friction holds (figure 22-48) depend solely on the friction of hands or feet against a relatively smooth surface with a shallow hold. They are difficult to use because they give a feeling of insecurity, which the inexperienced climber tries to correct by leaning close to the rock, thereby increasing the insecurity. They often serve well as intermediate holds, giving needed support while the climber moves over them; however, they would not hold if the climber decided to stop.

Figure 22-44. Lay-Back.

Figure 22-45. Underclings.

maximum steps up with each foot in turn, the longer foot sling should extend to about nose level and the other to several inches below the waist loop. Both prusiks should have loops just big enough for the foot. The chest sling is just long enough to keep the climber from toppling backwards when rigged. Standing upright with all the weight bearing on the foot slings, the climber lifts one foot and raises its unweighted knot. Now, stepping up and shifting weight on to this sling, it is repeated with the other foot. The climb is less tiring when the foot slings are of different lengths, allowing the climber to take equal steps with both feet. There is some difficulty in sliding the knots when the rope is wet or the slings are made from layed rope. Spinning is also a potential problem under overhangs where the wall cannot be touched to maintain stability, particularly when climbing a layed rope. While a rescuer (climber) is ascending via the prusik method, companions have little to do except guard the anchors and prepare to lift the climber over the edge.

22-11. Rappelling. The climber with a rope can descend quickly by sliding down a rope that has been doubled around such anchor points as a tree, a projecting rock, or several artificial anchors secured to one another with sling rope.

   a. Establishing a Rappel. In selecting the route, the climber should be sure that the rope reaches the bottom or a place from which further rappels or climbing can be safe. The rappel point should be carefully tested and inspected to ensure the rope will run around it when one end is pulled from below, and the area is clear of loose rocks. If a sling rope is used for a rappel point, it should be tied twice to form two separate loops. The first person down chooses a smooth route for the rope that is free of sharp rocks. Place loose rocks, which the rope might later dislodge, far enough back on ledges to be out of the way. The rappeller should ensure the rope runs freely around the rappel point when pulled from below. Each person down will give the signal "OFF RAPPEL," straighten the rope, and ensure the rope runs freely around its anchor. When silence is needed, a prearranged signal of pulling on the rope is substituted for the vocal signal. Recover the rope when the last person is down. The rope should be pulled smoothly to prevent the rising end from whipping around the rope. Climbers should stand clear of falling rope and rocks, which may be dislodged. The rope should be inspected frequently if a large number of people are rappelling. Rappellers should wear gloves during rappels to protect the palms from rope burns.

   b. Types of Rappel. The type of rappel is determined by the steepness of the terrain. The hasty rappel is used only on moderate-pitch slopes. The body rappel may be used on moderate to severely pitched slopes, but never

   c. Prusik Climbing. Prusiking is a method of ascension using a fixed rope (figure 22-49). This method does not require a belay. With two prusik slings, a chest prusik sling, and a carabiner, a person can climb the length of a fixed rope. The slings are attached with prusik knots that grip tightly when loaded yet slide when the load is removed. Prusiking is strenuous and requires the use of both hands and both feet, hence it is not a system for a badly injured person. Prusik slings are usually made from lengths of 7 mm Kernmantle or Mountain-Lay rope. Smaller diameter rope is less bulky and grips the climbing rope better, but in practice the knots are very difficult to work with gloved hands, and it is not recommended. To allow the climber to take

Figure 22-46. Mantleshelving.

Figure 22-47. Chimney Climbing.

Figure 22-48. Friction Holds.

Figure 22-49. Prusiking.

used on overhangs. The seat rappel is used on very steep pitches, including overhangs.

(1) The Hasty Rappel (figure 22-50). Facing slightly sideways to the anchor, the climber places the ropes across the back. The hand nearest the anchor is the guiding hand. To stop, the rappeller brings the braking hand across in front of the body, and at the same time turns to face the anchor point.

(2) The Body Rappel (figure 22-51). The climber faces the anchor point and straddles the rope, then pulls the rope from behind, runs it around either hip, diagonally across the chest, and back over the opposite shoulder. From there the rope runs to the braking hand that is on the same side of the hip that the rope crosses (for example, the right hip to the left shoulder to the right hand). The climber should lead with the braking hand down and should face slightly sideways. The foot corresponding to the braking hand should precede the other at all times. The guiding hand should be used only to guide and not to brake. To rappel, lean out at a

45-degree angle to the rock. Keep the legs well-spread and relatively straight for lateral stability, and the back straight since this reduces unnecessary friction. The collar should be turned up to prevent rope burns on the neck. Gloves should be worn, and any other articles of clothing may be used as padding for the shoulders and buttocks. To brake, lean back and face directly into the rock so the feet are flat on the rock.

(3) Four-Carabiner Seat Rappel. Seat rappels differ from the body and hasty rappels in that the friction is primarily absorbed by a carabiner in the sling-rope seat worn by the rappeller. The rappeller stands to one side of the rope (when braking with the right hand, on the left, and when braking with the left hand, on the right). Some slack between the carabiner and the anchor point is taken up and brought through two carabiners, which are attached to the harness and are horizontal and reversed and opposed (figure 22-52). Two additional carabiners are clipped from opposite sides, gates down, at 90 degrees to the original carabiners to form a friction-brake bar (figure 22-53).

c. Body Position. The body must be perpendicular to the face of the rock and the feet must be about shoulder width apart and flat on the rock (figure 22-54).

Figure 22-50. Hasty Rappel.

Figure 22-51. Body Rappel.

To descend, hold the brake out to the side to reduce friction. To brake, cinch down on the rope and move the brake hand to the small of the back.

22-12. Overland Snow Travel. Cold-weather operations conducted in snow-covered regions magnify the difficulties of reaching a survivor and effecting an extraction. Routes of travel surveyed from the air may not be possible from the ground. A straight line from the insertion point to the objective is the most desired route; however, inherent dangers (avalanches, collapsing cornices, etc.) may necessitate an alternate route entailing a longer trek. During all operations, safety, and not ease of travel, will be the primary concern.

   a. Snow Conditions. Travel time varies from hour to hour. Certain indicators may assist in the direction of travel; that is, the best snow condition is one that supports a person on or near the surface when wearing boots and the second best is a calf-deep snow condition. If possible, avoid traveling in thigh- or waist-deep snow. Snowshoes should be worn when conditions dictate.

      (1) South and west slopes offer hard surfaces late in the day after exposure to the sun and the surface is refrozen. East and north slopes tend to remain soft and unstable. Walking on one side of a ridge, gully, clump of

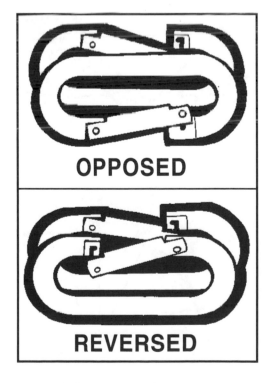

Figure 22-52. Seat Rappel (Carabiners).

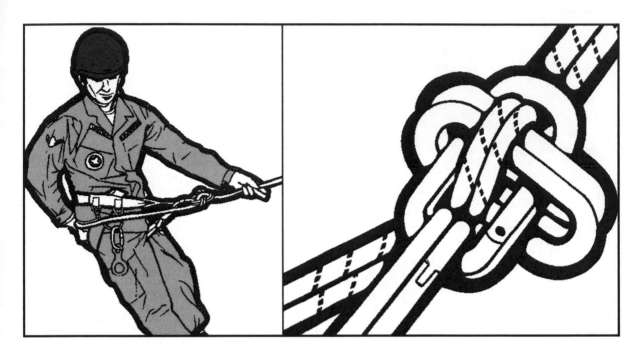

Figure 22-53. Four-Carabiner Brake System.

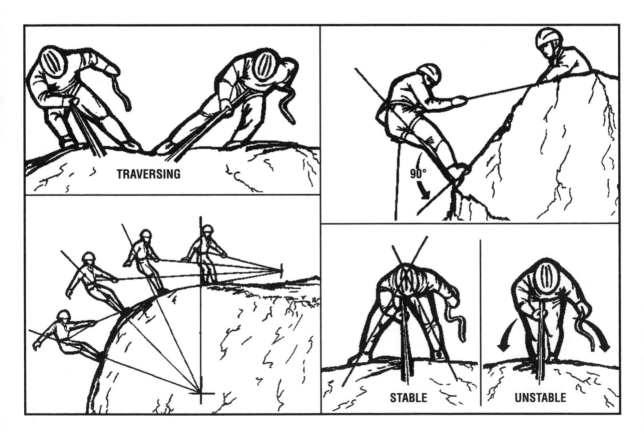

Figure 22-54. Body Position for Rappel.

trees, or large boulders is often more solid than the other side. Dirty snow absorbs more heat than clean snow; slopes darkened by rocks, dust, or uprooted vegetation usually provide more solid footing. Travel should be done in the early morning after a cold night to take advantage of stable snow conditions. Since sunlight affects the stability of snow, travel should be concentrated in shaded areas where footing should remain stable.

(2) In areas covered by early seasonal snowfall, travel between deep snow and clear ground must be done cautiously. Snow on slopes tends to slip away from rocks on the downhill side, forming openings. These openings, called moats, are filled by subsequent snowfalls. During the snow season, moats below large rocks or cliffs may become extremely wide and deep, presenting a hazard to the rescue team.

b. Travel Speed. An overzealous drive to reach an objective may be too fast for the endurance of the team. Fast starts at the point of insertion usually result in frequent stops for recuperation. The best way to reach an objective is to start with a steady pace and continue that pace throughout. Movement at reasonable speeds, with rest stops as required, will help prevent team "burnout." The following considerations further ensure steady advancement with minimal degradation to the team. A steady pace helps maintain an even rate of breathing. After the initial period of travel, that is, one-half hour, a shakedown rest should be initiated to adjust boots, snowshoes, crampons, packs, etc., or to remove or add layers of clothing.

c. Snowshoe Technique. A striding technique is used for movement with snowshoes. In taking a stride, the toe of the snowshoe is lifted upward to clear the snow, and thrust forward. Energy is conserved by lifting the snowshoe no higher than is necessary to clear the snow. If the front of the snowshoe catches, the foot is pulled back to free it and then lifted before proceeding with the stride. The best and least exertive method of travel is a loose-kneed, rocking gait in a normal rhythmic stride. Care should be taken not to step on or catch the other snowshoe.

(1) On gentle slopes, ascent is made by climbing straight upward. (Traction is generally very poor on hard-packed or crusty snow.) Steeper terrain is ascended by traversing and packing a trail similar to a shelf. When climbing, the snowshoe is placed horizontally in the snow. On hard snow, the snowshoe is placed flat on the surface with the toe of the upper one diagonally uphill to gain more traction. If the snow will support the weight of a person, it is better to remove the snowshoes and temporarily proceed on foot. In turning, the best method is to swing the leg up and turn the snowshoe in the new direction of travel.

(2) Obstacles such as logs, tree stumps, ditches, and small streams should be stepped over. Care must be taken not to place too much strain on the snowshoe ends by bridging a gap, since the frame may break. In shallow snow, there is danger of catching and tearing the webbing on tree stumps or snags. Wet snow will frequently ball up under the feet, making walking uncomfortable. This snow should be knocked off with a stick or pole.

(3) Generally, ski poles are not used in snowshoeing; however, one or two poles are desirable when carrying heavy loads, especially in mountainous terrain. The bindings must not be fastened too tightly or circulation will be impaired and frostbite can occur. During stops, bindings should be checked for fit and possible readjustment.

d. Uphill Travel. Maximum altitude may be obtained with less effort by traversing a slope. A zigzag or switchback route used to traverse steep slopes places body weight over the entire foot as opposed to the balls of the feet as in a straight-line uphill climb. An additional advantage to zigzagging or switchbacking is alternating the stress and strain placed on the feet, ankles, legs, and arms when a change in direction is made.

(1) When a change in direction is made, the body is temporarily out of balance. The proper method for turning on the steep slope is to pivot on the outside foot (the one away from the slope). With the upper slope on the right side, the left foot (pivot foot) is kicked directly into the slope. The body weight is transferred on to the left foot while pivoting toward the slope. The slope is then positioned on the left side and the right foot is on the outside.

(2) In soft snow on steep slopes, pit steps must be stamped in for solid footing. On hard snow, the surface is solid but slippery, and level pit steps must be made. In both cases, the steps are made by swinging the entire leg in toward the slope, not by merely pushing the boot into the snow. In hard snow, when one or two blows do not suffice, crampons should be used. Space steps evenly and close together to facilitate ease of travel and balance. Additionally, the lead climber must consider the other team members, especially those who have a shorter stride.

(3) The team should travel in single file when ascending, permitting the leader to establish the route. The physical exertion of the climbing leader is greater than that of any other team member. The climbing leader must remain alert to safeguard other team members, while choosing the best route of travel. The lead function should be changed frequently to prevent exhaustion of any one individual. Team members following the leader should use the same leg-swing technique to establish foot positions, improving each step as they climb. Each foot must be firmly kicked into place, securely positioning the boot in the step. In compact

snow, the kick should be somewhat low, shaving off snow during each step, thus enlarging the hole by deepening. In very soft snow, it is usually easier to bring the boot down from above, dragging a layer of snow into the step to strengthen and decrease the depth of it.

(4) When it is necessary to traverse a slope without an increase in elevation, the heels rather than the toes form the step. During the stride the climber twists the leading leg so the boot heel strikes the slope first, carrying most of the weight into the step. The toe is pointed up and out. Similar to the plunge step, the heel makes the platform secure by compacting the snow more effectively than the toe.

e. Descending. The route down a slope may be different from the route up a slope. Route variations may be required for descending different sides of a mountain or for moving just a few feet from icy shadows on to sun-softened slopes. A good surface-snow condition is ideal for descending rapidly since it yields comfortably underfoot. The primary techniques for descending snow-covered slopes are plunge stepping and descending step by step.

(1) The plunge step makes extensive use of the heels of the feet (figure 22-55) and is applicable on scree as well as snow. Ideally, the plunging route should be at an angle, one that is within the capabilities of the team and affords a safe descent. The angle at which the heel should enter the surface varies with the surface hardness. On soft-snow slopes, almost any angle suffices; however, if the person leans too far forward, there is a risk of lodging the foot in a rut and inflicting injuries. On hard snow, the heel will not penetrate the surface unless it has sufficient

force behind it. Failure to firmly drive the heel into the snow can cause a slip and subsequent slide. The quickest way to check a slip is to shift the weight on to the other heel, making several short, stiff-legged stomps. This technique is not intended to replace "the ice arrest" technique, which is usually more effective. When roped, plunging requires coordination and awareness of all team members' progress. Speed of the team must be limited to the slowest member. Plunging is unsatisfactory when wearing crampons due to the snow compacting and sticking to them.

(2) The technique of step-by-step descending is used when the terrain is extremely steep, snow significantly deep, or circumstances dictate a slower pace. On near-vertical walls, it is necessary to face the slope and cautiously lower oneself step by step, thrusting the toe of the boot into the snow while maintaining an anchor or handhold with the axe. Once the new foothold withstands the body's full weight, the technique is repeated. On moderately angled terrain, the team can face away from the slope and descend by step-kicking with the heels.

22-13. Snow and Ice Climbing Procedures and Techniques. Snow and ice climbing differs from rock climbing, yet many of the procedures and techniques are the same. Belay tie-in commands, principles of runner placement, straight-line climbing, and placement of protection are common to snow and ice as well as rock. As expected, however, there are major differences from rock climbing.

a. Ice Axe Techniques (figure 22-56). The axe is the most important tool a climber carries. It can be used for braking assistance when a climber begins sliding down a steep, snow-covered incline.

(1) Each rescuer (climber) must practice the self-arrest technique before venturing on to steep grades. Since the ice axe arrest requires the use of the ice axe, the climber must hang on to it at all times. The ice axe, whether sharp or not, is a lethal weapon when flying about on the attached cord. Physically, the climber rigs for arrest by rolling down shirtsleeves, putting on mittens,

Figure 22-55. Plunge Step.

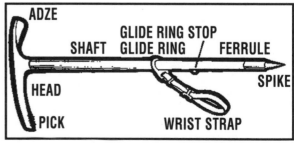

Figure 22-56. Ice Axe.

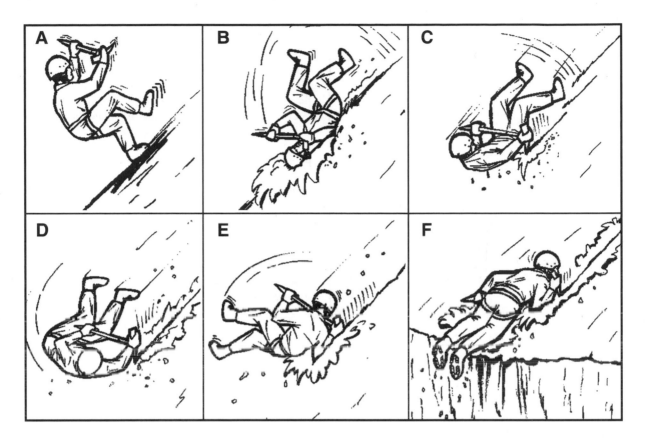

Figure 22-57. Ice Axe Self-Arrest.

securing loose gear, and most important of all, making certain the axe is held correctly. Mentally prepare by recognizing the importance of instantaneous application. A quick arrest, before the fall picks up speed, has a better chance of success than a slow arrest. Preparation for an ice-axe arrest should be taken when traveling on terrain that could result in a fall.

(2) The proper method of holding an ice axe for self-arrest (see figure 22-57) is to place one hand on the head of the axe with the thumb under the adze and fingers over the pick. The other hand is placed on the shaft next to the spike. The pick is pressed into the slope just above the shoulder so the adze is near the angle formed by the neck and shoulders. The shaft should cross the chest diagonally with the spike held firmly close to the opposite hip. A short axe is held in the same position, although the spike will not reach the opposite hip. Chest and shoulders should press strongly on the shaft, and the spine should be arched slightly to distribute weight primarily to the shoulders and toes. The legs should be stiff and spread apart, toes digging in (if wearing crampons, keep the toes off the surface until almost stopped)—and hang on to the axe!

b. Team Arrest. The team arrest is intermediate between self-arrest and belays. When there is doubt that a person could arrest a fall, such as on crevassed glaciers and steep snowfields, and conditions are not so extreme as to make belaying necessary, the party ropes up and travels in unison. If any member falls, arrest is made by two or three axes. The rope between the climbers must be fully extended, except for minimum slack carried by the second and subsequent persons to allow them to flip the rope out of the track (steps). This also allows easy compensation for pace variations. However, slack is minimized to bring the second and subsequent axes into action at the moment of need. A roped climber who falls should immediately yell "FALLING." It is not advisable to delay the alarm to see how "self-arrest" will develop because team members may hear the signal only after they have been pulled into their own falls, decreasing their ability to help. When roped climber(s) hear the cry "falling," they immediately drop into self-arrest position.

c. Boot Axe Belay (figure 22-58). The boot axe belay can be set up rapidly and is used when a team is moving together, and belaying is only required at a few spots. The boot axe belay should be practiced until a sweep

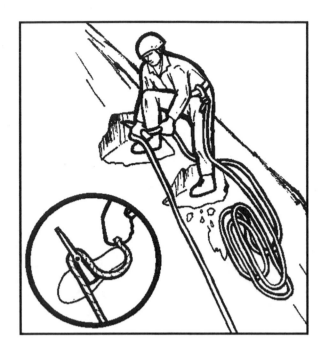

Figure 22-58. Boot Axe Belay.

and jab of the ice axe can set up the stance within a couple of seconds. The axe provides an anchor to the slope, and the slope and the boot braces the axe. Both give a friction surface over which the run of rope is controlled.

(1) To prepare a boot axe belay, a firm platform, large enough for the axe and uphill boot, is stamped out in the snow. The ice-axe shaft is jammed as deeply as possible, at a slight uphill angle (against the anticipated fall) into the snow at the rear of the platform. The pick is parallel to the fall line, pointing uphill, thus applying the strongest dimension of the shaft against the force of a fall. The length of the pick prevents the rope from escaping over the top of the shaft.

(2) The belayer stands below the axe, facing at a right angle to the fall line. The uphill boot is stamped into the slope against the downhill side of the shaft at a right angle to the fall line, bracing the shaft against downhill pull. The downhill boot is in a firmly compacted step below the uphill boot so the leg is straight, stiffly bracing the belayer. The uphill hand is on the axe head in arrest grasp, bracing the shaft against downhill and lateral stress. From below, the rope crosses the toe of the boot, preventing the rope from trenching into the snow. The rope bends around the uphill side of the shaft, then down across the instep of the bracing boot, and is controlled by the downhill hand. To apply braking through greater friction, the downhill, or braking, hand brings the rope uphill around the heel, forming an "S" bend.

d. Crampon Techniques:

(1) Donning. When attaching the harness, the buckles should be positioned to cinch on the outward sides of the boots. Special care must be taken to strap the crampons tightly to the boots, running the strap through each attachment prong or ring. If crampons do not have heel loops, ankle straps should be long enough to be crossed behind the boot before being secured, to prevent boots from sliding backward out of the crampons. Many crampons have been lost because this precaution was not taken. When trimming new straps, allowance must be made for gaiters, which sometimes cover the instep of the boot. Donning is best done by laying each crampon on the snow or ice with all rings and straps outward, then place the boot on the crampon and tighten the straps. Even modern neoprene-coated nylon straps should be checked from time to time to make sure they are tight, have not been cut, and are not trailing loop-strap ends, which could cause the wearer to trip.

(a) If it is believed crampons may be needed, they must be carried. Conditions change rapidly; an east-facing slope may be mushy enough for step-kicking during the morning, but can become a sheet of smooth white ice in the afternoon shade. Furthermore, cramponing may contribute directly to the team's safety by enabling it to negotiate stretches of ice faster and with less fatigue than having to chop steps. The decision of whether or not to wear crampons is determine by the situation. Wearing crampons should not be considered mandatory because of venturing on to a glacier; neither should a team attempt to save time by never wearing crampons on steep, exposed icy patches just because they are fairly short. Another important guideline is to don crampons before they are needed to avoid donning them while teetering in ice steps. On mixed rock-and-ice climbs, constant donning and removing of crampons takes so much time that the objective may be lost.

(b) Crampons should be worn throughout the entire climb if the terrain is 50 percent or more suitable for crampons (crampons may skid or be broken on rock surfaces). Crampons are not required if the snow or ice patches are fairly short, good belays are available, and rock predominates. These alternatives are suggestions and the team leader's decision must be based on the conditions at hand.

(c) Crampons should be taken off when the snow begins to ball up badly in them and no improvement in snow conditions is anticipated. On the ascent, it may be possible to clear away the soft surface snow and climb on the ice below, but this is usually impractical and futile on the descent. Occasionally the climber should kick the crampons free of accumulated snow. The timeworn practice of striking the ice-axe shaft against the crampons to knock out the snow is effective but hard on the axe and perhaps the ankle. In situations in which the crampons must be worn even though the snow balls up in them, shuffling the feet through the snow instead of stepping over the surface tends to force the accumulated snow through the back points. The normal kicking motion of the foot generally keeps the crampons snow-free on the ascent and while traversing.

(d) On the descent, drive the toe of the boot under the surface of the snow ahead of the heel, walking on the ball of the foot. Keep the weight well forward and use short, skating steps, allowing the foot to slide forward and penetrate the harder sublayers.

(2) Flat footing. Flat footing involves a logical and natural progression of coordinated body and ice-axe positions to allow the climber to move steadily and in balance while keeping all vertical points of the crampons biting into the ice. The weight is carried directly over the feet, the crampon points stamped firmly into the ice with each step, with the ankles and knees flexed to allow boot soles to remain parallel to the slope.

(a) On gentle slopes, the climber walks straight up the hill. Normally, the feet are naturally flat to the slope and the axe is used as a cane. If pointing the toes uphill becomes awkward, they may be turned outward in duck fashion. As the slope steepens, the body is turned to face across the slope rather than up it. The feet may also point across the slope, but additional flexibility and greater security are gained by pointing the lower foot downhill. The axe is used only to maintain balance and may be carried in the cane position or the arrest grasp with either the pick or point touching the slope. (Movement is diagonal rather than straight upward, and the climber takes advantage of terrain irregularities and graded slopes. Changes in direction are done as in step-kicking on snow, by planting the downhill foot, turning the body toward the slope to face the opposite direction, and stepping off with the new downhill foot.)

(b) On gentler slopes, the flat-footed approach is used throughout, but it is more secure and easier on steeper slopes to initiate the turn by kicking the front points and briefly front-pointing through the turn. At some point in the turn, the grip on the axe must be reversed. The exact moment for this depends on the climber and the specific situation. However, the climber's stance must be secure when the third point of support— the axe—is temporarily relinquished.

(c) On steep slopes, which approach the limit of practical use of this style of ascent, the climber relies on the axe for security of a hold as well as for balance. The axe is held in the arrest grasp with one hand on the head and the other on the shaft, above the point. The well-sharpened pick is planted firmly in the ice at about shoulder height to provide one point of suspension while a foot moves forward and the crampons are stamped in (figure 22-59).

(d) Descent follows the same general progression of foot and axe positions; descend the fall line, gradually turning the toes out as the slope gets steeper. As the slope steepens, widen the stance, flex the knees, and lean forward to keep weight over the feet, and finally, face sideways and descend with the support of the axe in the arrest position. On very steep or hard ice, it may be necessary to face the slope and front-point downward. When flat-footing downhill, all crampon points should be stamped firmly into the ice. It may be necessary to strive to take small steps, which allow the climber to maintain balance during moves; long steps require major weight shifts to adjust balance.

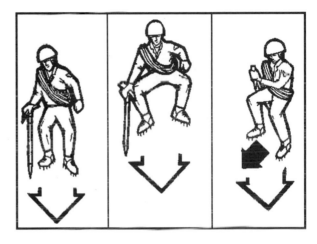

Figure 22-59. Descent with Crampons.

e. Anchors. Snow and ice conditions require the use of special devices for establishing belay anchors or placement of intermediate protection during a climb.

(1) Snow Pickets. Three- to four-foot lengths of aluminum "T" or tubular sections perform as long pitons and are suited for belaying. They must always be used in pairs or greater numbers, one anchoring the other (figure 22-60).

(2) Snow Fluke. A 12-inch piece of metal is buried with a runner coming to the surface; this can pull out if snow conditions are not just right. Better resistance to pull-out is gained with a large flat piece of metal driven into the snow surface at an angle, acting in the same manner as the fluke of an old-fashioned anchor. There is no danger of the runner being cut or weakened from wet conditions with the attachment of a wire cable. The softer the snow, the larger the size of the plate. When using flukes, it is very important that the proper angle with the surface is maintained; otherwise wire, instead of becoming stronger (going deeper) when pulled, will become weaker (surface). Additionally, the cable may

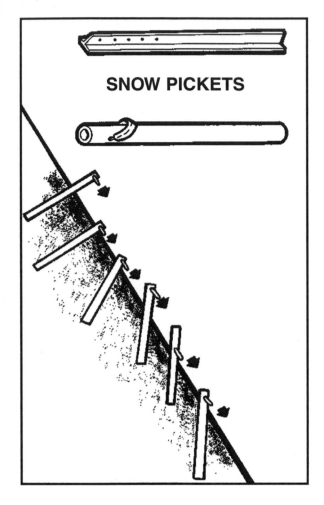

Figure 22-60. Snow Pickets.

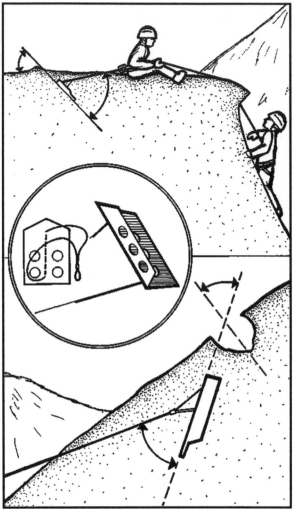

Figure 22-61. Snow Flukes.

act as a lever arm on hard snow, causing the fluke to pop out. This can be prevented by carefully cutting a channel in the snow for cables so the pull comes directly at the plate. If attention is paid to placement, snow flukes will provide great security as belay and rappel anchors. A properly placed fluke is secure for a sitting or standing hop belay on snow (figure 22-61).

    (3) Ice Screws:

        (a) Tubular screws are very strong and are the most reliable (figure 22-62). They are difficult to place in hard or water ice since they tend to clog and have a large cross section. Their main advantage is that they minimize "spalling" (a crater-like splintering of the ice around the shaft of the screw) by allowing the displaced ice to work itself out through the core of the screw. If the core of the ice remaining in the screw is frozen in place, it jams the screw in subsequent placements. The ice may be removed by pushing with a length of wire or by heating with a cigarette lighter. This type of screw

requires both hands for placement; however, once it is started, the pick of an ice hammer or axe inserted in the eye allows the climber to gain the advantage of leverage. Removal is easy and melt-out is slow due to the large cross section.

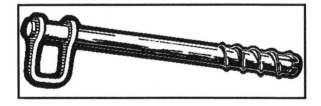

Figure 22-62. Tubular Ice Screw.

(b) Heavier "coathanger" type screws can be relied on to stop a fall (figure 22-63). They are easier to start in hard ice than tubular screws and can often be placed with one hand, although it may be necessary to tap them while twisting as they are started. Their holding power is less than tubular screws, as they tend to fracture hard ice and, under heavy loads, tend to shear through the ice because of their small cross section.

Figure 22-64. Solid Ice Screw.

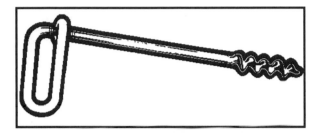

Figure 22-63. Coathanger Ice Screw.

(c) Developed as an attempt to make an easy-to-place and easy-to-remove screw, the solid screws are driven in like a piton and screwed out (figure 22-64). They offer excellent protection in water ice but are less effective in other ice forms. Melt-out is sometimes rapid because of limited thread displacement and, under load, they tend to shear through the ice, as do coathanger screws.

(d) Before placement of ice screws or pitons, any soft snow or loose ice should be scraped or chopped away until a hard and trustworthy surface is reached. A small starting hole punched out with the pick or spike of the axe or hammer facilitates a good grip for the starting threads, or teeth. The screw is pressed firmly into the ice and twisted in at the same time, angled slightly uphill against the anticipated direction of pull. Ice pitons are, of course, driven straight in, but must also be angled against the pull that would result from a fall (figure 22-65). If any spalling or splintering of the ice occurs, the screw should be removed and another placement tried 1 or 2 feet away. Some glacier ice will spall near the surface, but by continuing to place the screw and gently chopping out the shattered ice, a deep, safe placement may be obtained. As a general rule, short screws or pitons should be used in hard ice and long ones in softer ice. They should always be placed in the ice until the eye is flush with the surface. When removing ice hardware, take care not to bend it since this diminishes its effectiveness in future use.

(4) Ice/Snow Bollard. Although not a natural anchor in itself, an ice or snow bollard is easily made from natural materials. A semicircular trench is dug in the snow or ice. The trench should be 3 to 4 feet

across and 6 to 12 inches deep. Allow a larger size for poor snow or ice conditions. The rope can be positioned in the trench to provide a downward belay (figure 22-66).

f. Glissading. Glissading is a means of rapidly descending a slope. Consisting of two basic positions, glissading offers a speedy means of travel, with less energy exerted than using the descending step-by-step or plunging techniques.

(1) When snow conditions permit, the sitting glissade position is the easiest way to descend. The climber simply sits in the snow and slides down the slope while holding the axe in an arrest position (figure 22-67). Any tendency of the body to pivot head downward may be checked by running the spike of the axe rudder-like along the surface of the snow. Speed is increased by lying on the back to spread the body weight over a greater area and by lifting the feet in the air. Sitting back up and returning the feet to the snow surface reduces speed. On crusted or firmly consolidated snow, sit fairly erect with the heels drawn up against the buttocks and the boot soles skimming along the surface. Turns are nearly impossible in a sitting glissade; however, the spike, dragged as a rudder and assisted by body contortions, can effect a change in direction of several degrees. Obstructions on the slope are best avoided by rising into a standing glissade (figure 22-67) for the turn, and then returning to the sitting position. Speed is decreased by dragging the spike and increasing pressure on it. After the momentum has been checked by the spike, the heels are dug in for the final halt but not while sliding at a fast rate, as the result is likely to be a somersault. Emergency stops at high speeds are made by arresting.

(2) The standing glissade is similar to skiing. Positioned in a semi-crouch stance with the knees bent as if sitting in a chair (figure 22-67), the legs are spread laterally for stability, and one foot is advanced slightly to anticipate bumps and ruts. For additional stability, the spike of the axe can be skimmed along the surface, the shaft held alongside the knee in the arrest grasp, with the pick pointing down or to the outside away from the body. Stability is increased by widening the spread of

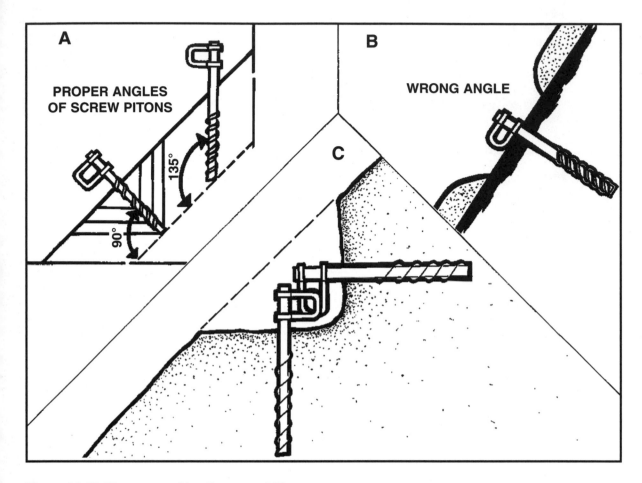

Figure 22-65. Placement of Ice Screw and Piton.

the legs, deepening the crouch, and putting more weight on the spike. A decrease in speed increases muscular strain and the technique becomes awkward and trying, although safe. Speed is increased by bringing the feet close together, reducing weight on the spike, and leaning forward until the boot soles are running flat along the surface like short skis. If the slide is too shallow, a long skating stride helps.

(3) A glissade should be made only when there is a safe runout. Unless a view of the entire run can be obtained beforehand, the first person down the run must use extreme caution, stopping frequently to study the terrain ahead. Equipment must be adjusted before beginning the descent. Crampons and other hardware must be properly stowed. Never attempt to glissade while wearing crampons, as it is extremely easy to snag a crampon and be thrown down the slope. Mittens or gloves are worn to protect the hands and to maintain control of the axe. Heavy waterproof pants provide protection to the buttocks. Gaiters are also helpful for all

glissading. Glissades should never be attempted in terrain where the axe safety cord is required. The hazards

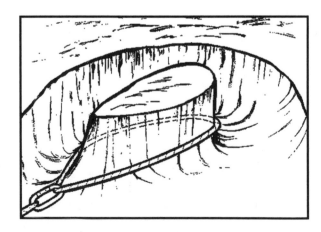

Figure 22-66. Ice/Snow Bollard.

Figure 22-67. Glissading.

of a flailing ice axe should never be risked during a glissade.

22-14. Glaciers and Glacial Travel:

a. Features. To cope with the problems that can arise in using glaciers as avenues of travel, it is important to understand something of the nature and composition of glaciers.

(1) A valley glacier is essentially a river of ice and it flows at a rate of speed that depends largely on its mass and the slope of its bed. A glacier consists of two parts:

(a) The lower glacier, which has an ice surface void of snow during the summer.

(b) The upper glacier, where the ice is covered even in summer with layers of accumulated snow that changes into glacier ice.

(2) To these two integral parts of a glacier may be added two others which, although not a part of the glacier proper, are generally adjacent to it and are of similar composition. These adjacent features—the ice and snow slopes—are immobile since they are anchored to underlying rock slopes. A large crevasse separates such slopes

from the glacier proper and defines the boundary between moving and anchored ice.

(3) Ice is plastic-like near the surface but not smooth enough to prevent cracking as the ice moves forward over irregularities in its bed. Fractures in a glacier surface, called crevasses, vary in width and depth from only a few inches to many feet. Crevasses form at right angles to the direction of greatest tension, and due to a limited area tension is usually in the same direction. Crevasses in any given area tend to be roughly parallel to each other. Generally, crevasses develop across a slope. Therefore, when traveling up the middle of a glacier, people usually encounter only transverse crevasses (crossing at right angles to the main direction of the glacier). Near the margins or edges of a glacier, the ice moves more slowly than it does in midstream. This speed differential causes the formation of crevasses diagonally upstream and away from the margins or sides. While crevasses are almost certain to be encountered along the margins of a glacier and in areas where a steepening in gradient occurs, the gentlest slopes may also contain crevasses.

(4) An icefall forms where an abrupt steeping of slope occurs in the course of a glacier. These stresses are set up in many directions. As a result, the icefall consists of a varied mass of iceblocks and troughs with no well-defined trend to the many crevasses.

(5) As a glacier moves forward, debris from the valley slopes on either side is deposited on its surface. Shrinkage of the glacier from subsequent melting causes this debris to be deposited along the receding margins of the glacier. Such ridges are called lateral (side) moraines. Where two glaciers join and flow as a single river of ice, the debris on the adjoining lateral margins of the glaciers also unites and flows with the major ice stream, forming a medial (middle) moraine. (By examining the lower part of a glacier, it is often possible to tell how many tributaries have joined to form the lower trunk of the glacier.) Terminal (end) moraine is usually found where the frontage of the glacier has pushed forward as far as it can go, that is, to the point at which the rate of melting equals the speed of advance of the ice mass. This moraine may be formed of debris pushed forward by the advancing edge, or it may be formed by a combination of this and other processes.

(a) Lateral and medial moraines may provide excellent avenues of travel. When the glacier is heavily crevassed, moraines may be the only practical routes. Ease of progress along moraines depends on the stability of the debris composition. If the material consists of small rocks, pebbles, and earth, the moraine is usually loose and unstable and the crest may break away at each footstep. If large blocks compose the moraine, they have probably settled into a compact mass and progress may be easy.

(b) In moraine travel, it is best either to proceed along the crest or, in the case of lateral moraines, to follow the trough that separates it from the mountainside. Since the slopes of moraines are usually unstable, there is a great risk of spraining an ankle on them. Medial moraines are usually less pronounced than lateral moraines because a large part of their material is transported within the ice. Travel on them is usually easy, but should not be relied on as routes for long distances since they may disappear beneath the glacier surface. Only rarely is it necessary for a party traveling along or across moraines to be roped together (figure 22-68).

(6) Glacial rivers are varied in type and present numerous problems to those who must cross or navigate them. Wherever mountains and highlands exist in the arctic regions, melting snows produce concentrations of water pouring downward in a series of falls and swift chutes. Rivers flowing from icecaps, hanging piedmonts (lake-like), or serpentine (winding or valley) glaciers are all notoriously treacherous. Northern glaciers may be vast in size and the heat of the summer sun can release vast quantities of water from them. Glacier ice is extremely unpredictable. An ice field may look innocent from above, but countless subglacial streams and water reservoirs may be under its smooth surface. These reservoirs are either draining or temporarily blocked. Mile-long lakes may lie under the upper snowfield, waiting only for a slight movement in the glacier to liberate them and send their waters into the valleys below. Because of variations in the amounts of water released by the sun's heat, all glacial rivers fluctuate in water level. The peak of the floodwater usually occurs in the afternoon as a result of the noonday heat of the sun on the ice. For some time after the peak has passed, rivers that drain glaciers may not be fordable or even navigable. However, by midnight or the following morning, the water may recede so fording is both safe and easy. When following a glacial river broken up into many shifting channels, choose routes next to the bank rather than taking a chance on getting caught between two dangerous channels.

(7) Glaciers from which torrents of water descend are called flooding glaciers. Two basic causes of such glaciers are the violent release of water that the glacier carried on its surface as lakes, or the violent release of large lakes that have been dammed up in tributary glaciers because of the blocking of the tributary valley by the main glacier. This release is caused by a crevasse or a break in the moving glacial dam, the water then roars down in an all-enveloping flood. Flooding glaciers can be recognized from above by the flood-swept character of the lower valleys. The influence of such glaciers is sometimes felt for many miles below.

Prospectors have lost their life while rafting otherwise safe rivers because a sudden flood entered by a side tributary and descended as a wall of white, rushing water.

(8) On those portions of a glacier where melting occurs, runoff water cuts deep channels in the ice surface and forms surface streams. Many such channels exceed 20 feet in depth and width. They usually have smooth sides and undercut banks. Many of these streams terminate at the margins of the glacier, where in summer they contribute to the torrent that constantly flows between the ice and the lateral moraine. Size increases greatly as the heat of the day moves to an end. The greatest caution must be taken in crossing a glacial surface stream since the bed and undercut banks are usually hard, smooth ice that offers no secure footing.

(9) Some streams disappear into crevasses or into round holes known as glacial mills, and then flow as subglacial streams. Glacial mills are cut into the ice by the churning action of water. They vary in diameter. Glacial mills differ from crevasses not only in shape but also in origin, since they do not develop as a result of moving ice. In places, the depth of a glacial mill may equal the thickness of the glacier.

b. Glacier Operations. The principal dangers and obstacles to operations in glacier areas are crevasses and icefalls. Hidden crevasses present unique problems and situations since their presence is often difficult to detect. When one is detected, often it is due to a team member having fallen through the unstable surface cover. The following techniques and procedures should be followed when performing glacier operations.

(1) Equipment Preparation. The prevention of hypothermia should be of primary importance when performing glacier operations. Sufficient protective clothing must be worn or carried to cover all climatic temperature variations. Climbers trapped in crevasses have died of hypothermia while their team members, helpless to assist from their position on the glacier surface above, were sweltering in the sunshine. Backpacks should be equipped with a lanyard consisting of a 6-foot piece of line with a figure-eight and nonlocking carabiner at one end. The free end of the lanyard is attached to the pack and the unlocking carabiner is snapped into the buttock strap of the seat harness. If the climber falls into a crevasse and is suspended upside down by the weight of the pack, the pack can be released with the lanyard and the person can return to an upright position.

(2) Team Composition. The first law of glacial travel is to rope-in during travel. The principal consideration is to avoid crevasses. When stepping on to a known glacier or on to a snowfield of unknown stability,

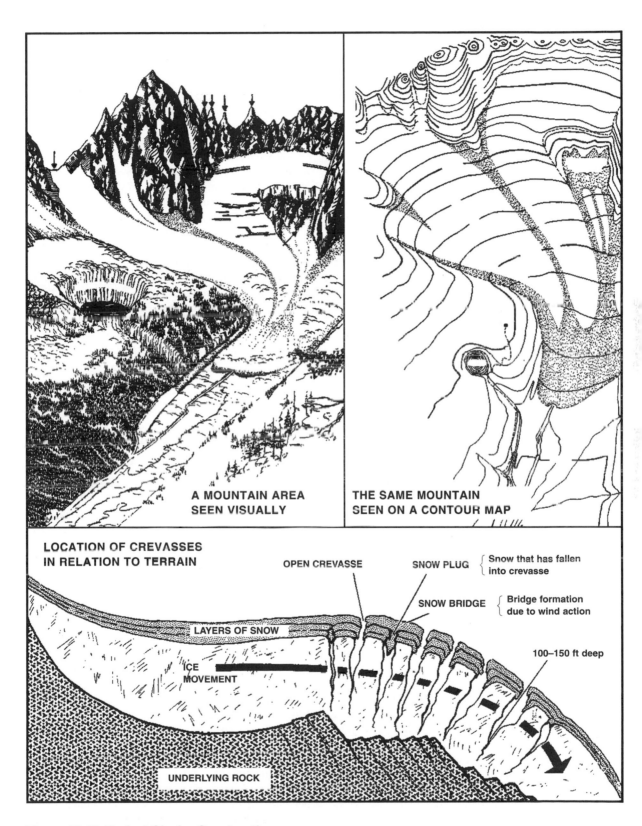

A MOUNTAIN AREA
SEEN VISUALLY

THE SAME MOUNTAIN
SEEN ON A CONTOUR MAP

LOCATION OF CREVASSES
IN RELATION TO TERRAIN

OPEN CREVASSE

SNOW PLUG { Snow that has fallen into crevasse

SNOW BRIDGE { Bridge formation due to wind action

LAYERS OF SNOW

ICE MOVEMENT

100—150 ft deep

UNDERLYING ROCK

Figure 22-68. Typical Glacier Construction.

whether crevasses are visible or not, the law of roping-in remains. The only variable to this law is when avalanches present a greater hazard than the threat of crevasses. The most experienced climber in glacial travel should be the lead climber; however, if crevasses are completely masked, the lightest climber may lead. During moderate climbs, three climbers tied in to a 165-foot rope is ideal. During severe climbs requiring belay, a 120-foot rope with only two climbers is recommended. If a two-person climbing team falls, the team must be arrested by a single axe. If a three-person climbing team is roped in, the rope is usually so shortened that if one climber falls, the others are often dragged in before they have time to react.

(3) Roping-In. Climbers are roped together by constructing figure-eight knots at the ends and middle of the rope. The rope is attached by passing a locking carabiner through the figure eight and the crotch strap of the seat harness. Associated climbing equipment such as ice axes, slings, and packs are donned. When completely roped in and prepared for travel, there should not be less than 50 feet of rope between each of the climbers. The more rope between the climbers, the better the chance for a successful arrest.

c. Glacier Travel. Due to the difficulty of crevasse rescue, two or more rope teams are recommended for glacier travel since a single team is sometimes pinned down in the arrest position, and members are unable to free themselves to begin rescue. Rope teams must travel close together to lend assistance to one another, however, not so close as to fall into the same crevasse. During extended periods on a glacier, skis and snowshoes are often of great value. This footgear will distribute the weight more widely than boots alone and place less strain on snow bridges. Neither skis nor snowshoes are substitutes for the rope, but may be used for easy travel.

(1) Operations in the mountains have certain limitations imposed by nature in glacial movement. Access to the end portion of a glacier may be difficult due to abruptness of the ice and possible presence of crevasses. Additional obstacles of mounting a glacier may be swift glacial streams or abrupt mountain terrain bordering the glacier ice. The same obstacles may also have to be negotiated when dismounting or mounting a valley glacier at any place along its course. Further considerations to movement on a glacier are steep sections, heavily crevassed regions, and icefalls. The use of up-to-date aerial photographs, when available, with aerial reconnaissance is a valuable means of gathering advance information about a particular glacier. The photos, however, only supplement and do not negate the advantages of surface reconnaissance conducted from available vantage points.

(2) Trail wands are used to mark the route and crevasses. The wands, especially essential to safety during periods of adverse weather, are placed every 150 feet along the route and can be used during day or night. A climbing team should not cluster close together during rest stops. If areas of safety cannot be found, the rope must be kept extended during rests just as during travel. A party establishing camp on a snow-covered glacier similarly remains roped-in for as long a period as required to safely inspect the area by stomping and probing the surface thoroughly before placing trust in the site. Hidden crevasses should always be assumed to exist in the area.

(3) Normally a team will travel in single file, stepping in the leader's footsteps or in echelon formation (figure 22-69). If a crevasse pinches out, an end run must be made (figure 22-70), even if it involves traveling half a mile to gain a few dozen feet of forward progress. The time taken to walk around is generally much less than in forcing a direct crossing. Important to remember in an end run is the possible hidden extension of a visible crevasse. A frequent error is aiming at the visible end. Unless the true or subsurface end is clearly visible during the approach, it is best to make a wide swing around the end.

(a) In late summer, the visible end is often the true end due to surface snow and ice having melted. When end runs are impractical because of the distance involved or because the end of one crevasse is adjacent to another, snow bridges may provide a crossing point. One kind consists of remnant snow cover sagging over an inner open space. Another kind, with a foundation that extends downward into the body of the glacier, is less a bridge than a solid area between two crevasses.

(b) Any bridge should be closely and completely examined before use. If overhanging snow obscures the bridge, the lead climber must explore at closer range by probing the depth and smashing at the sides while walking delicately, ready for an arrest or sudden drop. The second climber establishes a belay (figure 22-71) anchored by the third climber, who is also prepared to initiate rescue if the leader falls. An excessively narrow or weak bridge may be crossed by straddling or even slithering on the stomach, thereby lowering the center of gravity and distributing the weight over a broader area. When there is doubt about the integrity of a bridge, but it is the only possible route, the lightest climber in the team should be the first across, with the following climbers walking with light steps and taking care to step exactly in the same tracks.

(c) Bridges vary in strength with changes in temperatures. In the cold of winter or early morning, the thinnest and most fragile of bridges may have incredible structural strength. However, when the ice crystals melt

Figure 22-69. Echelon Formation.

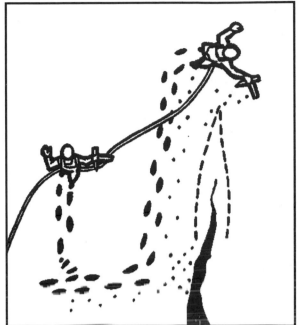

Figure 22-70. End Run.

in the afternoon heat, even the largest bridge may suddenly collapse. Each bridge must be tested with care, being neither abandoned nor trusted until its worth is determined (figure 22-71).

(d) Narrow cracks in a bridge can be stepped across, but wider crevasses require jumping. If the jump is so long that a run is required, the approach should be carefully packed. A running jump (figure 22-72) can carry the climber farther than a standing jump, although running jumps are often not practical. Most jumps are made with only two or three lead-up steps. In any case, care must be taken to locate the precise edge of the crevasse before any attempt is made to jump. Encumbering clothing and equipment must be removed before the jump, although the jumper must bear in mind the low temperature that often exists within crevasses.

d. Crevasse Rescue. Each climber must be able to effect a crevasse rescue if a team member falls into a crevasse. When a climber falls, the remaining team members must drop into a self-arrest position and stabilize their position. All climbers should never be dragged into the crevasse. If a climber falls, the remaining team members must support the weight until one of them can establish a reliable anchor point or until the second team arrives to help. If the fallen climber is able to assist in the recovery, self-extraction from the crevasse may be performed by using prusiks.

(1) A problem inherent to crevasse rescue is the imbedding of the rope (caused by the fallen climber's weight) in the ice and snow. Unless the rope is buffered with an ice axe during the climb out, it will tend to entrench itself deeper in the ice, eventually creating a deep groove from which it will be extremely difficult to use or retrieve, or it will freeze in place, rendering it useless. Corrective actions are to travel down along the rope, taking care not to drop debris on the climber, and free it from the ice. An additional method is to drop a spare rope down to the climber, who shifts weight off the imbedded rope until it can be freed.

(2) If the climber is using prusiks and the action of the climb seesaws the rope into the tip of the crevasse, it will be extremely difficult to ascend the few remaining feet. A procedure to overcome this situation is for the climber to tie in to the rope near the prusik. The climber then strikes the figure-eight knot from the harness and sends the end to the team above via a retrieving line.

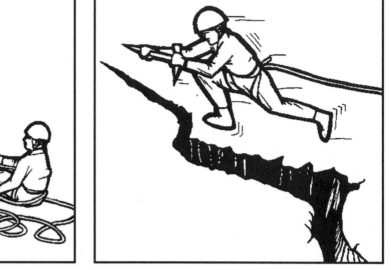

Figure 22-71. Crossing Bridged Crevasse.   **Figure 22-72. Jumping a Crevasse.**

Once firmly anchored in peace, the rope affords a viable route of ascent. Negotiating the final few feet of a crevasse is usually difficult due to the pressure of the rope against the lip or side of the wall. Prusiks tend to compress against, or gouge into, the wall, rendering them nonfunctional. In most cases, the final few feet are overcome by brute strength. If a second rope is available, an alternate method can be used (figure 22-73).

(3) If a fallen climber is unable to help in the recovery, another climber may be required to enter the crevasse. Before the team member is lowered, all assurances must be made that the assistance will enhance the outcome of the operation and not compound it. The rescuer should administer medical treatment as needed, paying special attention to preventing or treating cold-weather injuries, as the interior of the crevasse can become extremely cold. Warm protective clothing must be used if the medical situation does not permit immediate extraction.

22-15. Evacuation Principles and Techniques.
The performance of mountain rescue is not only physically demanding but also mentally challenging. Hard-and-fast procedures to fit all circumstances for mountain rescue cannot be established. The team's ability to innovate will, in most instances, allow the adaptation of the basic principles, techniques, and procedures into a system suitable to effect the rescue. A normal rescue system

will use anchors, belays, and various specialized systems. There are basically two methods of rescue: bringing the victim (patient) up to the rescuer's position, or evacuating the victim down from the position.

a. Safety. The establishment of rescue systems must be thoroughly tested prior to use. One missed step in setting up a rescue system may result in further injury of the victim and (or) injury to the rescuers.

b. Evacuations. Evacuation of a victim from the position in a downhill direction is an easier task than establishing a mechanical leverage for pulling a victim to the top of a hill. The victim's medical condition will dictate the method of evacuation and equipment used. The primary litter used is the Stokes litter. This is a tubular-frame litter with a wire basket. The tubular main bar provides a very strong framework for mountain operations. The patient may be secured in the litter by means of several cross-body straps, or by the interlacing of slings. When an evacuation team arrives on the scene, there are two activities that should take place at the same time: (1) The patient should be treated and prepared for transporting in the Stokes litter, and (2) the anchor and mechanical brake system must be established. Within obvious restraints of time and distance, low-angle evacuations are always preferable to high-angle evacuations. Low-angle work eliminates many hazards and requires a lesser degree of knowledge and skill. Using the Stokes litter eliminates the excessive knot tying and lashing essential with other litters.

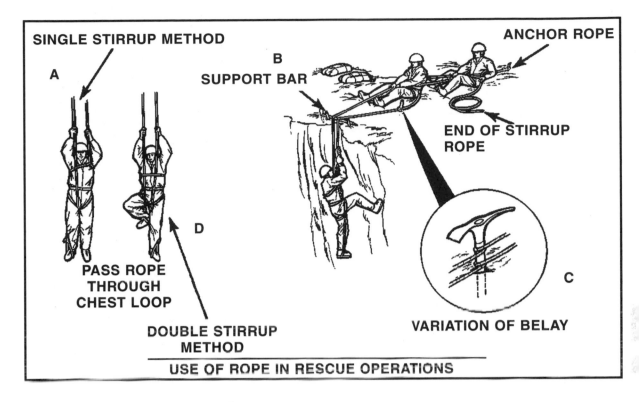

SINGLE STIRRUP METHOD

A

B
SUPPORT BAR

ANCHOR ROPE

END OF STIRRUP ROPE

PASS ROPE THROUGH CHEST LOOP

D

DOUBLE STIRRUP METHOD

C

VARIATION OF BELAY

USE OF ROPE IN RESCUE OPERATIONS

Figure 22-73. Two-Rope Crevasse Rescue.

c. Preparing the Braking System. The following describes the preparation of the brake system for a low-angle litter evacuation. Establish a very sound anchor, which will be in the direction of pull for the first pitch. If a sound anchor is not available, establish an anchor system. If the terrain does not allow the brake operator ample room to safely and effectively perform the required tasks, than an 11 mm (seven-sixteenth inch) rope sling or double-tape sling can be used to adjust for the distance from the anchor to the area where the brake operator will work. The mechanical braking device is securely attached to the anchor or sling. This braking device should either be a figure-eight rappel ring, or a four-carabiner brake system. The rope to be used for the litter descension should be backcoiled. If the rope is to be used for ascending, the rope should lay out along the route of ascension. The head of the litter will be attached to the figure eight at the end of the rope with a steel locking carabiner. If locking carabiners are not available, the rope should be tied directly to the litter using several round turns on the outer rail at the head of the Stokes litter, and tied off with a bowline and a safety knot. The rope is then properly locked into the mechanical braking system.

d. Preparing the Patient for Transport. While the rope and brake systems are being prepared, the litter and patient should also be prepared. The litter must be secured to prevent its loss or further injury to the patient. Additionally, the litter may be padded or insulated (blankets or foam pads) for protection. The ties for securing the feet and pelvis should be attached to the litter. Before evacuating, all emergency medical treatment appropriate to the situation should be performed (splinting fractures, maintaining an open airway, etc.). The patient should be insulated from environmental conditions such as cold, wind, or rain. The person in charge of the patient's medical condition should ensure that the patient's condition is stable enough for transporting. In mountainous terrain, the patient should be protected from further injury due to rockfall by wearing a helmet at all times. A litter team generally consists of four to six people. Fewer than six cannot withstand the fatigue of frequent or long trips while carrying an injured person.

e. Three- or Four-Man Lift. Three bearers take up positions on one side of the victim, one at the shoulder, one at the hip, and one at the knees. If one side is injured, the three bearers should be on the uninjured side. A fourth bearer, if available, takes a position on the opposite side, at the victim's hip.

(1) The bearers should kneel next to the victim. Then, simultaneously, the bearer at the victim's shoulder puts one arm under the victim's head, neck, and

Figure 22-74. Lifting the Patient.

shoulder, and the other under the upper part of the victim's back. Each bearer at the victim's hips places one arm under the victim's back and the other under the victim's thighs. The bearer at the victim's knees places one arm under the victim's knees and the other under the ankles (figure 22-74).

(2) The person at the victim's head gives all the commands. The command "Prepare to lift!" is followed by the command "Lift." Immediately, all the bearers lift simultaneously and place the victim in line on their knees. If the victim needs to be moved any distance to the litter, move as shown in figure 22-75.

(3) The fourth bearer, if available, places a stretcher under the victim and against the toes of the three kneeling bearers. The command "Prepare to lower!" is followed by the command "Lower!" and the victim is gently lowered to the litter. Once properly positioned in the litter, the victim must be secured in a manner to prevent further injury. The victim may be secured to the litter in a variety of ways depending on the evacuation route and the victim's condition.

f. Securing Patient in the Litter:

(1) The tape sling used to secure the feet is tied to the framework of the Stokes litter, which separates the

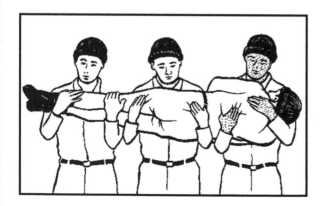

Figure 22-75. Moving the Patient.

legs near the groin area. The tape sling should be tied with a clove hitch in the middle of the tape in a manner to prevent the tape from sliding down to the feet when pulled tight. The feet are secured by running the tape across the legs to the window on the outside of the litter, then across the patient's legs to the feet. An overhand knot is made in each tape, which can be passed over the corresponding foot. When the feet are secured, there should be ample room to apply tension to the head if needed. The tape is then tied at the foot of the litter to a major support bar on the inside of the litter frame. The ties on a Stokes litter should never be made on any outside rail, as they are subject to abrasion. The tape slings should be tied off with a two-round turn and two half hitches. If the two-round turn does not hold tension, then a clove hitch can be used in its place.

(2) The tape used to secure the pelvis should be tied just above the tape used to secure the feet, and secured in the same fashion. Each end of the rope is passed over the leg to the larger upright cross-member of the Stokes litter between the outer rail and inner basket rail. This cross-member corresponds with the side of the hip. The tape is secured with a two-round turn and two half hitches, or a clove hitch and two half hitches. The ends of the tapes are then tied together at the middle of the patient's waistline with a square knot and two half hitches on either side of the knot.

(3) The upper torso is secured by placing the middle of the tape in the center of the patient's chest, and the two ends of the tape are secured to the large upright cross-member. The running ends of the tape are then passed diagonally across the patient to the cross-member that is next to the abdomen. The tape is secured again and the ends are tied at the midline of the body. The head is secured by running a tape sling over the helmet and securing the tape at the corresponding cross-members. The helmet can be used with a tape sling to provide traction; however, it is not a substitute for the neck collar (figure 22-76).

(4) Once the patient and the system are ready for the low-angle evacuation, the entire system must be double-checked. Once the litter and patient are prepared as described, ascent or descent is made through a team of litter bearers and a belay point. The minimum of essential members for a low-angle evacuation are five—a belayer and three litter bearers (figure 22-77).

g. Low-Angle Evacuation Descent. On the descent, the belay rope attached to the head of the litter runs through a mechanical belay brake system. One team member acts as the belayer. This secures the litter and aids in lowering. Another person may assist the belayer with the rope. Litter bearers take their positions on the litter, ideally three on each side. The chief medic and crew boss should be among these to ensure the patient

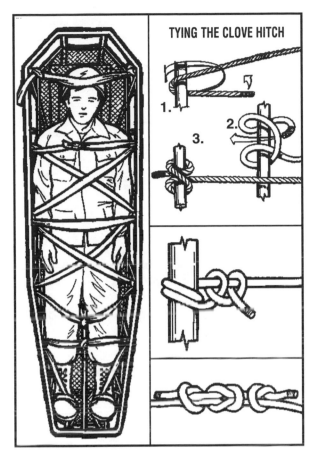

TYING THE CLOVE HITCH

1.

2.

3.

Figure 22-76. Tying the Patient into a Stokes Litter.

Figure 22-77. Carrying the Litter.

is monitored 100 percent of the time and that effective communication exists between the litter bearers and the belayer.

(1) In descending, the most direct practical passage that takes advantage of available trees and rocks for belay points should be used. Communication is made through a series of commands. As litter bearing is rapidly exhausting, team members should alternate roles. Additionally, a sling attached with a girth hitch to the litter may be used to transfer some of the weight from the arm to the skeletal system via the shoulder. It is also advantageous to use the belay-system brake by leaning forward, thus reducing the amount of lifting required.

(2) A scout may precede the team to pick a trail, make the passage more negotiable, or make a reconnaissance so the team need not retrace its course if an impasse is encountered. The scout can also select the site for and secure the next anchor. The scout must remember that the anchors and belay stations must be less than 140 feet apart (with 150-foot rope). Any time the route of descent or ascent changes course more than 90 degrees,

a new anchor and brake system must be established, or a runner used to change the direction of pull from the belay to the litter. In addition, if the rope is binding against vegetation or rock formations, a sling with a carabiner should be used to pull the rope away from the obstacle.

(3) Once all rescuers have been assigned positions and understand their responsibilities, the litter can be moved. The crew boss will count to three and give the command to lift. The crew boss should coordinate the litter bearers' activities while moving over rough terrain. Standardized commands between the crew boss and belayer are used to control the rate of descent. The following are the commands that should be used:

(a) "ON BELAY?"—Crew boss.

(b) "BELAY ON"—Belayer.

(c) "ROPE"—The crew boss is telling the belayer to feed the rope out in feet.

(d) "SLOW ROPE"—The crew boss is telling the belayer to feed the rope out in inches.

(e) "BRAKE"—Can be said by anyone to avoid a fall or obstacles.

(f) "LITTER SECURE"—The crew boss is telling the belayer that the litter will not be moving and it is attached to the anchor.

(g) "OFF BELAY"—The crew boss is telling the belayer to break the belay system and begin moving equipment to the next station.

(4) When ascending a steep slope, the procedures described in the descent are generally reversed. Additional manpower is required to pull the belay rope through a four-carabiner brake system or figure-eight clog while the litter bearers lift and slowly climb. One person is required to operate the brake system. As this procedure is considerably more fatiguing than descending, the litter bearers should not try to do all the work.

However, as they are ascending the slope, they should
assist and hold the litter off the ground.

h. The Buddy System. The buddy system is an
evacuation for a slightly injured patient or a patient
who is incapable of getting off a precipice. Equipment
required are two climbing ropes, four slings, chest and
seat harness for the rescuer, optional seat harness for the
patient, and sufficient equipment to construct an anchor
system. The following steps are necessary:

(1) An anchor system is set up with a belay device.
The first rope is backcoiled as a belay line. A figure-
eight knot is tied in the end of the rope with a fisherman's safety.

(2) The second rope is coiled with large enough
coils to fit around the shoulders of the patient and the
rescuer (figure 22-78). Once coiled, the loops are divided
into two groups so a figure eight is formed. The patient
steps into the divided coil so each leg is through one-half
of the figure eight. The knot securing the coil should be
in the small of the patient's back, and the coils should be
beneath the patient's arms.

(3) The rescuer then stands in front of the patient
and places each arm through the loops of the coil, half
over the right shoulder and the other half over the left
shoulder (figure 22-79).

Figure 22-79. Rescuer Donning Rope and Patient.

(4) A sling is passed around the back of the patient,
passing under the arms and over the shoulders of the rescuer. The sling is then wrapped around the coil as it passes over the front of the rescuer's shoulders (figure 22-80).
The working ends of the sling should pass over the top
of the loop formed in the wrap. Ensure the wrap is made
low enough so as not to cross the rescuer's neck or interfere with breathing. The two ends are tied together in the
center of the rescuer's chest using a square knot. The tails
are taken down to the rescuer's seat harness, one tail is
passed through each side, tied in a loop of the harness,
and back up to the square knot. Here they are secured
to the rope between the coil and square knot by a clove-
hitch knot on each tail (figure 22-81).

(5) To attach the party to the belay rope, the figure-
eight knot on the end of the belay rope is attached to the
carabiner on the crotch strap of the seat harness of the
rescuer. The two tie-end loops are secured with a carabiner sling; another sling is connected to the rescuer's
chest harness and the belay rope with a prusik knot (figure 22-82). This sling is used to support the additional

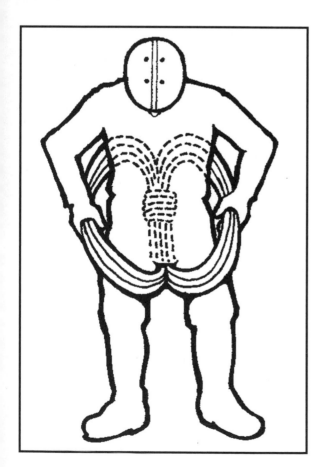

**Figure 22-78. Preparing Patient for Buddy
Evacuation System.**

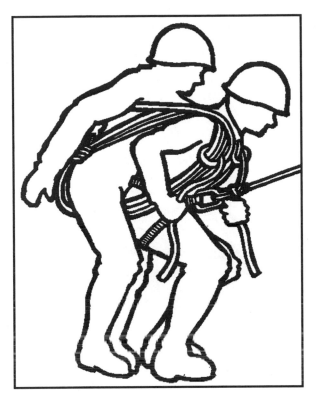

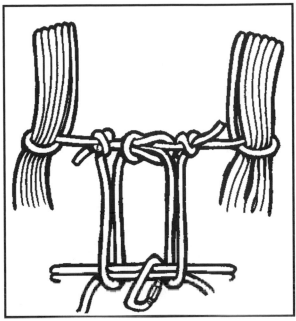

Figure 22-81. Tie-In Procedures.

Figure 22-80. Securing the Patient for Buddy Evacuation.

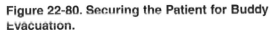

weight of the patient. This will enable the rescuer to remain perpendicular to the surface of the rock. Without this sling, the rescuer would fall over backward. The sling should be short enough so the rescuer can adjust the sling during the evacuation. The two people are then belayed down using a mechanical braking device (four-carabiner brake). (See figure 22-82.)

(6) The commands are:

(a) "ON BELAY?"—Given by rappeller.

(b) "BELAY ON"—Given by belayer.

(c) "BRAKE"—The belayer will stop the descent.

(d) "ROPE"—Belayer repeats "Rope." The belayer will feed the rope out in inches.

(e) "PATIENT SECURE"—Given by rappeller.

(f) "THANK YOU"—Given by belayer. The patient and the rescuer are in a safe location from rockfall and will not fall.

(g) "OFF BELAY"—Given by rappeller.

(h) "BELAY OFF"—Given by belayer. The rappeller is no longer secured by the belayer, and the belayer can disconnect from the system.

i. Vertical Litter Evacuation. The vertical litter evacuation is used in areas where a horizontal descent is not possible. The patient is secured to the litter in the same manner as for a horizontal evacuation. Two ropes are used for the vertical evacuation. The belay rope to the litter is configured with a figure-eight knot and a locking steel carabiner. A double-tape sling is passed between the two steel frame bars forming the main body at the head of the litter (figure 22-83). The sling should pass outside of the four joining bars between the frame bars. Both ends of the sling are connected into the belay-rope carabiners. Two carabiners are connected to the windows on one side of the litter. The second rope runs down the side of the litter through these carabiners to finish at the foot of the litter. A figure-eight knot is tied in the end of the rope. The seat harness of the barrelman attaches to this knot. A figure-eight knot is tied in the end of the sling and a carabiner clipped into the knot and connected to the foot of the litter (figure 22-84). A knot is added to the other end of the sling. A carabiner is added to the mariner's loop and connected to the barrelman's chest harness. As in horizontal evacuation, the commands are given by the barrelman. The commands are the same as in the high-angle horizontal litter evacuation (figure 22-85).

j. Horizontal High-Angle Litter Evacuation. The horizontal litter evacuation is the preferred position for an injured patient. This position allows for easier medical treatment and is required for shock prevention. The danger of this position is the patient's exposure to rockfall. Because of its complexity, it is also the most

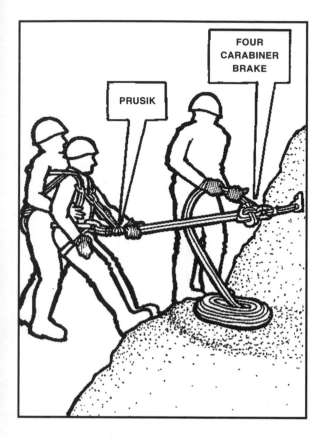

Figure 22-82. Hookup for Buddy Evacuation.

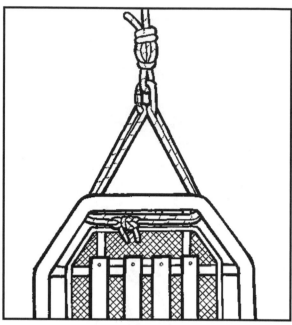

Figure 22-83. Attaching Rope to Stokes Litter.

potentially dangerous evacuation and should not be used unless no alternative exists.

(1) Team Composition. This method requires a minimum of three highly qualified team members. It also requires a great deal of equipment. It should not be attempted without the required equipment and qualified team members. Because the relative nature of the high-angle litter evacuation is potentially dangerous, the margin of safety must far exceed the stress on equipment. The safety factor is greatly increased by using equalizing anchors, two ropes for lowering a belayer, a rope handler, and a mechanical brake system. A high-angle evacuation on vertical walls is performed by only one litter bearer, with the litter fully supported by the lowering ropes. However, on walls that are not vertical, it requires a great deal of strength to pick up the litter and walk it down. When an injured victim is on a steep wall, a second person will administer first aid and then assist in loading the victim into the litter. In fact, this "second person" may perform the technically most difficult part of the evacuation in loading the injured person. In some cases, a "third person" can help load a severely injured patient. The job of the barrelman is to get the litter to the victim, to see that the victim is properly loaded and secured, to give the

victim a smooth evacuation by preventing the loaded litter from knocking against the rock, and to administer first aid to the victim if necessary. For first aid purposes, the litter should normally be horizontal. Proper operation of the brake controlling the litter is essential.

(2) Rigging the Litter. Two 7/16-inch Mountain-Lay ropes are used for lowering, each terminating in a figure-eight knot with a 4- to 5-foot tail attached to a large locking steel carabiner. Each tail will terminate with a figure-eight knot and safety knot. One end will be attached to the seat harness of the barrelman and the other will be attached to the patient. These ropes should be the same lengths for ease in rope changes, and ideally the ropes will also be matched in elasticity and of similar wear. The spiders are the nylon slings, between 30 to 36 inches long, which attach the litter to the rope. These spiders are attached to the litter with large locking carabiners. These large carabiners are clipped over the outer rail of the Stokes litter, and should have their gates facing in and locked (figure 22-86). If these carabiners are not available, the spiders can be attached to the major upright supports in the window of the litter with a double round turn and two or more half hitches. There are four spiders. One group of two spiders is attached to the windows at the sides of the litter at the head. The head group of two spiders is attached to the window at the sides of the foot of the litter. The spider groups should be connected with the key locking carabiner. This carabiner is attached to both of the ropes (figure 22-87). The spiders should be adjusted so the litter is horizontal, or with the head only

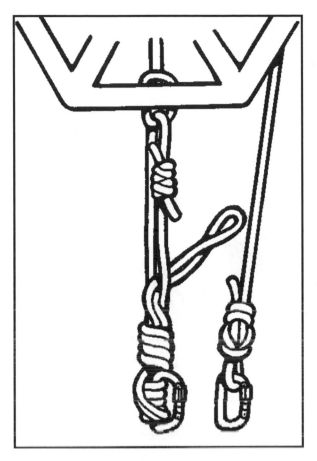

Figure 22-84. Tie-Ins for Foot of Litter.

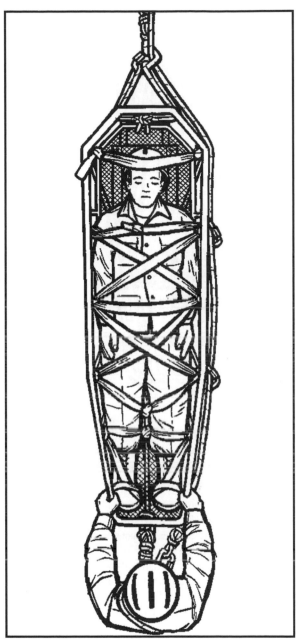

Figure 22-85. Horizontal Descent of Litter.

slightly elevated. The patient is tied into the litter in the same way as for the low-angle litter evacuation. However, the patient should have a chest or seat harness on. The figure-eight knot from one of the lowering ropes must be attached to the chest harness with a standard locking carabiner.

(3) The Barrelman. The barrelman is attached to the other lowering ropes by the figure-eight knot at the end of this rope. This rope is attached to the barrelman's seat harness with a standard locking carabiner. This is a safety line only. The majority of the barrelman's weight is supported with a 7 mm line that attaches to the barrelman's seat harness with a standard locking carabiner to the key carabiner. This line is secured with a prusik or Bachman's knot. The mariner's knot can be adjusted so the rescuer's feet are flat on the rock below the Stokes litter. The rescuer should be able to shield the patient with the upper body.

(4) Anchors. The anchor location should be directly above the victim, if possible. For an evacuation on a large face when visibility is good, a spotter on the ground with a plumb bob can direct an accurate

alignment of the anchor over the victim by communicating to the people on top. If the cliff is not vertical, the observer must observe the cliff "hang-on" and must be in the vertical plane containing the fall line through the victim. If the cliff is irregular, the plumb-bob technique may be unreliable, but the observer may still be the best source of information in locating the anchor. If the anchor is placed in such a way that the rope will have to pass over a sharp corner of rock at the lip of the face,

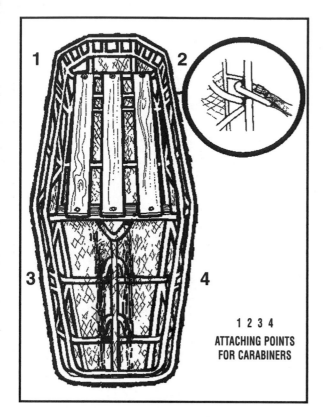

Figure 22-86. Attaching Points for Carabiners.

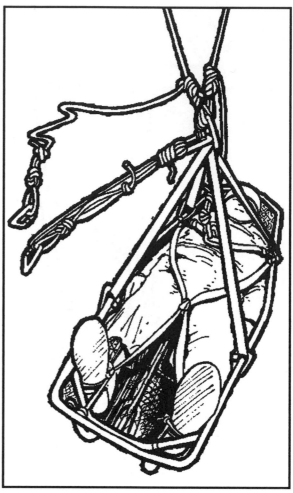

Figure 22-87. Rigging the Stokes Litter for Horizontal Evacuation.

the corner should be "softened" by breaking the edge with a hammer and by securing padding. If possible, an equalizing anchor system (figure 22-88) should be used for anchoring the top brake systems. This is the safest anchor system to use because it will adjust for a shift in position, and the shock load will be equally distributed if one of the individual anchors fail, the litter bearer falls, or the braking is of a fast descent.

(5) Braking System. The four-carabiner or figure-eight clog brake system can be used; however, the four-carabiner system is superior. This brake system will have a brake operator assigned. The brake operator obviously plays an important part in a successful evacuation, and should use gloves and avoid touching a hot braking device. To maintain full control while lowering, the brake operator should never let go of the ropes until tied off or "off belay." The operator should keep the descent smooth and steady, feeding the ropes equally into the brakes and never allowing a kink to form. A kink may jam in the brake and be difficult to release. Smoothness is easily produced if the loading by the litter is steady and has sufficient weight to require the brake operator to maintain a moderate grip on the rope and let the litter do the pulling. The litter should normally remain horizontal, thus the

two lines lowering the ropes should be fed equally into the brake. The ropes may be grasped together by one hand and fed equally, even with unbalanced loads. It is often helpful to the brake operator to have a rope handler pull the rope that is not feeding rapidly enough through the brakes. This is particularly useful when the litter is lightly loaded.

(6) Rope Management. The proper management of the rope by the team at the top is essential for efficient evacuations. In fact, the overall job on top usually involves more thought and skill than is required by the litter bearers. The operations on top are best done by at least two people. The brake operator directly controls the speed of descent. A rope handler provides slack rope with no kinks to the operator and assists on the brake when required. The rope handler's tasks are crucial since the evacuation will come to a rapid halt if the ropes become

snarled. The ideal situation on top is to be able to lay the lowering ropes out in a straight line for their entire length. The rope handler will then have an easy time assisting the operator. If space is limited, the ropes can be stacked separately and neatly. There may be a tendency for the rope to kink at the brake, in which case the rope handler will stay busy twisting and spinning the slack rope.

(7) Handling the Litter. The barrelman will always wear a hardhat, pack, and gloves. The reason for this is that in case small rocks fall toward the helpless victim, the barrelman can lean over the victim, providing a shield; and the packs and hardhats protect the barrelman. The victim will also wear a hardhat (when possible). A redundant barrelman tie-in is used for safety, and the locking of doubled carabiners should be used to prevent the barrelman tie-in from being accidentally released. Smooth handling of the litter by a skilled barrelman is the best way to ensure a safe ride down for the victim; banging the litter against the rock is very hard on the victim, and poor handling of the litter may result in injury to the barrelman. In starting the evacuation, the barrelman may have to carry the litter a short distance at the top of the face, taking care that ropes are positioned over a smooth rounded edge of the cliff and the spider carabiners have gates facing in. Once its weight is on the lowering ropes, the barrelman's main job is to hold the litter away from the rock. The feet are used against the rock, as they are in rappelling, to maintain footing. The barrelman's position on a nearly vertical wall is to be hanging from the tie-ins, which may be a comfortable seat sling. The barrelman (figure 22-88) is normally on the outside of the litter (away from the rock), gripping the outer or an underneath rail to hold the litter still farther off the rock to clear projections, etc., or to level the litter if loading is uneven. Occasionally, the rail at the cliff face is grasped, but care must be taken to avoid getting crushed hands.

(a) The descent is easy on flat, smooth, vertical walls. On a wall with obstructions, footwork techniques are the same as in rappelling. The legs should be perpendicular to the wall and spread comfortably apart. The prusik or Bachman's knot should be adjusted so the barrelman moves the litter to the edge of a vertical face, and may require readjusting so the space between the litter and the barrelman can be maintained. The barrelman will be bent at a 90-degree angle at the waist, which will enable the barrelman to bend over to protect the patient or administer first aid.

(b) The litter must be secured to an anchor to keep it from sliding and falling. The lowering ropes should be removed from the spiders and the knots untied to allow the ropes to be hauled back to the summit without a knot jamming. The spiders themselves should not be dragged up the cliff because of the danger of jamming. Normally, the ropes will be pulled up and carried down rather than dropped because of the

Figure 22-88. Body Position for Rescuer.

possibility of snagging or damage. After the evacuation is complete, the litter should be moved to an area sheltered from falling rock.

(8) Communication. The general goal of the brake operator is to lower the litter smoothly and safely as directed by the barrelman. The commands to be used are as follows:

(a) "ON BELAY?"—Has the usual meaning and is used before the evacuation begins.

(b) "ROPE"—Means the brake operator should feed the lowering ropes equally through the brake at a moderate speed.

(c) "SLOW ROPE"—Means the brake operator should feed the lowering ropes equally through the brakes at a slow speed.

(d) "BRAKE"—Means the brake operator should brake both lowering ropes simultaneously.

(e) "SECURE"—Means to maintain the brake until the litter is anchored, or in a position where it will not fall.

(f) "UP ROPE"—Means the litter is detached and all knots and slings have been removed. The rope can be pulled up without becoming caught on rock outcroppings or in cracks.

k. Horizontal Hauling Lines and Tyrolean Traverse. A suspension system used to transport a load off the ground along a taut (static) line may be useful when crossing streams, gorges, or other difficult terrain. These systems can be used for team movement or for transporting victims. The two-rope bridge is time consuming to set up, but in spite of the complexity the techniques can be useful in rescue. However, time should not be wasted waiting for the system to be set up. Sometimes the rigging may be done while the victim is being transported to the crossing site. Since system stresses are often great, the margin of safety is smaller than usual or desirable in rescue operations, and the rigging should be supervised by someone experienced in the techniques.

(1) Anchors. A major requirement for any suspension system is the availability of secure anchors at each end. Anchors have to be as strong for this use as in any mountain rescue. To withstand the high-line stress, the anchors may have to be a combination of several anchor points rigged to be self-equalizing. For reasonable safety, each anchor system should be able to hold at least 10 times the load to be transported.

(a) The anchors must be suitably placed in addition to being strong. To prevent the load from "bottoming out," or to minimize line tension by increasing sag, the anchors must be placed in a high location. If the rope contacts any sharp rock, the rock edges must be padded with suitable material, such as leather gloves, pack, tree branches, etc., since a taut rope can be easily cut. The anchors should be 6 to 10 feet back from the edge of the stream or cliff to provide for loading and unloading. The rope should be 3 feet or more off the ground at these positions, and if good anchors do not

allow this, A-frames may be improvised from sturdy tree trunks.

(b) If possible, two ropes or a double rope should be used to increase the safety margin. However, one rope will work, although the strength of the system is decreased. It is very important to understand the stresses involved with this system. For example, if the rope span is 40 feet and the sag in the middle of the rope is 1 foot with a 200-pound load, the rope tension is approximately 2,500 pounds. However, if the sag is 4 feet in a 40-foot span, then the tension is 650 pounds.

(2) Establishing the Two-Rope Bridge. Connect the center of the rope through the anchor. Do not tie, but simply pass through a carabiner. If the span requires two ropes to be tied together, a double figure eight should be used with single fisherman safeties. A carabiner must be placed in the double figure eight; if not, the double figure eight will be difficult to untie. One team member will cross the obstacle carrying the two ends of the rope, plus the hauling rope. Once across, an anchor is established and one end of the rope connected to it (figure 22-89). Remove as much slack as possible so the knot will be taut against a 4-inch pulley or carabiner on the opposite side when tied off to the anchor system. The other end of the rope is then put through a 3:1 mechanical advantage system.

(a) To use the mechanical advantage, a safety prusik is attached to the rope. This safety prusik must be attached to a stout anchor and be as long as practical

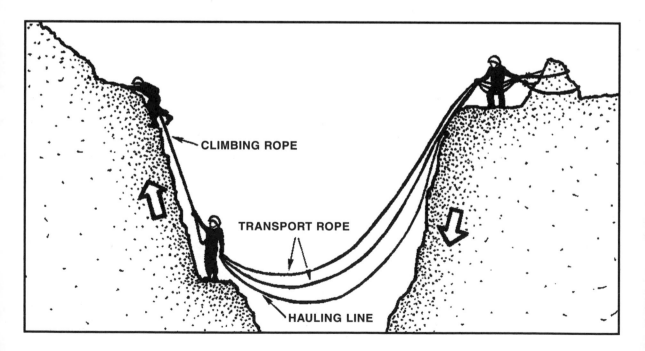

Figure 22-89. Horizontal Hand Evacuation System.

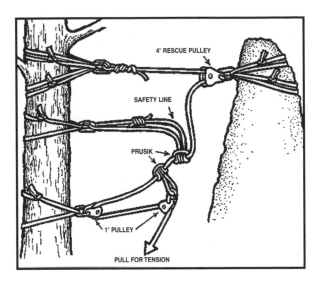

Figure 22-90. Rigging Hand Evacuation System.

toward the anchor. This is called a "three-to-one mechanical advantage" because there are three moving lines. The force of pull on the running end of the rope will be increased (ideally) three times. This does not account for the friction of the rope passing over the carabiners or pulleys.

(b) Using the mechanical advantage, the following steps should be taken to tighten the rope. Pull on the hauling line or the running end of the rope until the snap links come together. Have another person keep this safety prusik sliding out as far forward as possible. Then let the safety prusik take up the load, while slowly letting slack out of the running end of the rope. Then slide the second prusik out as far as possible, and repeat the entire maneuver. When the rope is tight enough, tie the rope off (figure 22-90).

(3) Crossing the Two-Rope Bridge. Half the team will now cross the rope by means of the Tyrolean traverse. This is a means of crossing a suspended rope while being secured to the rope. A pulley or carabiner is attached to the rope and clipped into the seat harness. The person then hangs under the rope and pulls hand over hand to the other side. The load is placed on the rope by pulleys or carabiners. The end of the second rope is tied to the load and used to pull the load across the rope to the other side (figure 22-91). Once the patient is secured in the litter, the slings to hold the litter horizontally can be attached. There are six slings in all, one set of three at the foot of the litter. In each of the

to hold the tension on the rope and to slide forward as the slack is pulled out of the rope. The rope is then passed through a carabiner that is attached to an anchor. A smaller prusik sling is tied on the rope behind the first prusik. A carabiner is attached to the prusik sling. The rope is then passed through this carabiner. The rope should resemble a "Z" when the running end is pulled

Figure 22-91. Rescuer Traversing.

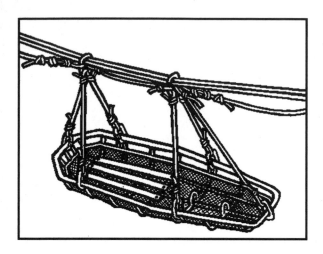

Figure 22-92. Rigging the Horizontal Hauling System.

sets, one is attached to the end of the litter and the other two to the sides of the litter. The slings are about 20 inches long, and attached to the inner rail of the Stokes litter with a round turn and two half hitches (figure 22-92). The remaining team then crosses the rope. Each set of slings is attached to a steel "D" locking carabiner. A short length of rope will join the two carabiners for equal pull during the traverse. An in-haul line should be attached to the litter to control the descent or to haul across the open expanse (figure 22-93). To retrieve the rope, the anchors are removed and the rope simply pulled across the obstacle. If an artificial anchor was used for the mid-rope anchor, it is not retrievable (figure 22-94).

(4) Crevasse or Ravine Recoveries. The ability of a team to recover a victim below them, as in a crevasse,

requires the use of a mechanical advantage. A mechanical advantage is the arrangement of ropes and use of pulleys to arrange a system that will allow a force directed on the system to produce a lifting power in multiples of the force. The primary use of a mechanical advantage is to perform vertical lifts, as in crevasse or cliff rescue, and to apply tension for tightening the system used in construction of hauling lines (figure 22-95). The mechanical advantage can be in different ratios depending on the number of pulleys and changes of direction. Pulleys should be used when a rope passes through a carabiner. If pulleys are not available, a carabiner can be used; however, a larger amount of friction will be present, causing a loss of some of the advantage in the system. Rope placement is important in the construction of a mechanical advantage. The ropes should run side by side with no twist around each other. Two basic types are illustrated. The Z-pulley method uses a portion of the load-bearing rope to gain the mechanical advantage and, therefore, is a one-rope system (figure 22-96). The second type uses a second rope to establish the mechanical advantage on the load-bearing rope (figure 22-97).

(5) Z-Pulley System. This system of mechanical advantage is used to lift a fallen climber or stranded individual out of a crevasse, up a cliff face, etc. If the fallen climber is attached to a rope, this rope may be used for the system. If the climber is not attached, it will be necessary for a member of the rescue team to rappel or be belayed down to the fallen climber. The rope should be connected to a seat harness and a chest harness. An improvised harness may be required (figure 22-97). Pull all rope slack up from the crevasse. Near the edge, anchor a prusik sling and attach it to the rope. This keeps the rope from slipping back down over the edge.

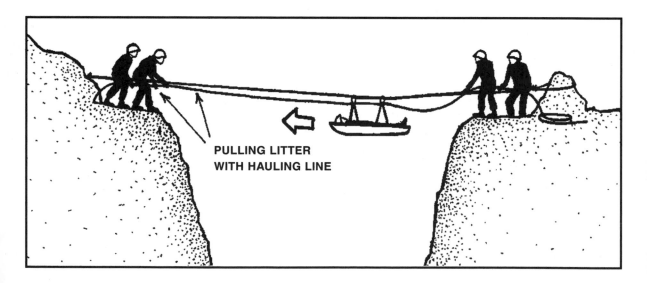

PULLING LITTER WITH HAULING LINE

Figure 22-93. Pulling the Litter.

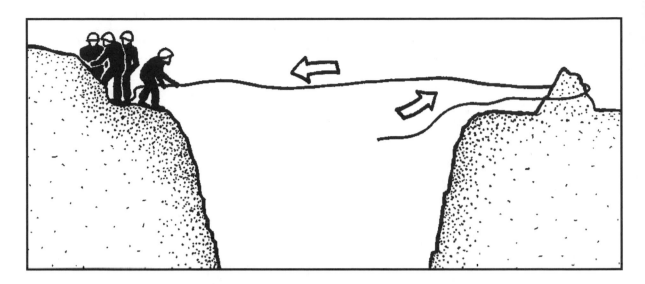

Figure 22-94. Retrieving the Rope.

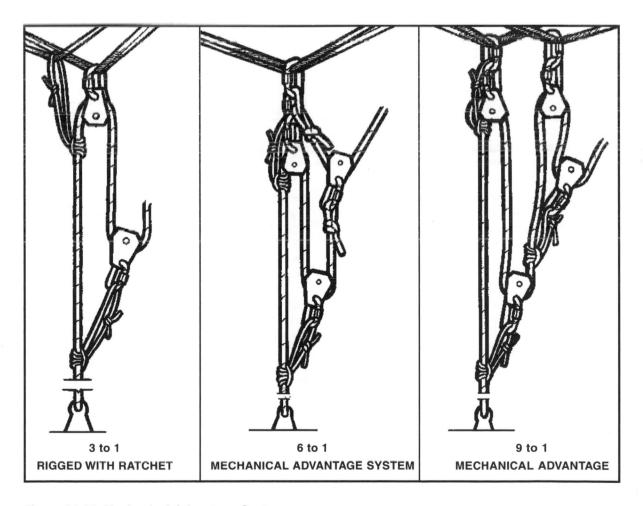

Figure 22-95. Mechanical Advantage Systems.

Figure 22-96. Two-Pulley System.

Figure 22-97. Second Rope Added.

An anchor system should be established back from the crevasse, and a pulley attached to the anchor. The rope is routed from the crevasse through this pulley. The rope comes out of the pulley and runs back to the crevasse edge. Here another prusik sling is added to the rope as it comes out of the crevasse, and a pulley attached to it. The rope from the anchor pulley is passed through this pulley. The system is now complete. The operation is conducted by pulling on the standing end of the rope until the second prusik sling is pulled to the anchor pulley. The prusik at the edge of the crevasse takes the load while the other prusik is reset in the original position. Continue with the lift until the climber is free of the crevasse.

(6) Added Rope Mechanical Advantage. This system requires an additional rope to form the system. As in the Z-pulley system, a rope from the fallen climber is required. Pull all rope slack up from the crevasse. Establish an anchor and connect a prusik sling to the anchor. Secure the belay rope with this prusik sling. Establish two separate anchors and anchor one end of a long sling to the anchor nearest the belay rope. Attach another prusik sling to the belay rope between the anchor prusik and crevasse edge. On the standing end of this sling, attach a carabiner using a figure-eight knot. Attach another carabiner through the body of the figure-eight knot. Place a pulley on each of these carabiners and on the remaining anchor point. Run the long sling through the pulley at the end of the sling, back through the pulley on the second anchor, and through the remaining pulley in the body of the figure-eight knot. The system is ready for operation (figure 22-96). Pull on the long sling end until the prusik with the two carabiners is pulled into the anchor system for the long sling. Allow the anchor prusik to hold the weight and reset the long sling in the initial way. Continue with this procedure until the climber is raised from the crevasse.

(7) Gorge Lift Rescue System. One rescuer rappels to the climber while the other goes around the crevasse or gorge to the other side. An anchor is established on both sides. The end of one rope is thrown across to the far side, where it is attached to the anchor. The center of the rope is allowed to drop in the crevasse, where the climber is attached to it by a pulley. The end of a second rope is

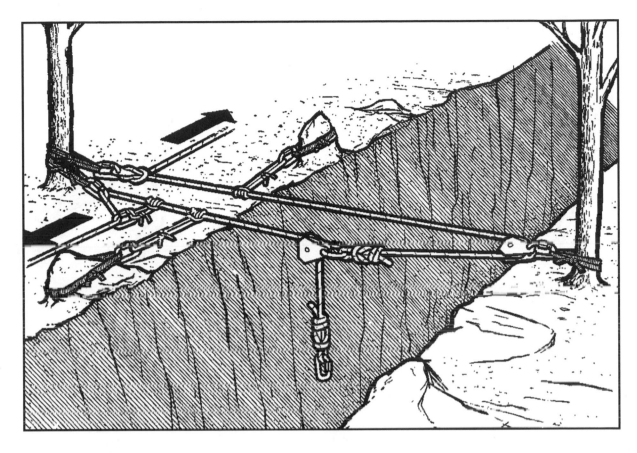

Figure 22-98. Gorge Lift.

lowered to the climber and anchored to the person.
The free end of the first rope is put into the mechanical
advantage system. As tension is applied to the rope, the
climber is lifted. When the rope is at an approximate 40-
degree angle, the climber is pulled toward the edge by the
ropes. As the climber nears the edge, the weight of the
climber will sharpen the angle of the rope. It will be necessary to apply more tension to the rope. Continue until
the climber is rescued (figure 22-98).

   (8) Rope/Knot Bypass While on Belay. When using
multiple ropes or adding extra ropes to a long descent, it
is necessary to pass the joining knots through the belay
system. The knots will not pass through the system, so
a method is used that will secure the rope while the belay
device is removed and the knot passed. As the knot
approaches the belay system, apply a brake and secure
the system. Place a prusik sling on the rope below the
belay device. Run this sling to the anchor and tie a
mariner's knot (figure 22-99). Take the brake off the
belay and allow enough rope through the belay until the

Figure 22-99. Multiple Ropes.

tension is taken by the mariner's knot. Replace the rope through the belay immediately behind the knot and secure the brake. Slowly remove the mariner's knot until the tension is again on the belay system (figure 22-100). Continue with the descent and repeat procedures as necessary.

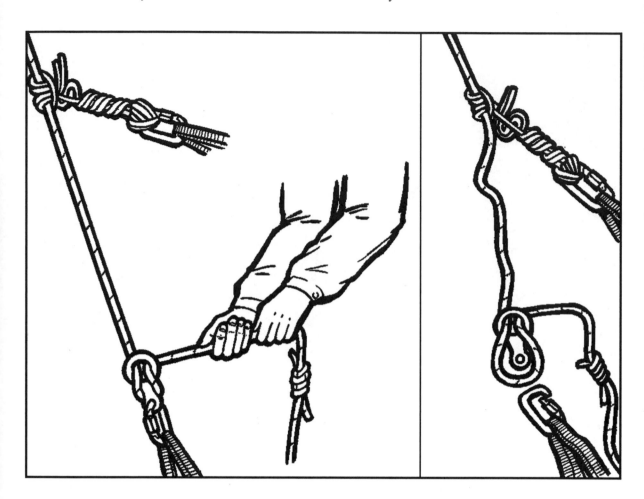

Figure 22-100. Rope Bypass.

Chapter 23

WATER TRAVEL

23-1. Introduction. In this chapter, the travel techniques that can be used on 71 percent of the Earth's surface—the water environment—will be addressed. The techniques of river and open-sea travel may be adapted to other water features, such as swamps and lakes. In the paragraphs concerning open-sea travel, the environmental factors of oceans of the world will be considered as they relate to travel. The techniques for individual water rescue and swimming are important for recovering injured survivors and equipment. Additionally, the problem of submersion must be considered, as well as how the anti-exposure suit, life preservers, and liferafts can extend a survivor's life expectancy. This is why a survivor must be familiar with the individual rafts and raft procedures. The ultimate goal of a survivor on the open seas is to be rescued; however, a second goal, if rescue is not made, would be to make it to land. A thorough knowledge of water-travel techniques will greatly increase the survivor's chances of reaching landfall. If done correctly, landfall can be reached with minimum loss of equipment or injury.

23-2. River Travel:

a. Rivers have been used for centuries in all environments as a safe means of travel and is the reason why most of the cities of the world are located on rivers. It is not uncommon for a river to flow at 4 or 5 knots per hour. A survivor could travel 20 to 25 miles in 5 hours of travel. This may contrast greatly with the rate of travel on land. The amount of energy required to carry survivors' equipment and other supplies, and to travel 20 to 25 miles on land, is much greater.

b. Each major continent has thousands of miles of navigable rivers. Some rivers, such as the Nile, Amazon, Mississippi, Lena, and Mackenzie, have hundreds of miles of navigable water with seldom a ripple. These navigable sections are generally found flowing through the flatlands, plains, tundras, and basins of the world. In these areas, only the temperatures of the water, and the plant and animal life may present hazards. In contrast, the headwaters of rivers like the Mackenzie, Yangtze, and Ganges are so rough that they would best be categorized as a threat to life. This would also be true of the Snake, Salmon, and Rogue Rivers of the northwestern United States. These rivers, although traveled by white-water rafters, pose an unreasonable hazard to survivors. Survivors must take into account individual or group skills, injuries, type and severity of rapids, the temperature of the water, and direction the river flows in making the decision to travel. Even if a portage of several miles is required, the energy saved by floating on a river might warrant river travel. However, once the energy expended

for portage exceeds the energy conserved by floating a section of a river, the river as a mode of travel should be abandoned.

c. In a nontactical situation, rivers will most likely carry survivors to indigenous people, who could aid the survivor in meeting the basic needs for sustaining life and effecting rescue. Even if some form of civilization is not encountered, the survivors would most likely reach a lakeshore or even the seacoast. These environments, particularly the seacoast, provide transition zones from land to water, which are rich in food and other survival resources. In these areas, the resources could improve the chances of survival. It is much easier to spot signs of survivors along a shore, as opposed to the interior of a landmass.

23-3. Using Safe Judgment and Rules for River Travel:

a. There are certain safety rules and guidelines that must be followed to reduce the dangers associated with river travel. Respect for these rules and guidelines is necessary to reduce the potential dangers.

b. The most important safety rule is personal preparation. Preparation should begin by thoroughly scouting the river. The conditions of the river will determine the intermittent stops. High edges along river edges provide needed visibility to plan each leg of travel. If there are numerous bends and poor lookout points to view the river, stops are frequent. Sound judgment must be used when planning routes. Patience in planning each leg of travel helps prevent disaster. All survivors must prepare, know the plans, and be able to handle the route safely, considering their skills and strength. Survivors should be aware of and avoid river hazards, and have alternate routes and communication signals in case flow conditions suddenly change, making the run more difficult. All rapids that cannot be seen clearly from the river should be scouted. The route should be discussed with the crew. The skills, knowledge, and abilities of the aircrew members must be considered, including swimming abilities and physical condition. Areas of high risk should not be attempted. Before reaching an area of suspected great difficulty, rafts should be beached and carried to the next point of travel; this is called portaging.

c. Before entering the raft, survivors should don life preservers and suitable clothing for adequate protection. The equipment should be tested to ensure it is serviceable. Bulkiness is not advisable due to the possibility of the raft capsizing and being weighted down with water. The antiexposure suit (if available) should be worn. Items that might absorb water should be packed in a waterproof container. The survivor should ensure that:

(1) All medical supplies, a repair kit, and a survival kit are in the raft.

(2) Survival kit is checked and inventoried.

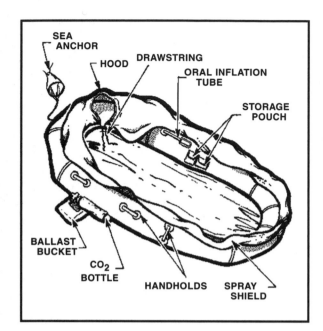

Figure 23-1. One-Man Raft.

(3) Extra efforts should be made to keep supplies and equipment in good condition.

(4) All items are secured to the raft to prevent loss and (or) injuries.

(5) Before use, the raft is checked for leaks and necessary repairs are made.

d. When using a one-man raft for river travel (figure 23-1), it may be advisable to tie or cut off the ballast bucket, fasten the spray shield in the opened position, and remove the sea anchor to prevent problems with swamping or entanglement with subsurface obstacles. Without the ballast bucket, the raft can be easily maneuvered by paddling with the backstroke, or for slight adjustment, with a front stroke. When using either the backstroke or the front stroke, the survivor will find it easier if the two underarm cells of the life-preserver underarms (LPU) are disconnected in front, and the cells placed behind the back (figure 23-2A). This gives the survivor a full range of motion. When rough water is encountered, the survivors should fasten the LPU and face downstream.

e. One of the primary methods of avoiding hazards on the river is to slow the speed of the raft and move across the river, avoiding a collision with the obstacle. This ferry position should be initiated early to avoid large rocks and reversals in the river. If the collision obstacles are to be avoided, point the bow of the raft toward them and backstroke against the current to slow the speed of the raft's downstream progress and move it across the river. Usually, the best angle is about 45 degrees to the current. The greater the angle, the quicker the movement across the river, but this also increases the downstream speed of the raft. Decreasing the angle will slow downstream speed, but movement across the width of the river will also be decreased (figure 23-2B). The raft will be more maneuverable if it is well inflated. If the raft should pass over a rock, arch the back up to prevent injury to the buttocks or back.

f. When using multiplace rafts, the boarding ladder and sea anchor should be removed to prevent entanglement. If available, about 50 feet of line should be tied on the bow and stern of the raft to be used for tie-offs. An additional 200 feet of line should be coiled and tucked away for emergency and rescue work; one end is secured to the raft while the other end has a fixed loop. An improvised suspension-line rope may be used for this; for example, three-strand braid can support about 1,000 to 1,500 pounds of pressure, or a two-strand twist strengthens the line to support about 700 to 900 pounds of pressure (figure 23-3).

g. Proper placement of equipment and personnel should equalize weight distribution to ensure stable control; overloading should be avoided. Assign personnel crew positions and responsibilities in the raft; captain (person in charge), stern paddler (maneuvers raft), and side paddlers. Twilight and night rafting should be avoided (nontactical), as poor visibility increases danger.

h. Two ways to steer a multiplace raft are:

(1) To steer a raft by using sweeps (long oar) and poles. A pole is more efficient in fairly shallow water, but a sweep is preferable in deep water. Poles and sweeps from both ends of the raft are used. The person in the bow (front) can see any obstructions ahead, and the one in the stern (rear) can follow directions for steering. Poles are also useful for pushing a raft in quiet water.

(2) Paddle techniques are used to maneuver the raft. When paddling, there are three possible body positions on a raft. The best way is to sit on the upper buoyancy tube with both legs angled to the inside of the raft. The body should be perpendicular to the sides of the raft, enabling the rafter to paddle. Another way is to sit cowboy style, straddling the upper buoyancy tube of the raft with one leg on either side, and folding at the knee with each leg back. However, the outside knee may collide with obstacles and cause injury. The third way is normally used in calmer waters because it consists of partially straddling the upper tube with legs comfortably extended. In a smaller raft, the survivor may be able to sit down inside the raft and reach over the buoyancy tube. The following strokes can be done from these positions. Knowing the parts of paddles will help in explaining the different paddling strokes (figure 23-4).

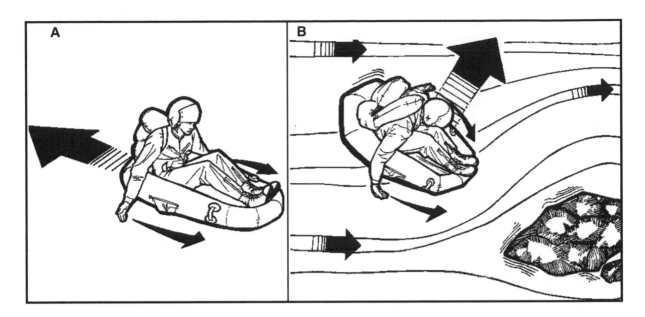

Figure 23-2. One-Man Raft (Paddling).

(a) One of the easiest is the forward stroke, which is done with smooth continuous movements using these techniques:

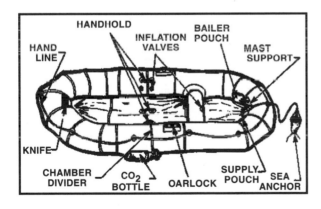

Figure 23-3. Seven-Man Raft.

-1. Thrust the blade of the paddle forward using the outboard arm, then momentarily keeping the outboard arm stiff and away from the raft, push the grip. The inboard hand is then moved forward to cut the blade deeply into the water. Continue the stroke by pushing on the grip and pulling on the shaft, keeping the blade at a 90-degree angle to the raft. Stop the motion as the blade comes slightly past the hip, because a full follow-through provides little forward power and wastes valuable energy. Slide the blade out of the water

by pushing down on the grip and swinging it toward the inboard hip, and turning the blade at a parallel angle to the water once it has cleared the water. By paralleling

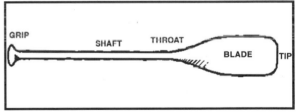

Figure 23-4. Paddle.

the blade, it cuts wind and wave resistance and saves time and energy. This cycle is repeated until the strokes are changed.

-2. In mild water, there is no need to over reach or excessively twist the upper trunk of the body. When extra speed is needed, lean deeply into the strokes, which brings the entire body into play. Position the inboard hand across the tip of the paddle grip and the outboard hand halfway to three-fourths of the way down the shaft.

(b) The opposite of the forward stroke is the backward stroke. The blade is thrust into the water just behind the hip, and pressure applied by simultaneously pushing forward on the shaft and pulling back on the grip. End the stroke where the forward stroke would

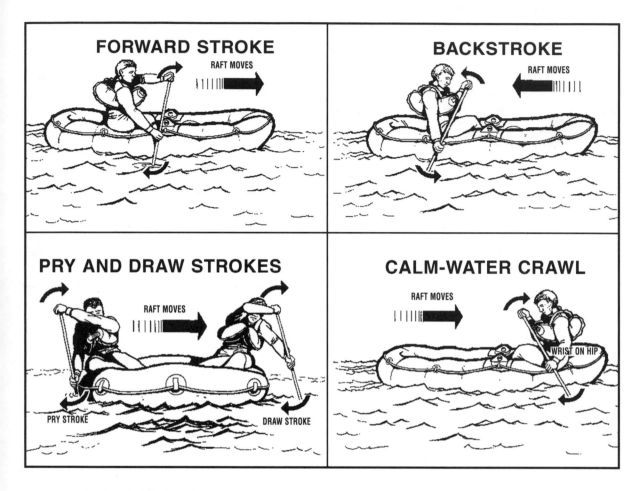

Figure 23-5. The Paddle Strokes.

begin, and again angle the blade out of the water back to the beginning of the backward stroke.

(c) The draw and pry strokes are opposite sideways strokes. These strokes are good for small sideways maneuvers and for turning the raft when used from the front or rear of the raft.

-1. Draw stroke. Reach out from the raft, dip the blade in parallel to the raft, and pull on the shaft while pushing on the grip. Pull the blade flat to the side of the raft. Pull the paddle out and repeat.

-2. Pry stroke. Dip the blade in close to the raft, and push out on the shaft while pulling in on the grip.

(d) The fifth stroke, called the calm water crawl, is used alternately with the forward stroke when paddling through long calms. Sit cowboy fashion while facing the stern and hold the paddle diagonally in front with the shaft that is held by the outboard hand against the outboard hip, and the grip held by the inboard hand in front of the inboard shoulder. Extend the inboard arm to swing

the blade behind, dip the blade in the water, and pull back on the grip, prying it forcefully, using the hip as a fulcrum (the point of support on which a lever works). Using the shoulder, hip, and hand as assisters, the crawl is easy yet powerful (figure 23-5).

(e) The ferry is a basic paddle-maneuvering technique used to navigate bends and to sidestep obstacles in swift currents. The ferry is essentially paddling upstream at an angle to move the raft sideways in the current. Paddle rafts can ferry either with the bow (front) angled upstream or downstream. The bow-upstream ferry is stronger because it uses the more powerful and easier forward stroke. It is carried out by placing the raft at a 45-degree angle to the current with the bow angled upstream and the side toward the desired direction. The bow-downstream ferry is weaker because it uses the less powerful backstroke, but it does offer certain advantages. It enables paddlers to look ahead without straining their necks, and makes it easy to put the bow into waves (figures 23-6 and 23-7). It is carried out by backstroking with the stern (back) angled upstream at a 45-degree angle and the side facing the desired direction.

(f) There may be times when the only way for a heavy raft to enter a small or violent eddy is with a reverse ferry (figure 23-8). The following steps may be used for an oar or paddle raft, except the paddle raft approaches the eddy bow-first and finishes in a bow-upstream position:

-1. Raft approaches sideways.

-2. Raft turns around to angle its bow down-stream.

-3. With careful timing, the captain should have the crew begin to pull powerfully on the paddles. The angle of the raft to the current can be close to 90 degrees, but is best at about 45 degrees.

-4. While aiming for the eddy, the crew should continue with the front stroke and gain momentum.

-5. With the crew still using the front stroke, the raft breaks through the eddy fence.

-6. With the bow in the upstream eddy current and the stern still in the downstream current, the raft is spun into a normal ferry angle. The crew continues with

the front stroke while making the necessary turn to bring the boat entirely into the eddy.

-7. The raft rides easily in the eddy. NOTE: The reverse ferry and eddy turns are not only used to enter eddies, but can also be used to dodge through tight places. The reverse ferry (or sometimes an extreme ferry) scoots the raft sideways, the eddy turn snaps the bow into a bow-downstream position, and the raft, rather than entering the eddy, rides the eddy fence past a major obstruction or hole.

(g) The straight forward paddle is used in calm and moderate waters where there is ample maneuvering time. Simply point the bow in the desired direction and follow the forward-stroke method of paddling. The back paddle is performed in the exact opposite manner of the forward paddle. Point the stern (back) in the desired direction and follow the backstroke method of paddling (figure 23-5).

(h) To make a left turn, the left side of the raft will backpaddle, while the right side paddles forward. It's just the opposite to make the raft turn right. The right side on the raft backpaddles while the left side paddles forward—both performing the paddling maneuvers at the same time (figure 23-9).

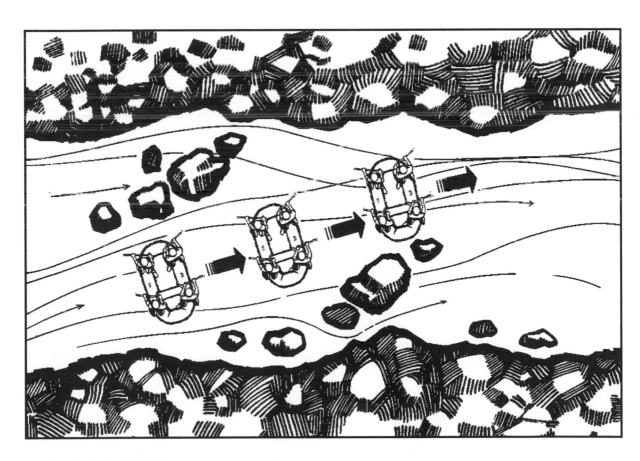

Figure 23-6. The Paddle Ferry.

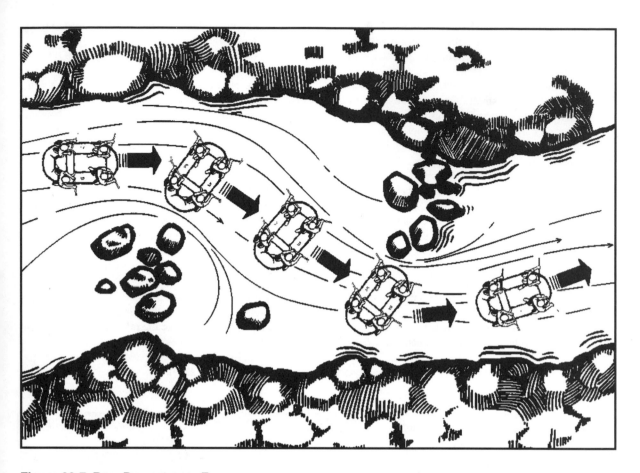

Figure 23-7. Bow Downstream Ferry.

(i) The pry and draw strokes are used to move the raft sideways (figure 23-10).

(j) Stern maneuvers are used to increase the maneuverability of the raft. The paddle used at the stern of the boat is basically a rudder, which controls direction. To turn right, the paddle blade is held to the right, square against the direction of the current. To turn left, the paddle is held to the left.

(k) Other strokes, such as a forward stroke or draw stroke, used at the stern of the raft, will cause it to turn or move faster. If stroking is done slightly to the side (either right or left) of the raft, it will help move the raft in the opposite direction (figure 23-11).

(l) River travel requires fast, decisive action. Therefore, a paddle raft needs a captain to coordinate the crew's actions by the use of commands or signals. Communications between captain and crew are crucial; all members must agree on a set of short, clear commands. The following are suggested commands:

CAPTAIN'S COMMANDS	CREW RESPONSE
Forward	Crew paddles forward.
Backpaddle	Crew does backstroke.
Turn right	Left side paddles forward, right side does the backstroke.
Turn left	Right side paddles forward, left side does the backstroke.
Draw right	Right side uses draw stroke, left side uses pry stroke.
Draw left	Left side uses draw stroke, right side uses pry stroke.
Stop	Paddlers relax.

-1. Commands must be carried out immediately, so the crew should practice until they can snap through all the commands without hesitation. The captain controls both the direction and speed of the raft with a specific tone of voice and commands. This control, and the captain's ability to anticipate how the water ahead will affect the raft, will help avoid undercompensation or overcompensation of maneuvers through obstacles. Good captains

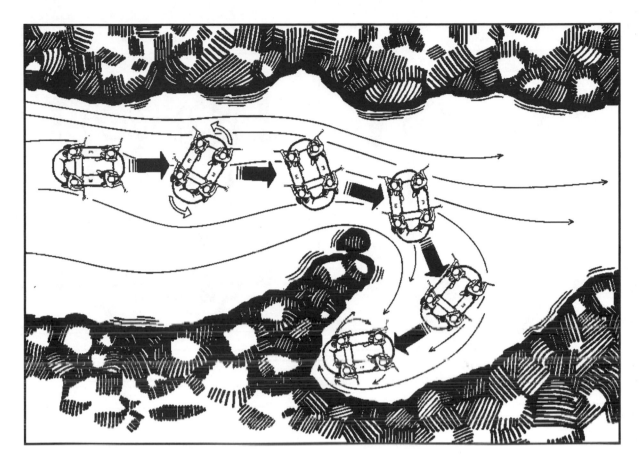

Figure 23-8. Entering an Eddy with a Reverse Ferry.

think well ahead and move with the river, issuing commands precisely and sparingly, working their crew as little as possible.

-2. When using commands and maneuvering the boat in harmony with the river's currents, paddling can be easy and effective, even fun. When instant action is necessary, the captain may say, "Paddle at will." When time permits, the captain should introduce commands with a preparatory statement, such as, "We're going to ferry to the right of that big rock. OK—[gives command]." This gives the crew time to prepare for the next response. If a command is not heard or understood, it should be repeated with zest until it is understood. If a raft member spots a better way through a rapid or channel, a fully extended arm is used to point it out. This signal, like the others agreed on, should be repeated until it is understood.

23-4. River Hydraulics. An understanding of river hydraulics is important to the survivor. Knowledge of the types of obstacles and why they should be avoided or overcome is necessary for a safe river journey.

a. Laminar Flow. The drag produced when moving water flows over or past various types of objects and surfaces is called a laminar flow (figure 23-12). The laminar-flow principle is that various layers or channels of water move at different speeds. The lower layer of the river moves more slowly than the top layer. This is due to the friction on the bottom and sides of the river, which is caused by soil, vegetation, or contours of the riverbank. The layers next to the bottom and sides are the slowest; each subsequent layer will increase in speed. The top layer of the river is affected only by the air. The fastest part of the flow on smooth, straight stretches of water will be between 5 and 15 percent of the river depth below the surface. Even straight-running riverbeds are not smooth; they have jutting and receding banks on the sides, which affect the laminar flow. The friction caused by the banks causes the sides of the flow to be slower than the midstream. The areas near the banks are shallower and have fewer layers. When the river travels at 4 to 5 knots, turbulence begins to develop, which interferes with the regular flow of the current. When this rate of river flow is achieved, the

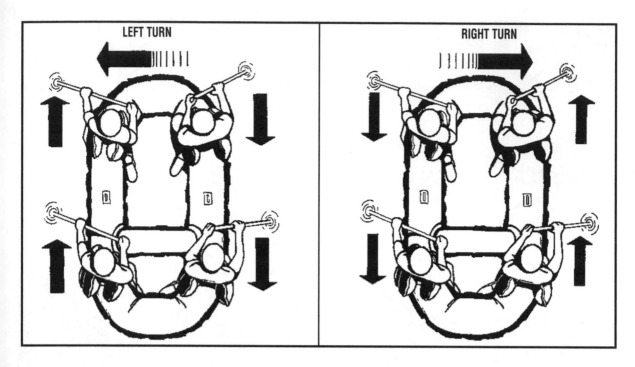

Figure 23-9. Turning the Raft to the Right or Left.

friction between the layers of water will cause whirling and spinning actions that agitate the smooth flow of water, creating more resistance.

b. River Currents:
(1) When the current of a river is deflected by obstructions, the overall downward flow of a river will

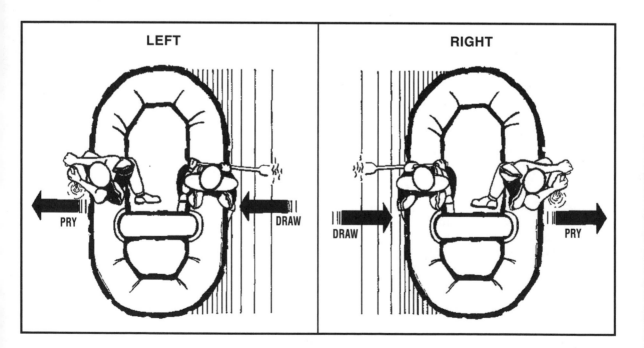

Figure 23-10. Making the Raft Move Left or Right with Pry and Draw Stroke.

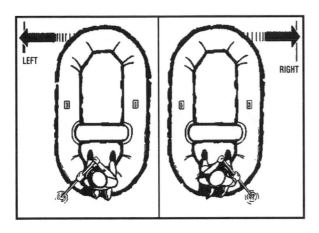

Figure 23-11. Stern Maneuvers.

respond. These responses vary from mild to radical deflection, creating direction and speed changes of water flow. These changes are called reflex current. The reflex current responds to an obstruction, such as bends or submerged rocks.

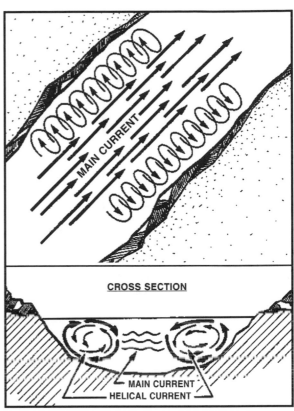

Figure 23-13. Helical Flow.

(2) One response to the laminar flow is a spiraling, coil-springing flow called a helical current (figure 23-13), which corkscrews as a result of friction with the riverbank. Going downriver, on the left side of the supposed straight-line river, the helical flow turns clockwise to the main current, and on the right side the helical flow is counterclockwise. This results from friction and drag caused by shallowing banks combined with the strong force of the main current flowing down. The helical flow and the mainstream create a circular, whirling secondary current that travels down along a line near the point of maximum flow. Helical current flow starts along the bottom of the river going out toward the riverbank, surfacing, and then spiraling back into the mainstream at a downward angle. This flow causes floating objects around the edges to be pulled into the mainstream and held there. By understanding where the fast water is and how to observe the characteristics that show the current, a survivor can maneuver the raft to take advantage of the faster water and increase the rate of travel. Even at the quietest edge of a flow, particles are still drawn into the strongest part of the current. Laminar and helical flows are always present in fast-flowing rivers.

(3) The main channel is the deepest part of the river and can wander from bank to bank. The turbulence

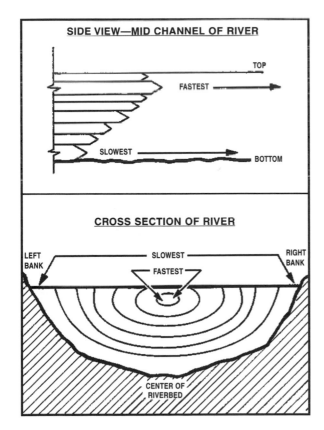

Figure 23-12. Laminar Flow.

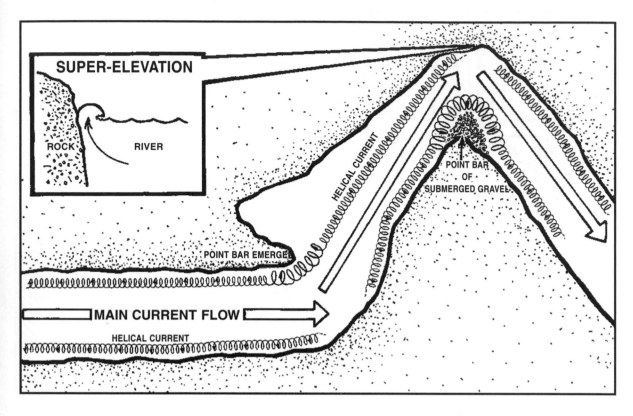

Figure 23-14. Macroturbulence.

caused by the wandering main current erodes wide curves into sharper, more defined bends, creating indirect courses.

(4) When the river makes sharp turns, the current is affected by centrifugal force, swinging it wide into the outside bank. The helical current diminishes, being smothered by the laminal flow, thereby increasing the corkscrew effect on the inside of the curve. The surface water is being whirled hard in the direction of the outside curve of the bank—the faster the water flow, the stronger the push. Floating objects are forced, with the surface water, to the outside of the curve and into the banks, usually getting lodged against and onto the shore.

(a) A powerful helical flow not only pushes the surface outward, but as it swirls up from the bottom it carries sediment with it. The sediment and other debris are deposited at the highest point of the inside bank of the bend. The sediment is then dropped during high water, and when the water recedes, a point bar (made of sand and gravel) is revealed. The point bar generally sticks out far enough to funnel floating objects into the swiftest part of the river during high waters, avoiding the sandbar.

(b) Super-elevation is a feature in which the water is being increased in volume, intensity, and height. When

both the stream volume and movement are high, centrifugal force exerts another type of influence on flow characteristics. The river surface water tends to curve in a dish shape toward the outer bend, like a banked turn on a racetrack. The dished inside curve is the easiest and safest route to travel through. If maneuvered correctly, the slight rise of the water and the force of the current around the curve will cause the raft to slip gently off the wave and into the quiet pools of water below. But if the raft was maneuvered across the line of the currents, the raft may either be sucked under by the dominant helical flow, or the power and force of the river on the outside of the dish on the curve could smash and pin any floating device against the outside bank.

(5) Macroturbulence is any extreme, unpredictable turbulence (figure 23-14). It is an especially dangerous phenomenon, caused by a drop or decline in the river bottom. The phenomenon also occurs when the water comes in contact with the river bend or rocks. The extreme amounts of froth created from the turbulence and the gravitational pull cause rafts to spin. Raft control is extremely limited because the lack of water viscosity causes resistance against paddles and a lack of buoyancy, making it difficult to float or maneuver. This type of white water can be impassable, depending on how

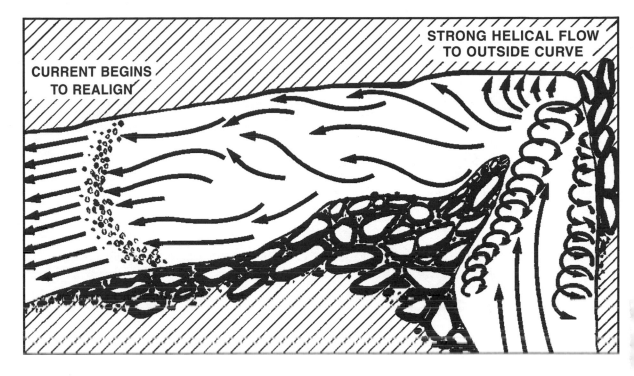

Figure 23-15. Anatomy of a River Bend.

extreme the dip and amount of water flow.

(6) Coming out of a sharp bend, the river currents are mixed, but the dangerous movement is still pulling. The result of the laminar flow shooting into a bank creates a helical-flow effect immediately below the turn, where the river is still trying to assume a natural "straight" flow. Being a liquid, water cannot resist stress, and it responds to a variety of obstacles (most common are submerged boulders) (figure 23-15).

(7) When water flows over obstructions, such as submerged boulders, the character of the laminar flow is changed. As the water flows over the top of the rock, the layers of the laminar flow increase in speed. This is known as a venturi effect. The hydraulic area is a type of "vacuum" formed as water flows around the rock. Created directly below the obstruction are confused and disordered currents that accelerate the layers of the laminar flow (figure 23-16).

(8) One type of hydraulic is the surge, which usually occurs when the current is slow and the water is deep. This hydraulic is formed downstream from an obstruction, with a surge in the water volume. When obstacles no longer have the ability to hold the water back, the pressure is released. Surges present few problems if the boulder or obstruction is covered with enough water flow to prevent contact when floating

over the top. Survivors should be aware of obstructions (known as "sleepers") if the water does not sufficiently cover them (figure 23-17). Failing to recognize a sleeper can result in raft destruction and severe bodily injuries. With large sleepers, the water flows over the top, creating a powerful current. This powerful, secondary current is trying to fill the vacuum created by the hydraulic downstream action.

(9) Another form of large sleeper is referred to as a dribbling fall. These are caused by minimal water flowing over submerged obstacles, with considerable drop below. This type of sleeper causes a bumpy ride, reducing speed, and can capsize the raft.

(10) Breaking holes occur where a large quantity of water flows over a sleeper, and the drop is not steep enough to create a suction hole (figure 23-18). A wave of standing water, much like an ocean breaker, is found downstream. This wave is stationary and can vary from 1 to 10 feet high, and even though it lacks the strong upriver flow of a suction hole, it can be a trap for rafts too small to climb up and over the crest. The size of the breaking hole and the survivor's seamanship must be considered before tackling this obstacle.

c. Suction Hole or Two-Dimensional Eddy:

(1) The vacuum created by a suction hole is strong enough to pull a survivor wearing a life preserver

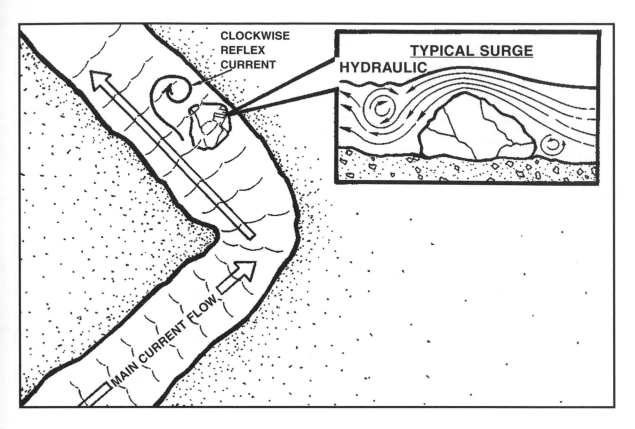

Figure 23-16. Response to a Submerged Rock.

beneath the surface. If pulled down into a suction hole, the survivor will normally be whirled to the surface downriver and returned to the suction hole by the upriver flow, to be pushed under once again. Objects too buoyant to sink usually remain trapped. It's often difficult to identify a suction hole because there is no frothing, no obvious curling water, and little noise. Extreme caution should be used when a large bulge appears in the water (figure 23-19). There are three possible ways of surviving and avoiding serious injury in suction holes. One way is to find the layer of water below the surface that is moving in the same desired direction. The second way is to reach down with a paddle or hand and feel for a current that is moving out of the hole. However, in a large suction hole the downstream flow will be too deep to reach. The survivor should attempt to cut across through the side of the eddy and into the water rushing by. The final and best solution is to scout ahead and try to identify the location of suction holes, and avoid them.

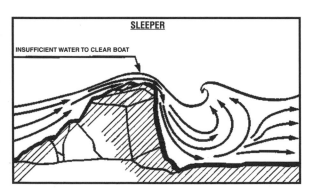

Figure 23-17. Sleeper.

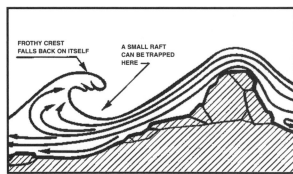

Figure 23-18. Breaking Hole.

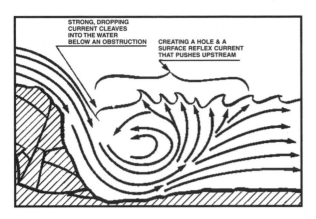

Figure 23-19. Suction Hole.

(2) An eddy (figure 23-20) is a reaction to an obstruction. The type of eddy that occurs next to the bank is caused by parts of the main current being deflected and forced to flow back upriver, where it again joins the mainstream. These areas are usually associated with quiet and slow-flowing water. They are also associated with areas where the river widens, or just above or below a bend in the river. An eddy has two distinct currents: the upstream current and the downstream current. The dividing line between the two is called an eddy fence. It is a line of small whirlpools spun off the upriver current by the power of the downstream current.

(3) A two-dimensional eddy occurs when the tip of an obstruction is slightly above the water and causes a two dimensional flow around the obstacle. Because of the speed and power of the current, the water is super-elevated and is significantly higher than the level of water directly behind the obstacle. This creates a hole that is filled by the flow of water around the obstacle (figure 23-21). Two-dimensional eddies that occur in midstream will create two eddy fences, one on each side of the obstacle. The water will enter the depression from both sides and will travel in a circular motion, clockwise from the right

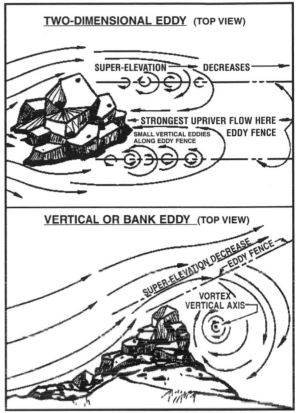

Figure 23-21. Two-Dimensional Eddies.

bank, counterclockwise from the left bank, and back upstream directly behind the obstacle. If the projection is large enough and a strong circular motion is created, it becomes a whirlpool. The outer reaches of the swirling water are super-elevated by centrifugal force and suction is created, similar to a drain in a bathtub. These vortexes are very rare and usually occur on huge rivers. Survivors may stop and rest where the eddies occur since there should be no strong swift currents in the eddy. If the obstacle is huge, it may be impossible to paddle fast enough to cross the eddy fence without being spun around in a pinwheeling manner.

d. Falls. In most falls, there are two reflex currents or suction holes forming, both whirling on crosscurrent axes into the falls (figure 23-22). One current falls behind the crashing main stream of water while the other falls in front. It's not too dangerous on a 2- or 3-foot drop, but when heights of 6 or more feet are present, it could be a death trap. The suction holes formed below these falls are inherently inescapable because of their power. The foam and froth formed at the bottom of falls

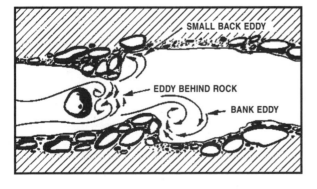

Figure 23-20. Eddies.

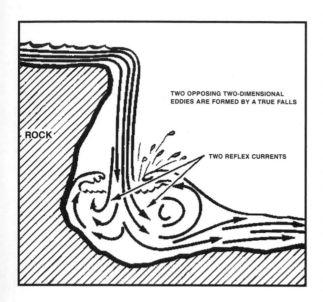

Figure 23-22. Falls.

can be dangerous. Jagged boulders and other hazards may be hidden.

e. Boils. A boil may occur below a fall or sleeper, downstream of the curling or reverse-current suction. This appears as a dome or mound-shaped water formation. Boils are the result of layers of flows hitting bottom, aimed upward, and reaching the surface and parting into a flower-like flow. The water billowing out in boils is super-oxygenated, taking away the resistance needed to push with the paddles or to suspend a survivor in a life preserver.

f. Rollers. Another difficulty found when traveling on fast rivers are rollers. Rollers are large, cresting waves caused by a variety of situations. A wave seen

below a breaking hole is one type of roller. Velocity waves are another type that occur on straight stretches of fast-dropping waters and are caused by the drag of sandy banks and submerged sandbars. They may be large enough to overturn a raft but are easily recognized, and are usually regular and easy to navigate by keeping the raft direction of travel in line with the crest of the wave (figure 23-23).

g. Tail Waves. Tail waves are quiet waves that are a reflex from the current hitting small rocks along the bed of the river and deflecting the current toward the surface. They are usually so calm that going over them is not noticeable.

h. Bank Rollers. Bank rollers are similar in appearance to crested tail waves and occur when there is a sharp bend in the riverbed that turns so sharply the water can't readily turn in the bend. The water slams into the outside bank and super-elevates, falling back upon itself. Small bank rollers cause few problems, but large ones cresting 5 feet or higher can capsize a raft (figure 23-24). There are three terms used to specify the severity or height of rollers; one being washboard. Washboard rollers are a series of swells that gently ripple and are safe and easy to ride. The next stage of rollers is called standing-water (cresting) rollers, in which the speed of contours on the bottom are such that the tops of the swells fall back onto themselves. The most dangerous and insurmountable rollers are referred to as haystacks, or roosters (figure 23-25). They resemble haystacks of water coming down from every direction—giant, frothing bulges. They may be so high a raft cannot go over them. It would be easy to become trapped in a deep trough (depression) and be buried under tons of frothing water. These troughs also hide dangers such as sharp rocks, which could tear a raft or break through the bottom of a boat.

i. Chutes. River chutes are good, logical travel routes, but they may harbor dangers. A chute (or tongue) is a

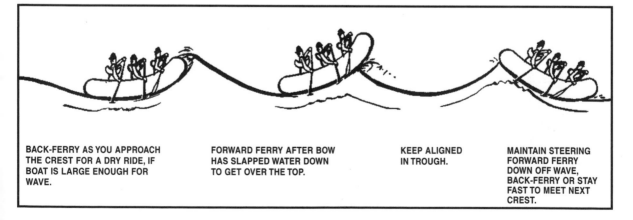

Figure 23-23. Tail Waves and Rollers.

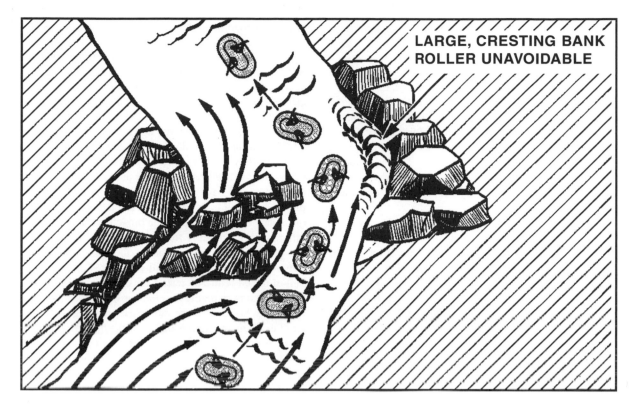

Figure 23-24. Bank Roller.

swift running, narrow passage between river obstructions caused by a damming effect. An example would be water forced between two large boulders. Because the water flow is restricted, it accelerates and a powerful current rushes through. Because of the water velocity, there may be a suction hole on either side of the chute (figure 23-26).

j. Logjams (Tongues of the Rapids). Logjams are

extremely dangerous. They consist of logs, brush, and debris collected from high waters that become lodged across the current. They remain stationary while the river flows through them. If a craft should be swept up against the stationary logs, it will be pinned in place by the current. Should the craft be tipped and swamped, it could be swept under the logjam. If the current is strong

Figure 23-25. Haystack.

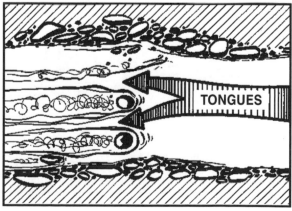

Figure 23-26. Tongues of the Rapids.

enough to do this, the occupants may also be pinned underneath the jam.

k. Sweepers. Sweepers can be the most dangerous obstructions on rivers. A sweeper is a large tree growing on a riverbank that has fallen over and is resting at or near the surface of the water. It may bounce up and down with the current. Survivors may be suddenly confronted with a sweeper that blocks the channel while rounding a bend in a river. The survivors are relatively helpless when they encounter sweepers in swift water. The only precautionary measure is to land above a bend in order to study the river ahead. (NOTE: Many people have met disaster by hitting sweepers.)

23-5. Emergency Situations:

a. Rock Collisions. Collisions with rocks above the surface of the water are common occurrences on a river. If the collision is unavoidable, survivors should spin the raft powerfully just before contact, or hit the rock bow-on. If the survivor is able to spin the raft, it will usually turn the raft off and around the rock. If they hit the rock bow-on, it will stop the raft momentarily, giving time to manipulate a spin-off with a few turn strokes. When the

bow-on method is used, occupants in the stern (aft) should move to the center of the tail before impact. This allows the stern to raise and the rushing current to slide under the raft. If the stern is low, the water will pile against it, causing the raft to be swamped. If survivors are colliding broadside with a rock, the entire crew must immediately jump to the side of the raft nearest the rock—always being the raft's downstream side (figure 23-27). This should be done before contact. If not, the river will flow over and suck down the raft's upstream tube. The raft will be flooded and the powerful force of the current will wrap it around the rock, possibly trapping some or all of the crew between the rock and the raft. Once it is unswamped, the broached (broadsided) raft on a rock is usually freed easily (figure 23-28). If two people push with both feet on the rock in the direction of the current, the raft will swing or slide into the pull of the current. The rest of the crew should shift to the end that is swinging into the current. These methods rarely fail; however, if the raft refuses to budge, it can be freed using the enormous power of the current. Large gear bags or sea anchors are securely tied to a long rope and secured to the end of the raft expected to swing downstream. The sea anchors (gear bags) are then tossed downstream into the current. (NOTE: A safety line should be used.)

b. Freeing a Wrapped Raft. Sometimes the powerful force of swift water may pin and wrap a raft around a rock. It is unusual for a raft to be equally balanced around the rock, so it will move more easily one way than it will in the other. The part of the raft with more weight and bulk should be moved toward the flow of the current. Lines are attached to at least two points on the raft so, when pulled, the force is equally distributed. One of these points is on the far end of the raft, around the tube, and is called the hauling line. (NOTE and CAUTION: A small hole may need to be cut in the floor of the raft to pass the line through and around the tube if

WRAPPED

HANDLING A BROADSIDE COLLISION WITH A ROCK

JUST BEFORE A RAFT BROADSIDES ON A ROCK, THE CREW SHOULD JUMP TO THE SIDE OF THE RAFT NEAREST THE ROCK. THIS LIFTS THE UPSTREAM TUBE AND ALLOWS THE CURRENT TO SLIDE EASILY UNDER THE RAFT. IF THE CREW IS NOT QUICK TO HOP TOWARD THE ROCK, SWIFT WATER WILL SPEEDILY SWAMP THE RAFT AND WRAP IT FLAT AROUND THE ROCK'S UPSTREAM FACE.

Figure 23-27. Handling Broadside Collision with Rock.

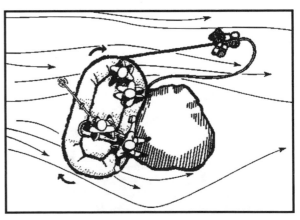

Figure 23-28. Freeing Unswamped Raft.

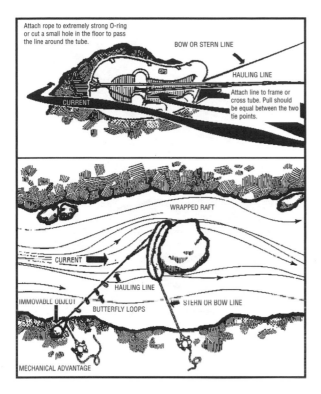

Attach rope to extremely strong O-ring or cut a small hole in the floor to pass the line around the tube.

BOW OR STERN LINE

HAULING LINE

CURRENT

Attach line to frame or cross tube. Pull should be equal between the two tie points.

WRAPPED RAFT

CURRENT

HAULING LINE

IMMOVABLE OBJECT

BUTTERFLY LOOPS

STERN OR BOW LINE

MECHANICAL ADVANTAGE

Figure 23-29. Freeing a Wrapped Raft.

there is no ring to pass it through.) The second tie-off point can be a cross tube. One person should hold the line attached to the stern of the raft. The raft should be moved over the rock into the pull of the current. Once it is freed, the person holding the stern line can move the raft to a safe position (figure 23-29).

c. Raft Flips. When a raft is about to flip, there is little time to react to the situation. If the raft is diving into a big hole, the primary danger is being violently thrown forward into a solid object in the raft. Survivors should protect themselves by dropping low and flattening themselves against the backside of the baggage or cross tube. If the raft is being upset by a rock, fallen tree, or other obstacle, members should jump clear of the raft to prevent being crushed against the obstruction, or struck by the falling raft. If the raft is pinned flat against an obstruction, the members should stay with the raft and try to safely climb up the obstruction.

d. Lining Unrunnable Rafts. Lining a raft through rapids is basically letting it run through rapids with a crew on shore controlling it by attached lines (figure 23-30). The raft should be moved slowly by maintaining tight control of the lines attached to the bow and stern. If no strong eddies or steep, narrow chutes are present, one member should walk along the shore and control the raft with a strong, long line. The lines running to shore should be long enough to allow the raft full travel through the rapids.

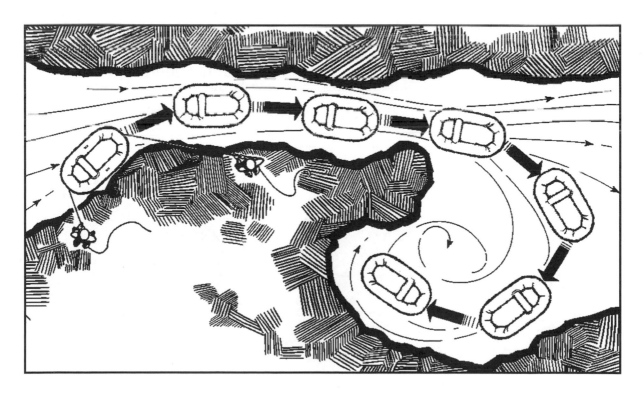

Figure 23-30. Lining a Chute.

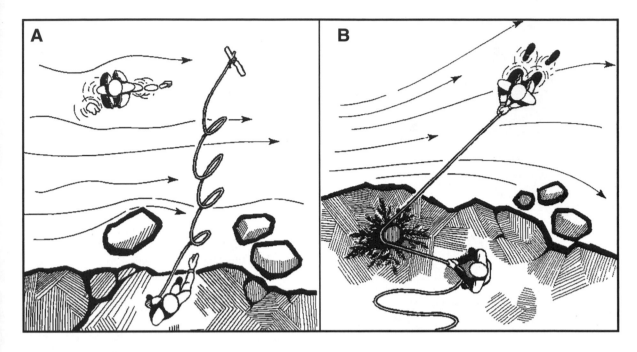

Figure 23-31. Making a Rescue.

e. Rescuing a Swimmer from Shore. The rescuer should carefully choose the right spot where the coil of rope thrown to the swimmer will not cross hazardous areas or obstacles, and yet be near a rock or tree that can be used for belaying (securing without being tied) the end of the line. The person throwing the line should make sure the line has a flotation device (life preserver) at the end before it is thrown to the swimmer. The rope should be coiled in a manner that will allow it to flow smoothly, without entanglement, to full extension. One hand holds one-half to one-third of the coil while the other throws the remainder of the coil out. The weighted coil should be thrown to a spot where the swimmer will drift, which is usually downstream, in front of the swimmer. As the rope travels out, all of the line, except the last 10 or 12 feet, should be uncoiled.

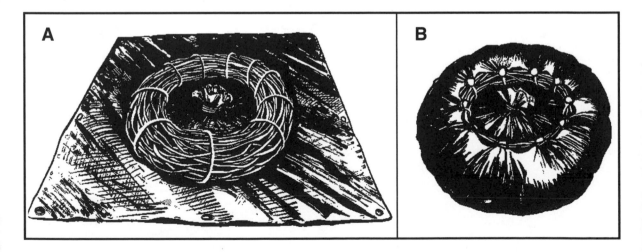

Figure 23-32. Poncho Ring Raft.

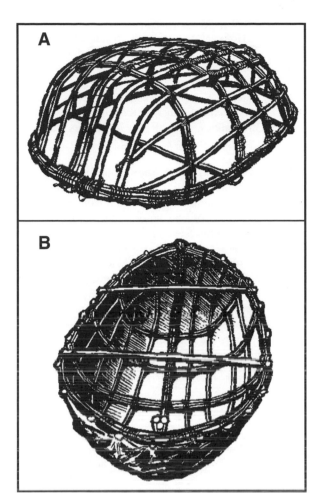

Figure 23-33. Bull Boat.

The member pulling the survivor should be braced, or have the line around a rock or tree to hold the swimmer once the line has been reached by the survivor.

f. The Swimmer's Responsibility. The swimmer should be aware of the rescuer's location and face downstream when waiting for the hauling line. When the line is thrown, the swimmer will usually be required to swim to reach it. Once holding the line, the swimmer should be prepared for a very strong pull from the current and line. The line should be held tightly, but not wrapped around a wrist or hand. Entanglement must be avoided. The swimmer should pull, hand over hand, until reaching shallow water, and then use the rope for steadiness while walking to shore (figure 23-31).

23-6. Improvised Rafts:

a. Types of Flotation Devices. There are various types of flotation devices that may be improvised and used as rafts for equipment and personnel.

(1) The Poncho (doughnut) raft (figure 23-32) can be used for transporting equipment but is not a good vehicle for people. The raft is constructed by using saplings or pliable willows and a waterproof cover. A hoop-shaped framework of saplings or pliable willow is constructed within a circle of stakes. The hoop is tied with cordage or suspension line and removed from the circle of stakes and placed on the waterproof cover to which it will be attached. Clothing and (or) equipment is then placed in the raft, and the survivor swims, pushing the raft.

(2) The bull boat (figure 23-33) is a shallow-draft skin boat shaped like a tub and formerly used by American Indians in the Great Plains area. Survivors should construct an oval frame, similar to a canoe, of willow or other pliable materials and cover the framework with waterproof material such as a signal paulin or skins. This makes a craft that is suitable for transporting equipment across a river with the survivor propelling it from behind.

(3) An emergency boat can be made by stretching a tarpaulin or light, canvas cover over a skillfully shaped framework of willows and adding a well-framed keel of green wood, such as slender pieces of spruce. Gunwales (sides) of slender saplings are attached at both ends and the spreaders or thwarts are attached as in a canoe. Ribs of strong willows are tied to the keel. The ends of the ribs are bent upward and tied to the gunwales. The inside of the frame is closely covered with willows to form a deck upon which to stand. Such a boat is easy to handle and is buoyant, but lacks the strength necessary for long journeys. This boat is entirely satisfactory for ferrying a group across a broad, quiet stretch of river. When such a boat has served its purpose, the cover should be removed for later use.

(4) The vegetation raft is built of small vegetation that will float, and is placed within clothing or a parachute to form a raft for a survivor and (or) equipment. Plants such as water hyacinth or cattail may be used (figure 23-34).

(5) A good floating device for the single survivor can be fabricated by using two balsa logs or other lightweight wood. The logs should be placed about 2 feet apart and tied together. The survivor sits on the lines and travels with the current (figure 23-35).

(6) A dugout canoe is good transportation but difficult to construct. One method is to build a long fire on the side to be dug out, and chop away the burned material when the fire is out. Repeat this procedure as often as necessary.

b. Building a Raft. The greatest problem in raft construction (figure 23-36) is being able to construct a craft strong enough to withstand the buffeting it may have to take from rocks and swift water. Even if 6- to 8-inch

Figure 23-34. Vegetation Bag.

spikes are available, they are not satisfactory since they pull or twist out easily. Rope quickly wears out from frequent, rough contact with rocks and gravel. Northern woodsmen have evolved a construction method (figure 23-37) that requires neither spikes nor rope, yet produces a raft superior in strength. The only materials required are logs, although rope is sometimes useful; the only tools needed are an axe and a sheath knife.

23-7. Fording Streams:

a. Survivors traveling on foot through wilderness areas may have to ford some streams. These can range from small, ankle-deep brooks to large rivers. Rivers are often so swift that a survivor can hear boulders on the bottom being crashed together by the current. If these streams are of glacial origin, the survivor should wait for them to decrease in strength during the night hours before attempting to ford.

b. Careful study is required to find a place to safely ford a stream. If there is a high vantage point beside the ever, the survivor should climb the rise and look over the river. Finding a safe crossing area may be easy if the river breaks into a number of small channels. The area on the opposite bank should be surveyed to make sure travel will be easier after crossing. When selecting a fording site, the survivor should:

(1) When possible, select a travel course that leads across the current at about a 45-degree angle downstream.

(2) Never attempt to ford a stream directly above, or close to, a deep or rapid waterfall or a deep channel. The stream should be crossed where the opposite side is comprised of shallow banks or sandbars.

(3) Avoid rocky places, since a fall may cause serious injury. However, an occasional rock that breaks the current may be of some assistance. The depth of the water is not necessarily a deterrent. Deep water may run more slowly and be safer than shallow water.

(4) Before entering the water, the survivors should have a plan of action for making the crossing. Use all possible precautions, and if the stream appears treacherous, take the steps shown in figure 23-38.

23-8. Traveling on Open Seas. Four-fifths of the Earth's surface is covered by open water. Although accounts of sea survival incidents are often gloomy, successful survival is possible; however, the raft is at the mercy of the currents and winds. There are many currents, both warm and cold, throughout the seas.

a. Currents. Sea currents flow in a clockwise direction in the Northern Hemisphere and counterclockwise in the Southern Hemisphere. This is caused by three factors: the sun's heat, the winds, and the Earth's rotation (Coriolis effect). Most sea currents travel at speeds of less than 5 miles per hour. Using currents as a mode of travel can be done by putting out the sea anchor and letting the current pull the raft along. Survivors should use

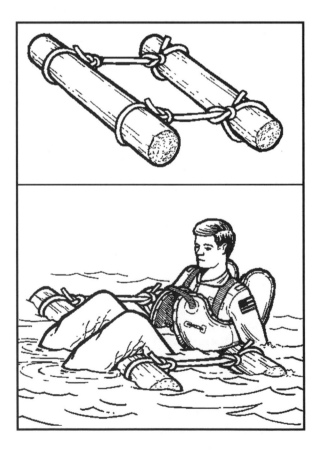

Figure 23-35. Log Flotation.

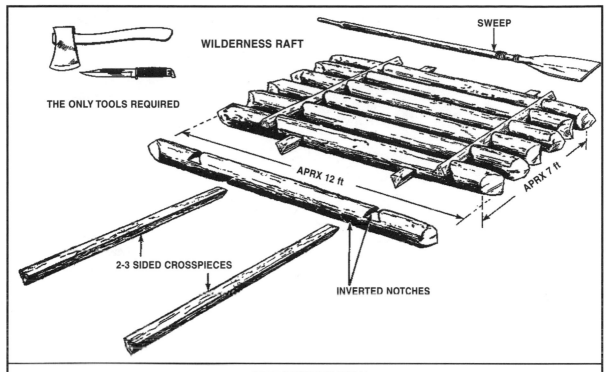

THE ONLY TOOLS REQUIRED

WILDERNESS RAFT

SWEEP

APRX 12 ft

APRX 7 ft

2-3 SIDED CROSSPIECES

INVERTED NOTCHES

RAFT CONSTRUCTION

A raft for three persons should be about 12 feet long and 6 feet wide, depending on the size of the logs used. The logs should be 12 to 14 inches in diameter and so well-matched in size that notches you make in them are level when crosspieces are driven into place.

Build the raft on two skid logs placed so they slope downward to the bank. Smooth the logs with an axe so the raft logs lie evenly on them. Cut two sets of slightly offset, inverted notches, one in the top and bottom of both ends of each log. Make the notches broader at the base than at the outer edge of the log, as shown in the illustration. Use small poles with straight edges or a string pulled taut to make the notches. A three-sided wooden crosspiece about a foot longer than the total width of the raft is to be driven through each end of the four sets of notches.

Complete the notches on all logs at the top of the logs. Turn the logs over and drive a 3-sided crosspiece through both sets of notches on the underside of the raft. Then complete the top set of notches, and drive through the two additional sets of crosspieces.

You can lash together the overhanging ends of the two crosspieces at each end of the raft to give it added strength; however, when the crosspieces are immersed in water they swell and tightly bind the raft logs together.

If the crosspieces fit too loosely, wedge them with thin, boardlike pieces of wood split from a dead log. When the raft is in water, the wood swells, and the crosspieces become very tight and strong.

Make a deck of light poles on top of the raft to keep packs and other gear dry.

Figure 23-36. Raft Construction.

caution when traveling through areas where warm and cold currents meet. It could be a storm-forming area with dense fog and high winds and waves.

b. Winds. Winds also aid raft travel. In tropical areas, the winds are easterly blowing (trade winds). In higher latitudes, they blow from the west (westerlies). To use the winds as a mode of travel, the sea anchor should be pulled into the raft and, if available, a sail should be erected.

c. Waves. Waves can be both an asset and a hazard to raft travel. Waves are normally formed by the wind. The severity of the wind determines the size of the waves. On open seas, waves range from a few inches to more than 100 feet in height. Under normal conditions, waves alone will move a liferaft only a few inches at a time; therefore, using waves as a mode of propulsion is not practical. Waves are a great help in finding land or shallow areas in the sea. Ocean waves always break when they enter shallow water or when they encounter an obstruction. The force of energy of the wave depends on how abruptly the water depth decreases and on the size of the waves. Breaking waves can be used as an aid to

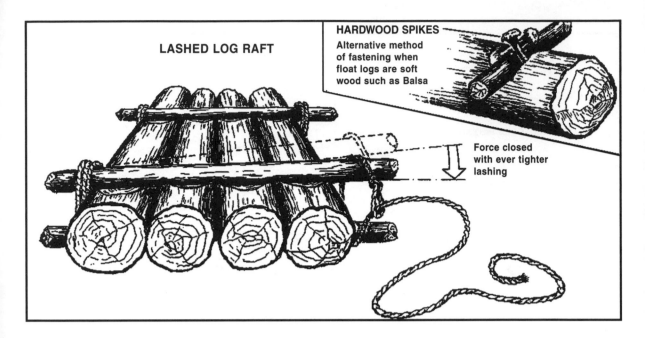

Figure 23-37. Lashed Log Raft.

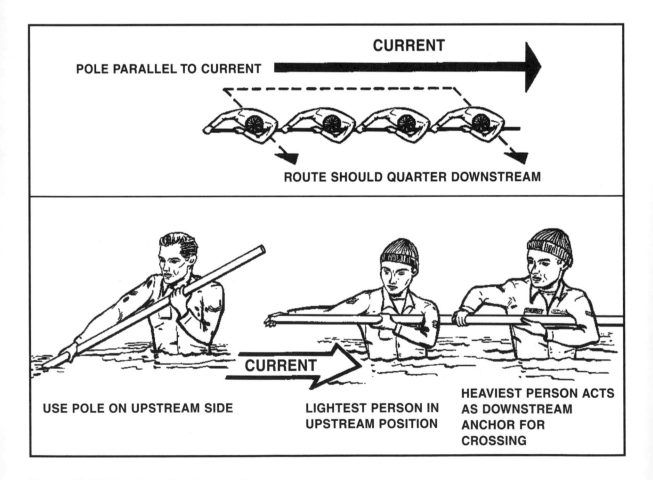

Figure 23-38. Fording a Trecherous Stream.

make a landfall. Storms at sea are probably the greatest hazard to survivors in rafts. Aside from the waves created by a storm, the wind and rain can make life in a raft very difficult. The waves and wind can capsize a raft, or throw a person out of the raft, and then will constantly fill the raft with water. Seasickness can result from gentle to severe wave action. Additionally, rescue efforts may be severely hampered by large waves.

d. Tides. Tides are another form of wave, but they are very predictable. They occur twice daily and usually cause no problems for anyone in a raft. Tides may range from one to 40 feet in height depending on the area of the world. They should be considered when planning a landing. When the tide is going in, it will help propel the raft to shore. The action of the water going away from shore when the tide is going out makes landing difficult.

e. Hazards. Certain marine life must be considered a hazard to survival. Sharks, jellyfish, eels, and most reef fishes can cause serious injuries if encountered. Waste materials should be disposed of when these creatures are not present.

(1) The survivor should be aware of the saltwater (estuarine) crocodile. It is found throughout the Southeast Asian shoreline. It is a well-known maneater, and is almost always found in salt or brackish water. It is more commonly found near river mouths and along the coasts; however, it has been known to swim as many as 40 miles out into the sea. Females with nests are likely to be vicious and aggressive. They can grow to a length of 30 feet, but most specimens are less than 15 feet in length. Survivors should watch for this reptile while landing their raft or fishing.

(2) The survivor will normally encounter reef fishes during the landing process by stepping on them. They may also be caught while fishing. Clothing and footgear must be worn at all times, whether landing raft or fishing.

(3) Coral is normally found in warm waters along the shores of islands and mainlands. There are many different types of coral. They should be avoided since all can destroy a raft or severely injure a survivor. It is best to stay in the raft when coral is encountered. If the survivor must wade to shore, footgear and pants should be worn for protection. Moving slowly and watching every step may prevent serious injuries. Coral does not exist where freshwater enters the sea.

(4) Ships can be a welcome sight to the survivor, but they can also be a hazard. Since the raft is a small object in a very large sea, the survivor must constantly be aware that at night or during inclement weather, the raft will be difficult to see and could be struck by a large ship.

f. Early Considerations. Survivors should stay upwind and clear of the aircraft (out of fuel-covered waters), but in the vicinity of the crash until the aircraft sinks. A search for survivors is usually activated around the entire area of and near the crash site. Missing personnel may be unconscious and floating low in the water. Rescue procedures are illustrated in figure 23-39.

(1) The best technique for rescuing aircrew members from the water is to throw them a line with a life preserver attached. The second is to send a swimmer (rescuer) from the raft with a line, using a flotation device that will support the weight of a rescuer. This will help to conserve energy while recovering the survivor. The least acceptable technique is to send an attached swimmer without floatable devices to retrieve a survivor. In all cases, the rescuer should wear a life preserver. The strength of a person in a state of panic in the water should not be underestimated. A careful approach can prevent injury to the rescuer.

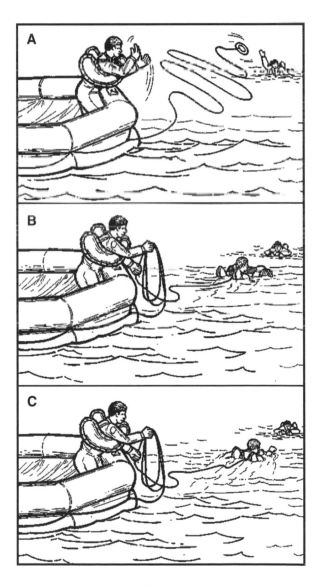

Figure 23-39. Rescue from Water.

(2) When the rescuer is approaching a survivor in trouble from behind, there is little danger of being kicked, scratched or grabbed. The rescuer should swim to a point directly behind the survivor and grasp the backstrap of the life preserver. A sidestroke may then be used to drag the survivor to the raft.

(3) All debris from the aircraft should be inspected and salvaged (rations, canteens, thermoses and other containers, parachutes, seat cushions, extra clothing, maps, etc.). Secure equipment to the raft to prevent loss. Special precaution should be taken with flashlights and signaling equipment to keep them dry so they will function when needed.

(4) Rafts should be checked for inflation. Leaks and points of possible chafing should be repaired, as required. All water should be removed from inside the raft. Care should be taken to avoid snagging the raft on shoes or sharp objects. Placing the sea anchor out will slow the rate of drift. If there is more than one raft, they should be connected with at least 25 feet of line. The lifeline attached to the outer periphery of the raft is to be used for the connection. Donning the antiexposure suit is essential in cold climates. Erecting windbreaks, spray shields, and canopies will protect survivors from the elements. Survivors should huddle together and exercise regularly to maintain body heat.

(5) Monitoring the physical condition of survivors and administering first aid to survivors is essential. If available, seasickness pills will help prevent vomiting and the resulting dehydration.

(6) Survivors should prepare all available signaling equipment for immediate use.

(7) Compasses, watches, matches, and lighters will become worthless unless they are kept dry.

(8) The raft repair plugs should be attached to the raft for easy access as soon as possible.

(9) All areas of the body should be protected from the sun. Precautions should be taken to prevent sunburn on the eyelids, under the chin, and on the backs of the ears. Sunburn cream and Chapstick will protect these areas.

(10) The leader should calmly analyze the situation and plan a course of action, to include duty assignments (watch duty, procuring and rationing food and water, etc.). All survivors, except those who are badly injured or completely exhausted, are expected to perform watch duty, which should not exceed 2 hours. The survivor on watch should be looking for signs of land, passing vessels or aircraft, wreckage, seaweed, schools of fish, birds, and signs of chafing or leaking of the raft.

(11) Food and water can be conserved by saving energy. Survivors should remain calm.

(12) Maintaining a sense of humor will help keep morale high.

(13) The survivor(s) should remember that rescue at sea is a cooperative effort. Search-aircraft contact is limited by the visibility of survivors based on the availability of visual or electronic signaling devices. Visual and electronic communications can be increased by using all available signaling devices (signal mirrors, radios, signal panels, dye marker, and other available devices) when an aircraft is in the area.

(14) A log should be maintained with a record of the navigator's last fix, time of ditching, names and physical condition of survivors, ration schedule, winds, weather, direction of swells, times of sunrise and sunset, and other navigation data.

23-9. Physical Considerations:

a. The greatest problem a survivor is faced with when submerged in cold water is death due to hypothermia. When a survivor is immersed in cold water, hypothermia occurs rapidly due to the decreased insulating quality of wet clothing and the fact that water displaces the layer of still air that normally surrounds the body. Water causes a rate of heat exchange approximately 25 times greater than air at the same temperature. The following lists life-expectancy times:

Temperature of Water	Time
70° - 60°	12 hours
60° - 50°	6 hours
50° - 40°	1 hour
40° - below	-1 hour

NOTE: These times may be increased with the wearing of an antiexposure suit.

b. The best protection for a survivor against the effects of cold water is to get into the liferaft, stay dry, and insulate the body from the cold surface of the bottom of the liferaft. If this is not possible, wearing of the antiexposure suit will extend a survivor's life expectancy considerably. It's important to keep the head and neck out of the water and well insulated from the cold-water effects when the temperature is below 66°F. The wearing of life preservers increases the predicted survival time just as the body position in the water increases the probability of survival. The following table shows predicted survival times for an average person in 50°F water:

Situation	Predicted Survival Time (Hours)
No Flotation	
Drown-proofing	1.5
Treading Water	2.0
With Flotation	
Swimming	2.0
Holding Still	2.7
Help	4.0
Huddle	4.0

Figure 23-40. HELP Position.

(1) HELP Body Position. Remaining still and assuming the fetal position, or heat escape lessening posture (HELP; figure 23-40), will increase the downed crewmember's survival time. About 50 percent of body heat is lost from the head. It is therefore important to keep the head out of the water. Other areas of high heat loss are the neck, the sides, and the groin.

Figure 23-41. Huddling for Temperature Conservation.

(2) Huddling. If there are several survivors in the water, huddle close, side to side in a circle, and body heat will be preserved (figure 23-41).

23-10. Life Preserver Use:
a. Survival Swimming Without a Life Preserver. A survivor who knows how to relax in the water is in little danger of drowning, especially in saltwater, where the body is of lower density than the water. Trapped air in clothing will help buoy the survivor in the water. If in the water for long periods, the survivor will have to rest from treading water. The survivor may best do this by floating on the back. If this is not possible, the following techniques should be used: Rest erect in the water and inhale; put the head face-down in the water and stroke with the arms; rest in this face-down position until there is a need to breathe again; raise the head and exhale; support the body by kicking arms and legs and inhaling; then repeat the cycle.

b. Swimming with a Life Preserver. The bulkiness of clothing, equipment, and (or) any personal injuries will necessitate the immediate need for flotation. Normally, a life preserver will be available for donning before entering the water.

(1) Proper inflation of the life preserver must be done after clearing the aircraft but, preferably, before entering the water. Upon entering the water, the two cells of the life preserver should be fastened together. Limited swimming may be done with the life preserver inflated by cupping the hands and taking strong strokes deep into the water. The life preserver may be slightly deflated to permit better arm movement.

(2) The backstroke should be used to conserve energy when traveling long distances. If aiding an injured or unconscious person, the sidestroke may have to be used. When approaching an object, it is best to use the breaststroke. If a group must swim, they should try to have the strongest swimmer in the lead, with any injured persons intermingled within the group. It is best to swim in single file.

23-11. Raft Procedures.
There are three needs that can be satisfied by most rafts; they are personal protection, mode of travel, and evasion and camouflage.

a. One-Man Raft:
(1) The one-man raft has a main cell inflation. If the CO_2 bottle should malfunction and not inflate the raft, or if the raft develops a leak, it can be inflated orally. The spray shield acts as a shelter from the cold, wind, and water. In some cases, this shield serves as insulation. The insulated bottom plays a significant role in the survivor's protection from hypothermia by limiting the conduction of the cold through the bottom of the raft (figure 23-42).

(2) Travel is more effectively made by inhaling or deflating the raft to take advantage of the wind or current.

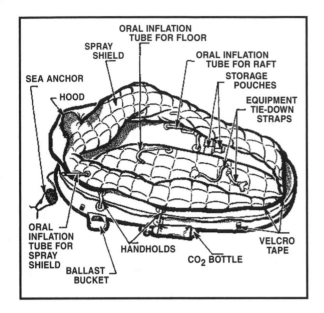

Figure 23-42. One-Man Raft with Spray Shield.

The spray shield can be used as a sail, while the ballast buckets serve to increase raft drag in the water. The last device that may be used to control the speed and direction of the raft is the sea anchor. (NOTE: The primary purpose of the sea anchor is to stabilize the raft.)

(3) Black rafts have been developed for use in tactical areas. These rafts blend in with the background of the sea. The raft can be further modified for evasion by partially deflating, which provides a low profile.

(4) The one-man raft is connected to the aircrew member by a lanyard parachuting to the water. Survivors should not swim to the raft, but pull it to their position

via the lanyard (figure 23-43). The parachute J-l releases should be closed and the life preserver separated before boarding the raft. The raft may hit the water upside down, but may be righted by approaching the bottle side and flipping it over. The spray shield must be in the raft to expose the boarding handles.

(5) If the survivor has an arm injury, boarding is best done by turning the back to the small end of the raft, pushing the raft under the buttocks, and lying back (figure 23-44). Another method of boarding is to push down on the small end until one knee is inside, and lie forward (figure 23-44).

(6) In rough seas, it may be easier for the survivor to grasp the small end of the raft and, in a prone position, kick and pull into the raft. Once in the rain, lying face down, the sea anchor should be deployed and adjusted. To sit upright in the raft, one side of the seat kit might have to be disconnected, and the survivor should roll to that side. The spray shield is then adjusted. There are two variations of the one-man raft, with the improved model incorporating an inflatable spray shield and floor for additional insulation. The spray shield is designed to help keep the survivor dry and warm in cold oceans, and protect them from the sun in the hot climates (figure 23-45).

(7) The sea anchor can be adjusted to either act as a drag by slowing down the rate of travel with the current, or as a means of traveling with the current. This is done by opening or closing the apex of the sea anchor. When opened, the sea anchor (figure 23-46) will act as a drag, and the survivor will stay in the general area. When the sea anchor is closed (figure 23-46), it will form a pocket for the current to strike and propel the raft in the direction of the current. Additionally, the sea anchor should be adjusted so when the raft is on the crest of a wave, the sea anchor is in the trough of the wave (figure 23-47).

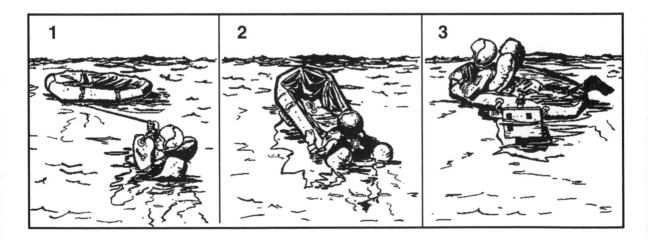

Figure 23-43. Boarding One-Man Raft.

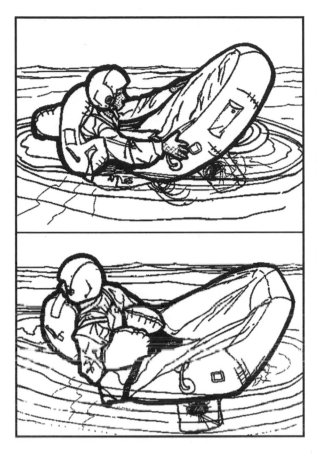

Figure 23-44. Boarding One-Man Raft (Other Methods).

Figure 23-45. One-Man Raft with Spray Shield Inflated.

b. Seven-Man Raft:

(1) The seven-man raft is found on some multiplace aircraft. It can also be found in the survival drop kit (MA-1 Kit, figure 23-48). This type of raft may inflate upside down and may, therefore, require the survivor to right the raft before boarding. The J-l releases should be closed before boarding. The survivor should always work from the bottle side to prevent injury if the raft turns over. Facing into the wind provides additional assistance in righting the raft. The handles on the inside bottom of the raft are used for boarding (figure 23-49).

(2) The boarding ladder is used to board if someone assists in holding down the opposite side. If no assistance is available, the survivor should again work from the bottle side with the wind at the back to help hold down the raft. The survivor should separate the life preserver, grasp an oarlock and boarding handle, kick the legs to get the body prone on the water, and then kick and pull into the raft. If the survivor is weak or injured, the raft may be partially deflated to make boarding easier (figure 23-50).

(3) Manual inflation can be done by using the pump to keep buoyancy chambers and cross-seat firm, but the raft should not be over-inflated. The buoyancy chambers and cross-seat should be rounded but not drum-tight. Hot air expands, so on hot days, some air may be released, while air may be added on cold days (figure 23-51).

c. Sailing a Raft into the Wind. Rafts are not equipped with keels, so they cannot be sailed into the wind. However, anyone can sail a raft downwind, and multiplace (except 20/25-man) rafts can be successfully sailed 10 degrees off from the direction of the wind. An attempt to sail the raft should not be made unless land is near. If the decision to sail is made and the wind is blowing toward a desired destination, survivors should fully inflate the raft, sit high, take in the sea anchor, rig a sail, and use an oar as a rudder, as shown in figure 23-52.

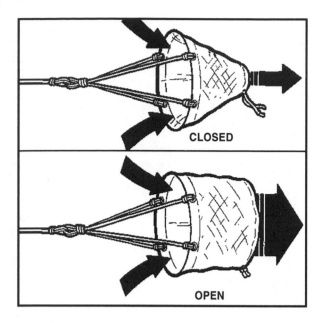

Figure 23-46. Sea Anchor.

d. Multiplace Raft. In a multiplace (except 20/25-man) raft, a square sail should be erected in the bow using oars, with their extensions as the mast and crossbar (figure 23-52). A waterproof tarpaulin or parachute material may be used for the sail. If the raft has no regular mast socket and step, the mast may be erected by tying it securely to the front cross-seat, using braces. The bottom of the mast must be padded to prevent it from chafing or punching a hole through the floor, whether or not a socket is provided. The heel of a shoe, with the toe wedged under the seat, makes a good improvised mast step. The corners of the lower edge of the sail should not be secured. The lines attached to the corners are held with the hands so a gust of wind will not rip the sail, break the mast, or capsize the raft. Every precaution must be taken to prevent the raft from

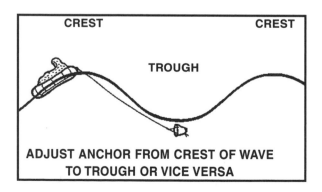

Figure 23-47. Deployment of the Sea Anchor.

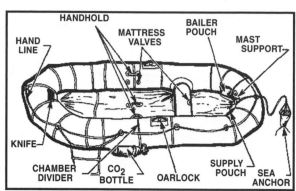

Figure 23-48. Seven-Man Raft.

Figure 23-49. Method of Righting Raft.

Figure 23-50. Method of Boarding Seven-Man Raft.

Figure 23-51. Inflating the Raft.

turning over. In rough weather, the sea anchor is kept out and away from the bow. The passengers should sit low in the raft, with their weight distributed to hold the upwind side down. They should also avoid sitting on the sides of the raft or standing up to prevent falling out. Sudden movements (without warning the other passengers) should be avoided. When the sea anchor is not in use, it should be tied to the raft and stowed in such a manner that it will hold immediately if the raft capsizes.

e. Twenty- to Twenty-Five-Man Rafts. The 20/25-man rafts may be found in multiplace aircraft (figures 23-53 and 23-54). They will be found in accessible areas of the fuselage, or in raft compartments. Some may be automatically deployed from the cockpit, while others may need manual deployment. No matter how the raft lands in the water, it's ready for boarding. The accessory kit is attached by a lanyard and is retrieved by hand. The center chamber must be inflated manually with the hand pump. The 20/25-man raft should be boarded from the aircraft if possible, if not, the following steps should be taken:

(1) Approach lower boarding ramp.

(2) Separate the life preserver.

(3) Grasp the boarding handles and kick the legs to put the body into a prone position on the water's surface; then kick and pull until inside the raft.

(4) If for any reason the raft is not completely inflated, boarding will be made easier by approaching the intersection of the raft and ramp, grasping the upper boarding handle, and swinging one leg on to the center of the ramp as in mounting a horse (figure 23-55).

(5) The equalizer tube should be clamped immediately upon entering the raft to prevent deflating the entire raft, in case of puncture (figure 23-56).

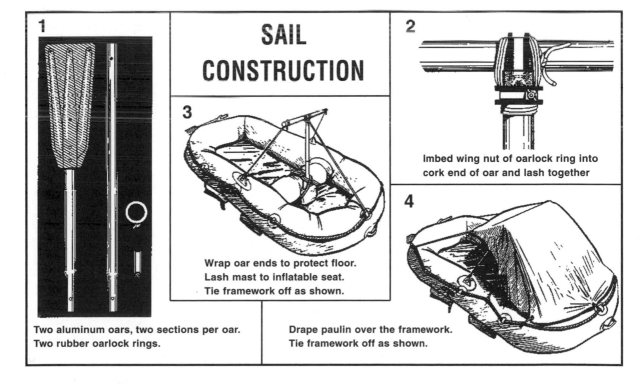

Figure 23-52. Sail Construction.

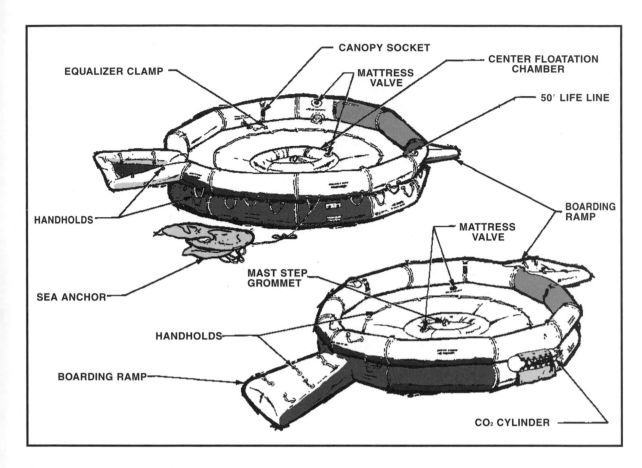

Figure 23-53. 20-Man Raft.

(6) The 20/25-man raft can be inflated by using the pump to keep the chambers and center ring firm. They should be well rounded but not drum-tight (figure 23-57).

23-12. Making a Landfall. The lookout should watch carefully for signs of land. Some indications of land are:

a. A fixed cumulus cloud in a clear sky, or in a sky where all other clouds are moving, often hovers over or slightly downwind from an island.

b. In the tropics, a greenish tint in the sky is often caused by the reflection of sunlight from shallow lagoons or shelves of coral reefs.

c. In the Arctic, ice fields or snow-covered land are often indicated by light-colored reflections on clouds, quite different from the darkish gray reflection caused by open water.

d. Deep water is dark green or dark blue. Lighter color indicates shallow water, which may mean land is near.

e. In fog, mist, rain, or at night, when drifting past a nearby shore, land may be detected by characteristic odors and sounds. The musty odor of mangrove swamps and mudflats and the smell of burning wood carries a long way. The roar of surf is heard long before the surf is seen. Continued cries of sea birds from one direction indicate their roosting place on nearby land.

f. Birds are usually more abundant near land than over the open sea. The direction from which flocks fly at dawn and to which they fly at dusk may indicate the direction of land. During the day, birds are searching for food, and the direction of flight has no significance unless there is a storm approaching.

g. Land may be detected by the pattern of the waves, which are refracted as they approach land. Figure 23-58 shows the form the waves assume. Land should be located by observing this pattern and turning parallel to the slightly turbulent area (marked "X" on the illustration), and following its direction.

23-13. Methods of Getting Ashore:

a. **Swimming Ashore.** This is a most difficult decision. It depends on many things. Some good swimmers

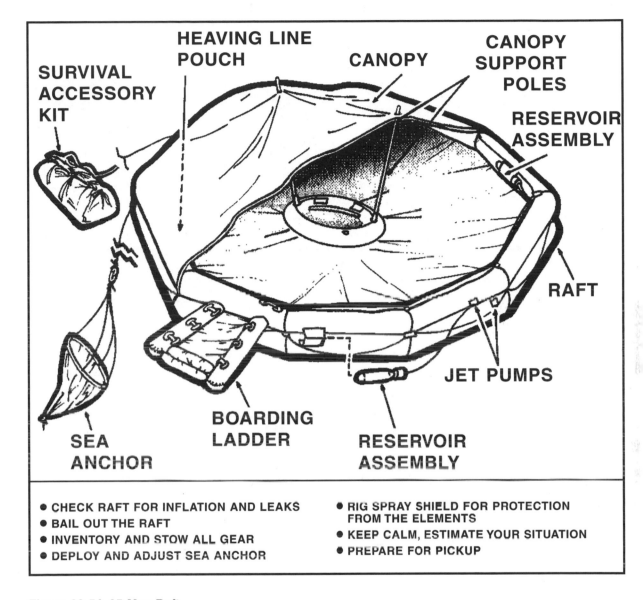

- CHECK RAFT FOR INFLATION AND LEAKS
- BAIL OUT THE RAFT
- INVENTORY AND STOW ALL GEAR
- DEPLOY AND ADJUST SEA ANCHOR
- RIG SPRAY SHIELD FOR PROTECTION FROM THE ELEMENTS
- KEEP CALM, ESTIMATE YOUR SITUATION
- PREPARE FOR PICKUP

Figure 23-54. 25-Man Raft.

have been able to swim eight-tenths of a mile in 50°F water before being overcome by hypothermia. Others have not been able to swim 100 yards. Furthermore, distances on the water are very deceptive. In most instances, staying with the raft is the best course of action. If the decision is made to swim, a life preserver or other flotation aid should be used. Shoes and at least one thickness of clothing should be worn. The side or breaststroke will help conserve strength.

(1) If surf is moderate, the survivor can ride in on the back of a small wave by swimming forward with it, and making a shallow dive to end the ride just before the

wave breaks. The swimmer should stay in the trough between waves in high surf, facing the seaward wave and submerging when the wave approaches. After the wave passes, the swimmer should work shoreward in the next trough.

(2) If the swimmer is caught in the undertow of a large wave, push off the bottom, swim to the surface, and proceed shoreward. A place where the waves rush up onto the rocks should be selected if it is necessary to land on rocky shores, but avoid places where the waves explode with a high, white spray. After selecting the landing point, the swimmer should advance behind a

Figure 23-55. Boarding 20-Man Raft.

large wave into the breakers. The swimmer should face shoreward and take a sitting position with the feet in front, 2 or 3 feet lower than the head, so the knees are bent and the feet will absorb shocks when landing or striking submerged boulders or reefs. If the shore is not reached the first time, the survivor should swim with hands and arms only. As the next wave approaches, the

sitting position with the feet forward should be repeated until a landing is made.

(3) Water is quieter in the lee of a heavy growth of seaweed. This growth can be very helpful. The swimmer should crawl over the top by grasping the vegetation with overhand movements.

(4) A rocky reef should be crossed in the same way as landing on a rocky shore. The feet should be close together with knees slightly bent in a relaxed sitting posture to cushion blows against coral.

b. Rafting Ashore. In most cases, the one-man raft can be used to make a shore landing with no danger. Going ashore in strong surf is dangerous. The time should be taken to sail around and look for a sloping beach where the surf is gentle. The landing point should be carefully selected. Landing when the sun is low and straight in front is not recommended. The survivor

1 TIGHTEN EQUALIZER CLAMP   **2 DEPLOY SEA ANCHOR**

3 DEPLOY LIFE LINE AND FORM CHAIN OF SURVIVORS STILL IN WATER

4 ERECT THE RAFT CONVOY

Figure 23-56. Immediate Action—Multiplace Raft.

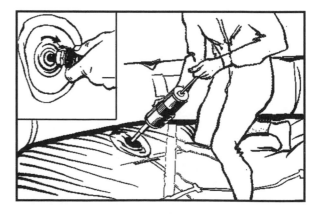

Figure 23-57. Inflating the 20-Man Raft.

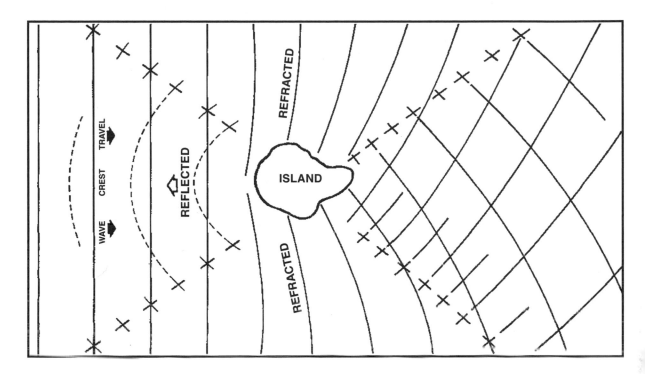

Figure 23-58. Diagram of Wave Patterns About an Island.

should look for gaps in the surf line and head for them, while avoiding coral reefs and rocky cliffs. These reefs don't occur near the mouths of freshwater streams. Avoid rip currents or strong tidal currents, which may carry the raft far out to sea.

    (1) When going through surf, the survivor should:

      (a) Take down the mast.

      (b) Don clothing and shoes to avoid injuries.

      (c) Adjust and fasten life preserver.

      (d) Stow equipment.

      (e) Use paddles to maintain control.

      (f) Ensure the sea anchor is deployed to help prevent the sea from throwing the stem of the raft around and capsizing it. CAUTION: The sea anchor should not be deployed when traveling through coral.

    (2) In medium surf with no wind, survivors should keep the raft from passing over a wave so rapidly that it drops suddenly after topping the crest. If the raft turns over in the surf, every effort should be made to grab hold.

    (3) The survivor should ride the crest of a large wave as the raft nears the beach, staying inside until it has grounded. If there is a choice, a night landing should not be attempted. If signs of people are noted, it might be advantageous to wait for assistance.

    (4) Sea-ice landings should be made on large, stable floes only. Icebergs, small floes, and disintegrating floes could cause serious problems. The edge of the ice can cut, and the raft can become deflated. Use paddles and hands to keep the raft away from the sharp edges of the iceberg. The raft should be stored a considerable distance from the ice edge. It should be fully inflated and ready for use in case the floe breaks up.

Part Eight

SIGNALING AND RECOVERY

Chapter 24

SIGNALING

24-1. Introduction:

a. Most successful recoveries have resulted primarily because survivors were able to assist in their own recovery. Many rescue efforts failed because survivors lacked the knowledge and ability necessary to assist. When needed, this knowledge and ability could have made the difference between life or death—freedom or captivity (figure 24-1).

Figure 24-1. Signaling and Recovery.

b. What can survivors do to assist in their own recovery? First, they need to know what is being done to find them. Next, they need to know how to operate the communications equipment in the survival kit, and when to put each item into use. Survivors should also be able to improvise signals to improve their chances of being sighted and to supplement the issued equipment.

c. It is not easy to spot one survivor, a group of survivors, or even an aircraft from the air, especially when visibility is limited. Emergency signaling equipment is designed to make a person easier to find. Emergency equipment may be used to provide rescue personnel with information about survivors' conditions, plans, position, or the availability of a rescue site where recovery vehicles might reach them (figure 24-2).

d. Part of a survivor's plan of action should be to visualize how emergencies will develop, recognize them, and, at the appropriate time, let friendly forces know about the problem. The length of time before survivors are rescued often depends on the effectiveness of emergency signals and the speed with which they can be used. Signal sites should be carefully selected. These sites should enhance the signal, and have natural or manufactured materials readily available for immediate use. Survivors should avoid using pyrotechnic signals wastefully, as they may be needed to enhance rescue efforts. Signals used correctly can hasten recovery and eliminate the possibility of a long, hard survival episode. Survivors should:

Figure 24-2. Signaling.

(1) Know how to use their emergency signals.

(2) Know when to use their signals.

(3) Be able to use their signals on short notice.

(4) Use signals in a manner that will not jeopardize individual safety.

e. The situation on the ground governs the type of information that survivors can furnish the rescue team, and will govern the type of signaling they should use. In nontactical survival situations, there are no limitations on the ways and means survivors may use to furnish information.

f. In hostile areas, limitations on the use of signals should be expected. The use of some signaling devices will pinpoint survivors to the enemy as well as to friendly personnel. Remember, the signal enhances the visibility of the survivors.

24-2. Manufactured Signals:
a. Electronic Signals:

(1) Current line-of-sight electronic signaling devices fall into two categories. One is the transceiver type; the other is the personal-locator-beacon type. The transceiver type is equipped for transmitting tone or voice and receiving tone or voice. The personal locator beacon is equipped to transmit tone only. The ranges of the different radios vary depending on the altitude of the receiving aircraft, terrain factors, forest density, weather, battery strength, type(s) of radios, and interference. Interference is a very important aspect of the use of these radios. If a personal locator beacon is transmitting, it will interfere with incoming and outgoing signals of the transceivers.

(2) Before using survival radios, a few basic precautions should be observed. These will help in obtaining maximum performance from the radios in survival situations.

(a) The survival radios are line-of-sight communication devices; therefore, the best transmission range will be obtained when operating in clear, unobstructed terrain.

(b) Extending from the top and bottom of the radio antenna is an area referred to as the "cone of silence." To avoid the "cone of silence" problem, keep the radio/beacon antenna orthogonal (at a right angle) to the path of the rescue aircraft.

(c) Since the radios have the capability of transmitting a tone (beacon) without being hand-held, they can be placed upright on a flat, elevated surface, allowing the operator to perform other tasks.

(d) Never allow the radio antenna to ground itself on clothing, body, foliage or the ground. This will severely decrease the effective range of the signals.

(e) Conserve battery power by turning the radio off when not in use. Do not transmit or receive constant-ly. Use the locator beacon to supplement the radio when transmitting is done. In tactical environments, the radio should be used as stated in the premission briefing.

(f) Survival radios are designed to operate in extreme heat or cold. The life expectancy of a battery decreases as the temperature drops below freezing, and exposure to extreme heat or shorting out of the battery can cause an explosion. During cold weather, the battery should be kept warm by placing it between layers of clothing to absorb body heat, or wrapped in some type of protective material when it is not being used.

(g) Survival radios are designed to be waterproof. However, precautions should be taken to keep them out of water.

(3) Presently, a satellite monitoring system has been developed to assist in locating survivors. To activate this system (SARSAT), the transmitter is "keyed" for a minimum of 30 seconds. In a nontactical situation, leave the beacon on until rescue forces are heard or sighted.

b. Pyrotechnics. Care should be used when operating around flammable materials.

(1) A device containing chemicals for producing smoke or light is known as a pyrotechnic. Hand-held flares are in this category. Survivors may be required to use a variety of flares. They must know the types of flares stored in their survival kits and (or) aircraft. Aircrew members should learn how to use each type of flare before they face an emergency. Flares are designed to be used during the day or night. Day flares produce a unique, bright-colored smoke, which stands out very clearly against most backgrounds. Night flares are extremely bright and may be seen for miles by air, ground, or naval recovery forces.

(2) The hand-held launched flares also fall in the pyrotechnic category. They were designed to overcome the problems of terrain-masking and climatic conditions. For example, a person may be faced with multilayer vegetation or atmospheric conditions known as an inversion, which keeps the smoke next to the ground.

(3) Flares must be fired at the right time to be of maximum use. Smoke flares, for example, take a second or two after activation before they produce a full volume of smoke. Therefore, the flare should be ignited just before the time it can be seen by rescue personnel. These signals should not be used in tactical environments unless directed to do so.

(4) Tracer ammunition is another pyrotechnic that may be issued to aircrew members. When fired, the projectile appears as an orange-red flash the size of a golf ball. According to specifications, these tracers have a range of 1,300 feet. Tracer bullets have been detected from a distance of 6 miles, but there is usually difficulty in pinpointing the survivor. A survivor should use

Figure 24-3. Sea Marker Dye.

this signaling device only when rescue forces can be seen or heard. Do not direct this device at the aircraft.

(5) Because of the rapidly changing technology in pyrotechnic signaling devices, an aircrew member should check regularly for new and improved models, making special note of the firing procedures and safety precautions necessary for their operation.

c. Sea Marker:

(1) Of the many dyes and metallic powders tested at various times for marking the sea, the most successful is the fluorescent, water-soluble orange powder. When released in the sea, a highly visible, light-green fluorescent cast is produced. Sea-marker dye has rapid dispersion power; a packet spreads into a slick about 150 feet in diameter that lasts an hour or more in calm weather. Rough seas will stream it into a long streak, which may disperse in 20 minutes (figure 24-3).

(2) Under ideal weather conditions, the dye can be sighted at 5 miles with the aircraft operating at 1,000 feet. The dye has also been spotted at 7 miles away from an aircraft operating at 2,000 feet.

(3) Sea-marker dye should be used in friendly areas during daytime, and only when there is a chance of being sighted (aircraft seen or heard in the immediate area). It is not effective in heavy fog, solid overcast, and storms with high winds and waves. The release tab on the packet of dye is pulled to open for use. In calm water, the dye can be dispersed more rapidly by stirring the water with paddles or hands.

(4) If left open in the raft, the escaping powder penetrates clothing, stains hands, face, and hair, and eventually may contaminate food and water. To avoid the inevitable messiness, some survivors have tied the sea-marker dye to the sea anchor. Others have dipped the packet over the side, letting it drain off the side into the sea. After using the dye, it should be rewrapped to conserve the remainder of the packet.

d. Paulin Signals. The paulin is a conventional signaling device used to send specific messages to aircraft. It may be packed with some sustenance kits and multiplace liferaft accessory kits. The paulin is constructed of rubberized nylon material, and is blue on one side and yellow on the other. These colors contrast against each other so when one side is folded over the other, the designs are easily distinguished (figure 24-4). The size is 7 feet by 11 feet, which is a disadvantage when folded because it makes a small signal. The paulin has numerous uses. It can be used as a camouflage cloth, sunshade, tent or sail, or it can be used to catch drinking water. The space blanket, used as a substitute for the sleeping bag in some survival kits, can be used in the same manner as the signal paulin because it is highly reflective (silver on one side and various colors on the other side).

e. Audio Signals. Sounds carry far over water under ideal conditions; however, they are easily distorted and deadened by wind, rain, or snow. On land, heavy foliage cuts down on the distance sound will travel. Shouting and whistling signals have been effective at short ranges for summoning rescue forces. Most contacts using these means were made at less than 200 yards, although a few reports claim success at ranges of up to a mile. A weapon can be used to attract attention by firing shots in a series of three. The number of available rounds determine whether this is practical. Survivors have used a multitude of devices to produce sound. Some examples are: striking two poles together, striking one pole against a hollow tree or log, and improvising whistles out of wood, metal, and grass.

f. Light Signals. When tested away from other manufactured lights, aircraft lights have been seen up to 85 miles. At night, a survivor should use any type of light to attract attention. A signal with a flashlight, or a light or fire in a parachute shelter, can be seen from a long distance. A flashing light (strobelight) is in most survival kits.

g. Signal Mirror.

(1) The signal mirror is probably the most underrated signaling device found in the survival kit. It is the most valuable daytime means of visual signaling. A mirror flash has been visible up to 100 miles under ideal conditions, but its value is significantly decreased unless it is used correctly. It also works on overcast days. Practice is the key to effective use of the signal mirror. Whether the mirror is factory manufactured or improvised, aim it so the beam of light reflected from its surface hits the overflying aircraft.

(2) The signal mirror's effectiveness is its greatest weakness if the survivor is in enemy territory. It is just as bright to the enemy as it is to the rescuer; use it wisely! Survivors should understand that even if the mirror flash is directly on the aircraft (especially if the aircraft is using terrain-masking techniques), that same flash may be visible to others (possibly the enemy) who are located at the proper angle in regard to the survivors' position.

(3) In a hostile environment, the exact location of the flash is extremely important. The signal mirror should be covered when not in use. One of the easiest

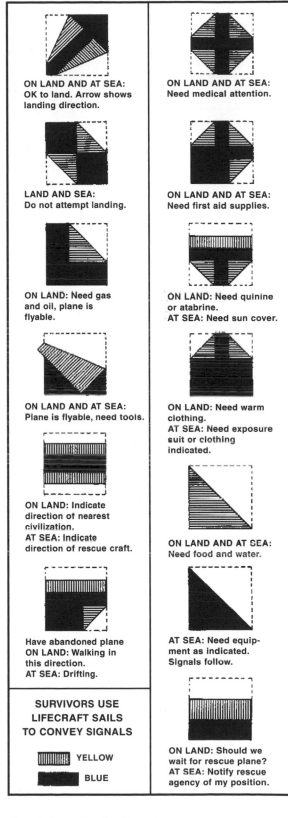

Figure 24-4. Paulin Signals.

Figure 24-5. Signal Mirrors.

methods is to tie the string from the mirror around the neck and tuck the mirror in the shirt or flight suit. When the mirror is removed from inside the clothing, the hand should be placed over the mirror surface to prevent accidental flashing. The covered mirror may then be raised toward the sky and the hand withdrawn. The flash can then be directed onto the free hand and the aiming indicator (sunspot) located. This minimizes the indiscriminate

flashing of surrounding terrain. When putting the mirror away, the survivor should remember to cover the mirror to prevent a flash.

h. Aiming Manufactured Mirrors. Instructions are printed on the back of the mirror. Survivors should:

(1) Reflect sunlight from the mirror onto a nearby surface—raft, hand, etc.

(2) Slowly bring the mirror up to eye level and look through the sighting hole, where a bright spot of light will be seen. This is the aim indicator.

(3) Hold mirror near the eye and slowly turn and manipulate it so the bright spot of light is on the target.

(4) In friendly areas, where rescue by friendly forces is anticipated, free use of the mirror is recommended. Survivors should continue to sweep the horizon even though no aircraft or ships are in sight (figure 24-5).

Figure 24-6. Aiming Signal Mirror.

Figure 24-7. Aiming Signal Mirror–Stationary Object.

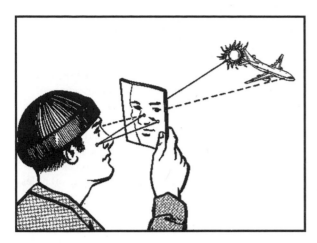

Figure 24-8. Aiming Signal Mirror–Double-Faced Mirror.

24-3. Improvised Signals:

a. Signal Mirrors. Improvised signal mirrors can be made from ration tins, parts from an aircraft, polished aluminum, glass, or the foil from rations or cigarette packs. However, the mirror must be accurately aimed if the reflection of the Sun in the mirror is to be seen by the pilot of a passing aircraft or the crew of a ship.

b. Aiming Improvised Mirrors:

(1) The simple way to aim an improvised mirror is to place one hand out in front of the mirror at arm's length and form a "V" with two fingers. With the target in the "V" the mirror can be manipulated so the majority of light reflected passes through the "V" (figure 24-6). This method can be used with all mirrors. Another method is to use an aiming stake as shown in figure 24-7. Any object 4 to 5 feet high can serve as the point of reference.

(2) Survivors should hold the mirror so they can sight along its upper edge. Changing their position until the top of the stick and target line up, they should adjust the angle of the mirror until the beam of reflected light hits the top of the stick. If stick and target are then kept in the sighting line, the reflection will be visible to the rescue vehicle.

(3) Another method is to improvise a double-faced mirror (shiny on both sides). A sighting hole can be made in the center of the mirror.

(a) When trying to attract the attention of a friendly rescue vehicle that is no more than 90 degrees from the Sun, proceed as shown in figure 24-8.

(b) The survivor's first step will be to hold the double-faced mirror about 3 to 6 inches away from the face, and sight at the rescue target through the hole in the center of the mirror. The light from the Sun shining through the hole will form a spot of light on the survivor's face. This spot will be reflected in the rear surface of the mirror. Then, aiming at the rescue vehicle through the hole, the survivor can adjust the angle of the mirror until the reflection of the spot on the face in the rear surface of the mirror lines up with, and disappears into, the sighting hole.

(c) When the reflected spot disappears and the rescue vehicle is still visible through the hole, the survivor can be sure the reflected light from the Sun is accurately aimed. The survivor may also "shimmer" the mirror by moving it rapidly over the target. This ensures that the part of the bright flash the rescuers see coincides with the position of the survivor. This "shimmering" is especially useful on a moving target.

(d) When the angle between the target and the Sun is more than 90 degrees (when the survivor is between the rescue vehicle and the Sun), a different method may be used for aiming. The survivor should adjust the angle of the mirror until the spot made by the

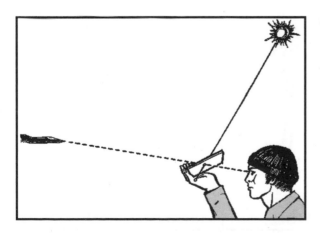

Figure 24-9. Aiming Signal Mirror–Angle Greater than 180 Degrees.

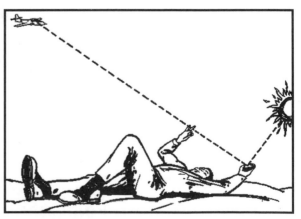

Figure 24-10. Aiming Signal Mirror–Angle Greater than 180 Degrees (Another Method).

Figure 24-11. Shelter as a Signal.

Sun's rays passing through the hole in the mirror lands on the hand instead of the face. The reflection in the back of the mirror that comes off the hand may then be manipulated in the same way (figure 24-9).

(e) Another method used when the angle is greater than 90 degrees is to lie on the ground in a large clearing

and aim the mirror using one of the methods previously discussed (figure 24-10).

c. Fire and Smoke Signals:

(1) Fire and smoke can be used to attract the attention of recovery forces. Three evenly spaced fires, 100 feet apart, arranged in a triangle or in a straight line, serve as an international distress signal. One signal fire will usually work for a survivor. During the night, the flames should be as bright as possible, and during the day, as much smoke as possible should be produced.

(2) Smoke signals are most effective on clear and calm days. They have been sighted from up to 50 miles away. High winds, rain, or snow tend to disperse the smoke and lessen the chances of it being seen. Smoke signals are not dependable when used in heavily wooded areas.

(3) The smoke produced should contrast with its background. Against snow, dark smoke is most effective. Likewise, against a dark background, white smoke is best. Smoke can be darkened with rags soaked in oil, pieces of rubber, matting, or electrical insulation, or plastic being added to the fire. Green leaves, moss, ferns or water produce white smoke.

(4) To increase its effectiveness, the signal fire must be prepared before the recovery vehicle enters the area.

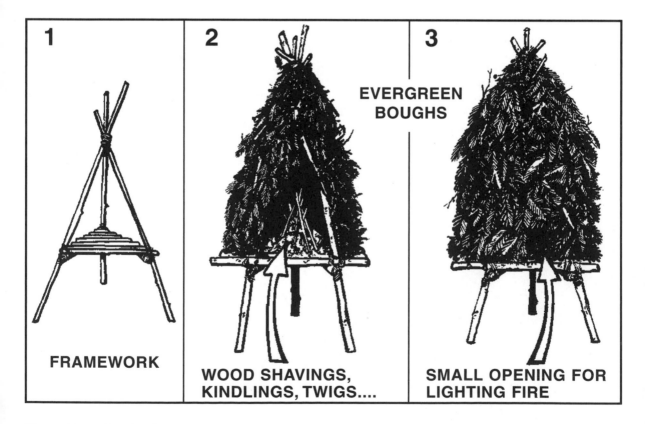

Figure 24-12. Smoke-Generator–Platform.

LOTS OF DEAD DRY TWIGS OR KINDLING FOR QUICK STARTING FAST-BURNING FIRE

EVERGREEN BOUGHS

SMALL OPENING FOR LIGHTING FIRE

Figure 24-13. Smoke Generator–Ground.

The fires used by survivors for heat and cooking may be used as signal fires as long as the necessary materials are available in the immediate vicinity. Survivors should supplement the fire to provide the desired signal (figure 24-11).

(5) Smoke Generators:

(a) Raised Platform Generator (figure 24-12). The survivor should:

-1. Build a raised platform above wet ground or snow.

-2. Place highly combustible materials on the platform.

-3. Then place smoke-producing materials over the platform, and light when search aircraft is in the immediate vicinity.

(b) Ground Smoke Generator (figure 24-13). The survivor should:

-1. Build a large, log cabin fire configuration on the ground. This provides good ventilation and supports the green boughs used for producing smoke.

-2. Place smoke-producing materials over the fire lay; ignite when a search aircraft is in the immediate vicinity.

(c) Tree Torch Smoke Generator (figure 24-14). To build this device, the survivor should:

-1. Locate a tree in a clearing, to prevent a forest fire hazard.

-2. Add additional smoke-producing materials.

-3. Add igniter.

-4. Light when a search aircraft is in the immediate vicinity.

(d) Fuel Smoke Generator. If survivors are with the aircraft, they can improvise a generator by burning aircraft fuels, lubricating oil, or a mixture of both. One to 2 inches of sand or fine gravel should be placed in the bottom of a container and saturated with fuel. Care should be used when lighting the fuel, as an explosion may occur initially. If there is no container available, a hole can be dug in the ground, filled with sand or gravel, saturated with fuel, and ignited. Care should be taken to protect the hands and face.

d. Pattern Signals. The construction and use of pattern signals must take many factors into account. Size, ratio, angularity, contrast, location, and meaning are each important if the survivors' signals are to be effective. The

Figure 24-14. Tree Torch.

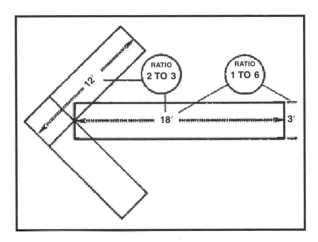

Figure 24-15. Pattern Signal Sizes.

type of signal constructed will depend on the material available to survivors. Not every crewmember will have a parachute, so ingenuity plays an important role in the construction of the signal. Survivors should remember to judge their signals from the standpoint of aircrew members who are flying over their location searching for them.

(1) Size. The signal should be as large as possible. To be most effective, the signal should have "lines" no less than 3 feet wide and 18 feet long (1:6). (See figure 24-15.)

(2) Ratio. Proper proportion should also be remembered. For example, if the baseline of an "L" is 18 feet long, then the vertical line of the "L" must be longer (27 feet), at a 2 to 3 ratio, to keep the letter in proper proportion.

(3) Angularity. Straight lines and square corners are not found in nature. For this reason, survivors should make all pattern signals with straight lines and square corners.

(4) Contrast. The signal should stand out sharply against the background. The idea is to make the signal look "larger." On snow, the fluorescent sea dye available in the liferaft accessory kit can be used to add contrast around the signal. The survivor should do everything

Figure 24-16. Contrast (Snow).

Figure 24-17. Contrast (Shadow).

Figure 24-18. Location.

NO.	MESSAGE	CODE SYMBOL
1	REQUIRE ASSISTANCE	V
2	REQUIRE MEDICAL ASSISTANCE	X
3	NO or NEGATIVE	N
4	YES or AFFIRMATIVE	Y
5	PROCEEDING IN THIS DIRECTION	↑

Figure 24-19. Signal Key.

possible to disturb the natural look of the ground.
In grass and scrubland, the grass should be stamped
down or turned over to allow the signal to be easily
seen from the air. A burned grass pattern is also effective. Survivors should use only one path to and from

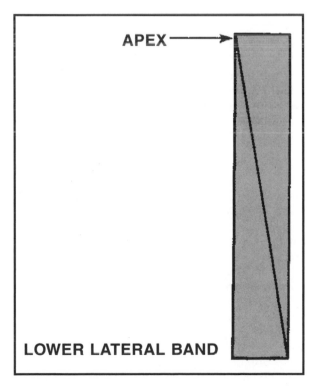

Figure 24-20. Parachute Strips.

APEX

LOWER LATERAL BAND

Figure 24-21. Parachute in Tree.

the signal to avoid disrupting the signal pattern. Avoid
using orange parachute material on a green or brown
background, as it has a tendency to blend in (figure
24-16). Contrast can be improved by outlining the
signal with green boughs, piling brush and rocks to
produce shadows, or raising the panel on sticks to cast
its own shadow (figure 24-17).

(5) Location. The signal should be located so it can
be seen from all directions. Survivors should make sure
the signal is located away from shadows and overhangs.
A large, high open area is preferable. It can serve a dual
function—one for signaling and the other for rescue
aircraft to land (figure 24-18).

(6) Meaning. If possible, the signal should tell the
rescue forces something pertaining to the situation. For
example: "require medical assistance," or a coded symbol
used during evasion, etc. Figure 24-19 shows the interna-
tionally accepted symbols.

e. Parachute Signals:

(1) Parachute material can be used effectively to
construct pattern signals. A rectangular section of parachute material can be formed as shown in figure 24-20.
When making a pattern signal, survivors should ensure
that edges are staked down so the wind will not blow
the panels away.

(2) A parachute caught in a tree will also serve
as a signal. Survivors should try to spread the material
over the tree to provide the maximum amount of signal
(figure 24-21).

Figure 24-22. Chute Over Trees or Streams.

Figure 24-23. Pennants and Banners.

(3) When open areas are not available, survivors should stretch the chute over low trees and brush or across small streams (figures 24-22 and 24-23).

f. Shadow Signals. If no other means are available, survivors may have to construct mounds that will use the Sun to cast shadows. These mounds should be constructed in one of the international distress patterns. Brush, foliage, rocks, or snow blocks may be used to cast shadows. To be effective, these shadow signals must be oriented to the Sun to produce the best shadow. In areas close to the Equator, a north-south line gives a shadow at any time except noon. Areas farther north or south require the use of an east-west line, or some point of the compass in between, to give the best results.

g. Acknowledgements:

(1) Rescue personnel will normally inform the survivors that they have been sighted by:

(a) Flying low with landing lights on (figure 24-24), and (or) rocking the wings.

(b) Emergency radio.

(2) Figure 24-25 depicts the standard body signals that can be used if electronic signaling devices are not available.

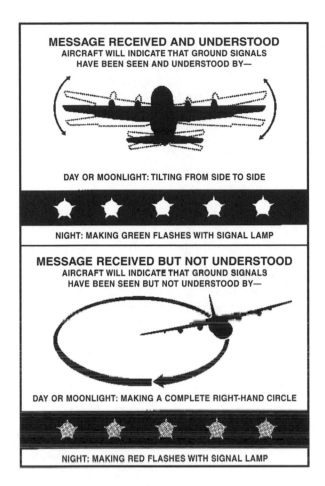

Figure 24-24. Standard Aircraft Acknowledgments.

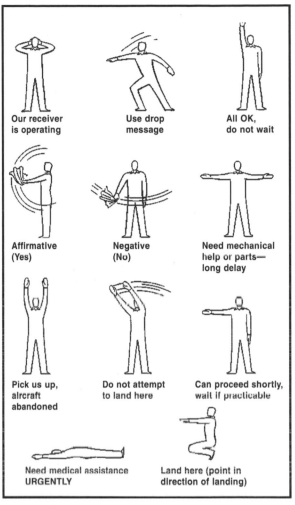

Figure 24-25. Close-In Visual Signals.

Chapter 25

RECOVERY PRINCIPLES

25-1. Introduction:

a. Receipt of a distress call sets a highly trained and well-equipped organization into operation; however, prompt and safe recovery is by no means assured. The success of the rescue effort depends on many factors. Such factors as the availability of rescue forces, the proximity of enemy forces, and weather conditions can affect the success of the rescue. Above all, the survivor's knowledge of what to do in the rescue effort may make the difference between success and failure (figure 25-1).

b. The role of survivors in effecting their rescue changes continuously as aircraft and rescue equipment become more sophisticated. The probability of a downed aircrew member applying long-term survival-training concepts under noncombat conditions continues to decrease, while increasing under combat conditions.

c. There are several independent organizations engaged in search-and-rescue (SAR) operations or influencing the SAR system. The organizations may be international, federal, state, county, or local governmental, commercial, or private organizations. Survivors are responsible for being familiar with procedures used by international SAR systems in order to assist in rescue efforts. Some international organizations are:

(1) International Civil Aviation Organization (ICAO).

(2) Intergovernmental Maritime Consultive Organization (IMCO).

(3) Automated Mutual-Assistance Vessel Rescue (AMVER) System.

25-2. National Search-and-Rescue (SAR) Plan:

a. The National SAR Plan is implemented the instant an aircraft is known to be down. There are three primary SAR regions; they are the Inland Region, the Maritime Region, and the Overseas Region.

b. The Air Force is the SAR coordinator for the Inland Region, which encompasses the continental United States. The Coast Guard is the SAR coordinator for the Maritime Region, which includes the Caribbean Area and Hawaii. The third National Region is the Overseas Region. The Secretary of Defense designates certain Defense Department officers as United Commanders of specified areas where US Forces are operating. Wherever such commands are established, the Unified Commander is the Regional SAR Coordinator. Overseas regions are normally served by the Joint Rescue Coordination Center, operated under the Unified Action Armed Forces. Under the terms of the National SAR Plan, the "inland" area of Alaska is considered part of the Overseas Region.

c. The National SAR Manual, Air Force Manual 64-2, provides a long-range rescue plan, which personnel should study for additional information.

25-3. Survivors' Responsibilities:

a. The survivors' responsibilities begin at the onset of the emergency, with the dispatching of an immediate radio message. The radio message should include position, course, altitude, groundspeed, and actions planned. This information is essential for initiating efficient recovery operations.

b. Once recovery operations have been initiated, survivors have a continuing responsibility to furnish information. Both ground and radio signals should be immediate considerations.

c. If a group of survivors should become separated, each group member should, when contacted by rescue forces, provide information surrounding the dispersal of the group.

d. The greatest responsibility of aircrew members is to follow all instructions to the letter. The intelligence officer will brief aircrew members on procedures for tactical situations. These instructions must be followed

Figure 25-1. Recovery.

explicitly since it could mean the difference between life and death. When rescue personnel tell the survivor to unhook from the raft—it should be done immediately! If instructions are not followed, survivors could be responsible for causing their own death and (or) the death of rescue personnel.

25-4. Recovery Site:

a. Consideration must be given to a recovery site. The survivor's major considerations are the type of recovery vehicle carrying out the recovery and the effects of the weather and terrain on the rescue aircraft, such as updrafts and downdrafts, heat, wind, etc. Survivors should try to pick the highest terrain possible in the immediate area for pickup. When locating this rescue site, they should watch for obstacles such as trees, cliffs, etc., which could limit the aircraft's ability to maneuver. Overhangs, cliffs, or the sides of steep slopes should be avoided. Such terrain features restrict the approach and maneuverability of the rescue vehicle, and require an increase in rescue time.

b. Even though survivors should select a recovery site, it is the ultimate responsibility of rescue personnel to decide whether the selected site is suitable.

25-5. Recovery Procedures:

a. **Knowing Current Procedures.** Since procedures involving recovery vary with changes in equipment and rescue capability, survivors must always know the current procedures and techniques. This is particularly

true of the procedures used for wartime recovery, which are in AFR 64-3.

(1) In deciding whether or not supplies should be dropped, rescue forces consider such factors as the relative locations of the distress site to rescue-unit bases, the lapse expected before rescue is initiated, and the danger of exposure. If a delay is expected, supplies are usually dropped to survivors to help sustain and protect them while they await rescue. The mobility of survivors on the land generally makes it possible to recover equipment dropped some distance away, but airdrops at sea must be accurate.

(2) Aircraft with internal aerial delivery systems, such as the HC-130, are the most suitable for delivery of supplies to survivors. Aircraft with bomb bays or exterior racks capable of carrying droppable containers or packages of survival requisites are the next most suitable for dropping supplies. However, these aircraft are not always available for supply-dropping operations, so aircraft not specifically designed for this function may have to be used.

b. **Rescue by Helicopter:**

(1) Helicopters make rescues by landing or hoisting. Landings are usually required at high altitudes due to limitations of helicopter power for maintaining a hover. Hoist recovery is the preferred method for effecting a water rescue. Helicopter landings are made for all rescues when a suitable landing site is available, and danger from enemy forces is not a problem. Hovering the helicopters and hoisting the survivor aboard requires

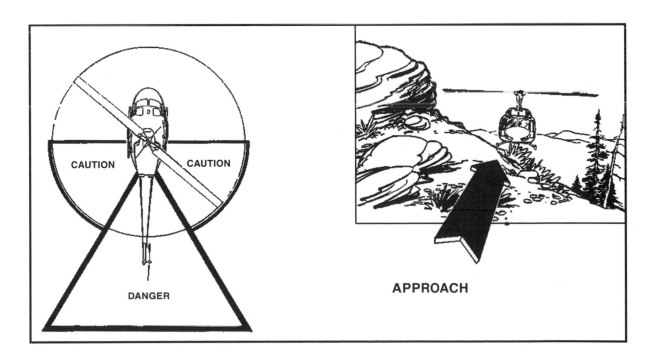

CAUTION   CAUTION

DANGER

APPROACH

Figure 25-2. Approaching Helicopter.

more helicopter power than landing, and presents a hazard to both the aircraft and the survivor. There is a danger if helicopters are operated close to collapsed parachutes. Parachute inflation by rotor downwash can cause the parachute to be sucked into the rotor blades of the helicopter.

(2) After landing, a crewmember will usually depart the aircraft. If for some reason this cannot be done, as in combat, the survivor should approach the helicopter from the 3 o'clock to 9 o'clock position relative to the nose of the helicopter, and follow instructions (figure 25-2).

c. Rescue by Fixed-Wing Aircraft on Land:

(1) The most significant role played by fixed-wing aircraft in rescue operations is providing immediate assistance to survivors, and serving as the "eyes" of approaching rescue units. This is done by pinpointing the survivors' position, orbiting the survivors, and dropping survival equipment. This type of operation improves the morale of the survivors, fixes the survivors' location to prevent additional searching, and saves valuable time in getting the pickup unit on the scene.

(2) The role of fixed-wing aircraft in actually performing a rescue is limited to instances in which there is a suitable runway near the survivor, or in which the aircraft is designed to operate from rough and improvised strips. Fixed-wing aircraft rescues have often been made in extremely cold climates where the aircraft have either used frozen lakes or rivers as runways or, when fitted with skis, have operated from snow-covered surfaces and glaciers. However, landing in unknown terrain under what appears to be ideal conditions is extremely hazardous.

d. Rescue by Ship:

(1) When a distress craft or survivors are a considerable distance from shore, rescue will normally be made by long-range ships (specialized SAR ships, warships, or merchant ships). The rescue methods used by these ships vary considerably according to their displacement and whether the rescue is made in mid-ocean or close to land. Weather, tides, currents, sea conditions, shallow water, reefs, daylight, or darkness may be important factors.

(2) Although it appears obvious that a marine craft should be used for rescue operations, it may be advisable to initiate an alternate method of recovery. For example, helicopters may be used to evacuate survivors picked up by marine craft in order to speed their delivery to an emergency-care center.

(3) Removal of survivors from the water, liferafts, lifeboats, or other vessels to the safety of the rescue-vessel deck may be the most difficult phase of a maritime search-and-rescue mission. In most cases, survivors will have to be assisted aboard. For this reason, all SAR vessels are usually equipped and prepared

to lift survivors from the water without help from the survivors. There are numerous methods for rescuing survivors that may be used by SAR vessels. The most commonly used methods are listed in this chapter, and are generally grouped as rescue of survivors in the water and rescue of survivors directly from their distressed vessel.

(a) When rescuing people from water, the following methods are generally used:
-1. Ship alongside/swimmer.
-2. Ship alongside/line thrower.
-3. Ship alongside/small boat.
-4. Ship circle/trail line.

(b) The most commonly used methods for rescuing personnel who are aboard distressed vessels are:
-1. Ship to ship/direct.
-2. Ship to ship/raft haul.
-3. Ship to ship/raft drift.
-4. Ship to ship/small boat.
-5. Ship to ship/haul-away line.

e. Rescue by Boat:

(1) When survivors are located on lakes, sheltered waters, rivers, or coastal areas, rescue will often be made by fast boats of limited range based close to the survivors, or by private boats operating in the vicinity.

(2) Rescue boats are usually small and may not be able to take all survivors onboard at one time; therefore, a sufficient number of boats to offset the rescue should be dispatched to the distress scene. When this is not possible, each boat should deploy its rafts so those survivors who cannot be taken aboard immediately can be towed ashore or kept afloat while they are waiting. The boat crew should make sure any survivors who must be left behind are made as secure as circumstances permit.

(3) Assistance to an aircraft that has crashed or ditched on the water will usually consist of transferring personnel from plane to boat, and picking up survivors from the water or liferafts. It may also include towing of an aircraft that is disabled on the water.

f. Coordinated Helicopter/Boat Rescues:

(1) Occasionally, boats and helicopters will be dispatched for a rescue operation. Generally, the first rescue unit to arrive in the vicinity of the survivors will attempt the first rescue.

(2) If the helicopter arrives first, the boat will take a position upwind of the helicopter in the 2 o'clock position at a safe distance, and stand by as a backup during the rescue attempt.

(3) If the helicopter must abort the rescue attempt, the pilot will depart the immediate area of the survivor and signal for the boat to move in and make its rescue attempt. Additionally, the helicopters may turn out the anti-collision rotating beacon to indicate they require boat assistance or are unable to complete the rescue. In certain operations in

which helicopter and boat coordinated rescue can be foreseen, specific signals should be prearranged.

(4) If the boat arrives first and makes the rescue, it will transfer the survivor to the helicopter to effect a rapid delivery to medical facilities.

25-6. Pickup Devices:

a. Assistance. When rescue forces are in the immediate area of survivors, they will, if conditions permit, deploy pararescue personnel to assist the survivors. Unfortunately, conditions may not always permit this, so survivors should know how to use different types of pickup devices.

b. Common Factors. Some common factors concerning all pickup devices are:

(1) The device should be allowed to ground to discharge static electricity before donning.

(2) To ensure stability, survivors should sit or kneel when donning a pickup device. Do not straddle the device.

(3) If no audio is available, survivors should visually signal the hoist operator when ready for lift-off—"thumbs up" or vigorously shake the cable from side to side.

(4) Most devices can be used as a sling (strop).

(5) Survivors must remember to follow all instructions provided by the rescue crew. When lifted to the door of the helicopter, survivors should not attempt to grab the door or assist the hoist operator in any way. They must not try to get out of the pickup device. The hoist operator will remove the device after the survivor is well inside the aircraft.

c. Rescue Sling. Before donning the rescue sling (strop), the survivor should face the drop cable and make sure the cable has touched the water or ground and has lost its charge of static electricity.

(1) The most commonly accepted method for donning the rescue sling (strop) is the same as putting on a coat. After connecting the ring to form the sling (strop), the survivor's arms should be inserted one by one into the sling (strop) as it swings behind. The sling (strop) loop should be against the survivor's back with an arm around each side of the strop. The webbing under the metal ring can be held until tension is put on the cable. The survivor's hands may then be interlocked and rested on the chest. This tends to lock the survivor into the sling (strop), as upward pressure is applied (figure 25-3).

(2) Another way to enter the strop is to grasp the strop with both hands and lift it over the head to bring it down under the arms and around the body. Regardless of the method used, the survivor should remember that the webbing and metal hardware of the device should be directly in front of the face.

d. Basket. If a basket is used, it will probably be accompanied to the water or ground by a member of the helicopter crew. The crewmember will assist survivors into the basket. There are two types of baskets: The litter type in which the person lies flat, and the seat type that survivors enter and sit down in as they would in a chair (figure 25-4).

e. Forest Penetrator:

(1) The forest-penetrator rescue seat is designed to make its way through interlacing tree branches and dense jungle growth. It can also be used in open terrain or over water. The device is equipped with three seats

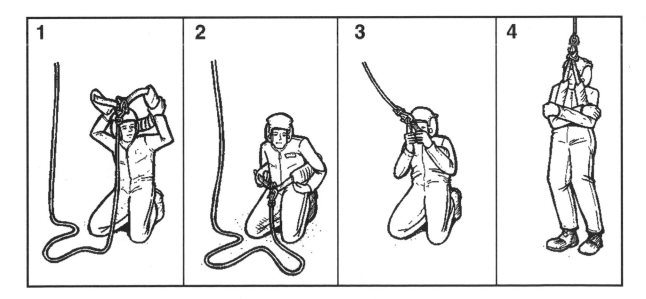

Figure 25-3. Horse Collar.

Figure 25-4. Basket.

that are spring-loaded in a folded position against the body or main shaft and must be pulled down to the locked position for use. On the main shaft of the tube, above the seats, there is a zippered fabric storage pouch for the safety (body) straps, which are stowed when lowered to the survivor for a land pickup. The penetrator

may also be equipped with a flotation collar. (NOTE: If the forest penetration is used for water pickup, it will be equipped with the flotation collar, which enables the device to float with the upper one-third [approximately] of the device protruding above the water. Additionally, for water pickups, one strap will be removed from the stowed position, and one seat will be locked in the down position to assist the survivor in using the penetrator.)

(2) The safety strap is pulled from the storage pouch and placed around the body to hold the person on the penetrator seat. The strap should not be unhooked unless there is no other way to fasten it around the body. The survivor must make certain the safety strap does not become fouled in the hoist cable. After the strap is in place, the seat should be pulled down sharply to engage the hook, which holds it in the extended position. The survivor can then place the seat between the legs. Then the survivor should pull the safety strap as tight as possible, ensuring the device fits snugly against the body. The survivor must always keep the arms down, elbows locked against the body, and not attempt to grab the cable or weighted snap link above the device. After making certain the body is not entangled in the hoist cable, the signal to be lifted can be given (figure 25-5).

(3) In a combat area, under fire, survivors may be lifted out of the area with the cable suspended before

1	2	3
PULL DOWN VELCRO FASTENER	PULL OUT STRAP, FASTEN AROUND BODY (UNDER ARMS), AND HOOK SNAP RING TO SNAP-RING BAR	FOLD DOWN SEAT
4	**5**	**6**
MOUNT SEAT AND TIGHTEN STRAP	GRASP CABLE AND SIGNAL WHEN READY	FOLD ARMS AROUND PENETRATOR— KEEP HEAD DOWN

Figure 25-5. Forest Penetrator.

being brought into the helicopter. It is important to be correctly and securely positioned on the pickup device. The seat should always be held tightly against the crotch to prevent injury when slack in the cable is taken up. The hands should be kept below and away from the swivel on the cable, with the arms around the body of the penetrator. Survivors should keep their head close to the body of the penetrator so tree branches or other obstructions will not come between the body and the hoist cable.

(4) When survivors reach a position level with the helicopter door, the hoist operator will turn them so they face away from the helicopter, and then pull them inside. The crewmember will disconnect the survivors from the penetrator once the device is safely inside the helicopter.

(5) The forest penetrator is designed to lift as many as three persons. When two or three survivors are picked up, heads should be kept tucked in, and each individual's safety strap drawn tight. The penetrator can be used to lower a paramedic or crewmember to assist injured personnel, and both (survivor and paramedic) can be hoisted to the helicopter. If the forest-penetrator

McGUIRE RIG

Figure 25-7. McGuire Rig.

seat blades have been lowered in a tree area, and if for any reason the pickup cannot be made, the blades should be returned to the folded position to prevent possible hangup on tree limbs or other objects while the device is being retracted.

(6) With all types of devices, it is necessary to watch the device as it is lowered. The devices weigh about 23 pounds. If the device were to hit a survivor, it could cause serious injury or death.

f. Other Devices. There are other devices that could be used to pick up survivors. Some of them are the Motley and McGuire rigs (figures 25-6 and 25-7), the Swiss Seat and Stabo rig (figures 25-8 and 25-9), and the Rope Ladder (figure 25-10).

(1) Motley and McGuire Rigs. These devices may be carried by Army helicopters either designated as the recovery aircraft in assault, or for use to insert or extract special ground forces. The device is normally packed in a weighted canvas container and dropped by rope. The device is dropped to the survivor, who is allowed time

MOTLEY RIG

Figure 25-6. Motley Rig.

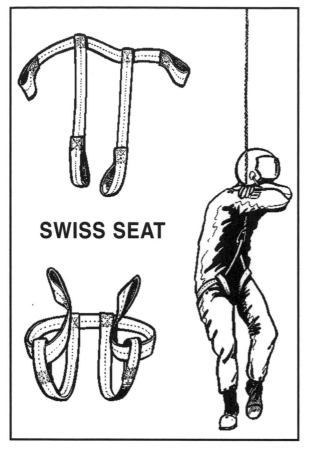

Figure 25-8. Swiss Seat.

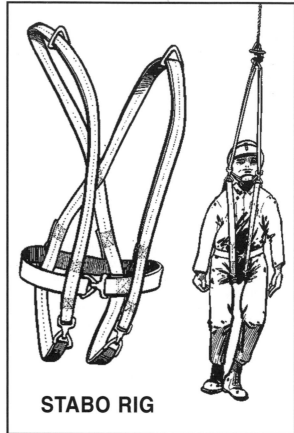

Figure 25-9. Stabo Rig.

for donning. The helicopter then returns trailing a rope, which is then fastened to the device for pickup. Generally, the survivor is not hoisted into the helicopter; therefore, all safety straps should be securely fastened.

(2) Swiss Seat and Stabo Rig (figures 25-8 and 25-9). These devices are carried by special ground forces that may require instant extraction by helicopter. Special ground forces put their devices on and wait for the helicopter to drop ropes, which are snapped into the devices for rapid extraction. Although not normally carried aboard the aircraft, the Army helicopter may supply one of these devices to the survivor. Again, the survivor would not be hoisted into the helicopter.

(3) Rope Ladder. This device is used primarily by the Army and special ground forces. If this device is used, it should be approached from the side and not the front. The survivor should climb up a few rungs, sit down on a rung, and inter-twine the body with rungs (figure 25-10). The survivor should not try to climb up the ladder and into the helicopter.

25-7. Preparations for Open Sea Recovery:

a. On sighting rescue craft approaching for pickup (boat, ship, conventional aircraft, or helicopter), survivors must quickly clear any lines (fishing lines, desalting-kit lines, etc.) or other gear that could cause entanglement during rescue. All loose items should be secured in the raft. Canopies and sails should be taken down to ensure a safer pickup. After all items are secure, the survivor should put on the helmet (if available). The life preserver should be fully inflated, with the oral valve-locking nut tight against the mouthpiece. Survivors should remain in the raft, unless otherwise instructed, and disengage all gear except the preservers. If possible, rescue personnel will be lowered into the water to assist survivors. The survivors should remember to follow all instructions given by rescue personnel.

b. If helicopter recovery is unassisted, the survivor will be expected to do the following before pickup:

(1) Secure all loose equipment in raft, accessory bag, or in pockets.

Figure 25-10. Rope Ladder.

    (2) Deploy sea anchor, stability bags, and accessory bag.

    (3) Partially deflate raft and fill with water.

    (4) Unsnap survival-kit container from parachute harness.

    (5) Grasp raft handhold and roll out of raft.

    (6) Allow recovery device and (or) cable to ground out on water surface.

    (7) Maintain handhold until recovery device is in the other hand.

    (8) Mount recovery device (avoid raft-lanyard entanglement).

    (9) Signal hoist operator for pickup.

Part Nine

EVASION

Chapter 26

LEGAL AND MORAL OBLIGATIONS

26-1. Introduction:

a. Aerial combat in the future, as in the past, will expose aircrews to possible ejection, bailout, or forced landings into enemy-controlled territory. The aircrews that encounter such traumatic circumstances must be prepared to survive and evade the enemy to return to friendly control. An active commitment to solving problems and to individual survival (the "will to survive") is essential. Aircrew members must be prepared to exert extreme effort, both mentally and physically, to successfully evade capture. In an excerpt from a debriefing, a survivor describes what it was like coming to terms with these problems: "I thought to myself . . . 'Well, I'm in a hell of a spot, what am I going to do about it?' The situation was such that I didn't do anything about it for quite a while. I just sat there in the rain where I had landed and stared at the ground. It became colder, and after what seemed like hours, I lifted my gaze and sat staring now into the forest, looking at it without even really seeing it. I started thinking about the chain of events that had preceded my landing in this god-forsaken forest so far from my lines. My thoughts were as jumbled and unreal as the fog into which I had dived. Soon, it started to snow and I realized that I was in danger of being discovered or of freezing to death where I sat, and so, with a huge effort of will, I forced myself to think coherently. I stood, and so began my journey." (See figure 26-1.)

b. This person had no survival training, but nevertheless managed to successfully evade for miles back to his own lines with little more than a strong will to survive and common sense. Potential evaders must understand that evading capture presents a difficult challenge. They will meet and have to overcome a succession of obstacles, both manmade and natural. Knowledge gained from the experience of others, their own training, and prior preparation and planning will help them to overcome these obstacles.

c. This part addresses covert survival—that is, evasion. Areas to be covered include:

(1) Evader's moral obligations and legal status.

(2) Principles and techniques of evasion, camouflage, and travel (assisted or unassisted).

(3) The special aspects of food and water procurement.

(4) Combat signaling and recovery.

26-2. Definitions:

a. Evader. An "evader" (JCS Pub 1) is "a person who, through training, preparation, and application of natural intelligence, avoids contact with, and capture by, hostiles, both military and civilian."

b. Evasion. Evasion refers to all the processes involved in living off the land, and at the same time, avoiding capture while returning to friendly control. As used here, it includes all the techniques of evasion employed by those on foot in enemy territory.

26-3. Military Drive:

a. A crewmember becomes an evader when isolated in hostile areas, unable to continue the assigned mission, and when prevented from rejoining friendly forces. Definitions are useful, but will not contribute to success in evading capture unless potential evaders understand what factors give direction and guidance to their efforts.

b. If the opportunity exists, evaders must be motivated to take advantage of it. The motivation to make a total effort to adhere to every evasion principle 24 hours a day may be personal or military. This strong central drive will give the survivor the necessary push to make these efforts.

c. Even if evasion is unsuccessful and the evader is captured, every hour spent eluding the enemy ties up enemy forces and lessens the evader's intelligence value.

d. In addition to the above reasons for evading capture, survivors also have moral and legal obligations to fulfill.

(1) Moral obligation is implied throughout the Articles of the Code of Conduct, specifically Article II. Article II states: "I will never surrender of my own free will. If in command, I will never surrender my men while they still have the means to resist." This Article of the Code should guide an evader's behavior during evasion just as it does in any other combat situation.

(2) The UCMJ continues to apply to evaders' conduct during evasion or captivity. Particularly applicable are Article 99, Misbehavior Before the Enemy, Article 104, Aiding the Enemy, and Article 92, Failure to Obey a Lawful Order. Thus, one can be tried for misconduct as a combatant or as a noncombatant. A combatant is defined in AFP 110-31 as "a person who engages in hostile acts in an armed conflict on behalf of a party to the conflict." The combatant must conform to the standards established under international law for combatants, be authorized by his or her country to so act, and be recognizable as a combatant by uniform, insignia, or

Figure 26-1. Evasion.

other sign. A noncombatant includes a wide variety of persons, including civilians, prisoners of war, sick and wounded persons, chaplains, medics, and other similar persons (AFP 110-31). Various countries around the world have developed written and unwritten laws of war. Four Geneva treaties were entered into by the United States and 60 other countries in 1949, and these treaties, as since amended, are in AFP 110-20.

(3) An "evader" is defined in international law as "any person who has become isolated in hostile or unfriendly territory and who eludes capture." An evader is a combatant, and retains this status as a fighting person under arms according to international law until captured. Evaders are considered instruments of their government, under orders to evade capture, and never to surrender of their own free will. The evader is still militarily effective and may take such steps as necessary, within the rules of warfare, to accomplish the mission, which includes returning after striking an enemy target. While in combatant status, the evader may continue to strike legitimate military targets and enemy troops without being held liable to prosecution after capture for violation of the local criminal law. To do so while evading capture is the legal function of a combatant.

(4) Once captured, an evader becomes a noncombatant and will occupy the status of a prisoner of war (PW). A PW who kills or wounds enemy personnel of the Detaining Power in an attempt to escape or evade may be tried and punished for the offense. International law provides certain rights to a PW, and requires issuance of an identity card showing name, rank, serial number, and date of birth. When questioned, a PW is bound to provide only this information. A PW who escapes remains a noncombatant until he or she rejoins the armed forces of his or her country, or the armed forces of a friendly power. Once escape is completed, combatant status is regained and no punishment may be imposed by the Detaining Power in the event of subsequent recapture. A PW who attempts to escape and is recaptured before rejoining his or her armed forces is liable under the Geneva Convention for disciplinary punishment only in respect of this act, even if it is a repeated offense. However, special surveillance may be imposed.

(5) Disguise is a lawful means of evading enemy forces so long as a means of military identification is retained on the person. However, an evader while in disguise may not participate in or commit hostile acts

involving destruction of life or property. A person involved in such an offense will be classified as an "unlawful combatant," will not be entitled to PW status, and may be tried by the enemy and sentenced to imprisonment or execution under certain circumstances. A disguise also cannot be used by a military person for the purpose of gathering enemy military information or for waging war, and to do so will result in loss of PW status if captured.

(6) Evasion can be classified as either assisted or unassisted. Assistance can be defined as any help that is offered to an American aircrew member by any person. This help may include food, clothing, medicine, shelter, money, and even so small an item as a shoelace. Evaders should, in fact, consider they have been assisted even if, while evading through hostile territory, their presence has been ignored by indigenous personnel. The term unassisted evasion would then relate to the situation in which the survivor, as an evader, must rely solely on his or her own knowledge and abilities to successfully emerge from an enemy-held or hostile area to areas under friendly control. The process of emerging may include aerial recovery, water recovery, or assisted evasion, but it is primarily an unassisted, individual effort. Certain principles apply to all evasion situations; certain procedures should be followed; and certain techniques have widespread

application. The beginning of an evasion experience is often the most critical phase. This is particularly true for a person who bails out over enemy territory during daylight hours and in sight of enemy personnel. The downed aircrew member can count on the enemy to make a determined effort to capture him if seen. Eluding the enemy is also a matter of effort and luck. Luck plays its part initially in establishing where a crewmember lands in relation to the location of any enemy personnel who may have seen the descent. For example, the downed crewmember who lands in a heavily populated area (city, military installation, or combat area) may be taken prisoner immediately. Here the problem becomes one of early escape rather than evasion.

(7) An evasion situation should not be categorized in terms of length. History has proven that predicting the length of any specific evasion situation is practically impossible. All crewmembers should be prepared to evade until rescued, no matter how long the evasion experience might last. Emphasizing the advantages of "short-term" evasion over "long-term" may cause an overly optimistic, possibly even foolhardy, attitude toward evasion planning. Or, evaders may decide that if they are not rescued in a "short" period of time, it is no longer worth the effort, thereby taking on a defeatist attitude.

Chapter 27

FACTORS OF SUCCESSFUL EVASION

27-1. Basic Principles. All potential evaders must have three things in their favor. These are the same three things needed by a potential escapee. The three factors that increase chances of successful evasion are preparation, opportunity, and motivation.

27-2. Preparation:

a. Preparation is one of the most important factors for successful evasion. The actions that crewmembers take before the evasion episode can make the difference between being able to evade or being captured. In a hostile area, the survivors should remember that evasion is an integral part of their mission, and plan accordingly. The enemy may make mistakes of every conceivable form and still not suffer more than indignation, anger, and fatigue. The evader, on the other hand, must constantly guard against mistakes of any sort. Being seen is the greatest mistake an evader can make. The evader must prepare for this task (figure 27-1).

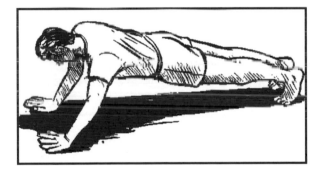

Figure 27-1. Preparation.

b. Three basic problems during evasion are:
  (1) Evading the enemy.
  (2) Surviving.
  (3) Returning to friendly control.

c. Chances for a successful evasion are improved if evaders:
  (1) Observe the elementary rules of movement, camouflage, and concealment.
  (2) Have a definite plan of action.
  (3) Be patient, especially while traveling. Hurrying increases fatigue and decreases alertness. Patience, preparation, and determination are key words in evasion.
  (4) Conserve food.
  (5) Conserve as much strength as possible for critical periods.
  (6) Rest and sleep as much as possible.

  (7) Maintain a highly developed "will to survive" and "can do" attitude. Evasion may require living off the land for extended periods of time and traveling on foot over difficult terrain, often during inclement weather.
  (8) Study the physical features of the land. Survivors should note the location of mountains, swamps, plains, deserts or forests, type of vegetation, and availability of water.
  (9) Consider the climate. Aircrew members should know the climatic characteristics and typical weather conditions of the area that may be flown over.
  (10) Study ethnic briefs, as well as survival, evasion, resistance, and escape (SERE) contingency guides before a mission and learn some of the customs and habits of the local people. Such knowledge will aid in planning missions and evasion plans of action. For example, it may give the evader the ability to avoid hostile people or groups, or to identify and deal with "friendlies." This knowledge may also allow for blending in to the local populace (figure 27-2).
  (11) Know the equipment well! One must know the location of each item in the kit, its operation, and its value. An evader must preplan which equipment should be retained and what should be left behind.

d. Once in the evasion situation, planning for travel will be a consideration for evaders. They must have a definite objective and be confident in their approach and ability to achieve it. They will normally have several options with variations to choose from in selecting a plan of action or destination. The enemy-force deployment, search procedures, terrain, population distribution, climate, distance, and environment (that is, NBC) will influence destination selection. Examples of options and destinations:
  (1) Await SAR forces.
  (2) Evade to a SAFE area.
  (3) Evade to a neutral country. (NOTE: Border areas not disrupted by combat may have a security system intact.)
  (4) If evaders are in the forward edge of a battle area (FEBA), and feel sure that friendly forces are moving in their direction, they should seek concealment and allow the FEBA to overrun their position. Evaders' attempts to penetrate the FEBA should be avoided. Evaders face stiff opposition from both sides.

e. The chances that one of these destinations might be nearby will be determined by many things, including the time and location of the bailout. Other determining factors are: the location and direction of movement of the FEBA, the presence or absence of willing assisters, and the knowledge of the evader's whereabouts possessed by rescue personnel. If the survivor does not land

Figure 27-2. Study Ethnic Briefs.

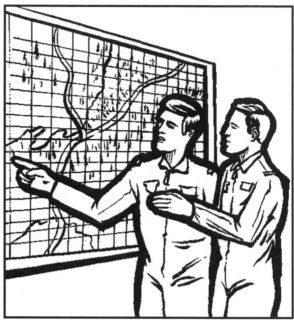

Figure 27-3. Planning for Travel.

close to one of the above areas, or if the previously mentioned factors do not favor immediate air pickup, the survivor may have to travel some distance to reach one of the destinations (figure 27-3).

f. One consideration in choosing destination and direction of travel after bailout is whether one of these suitable areas for pickup or contact with friendly forces exists, and, if so, its location. Some preplanning should have been done before the mission. Information upon which to base a decision is derived from command area briefings, area studies, SERE contingency guides, and premission intelligence briefings.

g. Another consideration is physical condition. One's physical condition is the responsibility of the potential evader and has a great effect on the evader's ability to survive. Once on the ground, it is too late to get in shape. Another aspect is an aircrew member's personal habits. On first consideration, personal grooming habits might not be considered an important premission briefing item. However, using aftershave lotion, hair dressing, or cologne could add to the problems of an evader. The odor can carry for great distances and give away the evader's presence.

27-3. Opportunity. Potential evaders must take advantage of any and all opportunities to evade. This starts in the aircraft when an emergency is declared. Following current, approved emergency in-flight procedures for the theater of operations (when ejection, bailout, or ditching appears imminent), the aircraft commander will attempt to establish radio contact by first calling on the secure frequency of the last contact; second, on an established common secure frequency; and third, on the international emergency frequency. When communication is established, the tactical call sign, type of aircraft, position, course, speed, altitude, nature of difficulties, and intentions will be transmitted. The identification of friend or foe (IFF) should be set to the emergency position. When possible, ejection or bailout should be attempted over or near a SAFE area, lifeguard station, or submarine pickup point. This minimizes threat involvement for evaders and SAR forces alike. After ejection or bailout and during descent, the aircrew member must remain alert, and steer the parachute away from potential threats (populated areas, gun emplacements, troop concentrations, etc.) or out to sea (feet wet). Once on the ground, the evader must be proficient in the use of the survival/evasion equipment to facilitate evasion (for example, use of the compass in conjunction with the survival radio to call in air strikes on enemy forces threatening the evader). In addition to the opportunity to evade, motivation is essential to the evader's success.

Figure 27-4. Avoiding Detection.

27-4. Motivation. A strong, central drive will give the evader the necessary push to succeed. It may be personal, ranging in nature from a frame of reference gained through training to a desire to return to family or loved ones. Motivation may be strictly military, involving one or all of the following reasons applicable to all military men:

a. To return and fight again.

b. To deny the enemy a source of military information.

c. To deny the enemy a source of propaganda.

d. To deny the enemy a source of forced labor.

e. To tie up enemy forces, transportation, and communications that otherwise might be committed to the war effort.

f. To return with intelligence information.

g. In addition, the Code of Conduct calls upon the military members not to surrender of their own free will.

h. Also, it is suggested that other personal reasons for being motivated to evade include: fear of death, pain, suffering, humiliation, degradation, disease, illness, torture, uncertainty, and fear of the unknown. From the evader's point of view, *evasion is far more desirable than captivity or death.*

27-5. Evasion Principles:

a. Besides the preparation, opportunity, and motivation factors important to evasion, there are other important principles. The evader should try to recall any previous briefings, standard operating procedures, or training. A course of action should then be chosen that has the greatest likelihood of resulting in the return to friendly forces.

b. Evader actions should be flexible. Flexibility is one of the most important keys to successful evasion. The evader, basically, must never be so firmly set in a course of action that a change is out of the question. The best thing an evader can do is to stay open to new ideas, suggestions, and changes of events. Having several backup plans of action can give the evader organized flexibility. If one plan of action is upset by enemy activity, the evader could rapidly switch to a backup plan without panic.

c. The evader is primarily interested in avoiding detection. Each evader should remember that people catch people. If the evader avoids detection, success is almost assured. Evaders should:

(1) Observe and listen for sounds of enemy fire and vehicle activity during parachute descent and move away from those enemy positions once on the ground. Fliers downed during daylight hours should assume they were seen during descent and expect a search to center on their likely point of landing.

(2) Be patient and determined while traveling.

(3) Use poor weather conditions as an aid in evading.

(4) Circumstances permitting, select time, routes, and methods of travel to avoid detection.

(5) Avoid lines of communication (waterways, roads, etc.; figure 27-4).

d. The evaders' main objective is immediate recovery. In hostile areas or situations, survivors must sanitize all evidence of presence and direction of travel (figure 27-5). Survivors may never be certain that rescue is imminent.

e. Although evaders would not normally move too far if rescue is imminent, in many situations they will have to leave the landing area quickly and travel as far as practical before selecting a hiding place. They should leave no sign that indicates the direction or presence of travel. All hiding places should be chosen with extreme care.

The time evaders will remain in the first location is governed by enemy activity in the area, their physical condition, availability of water and food, rescue capabilities, and patience. It is in this place of initial concealment that the evaders should regain strength, examine the current situation, and plan for the evasion problems ahead (figure 27-6).

f. Once in a place of concealment, evaders should make use of all available navigation aids to orient themselves. After finding their location, evaders should also select an ultimate destination and any necessary alternate destinations. The best possible route of travel should then be decided upon. When the time comes to move, they should have a primary plan and alternate plans for travel that cover eventualities they may encounter.

(1) Evasion in a forward area has one great advantage that is not present further to the rear: assistance may be close at hand. This assistance may come from several sources, each of which, under particular circumstances, may prove to be the most effective. These sources may be air cover by tactical fighter flights, helicopter recovery, and rescue by ground forces. Contact with friendly forces in forward areas requires extreme caution. Do not surprise them or move suddenly. They may mistake the evader for the enemy.

Figure 27-5. Sanitize Area.

Figure 27-6. Movement from the Area.

(2) The situation at the time of the emergency will determine the evaders' best course of action. High ground is normally the best position from which to await rescue; evaders may expect the best results from signaling devices, may observe the surrounding terrain, and may be kept under observation by friendly air cover. Whatever position is chosen, it must be clear of obstacles that would prevent a successful rescue.

(3) If not rescued immediately, the situation may compel evaders to move. Evaders must plan a course of action before leaving their position. When the evaders are certain that their position is known to friendly elements, they might expect ground forces to attempt a rescue. They should remember that their position might be detected as the enemy search parties approach. They must be prepared to evade to a new position.

(4) Evaders should remember that when traveling they are probably more vulnerable to capture. Once past the danger of an immediate search, evaders must avoid people. Inhabited areas should be bypassed rather than penetrated, even if it means miles of added travel. Many evaders have been captured because they followed the easiest and shortest route, or failed to employ simple techniques, such as scouting, patrolling, camouflage, and concealment. As a rule, the safest route avoids major roads and populated areas, even if it takes more time and energy. Unaccompanied evasion requires self-reliance and independent action (figure 27-7).

Figure 27-7. Avoiding Populated Areas.

Chapter 28

CAMOUFLAGE

28-1. Introduction. Presence of evaders in an area controlled by the enemy may require the evaders to adopt and maintain camouflage to avoid observation. Camouflage consists of those measures evaders use to conceal their presence from the enemy. Camouflage is a French word meaning disguise, and it is used to describe action taken to mislead the enemy by misrepresenting the true identity of an installation, an activity, an item of equipment, or an evader. As a tool for evasion, it enables evaders to carry out life-supporting activities, and to travel unseen, undetected, and free to return to friendly control. Camouflage allows them to see without being seen. They should try to blend in with the surrounding environment. Effective individual concealment often depends primarily on the choice of background and its proper use. Background is that portion of the surroundings against which an evader will be seen from the ground and the air. It may consist of a barren rocky desert, a farmyard, or a city street. It is the controlling element in individual camouflage and governs every concealment measure. At all times, camouflage is the responsibility of the individual evader. In the event of group evasion, the group leader and each individual are responsible for the camouflage of the group. Evaders should remember that camouflage is a continuous, never-ending process if they want to protect themselves from enemy observation and capture (figure 28-1).

28-2. Types of Observation:

a. Of the five senses, sight is by far the most useful to the enemy, hearing is second, while smell is of only occasional importance. But these same senses can be of equal value to the evader and observer.

b. How useful these senses are depends primarily on range. For this reason, basic camouflage stresses visual concealment that is relatively long-range. Most people are accustomed to looking from one position on the ground to another position on the ground.

c. Before evaders can conceal themselves from aerial observation, they should become familiar with what their activities look like from the air, both in an aerial photograph and from direct observation. The evaders must also have an understanding of the types of observation used by the enemy. There are two categories of observation—direct and indirect.

(1) Direct Observation:

(a) Direct observation refers to the process whereby the observer looks directly at the object itself without the use of telescopes, field glasses, or sniper scopes. Direct observation may be made from the ground or from the air. Direct aerial observation becomes more and more important because of the rapid changes in weapons and in tactical situations due to greater mobility of troops. Reconnaissance aircraft over enemy lines report locations of troops, vehicles, and

Figure 28-1. Camouflaging.

Figure 28-2. Direct Observation.

installations (or shelter areas) as seen from the air-to-ground control stations. Reported targets can be immediately fired upon, or troops can be sent in to investigate shelter areas or other suspicious areas (figure 28-2).

(b) The enemy may also use dogs, foot patrols, and mechanized units to patrol a given area. Such teams could physically search an area for signs of the passage of strangers, such as footprints, old campfires, discarded or lost equipment, and other "telltale" signs that would indicate that someone had been in the area.

(c) Observation by the local populace is also a possibility. Upon seeing an evader or "telltale" signs an evader left behind, they may contact the local authorities, which initiate organized searches.

(2) Indirect Observation:

(a) Indirect observation refers to the study of a photograph or an image of the subject via photography, radar, or television. This form of observation is becoming increasingly more varied and widespread, and may be used from either manned or unmanned positions.

(b) Views from the ground are familiar, but views from the air are usually quite unfamiliar. In modern warfare, the enemy may put emphasis on aerial photographs for information. It is important to become familiar with the "bird's-eye view" of the terrain as well as the ground view in order to learn how to guard against both kinds of observation.

Figure 28-3. Indirect Observation.

28-3. Comparison of Direct and Indirect Observation:

a. The main advantage of direct observation is that observers see movement of an evader without camouflage. An observation can be maintained for relatively long periods of time. The main disadvantage lies in human frailty. For example, the observer's attention may be diverted to another area, or the observer may be fatigued and unable to concentrate.

b. Indirect observation has many advantages. Indirect observation can be far-reaching, cover large areas, and be very accurate. It also produces a record of the area observed so the recorded picture can be studied in detail, compared, and evaluated. The principal disadvantage is that a photograph covers a very short period of time, making detection of movement difficult. This disadvantage can be partially overcome by taking pictures of the same area at different intervals and comparing them for changes (figure 28-3).

28-4. Preventing Recognition:

a. Recognition is the determination (through appearance, behavior, or movement of a hostile, or friendly, nature) of objects or persons. One objective of camouflage concealment is to prevent recognition. Another objective is to deceive or induce false recognition. This implies that camouflage is not always designed to be a "cloak of invisibility." In some instances, camouflage is used to allow deception. The camouflaged object or person is then seen as a natural feature of the landscape.

b. Recognition through appearance is the result of conclusions drawn by the observer from the position, shape, shadow, texture, or color of the objects or persons. Recognition through behavior or movement includes deductions made from the actual movements themselves, or from the record left by tracks of persons or vehicles or by other violations of camouflage discipline. Camouflage disciplines are those actions that contribute to an evader's ability to remain undetected. Proper use of camouflage discipline avoids any activity that changes an area or reveals objects to an enemy. Examples of common

Figure 28-4. Reflections.

breaches of camouflage discipline include reflections from brightly shining objects (watches, glasses, rings, etc.; figure 28-4), overcamouflaging, or using camouflage materials that are foreign to the area presently occupied by an evader. Evaders must also watch for signs that may reveal enemy camouflage efforts. Inadvertently walking into a camouflaged enemy position may result in capture.

28-5. Factors of Recognition. Regardless of the type of observation, there are certain factors that help to identify an object. They are called the factors of recognition, and are the elements that determine how quickly an object will be seen, or how long it will remain unobserved. The eight factors of recognition are position, shape, shadow, texture, color, tone, movement, and shine. These factors must be considered when camouflaging to ensure that one or more of these factors do not reveal the location of the evaders.

a. Position. Position is the relation of an object or person to its background. When choosing a position for concealment, a background should be chosen that will virtually absorb the evader (figure 28-5).

Figure 28-5. Position.

b. Shape. Shape is the outward or visible form of an object or person as distinguished from its surface characteristics and color. Shape refers to outline or form. Color or texture is not considered. At a distance, the forms or outlines of objects can be recognized before the observer can make out details in their appearance. For this reason, camouflage should disrupt the normal shape of an object or person (figure 28-6).

c. Shadow. A shadow may be more revealing than the object itself, especially when seen from the air. Objects such as factory chimneys, utility poles, vehicles, and tents (or people) have distinctive shadows. Conversely, shadows may sometimes assist in concealment. Objects in the shadow of another object are more likely to be overlooked. As with shape, it is more

Figure 28-6. Shape.

important to disrupt the shadow pattern than to totally conceal the object or person. The identifiable shadows can be broken up by the addition of natural vegetation at various points on the body. Wearing "shapeless" garments will also disrupt the outline. For example, a soft and shapeless field cap can be used instead of a helmet or flight cap (figures 28-7 and 28-8).

d. Texture. Texture is a term used to describe the relative characteristics of a surface, whether that surface is a part of an object or an area of terrain. Texture affects the tone and apparent coloration of things because of its absorption and scattering of light. Highly textured surfaces tend to appear dark and remain constant in tone regardless of the direction of view or lighting, whereas relatively smooth surfaces change from dark to light with a change in direction of view or lighting. The application of texture to an object often has the added quality of disrupting its shape and the shape of its shadow, making it more difficult to detect and identify as something foreign to the surroundings in which it exists. For example, one surface having the same color but with heavy "nap" or texture is tall grass. Each separate blade is capable of casting a shadow upon itself and its surroundings. The light-reflecting properties have been cut to a minimum. It will look and photograph dark gray. Looking straight down, the aerial observer sees all of the shadows, whereas a person on the ground may not. The textured surface may look light at ground level, but to the aerial observer the same surface produces an effect of relative darkness. The material used to conceal a person or an object must approximate the texture of the terrain in order to blend in with the terrain. Personnel walking, or vehicles moving, across the terrain will change the texture by mashing down the growth. Therefore, this will show up clearly from the air as vehicle tracks or footpaths.

e. Color:

(1) Pronounced color differences at close range distinguish one object from another. The contrast between the color of the object and the color of its background

Figure 28-7. Shadow.

Figure 28-8. Shadow Breakup.

Figure 28-9. Contrast.

because the textured surface now absorbs more light rays. Objects become identifiable as such because of contrasts between them and their background. Camouflage blending is the process of eliminating or reducing these contrasts. The principal contrast is that of tone; that is, the dark and light relationship existing between an object and its background. The two principal means available for reducing tone contrast are the application of matching or neutral coloration, and the use of texturing to form disruptive patterns. Poorly chosen, disruptive patterns tend to make the object more conspicuous instead of concealing it.

Figure 28-10. Three Half-Tone Blocks.

can be an aid to enemy observers. The greater the contrast in color, the more visible the object appears (figure 28-9).

(2) Color differences or differences in hue, such as red and green-yellow, become increasingly difficult to distinguish as the viewing range is increased. This happens because of atmospheric effects. Colors in nature, except for certain floral and tropical animal life, are not brilliant. The impression of the vividness of nature's colors results from the large areas of like colors involved, and contrast of these areas with one another. The principal contrast is in their dark and light qualities. However, the dark and light color contrast does not fade out quickly, and is distinguishable at greater distances. Therefore, as a first general principle, the camouflage should match the darker and lighter qualities of the background, and become increasingly concerned with the colors involved as the viewing range is decreased or the size of the object or installation becomes larger. A second general rule to follow is to avoid contrasts of hues. This is especially true in areas with heavy vegetation. Light-toned colors, such as leaf bottoms, should be avoided, as they tend to attract attention.

f. Tone. Tone is the amount of contrast between variations of the same color. It is the effect achieved by a combination of light, shade, and color. In a black-and-white photograph, the shades of gray in which an object appears is known as tone (figure 28-10). By adding texturing material to a smooth or shiny surface, the surface can be made to produce a darker tone in a photograph,

g. Movement. Of the eight factors of recognition, movement is the quickest and easiest to detect. The eye is very quick to notice any movement in an otherwise still scene. The aerial camera can record the fact that something has moved when two photographs of the same area are taken at different times. If an object has moved, the changed position is apparent when the two photographs are compared (figure 28-11).

h. Shine.

(1) Shine is a particularly revealing signal to an observer. In undisturbed, natural surroundings, there are comparatively few objects that cause a reflected shine. Skin, clean clothing, metallic insignia, rings, glasses, watches, buckles, identification bracelets, and similar items produce "shine." When light strikes smooth surfaces such as these, it may be reflected directly into the observer's eye, or the camera lens, with striking emphasis (figure 28-12).

(2) Such items must be neutralized by staining, covering, or removing to prevent their shine from revealing the location of evaders. This is especially true at night.

28-6. Principles and Methods of Camouflage:

a. No matter how applied, camouflage can be successful only by observing three fundamental principles. These basic principles of camouflage are choice of position, camouflage discipline, and camouflage construction.

b. When these factors have been considered, the evader is ready to begin application of various methods of deceiving the enemy. These methods are:

Figure 28-11. Movement.

(1) Hiding. The complete concealment of a person or object by physical screening.

(2) Disguising. Changing the physical characteristics of an object or person in such a manner as to fool the enemy.

(3) Blending. The arrangement of camouflage material on or about an object in such a manner as to make the object appear to be part of the background. To properly use these methods, three simple rules should be followed:

(a) First, the background should be changed as little as possible. When choosing a position to gain concealment, a background should be chosen that will visually absorb the elements of the position. Evaders should use a "natural" position, if available. They should look for an existing position that can be used almost as is, such as a cave or thicket, if there are many like it in the area. Isolated landmarks such as individual trees, haystacks, or houses should be avoided. They lend to attract attention and are likely to be searched first because they are so obvious. At times, by making use of background, complete concealment against visual and photographic detection may be gained with no construction. In terrain where natural cover is plentiful, this is a

Figure 28-12. Reflected Shine.

Figure 28-13. Background.

simple task. Even in areas where natural cover is scarce, concealment may be achieved through use of terrain irregularities. Regardless of the activity involved, evaders must always be mentally aware of their positions (figure 28-13).

(b) Second, the evader should use camouflage discipline. This means all of the factors of concealment are continuously applied.

-1. Daytime. Camouflage discipline is the avoidance of activity that changes the appearance of an area, or reveals military objects to the enemy. A well-camouflaged position is only secure as long as it is well maintained. Concealment is worthless if obvious tracks point like directional arrows to the heart of the location, or if signs of occupancy are permitted to appear in the vicinity. Tracks, debris, and terrain disturbances are the most common signs of activity. Therefore, natural lines in the terrain should be used. If practical, exposed tracks should be camouflaged by brushing or beating them out. If leaving tracks is unavoidable, they should be placed where they will be least noticed and are partially concealed (along logs, under bushes, in deep grass, in shadows, etc.). If tracks cannot be concealed, brushing them out will help them disintegrate quickly. Tying rags or brush to the feet will disguise boot pants and may help disguise them as refugee tracks (figure 28-14).

-2. Nighttime. Visual concealment at night is less necessary than in the daytime; however, noises at night are more noticeable. As simple an act as snoring may prove fatal. Calling to one another, talking, and even whispering should be kept to a minimum (figure 28-15).

Figure 28-14. Camouflaging Tracks.

Figure 28-15. Sound.

But by far, the most important aspect of night discipline is light discipline. Lights at night not only disclose the evaders' position, but also hinder the evaders' ability to detect the enemy. Even on the darkest nights, eyes grow accustomed to the lack of light in approximately 30 minutes. Every time a match is lit or a flashlight is used, the eyes must go through the complete process of getting adjusted to the darkness again. Smoking and lights should be prohibited at night in areas in close proximity to the enemy because the light is impossible to conceal. Additionally, a cigarette light aggravates the situation by creating a reflection that completely illuminates the face. The smell of the evader's foreign tobacco would stand out even if the enemy is smoking as well.

       -3. Evaders can lessen the effects of sound by simply taking precautions against sound production. They should avoid any sound-producing activity. Walking on hard surfaces should be avoided and full use should be made of soft ground for digging. Hand signals or signs should be used when possible during group travel. Individual equipment should be padded and fastened in such a manner as to prevent banging noises.

    c. The evader should consider the following points regarding the use of camouflage:

      (1) Take advantage of all natural concealment.

      (2) Don't over-camouflage. Too much is as obvious as too little.

      (3) When using natural camouflage, remember that it fades and wilts, so change it regularly.

      (4) If taking advantage of shadows and shade, remember they shift with the sun.

      (5) Above all, avoid unnecessary movement.

      (6) When moving, keep off the skyline; use the military crest (three-quarters' way up the hill).

      (7) Do not expose anything that may shine.

      (8) Break up outlines of manmade objects.

      (9) When observing an area, do so from a prone position, while in cover.

      (10) Match vegetation used as camouflage with that in the immediate locale, and when moving from position to position, change camouflage to blend with the new area's vegetation types.

28-7. Individual Camouflage:

a. At this point, with some of the general information about camouflage presented, it is time for a more detailed examination concerning individual camouflage.

b. Generally, individual camouflage is the personal concealment that evaders must use to deceive the enemy. Evaders must know how to use the terrain for effective concealment. Evaders must dress for the best concealment, and carefully select their routes to provide for as much concealment as possible. All of the methods and techniques of camouflage addressed in this section have been successfully used by past evaders. If this information is learned and practiced by today's aircrew members (tomorrow's possible evaders), they will be more successful in evading, and have a greater chance of returning to friendly forces.

c. Evaders should remember that in some areas they might have to engage in camouflage activities designed to deceive two types of enemy observation—ground and air. Many objects that are concealed from ground observation may be seen from the air. This means the evader should camouflage for both types of observation.

d. Form is basic shape (body outline) and height. Three things that give an evader away in terms of form are to reveal outline of head and shoulders, to present straight lines of sides, and to allow the inverted "V" of the crotch and legs to be distinguishable. If staying in shadows, blending in with the background, adopting body positions other than standing erect, and other behavioral procedures are inadequate. They can be camouflaged by using "add-ons" such as branches or twigs to break up the lines. This addition of vegetation will also help an evader blend in with the background.

e. Effective concealment of evaders depends largely on the choice and proper use of background. Background varies widely in appearance, and evaders may find themselves in a jungle setting, in a barren or desert area, in a farmyard, or on a city street. Each location will require individual treatment because location governs every concealment measure taken by the individual. Clothing that blends in with the predominant color of the background is desirable. There will be occasions when the uniform color must be altered to blend in with a specific background. The color of the skin must receive individual attention and be toned to blend in with the background.

f. There are certain general aspects of individual body and equipment camouflage techniques that apply almost anywhere. The evaders should take each of the following areas under consideration.

(1) Exposed Skin. The contrast in tone between the skin of face and hands and that of the surrounding foliage and other background must be reduced. The skin is to be made lighter or darker, as the case may be, to blend in with the surrounding natural tones. The shine areas are the forehead, cheekbones, nose, and chin.

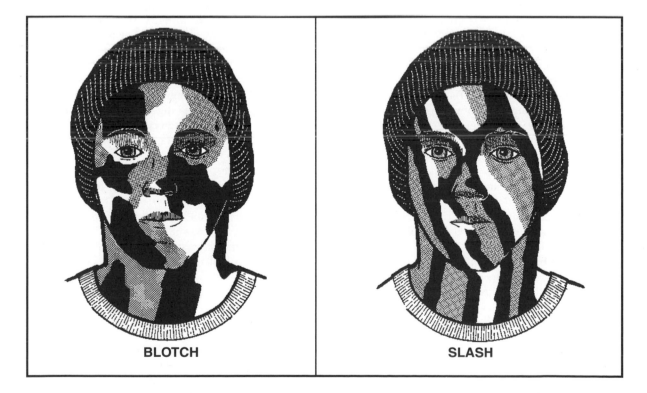

BLOTCH   SLASH

Figure 28-16. Camouflaged Faces.

These areas should have a dark color. The shadow areas, such as around the eyes, under the nose and under the chin, should have a light color. The hands, arms, and any other exposed areas of skin must also be toned down to blend in with the surroundings. Burnt cork, charcoal, lampblack, mud, camouflage sticks, berry stains, carbon paper, and green vegetation can all be used as toning materials.

(a) A mesh mosquito face net, properly toned down, is an effective method of breaking up the outlines of the face and ears.

(b) Two primary methods of facial camouflage have been found to be successful patterns. They are the "blotch" method for use in deciduous forests, and the "slash" method for use in coniferous forests (figure 28-16).

(c) Application of these two patterns is simply modified appropriately to whatever environment the evader is in. In the jungle, a broader slash method would be used to cover exposed skin; in the desert, a thinner slash; in barren snow, a wide blotch; and in grass areas, a thin-type slash. To further break up the outline of the facial features, a flop hat or other loosely fitting hat may help. A beard that is not neatly trimmed may also aid the evader.

(d) When toning down the skin, evaders should not neglect to pattern all of the skin; for example, the back of the neck, the insides and backs of the ears, and the eyelids. Covering these areas may help somewhat, especially if there is a lack of other material to tone down the skin. Vegetation hung from the hat, collar buttoned and turned up, a scarf, or even earflaps may help. To cover the hands, evaders may use flight gloves, mittens, or loose cloth if unable to tone down wrists, backs of hands, and between fingers sufficiently. Evaders should watch for protruding white undergarments, T-shirt, long-underwear sleeves, etc. They should also tone down these areas.

(e) Lack of hair or light-colored hair requires some type of camouflage. This could include those applied to the skin, or an appropriate hat, scarf, or mosquito netting.

(f) Odors in a natural environment stand out and may give evaders away. Americans are continually surrounded by artificial odors and are not usually aware of them. Human body odor would have to be very strong to be detected by ground troops (searchers) that have been in the field for long periods. The following odors should be of concern to the evader.

-1. Soaps and Shampoos. In combat areas, personnel should always use unscented toilet articles. Shaving cream, aftershave lotion, perfume, and other cosmetics are to be avoided. The potential evaders should also realize that insect repellent is scented. They should try to use headnets, but if forced to use a repellent,

the camouflage stick, which has repellent in it, is the least scented. Tobacco should not be used. The stain and odor should be removed from the body and clothing. Gum or candy may have strong or sweet smells—evaders should take care to rinse out their mouth after use. These odors, especially tobacco, can be detected at great distances.

-2. Smoke odors from campfires may permeate clothing, but if the potential searchers use fires for cooking and heat, they probably can't detect it on evaders.

(2) Clothing and Personal Items. These items require attention both before assignment to a combat mission and again if forced to evade. Prior preparation for the survivor may include:

(a) Ensuring that flight clothing does not smell of laundry products and is in good repair and not worn to the point where it shines or is faded.

(b) Checking zippers for shine and function.

(c) Checking rank insignia and patches for light-reflection and color. Remove name tag, branch of service tag, and rank (whether they are stripes or metal insignia, bright unit patches, etc.) from the uniforms, and place them either in the pack or in a secure pocket. Underwear should also be subdued for camouflage in the event that the outer layer is torn. Boots should be black but not shiny. Shiny eyelets should be repainted. Squeaky boots should be fixed or replaced. Sanitize pocket or wallet contents. Remove items that might aid in enemy exploitation attempts of the individual PW or PWs as a group; for example, credit cards, photographs, money, and addresses. Evaders should carry only those necessary pieces of identification that will prove a person is a US military member.

(3) Additional Clothing. Additional clothing may be desirable and located in flight clothing, such as hat, socks, scarf, and gloves. In an evasion environment, clothing and equipment need quick camouflaging attention. Anything to be discarded should be hidden at the initial landing site.

(4) Sanitizing Clothing. Clothing should be sanitized by removing anything bright or shiny. Evaders should consider camouflaging their clothing (including boots) just as they would their skin. This is detrimental to the insulative qualities of their clothes, but not as much as bullets or prison barbed wire if they are seen.

(5) Camouflaging Clothing:

(a) One principle of camouflage is to disrupt or conceal uniform color, straight lines, and squares—things rarely found in natural features. If ground-to-air signals are designed to exploit these visual characteristics, then evaders would certainly want to eliminate them.

(b) Evaders should reduce the tone of all equipment by smearing it with camouflage stick, mud, etc., or with whatever is available, in a mottled pattern. In some instances equipment will have to be lighter in tone, in others, darker.

(c) To maintain the functional capability of clothing and equipment, it must be kept clean. In some areas, however, these items may be the only natural camouflage material survivors have to work with (such as in desert regions). But, in areas where they have access to vegetation and the various dyes that can be made from vegetation, the vegetation should be used. In contrast to substances that soil the material and actually break down the fibers, dyes derived from grasses or plant sap (banana trees, ash trees, etc.) will offer the evader the toning material necessary to break up the solid green of a uniform, while leaving the fabric grit-free and still able to "breathe." The same saps that produce stains for cloth can be used to discolor metal objects. Banana-tree sap, when left on the metal blade of a knife, will produce a blue-black stain, which is a permanent discoloration. Trappers still boil their traps in ash-tree chips and water to produce the blue-black, rust-inhibiting coloration to the tools of their trade.

g. All principles and techniques for care and use of clothing and equipment cannot be forgotten or ignored in an evasion situation, although some modifications may be necessary. A number of variables will influence what changes or omissions will be necessary.

(1) All cutting tools must be kept sharp. Evaders should try to coincide these noisy—yet essential—tasks with natural noise in the area (a downpour of rain for instance), or in a protected, noise-dampening area.

(2) Clothing must be kept clean if it is to protect a survivor from a harsh environment. Dirt-clogged, perspiration-soaked fibers will not give the insulating qualities of clean cloth. Clothing can be washed during a downpour of rain or possibly under the cover of darkness in a stream. Convenient, secluded puddles of water may afford the opportunity a survivor needs to clean clothing and equipment properly.

(3) Cooking and eating utensils must be kept clean on the inside to prevent dysentery and diarrhea. Simultaneously, the outsides can be toned down with soil, mud, etc., to camouflage them.

(4) Where metal pieces come in contact with one another, there should be padding between them so they will not inadvertently "clank" together. Evaders should place all items needed for environmental protection in the top of the pack, where they will be most readily available. The rest of the gear can be used as padding around metal objects. In this manner, with everything stored inside the bundle (pack), it is secure from loss, damage, and enemy observation, as well as being readily available when needed. Evaders should also remove jewelry, watches, exposed pens, and glasses, if possible. If glasses are required, hat netting or masking may help reduce shine.

(5) An evader's pockets should be secured, and all equipment, including dog tags, arranged so no jingle or rattle sounds are made. This can be done with cloth, vegetation, padding, or tape.

(6) Evaders should minimize the sound of clothing brushing together when the body moves. Moving in a careful manner can decrease this sound. Evaders should remember that camouflaged clothing and equipment alone won't conceal, but it must be used intelligently in accordance with the other principles of camouflage and movement. For example, even if evaders are perfectly camouflaged for the Arctic, there could still be problems. Because snow country is not all white, shadows and dark objects appear darker than usual. A snowsuit cannot conceal the small patches of shadow caused by the human figure, but that is not necessary if the background contains numerous dark areas. If the background does not contain numerous dark areas, maximum use is to be made of snowdrifts and folds in the ground to aid in individual concealment (figure 28-17).

(7) The concept of blending in with the background is indeed an important one for the evader to understand. One major point in blending in with the background is not to show a body silhouette.

(8) Losing the body silhouette is done by making use of the shadows in the background. Evaders should be constantly aware of two factors—silhouette and shadow. From a concealment point of view, backgrounds consist of terrain, vegetation, artificial objects, sunlight, shadows, and color. The terrain may be flat and smooth, or it may be wrinkled with gullies, mounds, or rock outcroppings. Vegetation may be dense jungle growth, or no more than small patches of desert scrub growth. The size of artificial objects may range from a signpost to an entire city block. There may be many colors in a single background, and they may vary from the almost black of deep woods to the sand-pink of some desert valleys. Blending simply means matching with as many of these backgrounds as possible and avoiding contrast. If it is necessary for evaders to be positioned in front of a contrasting or fixed background, they must be aware of their position and take cover in the shortest possible time. The next point to which they will move for concealment must be selected in advance and reached as quickly as possible.

(9) As in the daytime, silhouette and background at night are still the vital elements in concealment (figure 28-18). A silhouette is always black against a night sky, and care must be taken at night to keep off the skyline. On moonlit nights, the same precautions must be taken as in daylight. It should be remembered that the position of the enemy observer, and not the topographic crest, fixes the skyline. At night, sound is an amplified, revealing signal. Movement must be careful, quiet, and

Figure 28-17. Arctic Travel.

Figure 28-18. Silhouetting.

Figure 28-19. Natural Materials.

close to the ground. If the pop of a flare is heard before the illuminating burst, evaders must drop to the ground instantly and remain motionless. If they are surprised by the light, they must freeze in place with their faces down.

28-8. Concealment in Various Geographic Areas:

a. When not otherwise specified, temperate zone terrain is to be assumed in this section. Desert, snow, and ice areas are mostly barren, and concealment may require considerable effort. Jungle and semi-tropical areas usually afford excellent concealment if the evader employs proper evasion techniques.

b. First, some general observations and rules regarding the addition of vegetation to the uniform and equipment. The cycles of the seasons bring marked changes in vegetation, coloring, and terrain pattern that requires corresponding changes in camouflage. Concealment that is provided in wooded areas during the summer is lost when leaves fall in the autumn. This will create a need for additional camouflage construction. Also, vegetation must be of the variety in the evaders' immediate location. It must be changed if it wilts or the evaders move into a different vegetation zone. Evidence of discarding the old and picking the new should be hidden. The vegetation should not be cut; this will give evidence of human presence.

c. Any type of material indigenous to the locality of the evaders may be classified as natural material.

Natural materials consist of foliage, grasses, debris, and earth. These materials match local colors and textures and when properly used are aids against both direct and indirect observation. The use of natural materials provides the best type of concealment. The chief disadvantage of natural foliage is that it cannot be prepared ahead of time, is not always available in usable types and quantities, wilts after gathering, and must be replaced periodically. Foliage of coniferous trees (evergreens) retains its camouflage qualities for considerable periods, but foliage that sheds leaves will wilt in a day or less, depending on the climate and type of vegetation (figure 28-19).

(1) The principal advantage in using live vegetation is its ability to reflect infrared waves and to blend in with surrounding terrain. When vegetation is used as garnishing or screening, it must be replaced with fresh materials before it has wilted sufficiently to change the color or texture. If vegetation is not maintained, it is ineffective. Thorn bushes, cacti, and other varieties of desert growth retain growing characteristics for long periods after being gathered.

(2) The arrangement of foliage is important. The upper sides of leaves are dark and waxy; the undersides are lighter. In camouflage, therefore, foliage must be placed as it appears in its natural growing state: top sides of leaves up, and tips of branches toward the outside of the leaves (figure 28-20).

Figure 28-20. Dark and Light Leaves.

(3) Foliage gathered by survivors must be matched to existing foliage. For example, foliage from trees that shed leaves must not be used in an area where only evergreens are growing. Foliage with leaves that feel leathery and tough should be chosen. Branches grow in irregular bunches and, when used for camouflage, must be placed in the same way. When branches are placed to break up the regular, straight lines of an object, only enough branches to do this should be used. The evader must adopt principles that apply, and know that the enemy also applies these principles.

(4) When vegetation is applied to the body or equipment of evaders, it must be secured to clothing or equipment in such a way that:

(a) Any inadvertent movement of the material will not attract attention.

(b) It appears to be part of the natural growth of the area; that is, when the evaders stop to hide, the light undersides of the vegetation are only visible from beneath. After evaders complete their camouflage, they should inspect it from the enemy's point of view. If it does not look natural, it should be rearranged or replaced.

(c) It does not fall off at the wrong moment and leave evaders exposed, or it should not show evidence of the evaders' passage through the area.

(5) Too much vegetation can give evaders away.

(6) If cloth material is used like vegetation to break up shape and outline and to help blend in with the environment, there are some points evaders should be aware of. Cloth can be used successfully when wrapped around equipment or designed into loose, irregular-shaped clothing or accessories (figure 28-21).

(a) Some of the materials evaders may use are:

-1. The colors (green, brown, white) of parachute materials, plus the harness.

-2. Excess clothing the evader may have prepacked (scarf, bandana, etc.).

-3. Burlap, when found. It is used in battle areas in the form of sandbags.

(b) Most artificial material is versatile, but it can have drawbacks:

-1. Parachute material, for example, tends to shine, and the unraveled edges may leave fine filaments of nylon on the ground as evidence of evaders in an area. Parachute material is very lightweight. A sudden breeze might cause it to move when movement is not desirable.

-2. A white suit of parachute material is excellent for winter evasion in snow and ice environments. The time required to fabricate this suit should be considered. (NOTE: Shadows cast on the snow cannot be

Figure 28-21. Breakup of Shape and Outline.

Figure 28-22. Concealment by Shadows.

Figure 28-23. Above Timberline.

camouflaged, they must be masked by terrain or other shadows.)

28-9. Concealment Factor for Areas Other than Temperate:

a. Desert. Lack of natural concealment, high visibility, and bright tone (smooth texture) all emphasize the need for careful selection of a position for a campsite. Deep shadows in the desert, strict observance of camouflage discipline, and the skillful use of deception and camouflage materials aid in concealing evaders in a desert area.

(1) Deserts are not always flat, single-toned areas. They are sometimes characterized by strong shadows with heavy, broken terrain lines, and sometimes by a mottled pattern. Each type of desert terrain presents its own problems. When evaders and their shelters are located in the desert, their shadows are inky-black and in strong contrast to their surroundings, and are extremely conspicuous. To minimize the effect of these shadows when possible, use concealment that is afforded by the shadows of deep gullies, scrub growth, and rocks (figure 28-22).

(2) Many objects that cannot be concealed from the air can be effectively viewed from the ground. Even though these objects are observed from the air, lack of reference points in the terrain will make them difficult to locate on a map.

b. Snow and Ice. From the air, snow-covered terrain is an irregular pattern of white spotted with dark tones produced by objects projecting above the snow, their shadows, and irregularities in the snow-covered surface, such as valleys, hummocks, ruts, and tracks. It is necessary, therefore, to make sure dark objects have dark backgrounds for concealment to control the making of tracks in the snow, and to maintain the snow cover on camouflaged objects.

(1) Mountain Areas Above Timberline and Arctic Areas. Common characteristics include an almost complete snow cover with a minimum of opportunities for concealment. Only a few dark objects protrude above the snow, except for rugged mountain peaks (figure 28-23).

(2) Mountain Areas Below the Timberline and Subarctic Areas. Common characteristics of these areas are forests, rivers, lakes, and artificial features such as trails and buildings. The appearance of the area is irregular in pattern and variable in tone and texture (figure 28-24).

Figure 28-24. Below Timberline.

(3) Areas Between the Subarctic Zone and the Southern Boundary of the Temperate Zone. These have the same characteristics as mountain areas below the timberline and subarctic areas.

c. Blending with Background in Snow and Ice Terrains. No practical artificial material has yet been developed that will reproduce the texture of snow well enough to be a protection against recognition by aerial observers. Concealment from direct ground observation is relatively successful with the use of white "snowsuits," white pants, and whitewash; these measures offer some protection against aerial detection. (White parachute cloth should also be considered.)

(1) People evading in snow-covered, frozen areas should wear a completely white camouflage outfit. A white poncho-like cape can be made easily from parachute material (figure 28-25).

(2) A pair of white pants will normally be sufficient in a heavily wooded area. However, following or during a heavy snowfall, when the trees are well covered with snow, the wearing of a completely white camouflage suit is necessary to blend in with the background. Other equipment, such as packs, should also be covered with white material.

d. Camouflage Checklist. The following checklist can be used to remember ideas concerning camouflage,

and to determine the completeness of individual camouflage application:

(1) Effective concealment entails protection from hostile observation from the ground as well as from the air.

(2) Natural terrain lines are to be used for help in concealing the evader, when possible.

(3) Every possible feature of the terrain should be used for concealment.

(4) A silhouette against the sky should be avoided.

(5) Every effort should be made to reduce tone contrast and eliminate shine.

(6) Evaders should be especially careful at night due to infrared and low-light detection equipment, which may be used by the enemy. Keeping close to the ground and using terrain masking for concealment provide the best protection.

28-10. Camouflage Techniques for Shelter Areas:

a. As used in this section, the word shelter refers to the concept of personal protection synonymous with the terms refuge, haven, or retreat. Readers should not visualize the word "shelter" when mentioned in this text to mean a dwelling traditionally occupied by survivors in nontactical situations, such as tents, cabins, or other such places of habitation. While it is true that these structures can and do provide safety and relief, it must

Figure 28-25. Evading in Snow-Covered Areas.

Figure 28-26. Shelter.

be understood that where an evasion situation is concerned, concealment, not personal comfort or convenience, will be the primary concern for the evader who seeks "shelter" or sanctuary (figure 28-26).

b. Besides resting or sleeping, there are a number of reasons survivors may need to conceal themselves for varying amounts of time. They may need to take care of problems concerning personal hygiene, adjustment of clothing, maintenance or alteration of camouflage, triangulation for position determination, food and water procurement, etc. Concealment cannot be overstressed in respect to the areas survivors (evaders) may select for shelter. One important reason evaders must be able to select secure areas for refuge is to avoid and prevent detection by the enemy. This is especially important if the haven selected is to be used for resting or sleeping. Anyone who is resting or sleeping will not be totally alert, and added precautions may be necessary to maintain security. Another factor to consider, since evaders are also survivors, is to protect themselves from the elements as much as possible.

c. At no time will evaders be able to safely assume they are free from the threat of either ground or aerial observation. Therefore, not only is the shelter area and type determined by the needs of the moment (enemy

Figure 28-27. Natural Shelter.

presence, etc.), but consideration must also be given to the terrain and climatic conditions of the area. Evaders must constantly be aware of how long they may have to remain in the area and, most important, of what type of enemy observation may be employed.

d. The shelters may be naturally present, or they may be those that are "assembled" and camouflaged by the evaders. Full use must be made of concealment and camouflage, no matter what types of shelter areas are selected. The use of the natural concealment afforded by darkness, wooded areas, trees, bushes, and terrain features are recommended; however, any method used for disguise or hiding from view will increase the chances for success. There is much for evaders to consider concerning the many facets of evasion shelter-site selection if they expect to establish and maintain the security of their area (figure 28-27).

e. Evaders should locate their shelter areas carefully. They should choose areas that are the least likely to be searched. They must be in the least obvious locations. The chosen areas should look typical of the whole environment at a distance. They should not be near prominent landmarks. Areas that look bland get a cursory glance. The areas should also be those least likely to be searched; for example, rough terrain and thickly vegetated areas. The shelter sites should also be situated such that in the event of impending discovery, the evaders will be able to depart the area via at least one concealed escape route. The shelter areas should never be in areas that may trap the evader if the enemy discovers the places of concealment.

f. Evaders should choose natural concealment areas— a "natural shelter." Examples include small, concealed caves, hollow logs, holes or depressions, clumps of trees, or other thick vegetation (tall grass, bamboo, etc.). The site should have as much natural camouflage as possible. There should be cover on all sides; this includes natural formations or vegetation, which can also protect evaders from aerial observations. The site should be as concealed as possible with a minimum of work. Sites chosen this way will make concealment easier and require less activity and movement. This is most important if the evaders are close to population centers or if the enemy is present.

g. The evaders should attempt to stay as high as possible, and to select concealment sites near the military crest of a hill if cover is available. Noises from ridge to ridge tend to dissipate. Whispers or other sounds made in a valley tend to magnify as listeners get further up a hill. Shelter areas located on a slope are subject to higher daytime and lower nighttime winds, thereby minimizing the chance of detection by sense of smell.

h. If possible, evaders should be in such a position in the shelter area that shadows will fall over the side of the area throughout the day. This can best be done in heavy brush and timber.

i. Evaders should try to locate alternate entrance and exit routes along small ridges or bumps, ditches, and rocks to keep the ground around the shelter area from becoming worn and forming "paths" to the site. They should avoid staying in one area so long that it develops the appearance of being "lived in." Evaders should try to stay away from and out of sight of any open areas; examples are roads and meadows. Several miles' distance from these may be desirable.

j. Waterways such as lakes, large rivers, and streams, especially at the junctions, are dangerous places. Power and fence lines or any prominent landmarks may indicate places where people may be. Evaders will want to stay clear of these areas. The enemy may patrol bridges frequently. Evaders should avoid any areas close to population centers. The evaders should be able to observe the enemy and their movements and the surrounding country from this hiding area, if at all possible. If any assemblage of camouflage is necessary at the shelter site, evaders should keep in mind that they should always "construct" to blend. They should match the shelter area with natural cover and foliage, remembering that over-camouflaging is as bad as no camouflage. Natural materials should be taken from areas of thick growth. Any place from which materials have been taken should be camouflaged. The following is an easy to remember acronym (BLISS) for evasion-shelter principles:

> B - Blend.
> L - Low silhouette.
> I - Irregular shape (outline).
> S - Small.
> S - Secluded location.

k. Other facilities evaders may use, such as latrines, caches, garbage pits, etc., must be located and camouflage in the same manner as the shelter sites. Evaders should avoid forming a line of installations that lead from point to point to their location. They should dogleg through ground cover to use concealment to its best advantage. A dry, level, sleeping spot is ideal, but the ideal spot to provide nonvisibility and comfort may be difficult to find. Evaders must have patience and perseverance to stay hidden until danger has passed, or until they are prepared and rested enough to safely move on. They must be constantly on the watch for shelter areas that need little or no improvement for camouflage, protection from the elements, or security.

28-11. Firecraft Under Evasion Conditions:

a. Whether or not to build a fire under evasion conditions is indeed a difficult decision evaders must make at times. Basically, fire should only be used when it is absolutely necessary in a life or death situation. Potential evaders must understand that the use of fire can greatly increase the probability of discovery and subsequent capture.

Figure 28-28. Fire.

b. If a fire is required, location, time, selection of tinder, kindling, and fuel, and construction should be major considerations.

(1) Evaders should keep the fire as inconspicuous as possible. The location of an evasion fire is of primary importance. All of the small evasion-type fires must be built in an area where the enemy is least likely to see them. If possible, in hilly terrain with cover, the fire should be built on the side of a ridge (military crest). No matter where the fire is built, it should be as small and smokeless as possible (figure 28-28).

(2) Fires are easier to disguise and will blend in better during dawn and dusk and in bad weather. At these times, there is a haze or vapor trap that hinges in and around hills and depressions and is prevalent on the horizon. Any smoke from the fire will be masked by this haze in the early morning and at sunset. This is the time when the local populace is most likely to have their cooking or heating fires lit. Another method of disguising the smoke from a fire is to build it under a tree. The smoke will tend to dissipate as it rises up through the branches, especially if there is thick growth or the boughs are low hanging. If this is not possible, it will be helpful to camouflage the fire with earthen walls, stone fences, bark, brush, or snow mounds to block the light rays and help disperse the smoke.

(3) The best wood to use on an evasion fire is dry, dead hardwood no larger than a pencil, with all the bark removed. This wood will produce more heat and burn cleaner with less smoke. Wood that is wet, heavy with pitch, or green will produce large amounts of smoke. When the wood has been gathered, evaders should select small pieces or make small pieces out of the wood collected. Small pieces of wood will burn more rapidly and cleanly, thus reducing the chance of smoldering and creating smoke. The wood selected should be stacked so the fire gets plenty of air, as ventilation will make the fire burn faster with less smoke.

c. One type of evasion fire that has the capability of being inconspicuous is called the Dakota Hole Fire (figure 28-29). After selecting a site for the fire hole, a "fireplace" must be prepared. This is done by digging two holes in the ground, one for air or ventilation, and the other to actually lay the fire in. These holes should be roughly 8 to 12 inches deep and about 12 inches apart with a wide tunnel dug to connect the bottoms of the holes. The depth of the holes depends on the intended use of the fire. Place dirt on a piece of cloth so it can be used to rapidly extinguish the fire and conceal the fire site. Evasion fires should be small. Evaders who

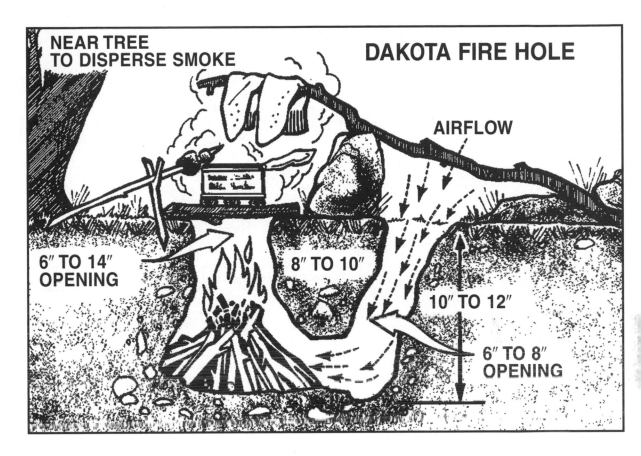

NEAR TREE TO DISPERSE SMOKE

DAKOTA FIRE HOLE

AIRFLOW

6″ TO 14″ OPENING

8″ TO 10″

10″ TO 12″

6″ TO 8″ OPENING

Figure 28-29. Dakota Hole.

build these fires should strive to keep the flame under the surface of the ground. Initially the fire may appear to be smoking due to the moisture in the ground. At night the area may glow, but there should be no visible flame. A fire built in a hole this way will burn fast, as all the heat is concentrated in a small area. This is a good type of fire for a single evader, as opposed to many persons taking turns cooking over this one hole. Everything about the evasion situation will have to be examined before deciding which fire configuration would be most useful if a fire must be built.

d. Another good evasion fire is the trench fire. This fire is built by digging below the earth's surface 8 to 12 inches in an elongated pattern. The length depends on how many people need to use it. This is a fire better suited to the needs of a small group. The fire should not crowd either end of the excavation, as it must be able to "draw" an adequate amount of air to help it burn hot and eliminate smoke (figure 28-30).

e. An evasion fire can also be built just below the ground cover (figure 28-31). Here the emphasis is on quick concealment of the area. It should be kept in mind, however, that some type of screen should be built

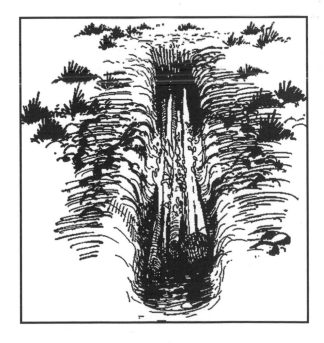

Figure 28-30. Trench Fire.

Figure 28-31. Fire Concealment.

to hide the flames. The small fire is built on the bare ground after a layer of sod or earth has been sliced and rolled back. After use, evaders simply scatter the fire remnants over the bare area and roll or fold the piece of ground back into place.

f. If evaders are in areas where holes can't be dug or sod can't be lifted, they will have to settle for some type of screen around the fire. They should also keep the fire small and finish using it quickly.

g. All traces of the fire should be removed. Unburned firewood should be buried. Holes should be totally filled in. Placing the soil on a holder, such as a map or piece of equipment, will aid in replacing it. This way there will be no leftover dirt patches on the ground after the holes are filled in. Once evaders feel all available measures have been taken to obliterate any leftover evidence, they should move out of the area if possible. Since evaders can never be positive the fire wasn't detected, they must assume it was spotted and take all necessary precautions.

28-12. Sustenance for Evasion:

a. As previously stated in this section, not only do evaders face the problem of remaining undetected by the enemy, they must also have the knowledge that will enable them to "live off the land" as they evade. They must be prepared to use a wide variety of both "wild" and domestic food sources, obtain water from different sources, and use many methods for preparing and preserving this food and water (figure 28-32).

b. No matter what the circumstances of the evasion situation are, evaders should never miss an opportunity to obtain food. Ordinarily food will be obtained from wild plant and animal sources. If possible, evaders should stay away from domestic plants (crops) and animals. Using wild animals and plant sources for food will reduce the probability of capture.

c. Animal foods are a prime source of sustenance for evaders, having more nutritional value for their weight than do plant foods. Evaders may obtain enough animal food in one place to last for several days while they travel.

d. There are several ways in which evaders may procure animal foods (trapping or snaring, fishing, hunting, and poisoning). A few modifications should be considered regarding the use of traps and snares in a tactical situation even though the same basic principles apply in both tactical and nontactical environments.

(1) Because small game is more abundant than large game in most areas, evaders should confine traps and snares to the pursuit of small game. There are other advantages to restricting trap and snare size. Evaders will find it is easier to conceal a small trap from the enemy. Small animals make less noise and create less disturbance of the area when caught.

(2) Conversely, there are also a few disadvantages pertaining to the use of snares or traps during evasion episodes. Two disadvantages are: Evaders must remain in one place while the snares are working, and there may be some disturbance of the area where materials have been removed.

e. Fishing is another effective means of procuring animal food. Fish are normally easy for evaders to catch, and they are easy to cook.

(1) There are several methods that evaders can use to catch fish. A simple hook and line is one of these methods; another is a "trotline." Evaders may construct a trotline by fixing numerous hooks to a pole, then

Figure 28-32. Sustenance for Evasion.

sliding it into the water from a place of concealment (figure 28-33).

(2) Nets and traps may also be used; however, they should be set below the waterline to avoid detection. Spearing is another option. Here again, exposure in open waterways can be very dangerous to the evader.

(3) "Tickling" the fish (figures 28-34, 28-35, and 28-36) is also effective, if evaders can remain concealed. This method requires no equipment to be successful. The main disadvantage to fishing is that people live by water bodies and travel on them. This greatly increases the chances of being detected. (NOTE: Caution should be used when tickling fish in areas with carnivorous fish or reptiles.)

f. Weapons may be used by evaders to procure animals. The best weapons are those that can be operated silently, such as a blowgun, slingshot, bow and arrow, rock, club or spear. These should be used primarily against small game. One major advantage of using weapons is that game can be taken while evaders travel. Because of noise, firearms should never be used in an evasion situation.

g. Plant foods are very abundant in many areas of the world. Evaders may be able to procure plant-food types that require no cooking. One advantage of procuring plant foods during evasion is that by collecting natural fruits and nuts, evaders can remain deep in unpopulated areas. In some areas, it is possible to find old garden plots where vegetables may be obtained. When possible, select foods that can be eaten raw. (Refer to Chapter 18—Food.)

(1) The disadvantage of plant-food procurement is that evaders may not be the only ones looking for food. The natives of the country could also be out looking for food. If natives know of a good area, they may visit that place many times. If evaders have been in the area, their presence could be discovered.

(2) Some other considerations and methods concerning plant-food procurement are as follows. Evaders should:

(a) Never take all the plants or fruits from one area.

(b) Pick only a few berries off of any one bush.

(c) When digging plants, take only one plant, and move on some distance before digging up another.

Figure 28-33. Fishing.

(d) When digging plants from old garden plots, make sure the plot is old. In many countries, people plant their crops and do not return to the plot until harvest time.

(e) Camouflage all signs of presence.

(3) Most domestic foods must be procured by theft, which is very dangerous. However, if proper methods are used and the opportunity presents itself, plants and animals may be stolen. The main reason thieves are captured is the boldness they display after several successful thefts. The basic rules of theft are:

(a) If at all possible, the theft should take place at night.

(b) Evaders should thoroughly observe the area of intended theft from a safe vantage point.

(c) Evaders should find the vantage points just before dusk and look the place over to make sure everything is the same as it was the last time a check was made.

(d) Evaders should check for dogs, which could be a big hazard. Barking draws attention; also, some dogs are vicious and can harm evaders. Besides dogs, other animals or fowl can alert the enemy to the evader's position.

(e) Evaders should never return twice to the scene of a theft.

(f) Every theft should be planned, and after its accomplishment, evaders should leave no evidence of either their presence or the theft itself (figure 28-37).

(g) Only small amounts should be taken (figure 28-38).

(4) When evaders find it necessary to take cultivated plant foods, they should never take the complete plant. Taking plants from the inside of the field, not the edge, and leaving the top of plants in place may help conceal the theft.

h. The rules of theft also apply to taking domesticated animals. The evader should concentrate on animals that don't make much noise. If a choice has to be made as to which animal to steal, evaders should take the smallest one.

i. Water is very essential, but it can be difficult to acquire (figure 28-39).

(1) When procuring water, evaders should try to find small springs or streams well away from populated areas. The enemy knows evaders need water and may check all known water sources. No matter where water is procured, evaders should try to remain completely

Figure 28-34. Fish Tickling - A.

Figure 28-35. Fish Tickling - B.

Figure 28-36. Fish Tickling - C.

concealed when doing so. The area around the water source should be observed to make sure it isn't patrolled or watched. While obtaining water and when leaving the area, evaders should conceal any evidence of their presence. Good sources of water include trapped rainwater in holes or depressions, and plants that contain water.

(2) The preparation and purification of food and water by cooking is a precaution that should be weighed against the possibility of capture. It might be necessary to eat raw plant and animal foods at times. Some plant foods will require cutting off thorns, peeling off the outer layer, or scraping off fuzz before consumption. Raw animal foods may contain parasites and micro-organisms that might not affect evaders for days, weeks, or months, by which time, hopefully, they will be under competent medical care. If the environment in which the parasites live is altered by cooking, cooling, or drying, the organisms may be killed. Meat can be dried and cooled by cutting it into thin strips and air-drying. Salting makes the meat more palatable. If the meat must be cooked, small pieces should be cooked over a small hot fire built in an unpopulated area. The best methods of preserving food during evasion are drying or freezing.

(3) In an evasion situation, boiling water for purification should not be used as a method except as a last resort. Iodine tablets are the best method of purification. If evaders do not have purification tablets, and the danger of the enemy detecting their fires is too great, evaders may have to forego purification. The only problem with this is if water is not purified, it may cause vomiting and (or) diarrhea. These ailments will slow down the evaders and make them susceptible to dehydration. Aeration and filtration may help to some degree and are better than nothing. If water cannot be purified, evaders should at least try to use water sources that are clean, cold, and clear. Rain, snow, or ice should be used if available.

28-13. Encampment Security Systems:

a. When evading for extended periods in enemy-held territories, it becomes essential for evaders to rest. To rest safely, especially if in a group, it is essential to devise and use some sort of early warning system to prevent detection and unexpected enemy infiltration. When establishing an evasion "shelter" area, there are certain things that should be done (day or night) for security purposes (figure 28-40).

b. Evaders should scout the area around their encampment for signs of people. They should pay particular attention to crushed grass, broken branches, footprints, cigarette butts, and other discarded trash. These signs may reveal identity, size, direction of travel, and time of passage of an enemy force. If large numbers of these signs are present, the evader should consider moving to a more secure area.

Figure 28-37. Stealing Vegetables.

Figure 28-38. Stealing Small Amounts of Food.

Figure 28-39. Procuring Water.

Figure 28-40. Resting Safely.

c. Once the camp area has been determined to be fairly secure, some type of alarm system must be devised. For a lone evader, this may consist of actually constructing wire or line with sound-producing devices attached. However, this system works for the enemy as well, and may prove to alert enemy forces to the evader's presence. A lone evader should use the natural alarm system available. Disturbances in animal life around an evader may indicate enemy activity in the area. Group situations may allow for more security. Two or more evaders may use a lookout(s) or "scout(s)" at observation posts strategically located around the encampment.

d. Readiness is another aspect of security. The evader should be aware that, at any time, the shelter area might be overrun, ambushed, or security compromised, making it necessary to vacate the area. If evaders are in a group and future group travel is desired, it is essential that everyone in the group know and memorize certain things, such as compass headings or direction of travel, routes of travel, destination descriptions, and rally points (locations where evaders regroup after separation). Alternate points must be designated in case the original cannot be reached or it is compromised by enemy activity.

(1) Once everyone in the evader's group has reached the final destination, alternate point, or rally point, a new emergency evacuation and rendezvous plan must be established.

(2) Evaders should always be aware of the next rally point, its location, and direction. These places, which provide concealment and cover, should be designated along the route in case an enemy raid or ambush scatters the group. There should be a rally point for every stage of the journey. Even when approaching the supposed "final destination" of the day, evaders should have an evacuation plan ready.

e. Maintaining silence is a very important aspect of security. It is essential to be able to communicate with individual group members and scouts so everyone is aware of what is going on at all times during evasion. Hand signals are the best method of communication during evasion as they are silent and easily understood. Instructions and commands that must be conveyed throughout the entire group are (see figure 28-41 for examples of hand signals):

(1) Freeze.	(5) Rally.
(2) Listen.	(6) All clear.
(3) Take cover.	(7) Right.
(4) Enemy in sight.	(8) Left.

28-14. Evasion Movement:

a. During evasion travel, the evader is probably most susceptible to capture. Many evaders have been captured as a direct result of their failure to use proper evasion-movement techniques. Evasion movement is the action of a person who, through training, preparation, and application of natural intelligence, avoids capture and contact with hostiles, both military and civilian. Not only is total avoidance of the enemy desirable for evaders, it is equally important for evaders not to have their presence in any enemy-controlled area even suspected. A fleeting shadow, an inopportune movement or sound, and an improper route selection are among a number of things that can compromise security, reveal the presence of evaders, and lead to capture.

b. One evasion situation will not be identical to another. There are, however, general rules that apply to most circumstances. These rules, carefully observed,

Figure 28-41. Hand Signals.

will enhance the evader's chances of returning to friendly control.

(1) Evasion begins even before a crewmember leaves an aircraft over enemy territory. Two factors that are essential to successful evasion and return are opportunity and motivation. Premission preparation and knowledge of areas of concern are very important to a crewmember. Premission knowledge gained must be based on the most current information available through command area briefings, area studies, and intelligence briefings.

(2) Some areas of interest to the prospective evader are:

(a) Topography and Terrain. An aircrew member should know the physical features and characteristics, possible barriers, best areas for travel, availability of rescue, and the type of air or ground recovery possible. A future evader should also know the requirements for long-term unassisted survival in the area of operation.

(b) Climate. The typical weather conditions and variations should be known to aid in evasion efforts.

(c) People. A very critical consideration may be knowledge of the people in the area. From ethnic and cultural briefs read before the mission, crewmembers should familiarize themselves with the behavior, character, customs, and habits of the people. It may, at times during the evasion episode, be necessary to emulate the natives in these respects (figure 28-42).

(d) Equipment. Aircrew members who would hope to be successful evaders should be thoroughly familiar with all of their equipment, and know where it is located. They should also preplan what equipment should be retained and what should be left behind under certain conditions of an evasion travel situation.

(3) Before addressing evasion-movement techniques, it is appropriate to go over some of the factors that influence an evader's decision to travel.

Figure 28-42. Knowing the People.

(a) The first few minutes after entering enemy or unfriendly territory is usually the most critical period for the evader. The evader must avoid panic and not take any action without thinking. In these circumstances, the evader must try to recall any previous briefings, standard operating procedures, or training, and choose a course of action that will most likely result in return to friendly territory.

(b) In those first few minutes, there is a great deal for the downed crewmember to think about. Quick consideration must be given to landmarks, bearings, and distance to friendly forces and from enemy forces, likely location for helicopter landing or pickup, and the initial direction to take for evasion. Knowledge of what to expect is important because when circumstances arise that have been considered in advance, they can be carried out more quickly and easily. Evaders should try to adapt this knowledge and any skills they have to their particular situation. Flexibility is most important, as there are no hard-and-fast rules governing what might happen in an evasion experience.

(c) In most evasion situations, evaders will be required to move if for no other reason than to leave the immediate landing area when pickup is not imminent. Because any movement has the inherent risk of exposure, some specific principles and practices must be observed. Periods of travel are the phases of evasion when evaders are most vulnerable. Many evaders have been captured because they followed the easier or shorter route, and failed to employ simple techniques such as watching and listening frequently, and seeking concealment sites.

28-15. Searching Terrain:

a. Evaders should visually survey the surrounding terrain from an area of concealment to determine whether the route of travel is a safe one. Evaders should first make a quick overall search for obvious signs of any presence such as unnatural colors, outlines, or movement. This can be done by first looking straight down the center of the area they are observing, starting just in front of their position, and then raising their eyes quickly to the maximum distance they wish to observe. If the area is a wide one, evaders may wish to subdivide it as shown in figure 28-43. Now all areas may be covered as follows: First, by searching the ground next to them. A strip about 6 feet deep should be looked at first. They may search it by looking from right to left parallel to their front. Second, by searching from left to right over a second strip farther out, but overlapping the first strip. Searching the terrain in this manner should continue until the entire area has been studied. When a suspicious spot has been located, evaders should stop and search it thoroughly (figure 28-44).

b. The evader must question the movement:

(1) Is the enemy searching for the evader?

Figure 28-43. Viewing Terrain - A.

Figure 28-44. Viewing Terrain - B.

(2) What is the evader's present location?

(3) Are chances for rescue better in some other place?

(4) What type of concealment can be afforded in the present location?

(5) Where is the enemy located relative to the evader's position?

c. Having considered the necessity and risks of travel, evaders must:

(1) Orient themselves.

(2) Select a destination, alternates, and the best route.

(3) Have an alternate plan to cover all foreseeable events.

d. Cautious execution of plans cannot be overemphasized since capture of evaders has generally been due to one or more of the following reasons:

(1) Unfamiliarity with emergency equipment.

(2) Walking on roads or paths.

(3) Inefficient or insufficient camouflage.

(4) Lack of patience when pinned down.

(5) Noise or movement or reflection of equipment.

(6) Failure to have plans if surprised by the enemy.

(7) Failure to read signs of enemy presence.

(8) Failure to check and recheck course.

(9) Failure to stop, look, and listen frequently.

(10) Neglecting safety measures when crossing roads, fences, and streams.

(11) Leaving tracking signs behind.

(12) Underestimating time required to cover distance under varying conditions.

e. Evaders should understand that progress on the ground is measured in stopover points reached. Speed and distance are of secondary importance. Evaders should not let failure to meet a precise schedule inhibit their use of a plan.

28-16. Movement Techniques That Limit the Potential for Detection of an Evader (Single).

a. Evaders should constantly be on the lookout for signs of enemy presence. They should look for signs of passage of groups, such as crushed grass, broken branches, footprints, and cigarette butts, or other discarded trash. These may reveal identity, size, direction of travel, and time of passage (figure 28-45).

(1) Workers in fields and normal activities in villages may indicate absence of the enemy.

(2) The absence of children in a village is an indication they may have been hidden to protect them from action that may be about to take place.

(3) The absence of young men in a village is an indication the village may be controlled by the enemy.

b. Knowledge of enemy signaling devices is very helpful. Those listed below are examples of some used by communist guerrillas in Southeast Asia.

(1) A farm cart moving at night shows one lantern to indicate no government troops are close behind.

(2) A worker in the fields stops to put on or take off a shirt. Either act can signal the approach of government troops. This is relayed by other informers.

(3) A villager fishing in a rice paddy holds a fishing pole out straight to signal all clear, up at an angle to signal the troops are approaching.

Figure 28-45. Signs of Passage.

c. The times evaders choose to travel are as critical as the routes they select. If possible, evaders should try to make use of the cover of darkness. Darkness provides concealment, and in some cases there is also less enemy traffic. If, out of military necessity, the enemy is active during the hours of darkness, evaders may then find it wise to move in the early morning or late afternoon. Night movement is slower, more demanding, and more detailed than daylight movement; but it can be done. The alternative to night movement might be capture, imprisonment, and death. Evaders should then consider traveling under the cover of darkness first. However, if the enemy knows the evaders' position, or if the other factors dictate (terrain, vegetation, navigation considerations, etc.), other choices may have to be made. If travel is to be done during darkness, the terrain to be traversed should be observed during daylight, if possible. While observing the area to be traveled, evaders should give attention to areas offering possible concealment, as well as the location of obstacles they may encounter on their route. If the evader has a map, a detailed study of it should be made. However, it should be remembered that natural terrain features change with time of day and time of year. Certain features (ditches, roads, burned-off areas, etc.) may not be on the evaders' map if they are new. Such pretravel reconnoitering of an area will give evaders a head start on knowing how to adapt their travel

movements and camouflage necessary from point to point (figure 28-46).

d. Evaders should try to memorize the routes they will take and the compass headings to their destinations. This information should not be written down on the map or on other pieces of material.

Figure 28-46. Pretravel Reconnoitering.

Figure 28-47. Military Crest.

e. If traveling through hilly country that provides cover, the military crest should be used, as it may be the safest route. An evader's route should avoid game trails and human paths on the tops of ridges. The chance of encountering the enemy is greater on the tops of ridges. The risk of silhouetting during both day and night is also increased (figure 28-47).

(1) When it is necessary to cross the skyline at a high point in the terrain, an evader should crawl to it and approach the crest slowly, using all natural conceal-ment possible. How the skyline is to be crossed depends on whether it is likely the skyline at that point is under hostile observation. Evaders may never be certain that any area is not under observation. When a choice of position is possible, the skyline is to be crossed at a point of irregular shapes, such as rocks, debris, bushes, fence lines, etc.

(2) Another important point about moving along the military crest is that it is easier to evade the enemy on the side of the ridge than it is to lose sight of the enemy on the top of a ridge or in the valley below. Evasion along the side of a hill will afford a better chance for evaders to reach sites that are suitable for air recovery.

f. Evaders should move slowly, stopping and listening every few paces. Additionally, they should not make noise and should take advantage of all cover to avoid revealing themselves. If spotted, they should leave the area quickly by moving in a zigzag route to their goal, if at all possible. If the enemy finds evidence of the evaders' passage, their route may help confuse the pur-suer as to direction and goal. Background noise can be either a help or hindrance to those who are trying to move quietly—both the evaders and the enemy. Sudden bird and animal cries, or their absence, may alert evaders to the presence of the enemy, but those same signals can also warn the enemy of an approaching or fleeing evader. Sudden movement of birds or animals is also something to look out for.

g. The following are some techniques for limiting or concealing evidence of travel. Evaders should:

(1) Avoid disturbing any vegetation above knee level. Evaders should not grab at or break off branches, leaves, or tall grass.

(2) Glide gently through tall grass or brush. Avoid using thrashing movements. A walking stick may be used to part the vegetation in front, and then it can be used

behind to push the vegetation back to its original position. The best time to move is when the wind is blowing the grass.

(3) Realize that grabbing small trees or brush as handholds may scuff the bark at eye level. The movement of trees can be spotted very easily from a distance. In snow country it may mark a path of snowless vegetation that can be spotted when tracks cannot be.

(4) Select firm footing and place the feet lightly but squarely on the surface, avoiding the following:

(a) Overturning duff, rocks, and sticks.

(b) Scuffing bark on logs and sticks.

(c) Making noise by breaking sticks.

(d) Slipping, which may make the noise of falling.

(e) Mangling of low grass and bushes that would normally spring back.

h. Evaders can mask their tracks in snow by:

(1) Using a zigzag route from one point of concealment to the next and, when possible, placing the unavoidable tracks in the shadows of trees, brush, and along downed logs and snowdrifts.

(2) Restricting movement before and during snowfall so tracks will fill in. This may be the only safe time to cross roads, frozen rivers, etc.

(3) Traveling during and after windy periods when wind blows clumps of snow from trees, creating a pockmarked surface that may hide footprints.

(4) Remembering that snowshoes leave bigger tracks, but they are shallower tracks that are more likely to fill in during heavy snowfall. Branches or bough snowshoes make less discernible prints.

i. Evaders' tracks in sand, dust, or loose soil should be avoided or else marked by:

(1) Using a zigzag route, making use of shadows, rocks, or vegetation to walk on to mask or prevent tracks.

(2) Wrapping cloth material loosely about the feet; this makes tracks less obvious.

j. By moving before or during wet or windy weather, evaders may find that their tracks are obliterated or worn down by the elements. Along roadways, evaders should be particularly cautious about leaving their tracks in the soft soil to the side of the road. They should step on sticks, rocks, or vegetation. They shouldn't leave tracks unless there are already tracks of natives on the road, and the evaders' tracks can be made to look like the existing ones (small bare feet, tire sandals, or enemy footgear). Rolling across the road is a method of avoiding tracks. Walking in wheel ruts with the toes parallel to the road will help conceal tracks. If the road surface is dry, sand, dust, or soil tracks may be "brushed" out to make them look old, or will help the wind erase them more quickly. This must be done lightly, however, so the tracks do not look as if they have been deliberately swept over. Mud will retain footprints unless the mud is shallow and there is a heavy rain. Evaders should try to go around these areas.

k. Many principles and techniques that work for the individual are also appropriate for use in a group of evaders. While it is not true that the only safe way to evade is individually, there is a certain danger when moving in a group.

(1) It is generally not advisable to travel in a group larger than three. If possible, the senior person should divide the group into pairs. Paired individuals must be compatible since any disagreement may prove fatal during the evasion process. Group travel can be advantageous because supplementary assistance is available in case of injury, in defense against hostile elements, in travel over rough terrain, and it provides moral support.

(2) In group travel, movement becomes critical. Movement attracts attention.

(3) The intervals and distances between individuals in the group should be made according to the terrain and the time of day. Intervals will probably be greater during the day. The natural but extremely dangerous tendency to "bunch up" is to be avoided when traveling in a group (figure 28-48).

(4) Under favorable conditions it is possible for the enemy to see 100 yards into open woods. If the undergrowth is light, the route must be further into the woods, and the interval between evaders must be greater. Added consideration must be taken in deciding whether to travel at night or during the day. The leader will direct and guide the group to and from the best positions. All communications within the group should be made with silent signals only. Security in group evasion is of paramount importance. All members should stay alert. Security posts, lookouts, or guards should be designated for periods of rest or stopping.

(5) Various formations are available for use by the evading group during periods of travel. The group must be flexible and able to adapt to changes in the conditions of the situation. The type of formation may also change with the route. In choosing a formation, the following points should be considered:

(a) Group control and intercommunications.

(b) Security.

(c) Terrain.

(d) Speed in movement.

(e) Visibility.

(f) Weather.

(g) Enemy placement.

(h) The need for dispersion.

(i) Flexibility in change in speed and direction of travel.

(6) A "formation" is merely the formal arrangement of individuals within a group. This formal arrangement is designed to give the greatest dispersion consistent with adequate intercommunication, ease and speed of movement, and flexibility to change direction and speed of travel at a moment's notice; that is, close control. Any arrangement that provides the above advantages

Figure 28-48. Group Travel.

is satisfactory. The Army, which constantly moves groups of men of various sizes on various missions, has found the squad file, squad column, and squad line to be satisfactory.

(a) In a squad file, the personnel are arranged in a single file, or one directly behind the other, at different distances. The distance will vary depending on need for security, terrain, visibility, group control, etc. It is primarily used when moving over terrain that is so restrictive that the squad cannot adopt any other formation. It is also used when visibility is poor and squad control becomes difficult. When people are in a squad file, it is easy to control the group and provide maximum observation of the flanks. This is a fast way to travel, especially in the snow.

-1. However, there are disadvantages to this type of movement. A major one is that this formation is visually eye-catching. All the noise of the group is also concentrated in one place. This type of formation is easily defined and infiltrated. Depending on how many people are in the group, the area they walk

through can become very packed down and easily detected by the enemy.

-2. If the group in the squad file is small, some members may have to double up on jobs; however, the "point" or lookout (scout) in the lead should perform only that job (figure 28-49).

(b) In the squad column, personnel are arranged in two files. The personnel are more closely controlled and yet maneuverable in any direction. There is greater dispersion with reasonable control for all-round security. It is used when terrain and visibility are less restrictive because it provides the best means of moving armed personnel into dispersed all-round security. It is easy to change into either the file or the line.

-1. There are many advantages to this type of formation, a major one being a greater dispersal of personnel. Visually, this way of moving is less eye-catching, and the sounds the group makes are less concentrated. With this formation, the rate of travel is reasonable, yet there is no well-trampled corridor for enemy trackers to follow.

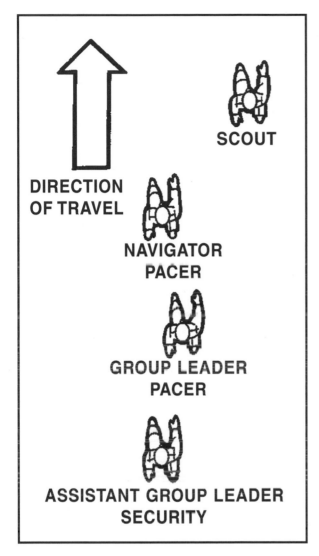

Figure 28-49. File Formation.

and small open spaces. It provides for tight control of individual movement while providing security for short-distance moves.

-2. The disadvantages are that there are extreme communication and control problems. Some personnel in this formation may be forced to traverse undesirable, rough terrain in contrast to the other two formations. Figure 28-51 illustrates the organization of personnel in this type of formation. The speed at which these formations will progress will vary with terrain, light, cover, enemy presence, health of personnel, etc.

l. Regardless of the formation used, the evader should pay particular attention to the technique used to travel. Knowing how to walk or crawl may mean the difference between success and failure.

(1) Correct walking techniques can provide safety and security to the evader. Solid footing can be maintained by keeping the weight totally on one foot when stepping, raising the other foot high to clear brush,

-2. There are a few disadvantages to movement in this manner. This formation of travel is hard to control in areas of dense vegetation with poor visibility, which makes straying from the group likely. Although the paths may be faint, this mode of travel will also form a large number of trails. The rate of travel will be slower overall when traveling this way. An example of how personnel may be dispersed using the squad-column method is shown in figure 28-50.

(c) In the squad line, all personnel are arranged in a line. This formation is used primarily by the Army as an assault formation because it is best for short, fast movements.

-1. The advantage of this formation is that it is the quickest way to cross such obstacles as roads, fences,

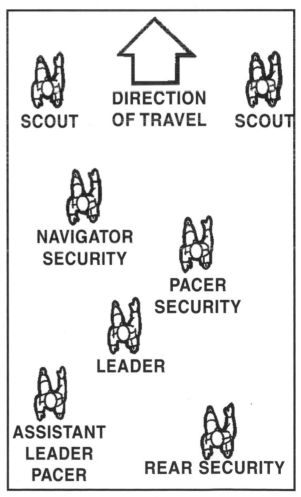

Figure 28-50. Squad Formation.

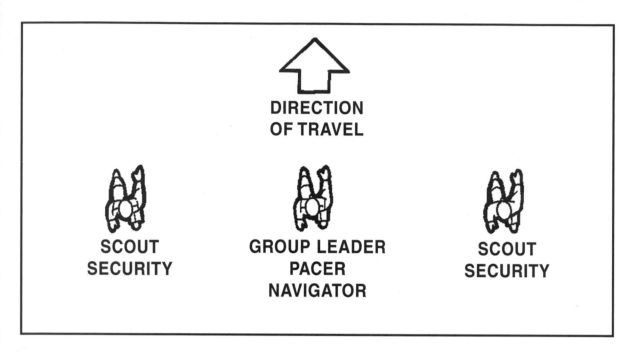

Figure 28-51. Squad Line.

grass, or other natural obstacles, and gently letting the foot down, toe first. Feel with the toe to pick a good spot—solid and free of noisy materials. Lower the heel only after finding a solid spot with the toe. Shift the weight and balance in front of the lowered foot and continue. Take short steps to avoid losing balance. When vision is impaired at night, a wand probe, or staff, should be used. By moving these aids in a figure-eight motion from near the ground to head height, obstructions may be felt.

(2) Another method of movement is by crawling. Crawling is useful when a low silhouette is required. There are times when evaders must move with their bodies close to the ground to avoid being seen and to penetrate some obstacles. There are three ways to do this: the low crawl, the high crawl, and the hands-and-knees position. Evaders should use the method best suited to the condition of visibility, ground cover, concealment available, and speed required.

(a) The Low Crawl. This can be done either on the stomach or back, depending on the requirement. The body is kept flat and movement is made by moving the arms and legs over the ground (figure 28-52).

(b) The High Crawl. This is a position of higher silhouette than the low-crawl position, but lower than the hands-and-knees position. The body is free of the ground, with the weight of the body resting on the forearms and lower legs. Movement is made by alternately advancing the right elbow and left knee, left elbow and right knee, elbows and knees laid flat on their inside surfaces (figure 28-53).

(c) Controlled Movement. The low crawl and high crawl are not always suitable when very near an enemy. They sometimes cause the evader to make a shuffling noise that is easily heard. On the other hand, carefully controlled movement can be made to be silent, and these techniques present the lowest possible silhouettes.

(d) The Hands-and-Knee Crawl. This position is used when near an enemy. Noise must be avoided, and a relatively high silhouette is permissible. It should be used only when there is enough ground cover and concealment to hide the higher silhouette involved (figure 28-54).

m. If evaders are moving as a group and are forced to disperse, being able to account for everyone after regrouping is important. Likely locations for rallying points are selected during map study or reconnaissance.

(1) Selecting a Rallying Point. The group leader must:

(a) Always select an initial rallying point. If a suitable area for this point is not found during map study or reconnaissance, the leader can select it by grid coordinates or in relation to terrain features.

(b) Select likely locations for rallying points en route.

(c) Plan for the selection and designation of additional rallying points en route as the patrol reaches suitable locations.

(d) Plan for the selection of rallying points on both near and far sides of danger areas that cannot be

Figure 28-52. Low Crawl.

bypassed, such as trails and streams. This may be done by planning that if good locations are not available, rallying points will be designated in relation to the danger area; for example, "...50 yards this side of the trail," or "...50 yards beyond the stream."

(2) Use of Rallying Points. If dispersed by enemy activity or through accidental separation, each evader in the group should be prepared to evade, individually, to the regrouping (rallying) point to arrive at a predesignated time. If this is not possible, the individual will become a "lone evader." The group should not make any effort to locate someone not reaching the rally point. The group should formulate a new plan with new rallying points and clear the area.

(a) Rallying points should be changed or updated as they are passed. Points should not be directly on the line (route) of travel. By selecting points off line, the job of searchers (trackers) is made more difficult, and the chance of being "headed off" or "blind stalked" by the enemy is reduced.

Figure 28-53. High Crawl.

Figure 28-54. Hands-and-Knees Crawl.

(b) If the group is dispersed between rallying points en route, the group rallies at the last rallying point or at the next selected rallying point. The group leader announces the decision at each rallying point as to at which point the group will rally.

(3) Actions at Rallying Points. Actions to be taken at rallying points must be planned in detail. Plans for actions at the initial rallying point and rallying points en route must provide for the continuation of the group as long as there is a reasonable chance to evade as a group. An example of a plan would be for the group to wait for a specified period, after which the senior person present will determine actions to be taken based on personnel and equipment present. Even during movement phases, it is important to be able to check on the presence and status of all group members. A low-toned, actual head count starting at the rear of the formation might be one way to do this; hand signals is another. One reason for keeping in touch with everyone is to establish new plans, or adjust old ones, while moving.

28-17. Barriers to Evasion Movement:

a. Obstacles. Throughout the evasion episode, many obstacles may be encountered that may impede evaders or influence the selection of travel routes. These barriers or obstacles can be divided into natural ones, such as rivers or mountains, and human ones, such as border guards or manmade fences or roads. Some of these obstacles might be helpful, while others might be a hindrance.

(1) Rivers and Streams. When crossing rivers and streams, bridges and ferries can seldom be used since the enemy normally establishes checkpoints at these locations. This leaves a choice of fording, swimming, crossing by boat, or using some improvised method (figure 28-55).

(2) Mountains. In mountainous areas, survival may be the primary concern. It may be necessary for evaders

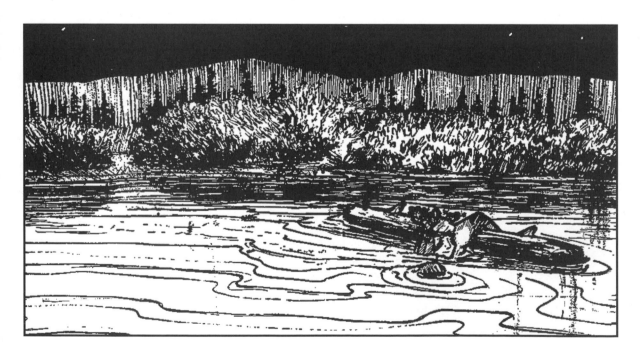

Figure 28-55. River Crossing.

Figure 28-56. Weather.

to remain in one location for an extended period of time, perhaps even waiting for the coming of spring before attempting travel. Many mountainous areas, however, are havens that afford cover, water, food, and low population densities. Also, the chances of receiving assistance from people in areas where homes and farms are separated by great distances are more likely. When traveling in mountainous regions, evaders should not forget to use the military crest if concealment is available. In plains areas, evaders should use depressions, drainages, or other low spots to conceal their movements. Route selection should be planned with the utmost care to avoid unnecessary delays caused by cliffs, large bodies of water, and flat areas.

(3) Vegetation. Some swamps, drainage areas, and thickets may be too thick for evaders to penetrate, and may require that a detour or alternate route be used. If the vegetation can be moved through, evaders should take care not to leave evidence of passage by disturbing the growth.

(4) Weather. Weather can sometimes be used to screen evaders from the enemy. Certain weather conditions mask the noise made by traveling (figure 28-56). Moving during a rainstorm may erase the footprints left by an evader; but after the rain the soft soil will leave definite signs of passage. Thunder may mask the sounds evaders make, but lightning may cause them to be seen. Snowstorms may be used to cover evaders' signs and sounds, but once the storm is over, evaders must use extreme care not to leave a trail.

b. Artificial Obstacles. Evaders may also encounter a wide variety of artificial obstacles while traveling within enemy territory or when attempting to leave a controlled area. As a general rule, evaders should not attempt to penetrate these obstacles if they can be bypassed. If an analysis of the situation reveals the obstacles cannot be bypassed, evaders must be skilled in the methods and techniques for dealing with specific artificial barriers to evasion. If possible, move to a less fortified (controlled) area or find a better, damaged area in the barrier.

Figure 28-57. Trip Wires.

(1) Evaders may encounter trip wires. These wires may be attached to pyrotechnics, booby traps, sensor devices, mines, etc. These wires are normally thin, olive green (or other colors that blend in with the environment), strong, and extremely difficult to see. A supple piece of wood can be improvised and used as a wand to detect these wires (figure 28-57).

(a) A trip wire may be set up to be from 1 inch to a number of feet above the ground, and to extend any number of feet from the device to which it is attached.

(b) The pressure necessary to activate a sensing device, mine, pyrotechnic device, or other trap-associated device to which the wire may be connected can vary from a few ounces to several pounds. This means that the evader must be very careful when attempting to determine the presence of these devices.

(c) Once a trip wire has been detected, evaders should move around the wire if possible. If not possible they should go either over or under it. They should not tamper with or cut the wire. If one device is discovered, be alert for "backup" devices.

(d) A number of devices activated by trip wires have a combination pressure-release arming mechanism. Cutting the wire or releasing the tension in the wire may activate the device. Some devices are electrically activated when there is a change in the current flowing through the attached wire—either because the wire is cut, in some

way grounded, or otherwise altered. Evaders should take extreme care to avoid touching trip wires, but if contact is made, they must try to avoid sufficient pressure for activation.

(2) Illumination flares may also present a problem to evaders. These, of course, can be activated by evaders themselves by a trip wire, by the enemy in the form of electronically activated ground flares, or by flares dropped from an overflying aircraft. Other overhead flares may be fired from mortars, rifles, artillery, or hand projectors.

(a) Illumination flares may burn as bright as 20,000 candlepower and illuminate up to a 300-foot radius in the case of a ground flare, or a much larger area in the case of an overhead flare, which is lofted or dropped and burns high above the ground.

(b) If evaders hear the launching burst of an overhead flare, they should, if possible, get down while it is rising, and then remain motionless. If caught in the light of a flare when they blend well with the background, they should freeze in position and not move until the flare goes out. The shadow of a tree will provide some protection. If caught in an open area, they may elect to crouch low or lie on the ground, and, as a general rule, should not move after the area is illuminated.

(c) However, if they are caught in the light of a ground flare, and their position is such that the risk of

Figure 28-58. Electrified Fences.

remaining is greater than that of moving, they should move quickly out of the area. If within a series of obstacles or an obstacle system, evaders must remember that running can be extremely dangerous because of the obstacles in the area and the fact that movement, especially fast movement, catches the eye of an observer. If it is determined that they cannot quickly move out of the area because of possible serious injuries due to existing obstacles or because they might be observed by the enemy, evaders should drop to the ground and conceal themselves as much as possible.

(d) Evaders should remember that the light of a flare (either ground or overhead bursts) is temporarily blinding, and the eyes should be covered to conserve night vision.

(e) If caught by a flare when actually penetrating an obstacle, such as a concertina wire, evaders should get as low as possible, stay still, and cover their eyes. The light of a flare can act to an evader's advantage because the searching enemy will lose its night vision, which may add to the evader's chance for success in departing the area after the flare has burned out. The light of a flare also creates very dark shadows that, under some circumstances, can afford good concealment from enemy observation.

(3) Various types of chain and wire fences may hinder the progress of evaders who are moving to the safety of friendly areas.

(a) For indications of electrical fences, evaders should watch for dead animals, insulators, flashes from wires during storms, and short-circuits causing sparks (figure 29-58). A quick, simple test can be conducted to determine whether a wire is electrified. This test is made by carefully approaching the wire holding a stem of grass or a small, damp stick on the wire. If the wire is charged, a mild shock will result but will not cause injury.

(b) Evaders should use a wand to check for booby traps between strands of multi-strand barbed wire. Generally, they should penetrate the fence under the wire

Figure 28-59. Penetrating Wire.

Figure 28-60. Penetrating Concertina Wire.

closest to the ground, with the body parallel or perpendicular to the wire, depending on circumstances (figure 28-59). They should lie flat on their backs both to project the lowest possible silhouette and to provide good visibility of the wire against the sky. The probe can sometimes be used to lift the wire. If the lowest wire is close to the ground and is tight, evaders may have to modify their approach to the problem.

(c) The apron fence is penetrated in the same manner as any multi-strand fence—one wire at a time. Evaders should check the area between wires before proceeding.

(d) Concertina wire is penetrated with the body perpendicular to the wire using a probe to lift the wire, if it is not secured to the ground (figure 28-60). If the wire is secured to the ground, the evader can crawl between the loops. If two loops are not separated enough, they may be tied apart using shoelaces, string, suspension line, or strips of cloth. After passing through, the ties should be removed for future use and to erase evidence of travel.

(e) Chain-link fences should be avoided completely if at all possible. These fences are usually found only in highly sensitive zones. This means the area is probably more highly guarded and patrolled than other areas. There also may be other traps or devices installed. The fence may also be electrified. If the fence must be penetrated, the evader should go under it if possible (figure 28-61). If digging is required, the soil should be placed on the opposite side so it can be replaced to remove

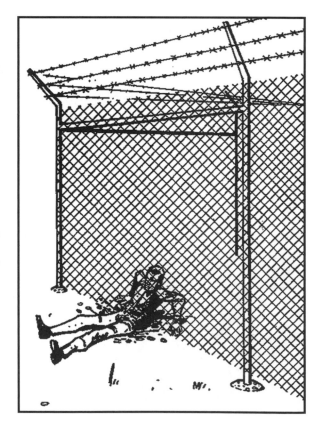

Figure 28-61. Penetrating Chain Link Fence.

Figure 28-62. Penetrating Log Fence.

evidence. Climbing the fence is recommended only as a last resort.

(f) Evaders may encounter rail and split rail fences while evading or escaping. These fences are penetrated by going under or between lower rails if possible. If not, evaders should go over at the lowest point, projecting as low a silhouette as possible (figure 28-62). They should check between the rails and on the other side of the fence to detect trip wires or booby traps. Firmness of the ground should be checked on both sides of the fence. The body should be parallel to the fence before penetration.

(4) Raked or plowed areas may be found in areas of both low and high-density security. If such an area is encountered, evaders should roll across the area—after making sure it is not a minefield—to avoid leaving footprints; or they may side-step, walk backwards, or brush out footprints. Any of the above may be done when it is a requirement not to leave clear-cut evidence of the direction of movement.

(5) Roads are common barriers to evasion and escape. When roads are encountered, evaders should closely observe the road from a place of concealment to determine enemy travel patterns (figure 28-63). Crossing from points offering best concealment, such as bushes, shadows, etc., is best (figure 28-64). Evaders should cross at straight stretches of road in open country, and on the inside of curves in hilly or wooded areas. This allows the evader to see in both directions so the chances of being spotted or surprised in the open is minimized. Avoid leaving tracks both in the road and on the shoulder of the road.

(6) Culverts and drains offer excellent means of crossing a road unobserved (figure 28-65).

(7) Railroad tracks often lie in the path of evaders. If so, evaders should use the same procedures for observation as for roads. If it is determined that tracks are patrolled, a check should be made for booby traps and trip wires between tracks. Aligning the body parallel to

Figure 28-63. Common Barriers.

Figure 28-64. Crossing Roads.

Figure 28-65. Crossing at Culverts.

the tracks with face down next to the first track, evaders should carefully move across the first track in a semi-pushup motion, repeating for the second track and subsequent set of tracks (figure 28-66). If there is a third rail, they should avoid touching it, as it could be electrified. Sound detectors can also be attached to the tracks and can reveal any crossing if a track is touched. If determined from observation that the tracks are not

Figure 28-66. Crawling Over Railroad Tracks.

Figure 28-67. Climbing Down Cliff.

patrolled, evaders should cross in a normal walking or hands-and-knees manner, attempting to attract as little attention as possible. Evaders should try to keep their hands and feet on the railroad ties to prevent leaving foot or handprints in the adjacent soil or gravel.

(8) Deep ditches (such as tank traps or natural drainages) may be obstacles with which evaders must deal. Ditches should be entered feet-first to avoid injury to the head or upper torso should there be large rocks, barbed wire, or other hazards at the bottom (figure 28-67). Using a wand to detect trip wires and booby traps in the ditch and on the sides is highly recommended. Maintaining a low silhouette upon exiting the ditch is imperative.

(9) Open terrain complicated by guard towers or walking patrols is a definite hazard to evaders. These areas should be avoided if possible. If it becomes necessary to traverse open terrain or come near guard towers, evaders should stay low to the ground and, when possible, travel at night or during inclement weather. Use terrain masking since night-vision devices may be used near border areas (figure 28-68).

(10) The problem of crossing areas that have been contaminated as a result of enemy or friendly NBC operations may arise. Chemical contamination should be suspected when the following are observed (figure 28-69):

(a) Shell craters with liquid in the bottom.
(b) Liquid droplets on vegetation.
(c) Water with "film" on the surface.
(d) Unexplained dead animals.
(e) Unseasonal discoloration of vegetation.
NOTE: Without protective clothing, mask, and accessories, evaders should bypass contaminated areas if possible.

(11) CAUTION: Stay away from borders unless absolutely necessary. The crossing of one or more borders presents a major problem. These areas may be located in any type of terrain.

(a) In areas where there is no well-defined terrain feature to indicate the border, artificial obstacles such as electrified or barbed-wire fences, augmented with trip wires, anti-personnel mines, or flares, may be encountered. Open areas may be patrolled by humans or dogs or both, particularly during the hours of darkness. The enemy may also use floodlights and plowed strips as aids to detecting evaders (figure 28-70).

(b) The plan to cross a border must be deliberate and must be designed to take advantage of unusually bad weather (as a major distraction to the enemy force), or areas where security forces are overextended. These areas are usually found where there are natural obstacles.

(c) Crossings should be made at night, when possible, in battle-damaged areas. If it is necessary to cross during daylight hours, evaders should select a crossing point that offers the best protection and cover. They should then keep the area under close observation for several days to determine:

-1. The number of guards in the area.
-2. The manner of their posting.
-3. Aerial patrols and their frequency.
-4. The limits of the areas they patrol.
-5. Location of mines, flares, or trip wires.

(12) A difficult task in any situation is the attempt to cross the forward edge of the battle area. If unable to determine the general direction to friendly lines, evaders should remain in position and observe the movement of enemy military forces or supplies, the noise and flashes of the battle area, or the orientation of enemy artillery. After arriving in the combat zone, evaders should select a concealed position from which as much of the battle area as possible may be observed. They can then select a route and critical terrain features that can guide them when infiltrating back to friendly positions under the cover of darkness. Several alternate

Figure 28-68. Open Terrain Near Border Areas.

Figure 28-69. Contaminated Areas.

Figure 28-70. Border Crossings.

routes should be selected, with care to avoid "easy" approaches to friendly lines, which are more likely to be covered by friendly fire and enemy patrols. If in uniform, select exposure time during daylight hours and be close enough to be easily recognized by friendly troops.

(13) Evaders should also watch out for friendly patrols. If a patrol is spotted, evaders should remain in position and allow the patrol to approach them. When the patrol is close enough to recognize them, evaders should have a white cloth displayed before the patrol gets close enough to see the movement. A patrol may fire at any movement. Shouting or calling out to them jeopardizes both the patrol and the evader. Evaders should stand silently with hands over their head and legs apart so their silhouettes are not threatening. If evaders elect not to make contact with a patrol, they should, if possible, observe their route and approach friendly lines at approximately the same location. This may enable them to avoid minefields and booby traps. (NOTE: The practice of following any patrol is extremely dangerous. The last persons in line are charged with security, and anyone following them would be considered hostile and be eliminated.)

(14) If unable to contact a friendly patrol, the only alternative for evaders may be to make a direct approach of front-line positions. This will require them to crawl through the enemy's forward position near forward friendly elements. This action should be done during the hours of darkness. Once near friendly lines, however, evaders should not attempt to make contact until there is sufficient light for them to be recognized.

(15) In dogs, the ability to perceive odors is much greater than that of humans.

(a) In this portion, the term dog is meant to describe only the animals specifically trained in the areas of patrolling, guarding, and searching. Since dogs are basically odor-seeking animals, anything developed to work against their odor-seeking capabilities is worthy of experimentation for survivors (evaders).

(b) A problem that must be considered is that evaders will not be working solely against a dog, but against a dog handler as well. There is no simple, sure method of evading a dog. Some possible means that could be tried by evaders are:

-1. Dogs detect the fatty acids in dead skin cells that humans shed by the thousands every day. Scenting dogs may be distracted by scattering an irritant such as pepper behind the evader or traveling across an asphalt road on a hot day.

-2. Dogs track better when the weather is humid and the air is still—there is less evaporation and dissipation of odor.

-3. If evaders know that they are being followed by dogs, either from the landing site or as the result of an escape, they should try to use water to conceal their tracks and eliminate their scent.

-4. If there is a choice of terrain and it is possible to travel on a hard surface, evaders should do so rather than travel on soft ground.

-5. Evaders should always attempt to move downwind of a dog. This should be attempted when they are traveling in open country, or penetrating obstacles such as dog guard posts or border areas. If penetration of obstacles or escape is planned (after careful location and study of guard and dog areas of responsibility and their methods), evaders should select a time for movement when noise will distract the dog to a point away from the planned maneuver.

-6. If evaders are physically capable, they should attempt to maintain the maximum distance possible from dogs. Moving fast through rugged terrain will slow and probably defeat the handlers of dogs. Here evaders must choose between making mistakes in travel techniques while evading, or being caught by dogs if they don't move fast enough.

28-18. Evasion Aids:

a. Survival Kits. Personnel may sometimes find it practical to devise and carry compact personal survival kits to complement issued survival equipment. If E&E kits are provided, potential evaders should be familiar with them, their uses, and their limitations.

b. Maps. Any maps of the area in an evader's possession should not be marked. A marked map in enemy hands can lead to the compromise of people and locations where assistance was given. Evaders should be wary of even accidentally marking a map; for example, soiled fingers will mark a map just as plainly as a pencil.

c. Pointee-Talkee. The "pointee-talkee" is a language aid that contains selected phrases in English on one side of the page and foreign language translations on the other side. To use it, evaders determine the question and statement to be used in the English text and then point to its foreign language counterpart. In reply, the natives will point to the applicable phrase in their own language; evaders then read the English translation.

(1) The major limitation of the "pointee-talkee" is in trying to communicate with illiterates. In many countries the illiteracy rate can be astoundingly high, and personnel have to resort to pantomime and sign language, which have been relatively effective in the past.

(2) "Pointee-talkee" phrases are presented under the following eight subheadings:

(a) Finding an interpreter.

(b) Courtesy phrases.

(c) Food and drink.

(d) Comfort and lodging.

(e) Communications.

(f) Injury.

(g) Hostile territory.

(h) Other military personnel.

d. Barter Kits. Barter kits may be available in some commands. If not, crewmembers may elect to develop their own. Items for consideration should be selected from area studies. Some items to consider might be rings, watches, knives, local currency, coins, and lighters. Items should have no markings of personal significance or military value. Military items packed in kits may be considered if not essential to the evader. Flashing large amounts of cash or valuables can have negative results in a depressed, war-torn area. Show only small amounts and drive a hard bargain.

e. Other Evasion Aids. Information on other evasion aids and tools is available from unit intelligence officers.

f. Assisted Evasion. There may be people in a hostile nation or in an enemy-occupied country who are dissatisfied with existing conditions.

(1) History has revealed that in every major conflict there are groups of people in every country who will aid a representative of their government's enemies. The motivating force behind these groups may vary from purely monetary considerations to idealistic concepts of government reform. In many cases, their real objective will be the political advancement of their particular group. During WW II, many underground or resistance groups and movements were established in occupied countries. One of the major purposes or functions of these groups was the aiding of downed allied aircrew personnel. In most cases, the driving force behind these movements was patriotism and desire for political recognition for their cause.

(2) These circumstances favor active resistance movements. One of the functions of such movements may be the operation of escape and evasion (E&E) systems for the purpose of returning evaders to friendly territory.

(3) US Special Forces may also organize and operate E&E mechanisms (figure 28-71).

(a) E&E Organizations. Assistance may range from that rendered by a sympathetic individual to elaborate E&E nets organized by local inhabitants. E&E organizations may be limited in nature, such as providing assistance to reach a national frontier, or they may be linked to larger organizations capable of returning to friendly control.

(b) Acts of Mercy. These are usually isolated events during which evaders may be provided food, shelter, or medical attention for a brief period of time. Local people may find an exhausted or incapacitated

Figure 28-71. E&E Organizations.

evader and provide that evader with limited sustenance. This type of assistance is frequently offered with reluctance or under fear of reprisal because an act of providing comfort to the enemy would mean punishment. Unless an evader is in immediate need of medical attention, acts of mercy may consist of only an offer of food followed by an urgent plea that the evader leave the area immediately. If an evader is physically able to depart with a reasonable chance of evading capture, he or she should do so. An evader should not insist on receiving additional aid other than what is offered by the person who renders assistance only through human impulses (figure 28-72).

(4) Evaders must understand that when dealing with any indigenous personnel while in enemy territory, their own actions will govern the treatment they will receive at the hands of these people. How evaders conduct themselves may also have much to do with getting back to their own forces should they fall into the hands of irregulars friendly to their (the evaders') own cause. The following list of suggestions may be a useful guide in dealing with these people or groups.

(a) Evaders should understand that failure to cooperate or obey might result in death.

(b) Evaders should try to avoid making any promises that they cannot personally keep.

(c) Evaders should remember that the four conditions called for by the rules of land warfare must be met before members of an irregular group can be recognized

as qualifying for PW status in the event of capture. These same rules will also apply to evaders who are involved with these groups.

(d) Prisoners of war, according to the current Geneva Convention, are persons belonging to one of

Figure 28-72. Receiving Aid.

the following categories, who have been captured by the enemy. Members of other militias and volunteer corps, including those of organized resistance movements, belonging to a conflicting Party and operating in any territory, even if this territory is occupied, provided that such groups fulfill the following four conditions of:

-1. Being commanded by a person responsible for subordinates.

-2. Having a fixed, distinctive sign recognizable at a distance.

-3. Carrying arms openly.

-4. Conducting their operations in accordance with the laws and customs of war.

(5) If evaders join such a unit, the closest thing to a guarantee of treatment as a military person will be their uniforms, if they are captured.

(6) If evaders have the opportunity to influence the group they are with, they should try to encourage them to abide by the four conditions mentioned.

(7) Evaders must avoid becoming associated in any way with atrocities that these groups may commit against civilians, prisoners, or enemy soldiers.

(8) Evaders should not become involved in their political or religious discussions, take sides in their arguments, or become involved with the opposite sex.

(9) Evaders should show consideration for being allowed to share food and supplies. It may also be helpful if evaders understand and show interest in the assister's customs and habits.

(10) The overall best and safest course for evaders to follow is to exercise self-discipline, display military courtesy, and be polite, sincere, and honest. Such qualities are recognized by any group of people throughout the world. The impression left can influence the aid provided to future evaders.

g. E&E Lines. An E&E line is a system of one or more secret nets organized to contact, secure, and when possible, evacuate friendly personnel. Well-organized and supported lines normally can be expected to provide the following assistance:

(1) Temporary shelter, food, and equipment for the next phase of the journey.

(2) Clothing and credentials acceptable in the area to be traveled.

(3) Information concerning enemy security measures along the evasion route.

(4) Local currency and transportation.

(5) Medical treatment.

(6) Available native guides.

h. Conduct of E&E Lines. The success of an E&E organization depends almost entirely upon its security. The organization of a line includes much planning and work carried out under dangerous conditions. The security of the system often depends on the evader's cooperation and working knowledge of how it functions, how

it may be contacted, and what rules of personal conduct are expected of the evader. The following paragraphs summarize the major aspects of the operation of an E&E line.

(1) Contacting the Line. Premission briefings may inform evaders where to go and what actions to take to make contact with an E&E mechanism. After being picked up by an evasion-and-escape mechanism, evaders will be moved under the control of this mechanism to territory under friendly control, or to a removal area, and arrangements may be made for air or sea rescue. The organizer of a line in friendly but enemy-occupied territory normally will have arranged a network of spotters who will be especially active when evaders are in the immediate area, but so will the enemy police and counter-intelligence organizations. For this reason, certain precautions must be observed when making contact.

(2) Approach. Help may be refused by a person simply because he or she thinks someone else has seen the evaders approach to seek assistance. If captured with a local helper, an evader will become a prisoner, but the helper and perhaps an entire family may be more severely punished.

(3) Making Contact. Contacts with the natives are discouraged, unless observation shows they are dissatisfied with the local governing authority, or previous intelligence has indicated the populace is friendly. Evaders should proceed to, and remain in, the nearest SAFE area where arrangements for contact can be developed. If the E&E system is operating successfully, the spotter will know evaders are present and will search the immediate area, making frequent visits to designated contact points. Identification signs and countersigns, if used, may be included in the preoperational briefing. It is seldom advisable to seek first contact in a village or town. Strangers are conspicuous by day, and there may be curfews or other security measures during the hours of darkness. The time of contact should be at the end of the daylight period or shortly thereafter. Darkness will add to the chance of escape if the contact proves to be unfriendly, and may be advantageous to the contact in providing further assistance.

(4) Procedure After Contact. If contact is made, evaders may be told to remain in the vicinity where spotted, or, more likely, they will be taken to a house or other structure used by the E&E net as a holding area. It must be decided at this time whether or not to trust the contact. If there is any doubt, plans should be made to leave at once. It is also possible that the house may not belong to the E&E organization, but rather to someone who will look after evaders until arrangements can be made for the line to identify and accept them in E&E net.

(5) Establishing Identity. Verification of identity will be required before anyone is accepted as a bona fide evader. The constant danger facing the operators of an

escape line is the penetration of the E&E system by enemy agents pretending to be evaders or escapees. Evaders should be prepared to furnish proof of identity or nationality. Since it may lead to later difficulties of iden-tification, they should never give a false name—just their name, grade, service number, and date of birth. It is best for them to avoid talking as much as possible.

(6) Awaiting Movement on the Line. Delays can be expected while proceeding along the escape line. If the period of waiting is prolonged, frustration and impa-tience may become unbearable, leading to a desire to leave the holding area. This must not be done, because if seen by other people, the lives of the assisting person-nel and the existence of the entire line itself may be endangered.

(a) Evaders should follow the orders of those assisting them. If kept indoors for any length of time they can keep fit by moderate physical exercise.

(b) The host should have a plan for rapid evacu-ation of the area if enemy personnel should raid the holding area, if not, evaders should have a personal plan that should include measures for removing all traces of having occupied the area. If the net is being overrun and capture is imminent, evaders must be pre-pared to fend for themselves. The evader is the only one who knows more than one part of the net. The assisters may attempt to eliminate the security risk to the net.

(7) Traveling the Line. It would be a grave breach of faith and security for evaders to discuss, at any point on the line, the earlier stages of the journey. Evaders might be tested to see if they are trustworthy—they should discuss nothing of the net. For security reasons and to protect the compartmentalization of the line, no information should be revealed. It is also useless to ask where a line leads or how it will eventually reach friendly territory. Evaders should not try to learn or memorize names and addresses and, above all, they shouldn't put these facts or any other information in writing. Evaders should give the impression of having received no assis-tance from local inhabitants.

(8) Fellow Evaders. Caution is required in the case of fellow evaders on an escape line unless they are personally known. Even when it has been satisfactorily determined that another person is a genuine evader, no information should be given.

(9) Travel with Guides. If under escort, this fact should not be apparent to outsiders. In a public vehicle, for example, evaders should never talk to their guide or appear to be associated with the person in any way unless told to do so. This will lessen the possibility of both the evader and the guide being apprehended if one should arouse suspicion. It should always be possible for the guide to disown an evader if the guide gets into difficulty. When evaders are escorted, they should follow the guide at a safe distance, rather then walk right next to the person, unless instructions indicate the latter action is required (figure 28-73).

(10) Speaking to Strangers. Evaders should never speak to a stranger if it can be avoided. As a last resort, they should pretend to be deaf and dumb or even half-witted. This technique has often been successful. To discourage conversation in a public conveyance, they can also pretend to read or sleep.

(11) Personal Articles and Habits. Evaders should not produce articles in public that might show their national origin. This pertains to items such as pipes, cigarettes, tobacco, matches, fountain pens, pencils, and wristwatches. Evaders should also ensure their personal habits do not give them away; for example, they should not hum or whistle popular tunes or utter involuntary oaths. Again, in restaurants, imitating local customs in the use of knives and forks and other table manners is advisable. Study of the area before the mission may help evaders avoid making mistakes.

(12) Payment to Helpers. On an escape line, evaders should not offer to pay for board, lodging, or other serv-ices rendered. These matters will be settled afterward by those who are directing and financing the line. If in pos-session of escape kits or survival packs, evaders should keep them as reserves for emergencies. If they have no

Figure 28-73. Travel with Guide.

food reserve, they should try to build up a small stock in case they are forced to abandon the line.

(13) Evaders Conduct:

(a) Be polite by local standards.

(b) Be patient and diplomatic.

(c) Avoid causing jealousy. Disregard the sex of the assisters.

(d) Avoid discussions of a religious or political nature.

(e) Eat and drink if asked, but don't over indulge or become intoxicated.

(f) Don't take sides in arguments between assisters.

(g) Don't become inquisitive or question any instruction.

(h) Help with menial tasks as directed.

(i) Write or say nothing about the other people or places in the net.

(j) Don't be a burden; care for self as much as possible.

(k) Follow all instructions quickly and accurately.

28-19. Combat Recovery:

a. Air recovery is one of several means of transportation for downed crewmembers in their quest for their final goal—returning to their own lines and units. How recovery will be effected will depend on a number of factors, among them the following:

(1) Terrain.

(2) Capability of the rescue vehicle.

(3) Condition of the survivor (evader).

(4) Enemy activity in the area.

(5) Weather conditions.

(6) High or priority mission.

b. Even though a maximum effort will be made to recover downed aircrew members, survivors can jeopardize the whole rescue operation and the lives of rescue personnel in a combat zone by not taking a responsible role in recovery operations. The responsibilities are many and varied but essential to a successful rescue mission. Evaders should recall that even though they may have little experience in participating in actual rescues, they must nonetheless be very proficient in their actions. Evaders should always remember that other people are endangering their lives in an attempt to retrieve them.

c. There are two basic types of air-recovery vehicles—rotary-wing and fixed-wing aircraft. The rotary-wing type can extract evaders from remote areas. With air-to-air refueling, the range of these aircraft has increased; they are limited only by the endurance of the crewmembers who operate them.

d. Survivors must be proficient in the operation of all the survival equipment at their disposal. For example, they must be able to switch off the beeper on the radio to receive verbal instructions from the rescue commander.

e. The initial contact with rescue or other combat aircraft in the area must be done as directed by authorities in that theater of operations; for example, in Southeast Asia, a contact method of transmitting 15-seconds beeper, 15-seconds voice (call sign), and 15-seconds monitoring, was used. The method an evader should use to establish contact with rescue will be briefed before the mission.

f. One important aspect of the rescue process is evader authentication by rescue personnel. Survivors must be able to authenticate their identity through the use of questions and answers, responding as directed before the missions. Authentication in a combat area changes rapidly to reduce the chances of compromising the rescue efforts. Survivors (evaders) must keep up with these changes so their rescue can be made without undue danger to rescue personnel.

g. The purpose of filling out and using a personal authenticator card is positive verification of an evader's identity, which is essential before risking search-and-rescue aircraft or the lives of assisters. The purpose of the photographs, descriptions, fingerprints, etc., on the card is to ensure all possible means are used for the proper identification of personnel. Intelligence personnel are responsible for ensuring that personnel fill out the card completely, and that they are aware of the purpose and use of the information. When filled in, the personal authenticator card becomes classified Confidential, and is reviewed at least semiannually by both the crewmember and intelligence personnel. (NOTE: See AFR 64-3, Wartime Search and Rescue [SAR] Procedures.)

h. When selecting a site for possible evasion recovery, there is much for evaders to keep in mind. The area they choose could well decide the success or failure of the mission.

(1) The potential recovery site should be observed for 24 hours, if possible, for signs of:

(a) Enemy or civilian activity.

(b) Roads or trails.

(c) Farming signs.

(d) Orchards.

(e) Tree plantations.

(f) Domesticated animals or droppings.

(g) Buildings or encampment areas.

(2) The rescue site should be observable by aircraft but unobservable from surrounding terrain, if possible. There should also be good hiding places around the area. The site should include several escape routes so evaders can avoid being trapped by the enemy if discovered. There should be a small open area for both signaling and recovery. It would be beneficial if the surrounding terrain provided a masking effect for rescue forces in order to avoid enemy observation and fire.

i. The size of evasion ground-to-air signals should be as large as possible but must be concealed from people passing by. Evaders should remember the six-to-one ratio. The contrast these signals make with the surrounding vegetation should be seen from the air only. Any signal displays should be arranged so they can be removed at a moment's notice, since enemy aircraft may also fly over the immediate area.

(1) All of the principles of regular (nontactical) signals should be followed by those building evasion signals. Crewmembers may be prebriefed as to the appearance of specially shaped signals (figure 28-74).

(2) Evasion signals, like all others, must be maintained to be effective. At times, evaders may be preinstructed to set out their signals according to a prearranged time schedule.

j. Whatever signal devices are available to evaders, they must be able to use them (mirror, flare, gyro-jet, etc.) with proficiency. These signals, like those used in nontactical situations, are to be used either to gain the attention of friendly aircraft or rescuers, or when directed by rescue. Extreme care must be taken to minimize or eliminate chances of enemy elements spotting the signals. For example, the strobe light and mirror can be directed and aimed instead of being used in an indiscriminate manner.

(1) If evaders are in or on the water, they should use lights, flares, dye, whistles, etc., with extreme care, as they are readily distinguishable over water at great distances.

(2) In addition to knowing how to use radios correctly to effect their rescue, evaders should also be familiar with the special points of evasion radio procedures; for example, ensuring the radio transmits continuously during a night-recovery operation using the electronic locator finder (ELF) system.

(3) It is also the evaders' responsibility to communicate all signs of enemy activity, such as:

(a) Locating anti-aircraft emplacements.

(b) Identifying when they are firing.

(c) Assisting strikes by spotting hits (high, low, or on target).

(d) Determining effectiveness of hits.

(e) Notifying personnel of changes in small-arms positions, etc.

(4) Downed evaders can normally expect to be hoisted to a helicopter by one of five methods: basket, Stokes litter, bell, horse collar, or forest penetrator. Other pickup devices that may be used are the McGuire Rig, Swiss Seat, Motley Rig, Stabo Rig, rope ladder, or rope. Another method of recovery that may be employed to extract evaders is the Surface-to-Air-Recovery (STAR) System (figure 28-75). Evaders should remember that whatever the pickup device, it should always be grounded before they grasp it.

(5) There is also fixed-wing capability of rescuing downed crewmembers. Evaders will be prebriefed as to which type of rescue vehicle and systems to expect in their areas of operation. No matter which type is used, evaders must be capable of mounting and riding the rescue devices in a minimum amount of

Figure 28-74. Evasion Signals.

Figure 28-75. Surface-to-Air Rescue.

time, allowing the rescue craft to effect the rescue and depart quickly.

(6) As downed crewmembers, whether as a result of enemy action or of mechanical failure, it is important that they all become familiar with the information that one day could prove instrumental in saving their lives. First, they must have prepared themselves to cope with the survival situation before an aircraft emergency. This

can be done by including, as part of their preflight planning, a thorough inspection of survival equipment to determine its availability and serviceability. Second, they must realize that there are many types of recovery vehicles available in the Air Force inventory, and knowledge of all recovery techniques is a necessity. They should request detailed briefings from area or local rescue personnel, who will explain the operational limitations and recovery potential for each. Third, evaders must be capable of fulfilling their part in the recovery operation. This can be done by knowing when, where, and how to initiate communications, and how to cooperate with the rescue aircraft crew.

(7) Other forms of assistance that may be available to evaders are Special Forces, Combat Control, RECON, SEALS, Riverine Operations, and Submarines. While aiding evaders is not the primary mission of these groups, as a secondary mission they travel in or near SAFE areas on their return journey to check for the presence of evaders.

Part Ten

INDUCED CONDITIONS—NUCLEAR, BIOLOGICAL, AND CHEMICAL (NBC)

Chapter 29

NUCLEAR CONDITIONS

29-1. Introduction. The possibility of "induced conditions" has served to intensify the difficulties of basic and combat survival because of the serious problems posed by nuclear, biological, and chemical warfare. Though the prescribed survival procedures recommended in other parts of this regulation are still applicable, a number of additional problems are created by the hazards of induced conditions (figure 29-1).

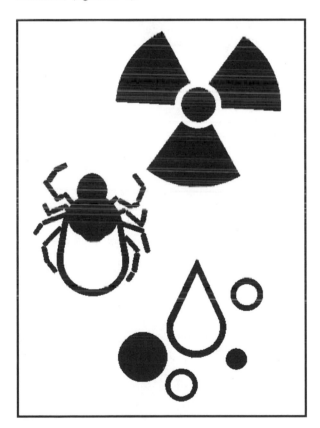

Figure 29-1. Induced Conditions.

29-2. Effects of Nuclear Weapons. Nuclear weapons cause casualties and material damage through the effects of blast, thermal radiation, and nuclear radiation. The degree of hazard from each of these effects depends on the type of weapon, height of the burst, distance from the detonation, hardness of the target, and explosive yield of the weapon.

a. Blast. The blast wave is the cause of most of the destruction accompanying a nuclear blast. After a nuclear detonation, a high-pressure wave develops and moves outward from the fireball. The front of the wave travels rapidly away from the fireball as a moving wall of highly compressed air. An example of the speed of the blast wave is: At 1 minute after the burst, when the fireball is no longer visible, the blast wave has traveled about 40 miles and is still moving slightly faster than the speed of sound. There are strong winds associated with the passage of the blast wave. These winds may have a peak velocity of several hundred miles an hour near ground zero. Ground zero is the point on the ground directly above or below the point of detonation. The overpressure, which is the pressure in excess of the normal atmospheric pressure, and the winds are major contributors to the casualty and damage-producing effects of the nuclear detonation. The overpressure can cause immediate death or injury to personnel, and damage to material by its crushing effect. The high-speed winds propel objects, such as tree limbs or debris, at great speed and turn them into potentially lethal missiles. These winds can also physically throw personnel who are not protected, resulting in casualties. People both inside and outside of a structure may be injured as a result of blast damage to that structure; those inside by the collapse of the structure and by fire; and those outside by the flying objects carried by the winds (figure 29-2).

b. Thermal Radiation:

(1) Heat. Within less than a millionth of a second of the detonation of a nuclear weapon, the extremely hot weapon residues radiate great amounts of energy. This leads to the formation of a hot and highly luminous, spherical mass of air and gaseous residue, which is the fireball. The heat radiated from the fireball contributes to the overall damage caused by a nuclear burst by igniting combustibles and thus starting fires in buildings and forests. These fires may spread rapidly among the debris produced by the blast. In addition, this intense heat can burn exposed personnel at great distances from ground zero where the effects of blast and initial nuclear radiation become insignificant. The degree of injury from thermal radiation becomes more marked with the increasing size of the weapon. The degree of injury from thermal radiation is also affected by weather and terrain. During periods of limited visibility, the heat effect will be reduced significantly. Additionally, since thermal

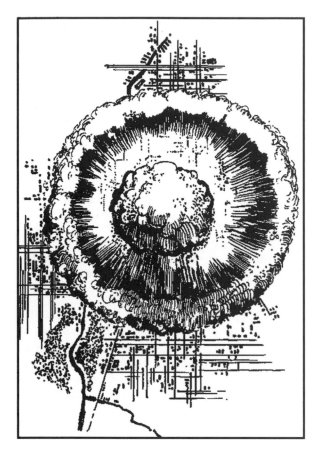

Figure 29-2. Blast.

Figure 29-3. Thermal Radiation.

radiation is primarily a line-of-sight phenomenon, terrain masking can help reduce its effects (figure 29-3).

(2) Light. The fireball formed at the instant of a nuclear detonation is a source of extremely bright light. To an observer 135 miles away from the explosion, the fireball of a 1-megaton weapon would appear to be many times more brilliant than the Sun at noon. The surface temperatures of the fireball, upon which the brightness depends, do not vary greatly with the size of the weapon. Consequently, the brightness of the fireball is roughly the same, regardless of the weapon yield. This light can cause injuries to personnel in the form of temporary or permanent blindness. Temporary blindness from a burst during daylight should be of very short duration and is not an important consideration for anyone other than air-crew members. At night, this loss of vision will last for longer periods because the eyes have been adapted to the dark. However, recovery should be complete within 15 minutes. The light flash can cause permanent injury to the eyes due to burns within the eye, but this is likely to occur only in personnel who happen to be looking direct-ly at the fireball at the instant of explosion (figure 29-4).

c. Nuclear Radiation:

(1) Initial nuclear radiation is the radiation emit-ted in the first minute after detonation. For practical

Figure 29-4. Light.

purposes, it consists primarily of neutrons and gamma rays. Both of these types of radiation, although different in character, can travel considerable distances through the air and can produce harmful effects in humans. Gamma rays are invisible rays similar to X rays. These penetrating rays interact with the human body and cause damage to tissues and the blood-forming cells. The effects of neutrons on the body resemble those of gamma rays. They are highly penetrating and are easily absorbed by human tissue. Neutron radiation can penetrate several inches of tissue. The neutron radiation produces extensive tissue damage within the body. The major problem in protecting against the effects of initial radiation is that a person may have received a lethal or incapacitating dose of radiation before taking any protective action (figure 29-5).

The contaminated areas created by fallout may be very small, or may extend over many thousands of square miles. The dose rate may vary from an insignificant level to an extremely dangerous one for all personnel not taking protective measures.

(b) A secondary hazard that may arise is the neutron-induced radioactivity on the Earth's surface in the immediate vicinity of ground zero. The intensity and extent of the induced radiation field depend on the type of soil in the area around ground zero, the height of the burst, and the type and yield of the weapon. The only significant source of residual radiation from an airburst weapon is induced activity in the soil of a limited circular pattern directly beneath the point of burst (figure 29-6).

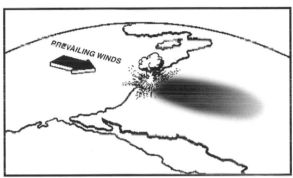

Figure 29-6. Residual Radiation.

29-3. Types of Nuclear Bursts. Nuclear bursts may be classified into three types according to the height of burst—airburst, surface bursts, and subsurface bursts (figure 29-7).

Figure 29-5. Nuclear Radiation.

(2) Residual nuclear radiation is that which lasts after the first minute, and consists primarily of fallout and neutron-induced radiation.

(a) The primary hazard of residual radiation results from the creation of fallout. Fallout is produced when material from the Earth is drawn into the fireball, vaporized, combined with radioactive material, and condensed to particles that then fall back to the Earth. The larger particles fall back immediately in the vicinity of ground zero. The smaller particles are carried by the winds until they gradually settle on the Earth's surface.

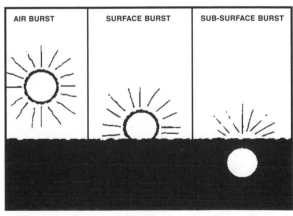

Figure 29-7. Types of Blasts.

a. Airburst. The detonation of a nuclear weapon at such a height that the fireball does not touch the surface of the Earth is called an airburst. Blast, thermal radiation, and initial radiation effects are increased in a low airburst. Fallout of radioactive material from an airburst is not of survival significance, unless rain or snow falls through the radioactive cloud and brings the material to Earth. Neutrons from the detonation will cause induced radiation in the soil around ground zero. Except for very high airbursts, neutron-induced radiation in the area of ground zero will be of concern to survivors who are required to go into or across the area. Radiological monitoring will be required as units pass through such an area so hazardous levels of radiation can be detected and avoided, if possible.

b. Surface Burst. The detonation of a nuclear weapon at such a height that the fireball touches the surface of the Earth or water is called a surface burst. Blast, thermal radiation, and initial nuclear radiation are not as widespread as from an airburst. Induced radiation is present but will be masked by residual radiation from fallout. The fallout produced by a surface burst is by far its most dangerous effect because the burst picks up a great deal more debris, and radioactivates this debris; and, depending on the prevailing winds, the fallout covers thousands of square miles with high levels of radioactivity.

c. Subsurface Burst. The detonation of a nuclear weapon so that the center of the fireball is beneath the surface of the Earth or water is called a subsurface burst. If a fireball of this type breaks through the surface, fallout will be produced. Thermal radiation will not be a significant hazard since it will be almost completely absorbed by the soil. Blast effects will also be significantly reduced. Shock waves passing through the ground or water will extend for a limited distance. The range of the initial nuclear radiation will be considerably less than from either of the other two types of bursts, because this will also be absorbed to a great extent by the soil. However, extremely hazardous residual radiation will occur in and around any crater. If the fireball does not break the surface, shock waves will pass through the ground and craters may result due to settling.

29-4. Injuries. The explosion of a nuclear bomb can cause three types of injures—blast, thermal radiation, and nuclear radiation. Many survivors receive a combination of two or all three of the above injuries. For example, an unprotected person could be killed by a piece of debris, could be burned to death, or could be killed by initial nuclear radiation if the person is within a few thousand yards from the center of the blast (figure 29-8).

Figure 29-8. Injuries.

a. Blast Injuries. Direct blast can cause damage to lungs, stomach, intestines, and eardrums, or can cause internal hemorrhaging. However, the direct blast is not considered a primary cause of injury because those close enough to suffer serious injury from the direct blast will probably die as a result of initial thermal radiation, or they will be crushed to death. The greatest number of blast injuries are received as an indirect result of the blast, from falling buildings, flying objects, and shattered glass.

b. Thermal Radiation Injuries. Burns are classified in degrees according to the depth to which the tissues are injured. In first-degree burns, the skin is reddened as in sunburn. In second-degree burns, the skin is blistered as from contact with boiling water or hot metal. In third-degree burns, the skin is destroyed or charred, and the injury extends through the outer skin to deeper tissues. The degree of burn received from thermal radiation depends on weather conditions, distance from the explosion, and available protection. Many thermal casualties are compounded by nuclear radiation and indirect-blast injury. This makes it difficult to attribute casualties to thermal radiation alone.

c. Nuclear Radiation Injuries. The injurious effects of nuclear radiation from a nuclear explosion represent a new threat, which is completely absent in conventional explosions. This does not infer that this source of injury is the most important in a nuclear explosion. Rays from radioactive material are not as great a hazard as people fear. The amount of danger from fallout depends on where and how the nuclear bomb explodes, and how well the person is protected. The greatest danger from residual radiation (fallout) comes from exposure for a long period of time to radioactive particles that are nearby, or from dust settling on the body or clothing. Since fallout (like X rays) can destroy living tissue, particularly in the blood-forming system, the exposure of persons working in a radioactive or "hot" area must be controlled so as not to exceed a safe limit. Although a person can become seriously ill and even die from breathing radioactive dust, there is less danger from this than when the whole body is exposed to fallout. Remember, all types of radiation are dangerous (nuclear, thermal, X ray, or even that from an infrared lamp).

Figure 29-9. Fallout.

29-5. Types of Residual Radiation. The radioactive debris deposited on the surface as fallout contains three types of nuclear radiation—alpha particle emission, beta particle emission, and gamma radiation (figure 29-9).

    a. Alpha Emitters. Alpha particles have low-penetrating abilities; therefore, survivors can easily shield themselves against these particles. Although alpha particles emitted (given off) from radioactive elements will not penetrate the skin, alpha emitters present a serious hazard if ingested, inhaled, or allowed to enter the body by any other means. From a survival standpoint, alpha particles present the least danger of the three radiation hazards.

    b. Beta Emitters. Basically, beta particles (radiation) are high-speed electrons, which are only slightly penetrating; therefore, survivors can easily shield themselves against beta radiation by wearing materials of moderate thickness, such as heavy shoes and clothing. Because serious skin burns may result from the direct contact of beta-emitting materials with the skin, survivors should take special care to brush themselves off, wash any previously unprotected areas of the body, or cover any exposed parts of the body. Since beta radiation is rapidly absorbed by the air, distance will provide a good form of protection; in fact, 6 to 7 feet of atmosphere will stop most of the beta radiation resulting from fallout. In addition to presenting an external hazard, beta radiation will also cause serious internal effects. By using care in decontaminating foods and water and practicing good hygiene, survivors can greatly decrease the seriousness of this hazard.

    c. Gamma Radiation. In contrast to either alpha or beta emitters, gamma radiation is highly penetrating. Gamma rays are similar to light rays and X rays, but are composed of shorter wavelengths and contain greater amounts of energy. Because of their penetrating abilities, they are the most hazardous of all types of external radiation.

    (1) Fortunately, for the survivor faced with the important problem of obtaining immediate protection against gamma radiation, the amount of shielding is not as great as that required for the initial gamma radiation emitted during the fireball stage of a nuclear detonation. In addition to shielding, other methods of protection against penetrating gamma radiation, later addressed in more detail, include using the factors of time and distance. Because of the low absorption coefficient of gamma rays, their relative internal hazard is greatly reduced, and, in this respect, they are far less dangerous than alpha and beta radiation. However, caution should be taken to ensure no radiation is absorbed—internally or externally.

    (2) Since the fallout dust or debris contains particles emitting gamma rays, survivors must be especially careful to decontaminate themselves and their shelter area. Though radioactive fallout sometimes has the appearance

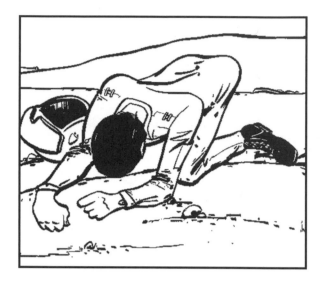

Figure 29-10. Effects.

of white ash or dust, it usually cannot be detected by the human senses. Survivors must assume any suspicious film of dust on water or plant life is radioactive, and they should apply decontamination procedures.

29-6. Effects of Fallout on the Individual. The most harmful effects of fallout result from the changing of the blood cells. Because of this change in cells, some of the tissues that are essential to normal functioning of the human body are damaged or destroyed. The cells are unable to rebuild, so normal cell replacement in the organism is stopped. In addition, the formed products act as poisons to the remaining cells. The extent of damage to the body cells depends on the dose received. Therefore, if the body receives a large dose of fallout radiation (gamma rays and possibly some beta particles), so many cells can be affected that survival is unlikely due to infection resulting from the loss of white blood cells (figure 29-10).

    a. An overdose of radiation from fallout could be received if the survivor stays in the open and doesn't seek shelter. It is also possible to receive an overexposure from the fallout that settles on the clothing or body. Clothing does not stop gamma rays from penetrating and seriously injuring body tissues. Overexposure can also occur from remaining in a highly radioactive area too long. A survivor should not leave a shelter area unless absolutely necessary.

    b. The first indication of an overdose of fallout radiation probably would not show up for several hours, or possibly days. The survivor would then most likely become ill and begin to vomit. The time elapsing before the illness would depend on how large a dose was received. When vomiting starts, it does not necessarily mean that death will follow. For a few days, a survivor

EXPOSURE TYPE	GENERAL SYMPTOMS	ONSET OF SYMPTOMS	ABLE TO WORK	MEDICAL CARE	DEATHS	REMARKS
50–200r	Nausea, weakness, fatigue, possible vomiting, possible radiation burns	Within 24 hours	Yes	No	5%	
200–450r	Nausea, weakness, fatigue, vomiting, diarrhea, loss of hair, radiation burns, easy to bleed	3–6 hours	No	Yes	Less than 50%	Sick a few days, 1–2 weeks temporary recovery, then serious symptoms
450–600r	Same as 200–450r, plus hemorrhaging and infection	Within 3 hours	No	Yes	More than 50%	1 week temporary recovery, death without medical aid
600r+	Same as 450–600r, plus severe, bloody diarrhea	Within 1 hour	No	Yes	100%	Death within 14 days
2,000r+	Complete incapacitation	Within minutes	No	Yes	100%	Death within hours
Internal Radiation	Varies in proportion to internal dose	Varies	Varies	Varies	Varies	
Mild Skin Injury	Itching and/or redness of skin, possible hair loss	1–2 days	Yes	No	None	
Severe Injury	Same as mild, plus radiation burns and hair loss	1–2 days	Yes	Yes	Varies	Death depends on burn area

Figure 29-11. Radiation Sickness Symptoms.

might feel far below par, but with proper medical care, complete recovery is possible (figure 29-11).

29-7. Radiation Detection. Since radiation is invisible and cannot be detected by the physical senses, detection instruments are used. This equipment includes devices for measuring the amount and intensity of area contamination, and devices for determining the radiation dosage a survivor has received while in a contaminated area. A single radiation measurement usually has limited operational significance, except to personnel in the immediate area, since it gives information at the point of the reading only. However, a number of individual measurements considered together can give a picture of the radiation pattern over an entire area. A number of readings made at the same point over a period of time are required to determine the rate of decay of the fallout. Several different points for taking readings may be required in varying types of terrain.

a. Measurement of Radiation. Instruments developed for the detection and measurement of radiation are called radiac instruments. Radiac instruments measure the absorption of radiation, either in terms of dose rate or dosage (figure 29-12).

(1) Dosage is the term applied to the total or accumulated amount of ionizing radiation (beta or gamma) received regardless of the time involved. Dosage is measured in terms of roentgens (r) or milliroentgens (mr). One milliroentgen is .001 of a roentgen. The total quantity of ionization received during a single radiation experience is called a dose. The radiation dose is also referred to as an exposure dose.

(2) When it is desirable to know how fast a certain dosage is being received, the term dose-rate is used to

Figure 29-12. Detection.

indicate this rate. Dose-rate is the amount of radiation being received per unit of time. Generally, the dose-rate is expressed in r/hr or mr/hr. The dose-rate is also used to indicate the level of radioactivity in a contaminated area (figure 29-l3).

b. Radioactive Decay. The concept of radioactive decay is of vital importance in obtaining protection against nuclear fallout, and in determining survival procedures. The debris from a nuclear explosion is made up of a mixture of radioactive materials of many kinds: unfissioned particles, fission products, and numerous other radioisotopes created by the neutron activation of inert material that takes place during the explosion. Fortunately, of the nearly 200 isotopes emerging, 70 percent are short-lived materials with a half-life of less than one day. Some elements, however, take a much longer time to decay; these are known as long half-life elements.

(1) Half-life is the time required for a radioactive substance to lose 50 percent of its activity through decay. Half-life for a mixture of isotopes (a term used to define precise species of elements) is not as simple as that described for a single radioactive isotope. The activity of the mixture diminishes very quickly after detonation, but as time passes, the longer-lived species become responsible for the major part of the radiation remaining, so that total radioactivity diminishes much more slowly.

(2) As a rule of thumb, radioactivity may be said to decrease in intensity by a factor of 10 for every sevenfold increase in time following the peak radiation level. Figure 29-13 illustrates the rapidity of the decay of radiation from fallout during the first 2 days after the nuclear explosion that produced it. Notice that it takes about seven times as long for the dose-rate to decay from 1,000 roentgens per hour (1,000 r/hr) to 10 r/hr (48 hours) as to decay from 1,000 r/fur to 100 r/hr (7 hours). Only in high-fallout areas would the dose-rate 1 hour after the explosion be as high as 1,000 roentgens per hour.

(3) If the dose-rate 1 hour after an explosion is 1,000 r/hr, it would take about 2 weeks for the dose-rate to be reduced to 1 r/fur solely as a result of radioactive decay. Weathering effects will reduce the dose-rate further; for example, rain can wash fallout particles from plants and buildings to lower positions on or closer to ground. Surrounding objects would reduce the radiation dose from these low-lying particles. Figure 29-13 also illustrates the fact that at a typical location where a given amount of fallout from an explosion is deposited later than 1 hour after the explosion, the highest dose-rate and the total dose received at that location are less than at a location where the same amount of fallout is deposited 1 hour after the explosion. The longer fallout particles have been airborne before reaching the ground, the less dangerous is their radiation.

(4) Two weeks after the last burst, the occupants of most shelters could begin working outside the shelters, increasing the number of hours each day. Exceptions would be in thermal-damaged areas or in areas of extremely heavy fallout, such as might occur downwind from important targets attacked with many weapons, especially missile sites and very large cities. To know when to come out safely, occupants would either need a reliable fallout meter to measure the changing radiation dangers, or they would need to receive information based on measurements made nearby with a reliable instrument, using the information in figure 29-14.

29-8. Body Reactions to Radiation. The effects of ionizing radiation on the human body may be divided into two broad categories—chronic effects and acute effects.

a. Chronic Effects. Chronic effects are defined as those occurring some years or generations after exposure to radiation. Included in this category are the cancer-producing and genetic effects. While of concern because of their possible damage to future generations, these effects are of minor significance insofar as they may affect immediate survival.

b. Acute Effects. Acute effects are of primary significance in survival. Some acute effects appear within a few hours after exposure to radiation. These effects are the result of direct physical damage to tissue caused by

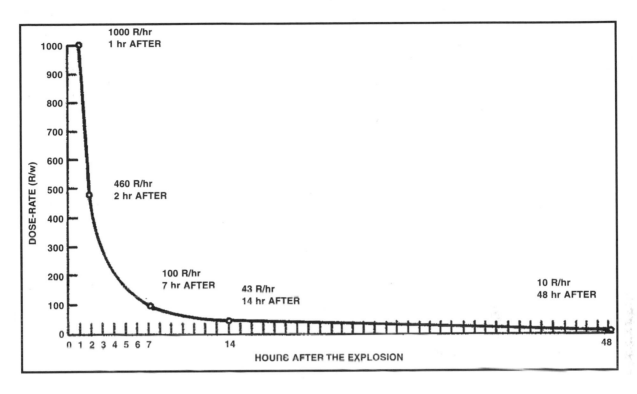

Figure 29-13. Decay Ratio.

radiation exposure; beta burns are examples of acute effects.

c. Body Damage. The extent of body damage depends to a large degree on the part and extent of the body exposed, and the ability of the body to recover from radiation damage. Certain parts of the body, such as the brain and the kidneys, have little ability to recover from injury, while other portions of the body, such as the skin and bone marrow, can repair damage. The extent of body exposure to radiation is of great importance in determining the chances of subsequent recovery. If a dose of 350 roentgens was applied to just a small portion of the body, such as the hands or face, there would probably be little effect on overall health. Serious damage would, of course, be created in these exposed parts.

d. Hazards from Residual Radiation. There are two main hazards resulting from residue hazard, one resulting from highly penetrating gamma radiation and less penetrating beta radiation, which causes burns; and an internal hazard, resulting from the entry of alpha- and beta-emitting particles into the body.

(1) The external hazards result in overall irradiation and serious beta burns, while the internal hazards result in irradiation of the critical organs, such as the gastrointestinal tract, thyroid gland, and bone.

(2) A very small amount of radioactive material can cause extensive damage in these and other parts of the body if allowed to enter the body by consumption of contaminated food or water, or by absorption by the bloodstream through cuts or abrasions in the skin. By comparison, the material gaining entry by breathing presents a lesser hazard. Fortunately, most of the fallout particles are filtered out by the upper-respiratory tract as seen in the observations of victims on the Marshall Islands, who were accidentally exposed to fallout during the 1954 nuclear test. By using good hygiene and careful decontamination of food and water, a survivor can further reduce the internal hazard.

e. Degree of Radiation Damage. Because of the sensitivity of the gastrointestinal tract to radiation, the severity and the speed with which vomiting and diarrhea appear after exposure are good indicators of the degree of radiation damage. Almost everyone confined to immediate-action shelters after a nuclear attack would be under stress and without clean surroundings. Many would also lack adequate water and food. Under these conditions, perhaps half the survivors who receive a whole-body dose of 200-450 roentgens would die within a few days. The human body can repair most radiation damage, if the daily radiation doses are not too large.

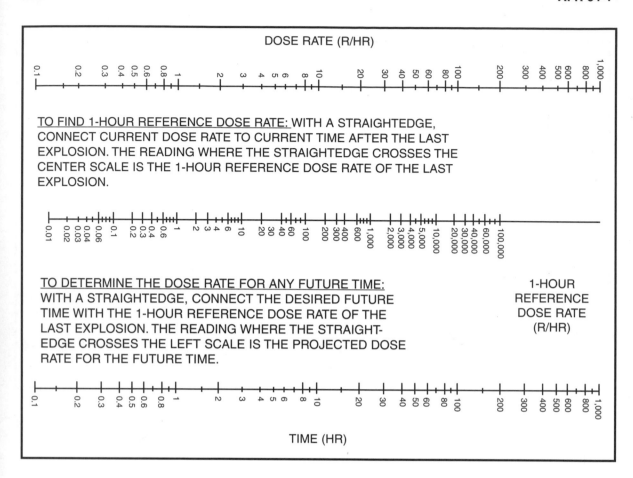

Figure 29-14. Dose Rate.

29-9 Countermeasures Against Penetrating External Radiation:

a. The means of protection from external radiation are threefold—time, distance, and shielding. By controlling the length of exposure time, controlling the distance between the individual and the source of radiation, and most important of all, by placing some absorbing or shielding material between the survivor and the source of radiation, the survivor's level of radiation can be significantly reduced, and can thereby increase the chances of survival.

(1) Time. The effect of time on radiation exposure is easy to understand. Take a simple example such as the following: Assume a person is in an area where the radiation level from penetrating external radiation is 100 roentgens per hour and the dose-rate is constant; then in 1 hour, that person would receive l00 roentgens (or more properly, roentgen absorbed dose (rad); in 2 hours, 200 rad; and in 8 hours, 800 rad. The important implication for a survivor is that, since radiation dosages received should be considered to be essentially cumulative, as little time as possible should be spent in an unprotected area, whether constructing a shelter or searching for water. Time is also of vital importance from another standpoint—that of radioactive decay. Knowledge of this characteristic can serve as one of the primary means of protection against radiation from fallout. The importance of time as a protective factor may be seen clearly by the following example. If survivors were to enter a high-intensity fallout area of 1,000 roentgens per hour in which the radiation intensity had just peaked, and remain there for 1 hour in the open, they would accumulate a biologically damaging dose of approximately 650 rad. This is a dose strong enough to kill 9 out of 10 people exposed to it. If, however, they entered this same area some 48 hours later and remained there for 1 hour, they would receive a dose of approximately 10 rad. Even if they were then required to remain in the open for 24 hours, their dose would only be 170 rad. Two weeks after the completed fallout, they would receive only 1 rad per hour in this same area. They could then remain in the area for several months, accumulating a dose of 180 rad, and not develop severe radiation sickness because of the rate at which the dosage was received (figure 29-15).

Figure 29-15. Countermeasures.

(2) Distance. Distance is effective as a means of protection against penetrating gamma radiation as shown by the inverse square law. This law states that the intensity of radiation decreases by the square of the distance from the source. In other words, survivors exposed to 1,000 units of penetrating external radiation at 1 foot from the source would receive only 250 units at 2 feet; when they double the distance, the radiation level is reduced by $(1/2)^2$, or one-fourth the amount. When the distance is tripled, the dose is reduced to $(5/8)^2$, or one-ninth of the original amount (or 111 units). At 10 feet, this is further reduced to $(^110)^2$, or one-hundredth of the radiation exposure at 1 foot. This relationship of distance to intensity of radiation exposure is dependent on the distribution pattern of the radiation source. As just seen, if the radiation is concentrated into a very small area (referred to as a point source), the intensity is decreased to about one-fourth the original amount each time the distance is doubled. A complicated relationship, however, is obtained when the radiation source is not concentrated in a point, but is spread around in random patterns, as in the case of surface contamination from fallout.

(3) Shielding. The third and most important method of protection against penetrating radiation is that of shielding. Since the damaging effects of penetrating radiation arise from the fact that the rays interact with electrons in the body, survivors must place dense material between themselves and the source of radiation. The more electrons there are in the makeup of the shielding material, the more gamma radiation will be stopped from

entering the body. Lead, iron, concrete, and water are all examples of good shielding materials. Of the three described countermeasures against penetrating external radiation, shielding gives the greatest protection, is the easiest to use under survival conditions, and is, therefore, the most desirable. If shielding is not possible, however, the other precautions should be rigorously followed. The degree of protection they afford is significant, and could provide the necessary margin of safety for survival (figure 29-16).

Figure 29-16. Shielding.

b. It is a common misconception that radioactive fallout can impart its radioactivity to the object with which it comes in contact, such as fruits and vegetables. The rays from radioactive fallout particles cannot make something radioactive merely by being in contact with it. For this reason, canned foods and smooth-skinned fruits and vegetables may be eaten once they have been decontaminated.

(1) Everyone entering a contaminated area should wear protective clothing (CW ensemble) to prevent exposure to radioactive dust. The main reason for this precaution is to shield the skin from beta particles. Beta particles won't penetrate far into the body, but they will harm the skin. The effect, a reddening and blistering of the skin, is called a "beta burn." Such damage may not appear until sometime after exposure since beta particles have a delayed action. Survivors may not know they have received skin burns from beta particles until it is too late. Therefore, protective clothing should be worn as instructed. Sleeves should be down and buttoned, wrist and ankle openings taped, and gloves and boots worn, if possible.

(2) Alpha particles are lowest in ability to penetrate, but they are most hazardous when taken into the body through contaminated food and drink. The best defense against alpha particles is to ensure all food is properly decontaminated before consumption.

(3) If at all possible, a survivor should always avoid "hot" areas. Sometimes, a survivor cannot avoid contact with this environment and, when entering a "hot" area, should keep radioactive dust off the body and out of the body's system. If survivors inhale radioactive dust, they can suffer serious internal injury. If there is radioactive dust in "hot" areas, a protective filter should be used over the mouth and nose.

c. The presence of fallout material in the area will require some minor modification of medical procedures. Wounds should be covered to prevent entry of active particles. Burns from beta activity are treated like any other burn, except the burned surface should be washed. Measures to prevent infection should be emphasized since the body will be especially sensitive to infections because of changes in the blood. Extra attention should be given to the prevention of colds and respiratory infections. The eyes should be covered to prevent entry of particles. Improvised goggles may be used for this purpose.

29-10. Shelter. A sufficient thickness of shielding material will reduce the level of radiation to negligible proportions. The thickness required to reduce gamma radiation from fallout is much less than that necessary against initial gamma radiation, because fallout radiation has much less energy than the initial radiation from a nuclear explosion. Thus, a comparatively small amount of shielding material can provide good protection against residual gamma radiation.

a. The following table illustrates the thicknesses of various materials required to reduce the penetration from residual fallout gamma sources by 50 percent:

Iron and steel	.7 inches
Concrete	2.2 inches
Brick	2.0 inches
Dirt	3.3 inches
Ice	6.8 inches
Wood (soft)	8.8 inches
Snow	20.3 inches

b. The principle of half-value layer thickness is useful in understanding the absorption of gamma radiation by various materials. According to this principle, if 2 inches of brick reduce the level of gamma radiation by one-half, adding another 2 inches of brick will reduce the intensity by another half, namely to one-fourth the original amount; 6 inches will reduce the level of fallout gamma radiation to one-eighth its original amount; 8 inches to one-sixteenth, and so on. Thus in a shelter protected by 3 feet of earth, a radiation intensity of 1,000 roentgens per

hour on the outside would be reduced to about one-half roentgen per hour inside the shelter.

c. Areas where terrain provides natural shielding and easy shelter construction, such as the sides of ditches, ravines, rocky outcroppings, hills, and riverbanks, are ideal locations for an emergency shelter (figure 29-17). In level areas lacking natural protection, dig a foxhole or slit trench.

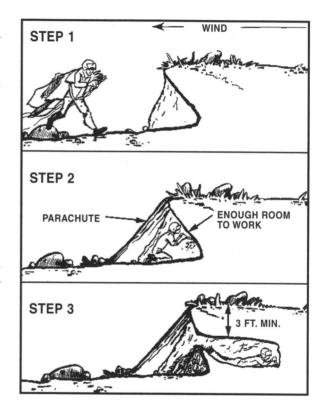

Figure 29-17. Immediate Shelter.

(1) In digging a trench, the survivor should work from the inside of the hole as soon as it is large enough to cover part of the body to prevent exposure of the entire body to radiation. In open country, an attempt should be made to dig the trench from a prone position, stacking the dirt carefully and evenly around the trench. On level ground, the dirt should be around the body for additional shielding. Depending on soil conditions, the time to construct a shelter will vary from a few minutes to a few hours. Rapid shelter construction will limit the dosage received. Building a shelter under a tree is not recommended because cutting or digging out the numerous roots will be very difficult. Another disadvantage in making a shelter under trees is that more of the gamma rays from fallout particles on the leaves and branches would reach and penetrate the shelter than if

these same particles were on the ground. Many gamma rays from fallout particles on the ground would be scattered or absorbed by striking rocks, clods of earth, tree trunks, or buildings before reaching a below-ground shelter (figure 29-18).

(2) While an underground shelter covered by 3 or more feet of earth would provide the best protection against fallout radiation, the following additional unoccupied structures (in the order listed) offer the next best protection:

(a) Caves and tunnels covered by more than 3 feet of earth.

(b) Storm or storage cellars.

(c) Culverts.

(d) Basements or cellars of abandoned buildings.

(e) Abandoned buildings made of stone or mud.

d. Building a roof on the shelter should not be required. A roof should only be added if the materials are readily available to the survivor, and will require only a brief exposure to the outside contamination. If the construction of a roof would require extended exposure to penetrating radiation, it would be wiser to leave the shelter roofless. The function of a roof is to lend distance from the source of fallout to the body, and unless

dense roofing is used, a roof provides only scant shielding. A simple roof can be constructed out of a piece of parachute anchored down by dirt, rocks, or other refuse from the shelter. Large particles of dirt and other debris may be removed from the top of this parachute canopy by beating it at frequent intervals. Such a parachute cover will not offer shielding from the radiation emitted by active fission products deposited on the outer surface, but it will increase the distance from the fallout source, and keep the shelter area covered from further contamination.

e. The primary criterion for locating and establishing a shelter site should be to obtain protection as rapidly as possible against the high-intensity radiation levels of early gamma fallout. Five minutes to locate the shelter site is an excellent guide. Speed in obtaining shelter is essential. Without shelter, the dosage received in the first few hours will exceed that received during the rest of the week in a contaminated area; the dosage received in this first week will exceed that accumulated during the rest of a lifetime spent in the same contaminated area.

(1) Several initial actions should be kept in mind while seeking a shelter location. The survivor should:

Figure 29-18. Terrain That Provides Natural Shielding.

(a) Where possible, attempt to remain with the aircraft until it is possible to eject or land in an area at least 20 miles upwind or crosswind from any known target. This action will ensure the landing area will have a minimum amount of fallout (figure 29-19).

(b) If it can be controlled, parachute opening should be delayed until a comparatively low altitude is reached in order to reduce the time of exposure to fallout during the descent.

-1. During the descent, select areas on the ground likely to provide good shelter.

-2. Immediately upon landing, take the parachute and survival kit and find cover and protective shelter.

-3. Take precautions to avoid detection and capture, but not at the expense of additional exposure to radiation, which will lessen the chance of survival.

(2) In selecting and establishing the shelter, the survivor should keep the following additional factors in mind in order to reduce the time of exposure and the dosage received:

(a) Where possible, seek a crude, existing shelter that can be improved. If none is available, dig a trench.

(b) Dig the shelter deep enough to obtain good protection, then enlarge it as required for comfort.

(c) Cover the top of the foxhole or trench with any readily available material and a thick layer of earth, if possible, without leaving the shelter.

(d) During the descent and while constructing a shelter, keep all clothing on, as well as a cap, scarf, and gloves to obtain protection against burns from beta radiation.

(3) The shelter location should be brushed or scraped clean of any surface deposits with a branch or some other object to be certain that contaminated materials are removed from the area to be occupied. The swept area should extend at least 5 feet beyond the area where the shelter is being dug. Any material brought into the shelter should be decontaminated. This includes grass or foliage that is being used as insulation or bedding material, outer clothing (especially footgear), and the parachute, if it is to be used. If weather permits and the parachute and outer clothing are heavily contaminated, the survivor may want to remove them and bury them under a foot of earth at the end of the shelter. These may later (after decay factor) be retrieved when leaving the shelter. If these materials are dry, decontamination may be done by beating or shaking them outside the entrance to the shelter to remove the radioactive dust (figure 29-20). Any body of water, even though it may contain contaminated

Figure 29-19. Route of Travel.

particles, may be used to rid materials of excess fallout particles by simply dipping the materials into the water, and shaking them to remove excess water. Do not wring out materials since this will trap the contaminated particles. If at all possible, wash the body thoroughly with soap and water without leaving the shelter. This washing will remove most of the harmful radioactive particles that are likely to cause beta burns or other damage. If water is not available, the face and any other exposed skin surfaces should be wiped to remove contaminated dust and dirt. This may be done with a clean piece of cloth, or a handful of uncontaminated dirt obtained by scraping off the top few inches of soil and using the "clean" dirt.

f. Upon completion of the shelter, the survivor should lie down, keep warm, sleep, and rest as much as possible during the time spent in the shelter. There is no need to panic if nausea and symptoms of radiation sickness are experienced. Even small doses of radiation can cause these symptoms, which may disappear in a short time. The following provides the time necessary to avoid serious dosage and still allow the opportunity to cope with survival problems:

(1) Complete isolation should be maintained from 4 to 6 days following delivery of the last weapon. A very brief exposure for procurement of water on the third day is permissible, but exposure should not exceed 30 minutes.

(2) On day 7, one exposure of not more than 30 minutes.

(3) On day 8, one exposure of not more than 1 hour.

(4) From day 9 through day 12, exposure of 2 to 4 hours per day.

(5) From day 13 on, normal operation, followed by rest in a protected shelter.

29-11. Water. In a fallout contaminated area, available water sources may be contaminated with radioactive particles. If possible, the survivor should wait at least 48 hours before drinking any water to allow radioactive decay to take place. Selecting the safest possible source of water will greatly diminish the danger of ingesting harmful amounts of radioactivity.

a. Although many factors, such as direction of wind, rainfall, and amount of particulate matter (clay, for example) in the water, will influence the choice in selecting water, the following priorities of water sources are

Figure 29-20. Decontamination.

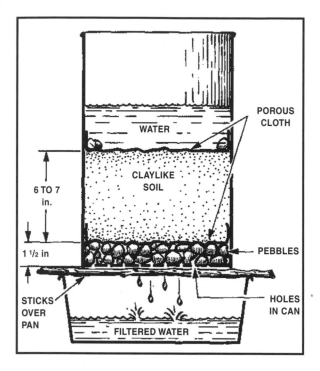

Figure 29-21. Filtering Water.

recommended (as an additional precaution against disease, all water sources should be treated with water-purification tablets from the survival kit, or boiled for at least 10 minutes):

(1) Water from springs, wells, or other underground sources having natural filtration will be the safest sources of water.

(2) Any water in the pipes or containers of abandoned houses or stores will also be free from radioactive

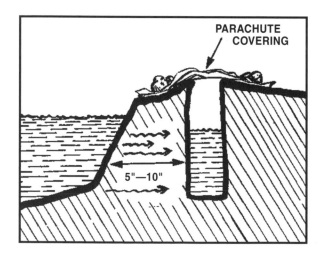

Figure 29-22. Settling Water.

particles and, therefore, safe to drink, although precautions will have to be taken against bacteria in the water.

(3) Snow taken from a level that was 6 or more inches below the surface during the fallout should be a safe source of water.

(4) Water from streams and rivers will be comparatively free from fallout within several days after the last nuclear explosion because of the dilution factor. If at all possible, such water should be filtered before drinking.

(5) Water from lakes, pools, ponds, and other standing sources is likely to be heavily contaminated, though most of the heavier, insoluble isotopes will settle to the bottom.

b. The degree of solubility of various isotopes varies. Some fission products are extremely water soluble, but most are relatively insoluble. Certain isotopes, in fact, have been found to be as much as 90 percent insoluble. The significance of this fact is that 99 percent of the radioactivity in water could be removed by filtering it through ordinary earth (figure 29-21). The best method of filtration is to dig sediment holes or seepage basins along the side of a water source. The water will seep laterally into the hole through the intervening soil, which will act as a filtering agent and remove the contaminated fallout particles that have settled on the original body of water. It is important that the hole be covered in some way (example, with a parachute) to prevent further contamination (figure 29-22).

c. Settling is one of the easiest methods used to remove most fallout particles from water. Furthermore, if the water to be used is muddy or murky, settling it before filtering will extend the life of the filter. The procedure is as follows:

(1) Fill a bucket or other deep container three-fourths full with the contaminated water.

(2) Dig dirt from a depth of 4 or more inches below the ground surface and stir it into the water. Use about a 1-inch depth of dirt for every 4 inches of water.

(3) The water is stirred until practically all of the dirt particles are suspended in the water.

(4) This mixture should be allowed to settle for at least 6 hours. The settling dirt particles will carry most of the suspended fallout particles to the bottom, and cover them. The clear water can then be dipped or siphoned out and purified (figure 29-23).

29-12. Food. Obtaining edible food in a radiation-contaminated area is possible. Survivors need to follow only a few special procedures in selecting and preparing rations and native foods for use. Since survival rations are protected by secure packaging, they will be perfectly acceptable for use after the ration containers are decontaminated, but survivors should supplement them with any food they can find on their trips away from the shelter. Any processed foods that may be found in abandoned buildings are acceptable for use once they are decontaminated. These include canned and packaged foods after the

Figure 29-23. Settling pool.

containers or wrappers are discarded or washed free of fallout particles, food stored in any closed container, and food stored in protected areas (such as cellars). The containers should be washed before opening. For purposes of discussion, native food sources may be divided into two categories: plant food and animal food.

a. Animal Food. In the category of native animal food, survivors must assume all animals, regardless of their habitat or living conditions, will be exposed to radiation. Since the effects of radiation on animals are similar to those in humans, most of the wild animals living in a fallout area are likely to become sick or die from radiation during the first month following the nuclear explosion. Even though animals may not be completely free from radioactive materials, it may be necessary to use them as a food source. With careful preparation and adherence to several important principles, animals can be safe sources of food.

(1) If an animal appears to be sick, it should not be eaten. The animal may have developed a bacterial infection as a result of a radiation dose. Contaminated meat could cause severe illness or death if eaten, even if thoroughly cooked.

(2) All animals should be carefully skinned to prevent any radioactive particles adhering to the outside of the skin from gaining entry into the body.

(3) Meat around the bones and joints should not be eaten. A large percentage of radioactivity in the body of animals is found in the skeleton. The remaining muscle tissue of the animal, however, will be safe to cut. Before cooking the animal, survivors should cut the meat away from the bone, leaving approximately one-eighth of an inch of meat on the bone. All internal organs such as the heart, liver, and kidneys, normally used as survival food, should be discarded (figure 29-24), since they tend to concentrate beta and gamma radioactivity. All meat should be cooked until it is very well done. To be sure the meat is well done, it should be cut into pieces less

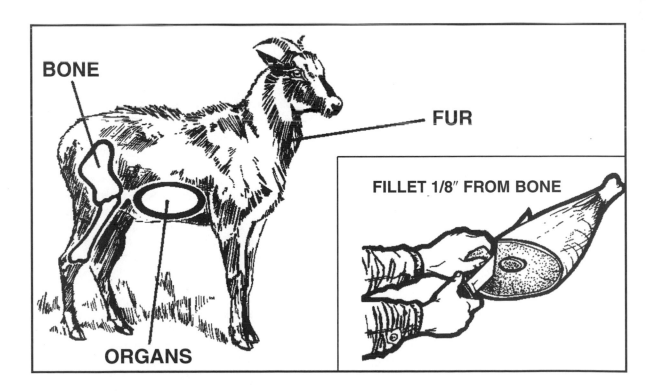

Figure 29-24. Meat Procurement.

Figure 29-25. Collecting Food.

than one-half inch thick before cooking. This precaution also reduces cooking time and saves fuel.

(4) The extent of contamination of fish and clams following nuclear tests in the Pacific was found to be much greater than that of the land animals. On the basis of these findings and those of other tests showing the high concentration of radioactivity in aquatic plants and animals, especially near coastal areas, it is recommended that aquatic food sources be used only in an extreme emergency.

(5) All eggs (excluding the shell), even if laid during the period of fallout, will be safe to eat. Because animals absorb large amounts of radioactive strontium from the plants upon which they graze, milk from any animals in a fallout area should be avoided.

b. Plant Food. Plant foods are contaminated by the accumulation of fallout on their outer surfaces, or by means of absorption through the roots.

(1) The survivor's first choice of plant food should be vegetables, such as potatoes, turnips, carrots, and other plants, whose edible portions grow underground. These are the safest to eat once they are scrubbed and the skins removed. Second in order of preference are those above-ground portions of the plant with edible parts that can be decontaminated by washing and peeling their outer surfaces; examples are bananas, apples, and other such fruits

and vegetables. Other smooth-skinned vegetables, fruits, or above-ground plants, which cannot be easily peeled or effectively decontaminated by washing the radioactive particles off their surfaces, will be the third choice of emergency food (figure 29-25).

(2) The effectiveness of decontamination by scrubbing is generally inversely proportional to the roughness of the surface of the fruit. After the Marshall Islands incident, smooth-surfaced fruits were found to lose 90 percent of their radioactivity after washing, while washing rough-surfaced plants removed only 50 percent of the contamination. Plant foods such as lettuce, having a very rough outer surface that cannot easily or effectively be decontaminated by peeling or washing, should be eaten only as a last resort. Other difficult foods to decontaminate by washing with water are dried fruits such as figs, prunes, peaches, apricots, pears, and soybeans.

(3) Generally speaking, any plant food ready for harvest can be used for food if it can be effectively decontaminated. Growing plants, however, can absorb some radioactive materials through their leaves as well as from soil, especially if rains have occurred during or after the fallout period. With some elements, such as strontium, which is extremely soluble in water, data has shown greater amounts were taken up by plants through

EQUIVALENT RESIDUAL DOSE (ERD): The body can repair 90% of a dose of radiation damage. The unrepaired damage for any given day is the radiation dose multiplied by the ERD factor of the number of days after the exposure. NOTE: Each subsequent dose ERD factor is based on the number of days after each subsequent dose.

DAY	FACTOR	DAY	FACTOR	DAY	FACTOR	DAY	FACTOR	DAY	FACTOR
1–4	1.000	10	.873	16	.764	22	.670	28	.590
5	.978	11	.854	17	.748	23	.656	29	.578
6	.956	12	.835	18	.731	24	.642	30	.566
7	.934	13	.817	19	.716	25	.629	31	.554
8	.913	14	.799	20	.700	26	.616	32	.543
9	.893	15	.781	21	.685	27	.602	33	.532

$$\binom{1st}{Dose}\binom{ERD}{Factor} + \binom{2nd}{Dose}\binom{2nd\ ERD}{FACTOR} + ...+ \binom{X}{Dose}\binom{X\ ERD}{Factor} = \binom{Dose}{Shelter}$$

WATER SOURCES: Uncontaminated water—solar still or snow 6 inches below contamination level. Delay as long as possible from drinking from contaminated sources. PRIORITY: (1) underground, (2) running, (3) stationary. Filter sources 2 and 3 through 12 inches of earth and add purification tablets before use. BOILING WILL NOT REMOVE RADIOACTIVITY.

ANIMAL FOOD: DO NOT BUTCHER SICK ANIMALS. Discard internal organs and meat next to bone. Hides/feathers may be heavily contaminated. Eggs OK to eat.

SEA FOOD: Ocean sources OK, others on a risk basis. Fish from stationary bodies of water probably contaminated.

PLANT FOOD: Plants with edible portions below ground first choice. Smooth plant food easy to wash. Rough-surfaced plants difficult to wash. Wash, pare, then wash again.

FIRST AID: NO FIRST AID FOR RADIATION SICKNESS. Infection, main danger. Personal hygiene important. Rest, avoid fatigue. Drink liquids.

BURNS: Normal first aid. Cool and cover burn. Do not use grease, etc. Treat for possible shock. Keep warm, lie with feet above head.

FRACTURES: Normal first aid. Immobilize and splint. Possible shock.

BLEEDING: Normal first aid. Apply pressure at break. Tight tourniquet can cause loss of limb—use only as last resort.

Figure 29-26. Equivalent Residual Dose Rate.

NUCLEAR EXPLOSIONS: FALL FLAT. Cover exposed body parts. Present minimal profile to direction of blast.

DO NOT LOOK AT FIREBALL. Remain prone until blast effects over.

SHELTER: Pick ASAP, 5 minutes unsheltered max. PRIORITY: (1) cave or tunnel covered with 3 or more feet of earth, (2) storm/storage cellars, (3) culverts, (4) basements, (5) abandoned stone/mud buildings, (6) foxhole 4 feet deep—remove topsoil within 2-foot radius of foxhole lip.

RADIATION SHIELDING EFFICIENCIES: One thickness reduces received radiation dose by one-half. Additional thickness added to any amount of thickness reduces received radiation dose by one half.

Iron/Steel	.7 in	Earth	3.3 in	Wood (Soft)	8.8 in
Brick	2.0 in	Cinder Block	5.3 in	Snow	20.3 in
Concrete	2.2 in	Ice	6.8 in		

SHELTER SURVIVAL: KEEP CONTAMINATED MATERIALS OUT OF SHELTER. Good weather, bury contaminated clothing outside of shelter—recover later. Bad weather, shake strongly or beat with branches. Rinse and (or) shake wet clothing—DO NOT WRING OUT.

PERSONAL HYGIENE: Wash entire body with soap and ANY water; give close attention to fingernails and hairy parts. No Water: Wipe all exposed skin surfaces with clean cloth or uncontaminated soil. Fallout/dusty conditions, keep entire body covered. Keep handkerchief/cloth over mouth and nose. Improvise goggles. DO NOT SMOKE.

EXIT PLANNING: Explosion time and a dose-rate known, use rate decay nomograph.

Explosion time unknown: (1) Take dose-rate readings every hour. (2) When any reading is ½ of any previous reading, multiply time difference between two readings by ¼. (3) Subtract the resultant from the time of first reading. (4) This new time is approximate time of explosion. Use new time with dose-rate nomograph to determine safe exit time.

No rate meter, complete isolation first 4–6 days after last explosion.

Days 3–7: Brief exposure, 30 minutes MAX.

Day 8: Brief exposure, 1 hour MAX.

Days 9–12: Exposure of 2–4 hours per day.

Day 13 on: Normal movement.

MAXIMUM SURVIVAL DOSE/ERD: 200 roentgens

Figure 29-27. Nuclear Explosions.

their leaves than through their roots. In the tests conducted in the Marshall Islands, high levels of radioactivity were found in the water and on the external surfaces of the plants early after the detonation, but only small amounts of beta activity and no alpha activity were detected in the edible portions of the fruits and vegetables. High levels of contamination, similar in activity to that of the contaminated water nearby, were found in the sap of the coconut tree. From these and other tests, we know that small amounts of fission products may be immediately absorbed by plants growing on soil contaminated by fallout material.

(4) If the countermeasures recommended for obtained protection against penetrating external gamma radiation and beta radiation are taken immediately, and the rules for constructing a shelter and selecting food and water are applied, a survivor's chances of surviving a nuclear attack are excellent. Figures 29-26 and 29-27 contain some summarized information applicable to a radioactive environment.

Chapter 30

BIOLOGICAL CONDITIONS

30-1. Introduction. Biological agents are viruses and micro-organisms, or their products, which are used to cause disease, injury, or death to people, animals, or plants (and, to a lesser extent, deterioration of material). Their use is an attempt to produce disease on a large scale. During war, these agents will probably be used primarily in a strategic role to attack rear area bases.

a. Most micro-organisms are harmless to humans, animals, or plants, and a few are helpful. Yeast is used in bread, beer, and cheese production. Some micro-organisms, however, do produce disease. Biological agents could consist of:

(1) Fungi - mold, mildew, athlete's foot; histoplasmosis; and other pathogenic fungi.

(2) Bacteria - plague, tularemia, anthrax.

(3) Rickettsiae - Rocky Mountain spotted fever, typhus.

(4) Viruses - yellow fever, smallpox.

(5) Biotoxins - mushroom, algae, and bacterial poisons.

b. Many biological agents are living, and require moisture, food, and certain temperature limits for life and growth. They are killed by simple acts such as boiling water, adding purification tablets to water, cooking food, exposing them to sunlight, and (or) using germicides. Biological agents enter the body through the nose, mouth, or skin; however, most will not penetrate intact skin. By preventing their entry into the body, a survivor is safe from biological agents.

30-2. Detection. There is no simple method of detecting biological agents. A person cannot see, feel, or taste these agents in a biological attack, whether spread through conventional means or sabotage. Additionally, a person cannot taste the toxins in food. The basic methods of disseminating the agents are through the generation of an aerosol, the use of disease-carrying vectors, and food and water.

a. Aerosols are particles composed of many organisms or a single organism, which are dispersed into the air and transported by air currents. Effective transmission as an aerosol requires that biological agents reach the target area with an effective percentage remaining alive and capable of causing disease. The appearance of certain clues may warn a survivor of an aerosol biological attack. They are:

(1) Aircraft dropping objects or spraying. Both enemy aircraft and friendly aircraft could be engaged in neutralizing or destroying opposing forces.

(2) Breakable containers or new and unusual types of shells and bombs, particularly those which burst with little or no blast.

(3) Smokes and mists of unknown origin.

(4) Unusual substances on the ground or on vegetation. Sick, dead, or dying animals.

b. Vectors such as mosquitoes and ticks, which carry disease, can be delivered to the target area in containers

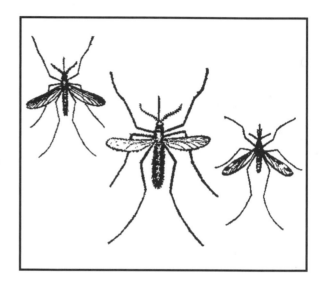

Figure 30-1. Vectors.

by aircraft or missiles. The containers rupture on impact and release the vectors (figure 30-1.). Some vectors need a "host" or carrier that can transmit the disease organism through the skin; others may be inanimate objects, such as contaminated food and water. Because disease organisms can infect personnel, the use of a "protective" mask may not help protect against viruses. The vectors can produce disease throughout their entire life spans, regardless of how far or where they travel. Furthermore, they may pass the disease on to successive generations. Therefore, survivors must be extremely cautious when skinning wild game by wearing gloves and other protective clothing, as the game may host fleas, which carry many diseases. (NOTE: Contact with animals should be avoided unless they are to be used as food.

30-3. Climate. The various characteristics of aerosols and vectors will affect their utility in varying climates. While a survivor may encounter any form of biological agent once their use in combat occurs, the following factors may help a survivor assess the relative risk of various types of biological agents.

a. Aerosols are generally much more controllable in the area of application than are vector-borne agents.

However, most aerosol agents deteriorate to some degree when exposed to direct sunlight. They are more suitable for use against small-area targets.

b. Strong winds are necessary to ensure maximum area coverage of both aerosols and vectors. However, the density of coverage may be decreased by extremely high winds.

c. Vectors are less controllable than aerosols in the area of application following release. They are more suited for use against broad-area targets. Although vectors tend to last longer in humid climates, many potential vectors (flies, fleas, mosquitoes, lice, etc.) will thrive in virtually any environmental area.

d. Generally, nighttime (1 hour before sunset to 1 hour after sunrise) is the best period to dispense vectors. Vector movement and activity is usually greater during the cooler hours of darkness.

30-4. Terrain. Biological-agent aerosols tend to follow the contour of rolling terrain and valleys very much like airborne particles, such as fog. Vegetation can slow the downwind travel of agents by removing some particles from the air. Due to a lack of sunlight, densely vegetated areas, such as a jungle (warm, humid), allow some agents to thrive for extended periods of time. Because most of the biological agents are more hazardous when inhaled than when directly exposed to the skin, contamination of the ground following an attack is less dangerous to the survivor than exposure to an aerosol attack.

30-5. Protection. Defense against biological warfare is neither simple nor easy. The best defense against these agents is the natural resistance of the survivor's body, a high standard of personal cleanliness, careful attention to sanitation, good nutrition, up-to-date immunization status, proper use of drugs, and immediate self-aid to any break in the skin or a puncture wound. Germs must actually get inside the body to cause disease.

a. A protective mask, properly fitted and in good condition, will greatly reduce the danger of inhaling germs. If a mask is not available, a handkerchief or parachute material over the mouth and nose will suffice to provide protection. Since survivors cannot detect the presence of biological agents, they should wear the mask or some other protective equipment over the mouth and nose until they are rescued, if possible.

b. Cuts and sores, in addition to the nose and mouth, are open doors to germs trying to enter the body. Wounds must be kept clean and protected with a bandage. Any type of clothing will give some protection. Fasten the shirt or jacket collar, roll the sleeves down and button the cuffs, wrap the trouser legs or tuck them inside the boots, and tie down all other clothes to stop the entry of germs that may be in the air or on the ground. If the survivor has a uniform used for protection against chemical agents, it should be worn because it gives a greater degree of protection against germs than ordinary clothing.

c. Survivors should always be careful about eating and drinking during and after a biological attack. One of the easiest ways for germs to enter the body is through food and water.

30-6. Tips for a Survivor:
a. Keep the body and living quarters clean.

b. Don't neglect preventive medicine. Keep the shot record up to date.

c. Keep alert for signs of biological attack.

d. Keep the nose, mouth, and skin covered.

e. Protect food and water. Bottled or canned foods are safe if sealed. If in doubt, boil the food and water for 10 minutes.

f. Build a shelter, if possible. Shelters should be located and constructed to minimize vector and aerosol access to the survivor; for example, shelter enclosed entrance 90 degrees to the prevailing wind.

g. If traveling, travel crosswind or upwind.

Chapter 31

CHEMICAL CONDITIONS

31-1. Introduction. Chemical agents may be solid particles, liquids, or gases that are toxic (poisonous) chemicals. These agents produce poisonous gas, fire, and smoke. The poisonous agents may produce casualties or irritating effects, and render material or areas unusable. The body is attacked by chemical agents, which produce specific damage depending on the type and concentration exposure of the agent used. Survivors must have a thorough knowledge of how each of these agents affects the body.

31-2. Chemical Groups. Chemical agents are divided into seven groups—nerve, blood, blister, choking, vomiting, incapacitating, and riot control. These agents can be dispersed by artillery shells, mortar shells, rockets, aircraft spraying, and bombs (figure 31-1).

   a. Nerve Agents. The nerve agents are among the deadliest of all chemical agents. They directly affect the nervous system and are highly toxic in both liquid and vapor forms. Examples of G-agents are tabun (GA), sarin (GB), and soman (GD). These nerve agents may be absorbed through any body surface. When dispersed as a vapor, they are absorbed through the respiratory tract or the eyes, but as liquid nerve agents, they can be absorbed through the skin. They are usually quick-acting casualty agents. Symptoms accompanying very small doses are headaches, dizziness, dimmed vision, and nausea. Large doses of nerve agents can interfere with breathing and may cause tightness in the chest, or convulsions, paralysis, and death. Symptoms of large doses of nerve agents are unpredictable, and circulatory collapse can occur without warning. The first effect of eye exposure to agents will probably be a dimming of vision, caused by contraction of the eye pupils to pinpoint size. The pinpoint pupils will more noticeably affect vision in dim light. If the nerve agent contaminates the skin only, the pupils may remain normal or be only slightly reduced in size, with other symptoms being first to appear. The injuries caused by nerve agents may range from mild disability to death, depending on the dose received and the adequacy and speed of self-aid treatment. Nerve agents are odorless, unlike most chemical agents. A survivor must rely on observation of living things and detection devices to identify their presence.

   b. Blood Agents (Cyanides). Blood agents produce their effects by interfering with some vital process within the body. The usual route of entry is inhalation. They prevent the body cells from using oxygen. Hydrocyanic acid (AC) and cyanogen chloride (CK) are the important agents in the group. CK also acts as a choking agent (figure 31-2).

Figure 31-1. Chemical Conditions.

   (1) Symptoms associated with blood-agent contamination vary. One type of blood agent causes a marked increase in the breathing rate; whereas another type

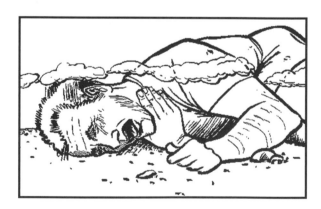

Figure 31-2. Symptoms.

causes a slow breathing rate, a choking effect, and a strong irritating effect. A slight exposure to still another type of blood agent causes headaches and uneasiness.

   (2) Blood agents cause the skin to have a cherry-red color similar to that seen in carbon monoxide poisoning. This symptom, by itself, may help identify the blood agents' poisoning. The symptoms produced by

blood agents also depend on the concentration of the agent and the duration of exposure. If irreversible damage has not occurred, removal from exposure to the agent may enhance recovery. Blood agents are used as quick-acting casualty agents. The speed in donning a protective mask is critical to survival in a blood-agent attack.

c. Blister Agents (Vesicants). Blister agents were developed during World War I to circumvent the protective mask that had made chlorine gas obsolete. These agents are primarily designed to attack the body through the skin and eyes. They can also attack through the respiratory or digestive tracts and cause inflammation, blisters, and general destruction of tissue. Some examples of blister agents are mustard (HD), nitrogen mustards (HN), lewisite (L), and other arsenicals, mixtures of mustards and arsenicals, and phosgene oxime (CX). They are effective even in small quantities, and produce delayed casualties.

(1) A drop of a mustard-type agent the size of a pinhead can produce a blister the size of a quarter. Blister agents are more effective in hot weather than in cold weather. Vapors first affect the moist parts of the body (joints of arms and knees, armpits, and crotch). People who are sweating are especially sensitive to the agents. Blister agents are quickly absorbed through the skin. Reddening of the affected area may appear any time up to about 12 hours after exposure, depending on the degree of contamination and the weather conditions. Blisters may appear in a day or less following the reddening. Healing time varies from about 6 days to as many as 8 weeks in severe cases. Since the damage is done during the first few minutes of exposure, speed in administering self-aid is essential.

(2) Damage to the eyes may be worse than the effects on the skin. Even as a liquid, the agent may only mildly irritate the eyes at first, or there may be no pain at all. In a few hours, however, the eyes hurt, become inflamed, and are sensitive to light. Tears and great pain follow, and permanent injury can result. Some blister agents cause immediate pain in the eyes.

(3) When inhaled, blister agents inflame the throat and windpipe and cause a harsh cough. Cases of serious exposure may result in pneumonia and death. Immediate detection of blister agents and prompt protection against entry into the eyes and lungs and on the skin is vital.

(4) Blister agents may be absorbed by any material (wood, concrete, clothing, metal, plastics, or rubber. Direct skin contact with these objects can cause blistering. Liquid blister agents will eventually penetrate gloves and other garments. Immediate decontamination after exposure is essential to prevent delayed absorption.

d. Choking Agents (Lung Irritants). These agents cause irritation and inflammation of bronchial tubes and lungs, but do not harm the skin or digestive system. They are usually disseminated as gases and inhaled into the body. The best known of these agents is phosgene. During and immediately after exposure, symptoms include coughing, choking, and a feeling of tightness in the chest, nausea, and occasionally vomiting, headache, and crying. If large amounts enter the lungs, they will fill with liquid, causing death from lack of oxygen. A properly operating and well-fitted mask protects against all choking agents.

e. Vomiting Agents. These agents produce strong pepper-like irritation in the upper respiratory tract. Other symptoms include irritation of the eyes and uncontrollable tearing. Symptoms of these agents include a very stuffy nose, severe headache, intense burning in the throat, and tightness and pain in the chest. These are followed by uncontrollable sneezing, coughing, nausea, vomiting, and a general feeling of bodily discomfort. These agents are dispersed as aerosols, and produce their effects by inhalation or by direct action on the eyes. If survivors inhale a vomiting agent before donning their masks, they may become ill after the respirator is on. As long as a vomiting agent is present, however, mask-wear is essential. The mask should be pulled away from the chin during actual vomiting, but not removed. If the survivor has vomited in the mask, caution should be taken to avoid inhaling or ingesting the vomit. Vomiting agents are not considered a major threat because of the comparative ease of protection against them, and their lower toxicity unless used with other agents. Vomiting agents alone seldom result in death (figure 31-3).

Figure 31-3. Avoiding Agents.

f. Incapacitating Agents. An incapacitating agent is any chemical that produces a temporary disabling condition that persists for hours to days after exposure to the agent has ceased (unlike that produced by riot-control agents), and for which medical treatment, while not required, facilitates a more rapid recovery.

(NOTE: Symptoms of riot-control agents may not be distinguished from other lethal agents; therefore, the survivor must be prepared to provide treatment for lethal agents.) In actual usage, the term "incapacitating agent" has come to refer primarily to those agents that:

(1) Produce their effects mainly by altering or disrupting the higher regulatory activity of the central nervous system (figure 31-4).

(2) Last for hours or days rather than the very short duration of riot-control agents.

(3) Do not seriously endanger life (except when large doses are received), and produce no permanent injury.

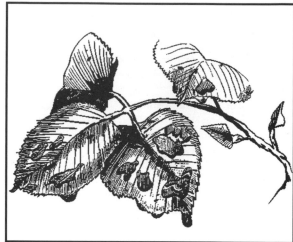

Figure 31-5. Detection.

Figure 31-4. Effects.

g. Riot Control (RC) Agents (Irritant Agents). RC agents are the least poisonous of the seven groups of chemical agents. They act primarily on the eyes, causing intense pain and tearing. Higher concentrations irritate the upper respiratory tract and the skin, and sometimes cause nausea and vomiting. These agents may be dispersed as smoke, or in solutions as droplet aerosols. Although they are used primarily in training and in riot control, some agents may be used in combat. When an unmasked person comes in contact with riot-control agents, the effects are felt almost immediately. The effects begin in 20 to 60 seconds, depending on agent concentration. Duration of effects lasts 5 to 10 minutes after removal to fresh air. There is usually no permanent damage to the eyes. For a short time (minutes), a person may be unable to see. If the mask is used before RC agents enter the eyes, increased protection is afforded.

31-3. Detection:
  a. General Indications. Detection of a chemical agent requires the recognition of evidence gathered by direct or indirect means. Therefore, every survivor must be alert and able to detect any clues indicating chemical warfare is being used. General symptoms of chemical

agents are tears, difficulty breathing, choking, itching, coughing, and dizziness. In the presence of agents that are very hard to detect without the use of detection devices, survivors must watch their fellow aircrew members constantly for symptoms. Additionally, a survivor's surroundings may provide valuable clues to the presence of chemical agents; for example, dead animals, sick people, or people displaying abnormal behavior.

  b. Smell. Survivors cannot rely on the nose as a foolproof means of detecting chemical agents. Although some agents do have a characteristic odor, many others have little or no odor at all. An agent may smell quite differently to different individuals. A mixture of agents will have a different smell than any one agent by itself.

  c. Sight. Since chemical agents are in one of three physical states—solid, liquid, or vapors—the sense of sight may help detect their presence. Most chemical agents in the solid or liquid state have some color. In the vapor state, some chemical agents can be seen as mist or thin fog immediately after bomb or shell bursts. Nerve agents are either a colorless liquid, or a colorless vapor. Although survivors can't see nerve agents, their eyes may help by detecting the methods used to dispense the agents. Mustard gas, unless purified, is dark brown in its liquid state. As a liquid, it is easy to detect and would appear as oily, dark patches on leaves and buildings. However, liquid mustard changes slowly to a colorless gas. As a gas, it is still very toxic, but now the eyes will not be an effective aid to detection (figure 31-5).

  d. Hearing. If survivors know the methods being used by the enemy to spread chemical agents, they can detect the sounds of the enemy's chemical munitions. For example, a bomb filled with an agent would probably cause only a muffled explosion; however, aircrew members untrained in ordnance may have difficulty in making this distinction.

e. Feel and Taste. Irritation in the nose or eyes or on the skin is an urgent warning to protect the body from chemical agents. Additionally, a strange taste in food, water, or cigarettes may serve as a warning that they have been contaminated.

31-4. Protection:

a. Protective Actions. Survivors should use the following steps, listed in the order of importance, to protect themselves from chemical attack:

(1) Use protective equipment.

(2) Give quick and correct self-aid when contaminated.

(3) Avoid the areas where chemical agents exist.

(4) Decontaminate equipment and the body as soon as possible.

b. Equipment. Survivors' masks are as vital to them as lifejackets are to sailors or as parachutes are to fliers. If properly adjusted, they protect the face, eyes, and lungs from chemical agents. Survivors are responsible for proper care of the mask and should inspect the masks frequently to ensure they are free from damage and in good condition. Aircrew members located in areas of potential contamination are issued protective clothing.

c. Self-Aid. Survivors must apply self-aid skillfully and promptly after exposure. Not only is it important for them to know what to do, but they must also know what not to do. It is evident from previous information in this chapter that each type of chemical agent produces certain conditions that require special treatment. However, there are certain essentials of self-aid that, if applied soon enough, give some relief and may prevent serious injury.

(1) Since there are definite time limits after which self-aid becomes useless, immediate self-aid or personal decontamination is all-important if survivors are exposed to liquid nerve or blister agents. Since they may not know whether the contamination is by liquid nerve agent or liquid blister agent, the following procedures are recommended:

(a) Don the mask and clear it.

(b) Contact with thickened (persistent) nerve or mustard agents requires the use of a decontamination kit, or if not available, tear away the contaminated area of clothing and rinse immediately with water.

(c) Rinse contaminated areas with water (removes nerve agents).

(d) If effects of nerve agents become apparent, then and only then, use an antidote, realizing that the antidote provides protection only from nerve agents (GA, GB, GD). It is also incapacitating and is not effective against vector agents.

(2) Use the self-aid procedure given in the following paragraphs for specific agents, if the agent has been identified.

(a) Nerve Agents. The protective mask and hood, if available, must be donned immediately at the first sign of a nerve agent in the air. Stop breathing until the mask is on and the face piece cleared and checked. The mask should be worn constantly until the absence of the nerve agent in the air is indicated, or the individual moves into a clean area (where there are live animals etc.). If, after masking, the survivor has an unexplained runny nose, tightness of the chest, dimness of vision, or breathing difficulty, the use of the antidote should be considered. (Use of the antidote is moderately to severely incapacitating. The survivor should consider the severity of symptoms, availability of the buddy-care system, and requirements for rescue before injecting the antidote.) Exposure to high concentrations of a nerve agent may bring on incoordination, mental confusion, and collapse so rapidly that the survivor cannot perform self-aid. If this happens, a fellow aircrew member must administer first aid. Severe nerve-agent exposure may rapidly cause unconsciousness, muscular paralysis, and breathing stoppage. Any remaining survivors should keep their masks on and move out of the area as soon as possible. The following precautions should be used when applying self-aid for nerve agents:

-1. Antidote should not be used until certain it is needed. Pinpointing of the eye pupils or blurred vision, tightness in the chest, and difficulty in breathing are signs it is needed. If certain nerve agents are inhaled, the antidote counteracts them and makes the survivor feel better.

-2. If survivors have inhaled a very large dose of nerve-agent vapor, they may need more than one injection of the antidote to relieve their symptoms. If the symptoms are steadily becoming worse, and the first injection does not relieve them, or if their mouth does not become dry, it may be necessary to use an extra dose. Inject the second dose in a different muscle. (NOTE: Do not use your own combo pen to inject a victim, use theirs. Additionally, if you find a deceased aircrew member, remove the combo pens from the deceased and take them with you.)

-3. If the difficulty in breathing is not relieved by the second injection, one more dose may be administered. Dryness of the mouth is a good sign. It means that they have had enough antidote to overcome the dangerous effects of the nerve agent.

-4. If a drop or splash of liquid nerve agent gets in the eye, instant action is necessary to avoid serious effects. The eye should be irrigated immediately with water by tilting the head back, and looking straight upward. Slowly pour water into the contaminated eye. Hold the eye open with the finger if necessary. Pour the water slowly so irrigation will last at least 30 seconds. Survivors must irrigate in spite of the danger of breathing nerve-agent vapor. Don the mask quickly after completing the irrigation. The pupil of the contaminated eye should be observed during the next minute, in a mirror if one is available, or by someone nearby. If the pupil rapidly gets smaller, inject antidote intramuscularly at once. If the pupil does not get smaller, antidote is not needed.

-5. If liquid nerve agent gets on the skin or clothing, it should be removed. The liquid should be blotted off the skin with a handkerchief, a piece of cloth torn from the outer clothing, or personal decontaminating kits. Pinch-blotting the liquid won't spread the contamination. Contaminated clothing must be quickly removed, and the skin washed with soap and water. In an emergency, the contaminated portion of the clothing can be cut away, and the contaminated skin area flushed with water. The muscles under the contaminated area should be observed for any signs of twitching. If none develops in the next half an hour, and the survivor has no tightness in the chest, the decontamination was successful. If twitching of the muscles under the area of contaminated skin does develop, the antidote should be administered at once.

-6. Food and water that may be contaminated with nerve agents must be avoided. If a survivor has swallowed contaminated food or water and all of these symptoms occur—nausea, pains in the stomach, increased flow of saliva, and tightness in the chest—the antidote must be administered.

(b) Blood Agents. If, during any chemical attack, a sudden stimulation of breathing, or an odor like bitter almonds is noticed, the survivor must don the mask as quickly as possible. Speed is absolutely essential; this agent acts so rapidly that within a few seconds its effects will make it impossible for survivors to don the mask by themselves. The breath should be held until the mask is on the face, if at all possible. This may be very difficult since the agent strongly stimulates respiration.

(c) Blister Agents. The protective mask, hood, and clothing must be worn when liquid or vaporized blister agents are known to be present. There are two groups of blister agents, one called mustards and the other called arsenicals. Self-aid against mustards and arsenicals is the same. A liquid mustard in the eye will not hurt immediately. A liquid arsenical in the eye will sting and hurt severely.

-1. To remove a liquid blister agent from the eye, the eye is irrigated using the same procedure as for removing nerve agents. Speed in decontaminating the eye is absolutely essential. The self-aid procedure is very effective for mustard within the first few seconds after exposure, but is of little value after 2 minutes.

-2. Generally, for any liquid blister agent on the skin, the survivor should:

-a. Pinch-blot to prevent the liquid from spreading, using cloth or any other absorbent material at

hand. The used cloth or absorbent material should then be discarded.

-b. Scrub the skin with soap and water, and rinse thoroughly with clean water. When scrubbing, special attention must be paid to areas not covered by clothing (neck and ears).

-3. Survivors must decontaminate or remove any clothing that is contaminated with a liquid blister agent. Small areas on the clothing can be decontaminated by using soap and water. The contaminated parts of the clothing can also be cut out, thus making the clothing safe to wear.

-4. When mustard vapor is detected, a survivor must put on the mask and leave it on until clear of the area. There is no decontamination procedure of any value when a mustard vapor agent is used. The damage is done as soon as the mustard vapor strikes the eyes, although the full extent of the injury may not appear for several hours.

(d) Choking Agents. The protective mask should be donned immediately upon detection of any choking agents in the air by odor (like cut green corn or grass), irritation of the eyes, or change in the taste of a cigarette (smoking may become tasteless or offensive in taste). Survivors should hold their breath while masking. If an agent has been inhaled, normal survival duties should be continued unless there is difficulty in breathing, nausea and vomiting, or more than the usual shortness of breath from exertion. If any of the above symptoms occur, survivors should rest.

(e) Incapacitating Agents. The mask should be donned immediately. Complete cleaning of the skin with soap and water should be done at the earliest opportunity. The eyes should be flushed with clear water only. Survivors should shake or brush clothing, and when conditions permit, change clothing and thoroughly wash the contaminated clothing.

(f) Vomiting Agents. The protective mask must be worn in spite of coughing, sneezing, salivation, and nausea. The mask can be lifted from the face briefly if necessary to permit vomiting or to drain saliva from the face piece. Carrying on duties as vigorously as possible will help to lessen and to shorten the symptoms. Survival duties can usually be performed despite the effects of vomiting agents.

(g) Riot Control Agents. After the protective mask has been donned and cleared, the eyes should be kept open as much as possible. When vision clears, activities can continue. The eyes should not be rubbed. If drops or particles have entered the eyes, the eyes can be flushed with water. Chest discomfort can usually be relieved merely by talking.

31-5. Avoiding Chemical Agents. If survivors are hit by a chemical attack, they may have to remain in a contaminated area. After an attack, they should not expose themselves to other enemy weapons, and must seek areas that are less contaminated. If the attack is on a very small scale, they might seek an upwind area. Depending on the area and weather conditions, crosswind movement may be best. Chemical-agent attacks may cover too large an area to permit area avoidance. Selecting routes on high ground may be advisable because gas is usually heavier than air and tends to settle in low places. Cellars, trenches, gullies, and valleys are examples of places to avoid when possible. Woods, tall grass, and bushes tend to hold chemical agent vapors. (NOTE: Survivors have a better chance to avoid chemicals if they are familiar with the attack areas, and if they know their personal location.)

31-6. Decontamination of Chemical Agents.

Decontamination is removing, neutralizing, or destroying the agents. The purpose of personal decontamination is to remove agents from the body or personal equipment before serious injury occurs. An example of decontamination by removal is pinch-blotting the chemical agent from the skin. Neutralization makes the agent harmless. The contaminated cloth could also be buried. Common sense and quick thinking play a big role in personal decontamination. Survivors may have to rely on whatever they have at hand to remove chemical agents from the skin, eyes, or equipment. If liquid nerve or blister agents touch any part of the body, they must be removed as quickly as possible. If survivors are caught without soap and water, then anything that can dilute or remove the agents will have to be used; it may be mud, dirt, or urine. A crude remover may remove only two-thirds of the agent, but it is far better than leaving the agent in full concentration. It must be remembered that nerve and blister agents penetrate very rapidly. (NOTE: Use a scraping action to avoid pressing the agent into the skin.)

a. Soap is excellent for removing chemical agents. Cold water, while not as good as warm water, does dilute or weaken chemical agents. Hot, soapy water removes agents quickly. If the operational situation permits, a bath or shower should be taken. The mask should be left on until after survivors have washed their hair and thoroughly scrubbed themselves, while avoiding wetting the canister or cheek pads. Exposed areas and hairy regions of the body should be given extra attention.

b. When clothes have been exposed only to chemical agent vapors, airing usually decontaminates them (with the exception of mustard vapors, which will absorb into the garment and require washing for removal). If chemical-agent droplets or liquid splashes are present, survivors will need detergent or soap and water. Wool clothes are best washed in soapy, lukewarm water. Cotton clothes can be boiled.

c. Boots or shoes can be scrubbed with soap and water and rinsed at least twice. If the survivor's choice is wearing contaminated clothes and shoes or nothing, decontamination of the material must be done the best way possible. Almost any effort will help the survival situation.

31-7. Tips for a Survivor:

a. Keep the body and living area clean.

b. Keep the nose, mouth, and skin covered.

c. If a protective mask is needed but unavailable, improvise. The charcoal cloth from the CD suit or undergarment makes a moderately effective mask for short-term agent exposure. The use of the aircrew helmet, visor, and charcoal mask may provide a higher level of protection for the eyes and respiratory system.

d. Build a shelter or rest area in a clearing away from vegetation. Decontaminate the ground by removing the topsoil. Keep the entrance closed and 90 degrees to the prevailing wind.

e. Do not use wood or vegetation from a contaminated area for a fire.

f. Look at the area around a water source and check for foreign odors (garlic, mustard, geranium, bitter almond), oily spots on the surface or nearby, and the presence of dead fish and animals. If they are present, do not use the water.

g. Keep food and water protected. Bottled or canned foods and water are safe if sealed, and the cans are decontaminated before opening.

h. If possible, obtain water from a closed source, precipitation (if there is no evidence of agent vapor in the air), or from a slow-moving stream after it has been filtered.

i. Do not use plants for food, or water from a contaminated area.

j. Do not use sick animals as a food source. When skinning animals in a contaminated area, use protective clothing (gloves).

k. If traveling, travel crosswind or upwind.

BY ORDER OF THE SECRETARY OF THE AIR FORCE

OFFICIAL

CHARLES A. GABRIEL, General, USAF
Chief of Staff

JAMES H. DELANEY, Colonel, USAF
Director of Administration

SUMMARY OF CHANGES

This regulation [originally dated 15 July 1985, supersedes AFM 64-3, 15 August 1969 and] has been rewritten to provide sufficient information for aircrew members, and instructors who teach survival training. The information has been changed from environmental areas to subject-matter sections. Color illustrations have been added to provide realistic depiction of navigational aids, climatic areas, and fauna in its appropriate environment. Information has been expanded in all areas of the regulation. The following new information has been added: Weather, Geographic Principles, Chemical, and Biological Conditions.

Appendix: Poisonous Plants and Animals

The following is a guide to poisonous or venomous plants and animals that might be encountered in nature. It is by no means an exhaustive list and does not include snakes or mushrooms, which, as a general rule, should be avoided (for more information on sustenance see *Chapter 18*).

The animal category is almost entirely made up of sea-life because these are a less-obvious and often less-familiar source of danger than are other members of the animal kingdom. The sea-life section is divided into vertebrates and invertebrates.

Within the plant section, there are two types: those that cause external skin irritation and those that cause poisoning internally. PLEASE NOTE: If you are at all in doubt as to a plant's identity, DON'T EAT IT! unless you have first determined the plant's edibility by using the guidelines on page 235.

ANIMALS

Venomous Fish—Vertebrates (Fish That Sting)

BLUE-SPOTTED FANTAIL RAY, A.K.A. BLUE-SPOTTED RIBBONTAIL RAY *(Taeniura lymma)*

- Disc shape is an elongated oval with broadly rounded corners
- Width of disc does not normally exceed one foot
- Highly colorful, with bright blue spots covering the upper side of disc
- Underside of disc is whitish
- Rounded snout, sometimes raised to lure prey underneath
- Thick, tapering tail is longer than body and has bright blue side stripes
- Venomous spine located roughly halfway back on tail

Warning:
Like other stingrays, the blue-spotted stingray, when stepped on, can pierce the skin of an intruder with the spine on its tail and inject a painful venom. The sting is unlikely to cause long-term injury to humans.

Habitat and Range:
Often found in shallow sandy waters during high tides, the blue-spotted stingray retreats to the shelter of coral reefs during low tide. It ranges through the Western Pacific and Indian oceans, from southern Japan to northern Australia, and from the Red Sea to southeastern Africa.

BROWN BULLHEAD CATFISH
(*Ameiurus nebulosus*)

- Adults can attain one and one-half feet in length and weigh five and one-half pounds
- Elongated but full, heavy body, rounding towards the rear
- Medium brown to black on its back, with sides somewhat lighter
- Speckled with brownish patches on its upper parts; whitish on underside
- Wide, flattened mouth with slight overhang to upper jaw
- Four pairs of sensitive barbels project from its face, those on the upper jaw being especially long
- Pectoral fin has a spine with serrations on the rear edge
- Lengthy anal fin with twenty or more rays

Warning:
Many catfish, including the brown bullhead, have spines near their dorsal and pectoral fins connected to small poison glands. Often the pain caused by the sharp spine is worse than that of the fairly mild poison, which is not life threatening to humans.

Habitat and Range:
The brown bullhead's natural range covers much of eastern North America, as far north and west as Saskatchewan, though it has been widely introduced elsewhere. It prefers the bottoms of deep, still pools with submerged vegetation.

GULF TOADFISH
(*Opsanus beta*)

- Adult length between seven and ten inches
- Dull coloring provides camouflage on ocean floor
- Large flattened head
- Wide mouth with fleshy lips and blunt teeth
- Fanlike pectoral fins
- Dark brown bars on rounded caudal fin
- Roughly twenty-six dorsal fin rays with poisonous spines

Warning:
The dorsal spines of the toadfish are extremely sharp and poisonous. Since the fish often bury themselves in the sand, they can easily be stepped on. Ingesting the flesh of the gulf toadfish can also result in serious poisoning.

Habitat and Range:
The gulf toadfish is common in coastal shallows, bays, and lagoons throughout the Caribbean, as well as the coast of Florida and the Bahamas. It often lurks among seagrass or on rocky surfaces.

RABBITFISH *(Siganus spp.)*

- Adults are nine to twelve inches in length
- Gently curving form from head to tail is moderately elongated
- Rabbitlike facial features, including mouth and large dark eyes
- Dorsal fins make a prominent ridge with thirteen spines
- Rabbitfish vary widely in coloration; some can change color and pattern to blend in with surroundings
- Narrow, irregular dark bands cover the scales and continue onto fins

Warning:
The dorsal fin spines of the rabbitfish are very sharp and can inject venom into intruders, including humans who accidentally step on the fish in the sand. The wounds can be intensely painful and slow to heal.

Habitat and Range:
Rabbitfish are reef-dwelling creatures that hide among coral and rocks in the tropical waters of the Pacific and Indian oceans. Concentrations of rabbitfish can be found in the Great Barrier Reef and in reefs and lagoons in New Guinea, Indonesia, and the western Philippines.

RATFISH *(Hydrolagus colliei)*

- Maximum length is about three feet
- Elongated body tapers to a long, thin caudal fin
- Head is large relative to its body, with a rounded "rat-like" snout
- Upper body is a metallic bronze color with many white spots
- Lower parts are silvery; eyes are large and green
- Tall, spiky spine at the base of the front dorsal fin

Warning:
Although edible, ratfish are unappetizing to taste and are potentially dangerous to handle. Their teeth are very sharp, as are the retractable claspers in front of their pelvic fins. The first dorsal fin is preceded by a long venomous spine, which can cause sharp pain in humans.

Habitat and Range:
Ratfish live in cool eastern Pacific waters from Baja California to southern Alaska. They dwell most often on the ocean floor, ranging from coastal shallows to deep sea.

SMOOTH STARGAZER
(*Kathetostoma averruncus*)

- Maximum length is slightly more than one foot
- Characterized by enormous head and wide mouth used to gulp prey
- Eyes are placed on top of head, watching for prey as fish lies motionless
- Body is flattened and tapers to a laterally flattened tail
- Upper surface is brownish-gray with small white spots; underside is whitish
- Venomous spine above pectoral fin

Warning:

Stargazers can inject a mild venom from the spines located above their pectoral fins. Human divers may be injured if they step on or attempt to handle the fish.

Habitat and Range:

Smooth stargazers are bottom-dwellers of moderately deep waters. They inhabit the eastern Pacific from California to Peru and the Galapagos Islands.

SOUTHERN STINGRAY
(*Dasyatis americana*)

- Maximum width of disc (from "wing tip to wing tip") is about five feet
- Body is roughly rhombic in shape, with rounded outer corners
- Top surface is light brown, greenish, or gray, changing with surroundings
- Bottom surface is mostly white with darker edges
- No dorsal fin on flattened upper surface
- Long, tapered tail resembling a whip can reach twice the length of the rest of its body
- Tail has a dark ridge along its length and a sharp spine near its base

Warning:

If stepped on, the stingray will lash its poisonous spine into the intruder, usually striking the lower leg of a human wader. Shuffling one's feet underwater warns the rays of one's approach; they normally swim harmlessly away. The injury from a stingray's spine is characterized by intense pain and respiratory difficulty, and can occasionally prove life-threatening without immediate medical treatment.

Habitat and Range:

Shallow soft-bottomed waters of bays and coastal ocean from Brazil to the American Mid-Atlantic, as well as the Gulf of Mexico and Caribbean islands.

Spiny dogfish
(*Squalus acanthias*)

- Slim, elongated form, with an adult length of three to four feet
- Narrow, pointed snout characteristic of the shark family
- Grey upper body with white underside
- White spots on back and flanks are typical, though not always present
- Two dorsal fins are each preceded by a sharp spine
- Forked tail is extended on top half, culminating in a blunt tip

Warning:

The spines preceding each dorsal fin release a toxin used to ward off attackers. Although ordinarily harmless to humans in the wild, a spiny dogfish when caught may whip its tail around to inflict wounds with its spines, causing mild irritations and occasionally strong allergic reactions.

Habitat and Range:

Spiny dogfish are abundant in temperate and subarctic waters of the North Atlantic and North Pacific. Some specimens have been found as far south as Argentina, as well as in the Black and Mediterranean seas.

Surgeonfish
(*Zebrasoma spp.*)

- Averages eight to ten inches in length
- Body is tall and slender with a fanning tail
- Head slopes down to a small beaklike mouth
- Numerous small teeth for feeding on algae
- Often brightly colored, with a covering of small dark spots
- Scalpel-like spines (hence the fish's name) at the base of the tail

Warning:

Spines alongside the tail, originating at the base, are extremely sharp and release a very painful venom. The fish are nonaggressive and unlikely to cause injury unless handled. Ingesting the flesh of the surgeonfish may also cause ciguatera poisoning.

Habitat and Range:

Surgeonfish prefer shallower waters near abundant sources of food and shelter, especially coral reefs and rocky sea bottoms. They are native to warmer waters of the Atlantic, the Gulf of Mexico, the Pacific, and the Indian Ocean.

YELLOW STINGRAY
(Urolophus jamaicensis)

- Average size is eight to fifteen inches
- Body form is a flattened disc with strongly rounded corners
- Top surface is yellow-brown, with dark and pale markings forming a variety of patterns
- Bottom surface is yellowish, off-white, or pale green
- No dorsal fin, but a prominent caudal fin at the tip of the tail
- Tail is thick and shorter than length of body disc
- Serrated spine located far out on the tail, near the base of the caudal fin

Warning:
The spine on the yellow stingray's tail cuts into flesh and releases a potent venom if the animal is stepped on or handled. Its sting is rarely lethal but can cause severe pain.

Habitat and Range:
The yellow stingray inhabits shallow tropical and subtropical waters, especially in sandy areas around reefs, from the coast of North Carolina to northern South America and throughout the Caribbean.

Poisonous and Venomous Marine Animals—Invertebrates

EGG-YOLK JELLY
(Phacellophora camtschatica)

- Common name derives from yellow mass of gonad tissue surrounded by whitish and transparent parts of bell
- Bell can reach two feet in diameter
- Bell does not hold shape well and is easily contorted by water currents
- Sixteen clusters of tentacles extend from edges of bell
- Tentacles can number in the hundreds and reach twenty feet in length
- Oral arms beneath bell are folded and fairly short
- Tentacles and oral arms are sticky to trap other jellies as prey

Warning:
The tentacles of egg-yolk jellies contain thousands of stinging cells that release a relatively mild toxin. Casual contact can produce burning pain and possibly allergic reactions, but is not life threatening.

Habitat and Range:
The egg-yolk jelly moves about through open ocean in all the world's temperate regions.

HAIRY JELLYFISH, A.K.A. SEA BLUBBER, LION'S MANE JELLYFISH (*Cyanea capillata*)

- Bell is flattened or plate-like
- Bell of cold-water Atlantic variety can reach eight feet in diameter
- Up to one foot across in Australian Pacific variety
- Younger specimens are pink or red; older ones become brownish purple
- Delicate hairy tentacles can number in the thousands
- Tentacles of Atlantic variety can attain lengths of two hundred feet
- Tentacles contain perhaps millions of nematocysts, stinging cells which utilize a coiled harpoon mechanism to inject venom upon contact

Warning:
Stings from the tentacles of hairy jellyfish can cause intense pain lasting for an hour or more, and there are reports of severe stings being fatal. Other effects may include respiratory trouble, nausea, sweating, and muscle cramps.

Habitat and Range:
Hairy jellyfish are widely distributed through the Atlantic and Pacific, from the tropics to the Arctic.

PORTUGUESE MAN-O-WAR (*Physalia physalis*)

- Not considered a true jellyfish, a man-o-war actually comprises a colony of several symbiotic organisms
- The float is bluish purple and symmetrical around a central axis
- Float is filled with a light gas for buoyancy
- Float length is typically five to twelve inches
- Long central "feeding" tentacle is surrounded by more delicate tentacles
- Tentacles can reach sixty-five feet in length
- Feeding polyps, separate but dependent organisms, hang beneath float

Warning:
The immensely long tentacles of the Portuguese man-o-war are covered with stinging cells and can inflict extremely painful wounds. The stings can be cause temporary muscle paralysis but are rarely fatal.

Habitat and Range:
The Portuguese man-o-war is most abundant in tropical regions of the Atlantic and Pacific. Currents can carry it as far north as Europe and as far south as Australia.

SEA NETTLES
(Chrysaora fuscescens)

- Rounded bell of west coast sea nettle grows to roughly one foot in diameter
- Bell has a reddish brown or golden hue
- Twenty-four tentacles extend from bell
- Tentacles may stretch for more than eight feet
- Tentacles are very thin and dark red in color; oral arms are whitish
- East coast sea nettles grow to half the size and lack the reddish color

Warning:

Both Pacific and Atlantic sea nettles are common along shorelines, sometimes swarming together in groups of several thousand. This leads to frequent run-ins with unfortunate bathers, who can suffer very painful stings. The toxin is mild, however, and unlikely to cause serious harm.

Habitat and Range:

Sea nettles are open-sea swimmers of the temperate Pacific and Atlantic oceans. They frequently congregate in waters close to shore.

STINGING OR FIRE CORAL
(Millepora complanata)

- Huge colonies comprising thousands of individuals form on or near reefs
- Individual animals are three-quarters to one and a quarter inches in length
- Tough calcareous external skeleton
- Fire corals are very diverse in shape: can be plate-, lobe-, or antlerlike
- Most are brownish tan, due to the coloration of a symbiotic algae, zooxanthellae, living inside their tissues
- White hairlike tentacles are extended for feeding during the day

Warning:

Fire coral animals have nematocysts, or stinging cells, used to stun prey and to ward off predators. Contact with the coral, usually by scuba divers, can be very painful and cause an irritating rash.

Habitat and Range:

Fire coral build their colonies along reefs and rock surfaces in all warm-water regions of the world's oceans.

PLANTS

External (Can cause primarily skin irritation and/or rash)

POISON IVY
(Toxicondendron radicans)

- May grow self-standing, or as a climbing or trailing shrub of variable size
- Leaves are two to four inches long, glossy or dull green, and grow in clusters of three
- Clusters of small yellowish flowers appear in late spring and summer
- White berrylike fruits (up to one-quarter inch wide) appear from late summer into winter
- Older stems acquire a hairy look due to accumulation of root fibers

Warning.
Oil found throughout the poison ivy plant contains the allergen urushiol, which can cause skin rash, inflammation, blisters, and itching upon skin contact or inhalation of burning plants. The berries are also poisonous to eat.

Habitat and Range:
Poison ivy is widely distributed through the East, growing in thickets and open woodland and along roadsides and fences. It is found from Nova Scotia to Florida, and westward as far as Texas and Minnesota.

POISON OAK *(Toxicondendron diversiloba)*

- Eastern U.S. variety grows as a small shrub; western variety can grow as a shrub (sometimes quite large) or as a vine crawling over other plants
- Leaves grow in clusters of three, like poison ivy, with the center leaf on a short stalk
- Leaves are rounded with blunt serrations, and light to medium green in color
- Loose bunches of small greenish flowers appear in late spring
- In summer and fall, female specimens produce small clusters of white fruits, each encased in a paperlike outer sheath
- Cut stems exude a sticky liquid which turns into a black, shiny lacquer

Warning:
Canals in the stems of poison oak contain an allergen called urushiol, which is not present on the surface of poison oak unless it has oozed out due to breakage or animal bites. Urushiol can affect humans by contact with the skin or through inhaling particles released by burning plants. Poison oak dermatitis is marked by redness, blistering, and severe itch that can last for several days or even weeks. Inhaled particles can cause inflammation of the respiratory mucous membranes.

Habitat and Range:
Western poison oak occurs from the Rockies to the Pacific coast. In valleys and shaded forests it commonly takes the form of a clinging vine, while in open coastal scrubland it forms dense thickets in shrub form. Eastern poison oak flourishes in sandy soil of the southeastern United States.

Poison sumac *(Toxicondedron vernix)*

- Rounded shrub or small tree attaining a height of twenty-five feet and a diameter of six inches
- Pinnate leaf clusters, seven to twelve inches long, are paired except at the end
- Oval or elliptical leaves, about three inches long, lack teeth and culminate in a point
- Leaves are shiny and dark green on upper surface, pale and fuzzy on underside, but turn brilliant red or orange in autumn
- Bark generally smooth and gray to black in color
- Very small greenish flowers, with five petals, appear in long clusters in late spring/early summer
- Waxy white berries, one-quarter inch in diameter, appear in autumn and winter; they are edible and prized by birds and other species

Warning:

Like poison oak and poison ivy, poison sumac has reserves of allergenic oil beneath its bark. When released to the surface, this toxic urushiol turns black and can adhere to human skin or clothing brushing against it. Contact with skin can cause severe skin rash, inflammation, and itching.

Habitat and Range:

Poison sumac prefers damp shady areas, such as swamps, flood lands, and sheltered hardwood forests. It can be found throughout much of the eastern United States and southern Canada, but most commonly along the coasts of the Atlantic, the Gulf of Mexico, and the Great Lakes.

Internal (Can cause internal poisoning)

Belladonna or deadly nightshade *(Atropa belladonna)*

- Shrublike, many-branched herb averaging three feet tall
- Smooth, purple-green stems
- Dark green leaves are smooth and oval-shaped, with conspicuous veins
- Bell-shaped, drooping purple blossoms with five petals
- Flowers one to one and one-half inches in length
- Shiny black fruit resembling a cherry

Warning:

All parts of the belladonna contain poisons, primarily the alkaloid atropine. Ingestion can cause dizziness, nausea, visual and respiratory impairments, increased heart rate, and nervousness turning to intense sleepiness. In some cases, eating the plant can be fatal to humans.

Habitat and Range:

Belladonna originated in moist regions of southern Europe and Asia, but has become widespread in the Americas as an ornamental or landscaping plant.

BLEEDING HEART
(*Dicentra spp.*)

- White to rose-pink heart-shaped flowers with four petals
- Two outer petals are fused at the base and free at the ends and form the heart shape, while two inner petals protrude slightly suggesting a drop of blood
- Flowers hang on an elongated, leafless, slender stalk
- Leaves are pinnately divided and fernlike
- Fruit is a capsule with many seeds

Warning:
The leaves and roots of the bleeding heart plant are poisonous if ingested in large quantities. Repeated contact with the sap can cause minor skin irritation.

Habitat and Range:
Bleeding heart is found in damp, shady areas and in dry or moist woods. It is widely cultivated as a house or garden plant. Members of the genus Dicentra are native to Asia and, in North America, occur from British Columbia to central California, from Michigan east to Massachusetts, south to Georgia, and northwest to Tennessee and Illinois.

COMMON VETCH (*Vicia sativa*)

- Leaves are pinnately compound (arranged oppositely from each other) with eight to twelve leaflets, each one half to one and one-half inches long, on a midrib that extends to a terminal, snaky tendril
- Leaflets are oblong to elliptic, either smooth or with some short hairs
- Stems climb on other plants or trail on the ground and are either smooth or have some short hairs
- Flowers measure roughly one inch long and range from pinkish to violet to bluish pea-flowers with a five-toothed calyx; grow in a cluster of one to three on a short stalk at the leaf axils
- Fruit is a one- to two-chambered flat pod measuring two or three inches long and containing one to several seeds

Warning:
The seeds of some varieties of the genus Vicia contain levels of cyanide that are toxic to humans when ingested.

Habitat and Range:
Vetches are commonly found in open, disturbed areas, on roadsides, and in pastures. Common vetch is native to Europe but is now distributed over temperate regions of the Northern Hemisphere and South America. It is also commonly cultivated in North America as a cover crop.

Corn cockle (*Agrostemma githago*)

- Thin, stiffly upright plant averages two feet in height
- Forked branches covered with fine hair
- Light green leaves are two to six inches long
- Leaves grow oppositely and erect against stem
- Pink to reddish purple flowers, one to two inches wide, appear in summer
- Flower has five sepals, each one inch long with a narrow tip, and five petals shorter than sepals
- Capsule-like fruit containing many seeds

Warning:
The seeds of the corn cockle contain toxic substances called saponins, which can poison humans when eaten directly or when transported by breezes onto other plants. Children are most seriously affected due to their small size, though ingestion can cause pain and nausea in adults as well.

Habitat and Range:
Widely distributed through the United States and southern Canada, corn cockle is found in fields, along roads, and in disturbed areas such as waste dumps.

Daphne (*Daphne spp.*)

- A low, deciduous or evergreen shrub
- Leaves are alternate, smooth-edged and lance-shaped and grow in clusters at the ends of the stems
- Very fragrant flowers may be lavender to rosy-purple to white, or, depending on the species, pale greenish yellow; flowers grow in clusters of two to three and develop before the leaves
- Fruit is an oval-shaped drupe with a single seed that grows in clusters
- Berries may be scarlet to yellow or bluish black, depending on the species

Warning:
The fruits and leaves of the genus Daphne are highly toxic and may be fatal if ingested. Contact with the leaves may result in minor skin irritation with blisters.

Habitat and Range:
Daphne grows on plains and mountains up to sixty-five hundred feet. It is native to Near East Asia and is widely cultivated as an ornamental shrub.

DEATH CAMAS (*Zigadenus elegans*)

- Grows to a maximum of fifteen inches tall
- Grasslike leaves in groups of three
- Grows from a deeply buried bulb, similar to an onion, but with no onion scent
- Yellowish-white flowers grow on stalks above the leaves
- Flower petals have a green, heart-shaped mark

Warning:
Eating the bulbs of the death camas can be fatal, but all parts of the plant—including the pollen—are poisonous if ingested.

Habitat and Range:
Death camas is found in wet, open, sunny areas, although some species favor dry, rocky slopes. They are common in the western United States. Some species are found in the eastern states and in parts the subarctic regions of North America and eastern Siberia.

DELPHINIUM AND LARKSPUR (*Delphinium spp.*)

- Height ranges from one to three feet
- Leaves alternate and are palmately divided and lobed or toothed; they may be narrow or broad, depending on the species
- Spurred flowers are purple to blue or white
- Flowers grow in terminal clusters on a spire
- Fruit is dry, with many seeds

Warning:
All parts of the plants of the genus Delphinium are poisonous if ingested. Symptoms include a burning sensation on the lips and mouth, followed by intense vomiting and diarrhea, muscular weakness and spasms, weak pulse, paralysis of the respiratory system, and convulsions.

Habitat and Range:
Larkspurs are divided into two groups, based on their height and habitat: Tall larkspurs and low larkspurs. Tall larkspurs are found primarily at high altitudes (above six thousand feet) in moist areas, such as mountain meadows, stream banks, and near or under stands of aspen. Low larkspurs grow at lower elevations in varied habitats ranging from dry or moist sagebrush areas to dry, grassy hillsides or open meadows. Larkspurs range throughout western North America. Many species of larkspur are cultivated as ornamental plants.

DOGBANE *(Apocynum spp.)*

- One to four feet tall, with bushy, spreading branches
- Smooth, oval, bluish-green leaves, two to four inches long
- Clustered bell-shaped flowers about one-third inch wide each
- Flowers are fragrant and pink or white; darker hued inside
- Milky toxic fluid oozes from broken or bitten stems and leaves
- Two thin, hanging seed pods, three to eight inches in length
- Seeds end in a hairy tuft

Warning:

Toxins in the stems and leaves of dogbane can be fatal for animals and, in some cases, humans to ingest. Children are most at risk because of their smaller size, but adults may also suffer skin blisters from contact with the sap or severe sickness from eating the plant.

Habitat and Range:

Dogbane is present throughout much of North America, from southern Canada (east and west) to the southeastern United States and California. It thrives in moist temperate areas, such as fields, roadsides, brushy thickets, and the edges of woods.

ERGOT
(Claviceps purpurea)

- Parasitic fungus of cereal crops and grasses that congregates on seed heads and flowers of host species
- Ergot "grains" form on infected hosts, looking like greatly enlarged discolored seeds
- Grains are purplish to black
- Tough, hornlike outer surfaces
- Sticky oozing substance called "honeydew" containing many ergot spores, may form on infected plants

Warning:

Eating ergot-infected grains, or even animals that have eaten them, can lead to severe illness and death. Ergot fungi contain toxic alkaloids produced by purple spores called sclerotia. The toxins in them can cause intense burning pain, convulsions, hallucinations, and miscarriage.

Habitat and Range:

Ergot appears worldwide in regions with ample grasses or grains, including open fields, roadsides, meadows, and woodlands.

FOXGLOVE (*Digitalis purpurea*)

- Plants grow to be two to seven feet tall
- Numerous tubular, deep purple to white flowers, about two inches long
- Flowers arranged in elongated clusters on a spike
- Flowers hang down and have conspicuous reddish speckles in their interiors
- Lance- or egg-shaped leaves are soft, fuzzy, and toothed, and grow in a basal rosette

Warning:

All parts of the foxglove plant are poisonous if ingested. The upper leaves tend to be more toxic than the lower leaves, and it is most poisonous just before the seeds ripen. Symptoms include stomach pain (vomiting and diarrhea), headache, erratic pulse, and convulsions.

Habitat and Range:

Foxglove is a native of Europe, but has been naturalized in western North America, occurring from British Columbia south to central California. It is also widely cultivated as a garden plant. Foxglove needs little soil to survive and is found in rocky areas or in cracks in walls. It is also occurs along roadsides and on wooded slopes.

GREAT LOBELIA AND CARDINAL FLOWER (*Lobelia spp.*)

- Heights range from one to four feet
- Erect leafy stems, often in clusters
- Thin, lance-like leaves, two to six inches long, often toothed (less so, if at all, in great lobelia)
- Great lobelia has one-inch-long bell-shaped blue flowers with white stripes, and hairy calyx
- Cardinal flower has bright red flowers, one inch long, with five petals in the shape of slender spires
- Stamens in both species form a tube around the style, extending beyond the corolla in the cardinal flower

Warning:

Many bellflowers of the genus *Lobelia*, including the cardinal flower, great lobelia, and Indian tobacco, contain poisons in their leaves, seeds, and roots. Ingesting any part of these plants can cause severe sickness, vomiting, and, in some cases, death.

Habitat and Range:

Lobelias grow in wet soil along streams, ponds, shores, and in swamps. They are also cultivated as ornamental garden plants. The great lobelia is native to the eastern half of the United States, from the highlands of the southeast to western New England, and as far west as Kansas. The cardinal flower is found throughout the eastern United States, the southwest as far as California, and northward to southeastern Canada.

HORSE CHESTNUT AND BUCKEYE (*Aesculus spp.*)

- Heights range from twenty-five to ninety feet
- Rounded canopy of sturdy, spreading branches
- Leaf clusters contain five to seven leaves arranged palmately
- Leaves are dark green above, paler underneath
- Leaves three to ten inches long, fingerlike, saw-toothed, short-stalked or stalkless
- Gray or brown bark is thin, usually smooth
- Clusters of bell-shaped white, yellow, or pink flowers, about one inch long, with four or five petals
- Tough brown seed cases, about two inches long, are spiny in the horse chestnut, smooth in most buckeyes
- Seed cases contain one or two (occasionally three) large, glossy brown seeds

Warning:

The shiny brown seeds common to species of the genus *Aesculus* are poisonous to eat, and can be fatal to both animals and humans. The bark and young leaves can also be poisonous to ingest.

Range and Habitat:

Horse chestnuts and buckeyes thrive in rich moist soil, often in dense woodland. With the exception of the California buckeye, they are natives of the eastern United States, extending as far west as Texas and northward to southern Canada. The horse chestnut is a European species introduced as a shade tree and now widely distributed in eastern North America.

JIMSONWEED, DOWNY THORNAPPLE, DEVIL'S TRUMPET, ANGEL'S TRUMPET, STINKWEED (*Datura spp.*)

- Height ranges from one to five feet
- Stem is greenish or purplish
- Large white to violet flowers, which are five-pointed, trumpet-shaped corollas measuring three or four inches wide and set on short stalks
- Calyx is tubular and green, about half the length of the corolla
- Leaves are ovate, irregularly toothed, and grow three to eight inches long; they emit an unpleasant odor when touched
- Fruit is a prickly, egg-shaped capsule about two inches in diameter that splits into four segments when ripe

Warning:

All parts of the jimsonweed are toxic if a large quantity is ingested, including the nectar. Symptoms include flushed skin, hallucinations, pupil dilation, headache, and possibly coma. Skin irritation may result from touching the leaves or flowers.

Habitat and Range:

Jimsonweed often grows as a weed in cultivated fields and is common in barnyards, pastures, wasteland, and along roadsides. It is found worldwide.

LILY OF THE VALLEY
(*Convallaria majalis*)

- Adult height averages eleven or twelve inches
- Two four-inch long leaves grow from low on stem
- Leaves are lance-shaped and smooth, with veins extending from base to tip
- Cluster of flowers on upper part of erect central stem
- Blossoms are small, white, and fragrant
- Bright red strawberry-like fruit

Warning:

The various parts of the lily of the valley, or May lily, contain an array of poisons, including glycosides and saponins. Ingestion of any part can cause nausea and vomiting, headaches, slowed pulse, and urination problems.

Habitat and Range:

A European native, the lily of the valley has been widely cultivated for its beautiful flowers, and is now at home in wet woodland areas throughout North America and Asia, as well as Europe.

MAYAPPLE OR MANDRAKE
(*Podophyllum peltatum*)

- Short-stalked herb often appearing in clusters of many plants
- Umbrella of five to nine leaves radiating from central stem
- Leaves are lobed and two to four inches long
- Single white flower with up to nine petals and six sepals
- Large, yellow berry
- Flower and berry emerge only slightly from fork between leaves

Warning:

All parts of the mayapple contain a mixture of toxins called podophyllin. The berry is only mildly toxic and is sometimes eaten for its laxative effect. The stems and leaves are more poisonous, and if ingested in sufficient quantities can cause vomiting and possibly death.

Habitat & Range:

Clusters of mayapple appear in rich soil on forest floors and in open fields. Its range extends from southeastern Canada and Minnesota to Florida and Texas in the south.

MOONSEED
(Menispermum canadense)

- Woody, deciduous, twining vine without tendrils
- Leaves alternate, simple, long-stalked, and palmately veined, with fewer than ten shallow, rounded lobes
- Small, greenish-white flowers in axillary clusters
- Fruit grows in grape-like clusters and is a blue or bluish-black drupe
- Each fruit has a single disclike seed

Warning:
All parts of the Moonseed plant are poisonous. Life-threatening convulsions may ensue after ingestion. Not to be confused with the edible wild grape, the fruit of which contains several seeds.

Habitat and Range:
Moonseed is found in thickets and fencerows and along streams in moist woods. Its range extends from Canada south to Georgia and Arkansas.

POISON HEMLOCK
(Conium maculatum)

- Maximum height of about ten feet
- Abundant branches with purple-spotted stems
- Leaves resemble parsley, with a triangular shape and deep lobes
- Conspicuous veins extend to tips of teeth
- Small umbels of multiple white flowers, each one-sixteenth inch long
- Flowers have five petals, no sepals
- Small, rough seedlike fruit

Warning:
All parts of the poison hemlock contain coniine, a toxic alkaloid that can be deadly to ingest. Sufficient intake of the poison can lead to respiratory paralysis and fatal suffocation.

Habitat and Range:
Originally a European species, the poison hemlock has spread worldwide—including almost all of the United States and southern Canada. It commonly grows in weedy fields and along the edges of woods.

STAR OF BETHLEHEM
(Ornithogalum umbellatum)

- Grows to four to twelve inches tall
- Shiny, dark-green leaves are stiff and grasslike with a white midrib
- Leaves are four to twelve inches long and are hollow in cross section
- White, star-shaped flowers with six petals are marked with green on the back and occur on a leafless stem
- Emerges from an onionlike, oval-shaped bulb
- Fruit is a three-lobed capsule with several oval black seeds.

Warning:
All parts, especially the bulbs, of the Star of Bethlehem are poisonous if ingested. Burning and swelling of parts of the mouth, vomiting, diarrhea, and shortness of breath are indications of poisoning.

Habitat and Range:
The Star of Bethlehem is native to the Mediterranean region but escaped cultivation and is now naturalized in North America and grown in gardens. It is found along roads and in fields and other grassy areas and prefers shade. This plant resembles wild garlic, but lacks any onion or garlic odor.

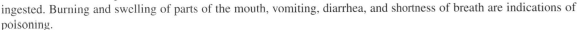

WATER HEMLOCK OR SPOTTED COWBANE
(Cicuta maculata)

- Grows to a height of three to six inches
- Highly branched with three inch-wide umbrella-like clusters of small, white flowers
- Stem is hollow, sectioned off like bamboo, and may be streaked with purple
- Lower leaves are one inch long, doubly divided, sharp-pointed and toothed; side veins end at notches between the teeth, not to tips at the outer margin
- Thick tubers and slender roots grow from the bottom of the rootstalk, which has hollow air chambers and produces a highly toxic brownish liquid when cut
- Fruit is flat and round, with thick ridges

Warning:
Water hemlock is violently poisonous if ingested. It contains a poisonous alcohol with a strong carrotlike odor. This toxin in concentrated in the tubers, but is present in all parts of the plant. Symptoms include dilated pupils, vomiting, muscle spasms, and convulsions. Water hemlock can be mistaken for wild parsnip or other medicinal herbs.

Habitat and Range:
Water hemlock is a wetland plant that grows in wet meadows and pastures, and in streams, ponds, and freshwater swamps. It is found throughout North America.

WHITE SNAKEROOT
(Eupatorium rugosum)

- Height is normally two or three feet
- Stems may be single or present in bunches
- Triangular or spade-shaped green leaves, three to seven inches long, are often sharply toothed
- Rounded clusters of very small white flowers
- White, bristly seeds

Warning:
All parts of the white snakeroot are poisonous to ingest, although illness usually occurs through consuming dairy products from cattle that have eaten the plants. Effects of poisoning include nausea, delirium, and coma.

Habitat and Range:
Widespread throughout the eastern United States and southeastern Canada, the snakeroot inhabits dry shady woodlands, fields, and roadsides.

WISTERIA *(Wisteria spp.)*

- Woody vines that can reach a height of sixty feet
- Clusters of scented, light blue to purple flowers; some varieties have pink or white flowers
- Flowers in the spring and late summer
- Fruits are oblong, velvety pods
- Eleven to thirteen leaflets radiate from a stem
- Leaflets measure about one and one-half inches in length

Warning:
Wisteria seeds are poisonous if ingested and will cause nausea, vomiting, and diarrhea.

Habitat and Range:
Wisteria is widely cultivated in temperate regions as an ornamental vine to cover walls, pergolas, and balconies. It is native to the southern United States, Japan, and China.

BIBLIOGRAPHY

Standardized Publications:
Manuals:
AFM 51-40, Air Navigation. Washington, D.C.: Department of the Air Force.

AFM 64-2, National Search and Rescue Manual. Washington, D.C.: Department of the Air Force, 1 July 1973.

AFM 64-5, Search and Rescue Survival. Washington, D.C.: Department of the Air Force, 1969.

AFM 64-6, Aircraft Emergency Procedures Over Water. Washington, D.C.: Department of the Air Force, 1955.

AFM 161-10, Field Hygiene and Sanitation. Washington, D.C.: Department of the Air Force, July 1970.

AFM 200-3, Joint Worldwide Evasion and Escape Manual. Washington, D.C.: Department of the Air Force, August 1961.

AARSM 55-1, Rescue and Recovery Operations. Washington D.C.: Department of the Air Force, 3 November 1978.

Army FM 5-20, Camouflage. Washington D.C.: Department of the Army, 1968.

Army FM 21-15, Care and Use of Individual Clothing and Equipment. Washington D.C.: Department of the Army, 1977.

Army FM 21-26, Map Reading. Washington, D.C.: Department of the Army, January 1969.

Army FM 21-26-1, Map Reading. Washington, D.C.: Department of the Army, 30 May 1975.

Army FM 21-31, Topographic Symbols. Washington, D.C., June 1961.

Army FM 21-40, NBC (Nuclear, Biological, and Chemical) Defense. Washington, D.C.: Department of the Army, 14 October 1977.

Army FM 21-60, Visual Signals. Washington, D.C.: Department of the Army, December 1974.

Army FM 21-75, Combat Training of the Individual Soldier and Patrolling. Washington, D.C.: Department of the Army, 10 July 1967.

Army FM 21-76, Survival Evasion And Escape. Washington, D.C.: Department of the Army, March 1969.

Army FM 21-78, Prisoner of War Resistance. Washington, D.C.: Department of the Army, December 1981.

Army FM 27-10, The Law of Land Warfare. Washington, D.C.: Department of the Army, July 1956.

Army FM 31-35, Jungle Operations. Washington D.C. Department of the Army, September 1969.

Army FM 31-70, Basic Cold Weather Manual. Washington, D.C.: Department of the Army, 1968.

Army FM 31-71, Northern Operations. Washington, D.C.: Department of the Army, 21 June 1971.

Army FM 90-3, Desert Operations, Washington, D.C.: Department of the Army, 19 August 1977.

Army FM 90-6, Mountain Operations. Washington, D.C.: Department of the Army, 30 June 1980.

Army FM 90-13, River Crossing Operations. Washington, D.C.: Department of the Army, 1 May 1978.

Army TC 21-3, The Soldier's Handbook for Individual Operations and Survival in Cold Weather. Washington, D.C.: Department of the Army, 30 September 1974.

Army TC 90-6-1, Military Mountaineering. Washington, D.C.: Department of the army, 30 September 1976.

Regulations:
AFR 64-3, Wartime Search and Rescue (SAR) Procedures. Washington D.C.: Department of the Air Force, 30 November 1971.

ARRSR 55-11, Pararescue Operational Regulation. HQ ARRS (MAC), Scott Air Force Base, Illinois 62225.

USAFER 64-3, Wartime Search and Rescue (SAR) Procedures (Europe). Washington, D.C.: Department of the Air Force, 7 October 1975.

Pamphlets:
AFP 6415, Survival and Emergency Uses of the Parachute. Washington, D.C.: Department of the Air Force, 1 June 1983.

AFP 161-43, Venomous Arthropod Handbook. Washington, D.C.: Department of the Air Force, 1977.

Army Pam 21-52, Cold Facts for Keeping Warm. Washington, D.C. Department of the Army, 1963.

Commercial Publications:
Abel, Michael, *Backpacking Made Easy*. Happ Camp CA: Naturegraph Publishers Inc., 1975.

Aleith, R.C., *Basic Rock Climbing*. New York: Charles Scribner's Sons, 1975.

Baird, P.D., *The Polar World*. New York: John Wiley and Son's Inc., 1964.

Benton, Allen and William Werner, *Field Biology and Ecology*. New York: McGraw, Hill Book Company, 1966.

Bergamini, David, *The Universe*. New York: Life Nature Library, Time Inc., 1966.

Brower, Kenneth, "A Galaxy of Life Fills the Night." National Geographic, Vol 160 No. 6 (December 1981), 834-847.

Bruemmer, F., *Encounter with Arctic Animals.* Toronto, Canada: McGraw-Hill Reyerson, Ltd Clark PH.D., Eugenie. "The Strangest Sea," National Geographic, Vol. 148 No. 3, (September 1975), 388-365.

Cousteau, Jacques-Yves, "The Ocean," National Geographic, Vol 160 No. 6 (December 1981), 780-791.

Darvill, Fred T., Jr., M.P., *Mountaineering Medicine.* Skagit Mountain Rescue Unit, Inc., 1969.

Dodge, Natt N., *Poisonous Dwellers of the Desert.* Arizona: Southwest Parks and Monuments Association, 1974.

Engel, Leonard, *The Sea.* New York: Life Nature Library, Time Inc., 1963.

Faub, P., *Ecology.* New York: Life Nature Library, Time Inc., 1963.

Fear, G., *Surviving the Unexpected Wilderness Emergency.* Tacoma, Washington: Survival Education Association, 1972.

Fear, Gene, *Wilderness Emergency,* Tacoma, Washington, Survival Education Association, 1975.

Fear, G. and J. Mitchel, *Fundamentals of Outdoor Enjoyment.* Tacoma, Washington: Survival Education Association, 1977.

Ferber, P., Mountaineering, *The Freedom of the Hills,* Third Edition. Seattle, Washington: The Mountaineers, 1977.

Freeman, Otio W. and H.F. Ranp, *Essentials of Geography.* New York: McGraw-Hill Book Company, 1959.

Gibson, C.E., *Handbook of Knots and Splices.* Emerson Books, 1972.

Glasstone, Samuel and Dolan, Philip J., *The Effects of Nuclear Weapons.* United States Department of Defense and the United States Department of Energy, 1977.

Gore, Rick, *A Bad Time to be a Crocodile.* National Geographic, Vol 153, No. 1 (January 1978), 90-115.

Halstead, B.W., *Dangerous Marine Animals.* Cornell Maritime Press, 1959.

Halstead, Bruce W., *Poisonous and Venomous Marine Animals of the World.* Princeton, New Jersey: Darwin Press, Inc., 1978.

Hanuritz, Bernard and Austin, James, *Climatology.* New York: McGraw-Hill Book Company, 1944.

Kaplan, M.D. Harold I., Freedman, M.D. Alfred M., Sadock, M.D. Benjamin J., *Comprehensive Textbook of Psychiatry/III.* Baltimore, Maryland: Williams and Wilkins Company, 1980.

Kearny, Cresson H., *Nuclear War Survival.* Oregon: NWS Research Bureau Coos bay.

Kjellstorm, Bjorn, *Map and Compass, The Orienteering Handbook.* American Orienteering Service, New York, 1955.

Kuhue, Cecil, *River Rafting.* World Publication Inc., Mt View, California 1979.

Lathrop, Theodore, M.D., *Hypothermia: The Killer of the Unexpected.* Portland, Oregon, 1972.

Ley, Willy, *The Poles.* New York: Life Nature Library, Time Inc., 1962.

Leopold, Starker A., *The Desert.* New York: Life Nature Library, Time Inc., 1962.

Lounsbury, John F. and Lawrence Ogden, *Earth Science.* New York: Harper and Row, 1969.

Matthews, Samuel W., "New World of the Ocean." National Geographic, Vol 160 No. 6 (December 1981), 792-833.

May, W., *Mountain Search and Rescue Techniques.* Bolder CO: Rocky Mountain Rescue Group, Inc., 1973.

McGinnis, William, *White Water Rafting.* New York: Time Books, 1979.

Nickelsbury, Janet, *Ecology: Habitats, Niches, and Food Chain.* New York: J.B. Lippincott Company, 1969.

Ormond, C., *Complete Book of Outdoor Lore, Outdoor Life.* New York: Harper and Row, 1964.

Peterson, Roger Tory, *The Birds.* New York Life Nature Library, Time Inc., 1963.

Shanks, Bernard, *Wilderness Survival.* New York: Universe Books, 1980.

Stefansson, V., *Arctic Manual.* New York: Greenwood Press, Publishers, Reprinted, 1974.

Stefansson, Y., *The Friendly Arctic.* New York: Greenwood Press, Publishers, Reprint of 1943 Edition.

Strahler, Arthur N., *Physical Geography.* New York: John Wiley and Sons, Inc., Third Edition 1969.

Strahler, Arthur N., *Introduction to Physical Geography.* New York: John Wiley and Sons, Inc., 1973.

Stuung, Norman: Curtis, Sl, Perry E., *White Water.* New York: Collier, MacMillan Publishers, 1976.

Van Dorn, William G., *Oceanography and Seamanship.* New York: Dodd, Mead, and Company 1974, 79-94, and 111-128.

Washburn, Bradford, *Frostbite.* Boston: Museum of Science, 1978.

Watson, Peter, *War on the Mind.* New York: Basic Books, Inc., Publishers, 1978.

Weiner, Michael A., *Earth Medicine - Earth Food.* New York: MacMillan Publishing Co., Inc., 1980.

West, James E. and Hillcomt, William, *Scout Field Book.* New Jersey: Boy Scouts of America, 1958.

Wirth, Eve R., *Survival Sense Emergency* (May 1982) 38 and 66.

Wolf, A.V., *Thirst.* Springfield, IL: C.C. Thomas, 1958.

American Wilderness, Time Life Books, Time Life Inc., 1972.

Encyclopedia Britannica, Inc., Encyclopedia, William Benton, Publisher, 1972.

National School of Conservation, Conservation of Natural Resources, Vol 1, "Tools and Techniques of Resource Management, Lesson 5." National School of Conservation Inc., Washington, D.C., 1973.

Publication No. 40, "Wild, Edible and Poisonous Plants of Alaska." Fairbanks AK: Cooperative Extension Service, University of Alaska.

Special Scientific Reports, Project Mint Julep Part II. Maxwell Air Force Base, Alabama. Research Studies Institute, May 1955.

TC 61-23, Private Pilot's Handbook of Aeronautical Knowledge. Washington D.C., Federal Aviation Agency, Flight Standard Service, 1965.

Other Selected References:

AALTDR 64-23, Project Cold Case, AD 462767, February 1965.

AALTN 57-16, Emergency Food Value of Alaskan Wild Plants. AD 293-31, July 1957.

ADTIC Publication A-103, Down in the North. Maxwell Air Force Base, Alabama. Research Studies Institute, 1976.

ADTIC Publication A-105, Glossary of Arctic and Subarctic Terms. Maxwell Air Force Base, Alabama. Air University, 1955.

ADTIC Publication A-107, Man in the Arctic. Maxwell Air Force Base, Alabama. Research Studies Institute, January 1962.

ADTIC Publication D-100 Afoot in the Desert. Maxwell Air Force Base, Alabama. Research Studies Institute, October 1980.

ADTIC Publication D-102, Sun, Sand and Survival. Maxwell Air Force Base, Alabama. Research Studies Institute, 1974.

AGARD Report No. 620, The Physiology of Cold Weather Survival. AD 784-268, April 1973.

Air Force CDC 20450 Intelligence Operations Specialist Vol 2, Maps and Charts. Extension Course Institute, Air Training Command, Gunter AFS, Alabama 36118, March 1979.

EID Bulletin No. 1., Sharks. Maxwell Air Force Base, Alabama. Aerospace Studies Institute.

EID Bulletin No. 2., Poisonous Snakes of North America. Maxwell Air Force Base, Alabama. Aerospace Studies Institute.

EID Bulletin No. 3, Poisonous Snakes of Central and South America. Maxwell Air Force Base, Alabama. Aerospace Studies Institute.

EID Bulletin No. 7, Water Resources. Maxwell Air Force Base, Alabama. Environmental Information Division, July 1969.

EID Bulletin No. 7a, Plant Sources of Water in Southern Asia. Maxwell Air Force Base, Alabama. Aerospace Studies Institute, August 1969.

EID Bulletin No. 8, Survival Nutrition. Maxwell Air Force Base, Alabama. Environmental Information Division.

EID Bulletin No. 13, Edible And Hazardous Marine Life. Maxwell Air Force Base, Alabama. Aerospace Studies Institute, April 1976.

EID Publication G-104, Airman Against the Sea. Maxwell Air Force Base, Alabama. Aerospace Studies Institute.

EID Publication G-105, Analysis of Survival Equipment. Maxwell Air Force Base, Alabama. Aerospace Studies Institute, 1957.

EID Publication G-107, Water Survival Field Tests. Maxwell Air Force Base, Alabama. Aerospace Studies Institute, June 1958.

EID Publication T-100, 999 Survived. AD 727-726, Maxwell Air Force Base, Alabama. Aerospace Studies Institute.

Know Your Knots, Missile Hazard Control Section ATC, Sheppard Air Force Base, Texas.Laboratory Note CRL-LN-55-211, The Will to Survive. Reno, USAF Survival Training School, 1955.

Synopsis of Survival Medicine, Fairchild Air Force Base, Washington, USAF Survival School, 1969.

INDEX

PHOTO CREDITS